Study Guide to accompany Atkins and Beran

GENERAL CHEMISTRY

Study Guide to accompany Atkins and Beran

GENERAL CHEMISTRY

SECOND EDITION

David Becker
Oakland Community College/Oakland University

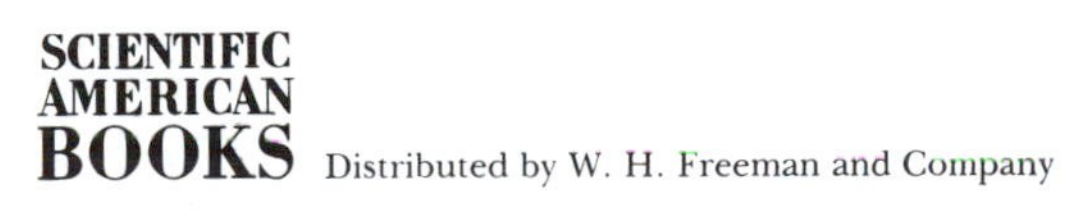

Cover image: Photograph by C. Bryan Jones © 1990

ISBN: 7167-2274-7

Printed in the United States of America.

Scientific American Books is a subsidiary of Scientific America Inc.
Distributed by W. H. Freeman and Company.
41 Madison Avenue, New York, New York 10010.

3 4 5 6 7 8 9 0 KP 9 9 8 7 6 5 4

CONTENTS

PREFACE

This *Study Guide* is designed to accompany the second edition of *General Chemistry* by P. W. Atkins and J. A. Beran. The positive response to the first edition, especially by chemistry students, has been heartening. In preparing the second edition I have, therefore, followed the format and style of the first edition. I also have tried to anticipate and elaborate the fundamental and critical information that is required by a student in learning the textbook material. Rather than being comprehensive, I have chosen to cover what I consider to be the most important topics in depth so that the *Study Guide* will be helpful not only to the student who requires basics to build on but also to the student who desires some elaboration of topics. (Even so, approximately 95 percent of the textbook material is covered in some form or other.) Because I believe that it is largely through working problems that a student develops a full appreciation of chemical principles, the focus of the *Study Guide* is decidedly on problem solving. With the addition of more questions to the Self-Test Exercises, the second edition now provides the student with more than 950 such exercises. The symbols, concepts, and formalism in the *Study Guide* are identical to those in the Atkins/Beran text, so students should have little difficulty in moving from one to the other. An unusually large number of graphs, figures, and chemical structures amplify the *Study Guide* material.

Each *Study Guide* chapter parallels a chapter in Atkins/Beran *General Chemistry*, second edition. A list of **Key Words** is given for each of the three or four major sections within a chapter. The key ideas of each textbook chapter are then stated, and sometimes elaborated, in **Key Concept** sections. The Key Concepts introduce no new material but restate textbook material in different words. Each Key Concept is accompanied by at least one example that is worked out in detail and a parallel exercise that has an answer but no detailed solution. A **Pitfalls** section sometimes points out common errors or misconceptions; at other times it offers insights on concepts or problem-solving techniques. The descriptive chemistry that appears in Chapters 1 through 17 and 22 of the text is reinforced at the end of the *Study Guide* chapters by **Descriptive Chemistry to Remember** and **Chemical Equations to Know.** The choice of what to include in these sections is, by necessity, very subjective; individual instructors may wish to modify them according to personal taste and emphases. Also included is a **Mathematical Equations to Know and Understand** section. Finally, numerous multiple-choice questions appear as **Self-Test Exercises** for each chapter. These questions vary in difficulty from remedial to very challenging, although basic single-concept problems have been emphasized.

How helpful this *Study Guide* is to students depends on its accuracy and clarity. If you, the reader or instructor, discover any errors or have ideas regarding how to say something better than what is now present, please take the time to drop me a note at: David Becker, Department of Chemistry, Oakland University, Rochester, MI 48309; Fax 313-370-2321.

TO THE STUDENT

You have registered for a General Chemistry course with a goal in mind, and possibly a long-term career plan that requires a knowledge of chemistry. If you have been permitted by your college or university to sign up for a chemistry course, you can succeed in it; but it will take time, discipline, and perseverence. As a rule of thumb, a four-credit chemistry course requires a minimum of 10 to 15 hours of home study per week. It would be wise for you to set up a weekly schedule that includes specific times for home study and stick to it! (A blank weekly calendar follows this section.)

It would be foolhardy to try to write a rigid prescription for studying chemistry; every one of us is different. Nevertheless, the following steps have proved successful for many students (including many of my own). I suspect that any successful plan of study will not deviate too much from them. Do the following for each chapter in your textbook:

1. Read the chapter once lightly. Try to get an overview of the material without taking detailed notes or underlining. Take a break.
2. Read the chapter (or required sections) very carefully. Underline or take notes. Work out derivations. Try to work all example problems and exercises within the chapter. When you finish each textbook section, read the relevant Key Concepts in the *Study Guide* and try to work out the *Study Guide* examples and exercises that accompany the Key Concepts. Write the definitions of all the Key Words in the *Study Guide* (writing the definitions of all the boldface words in the text narrative would be even better.)
3. Try to do the assigned homework problems. It is most important to write *all* the answers to the problems in an organized way. At this stage you will probably not be able to do all the assigned problems; don't spend more than 15 minutes on any one problem.
4. Attend the lecture. Take notes and listen. But don't let taking notes deter you from listening to your instructor. Don't be afraid to ask questions in the lecture. (I have found in my own lectures that when one student asks a question, many other students have the same question in mind. There is no such thing as a "dumb question.")
5. Complete the homework problems and exercises from the text and *Study Guide* that you could not do before the lecture. If you are unable to do any problems at this stage with the textbook and *Study Guide* alone, get help from your instructor, college tutors, or fellow students. In many colleges, computer-aided instruction (CAI) is available. (You don't have to know how to use a computer to use CAI; it's a marvelous opportunity for self-paced learning.)
6. At this point you will understand most of the material in the chapter. Reread the chapter carefully to try to put all the material together into a unified whole.
7. Do the Self-Test Exercises in the *Study Guide* as a test of your knowledge. If you can't work any of these exercises, get help.
8. Before an exam, go over your text underlinings, your notes, and any *Study Guide* material that your instructor has assigned (such as the Descriptive Chemistry to Remember, Chemical Equations to Know, and Mathematical Equations to Know and Understand).
9. Try to pace your studying so that you have a few hours to relax and get a good night's sleep before the exam. Last minute cramming is notoriously unsuccessful in chemistry.

It is most important to set up a disciplined study schedule that you will follow each week. After consideration of your schedule of classes and job and family obligations, block out, on the following calendar, at least 10 hours a week (more if your instructor suggests) when you can faithfully study chemistry. If possible, reserve one day (or half a day) a week for recreation and relaxation; try to avoid doing any school work during this time.

	Mon	Tues	Wed	Thurs	Fri	Sat	Sun
7 A.M.							
8							
9							
10							
11							
Noon							
1 P.M.							
2							
3							
4							
5							
6							
7							
8							
9							
10							
11							

ACKNOWLEDGMENTS

Any author who writes a scholarly text eventually discovers that, to be successful, any such endeavor must be a group effort. Many people contributed to both editions of this *Study Guide* and helped to improve it in countless ways.

George W. Eastland, Jr., of Saginaw Valley State University, University Center, Michigan, suggested a number of insightful improvements to the second edition; Margot E. Getman and Georgia Lee Hadler of Scientific American Books provided outstanding editorial support with remarkable good humor. The copy editor, Yvonne Howell, also made substantial and thoughtful contributions to the entire text.

A large number of individuals carefully reviewed the first edition and suggested many important changes. Among them are Judith R. Fish, Oakland University, Rochester, Michigan; John Forsberg, Saint Louis University; Forrest C. Hentz, Jr., North Carolina State University; Jeffrey A. Hurlburt, Metropolitan State College, Denver, Colorado; Earl S. Huyser, University of Kansas; Wayne Lowder, Department of Energy, New York; Richard A. Potts, University of Michigan, Dearborn Campus; Don Roach, Miami Dade Community College; and Robert L. Stern, Oakland University. Peter W. Atkins of Oxford University offered invaluable ideas and moral support at many stages of the writing.

My wife, Rita, and sons, Shanan, Seth, and Daniel, again deserve thanks for their patience and support.

Study Guide to accompany Atkins and Beran

GENERAL CHEMISTRY

1 PROPERTIES, MEASUREMENTS, AND UNITS

Matter, anything that has mass and takes up space, is the material, or "stuff," that makes up the universe. **Chemistry** is the study of matter and the changes it undergoes. A **substance** is a single kind of matter.

THE PROPERTIES OF SUBSTANCES

KEY WORDS Define or explain each of the following terms in a written sentence or two.

atom
change of state
chemical change
chromatography
distillation
filtration
molecule
mixture
physical change
solution

1.1 PHYSICAL AND CHEMICAL PROPERTIES

KEY CONCEPT **Physical and chemical properties**

The properties of a substance are characteristics of the substance that distinguish it from other substances. There are two classes of properties, **physical properties** and **chemical properties.** A physical property is one that is characteristic of the substance itself, such as its color, smell, ability to conduct electricity, or physical state. A chemical property describes how a substance changes into a new substance, for instance, how candle wax burns into water and carbon dioxide. To illustrate the nature of physical and chemical properties, let us consider the familiar substance water.

Some physical properties of water	Some chemical properties of water
liquid at room temperature	not flammable; will not react with oxygen
colorless, odorless	forms hydrogen and oxygen gas when electrolyzed
melts at 0°C, boils at 100°C	

Notice that the physical properties are all properties of the water itself and do not depend on the presence of any other substance. The chemical properties all involve the conversion of water to some new substance.

EXAMPLE **Physical and chemical properties**

State three physical properties and one chemical property of iron.

SOLUTION We must rely on our observations of familiar objects to answer this question. For the physical properties, we need only recall some distinguishing features of iron. For example, at room temperature it is a gray solid and an excellent conductor of electricity. These three properties (gray, solid, good conductor) suffice to answer the question, but we may also note that it is odorless. A chemical property involves the conversion of one substance into another. We can observe such a conversion when iron rusts, which converts the iron to an oxide.

EXERCISE State one physical property and one chemical property of gasoline.

ANSWER: Gasoline is a liquid (physical property) that is flammable (chemical property).

1.2 SUBSTANCES AND MIXTURES

KEY CONCEPT A Substances and mixtures

A substance is matter uncontaminated by the presence of any other substance. If we take two or more substances and mix them together (without permitting complete conversion to a new substance), the result is a **mixture.** A mixture has a combination of physical properties that originate with the different substances in the mixture. It can be separated into its various substances by physical methods, which are methods that take advantage of the differences in certain physical properties of the substances. A substance cannot be separated into different substances by physical methods because separation by a physical method requires the presence of at least two substances with two different sets of physical properties. A substance has only a single set of physical properties.

EXAMPLE Using physical properties

Develop a method to separate the components in a mixture of sugar and sand.

SOLUTION We must find a physical property of sugar and of sand that can be exploited to separate the two. For the separation to be successful, however, the property must not only be different for the two substances, but a technique must be available to exploit the difference. Because sugar is very soluble in water and sand is insoluble, we could accomplish a separation by adding the mixture of sugar and sand to a beaker of water and stirring thoroughly. The sugar dissolves, but the sand does not. If we now pour the contents of the beaker into a piece of filter paper (such as a coffee filter), the sand would stay behind in the filter paper and the dissolved sugar would pass through the filter paper with the water. The water can now be evaporated or *carefully* boiled off to leave the pure sugar.

EXERCISE You purchase a pound of table salt (NaCl, sodium chloride) from one company and mix it with a pound of table salt purchased from another company. Would it be possible to separate the table salt from the two different sources by a physical method?

ANSWER: No; only one substance is present.

KEY CONCEPT B Solutions

A **solution** is a **homogeneous** mixture. It is homogeneous because a visual inspection of a solution does not reveal its individual components. You can try this yourself: thoroughly dissolve $\frac{1}{4}$ teaspoon of table salt in a cup of water to make an NaCl solution and examine it carefully with a magnifying lens. The presence of the salt is not detectable. In a solution, the **solvent** is the component in which the **solute** is dissolved. Usually much more solvent is present than solute. Aqueous solutions are *always clear,* but they can also be colored. To understand what the **solubility** of a solute is, imagine trying to dissolve 5 pounds (lb) of sugar in a cup of water. After you dissolve a certain amount of sugar, the solution will contain as much sugar as it can possibly hold; it will be impossible to dissolve any more. The maximum amount that dissolves is called the solubility of the sugar; in our example, the solubility is approximately 1 lb/cup. It is possible for a solution to momentarily contain too much of a solid solute, in which case the extra solute will spontaneously form a solid. When the extra solute forms a solid slowly, large crystals tend to form. This process is called **crystallization.** When the extra solute forms a solid rapidly, a fine powder tends to form. The formation of the solid is called **precipitation.**

EXAMPLE Solutions

When we mix a small amount of sodium metal with excess liquid ammonia (which is colorless), the sodium dissolves, resulting in a homogeneous, clear, blue mixture. Is this a solution? If it is, identify the solute and solvent and state whether it is aqueous or nonaqueous.

SOLUTION A homogeneous mixture is a solution; hence the mixture described in the problem is a solution. (Solutions may be colored.) The component in excess, in this case ammonia, is the solvent; and the dissolved substance, in this case sodium, is the solute. Because the solvent is not water, the solution is nonaqueous.

EXERCISE When we mix a colorless aqueous solution in a test tube with a light blue aqueous solution in a beaker, the solution in the beaker becomes cloudy. What has happened?

ANSWER: A chemical reaction has occurred and a precipitate has formed.

MEASUREMENTS AND UNITS

KEY WORDS Define or explain each of the following terms in a written sentence or two.

absolute zero
base units
density
derived units
extensive property
intensive property
International System of Units
kelvin
metric system

1.3 THE INTERNATIONAL SYSTEM OF UNITS

KEY CONCEPT A The SI system of units: Mass, length, and time

To facilitate the sharing of information among scientists, a system of units called the **International System of Units** (SI system of units) has been developed. With such a system, a measured value reported by a scientist in one laboratory can immediately be understood and used by a scientist in a distant laboratory because the units used for the measurement will be familiar to both. The SI base unit of mass is the **kilogram** (kg), which is defined by a standard mass maintained in France. The **gram** (g) and **milligram** (mg) are other mass units which are related to the kilogram in a simple way and which are more convenient to use in certain situations. The SI base unit of length is the **meter** (m). An often convenient unit of length that is simply related to the meter is the **centimeter** (cm). The SI unit of time, the **second** (s), is the familiar unit of time used in everyday activities. Different size units enable us to select convenient units for a particular application. As an example, the mass of a sugar cube written as 4.5 g seems more natural than 0.0045 kg (too many zeros before the first nonzero digit) or 4500 mg (too many digits before the decimal), even though all represent the same mass. Units of different sizes are defined as a matter of convenience, not of necessity.

EXAMPLE Selecting convenient units

A first-class letter with a mass under 0.028 kg requires 29 cents postage. What are the most convenient units in which to express this mass, and what does 0.028 kg equal in these units?

SOLUTION We want a unit that expresses the mass of a letter without too many digits or too many zeros. Let's try a few examples of units, using the information given in Table 1.6 of the text.

Unit	Mass of letter	Comment
kg	0.028 kg	too many zeros before first nonzero digit
Mg	0.000028 Mg	too many zeros before first nonzero digit
μg	28,000,000 μg	too many digits before the decimal
g	28 g	best choice of units to use
mg	28,000 mg	too many digits before the decimal

EXERCISE The diameter of a red blood cell is approximately 0.000007 m. What is this diameter in a more convenient unit?

ANSWER: 7 μm

KEY CONCEPT B Powers of ten and scientific notation: Numbers > 10

Scientists must often deal with very large numbers (such as 300,000,000 m/s, the speed of light) or very small numbers (such as 0.000 000 000 05 m, the radius of a hydrogen atom). Writing numbers in **scientific notation** enables us to write very large numbers or very small numbers in a particularly convenient way. To write a number in scientific notation, we must use powers of 10, so we start with a review of this topic. When any number is raised to some power, the number is multiplied by itself the number of times given by the power:

$$2^3 = 2 \times 2 \times 2 = 8$$
$$2.4^2 = 2.4 \times 2.4 = 5.76$$
$$10^5 = 10 \times 10 \times 10 \times 10 \times 10 = 100{,}000$$

If we want to write a large number in scientific notation, we first factor the large number into two parts: a number between one and ten and a second number, such as 100 or 1000 or 10,000 or any similar multiple of ten. For the speed of light, for instance, this gives

$$300{,}000{,}000 \text{ m/s} = 3.00 \times 100{,}000{,}000 \text{ m/s}$$

Number between 1 and 10 (3.00); Multiple of 10 (100,000,000)

The multiple of ten can be written as a power of ten and substituted into the number:

$$100{,}000{,}000 = 10 \times 10 \times 10 \times 10 \times 10 \times 10 \times 10 \times 10 = 10^8$$
$$300{,}000{,}000 \text{ m/s} = 3.00 \times 10^8 \text{ m/s}$$

In this case, 3.00×10^8 is the speed of light, in meters per second, written in scientific notation. The reason for writing 3.00 instead of 3 as the number between 1 and 10 will be discussed later. To enter this number into a scientific calculator, use the following steps, which work for the common calculators that use arithmetic notation:

Step	Display shows
1. enter 3	3.
2. press EE or EXP key	3. 00
3. enter 8	3. 08

The 3. 08 now displayed is your calculator's way of indicating the number 3×10^8.

EXAMPLE Writing numbers larger than 10 in scientific notation

The national debt is approximately \$2,300,000,000,000. Express this in scientific notation.

SOLUTION We first factor \$2,300,000,000,000 into a number between 1 and 10 and a multiple of 10. Usually, in writing the number between 1 and 10, we drop all extra zeros:

$$\$2{,}300{,}000{,}000{,}000 = \$2.3 \times 1{,}000{,}000{,}000{,}000$$

We then decide how to write the multiple of ten as a power of 10:

$$1{,}000{,}000{,}000{,}000 = 10 \times 10 \times 10 \times 10 \times 10 \times 10 \times 10 \times 10 \times 10 \times 10 \times 10 \times 10 = 10^{12}$$

Putting together the two results, we get

$$\$2{,}300{,}000{,}000{,}000 = \$2.3 \times 10^{12}$$

EXERCISE Approximately 12,500,000 students are enrolled in colleges and universities in the United States. Write this number in scientific notation.

ANSWER: 1.25×10^7

KEY CONCEPT C Powers of ten and scientific notation: Numbers < 1

To write small numbers in scientific notation, we must deal with negative exponents. The meaning of a negative exponent is defined by the relationship

$$Y^{-n} = \frac{1}{Y^n} = \frac{1}{Y \times Y \times Y \times Y \quad (n \text{ times})}$$

For powers of 10, for example,

$$10^{-5} = \frac{1}{10^{+5}} = \frac{1}{10 \times 10 \times 10 \times 10 \times 10} = \frac{1}{100{,}000} = 0.00001$$

To write a number less than 1 in scientific notation, we first factor the number into a number between 1 and 10 and a multiple of 10, and then substitute in a power of 10 for the multiple of 10. For instance,

$$0.000567 = 5.67 \times 0.0001$$

$$0.0001 = \frac{1}{10000} = 10^{-4}$$

$$0.000567 = 5.67 \times 10^{-4}$$

To enter this number into your calculator requires the following steps, which will work for most scientific calculators:

Step	Display shows
1. enter 5.67	5.67
2. press EE or EXP	5.67 00
3. enter 4	5.67 04
4. press the +/− key or change-sign key	5.67 − 04

The display 5.67 − 04 is your calculator's way of indicating the number 5.67×10^{-4}.

EXAMPLE Using scientific notation

Rhode Island, the smallest state, comprises 0.0033% of the total area of the United States. Write this number in scientific notation.

SOLUTION Because 0.0033 is less than 1, we recognize that the power of 10 that results must be negative if we write the number in standard scientific notation. We first factor the number into a number between 1 and 10 and a multiple of 10:

$$0.0033 = 3.3 \times 0.001$$

We then convert 0.001 into a power of 10:

$$0.001 = \frac{1}{1000} = 10^{-3}$$

Finally, we substitute 10^{-3} for 0.001 in the original factored number:

$$0.0033 = 3.3 \times 10^{-3}$$

EXERCISE An averge chemistry class of 25 students contains approximately 0.000 000 61% of the world's population. Express this number in scientific notation.

ANSWER: 6.1×10^{-7}

PITFALL Entering negative exponents on a calculator

On many calculators, you cannot enter a negative exponent by using the minus operation key, −. Only the change-sign key, +/−, or its equivalent works.

KEY CONCEPT D The three common temperature scales

Three temperature scales are in use throughout the world today: the Celsius (°C) scale, the Fahrenheit (°F) scale, and the Kelvin (K) scale. To convert from one of these scales to another, we use the following relationships:

From °C to °F: $°\text{F} = 32 + \frac{9}{5}(°\text{C})$

From °F to °C: $°\text{C} = \frac{5}{9} \times (°\text{F} - 32)$

From °C to K: $\text{K} = 273.15 + °\text{C} = 273 + °\text{C}$ (for everyday use, this form has enough precision)

From K to °C: $°\text{C} = \text{K} - 273.15 = \text{K} - 273$

It is helpful to recognize that the two equations that relate degrees Celsius to degrees Fahrenheit are equivalent. It is necessary to learn only one of them since it is straightforward to algebraically convert to the other.

EXAMPLE Converting between Celsius and Fahrenheit temperatures

The highly **volatile** (low boiling point) liquid ether boils at 34.5°C. What is this temperature on the Kelvin and Fahrenheit scales?

SOLUTION To convert 34.5°C to the Kelvin scale, we use the relationship in which the Kelvin temperature is expressed on one side of the equation by itself, because this is the unknown, and everything else, including degrees Celsius, on the other side: $\text{K} = 273.15 + °\text{C}$. Because 34.5°C is expressed to a tenth of a degree, we use 273.15 in the conversion formula rather than 273:

$$\text{K} = 273.15 + 34.5 = 307.65 = 307.7$$

The answer is rounded to a tenth of a degree to agree with the tenth of a degree precision in the number 34.5. For conversion to the Fahrenheit scale, we use the equation in which degrees Fahrenheit (the unknown) is on one side and everything else, including degrees Celsius (the known), on the other side:

$$°F = 32 + \frac{9}{5}(°C)$$

Because $°C = 34.5$,

$$°F = 32 + 62.1 = 94.1°$$

Because the number 32 is an exact number, we express the answer to a tenth of a degree precision, the original precision in the number 34.5.

EXERCISE The coldest (official) temperature ever recorded in the United States was $-79.8°F$ at Prospect Creek Camp, Alaska, on January 23, 1971. What is this temperature in degrees Celsius and in kelvins?

ANSWER: $-62.1°C$, 211.1 K

PITFALL Negative temperatures in temperature conversions

As we have seen, negative temperatures are possible on the Celsius and Fahrenheit temperature scales. In doing a conversion with a negative temperature, care must be taken to properly subtract and add signed numbers. Consider the conversion of $-22°F$ to degrees Celsius. The equation to use is

$$°C = \frac{5}{9} \times (°F - 32)$$

Substituting in °F gives

$$°C = \frac{5}{9} \times (-22 - 32)$$

The subtraction in the parentheses involves signed numbers. Here we note that $-22 - 32 = -54$. The final answer for the Celsius temperature is $\frac{5}{9} \times (-54) = -30°C$.

KEY CONCEPT E Derived units: Volume

The sciences use a vast number of **derived units,** or units that are a combination of two or more base units. For instance, consider the unit associated with volume. If we calculate the volume of any object, for example, a rectangular solid, a sphere, a pyramid, or even an irregularly shaped object (such as a banana), the answer will always be in units of $(\text{length})^3$. The **liter** (L) is a common unit of volume. It is not, specifically, a $(\text{length})^3$ but is related to the cubic centimeter through the relationship that $1\ L = 1000\ cm^3$ (exactly).

EXAMPLE Units of volume

A quart is equal to 0.946 L (Table 1.5 in the text). How many milliliters are there in a quart?

SOLUTION From Table 1.6 of the text we conclude that 1 L = 1000 mL, or equivalently 0.001 L = 1 mL. Using the method of ratios or unit analysis, we conclude that 0.946 L = 946 mL; so 1 qt = 946 mL:

$$0.946\ \cancel{L} \times \frac{1000\ mL}{1\ \cancel{L}} = 946\ mL$$

EXERCISE How many milliliters are there in a cup? (4 cup = 1 qt)

ANSWER: 237

KEY CONCEPT F Derived units: Density

The **density** (d) of a substance is a measure of how much mass can be contained in a given volume of it. A substance that has a large mass packed in a small volume has a high density; a substance that packs a small mass into a large volume has a low density. For example, a brick has a relatively large mass in a relatively small volume and, therefore, has a high density; a feather, on the other hand, uses a large volume for a relatively small mass and, therefore, has a low density. To calculate the density of a sample, we divide the mass of the sample by its volume:

$$d = \frac{\text{mass}}{\text{volume}} = \frac{m}{V}$$

The units of density must agree with the definition of density as a mass divided by a volume; units such as gram per milliliter (g/mL), gram per cubic centimeter (g/cm^3), and gram per liter (g/L) are common. The density depends on both temperature and pressure.

EXAMPLE 1 Calculating the density of a sample

What is the density of a brick with a mass of 1600 g and a volume of 1100 cm^3?

SOLUTION To calculate the density, we use the defining equation

$$d = \frac{m}{V}$$

Inserting $m = 1600$ g and $V = 1100$ cm^3 into the equation gives us

$$d = \frac{1600 \text{ g}}{1100 \text{ cm}^3} = 1.500 \text{ g/cm}^3$$

EXERCISE A helium balloon with a volume of 2.0 L contains 0.36 g of helium. What is the density of helium?

ANSWER: 0.18 g/L

EXAMPLE 2 Calculating the volume of a sample of known density and mass

One ounce of gold cost approximately $450 in 1988. What is the volume of 1.00 oz (28.4 g) of gold? Its density is 19.3 g/cm^3.

SOLUTION The formula for calculating density is

$$d = \frac{m}{V}$$

Because the mass and density are given and the volume requested, we isolate the volume on one side of the equation by multiplying both sides of the equation by V and dividing both sides by d:

$$V = \frac{m}{d}$$

Substituting in $m = 28.4$ g and $d = 19.3$ g/cm^3 gives us

$$V = \frac{28.4 \text{ g}}{19.3 \text{ g/cm}} = 1.47 \text{ cm}^3$$

Thus 1.00 oz of gold occupies 1.47 cm^3 (about $\frac{1}{2}$ teaspoon).

EXERCISE Calculate the volume of 1.00 oz of nitrogen, which has a density of 1.25 g/L.

ANSWER: 22.7 L

EXAMPLE 3 Calculating the mass of a sample of known density and volume

What is the mass of 10.0 L of oxygen gas, which has a density of 1.43 g/L?

SOLUTION The formula for calculating density is

$$d = \frac{m}{V}$$

Because the volume and density are given and the mass requested, we isolate the mass on one side by multiplying both sides of the equation by V:

$$m = d \times V$$

Substituting $d = 1.43$ g/L and $V = 10.0$ L gives

$$m = (1.43 \text{ g/L})(10.0 \text{ L}) = 14.3 \text{ g}$$

EXERCISE What is the mass of an ice cube 3.0 cm on a side? The density of ice is 0.92 g/cm^3.

ANSWER: 25 g

1.4 EXTENSIVE AND INTENSIVE PROPERTIES

KEY CONCEPT Extensive and intensive properties

An **extensive property** of a sample is a property that depends on how much of the sample is present, that is, the extent of the sample. Imagine that we have three samples of water and that we tabulate the volume, mass, and density of each sample.

Sample number	Relative amount present	Mass (g)	Volume (mL)	Density (g/mL)
I	little (about 20 drops)	1.00	1.00 mL	1.00
II	more (1 cup)	237	237 mL	1.00
III	most (1 quart)	946	946 mL	1.00

Clearly, both the volume and the mass of the water depend on how much water is present; the volume and mass are, therefore, extensive properties. The density, however, is an **intensive property,** a property that is independent of the size of the sample. The density of water is 1.00 g/mL regardless of the size of the sample. Many intensive properties are defined by dividing one extensive property by another, as we define density.

EXAMPLE Extensive and intensive properties

Examine a sample consisting of a teaspoon of NaCl (sodium chloride or table salt). Assume that your sample has a mass of 11 g and a volume of 5 mL. Give three intensive properties and two extensive properties of the sample.

SOLUTION An intensive property is one that does not depend on the size of the sample; an extensive property is one that does. To answer the question, we imagine what would happen to various properties if we enlarge or shrink the size of the sample (either will do). We can easily obtain a larger sample of table salt, for example by pouring 1 cup of NaCl (approximately 237 mL, or 513 g) out of its container. We compare this sample with the smaller sample and list some properties that are different in the two samples and properties that are not.

Property	Sample 1	Sample 2	Intensive or extensive
mass	11 g	513 g	extensive
volume	5 mL	237 mL	extensive
color	white	white	intensive
density	2.2 g/mL	2.2 g/mL	intensive
taste	salty	salty	intensive

EXERCISE Is the electrical conductivity of copper wire (the ease of electrical conduction per meter of wire) an extensive or intensive property?

ANSWER: Intensive

USING MEASUREMENTS

KEY WORDS Define or explain each of the following terms in a written sentence or two.

accurate
conversion factor
mass percentage
precise
random error
significant figures
systematic error
unit analysis

1.5 CONVERSION FACTORS

KEY CONCEPT A Unit analysis and conversion factors

A working scientist frequently has to change a measured or calculated number from one set of units to another. As an example, if we are asked to fill a dime wrapper for a bank, the only information given on the wrapper is that it is meant to hold \$5; we must decide how many dimes are equivalent to \$5. This procedure is actually a unit conversion, from a certain number of dollars to a certain number of dimes:

$$5 \text{ dollar} \longrightarrow ? \text{ dime}$$

Most of us can solve this problem easily because we are so familiar with our money system, but many chemical problems are similar. A powerful calculational technique called **unit analysis** can also be used to solve such a problem; with this method we use the formulation

$$(\text{Answer with new units}) = (\text{given number with old units}) \times (\text{conversion factor})$$

For the problem of converting 5 dollars to the equivalent number of dimes, we have

$$? \text{ dime} = 5 \text{ dollar} \times (\text{conversion factor})$$

In this problem, the old units are "dollar" and the new units "dime." The next step is to find the proper **conversion factor,** a factor that expresses a known relationship between the old units and the new units. For our example, we use the fact that 1 dollar = 10 dime to derive two conversion factors

$$\frac{1 \text{ dollar}}{10 \text{ dime}} \quad \text{and} \quad \frac{10 \text{ dime}}{1 \text{ dollar}}$$

The first conversion factor gives the number of dollars per dime ($\frac{1}{10}$), the second, the number of dimes per dollar (10). It is most important to remember that conversion factors always come in pairs, as just indicated, with one factor the inverse of the other. The correct conversion factor for any calculation will result in cancellation of the old units; the remaining units are the correct new units for the desired answer:

$$50 \text{ dime} = 5 \cancel{\text{dollar}} \times \frac{10 \text{ dime}}{1 \cancel{\text{dollar}}}$$

When we use the correct conversion factor, the dollar units on top of the fraction cancel the dollar units on the bottom, leaving the expected units of dime for the answer. If we try the second (incorrect) conversion factor of the pair, we get

$$0.50\,\frac{(\text{dollar})^2}{\text{dime}} = 5\ \text{dollar} \times \frac{1\ \text{dollar}}{10\ \text{dime}} \quad \text{(incorrect answer, wrong units)}$$

The second conversion factor does not permit cancellation of units and thus is the wrong one to use. We use five steps in a unit conversion problem:

1. Analyze the problem. Decide which number with old units is to be converted into a new number with new units. Write your analysis in the following form:

   Given number with old units → answer with new units

2. Find a known relationship between the old and the new units.
3. Set up the two conversion factors that result from the relationship between the units.
4. Set up the formulation

   (Answer with new units) = (given numbers with old units) × (conversion factor)

5. Do the indicated calculation, taking care that units cancel properly.

EXAMPLE Unit conversions

What is the mass, in kilograms, of a 16-lb sledge hammer?

SOLUTION We follow steps 1 through 5.

1. Analyze the problem. Decide which number with old units is to be converted into a new number with new units. Write your analysis in the form "given number with old units → answer with new units."

$$16\ \text{lb} \rightarrow ?\ \text{kg}$$

2. Find a known relationship between the old and the new units:

$$2.205\ \text{lb} = 1\ \text{kg} \quad \text{(refer to inside back cover of the text)}$$

3. Set up the two conversion factors that result from the relationship between the units:

$$\frac{2.205\ \text{lb}}{1\ \text{kg}} \quad \text{and} \quad \frac{1\ \text{kg}}{2.205\ \text{lb}}$$

4. Set up the formulation (given number with old units) × (conversion factor) = (answer with new units):

$$16\ \text{lb} \times \frac{1\ \text{kg}}{2.205\ \text{lb}} = ?$$

5. Do the indicated calculation, taking care that units cancel properly:

$$16\ \cancel{\text{lb}} \times \frac{1\ \text{kg}}{2.205\ \cancel{\text{lb}}} = 7.3\ \text{kg}$$

EXERCISE How many minutes are there in 245 seconds?

ANSWER: 4.08 min

KEY CONCEPT B Converting units in a denominator

Unit conversions can be used to convert a unit in the denominator of a number. For example, if we know the density of a substance in grams per milliliter and want to calculate the density in grams per liter, we must convert milliliters (mL) in the denominator of the original unit to liters (L). The principle involved is the same as in other unit conversions: the conversion factor must be such that the old units cancel and leave the correct new units.

EXAMPLE Converting units in the denominator of a complex unit

The density of $CHCl_3$ (chloroform) is 1480 g/L. What is the density in grams per milliliter?

SOLUTION We apply the five steps used earlier.

1. Analyze the problem. Decide which number with old units is to be converted into a new number with new units. Write your analysis in the form "given number with old units" → "answer with new units." In this case, we must change the unit L in g/L to mL:

$$1480 \text{ g/L} \rightarrow ? \text{ g/mL}$$

2. Find a known relationship between the old and the new units. Table 1.6 of the text gives us the clue to the relationship between liters and milliliters:

$$1000 \text{ mL} = 1 \text{ L}$$

3. Set up the two conversion factors that result from the relationship between the units:

$$\frac{1000 \text{ mL}}{1 \text{ L}} \quad \text{and} \quad \frac{1 \text{ L}}{1000 \text{ mL}}$$

4. Set up the formulation (answer with new units) = (given number with old units) × (conversion factor):

$$1480 \frac{\text{g}}{\text{L}} \times \frac{1 \text{ L}}{1000 \text{ mL}} = ?$$

5. Do the indicated calculation, taking care that units cancel properly:

$$1480 \frac{\text{g}}{\cancel{\text{L}}} \times \frac{1 \cancel{\text{L}}}{1000 \text{ mL}} = 1.480 \frac{\text{g}}{\text{mL}}$$

EXERCISE A leaky faucet drips water at the rate of 0.055 mL/s. How many milliliters per hour (h = hour) does this correspond to?

ANSWER: 2.0×10^2 mL/h

PITFALL Choosing the correct conversion factor

A common mistake made in unit conversion calculations is choosing the wrong conversion factor from the pair of possible factors. To avoid this error, make a quick, rough mental calculation of the answer you expect and compare your estimate with the answer you get from a detailed calculation. At the very least, estimate whether the answer should be larger or smaller than the given number. For example, in a problem that asks for the number of centimeters in a mile, we should note that a centimeter is much smaller than a mile, so there should be a lot of centimeters in 1 mile. If we get an answer such as 161 cm or 1.609×10^{-5} cm (0.00001609 cm) for the number of centimeters in 1 mile, we would immediately recognize that something was wrong and look for an error somewhere in the problem solution. The moral: *look carefully at your answer!*

1.6 THE RELIABILITY OF MEASUREMENTS AND CALCULATIONS

KEY CONCEPT A The uncertainty in a measurement

Every measured number must be somewhat uncertain. For instance, the diameter of a nickel, measured with a vernier calipers, is 2.115 cm. When we use the calipers, we must estimate the last digit (the 5); thus, there is some uncertainty in this digit. For most of the equipment we encounter in a laboratory, similar estimates are required and there is an uncertainty of ± 1 in the last digit read. For our measurement, reporting the diameter as 2.115 cm means that the actual diameter is between 2.114 cm (2.115 − 0.001, corresponding to −1 in the last digit) and 2.116 cm (2.115 + 0.001, corresponding to +1 in the last digit). In summary:

If we report a measurement as 2.115,	⟶	we assume there is an uncertainty of ±1 in the the last digit,	⟶	which means the actual value is between 2.114 and 2.116

EXAMPLE Uncertainty in measured values

The mass of a beaker is reported as 22.56 g. Interpret this number with regard to its uncertainty.

SOLUTION Unless a statement is made to the contrary, it is safe to assume that the last digit in any measured number is reliable only to ±1. For the value 22.56 g, the 6 at the end of the number could be a 7 (6 + 1) or a 5 (6 − 1). Thus, the actual value of the mass is between 22.55 g and 22.57 g. That is, it could be 22.550 g, 22.551 g, 22.552 g, 22.553 g, . . . , 22.567 g, 22.568 g, 22.569 g, or 22.570 g.

EXERCISE The volume of a industrial reaction vessel is determined to be 1346 L. What is the uncertainty associated with this value?

ANSWER: It is uncertain by ±1 L, so the volume is between 1345 L and 1347 L.

KEY CONCEPT B Significant figures

A subtle but direct interplay exists between the uncertainty of a number and the number of significant figures in the number. The word "significant" in the term gives a clue to its meaning: **significant figures** (written sf) are the meaningful digits in a reported number. The last digit in the number, which is known only to ±1, is counted as a significant digit. When we write a number, we must be certain to use the proper number of significant figures, neither too few nor too many. As an example, assume we determine the mass of a sample on a balance that is calibrated to 0.01 g, and get 4.33 g. If we write 4.3 g (instead of the correct 4.33 g), we have used too few significant figures because our balance is capable of better precision. On the other hand, if we write 4.331 g (instead of the correct 4.33 g), we have used too many significant figures because our balance is not calibrated to 0.001 g and is incapable of the precision expressed in 4.331 g. When no zeros are present, the number of significant figures is the same as the total number of digits in the number. For our example,

4.3 g	2 sf	too few for the balance used
4.33 g	3 sf	correct number for the balance used
4.331 g	4 sf	too many for the balance used

Whether a zero is a significant figure or not depends on its role in a number. If a zero sets the position of the decimal point or is written by itself in front of a decimal (such as in 0.866), it is not significant. All other zeros are significant. The following examples illustrate the rules for handling zeros and the alternate technique of writing a number in standard scientific notation in order to count the significant figures in the number. The digits in boldface are significant:

3.045	**3.045** $\times 10^{0}$	4 sf
0.0**33**	**3.3** $\times 10^{-2}$	2 sf
83.670	**8.3670** $\times 10^{1}$	5 sf
0.00**802**	**8.02** $\times 10^{-3}$	3 sf

Numbers such as 360 present one last ambiguity regarding the handling of zeros. The zero in 360 may simply set the position of the decimal point; in this case, the 6 in the number is known to ±1, and the number contains 2 significant figures. Or the zero may be a measured number, one that has been determined to ±1; in this case, 3 significant numbers are present. We shall resolve this problem as is done in the text: All zeros at the end of numbers are significant unless otherwise stated. A number such as 360 that has two significant figures will always be written in scientific notation.

EXAMPLE Counting significant figures

How many significant figures are there in each of the following numbers? 33.4, 0.6600, 0.044×10^{-2}

SOLUTION We have two ways to solve this problem. We may use the rules for determining the significance of zeros, or we may write the numbers in standard scientific notation and count digits. The correct answers are

33.4	$\mathbf{3.34} \times 10^{1}$	3 sf
0.**6600**	$\mathbf{6.600} \times 10^{-1}$	4 sf
$0.\mathbf{044} \times 10^{-2}$	$\mathbf{4.4} \times 10^{-4}$	2 sf

Note, in the last answer, that a number must be written in *standard* scientific notation (as a number between one and ten times a power of 10) to directly give the correct number of significant figures; 0.044×10^{-2} is not in standard scientific notation because 0.044 is not between 1 and 10.

EXERCISE How many significant figures are there in each of the following numbers? 350, 0.00431, 3.005

ANSWER: 3, 3, and 4 sf

KEY CONCEPT C Precision

The **precision** of a number is a measure of its uncertainty. A number with a large relative uncertainty is said to be imprecise; one with a small relative uncertainty is said to be precise—that is, it has been "well characterized." A number with high precision contains a large number of significant figures, one with low precision a small number of significant figures. Assume we measure the volume of a sample of water with two different instruments. One instrument, calibrated in 1-mL increments, gives us 21 mL for the volume; the other, calibrated in 0.01-mL increments, gives us 20.88 mL. The uncertainty in the first measurement is ± 1 mL whereas the uncertainty in the second is ± 0.01 mL. The second measurement has a smaller uncertainty and is therefore more precise.

EXAMPLE Deciding which of two numbers is more precise

The diameter of an iron nail is measured with vernier calipers and with a micrometer to be 2.2 mm and 2.167 mm, respectively. Which measurement is more precise? Do the measurements agree or disagree?

SOLUTION In general, the number with higher precision has more significant figures. On this basis, 2.167 mm is a more precise measurement than 2.2 mm. In agreement with this idea, we note that the value 2.2 mm has an uncertainty of ± 0.1 mm and the value 2.167 mm an uncertainty of ± 0.001, indicating that the value 2.167 mm is much more precise than the value 2.2 mm. Regarding the agreement of the two values, we note that a reported value of 2.2 mm means that the correct value lies between 2.1 mm and 2.3 mm. The more precise measurement is also between 2.1 mm and 2.3 mm, so the two measurements agree nicely.

EXERCISE Which of the two measured values for the length of a pencil, 2×10^{1} cm and 15 cm, is more precise? Do the measurements agree?

ANSWER: 15 cm; yes

KEY CONCEPT D Accuracy

The **accuracy** of a measured number is determined solely by how close the measured value is to the correct value of the property. Accuracy and precision are different qualities of a measured value and should not be confused. The precision of a number relates to its uncertainty and is given by the number of significant figures, whereas the accuracy is concerned with how correct the measurement is. It is worth noting that a working scientist generally cannot determine the accuracy of a measurement at the time it is taken because the correct value is not usually known in advance of the experiment. In student laboratories, the correct answer is often known by an instructor, who can inform a student of the accuracy of the day's laboratory work.

EXAMPLE Determining accuracy

The actual mass of a piece of brass is known to be 16.3355 g. Of the three determinations—16.3336 g, 16.3387 g, and 16.3379 g—which is the least accurate?

SOLUTION The determination that is furthest from the correct value is the least accurate determination. We analyze the results by setting up a table showing how far each is from the correct mass.

Determination (g)	Deviation from correct value (g)
16.3336	0.0019 too low
16.3387	0.0032 too high
16.3379	0.0024 too high

The second determination, 16.3387 g, is the furthest from the correct value and is, therefore, the least accurate.

EXERCISE The distance between two lakes is 3.22 km. Of the three independent measurements for this distance—1.89 mile, 3.31 km, and 3160 m—which is the most accurate?

ANSWER: 3160 m

1.7 SIGNIFICANT FIGURES IN CALCULATIONS

KEY CONCEPT A Significant figures in addition and subtraction

Measured properties are frequently used to calculate other properties. How does the result of a calculation reflect the precision of the data used in the calculation? As we expect, the number of significant figures in the result depends on the significant figures in the data used in the calculation. For **addition** and **subtraction,** we express only the place values that are known for every number used in the calculation. If a place value is unknown for any number in the calculation, that place value cannot be known for the answer. For example, let us find the sum 2.233 + 10.11 + 5.6. This sum is indicated in the following standard manner except that unknown place values in the addends are denoted by question marks:

```
  2.233
 10.11?
  5.6??
 ------
 17.943   (sum before correct rounding off)
 17.9     (correct number of significant figures)
```

The question marks indicate that the thousandths place value is unknown for the number 10.11 and that the thousandths and hundredths are unknown for 5.6. Therefore, the hundredths and thousandths place values cannot be known in the answer, and the last two digits (43) must be rounded off. The correct answer is 17.9. Notice, in this case, that the answer has three significant figures, more than the two significant figures in 5.6 and less than the significant figures in either 2.233 or 10.11.

EXAMPLE Significant figures in a subtraction problem

Express 863 − 20 to the correct number of significant figures, given that 20 has one significant figure.

SOLUTION On the left we write the subtraction in the normal way. However, the number 20 has only one significant figure, so the zero in 20 represents an unknown place value. The subtraction is rewritten on the right, with a question mark in the units place value for 20 to emphasize that it is unknown.

```
  863                   863
 − 20                  − 2?
 ----                  ----
  843  (incorrect)      84?  (which rounds to 840)
```

Because the units place value is unknown for the number 20, we cannot express the units place value in the answer; so the 3 must be rounded off. This leaves 840 (or better, 8.4×10^2) as the correct answer.

EXERCISE Express 863 − 20 (with 2 significant figures in 20) to the correct number of significant figures.

ANSWER: 843

KEY CONCEPT B Significant figures in multiplication and division

The correct number of significant figures in the result of a **multiplication** or **division** problem is the same as the smallest number of significant figures present in the data used in the calculation. In the following calculation, the number of sigificant figures is indicated for each number used in the calculation. The smallest number of significant figures used is two (for 4.2) so the answer must be expressed to two significant figures.

$$\frac{\underset{(3\text{ sf})}{6.78} \times \underset{(2\text{ sf})}{4.2} \times \underset{(4\text{ sf})}{5.115}}{\underset{(4\text{ sf})}{10.22} \times \underset{(3\text{ sf})}{5.58}} = 2.4572296 \quad \text{(calculator answer)}$$

$$= 2.5 \quad \text{(only two significant figures allowed)}$$

EXAMPLE Significant figures in a combined problem

Give the answer to the correct number of significant figures: $(61.82 \times 0.0212)/1.5 = ?$

SOLUTION The number of significant figures allowed in the answer is equal to the smallest number of significant figures in the numbers used in the calculation. Thus, we count the significant figures in each of the numbers: 61.82 (4 sf), 0.0212 (3 sf), and 1.5 (2 sf). The smallest number of significant figures used in the calculation is two, for the number 1.5. Therefore, the answer should be expressed to two significant figures.

$$\frac{61.82 \times 0.0212}{1.5} = 0.873723 \quad \text{(calculator answer)}$$

$$= 0.87 \quad \text{(two significant figures allowed)}$$

EXERCISE What is $(3.50 \times 1.234)/113.3$?

ANSWER: 0.0381

1.8 MASS PERCENTAGE COMPOSITION

KEY CONCEPT Mass percentage

Before discussing mass percentage, we shall briefly review how percentages work. Let's assume we have a combination of three components A, B, and C. We can represent this combination with a pie chart:

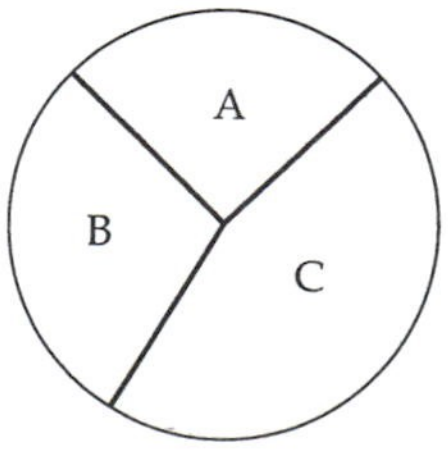

A

B

C

By definition, the percentage of any component is given by

$$\text{Percentage} = \frac{\text{piece of pie}}{\text{whole pie}} \times 100$$

For any component, A, for instance,

$$\text{Percentage A} = \frac{\text{amount of A}}{\text{amount of A} + \text{amount of B} + \text{amount of C}} \times 100$$

Chemists usually base percentage calculations on the mass of each component. For example, a typical piece of 14-carat gold jewelry with a mass of 3.672 g might contain 2.140 g gold (Au), 0.661 g copper (Cu), and 0.871 g silver (Ag). The following pie chart represents the composition of this item of jewelry:

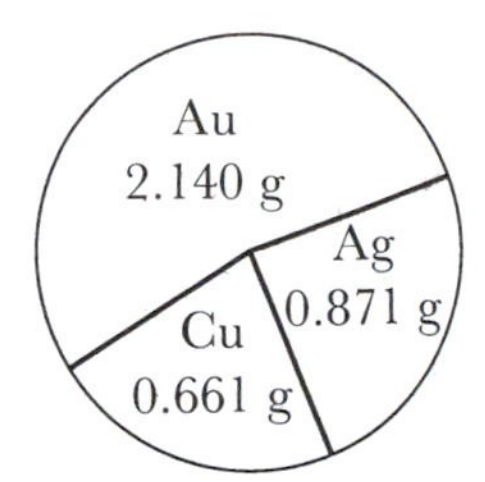

The mass percentage of each component, which means the percentage of each based on the mass present, is calculated by the formulation given earlier:

$$\text{Mass percentage gold} = \frac{\text{mass gold}}{\text{total mass}} \times 100 = \frac{2.140\ \cancel{g}}{3.672\ \cancel{g}} \times 100 = 58.28\%$$

$$\text{Mass percentage copper} = \frac{\text{mass copper}}{\text{total mass}} \times 100 = \frac{0.661\ \cancel{g}}{3.672\ \cancel{g}} \times 100 = 18.0\%$$

$$\text{Mass percentage silver} = \frac{\text{mass silver}}{\text{total mass}} \times 100 = \frac{0.871\ \cancel{g}}{3.672\ \cancel{g}} \times 100 = 23.7\%$$

▼ EXAMPLE 1 Calculating the percentage composition of a mixture

A sample of dry air is analyzed, with the following results:

nitrogen	19.5 g
oxygen	5.2 g
other	0.3 g

Calculate the mass percentage composition of air.

SOLUTION The mass percentage is calculated by taking the "piece over the whole" and multiplying by 100. More specifically, we take the mass of each component, divide it by the total mass of the sample, and multiply the result by 100. The total mass of the sample is 19.5 + 5.2 + 0.3 = 25.0 g. Thus, the mass percentage of each component of air is

$$\text{Mass percentage nitrogen} = \frac{\text{mass nitrogen}}{\text{total mass of sample}} \times 100 = \frac{19.5\ \cancel{g}}{25.0\ \cancel{g}} \times 100 = 78.0\%$$

$$\text{Mass percentage oxygen} = \frac{\text{mass oxygen}}{\text{total mass of sample}} \times 100 = \frac{5.2\ \cancel{g}}{25.0\ \cancel{g}} \times 100 = 21\%$$

$$\text{Mass percentage other} = \frac{\text{mass of other gases}}{\text{total mass of sample}} \times 100 = \frac{0.3\ \cancel{g}}{25.0\ \cancel{g}} \times 100 = 1\%$$

EXERCISE A sample of the solids in whole milk is analyzed and found to contain 3.31 g of protein, 4.09 g of fat, and 4.92 g of carbohydrate. What is the mass percentage composition of the solids in whole milk?

ANSWER: 26.9% protein, 33.2% fat, 39.9% carbohydrate

EXAMPLE 2 Calculating the mass of a sample component

A sample of solder containing 3.5 g of lead contains 25% lead. What is the mass of the sample of solder?

SOLUTION The percentage lead in the solder is given by

$$\text{Percentage lead} = \frac{\text{mass lead}}{\text{mass solder}} \times 100$$

Substituting mass lead = 3.5 g and percentage lead = 25% gives

$$25 = \frac{3.5\text{ g}}{\text{mass solder}} \times 100$$

Multiplying both sides of the equation by the mass solder and dividing both sides by 25 isolates the unknown on one side of the equation:

$$\text{Mass solder} = \frac{3.5\text{ g}}{25} \times 100 = 14\text{ g}$$

EXERCISE A sample of iron ore containing 26.2 g of iron is 31.2% iron. What is the mass of the sample of ore?

ANSWER: 84.0 g

DESCRIPTIVE CHEMISTRY TO REMEMBER

- The flammable gas **methane** is the main component of natural gas.
- The only elements that are liquid at room temperature are **mercury** and **bromine.**
- Seawater contains **sodium chloride.**
- Air is a mixture of **nitrogen, oxygen, carbon dioxide,** and other colorless gases.
- Helium, neon and argon are three **noble gases.**
- At room temperature and atmospheric pressure, solid carbon dioxide ("dry ice") undergoes **sublimation.**
- **Electrolysis** is the only method available for producing the gas fluorine.
- **Water** freezes at 0°C and boils at 100°C.

MATHEMATICAL EQUATIONS TO KNOW AND UNDERSTAND

$$°\text{F} = 32 + \left(\frac{9}{5}\right)°\text{C}$$ conversion from °C to °F

$$°\text{C} = \left(\frac{5}{9}\right)(°\text{F} - 32)$$ conversion from °F to °C

$$\text{K} = 273.15 + °\text{C}$$ relation between kelvins and °C

$$d = \frac{m}{V}$$ definition of density

$$\text{Mass \% A} = \frac{\text{mass of A in a sample}}{\text{total mass of sample}} \times 100$$ mass percentage component in a mixture

SELF-TEST EXERCISES

The properties of substances

1. Which of the following is a physical property of water? (All statements are correct.)
(a) Water reacts violently with potassium.
(b) Water is a solid at $-20°C$.
(c) Water is necessary for life.
(d) Water does not react with xenon.

2. Which of the following describes a solid?
(a) fluid
(b) takes the shape of the part of the container it fills
(c) easily compressed
(d) fills the container it occupies
(e) shape is independent of the container it is in

3. Which of the following states a chemical property of methane? (All statements are correct.)
(a) Methane is flammable.
(b) Methane is a gas at room temperature.
(c) Methane is colorless.
(d) Methane is insoluble in water.

4. One of the following statements regarding the melting and boiling of substances is incorrect. Which one?
(a) The melting temperature of a substance is characteristic of the substance.
(b) The temperature remains constant during boiling.
(c) The change from solid to liquid occurs over a wide temperature range.
(d) The melting temperature and the freezing temperature of a substance are identical.

5. Chromatography is used to separate substance A into substances B and C. Electrolysis is used to separate substance B into substances D and E. All attempts to chemically separate D into simpler substances fail, but E is successfully decomposed into simpler substances by heating to 328°C. Which of the following is correct?
(a) A is a mixture, B and E are compounds, and D is an element.
(b) A is a compound, B and E are compounds, and D is an element.
(c) A, B, and E are compounds; C and D are elements.
(d) A is an element, B and E are compounds, and C and D are elements.
(e) A, B, and C are compounds and D and E are elements.

6. Which of the following is a substance?
(a) milk (b) air (c) concrete (d) steel (e) water

7. Which of the following is a mixture?
(a) table salt (NaCl) (b) sugar (c) aluminum (d) vinegar (e) oxygen

8. Which of the following is a homogenous mixture?
(a) blood (b) sugar solution (c) concrete (d) granite (e) milk

9. Which method of separation depends on differences in volatility?
(a) filtration (b) centrifugation (c) chromatography
(d) distillation (e) recrystallization

Measurements and units

10. Which of the following describes the quantity of matter in a sample?
(a) weight (b) volume (c) mass (d) temperature (e) length

11. How many kilograms are in 675 g?
(a) 6.75 (b) 67.5 (c) 0.00675 (d) 0.675 (e) 675,000

12. How many milligrams are in 1.25 g?
(a) 1250 (b) 0.00125 (c) 0.125 (d) 125 (e) 12.5

13. What unit of length is most convenient for measuring the height of a human?
(a) millimeter (b) centimeter (c) meter (d) kilometer (e) megameter

14. The prefix "nano" corresponds to what multiplier?
(a) 10^{-3} (b) 10^{3} (c) 10^{-2} (d) 10^{-6} (e) 10^{-9}

15. Convert 36°F to degrees Celsius.
(a) −12°C (b) 2.2°C (c) 7.2°C (d) 61°C (e) 99°C

16. Convert −8°C to degrees Fahrenheit.
(a) −22°F (b) 28°F (c) −18°F (d) 18°F (e) −13°F

17. Convert 31°C to kelvins.
(a) 304 K (b) 88 K (c) −0.55 K (d) 63 K (e) 242 K

18. A volatile substance
(a) has a low boiling point. (b) is very flammable. (c) has a strange odor.
(d) is insoluble in water. (e) is toxic.

19. What is the volume (in cubic meters) of a rectangular solid 1.2 m long, 158 mm deep, and 62 cm wide?
(a) 1.2×10^{4} m^3 (b) 1.2 m^3 (c) 0.12 m^3 (d) 1.2×10^{3} m^3 (e) 12 m^3

20. How many milliliters are there in 0.35 L?
(a) 0.00035 mL (b) 35 mL (c) 0.0035 mL (d) 0.035 mL (e) 3.5×10^{2} mL

21. A sample of metal has a mass of 63.22 g and a volume of 8.89 cm^3. What is its density?
(a) 562 g/cm^3 (b) 14.1 g/cm^3 (c) 0.00178 g/cm^3
(d) 7.11 g/cm^3 (e) 0.141 g/cm^3

22. What is the volume of a 15.65-g piece of wood with a density of 0.857 g/cm^3?
(a) 7.89 cm^3 (b) 18.3 cm^3 (c) 16.5 cm^3 (d) 12.5 cm^3 (e) 54.6 cm^3

23. The density of aluminum is 2.70 g/cm^3. What is the mass of a 252-cm^3 sample of aluminum?
(a) 10.7 g (b) 680 g (c) 392 g (d) 1.47 g (e) 93.3 g

24. Which of the following properties of a sample of lead is an extensive property?
(a) Its density is 11.4 g/cm^3. (b) It is gray. (c) Its mass is 65.32 g.
(d) It is an electrical conductor. (e) It is pliable and soft.

Using measurements

25. 15.2 oz =
(a) 2.32 g (b) 1.87 g (c) 0.536 g (d) 477 g (e) 431 g

26. 52 km =
(a) 62 mile (b) 84 mile (c) 3.2×10^{4} mile (d) 32 mile (e) 8.4×10^{4} mile

27. Convert 8.2 gal to microliters
(a) 3.2×10^{-7} μl (b) 3.2×10^{-8} μl (c) 3.1×10^{-6} μl (d) 3.1×10^{7} μl
(e) 3.1×10^{6} μl

28. Convert 0.00421 mi to centimeters.
(a) 6.77 cm (b) 678 cm (c) 382 cm (d) 1.48 cm (e) 148 cm

29. How many cubic kilometers of water are there in a lake that contains 2.0 mi^3 of water?
(a) 8.3 km^3 (b) 66 km^3 (c) 3.2 km^3 (d) 33 km^3 (e) 5.0 km^3

30. What is the density of iron (7.86 g/cm^3) in grams per tablespoon (g/tbl), where 1 tbl = 4.93 mL?
(a) 0.626 g/tbl (b) 24.1 g/tbl (c) 1.60 g/tbl (d) 0.0258 g/tbl (e) 38.8 g/tbl

31. Convert 30 mi/hr to feet per second.
(a) 20 ft/s (b) 57 ft/s (c) 44 ft/s (d) 30 ft/s (e) 81 ft/s

32. Which of the following values for the mass of a dime is most precise?
(a) 2 g (b) 2.3 g (c) 2.28 g (d) 2.278 g (e) 2.2777 g

33. A digital watch shows the time to be 8:05.22 A.M. when it is actually 8:10.22 A.M. An analog watch gives the time as 8:09 A.M. Which of the following is correct?
(a) analog more accurate and precise
(b) analog more precise, digital more accurate
(c) analog more accurate, digital more precise
(d) digital more accurate and precise

34. Which of the following describes a systematic error in an experiment?
(a) Temperature fluctuations cause random differences in a balance reading.
(b) Vibrations in a building cause a meter to jiggle around the correct reading.
(c) A miscalibrated thermometer consistently reads 2°C low.
(d) Human judgment in estimating a measurement causes some uncertainty in the last digit.

35. How many significant figures are in the number 9.225?
(a) 4 (b) 3 (c) 2 (d) 1 (e) 0

36. How many significant figures are in the number 1680?
(a) 4 (b) 3 (c) 2 (d) 1 (e) 0

37. How many significant figures are in the number 0.0552?
(a) 3 (b) 4 (c) 0 (d) 5 (e) 2

38. What is 0.00950 in standard scientific notation?
(a) 950 (b) 0.950×10^{-2} (c) 9.50×10^{3} (d) 9.5×10^{-3} (e) 9.50×10^{-3}

39. What is 6.200×10^{2} in decimal notation?
(a) 62 (b) 62,000 (c) 620 (d) 6200 (e) 620.0

40. 2.235 + .01 =
(a) 2 (b) 2.245 (c) 2.250 (d) 2.25 (e) 2.24

41. 2.235 × .01 =
(a) 0.02 (b) 0.022 (c) 0.0223 (d) 0.0224 (e) 0.02235

42. (10.24 − 0.11) × 6.55 =
(a) 66.3515 (b) 67 (c) 66.3 (d) 66.4 (e) 66

43. It is possible to dissolve 35.9 g of sodium chloride (NaCl) in 100 g of water at 30 °C; no more will go into solution. What is the mass percentage composition of this solution?
(a) 56.0% NaCl, 44.0% H_2O
(b) 35.9% NaCl, 100% H_2O
(c) 26.4% NaCl, 73.6% H_2O
(d) 35.9% NaCl, 64.1% H_2O
(e) 0% NaCl, 100% H_2O

44. An unknown mixture in an experiment is found to contain 0.653 g magnesium, 0.151 g tin and 0.448 g aluminum. What is the mass percentage of tin in the sample?
(a) 25.2% (b) 12.1% (c) 15.1% (d) 33.4% (e) 13.7%

45. A gold ore is analyzed and found to contain 0.0021% gold. How much gold can be extracted from 1000 kg of the ore?
(a) 2.1 kg (b) 21 g (c) 0.21 kg (d) 210 g (e) 0.021 g

46. A certain breakfast cereal is known to contain 32% sugar. How many grams of sugar are present in an average serving of 28 g of the cereal? (The answer corresponds to about a teaspoon of sugar per serving.)
(a) 2.8 g (b) 4.7 g (c) 14 g (d) 9.0 g (e) 3.2 g

47. A sample of solder containing 35% tin is analyzed and found to contain 2.6 g tin. What is the mass of the sample?
(a) 13.5 g (b) 2.6 g (c) 7.4 g (d) 260 g (e) 6.5 g

Descriptive chemistry

48. What is the principal component of natural gas?
(a) methane (b) oxygen (c) nitrogen (d) acetylene (e) carbon dioxide

49. What is the principal component of air?
(a) oxygen (b) carbon dioxide (c) methane (d) nitrogen (e) water vapor

2 THE COMPOSITION OF MATTER

Experience shows that all matter is made from approximately 100 basic building blocks called elements. These elements can combine to form the millions of substances known to science, but an element does not change to another element in a normal chemical process.

ELEMENTS

KEY WORDS

Define or explain each of the following terms in a written sentence or two.

alkali metals
alkaline earths
coinage metals
congeners
halogens
metal
noble gases
nonmetal
periods
transition metals

2.1 THE NAMES AND SYMBOLS OF THE ELEMENTS

KEY CONCEPT Names and symbols of the elements

Chemists (and chemistry students) must be able to write the **symbols** of the elements in their day-to-day work. When writing the symbols that have two letters, remember that the first letter of an elemental symbol is always uppercase and the second lowercase. Thus, for mercury, Hg is the only correct symbol. In some cases, the use of only uppercase letters can lead to an erroneous representation, because a symbol improperly written in uppercase can be confused with the symbols for two elements.

EXAMPLE Writing elemental symbols properly

Find four examples in which writing an elemental symbol in uppercase letters could lead to confusion.

SOLUTION Chemists expect an elemental symbol to be an uppercase letter if the symbol has only one letter and to be an uppercase followed by a lowercase letter if the symbol has two letters. Thus, two uppercase letters look like adjacent symbols for two elements. To answer the question, we must look at the periodic table for symbols that have two letters, each of which corresponds to the symbol for an element.

Symbol for one element	Possible confusing misrepresentation
Sc scandium	SC sulfur, carbon
Hf hafnium	HF hydrogen, fluorine
Ni nickel	NI nitrogen, iodine
Pu plutonium	PU phosphorus, uranium

EXERCISE Formulate a situation in which writing the symbol of an element in lowercase letters may cause confusion.

ANSWER: NO intended (combination of nitrogen and oxygen), but No written (element nobelium).

2.2 THE PERIODIC TABLE

KEY CONCEPT A Groups and periods in the periodic table

The periodic table is a map of the elements, arranged, among other things (see Chapter 7), according to certain of their properties. It is divided into major subdivisions called groups and periods. A **period** is a horizontal row of elements; there are seven periods in the periodic table. Some of the elements in the sixth and seventh periods are placed, for convenience, under the main part of the table instead of in the main part where they actually belong. For instance, tracing across the sixth period, we start at cesium (Cs, $Z = 55$), move across to barium (Ba, $Z = 56$), down to lanthanum (La, $Z = 57$), across to Ytterbium (Yb, $Z = 70$), up to lutetium (Lu, $Z = 71$), and finally, across to the end of the period at xenon (Xe, $Z = 54$). The **groups** (also called **families**) are the vertical columns in the periodic table. Although there are many groups, we are concerned principally with the eight groups labeled I through VIII on the periodic table in the text. The elements in a group, called **congeners,** frequently show similarities in chemical and physical properties. In addition, it is common to observe a smooth change in properties as we scan down a group. This smooth change is evident in the following examples (two entries are left blank intentionally; hydrogen is not included because it is a nonmetal).

Group I	Melting point (K)	Group VIII	Boiling point (K)
Li	454	He	4
Na	371	Ne	27
K	?	Ar	87
Rb	337	Kr	?
Cs	312	Xe	166
Fr	302	Rn	211

Note the smooth change in melting points as we go down Group I and in boiling points as we go down Group VIII. However, the smoothness is *not inviolate;* it is merely a tendency, not a certainty.

EXAMPLE Using the gradation in properties of a group to estimate a property

Estimate the melting point of potassium (K).

SOLUTION To make the estimate, we depend on the regularity of changes in properties as we go down the elements in a group (and hope we have not encountered an exception). From the table, the melting point of potassium must lie between 371 K and 337 K. Without further information, the most reasonable approach would be to assume that the melting point is approximately halfway between 371 K and 337 K, or 354 K [$(371 + 337)/2 = 354$]. The actual melting point of potassium is 366 K. The estimate is remarkably good.

EXERCISE Estimate the boiling point of krypton (Kr).

ANSWER: 126 K estimate, 116 K actual

KEY CONCEPT B Metals and nonmetals in the periodic table

Most of the elements in the periodic table can be classified as either metals or nonmetals. Common knowledge tells us much about the differences between a metal and nonmetal. The following table illustrates a few of these differences.

Metals	Nonmetals
good electrical conductors	poor conductors or nonconductors
good heat conductors	poor heat conductors
lustrous (shiny)	dull in appearance
malleable, ductile	brittle
mostly solids	many gases

As an example, we compare a copper wire (a metal) with graphite (a nonmetal that is the main component of pencil lead). Copper possesses all of the properties of a metal, including being malleable (able to be shaped by hammering or bending) and ductile (able to be drawn into wire or hammered thin). Graphite posseses most of the typical properties of a nonmetal. Nonmetals are in the upper right of the periodic table; metals are lower and to the left (see Fig. 2.9 of the text).

EXAMPLE The location of metals and nonmetals in the periodic table

Use the periodic table to decide whether the elements oxygen (O) and iron (Fe) are metals or nonmetals. List three properties of each to verify your conclusion.

SOLUTION The position of an element in the periodic table indicates whether it is a metal or a nonmetal. Oxygen is high and toward the right, so we conclude it is a nonmetal. Iron is lower and in the center, but far to the left of the nonmetals, so we conclude it is a metal. A list of properties verifies this conclusion.

Oxygen	Iron
colorless	shiny
electrically nonconducting	good electrical conductor
gas	workable solid (can be bent, hammered)

EXERCISE Give the number of a group that has a nonmetal at the top of the group and a metal at the bottom.

ANSWER: IV, carbon (nonmetal) and lead (metal)

KEY CONCEPT C Dalton's atomic hypothesis

Imagine cutting a piece of copper in half, and then cutting it in half again and again until we get the smallest piece of matter that is still copper. This smallest piece is an atom of copper. Any further division of the atom into smaller pieces would result in matter that is no longer copper. John Dalton's atomic hypothesis starts with this concept of the atom, but its proposals regarding the nature of matter are much more specific.

1. All matter is composed of atoms.
2. All the atoms of a given element are identical. Embodied in this statement is the idea that atoms of one element are different from the atoms of any other element.
3. The atoms of different elements have different masses.
4. A compound is a specific combination of atoms of more than one element. (For instance, hydrogen chloride is a combination of hydrogen atoms and chlorine atoms, with a ratio of one hydrogen to each chlorine.)
5. In a chemical reaction, atoms are neither created nor destroyed but undergo a rearrangement to form new substances.

EXAMPLE Applying Dalton's atomic hypothesis

A sample of a type of atom is found to be magnetic; that is, the atoms respond to the presence of a nearby magnet. Another sample of atoms, under identical conditions, is not magnetic. Are these samples made up of atoms of the same element?

SOLUTION No. According to Dalton's hypothesis, all the atoms of a given element are identical. Because the two samples respond differently to a magnetic field, the atoms are not identical and, therefore, not of the same element.

EXERCISE The ratio of bromine atoms to calcium atoms in a 20-g sample of calcium bromide is found to be 2/1. What is the ratio in a 40-g sample of calcium bromide?

ANSWER: 2/1 (the same)

ATOMS

KEY WORDS Define or explain each of the following terms in a written sentence or two.

atomic mass unit
Avogadro's number
electron
isotope
mass number
molar mass
mole
neutron
nucleon
proton

2.3 THE NUCLEAR ATOM

KEY CONCEPT A The nuclear atom: Electrons, protons, and neutrons

The three **subatomic particles** in the atom are the electron, proton, and neutron. The **electron,** discovered by J. J. Thomson, is a negatively charged particle that is a substituent of all matter. Because all atoms have zero electrical charge, some positively charged particle must be present in matter to offset the negatively charged electron. The positive particle is called a **proton.** The electron has a charge of -1 and the proton $+1$; thus, the number of protons in an atom must be the same as the number of electrons for the atom to be electrically neutral. Rutherford established the location of the electrons and protons in the atom with a scattering experiment. The experiment consists of firing α particles at gold atoms and measuring how the particles deflect as they interact with the atoms. The results Rutherford observed were:

1. Most of the α particles went through the gold foil undeflected.
2. A few of the α particles were deflected a bit.
3. Very few α particles (about 1 in 20,000) were strongly deflected; some even bounced straight back.

We explain these results by assuming that a gold atom consists of an extremely small, heavy, positively charged nucleus surrounded by tenuous, light electrons. The accompanying figure shows the experiment schematically: the small black dots represent the massive, positively charged gold nuclei. α particles (which are positively charged) are repulsed by the positive charge of the nucleus.

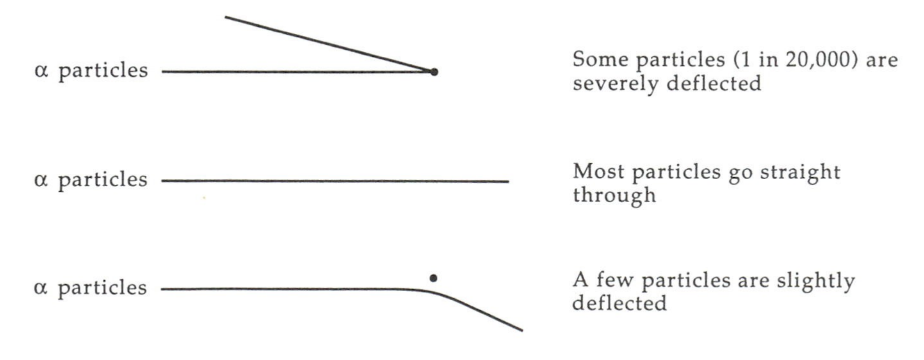

The **neutron,** which has zero charge and a mass close to that of the proton, completes the model of the **nuclear atom.** In this model, the positively charged protons and the electrically neutral neutrons, which are both massive, form the tiny nucleus. The nucleus is positively charged because the proton carries a positive charge. The electron which is a light particle, moves about the nucleus in such a way that most of the volume of the atom is empty space; the negative charge of the electrons offset the positive charge of the protons.

EXAMPLE The nuclear atom

The text states that the nucleus of a gold atom is a dense sphere about 10^{-14} m in diameter and that the diameter of the whole atom is about 10^{-10} m in diameter (10,000 times greater). Use the formula for the volume of a sphere, $V = \frac{4}{3}\pi r^3$, to calculate the percentage of the volume of the gold atom taken up by the nucleus. The r in the formula is the radius of the sphere.

SOLUTION We expect the nucleus to have a much smaller volume than the whole atom because its diameter is so much smaller than that of the atom. To get the percentage of the volume of the atom taken up by the nucleus, we use the following formula:

$$\text{Percentage volume occupied by nucleus} = \frac{\text{volume of nucleus}}{\text{volume of atom}} \times 100$$

For the nucleus, $r = 10^{-14}/2 \text{ m} = 5 \times 10^{-15}$ m, and

$$V = \frac{4}{3}\pi r^3 = \frac{4}{3}\pi(5 \times 10^{-15} \text{ m})^3 = \frac{4}{3}\pi(125 \times 10^{-45} \text{ m}^3) = 5 \times 10^{-43} \text{ m}^3$$

For the whole atom, $r = 10^{-10}/2 \text{ m} = 5 \times 10^{-11}$ m, and

$$V = \frac{4}{3}\pi r^3 = \frac{4}{3}\pi(5 \times 10^{-11} \text{ m})^3 = \frac{4}{3}\pi(125 \times 10^{-33} \text{ m}^3) = 5 \times 10^{-31} \text{ m}^3$$

Now, using the volumes just calculated, we can answer the question:

$$\text{Percentage volume occupied by nucleus} = \frac{5 \times 10^{-43} \cancel{\text{m}^3}}{5 \times 10^{-31} \cancel{\text{m}^3}} \times 100 = 1 \times 10^{-10}\ \%$$

This is 0.000 000 000 1% of the volume of the atom and confirms that only a tiny bit of the volume of the atom is taken up by the nucleus.

EXERCISE A gold atom has 79 protons and 79 electrons. Assume it has 118 neutrons. Use the data in Table 2.2 of the text to calculate the percentage of the mass of the atom that is in the nucleus. Is it correct to say that most of the mass of the atom is in the nucleus?

ANSWER: 99.98%; yes

KEY CONCEPT B Atomic number

The number of protons in an atom is called the **atomic number** of the atom. Because all atoms are electrically neutral (have zero electrical charge), the number of electrons in the atom equals the number of protons, and the atomic number equals the number of electrons for a neutral atom. The atomic number specifies the element that any atom corresponds to. For instance, every oxygen atom has an atomic number of 8, and any atom with an atomic number of 8 must be an oxygen atom. On the periodic table, the atomic number is the integer that accompanies each elemental symbol: 1 for hydrogen, 2 for helium, 92 for uranium. The atomic number is often symbolized by Z.

EXAMPLE 1 The protons and electrons in an atom

How many protons and how many electrons are present in a single atom of cadmium (Cd)?

SOLUTION Both the number of protons and the number of electrons in an atom are given by the atomic number. Cadmium is a transition metal in the fifth period; its atomic number is 48. Cadmium, therefore, has 48 protons and 48 electrons.

EXERCISE Give the name and symbol of the element that has 22 electrons? Is there more than one element with 22 electrons?

ANSWER: titanium, Ti; no

EXAMPLE 2 Using the atomic number

A 15-g crystal of sodium chloride contains 1.5×10^{23} atoms of sodium and an equal number of chlorine atoms. What is the mass of the electrons in the sample? Is it a large or small percentage of the total mass of the crystal?

SOLUTION The atomic numbers of sodium and chlorine are 11 and 17, respectively; so each sodium atom must contribute 11 electrons to the crystal, and each chlorine must contribute 17 electrons. Thus, we can calculate the total number of electrons in the crystal:

$$\text{Electrons from Na} = 1.5 \times 10^{23}\ \cancel{\text{atom Na}} \times \frac{11\ \text{electrons}}{1\ \cancel{\text{atom Na}}} = 1.7 \times 10^{24}\ \text{electrons}$$

$$\text{Electrons from Cl} = 1.5 \times 10^{23}\ \cancel{\text{atom Cl}} \times \frac{17\ \text{electrons}}{1\ \cancel{\text{atom Cl}}} = 2.6 \times 10^{24}\ \text{electrons}$$

$$\text{Total number of electrons} = 1.7 \times 10^{24}\ \text{electrons} + 2.6 \times 10^{24}\ \text{electrons} = 4.3 \times 10^{24}\ \text{electrons}$$

The mass of each electron is 9.11×10^{-28} g (Table 2.2 of the text), so

$$\text{Total mass of electrons} = 4.3 \times 10^{24}\ \cancel{\text{electrons}} \times \frac{9.11 \times 10^{-28}}{1\ \cancel{\text{electron}}} = 3.9 \times 10^{-3}\ \text{g}$$

The percentage of the mass of the crystal due to the electrons is

$$\text{Percentage} = \frac{\text{mass electrons} \times 100}{\text{total mass}} = \frac{3.9 \times 10^{-3}\ \text{g}}{15.0\ \text{g}} \times 100 = 2.6 \times 10^{-2}\%$$

The electrons, as expected, comprise a small percentage of the mass of the crystal.

EXERCISE A cup containing 250 g of water contains approximately 8.4×10^{24} atoms of oxygen and 1.7×10^{25} atoms of hydrogen. What is the mass of the electrons in the water? What percentage of the mass of the water is contributed by the electrons?

ANSWER: 7.7×10^{-2} g; 0.031%

2.4 THE MASSES OF ATOMS

KEY CONCEPT A Isotopes

The development of an instrument called a mass spectrometer enabled scientists to obtain very precise measurements of the masses of atoms. In a mass spectrometer, an atom is converted to a positive ion by removal of an electron. (A chemical species with a positive or negative charge is called in ion.) The ions are then separated according to their masses. The precise measurement of masses resulted in a completely unexpected discovery. For almost every element, all the atoms of the element do not have the same mass (contrary to Dalton's expectations). A variety of atoms exist with slightly different masses. For instance, with the element neon, three types of neon atoms can be characterized as shown in the table that follows:

Element	Number of protons	Mass (g)	Mass relative to hydrogen
neon-20	10	3.32×10^{-23}	20
neon-21	10	3.49×10^{-23}	21
neon-22	10	3.65×10^{-23}	22

Each of the three atoms shown has 10 protons, so each is a form of neon. Therefore, all are chemically equivalent and all are represented by the symbol Ne on the periodic table. Each has a slightly different mass, however. When the same element exists with different masses, the different forms are called **isotopes** of the element. In this case, three isotopes of neon exist: neon-20, neon-21, and neon-22.

EXAMPLE How much of the total mass in a neon-20 atom can be accounted for by the 10 protons and 10 electrons in the atom? Is there a significant mass left unaccounted for?

SOLUTION We know the total mass of a neon-20 atom and the individual masses of an electron and a proton (Table 2.2 of the text). Because we also know the number of electrons and protons in the atom, we can calculate how much mass they account for by multiplying the number of each by its mass. The mass unaccounted for is obtained by subtracting the total mass of the protons and electrons from the mass of the atom:

$$\text{Total mass of protons} = 10\ \cancel{\text{protons}} \times \frac{1.67 \times 10^{-24}\ \text{g}}{1\ \cancel{\text{proton}}} = 1.67 \times 10^{-23}\ \text{g}$$

$$\text{Total mass of electrons} = 10\ \cancel{\text{electrons}} \times \frac{9.11 \times 10^{-28}\ \text{g}}{1\ \cancel{\text{electron}}} = 9.11 \times 10^{-27}\ \text{g}$$

$$\text{Total mass accounted for} = 1.67 \times 10^{-23}\ \text{g} + 9.11 \times 10^{-27}\ \text{g} = 1.67 \times 10^{-23}\ \text{g}$$

The total mass unaccounted for equals the mass of the atom minus the mass of the electrons and protons:

$$\begin{aligned}\text{Unaccounted mass} &= (\text{mass of neon-20}) - (\text{mass of protons + electrons})\\ &= 3.32 \times 10^{-23}\ \text{g} - 1.67 \times 10^{-23}\ \text{g} = 1.65 \times 10^{-23}\ \text{g}\end{aligned}$$

A significant amount of mass, 49.7% of the mass of the atom, is left unaccounted for:

$$\frac{1.65 \times 10^{-23}\ \text{g}}{3.32 \times 10^{-23}\ \text{g}} \times 100 = 49.7\%$$

(Owing to effects to be described in Chapter 22, this calculation is an approximation.)

EXERCISE Assume that the difference in the masses of neon-20 and neon-21 is due to a single extra subatomic particle in the neon-21. What is the mass of the particle? Could it be any of the particles shown in Table 2.2 of the text?

ANSWER: 1.7×10^{-24} g; neutron (could not be proton because proton would change identity of element)

KEY CONCEPT B Atomic mass units

It would be inconvenient indeed to regularly use grams as the units of mass for atoms because all atoms have very small masses: we would be forced to use very small powers of 10 all the time. We solve this problem by defining a mass unit that results in convenient numbers for the masses of atoms, the atomic mass unit (u). Because the neutron and proton each have a mass of approximately 1 u, the mass number equals the approximate mass of an atom in atomic mass units. For example, chlorine-35 with mass number 35 has a mass of 34.97 u. As a final comment, we note from the definition of the atomic mass unit that

$$1\ \text{u} = 1.6605 \times 10^{-24}\ \text{g}$$

EXAMPLE Calculating the mass of an atom

The mass, in grams, of an atom of chlorine-35 is 5.807×10^{-23} g. Confirm that this mass corresponds to 34.97 u.

SOLUTION We have a relationship between atomic mass unit and gram, so it is possible to use the unit analysis method to solve the problem. From the relationship,

$$1 \text{ u} = 1.6605 \times 10^{-24} \text{ g}$$

we get two conversion factors,

$$\frac{1 \text{ u}}{1.6605 \times 10^{-24} \text{ g}} \quad \text{and} \quad \frac{1.6605 \times 10^{-24} \text{ g}}{1 \text{ u}}$$

We start with 5.807×10^{-23} g and multiply by the first conversion factor:

$$5.807 \times 10^{-23} \text{ g} \times \frac{1 \text{ u}}{1.6605 \times 10^{-24} \text{ g}} = 34.97 \text{ u}$$

EXERCISE What is the mass in atomic mass units of an atom of copper-62 with mass 1.03×10^{-22} g.

ANSWER: 62.0 u

PITFALL Mass number and atomic mass

Because the masses of a proton and neutron are each close to 1 u, the mass number and the atomic mass of an atom are nearly the same. However, the mass number is an integer (based on counting nucleons) and, therefore, has an unlimited number of significant figures; it also has no dimensions. The mass itself has dimensions of atomic mass units and is expressed with a specific number of significant figures.

KEY CONCEPT C Average mass of an element

An element is usually found in nature as a mixture of isotopes; so when a chemist deals with an element in a laboratory, more than one isotope is involved. The mass of an element as shown on the periodic table is the average of the atomic masses of the naturally occurring isotopes, based on the natural abundance of each isotope. To calculate the mass as it appears on the periodic table, we multiply the atomic mass of each isotope by the natural abundance of the isotope divided by 100 (natural abundance/100) and sum the results for all of the isotopes. This calculation is illustrated for chlorine, which consists of two naturally occurring isotopes, chlorine-35 and chlorine-37:

Isotope	Mass (u)	Natural abundance (%)
chlorine-35	34.97	75.77
chlorine-37	36.97	24.23

$$\text{Average mass of chlorine} = (\text{mass of chlorine-35}) \times \frac{(\text{natural abundance of chlorine-35})}{100} + (\text{mass of chlorine-37}) \times \frac{(\text{natural abundance of chlorine-37})}{100}$$

$$\text{Average mass of chlorine} = (34.97 \text{ u}) \times \frac{(75.77)}{100} + (36.97 \text{ u}) \times \frac{(24.23)}{100}$$

$$= (26.50 \text{ u}) + (8.96 \text{ u}) = 35.46 \text{ u}$$

The 100 is an exact number and so does not determine the number of significant figures in the answer. The number 35.453 u, the actual atomic weight of chlorine, agrees nicely with the 35.46 u we just calculated. Similar numbers in the periodic table indicate the atomic weights of all of the elements. These are the atomic weights that chemists use in day-to-day calculations in the laboratory since chemists generally deal with a natural mixture of isotopes.

EXAMPLE 1 Calculating the average mass of an element

Calculate the average mass of lithium from the data shown.

Isotope	Mass (u)	Natural abundance (%)
lithium-6	6.0151	7.42
lithium-7	7.0160	92.58

SOLUTION The average mass of an element is the average of the masses of the naturally occurring isotopes, averaged according to their natural abundances:

$$\begin{aligned}\text{Average mass of lithium} &= (\text{mass of lithium-6}) \times \frac{(\text{natural abundance of lithium-6})}{100} \\ &\quad + (\text{mass of lithium-7}) \times \frac{(\text{natural abundance of lithium-7})}{100} \\ &= (6.0151\ \text{u}) \times \frac{(7.42)}{100} + (7.0160\ \text{u}) \times \frac{(92.58)}{100} \\ &= 0.446\ \text{u} + 6.495\ \text{u} = 6.941\ \text{u}\end{aligned}$$

EXERCISE Calculate the average mass of boron from the data shown.

Isotope	Mass (u)	Natural abundance (%)
boron-10	10.0129	19.7
boron-11	11.0093	80.3

ANSWER: 10.8 u

EXAMPLE 2 Calculating the number of atoms in a sample

A 15.00-g sample of ammonia contains 2.663 g of hydrogen. How many hydrogen atoms are present in the sample?

SOLUTION The periodic table tells us that one hydrogen atom has mass 1.008 u. Because $1\ \text{u} = 1.6605 \times 10^{-24}\ \text{g}$, the logic flow of the problem is

$$2.663\ \text{g H} \rightarrow ?\ \text{u} \rightarrow \text{number of hydrogen atoms}$$

The mass of the hydrogen in atomic mass units is calculated first:

$$2.663\ \cancel{\text{g}} \times \frac{1\ \text{u}}{1.6605 \times 10^{-24}\ \cancel{\text{g}}} = 1.604 \times 10^{24}\ \text{u}$$

Because 1 hydrogen atom has a mass of 1.008 u, we can now determine the number of hydrogen atoms that must be present to give a mass of 1.604×10^{24} u:

$$1.604 \times 10^{24}\ \cancel{\text{u}} \times \frac{1\ \text{hydrogen atom}}{1.008\ \cancel{\text{u}}} = 1.591 \times 10^{24}\ \text{atoms H}$$

EXERCISE The same sample of ammonia contains 12.337 g of nitrogen. How many nitrogen atoms are present? What is the ratio of hydrogen to nitrogen in ammonia?

ANSWER: 5.303×10^{23} atoms N; 3:1

2.5 MOLES AND MOLAR MASS

KEY CONCEPT A The mole concept

The **mole** (mol) is formally defined as the number of atoms in exactly 12 g of carbon-12. This definition is rarely used, except to arrive at the conclusion that 1 mol of anything is 6.022×10^{23} of those things. The mole is often called the "chemist's dozen," and the idea behind it is exactly like that behind the dozen. A dozen is a certain number of objects (12) and a mole is a certain number of objects (6.022×10^{23}). The number 6.022×10^{23} is called **Avogadro's number.** The following comparisons illustrate the idea behind the mole:

$$1 \text{ dozen eggs} = 12 \text{ eggs}$$
$$1 \text{ gross pencils} = 144 \text{ pencils}$$
$$1 \text{ ream paper} = 500 \text{ sheets paper}$$
$$1 \text{ mol atoms} = 6.022 \times 10^{23} \text{ atoms}$$

The mole is a frequently used quantity, so it will be to your advantage to learn and understand how to use it. Avogadro's number, $N_A = 6.022 \times 10^{23}$, should be memorized.

EXAMPLE Using the mole

How many bromine atoms are there in 2.44 mol Br?

SOLUTION The relation between the number of bromine atoms and the moles of bromine atoms is given by the definition of the mole:

$$1 \text{ mol Br} = 6.022 \times 10^{23} \text{ Br atoms}$$

From this relationship, we can write two conversion factors:

$$\frac{1 \text{ mol Br}}{6.022 \times 10^{23} \text{ Br atoms}} \quad \text{and} \quad \frac{6.022 \times 10^{23} \text{ Br atoms}}{1 \text{ mol Br}}$$

The question requires that 2.44 mol of Br be converted to number of Br atoms. We use the second conversion factor so that the units "mol Br" cancel properly, leaving us with the desired units "Br atom":

$$2.44 \cancel{\text{mol Br}} \times \frac{6.022 \times 10^{23} \text{ Br atoms}}{1 \cancel{\text{mol Br}}} = 14.7 \times 10^{23} \text{ Br atoms} = 1.47 \times 10^{24} \text{ Br atoms}$$

One thing to look for in a problem such as this is thc reasonableness of the answer. Whenever we consider the number of atoms in a sample, we are likely to obtain a very, very large number: if we get a small number of atoms or a fraction of one atom, we can be certain that we have solved the problem incorrectly.

EXERCISE How many moles of mercury are present in 8.66×10^{22} Hg atoms?

ANSWER: 0.144 mol Hg

KEY CONCEPT B Molar mass

The molar mass of an atom is the mass of a sample of the atoms that contains one mole of atoms. Because of the way the mole is defined, the molar mass of any atom is the same number as the average mass of the atom but is expressed in grams rather than atomic mass units. The following table illustrates these relationships:

Element	Average mass (u)	Molar mass (g/mol)	Interpretation of molar mass
Fe	55.85	55.85	55.85 g Fe = 1 mol Fe = 6.022×10^{23} Fe atoms
P	30.97	30.97	30.97 g P = 1 mol P = 6.022×10^{23} P atoms
Au	197.0	197.0	197.0 g Au = 1 mol Au = 6.022×10^{23} Au atoms

The relationships expressed in this table allow for a number of types of calculations, as the following examples illustrate. These are some of the most important calculations you will do in this course and merit close attention.

EXAMPLE 1 Calculating the moles in a given mass

How many moles of iron are equivalent to 122 g Fe?

SOLUTION The molar mass of Fe is 55.85 g/mol. This tells us that 55.85 g Fe is equivalent to 1 mol Fe:

$$1 \text{ mol Fe} = 55.85 \text{ g Fe}$$

This relationship can be used to determine the moles of iron in any mass of iron. For our problem,

$$122 \not{g} \text{ Fe} \times \frac{1 \text{ mol Fe}}{55.85 \not{g}} = 2.18 \text{ mol Fe}$$

EXERCISE How many moles of argon atoms are there in 35.8 g Ar?

ANSWER: 0.896 mol Ar

EXAMPLE 2 Calculating the mass in a certain number of moles

How many kilograms of cadmium are present in 1.00×10^2 mol Cd?

SOLUTION Cadmium has an atomic mass of 112.4 u; its molar mass is 112.4 g/mol:

$$1 \text{ mol Cd} = 112.4 \text{ g}$$

Using unit analysis so that mol Cd cancels and g Cd remain as the units in the answer gives

$$1.00 \times 10^2 \cancel{\text{mol Cd}} \times \frac{112.4 \text{ g}}{1 \cancel{\text{mol Cd}}} = 1.12 \times 10^4 \text{ g}$$

The answer is converted to kilograms (1 kg = 1000 g):

$$1.12 \times 10^4 \not{g} \times \frac{1 \text{ kg}}{1000 \not{g}} = 11.2 \text{ kg Cd}$$

EXERCISE How many kilograms of iridium are present in 325 mol Ir?

ANSWER: 62.5 kg

EXAMPLE 3 Calculating the number of atoms in a given mass

How many atoms of nickel are present in 25.0 g Ni?

SOLUTION We have no direct connection between the number of atoms in a sample and the grams of sample. However, the moles of nickel can be calculated from the grams, and the number of nickel atoms from the moles:

$$\text{g Ni} \rightarrow \text{mol Ni} \rightarrow \text{number of nickel atoms}$$

We start by calculating the moles of nickel, using the molar mass of 58.71 g/mol:

$$25.0 \text{ g Ni} \times \frac{1 \text{ mol Ni}}{58.71 \text{ g Ni}} = 0.426 \text{ mol Ni}$$

The number of atoms of nickel can be calculated from the fact that 1 mol Ni = 6.022×10^{23} atoms Ni:

$$0.426 \text{ mol Ni} \times \frac{6.022 \times 10^{23} \text{ atoms Ni}}{1 \text{ mol Ni}} = 2.57 \times 10^{23} \text{ atoms Ni}$$

EXERCISE How many neon atoms are present in 58.2 g Ne?

ANSWER: 1.74×10^{24} atoms Ne

COMPOUNDS

KEY WORDS

Define or explain each of the following terms in a written sentence or two.

acid
anion
cation
formula unit
ionic compound
molecular compound
molecule
polyatomic ion

2.6 MOLECULES AND MOLECULAR COMPOUNDS

KEY CONCEPT A Ionic and molecular compounds

Most compounds can be classified as either molecular or ionic. A molecule is a unit of matter consisting of two or more atoms bonded together. A **molecular compound** is a compound made up of molecules. For instance, a water molecule consists of two hydrogen atoms bonded to one oxygen atom; it is a distinct, definite group of bonded atoms. A collection of water molecules (a glass of water, for instance) is like a bag of marbles; each water molecule is a distinct entity, just as each marble is. An **ion** is a charged chemical species. An **ionic compound** is a compound made up of a collection of oppositely charged ions positioned in a huge lattice. No ions are present in a molecular compound; molecules are not the principal structural unit in an ionic compound.

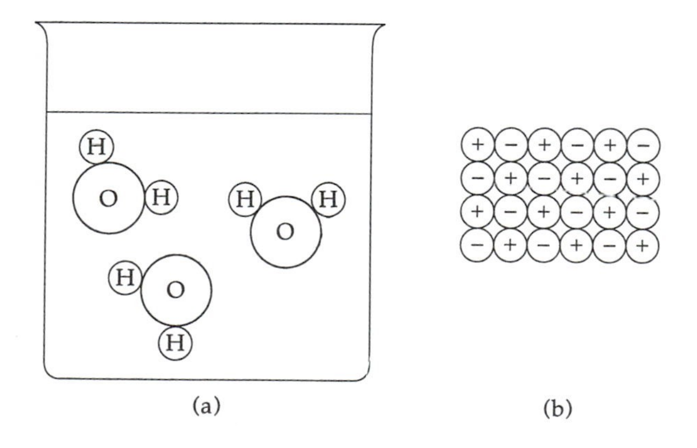

In the figure, (a) shows three discrete water molecules in a beaker of water; (b) shows a cross section of an ionic crystal, with a 1:1 ratio of positive to negatives ions. No distinct molecules are present in the ionic crystal lattice.

EXAMPLE Understanding the nature of a molecule

A picture of a model of ethanol is shown in the margin of the text on page 62. How many atoms of each kind are bound together to make up one ethanol molecule? Describe, in writing, which atoms are bonded to each other. What is the charge on this molecule?

SOLUTION The picture of ethanol in the margin shows two black spheres (representing two carbon atoms), six white spheres (representing six hydrogen atoms), and one red sphere (representing one oxygen atom). Thus, there are two carbons, six hydrogens, and one oxygen in ethanol. In the molecule, two carbons are bonded to each other, and the single oxygen is bonded to one of the carbons; five hydrogens are each bonded to carbon, and one hydrogen is bonded to oxygen. None of the hydrogens is bonded to another hydrogen. The molecule is made up of neutral atoms bound together, so it has zero charge.

EXERCISE A picture of a model of ammonia is shown in the margin of the text. Perform the same kind of analysis for ammonia as was just done for ethanol.

ANSWER: three hydrogens, one nitrogen present; the three hydrogens are all bonded to nitrogen, zero charge.

KEY CONCEPT B Molecules

A **molecular formula** symbolizes how many of each kind of atom join together to make a molecule. A water molecule is made up of two hydrogen atoms and one oxygen joined together; hence, the molecular formula of water is H_2O. The subscript 2 indicates two hydrogen atoms. When only one atom is present in a molecule, the subscript 1 is omitted from the formula; so water is written H_2O rather than H_2O_1. We should note that although a molecular formula gives the number of atoms of each kind that join together to form a molecule, it does not tell us how the atoms are linked.

EXAMPLE Interpreting a molecular formula

How many atoms of each kind are present in one molecule of sucrose, $C_{12}H_{22}O_{11}$?

SOLUTION The subscripts after each element symbol tell us how many atoms of that element are present in the formula. In this case, the subscript 12 following carbon tells us there are 12 carbon atoms; the subscript 22 after hydrogen indicates 22 hydrogen atoms; and the 11 after oxygen, 11 oxygen atoms.

EXERCISE How many atoms of each kind are present in one molecule of the Freon $C_2Cl_3F_3$?

ANSWER: 2 carbon, 3 chlorine, and 3 fluorine atoms

PITFALL Similar symbols in an element and a compound

Even though they may look the same, the H_2 symbol for the hydrogen molecule and the H_2 in H_2O do not mean exactly the same thing. In the hydrogen molecule, the two hydrogen atoms are joined to each other. In water, two hydrogen atoms are present, but they are not joined to each other. Each is linked to the centrally located oxygen atom. For any molecule, we must remember that the formula tells us which atoms are present but not how they are linked, and the same symbols in two different molecules may not represent exactly the same thing.

KEY CONCEPT C Empirical formulas

The **empirical formula** of a compound is a chemical formula that gives the simplest whole-number ratio of atoms in the compound. If we know the molecular formula or formula unit for a compound, we can determine the empirical formula by taking the relative numbers of atoms in the compound and reducing the relative numbers to the simplest whole-number ratio.

Different compounds can have the same empirical formula; for example, benzene (C_6H_6), acetylene (C_2H_2), and cyclobutadiene (C_4H_4) all have the empirical formula CH. Also, the correct molecular formula or formula unit for a compound may be the same as the empirical formula, as for water (H_2O) and calcium sulfate ($CaSO_4$).

EXAMPLE 1 Understanding empirical formulas

Give the empirical formulas for the following compounds: (a) anise alcohol, $C_8H_{10}O_2$; (b) aluminum chloride, $AlCl_3$.

SOLUTION (a) Here we are given the whole-number ratio of atoms in the compound, 8:10:2. To determine the empirical formula, we need the simplest whole-number ratio; we get it by dividing each of the numbers in the ratio 8:10:2 by the smallest of the numbers, 2, which gives 4:5:1. The empirical formula is therefore C_4H_5O. (b) The ratio of atoms in the formula unit (1:3) is already the simplest whole-number ratio, so the empirical formula is the same as the formula unit, $AlCl_3$.

EXERCISE Give the empirical formula of the following compounds: (a) ethylene glycol, $C_2H_6O_2$; (b) propylene glycol, $C_3H_8O_2$.

ANSWER: (a) CH_3O; (b) $C_3H_8O_2$

EXAMPLE 2 Determining an empirical formula from percentage composition

Tetroquinone has the following percentage composition: C, 41.87%; H, 2.340%; O, 55.78%. What is the empirical formula of tetroquinone?

SOLUTION The percentage composition gives the relative mass of each element present; we must convert to the relative moles of each element present. First, assume we are dealing with exactly 100 g of compound. (Any amount can be used, but using 100 g makes the calculation simpler because the mass of each element then equals the percentage of each.)

$$\text{g C} = 0.4187 \times 100 \text{ g} = 41.87 \text{ g}$$
$$\text{g H} = 0.02340 \times 100 \text{ g} = 2.340 \text{ g}$$
$$\text{g O} = 0.5578 \times 100 \text{ g} = 55.78 \text{ g}$$

We next convert the mass of each element to moles of each, using the molar mass of each element:

$$\text{mol C} = 41.87 \text{ g} \times \frac{1 \text{ mol C}}{12.01 \text{ g}} = 3.486 \text{ mol C}$$
$$\text{mol H} = 2.340 \text{ g} \times \frac{1 \text{ mol H}}{1.008 \text{ g}} = 2.321 \text{ mol H}$$
$$\text{mol O} = 55.78 \text{ g} \times \frac{1 \text{ mol O}}{16.00 \text{ g}} = 3.486 \text{ mol O}$$

The numbers 3.486, 2.321, and 3.486 give the ratio of moles of atoms. The empirical formula is the simplest whole-number ratio of moles of atoms, so the ratio 3.486:2.321:3.486 must be converted to a ratio of simple whole numbers. To do this, we divide by the smallest of the numbers in the ratio:

$$\text{C:} \quad \frac{3.486}{2.321} = 1.502$$
$$\text{H:} \quad \frac{2.321}{2.321} = 1.000$$
$$\text{O:} \quad \frac{3.486}{2.321} = 1.502$$

Multiplying each number in the ratio by 2 results in the desired simplest whole-number ratio of 3:2:3.

$$\begin{aligned} C&: \ 2 \times 1.502 = 3.004 \\ H&: \ 2 \times 1.000 = 2.000 \\ O&: \ 2 \times 1.502 = 3.004 \end{aligned}$$

The empirical formula of the compound is $C_3H_2O_3$.

EXERCISE The percentage composition of dibromoanthracene is C, 50.04%; H, 2.400%; Br, 47.56%. What is the empirical formula of dibromoanthracene?

ANSWER: C_7H_4Br

KEY CONCEPT D The molar mass of compounds

The **molar mass** is the mass in 1 mol of a compound. It is equal to the sum of the molar masses of the atoms in the compound. (For elements that exist as molecules, the molar mass is the sum of the molar masses in a molecule of the element.) The molar mass is calculated by adding the molar masses of all the atoms in the molecule. For instance, C_2H_6O (ethanol) contains 2 carbon atoms, 6 hydrogen atoms, and 1 oxygen atom. Its molar mass is the sum of the molar masses of these atoms:

$$\begin{aligned} 2\ C&: \ 2 \times 12.01 \text{ g/mol} = 24.02 \text{ g/mol} \\ 6\ H&: \ 6 \times 1.008 \text{ g/mol} = 6.048 \text{ g/mol} \\ 1\ O&: \ 1 \times 16.00 \text{ g/mol} = \underline{16.00 \text{ g/mol}} \\ & \qquad 46.068 \text{ g/mol} = 46.07 \text{ g/mol} \end{aligned}$$

EXAMPLE Calculating and interpreting molar masses

Alanine, an important biological compound, has the molecular formula $C_3H_7NO_2$. What is the molar mass of alanine? How many molecules are present in a molar mass of alanine?

SOLUTION To calculate the molar mass of a compound, we must sum the molar masses of all of the atoms in the compound. Alanine contains 3 carbon atoms, 7 hydrogen atoms, 1 nitrogen atom, and 2 oxygen atoms. Thus, we obtain

$$\begin{aligned} 3\ C&: \ 3 \times 12.01 \text{ g/mol} = 36.03 \text{ g/mol} \\ 7\ H&: \ 7 \times 1.008 \text{ g/mol} = 7.056 \text{ g/mol} \\ 1\ N&: \ 1 \times 14.01 \text{ g/mol} = 14.01 \text{ g/mol} \\ 2\ O&: \ 2 \times 16.00 \text{ g/mol} = \underline{32.00 \text{ g/mol}} \\ & \qquad 89.10 \text{ g/mol} \end{aligned}$$

The molar mass of a compound is the mass of 1 mol. Thus, 89.10 g of alanine contains 6.022×10^{23} molecules of alanine.

EXERCISE Calculate the molar mass of each of the following and state the number of molecules of each in one molar mass: (a) HNO_3 (nitric acid); (b) O_2 (oxygen).

ANSWER: (a) 63.01 g/mol, 6.022×10^{23} HNO_3 molecules; (b) 32.00 g/mol, 6.022×10^{23} O_2 molecules

KEY CONCEPT E Empirical formulas to molecular formulas

The molecular formula of a compound can be determined if its molar mass and empirical formula are both known. For instance, consider compounds with the empirical formula CH_2O. The simplest compound with the empirical formula CH_2O has the molecular formula CH_2O; the molar mass equals the empirical formula mass, 30.02 g/mol. The next compound with the same empirical formula contains the equivalent of not one, but two empirical formula units. By this we do not mean that two empirical units are bound together, but rather that the compound has twice the number of each atom as the empirical formula unit has. The compound, therefore, has 2 carbon atoms, 4 hydrogen atoms, and 2 oxygen atoms bound together. Its molecular formula is $C_2H_4O_2$, because the equivalent of two empirical formula units is present, its molar mass (60.04 g/mol) must be twice the empirical formula mass (32.02 g/mol). This result can be extended to compounds with three, four, or more empirical formula units. It becomes evident that since the molecular formula contains an integer number of empirical formula units, the molecular mass must be an integer multiple of the empirical formula mass. The table that follows summarizes the results for the empirical formula CH_2O up to the molecule $C_5H_{10}O_5$.

Molecular formula	Molar mass (g/mol)	Molar mass/empirical formula mass	Number of empirical formulas in the molecular formula
CH_2O	30.02	1	1
$C_2H_4O_2$	60.04	2	2
$C_3H_6O_3$	90.06	3	3
$C_4H_8O_4$	120.08	4	4
$C_5H_{10}O_5$	150.10	5	5

We see that by dividing the molar mass by the empirical formula mass, we find the number of empirical formulas that are contained in the molecular formula. Thus, if we know the empirical formula and the molar mass of a compound, we can determine the molecular formula.

EXAMPLE Determining the molecular formula from the empirical formula and the molar mass

A compound with the empirical formula C_5H_5O is found to have a molar mass of 324.36 g/mol. What is the molecular formula of the compound?

SOLUTION We must determine how many empirical formula units are contained in the molecular formula. The empirical formula mass of C_5H_5O is 81.09 g/mol. We divide the molar mass by the empirical formula mass to find the number of empirical units in the molecular formula:

$$\frac{324.36 \text{ g/mol}}{81.09 \text{ g/mol}} = 4.000 = 4 \quad \text{(remember, the answer must be an integer)}$$

The molecular formula contains four of the C_5H_5O empirical formula units, so the molecular formula must be $C_{20}H_{20}O_4$. A check shows the molar mass of $C_{20}H_{20}O_4$ to be 324.36 g/mol.

EXERCISE A compound with the empirical formula C_2H_2O has a molar mass of 210.18 g/mol. What is the molecular formula of the compound?

ANSWER: $C_{10}H_{10}O_5$

KEY CONCEPT F Using molar masses

The molar mass of any species relates the mass of a sample to the number of moles in the sample. Thus, we can use the molar mass to calculate the number of moles in a sample if we know the mass, or the mass in the sample if we know the number of moles. Because one mole of any sample contains 6.022×10^{23} of the species involved, the molar mass also provides an eventual connection between the mass of the sample and the number of molecules present:

Mass	$\xleftrightarrow{\text{molar mass}}$	Moles	$\xleftrightarrow{N_A = 6.02 \times 10^{23}}$	Number of molecules

The calculations here are the same as those we did earlier with the molar masses of atoms, except that molecules rather than atoms are involved.

EXAMPLE 1 Using the molar mass to calculate the moles and number of molecules

How many grams of C_3H_8O (isopropyl alcohol, the main component of rubbing alcohol) are present in 2.85 mol C_3H_8O? How many C_3H_8O molecules are present?

SOLUTION The problem asks for the number of grams in a given number of moles of a compound. The connection between the moles of a compound and the grams of a compound is given by the molar mass of the compound. For C_3H_8O, the molar mass is

$$\begin{array}{ll} 3\text{ C:} & 3 \times 12.01 = 36.03 \\ 8\text{ H:} & 8 \times 1.01 = 8.08 \\ 1\text{ O:} & 1 \times 16.00 = \underline{16.00} \\ & 60.11 \text{ g/mol} \end{array}$$

$$60.11 \text{ g } C_3H_8O = 1 \text{ mol } C_3H_8O$$

Two conversion factors result:

$$\frac{1 \text{ mol } C_3H_8O}{60.11 \text{ g } C_3H_8O} \quad \text{and} \quad \frac{60.11 \text{ g } C_3H_8O}{1 \text{ mol } C_3H_8O}$$

We multiply the given moles of by the correct conversion factor to get the mass:

$$2.85 \cancel{\text{mol } C_3H_8O} \times \frac{60.11 \text{ g } C_3H_8O}{1 \cancel{\text{mol } C_3H_8O}} = 171 \text{ g } C_3H_8O$$

For the second part of the problem, we use the relationship between moles of molecules and number of molecules:

$$1 \text{ mol } C_3H_8O = 6.022 \times 10^{23} \ C_3H_8O \text{ molecules}$$

Two conversion factors result:

$$\frac{1 \text{ mol } C_3H_8O}{6.022 \times 10^{23} \text{ molecules } C_3H_8O} \quad \text{and} \quad \frac{6.022 \times 10^{23} \text{ molecules } C_3H_8O}{1 \text{ mol } C_3H_8O}$$

Finally, we multiply the given number of moles by the correct conversion factor:

$$2.85 \cancel{\text{mol } C_3H_8O} \times \frac{6.022 \times 10^{23} \ C_3H_8O \text{ molecules}}{1 \cancel{\text{mol } C_3H_8O}} = 1.72 \times 10^{24} \text{ molecules } C_3H_8O$$

(Remember that $17.2 \times 10^{23} = 1.72 \times 10^{24}$.)

EXERCISE How many moles of $C_6H_4Cl_2$ (dichlorobenzene) are present in 125 g of $C_6H_4Cl_2$? How many $C_6H_4Cl_2$ molecules are present?

ANSWER: 0.850 mol $C_6H_4Cl_2$; 5.12×10^{23} $C_6H_4Cl_2$ molecules

EXAMPLE 2 Calculating the mass of a compound from the number of molecules

What is the mass of 5.63×10^{24} molecules HNO_3?

SOLUTION The calculational outline for this problem is

$$\text{Molecules } HNO_3 \rightarrow \text{mol } HNO_3 \rightarrow \text{g } HNO_3$$

The first step uses the definition of the mole (1 mol HNO_3 = 6.022×10^{23} molecules HNO_3).

$$5.63 \times 10^{24} \cancel{\text{molecules } HNO_3} \times \frac{1 \text{ mol } HNO_3}{6.022 \times 10^{23} \cancel{\text{molecules } HNO_3}} = 9.35 \text{ mol } HNO_3$$

The molar mass of HNO_3 is 63.02 g/mol:

$$\begin{array}{lll} 1 \text{ H:} & 1 \times 1.008 = & 1.008 \\ 1 \text{ N:} & 1 \times 14.01 = & 14.01 \\ 3 \text{ O:} & 3 \times 16.00 = & \underline{48.00} \\ & & 63.018 = 63.02 \text{ g/mol} \end{array}$$

This value enables us to convert from moles HNO_3 to grams HNO_3:

$$9.35 \cancel{\text{mol } HNO_3} \times \frac{63.02 \text{ g } HNO_3}{1 \cancel{\text{mol } HNO_3}} = 589 \text{ g } HNO_3$$

EXERCISE What is the mass (in grams) of 1.56×10^{22} molecules CCl_4?

ANSWER: 3.99 g

PITFALL Numerical errors in calculating molar masses

Chemistry students are frequently asked to calculate molar masses, and doing so will become almost second nature. As we have seen, determining the correct molar mass amounts to finding the sum of a few numbers; but as any accountant or bookkeeper knows, mistakes are often made in simple procedures. To avoid silly errors in the calculation of molar masses, *develop a formal written procedure for the calculation.* Do not do the sum in your head or just on a calculator: write it down! The text suggests one way of writing a molecular weight calculation:

$$\begin{aligned} \text{Molar mass } (C_2H_6O) &= [2 \times \text{molar mass (C)}] + [6 \times \text{molar mass (H)}] + [1 \times \text{molar mass (O)}] \\ &= (2 \times 12.01) + (6 \times 1.01) + 16.00 = 46.08 \text{ g/mol} \end{aligned}$$

This study guide has suggested an alternative:

$$\begin{array}{lll} 2 \text{ C:} & 2 \times 12.01 \text{ g/mol} = & 24.02 \text{ g/mol} \\ 6 \text{ H:} & 6 \times 1.008 \text{ g/mol} = & 6.048 \text{ g/mol} \\ 1 \text{ O:} & 1 \times 16.00 \text{ g/mol} = & \underline{16.00 \text{ g/mol}} \\ & & 46.068 \text{ g/mol} = 46.07 \text{ g/mol} \end{array}$$

Either of these methods, or one of your own design, may be used. Whatever method you choose, experience shows that writing down the steps in this calculation helps you to avoid mistakes.

2.7 IONS AND IONIC COMPOUNDS

KEY CONCEPT A Cations

A **cation** is an ion with a positive charge. Before proceeding, we make certain the origin of the positive charge is clear by examining the familiar sodium ion (Na^+). A sodium ion is formed by removing an electron from a sodium atom. If we account for the charges on the sodium atom before and after removal of the electron, we obtain the following:

Sodium atom (Na)			**Sodium ion (Na^+)**	
11 protons:	+11		11 protons:	+11
11 electrons:	−11	remove 1 electron →	10 electrons:	−10
net charge on atom:	0		net charge on ion:	+1

All positive ions, whether monatomic or not, have a positive charge for the same reason: they have an excess of protons over electrons. Table 2.4 of the text lists some important cations. From this table, we note the following:

1. For Groups I and II, the charge number of a cation is the same as the group number: for example, K^+ in Group I and Mg^{2+} in Group II.
2. For Groups III and IV, the maximum charge number of the cation is the same as the group number: for example, Tl^{3+} in Group III and Pb^{4+} in Group IV. Also, Al^{3+} and Ga^{3+} are the only common cations that Al and Ga form.
3. Some atoms, especially the transition metals, are able to form differently charged cations: for example, copper forms Cu^+ and Cu^{2+}.

EXAMPLE The charge on a cation

What is the charge number of the ion formed by radium, element number 88? State how many electrons the ion has.

SOLUTION Radium is a Group II element (check your periodic table.) We know that the Group II elements form ions with charge number +2 only, so the ion is Ra^{2+}. The ion is formed by loss of two electrons from the element, so it has 86 electrons ($88 - 2 = 86$).

EXERCISE Write the symbols of the two ions formed by tin, and state how many electrons each has.

ANSWER: Sn^{2+}, 48 electrons; Sn^{4+}, 46 electrons

KEY CONCEPT B Anions

An **anion** is a negatively charged ion. An anion is formed when an atom gains one or more electrons. Let us reconcile the charge on the chloride ion (Cl^-) as we did earlier for the sodium ion:

Chlorine atom (Cl)			**Chloride ion (Cl^-)**	
17 protons:	+17		17 protons:	+17
17 electrons:	−17	gain 1 electron →	18 electrons:	−18
net charge on atom:	0		net charge on ion:	−1

All anions have negative charge because they have an excess of electrons over protons. Table 2.5 and Appendix 3A of the text list some important anions; we note two facts from these tables:

1. Most of the simple anion names end with the suffix *-ide*. For some of the more common ions (carbide, nitride, phosphide, oxide, sulfide, fluoride, chloride, bromide, and iodide), this suffix is added to the first part of the name (the stem) of the element. For example, bromide ion is *brom* (first part of bromine) + *ide*.
2. The charge number of many of the simple anions (carbide, nitride, phosphide, oxide, sulfide, fluoride, chloride, bromide, and iodide) is equal to (group number) − 8. For example, the charge number on phosphide, a Group V ion, is $5 - 8 = -3$.

EXAMPLE The charge on an anion

What is the charge of the anion formed by selenium, number 34? How many electrons does the ion have?

SOLUTION We use the periodic table as a guide to determine the charge of the ion. Selenium is a Group VI atom. Thus, using the rule that the charge number on a simple anion equals (group number) −8, we expect the charge number on the selenium anion to be $6 - 8 = -2$. The ion is formed by addition of two electrons to a selenium ion, so it has 36 electrons ($34 + 2 = 36$).

EXERCISE Refer only to the periodic table to predict the charge number on the simple anion formed by nitrogen. Name the ion, calculate the number of electrons on it, and write its symbol.

ANSWER: -3; nitride; 10 electrons; N^{3-}

KEY CONCEPT C Polyatomic ions

Let us envision a group of atoms joined together to form a unit, but with extra electrons or too few electrons. Such a unit, which would have a negative charge (extra electrons) or positive charge (too few electrons), is called a **polyatomic ion.** A large number of common polyatomic anions contain one or more oxygen atoms and are, therefore, called **oxoanions.** Table 2.7 of the text shows some oxoanions.

EXAMPLE The origin of the charge on a polyatomic ion

What is the origin of the charge on the ammonium ion, NH_4^+?

SOLUTION The charge on an ion originates with an imbalance between the number of electrons and the number of protons in the ion. Nitrogen (atomic number 7) contributes 7 protons to the ion and each of the four hydrogen atoms (atomic number 1) contributes 1 proton; so 11 protons are present. For a overall charge of $+1$, 10 electrons must be present.

EXERCISE Account for the charge on the sulfate ion, SO_4^{2-}.

ANSWER: 48 protons $+50$ electrons result in -2 charge.

PITFALL The charge number on polyatomic ions

Some confusion occasionally arises regarding the charge number of the various oxoanions described in the text. In an oxoanion such as SO_4^{2-}, the subscript 4 has nothing directly to do with the charge on the anion; it refers to the number of oxygen atoms in the anion. The $2-$ superscript gives the correct charge number. In a similar vein, PO_4^{3-} has a -3 charge number, NO_3^- a -1 charge number, and ClO_2^- a -1 charge number.

KEY CONCEPT D The formula unit

When ions come together to form an ionic compound, molecules do not form. In an ionic crystal, the positive and negative ions form a huge, repeating lattice in which there are no distinct molecules. The **formula unit** is the formula that expresses the relative number of positive ions to negative ions in the lattice, usually in a form that gives the simplest ratio of positive to negative ions. We can derive the formula unit by using the fact that all salts are electrically neutral. We assure the electrical neutrality by writing a formula unit with zero charge, one in which the sum of the charges of all the ions equals zero. Consider magnesium chloride as an example. This ionic salt is made up of Mg^{2+} ions combined with Cl^- ions. To construct a formula unit with zero charge, we must combine two Cl^- ions with one Mg^{2+} ion to give the formula unit $MgCl_2$. The two chlorides contribute a charge of -2 and the magnesium a charge of $+2$, so the sum of charges is $-2 + 2 = 0$, as required by electrical neutrality.

EXAMPLE Predicting a formula unit

What is the correct formula for the formula unit of the salt composed of Na^+ (sodium ion) and CO_3^{2-} (carbonate ion)?

SOLUTION The criterion for writing a correct formula unit is that the charges of all the ions must sum to zero. The sodium ion has a +1 charge and carbonate ion a −2 charge. It takes two sodium ions, which together have a total charge number of +2 to cancel the −2 charge on the carbonate ion. The correct formula is Na_2CO_3, indicating that two Na^+ ions combine with one CO_3^{2-} ion to form the Na_2CO_3 formula unit.

EXERCISE What is the formula unit for the salt formed from the combination of Mg^{2+} ion with PO_4^{3-} ion?

ANSWER: $Mg_3(PO_4)_2$

PITFALL Symbols for elements and compounds

In a formula unit such as $MgCl_2$, the Cl_2 has a different meaning than the Cl_2 used as the symbol for the diatomic chlorine molecule: the Cl_2 in $MgCl_2$ indicates two Cl^- ions, which are part of the formula unit. In the diatomic Cl_2 molecule, there are no Cl^- ions; two chlorine atoms join together to form a neutral molecule.

KEY CONCEPT E Molar mass of an ionic compound

The **molar mass** of an ionic compound is the sum of the molar masses of the atoms in the formula unit. It is the mass of compound that contains 1 mol of formula units. For instance, for the molar mass of calcium sulfate ($CaSO_4$), we get

$$\begin{aligned}
1\ \text{Ca:}\quad & 1 \times 40.08\ \text{g/mol} = 40.08\ \text{g/mol}\\
1\ \text{S:}\quad & 1 \times 32.06\ \text{g/mol} = 32.06\ \text{g/mol}\\
4\ \text{O:}\quad & 4 \times 16.00\ \text{g/mol} = 64.00\ \text{g/mol}\\
& \overline{\quad 136.14\ \text{g/mol}}
\end{aligned}$$

The following table gives the interpretation of the molar mass for ionic compounds.

Compound	Molar mass (g/mol)	Interpretation of molar mass
$BaCO_3$	197.35	197.35 g $BaCO_3$ = 1 mol $BaCO_3$ = 6.022×10^{23} $BaCO_3$ formula units
$Ca(NO_3)_2$	164.09	164.09 g $Ca(NO_3)_2$ = 1 mol $Ca(NO_3)_2$ = 6.022×10^{23} $Ca(NO_3)_2$ formula units
$Fe_2(Cr_2O_7)_3$	759.66	759.66 g $Fe_2(Cr_2O_7)_3$ = $Fe_2(Cr_2O_7)_3$ = 6.022×10^{23} $Fe_2(Cr_2O_7)_3$ formula units

EXAMPLE 1 Calculating moles of an ionic compound from the mass

How many moles of $Sr(HSO_4)_2$ (strontium hydrogen sulfate) are present in 500 g? How many formula units are present?

SOLUTION The question asks for the moles of formula units in a given mass of a sample. The molar mass relates the mass of a sample to the number of moles of sample, so the first task is to calculate the molar mass of $Sr(HSO_4)_2$. We must remember that the subscript 2 after the parentheses around the HSO_4^- ion indicates that two of the ions are present in the formula unit, contributing two hydrogen atoms, two sulfur atoms, and eight oxygen atoms to the formula unit:

$$\begin{aligned}
1\ \text{Sr:}\quad & 1 \times 87.62 = 87.62\\
2\ \text{H:}\quad & 2 \times 1.01 = 2.02\\
2\ \text{S:}\quad & 2 \times 32.06 = 64.12\\
8\ \text{O:}\quad & 8 \times 16.00 = 128.00\\
& \overline{\quad 281.76\ \text{g/mol}\ Sr(HSO_4)_2}
\end{aligned}$$

We use the relationship that 281.76 g $Sr(HSO_4)_2$ = 1 mol $Sr(HSO_4)_2$ to get the conversion factor for the conversion from 500 g $Sr(HSO_4)_2$ to moles:

$$500\ \text{g Sr(HSO}_4)_2 \times \frac{1\ \text{mol Sr(HSO}_4)_2}{281.76\ \text{g Sr(HSO}_4)_2} = 1.77\ \text{mol Sr(HSO}_4)_2$$

For the second part of the question, we use the definition of the mole:

$$1\ \text{mol Sr(HSO}_4)_2 = 6.022 \times 10^{23}\ \text{Sr(HSO}_4)_2\ \text{formula units}$$

$$1.77\ \text{mol Sr(HSO}_4)_2 \times \frac{6.022 \times 10^{23}\ \text{Sr(HSO}_4)_2\ \text{formula units}}{1\ \text{mol Sr(HSO}_4)_2} = 1.07 \times 10^{24}\ \text{Sr(HSO}_4)_2\ \text{formula units}$$

EXERCISE How many moles of $CaCl_2$ are present in 0.455 g $CaCl_2$? How many formula units are present?

ANSWER: 4.10×10^{-3} mol; 2.47×10^{21} formula units

EXAMPLE 2 Calculating the mass of an ionic compound from the moles

What mass (in grams) of $Ba(NO_3)_2$ is present in 0.335 mol? How many formula units are present?

SOLUTION The relationship between mass and moles is given by the molar mass:

1 Ba:	1 × 137.3 g/mol =	137.3 g/mol
2 N:	2 × 14.01 g/mol =	28.02 g/mol
6 O:	6 × 16.00 g/mol =	96.00 g/mol
		261.32 g/mol = 261.3 g/mol

It is now possible to directly calculate the grams $Ba(NO_3)_2$ in 0.335 moles:

$$0.335\ \text{mol Ba(NO}_3)_2 \times \frac{261.3\ \text{g}}{1\ \text{mol Ba(NO}_3)_2} = 87.5\ \text{g Ba(NO}_3)_2$$

The number of formula units is calculated with the definition of the mole, 1 mol $Ba(NO_3)_2$ = 6.022×10^{23} formula units $Ba(NO_3)_2$:

$$0.335\ \text{mol Ba(NO}_3)_2 \times \frac{6.022 \times 10^{23}\ \text{formula units}}{1\ \text{mol Ba(NO}_3)_2} = 2.02 \times 10^{23}\ \text{formula units Ba(NO}_3)_2$$

EXERCISE How many grams of $Sn(CrO_4)_2$ are present in 0.495 mol $Sn(CrO_4)_2$? How many formula units are present?

ANSWER: 174 g; 2.98×10^{23} $Sn(CrO_4)_2$ formula units

2.8 CHEMICAL NOMENCLATURE

KEY CONCEPT A Naming cations and anions

The name of a cation depends on whether the atom is able to form ions of different charges; if not, the name of the cation is the same as the name of the element. The most commonly encountered cations of this type are the group I and group II cations and Zn^{2+}, Ag^+, Cd^{2+}, Al^{3+}, and Ga^{3+}. Some examples are

Li^+ lithium ion Ca^{2+} calcium ion Cd^{2+} cadmium ion

However, if the atom forms ions of different charges, we add to the name of the element the charge on the ion expressed as roman numerals in parentheses. Some examples are

Cu^+ copper(I) ion	Fe^{2+} iron(II) ion	Sn^{2+} tin(II) ion
Cu^{2+} copper(II) ion	Fe^{3+} iron(III) ion	Sn^{4+} tin(IV) ion

An older system for naming cations, which is still occasionally encountered, uses the stem of the name of the element with the suffix *-ous* for the lower charged ion or *-ic* for the higher charged ion. If the name of the element originates with a Latin name, the stem is also from the Latin. This two-suffix system can handle only the two most common ions formed by an element.

Monatomic anions are named by adding the suffix-*ide* to the stem name of the element. For instance,

Br^- bromide S^{2-} sulfide N^{3-} nitride

As described in the text, oxoanions are named according to the oxygen content of the ion:

CO_3^{2-} carbonate	SO_4^{2-} sulfate	NO_3^- nitrate	ClO^- hypochlorite
	SO_3^{2-} sulfite	NO_2^- nitrite	ClO_2^- chlorite
			ClO_3^- chlorate
			ClO_4^- perchlorate

Many other variations are shown in Table 2.7 of the text. The names and formulas of these oxoanions must be learned because so much of the chemistry that follows depends on recognizing them.

EXAMPLE Naming cations and anions

Without consulting any references, name the following ions: Rb^+, Mg^{2+}, Au^{3+} (two names), Tl^{3+}, O^{2-}, and Br^-.

SOLUTION

Rb^+	rubidium ion	Mg^{2+}	magnesium ion
Au^{3+}	gold(III) ion, auric ion	Tl^{3+}	thallium(III) ion
O^{2-}	oxide ion	Br^-	bromide ion

EXERCISE Without consulting any references, name these ions: SO_3^{2-}, NO_3^-, OH^-, CrO_4^{2-}, $Cr_2O_7^{2-}$, and HCO_3^- (two names).

ANSWER: SO_3^{2-} sulfite ion; NO_3^- nitrate ion; OH^- hydroxide ion; CrO_4^{2-} chromate ion; $Cr_2O_7^{2-}$ dichromate ion; HCO_3^- hydrogen carbonate ion, bicarbonate ion

PITFALL The ions formed by mercury

The mercury(I) ion exists not as a single Hg^+ ion but as a dimer. It may be thought of as two Hg^+ ions joined together. Its correct formula is Hg_2^{2+}; it should not be confused with the mercury(II) ion, Hg^{2+}:

mercury(I) Hg_2^{2+} mercury(II) Hg^{2+}

KEY CONCEPT B Naming ionic compounds

In an ionic compound, the first name is the name of the positive ion; the last name, that of the negative ion. The name of the positive ion includes the charge number in roman numerals in parentheses if appropriate. To name an ionic compound, we must

1. Identify what ions are present.
2. Determine the correct name for each of the two ions.
3. Name the compound as the positive ion name followed by the negative ion name.

In certain ionic compounds, individual water molecules are incorporated into the crystal lattice. This water can be driven off by heating the crystal. When the water is present, the compound is called a **hydrate;** when the water is absent, the compound is said to be **anhydrous.** The formula of the hydrate shows the presence of water in the crystal by use of a centered dot followed by the number of water molecules per formula unit: for instance, $MgCO_3 \cdot 3H_2O$ and $Na_3PO_4 \cdot 12H_2O$. This water can be driven off by heating the crystal, as the following equation indicates:

$$\text{Hydrate} \xrightarrow{\text{heat}} \text{anhydrous form} + \text{water}$$

$$MgCO_3 \cdot 3H_2O \longrightarrow MgCO_3 + 3H_2O$$

The Greek prefixes shown in Table 2.8 of the text are used in naming ionic compounds to indicate the number of waters of hydration in the formula unit. This is now the fourth rule for naming ionic compounds.

4. If the compound is a hydrate, indicate the number of waters of hydration by using a Greek prefix plus the term "hydrate" as the last part of the name.

EXAMPLE 1 Naming ionic compounds

Name the compounds Cu_2SO_4, $Na_2MnO_4 \cdot 10H_2O$, and $AuBr_3$.

SOLUTION

Cu_2SO_4: We recognize the $SO_4{}^{2-}$ as the sulfate ion. To balance the -2 charge number on $SO_4{}^{2-}$, each of the two copper ions must have a $+1$ charge: each is a Cu^+ ion. Because copper forms different charged ions, this ion is the copper(I) ion (rather than "copper ion") and Cu_2SO_4 is copper(I) sulfate.

$Na_2MnO_4 \cdot 10H_2O$: The ions present are sodium ion and manganate ion. Sodium forms Na^+ and does not form different charged ions, so its name is "sodium." The 10 waters of hydration are indicated by using the term "decahydrate" at the end of the name, so $Na_2MnO_4 \cdot 10H_2O$ is sodium manganate decahydrate.

$AuBr_3$: The anion in this formula is a bromide, Br^-, ion with charge number -1. The three bromides contribute a total charge of -3 to the formula unit, so the gold is present as a Au^{3+} ion. Because gold forms different charged ions, the name of Au^{3+} is gold(III) ion and $AuBr_3$ is gold(III) bromide.

EXERCISE Name $Mg(NO_3)_2$, $Pb(N_3)_2$, and $LiClO_4 \cdot 3H_2O$

ANSWER: magnesium nitrate, lead(II) azide, lithium perchlorate trihydrate

EXAMPLE 2 Writing the formulas of ionic compounds from their names

Write the formulas of cobalt(II) chlorate hexahydrate, mercury(I) bromide, and strontium peroxide

SOLUTION

Cobalt(II) chlorate hexahydrate: The cobalt(II) ion is Co^{2+} and the chlorate ion is $ClO_3{}^-$. For a formula unit with zero charge, we must combine two chlorate ions (total charge -2) with one cobalt(II) ion (charge $+2$). Hexahydrate means "six waters" of hydration, so the complete formula of cobalt(II) chlorate hexahydrate is $Co(ClO_3)_2 \cdot 6H_2O$.

Mercury(I) bromide: The mercury(I) ion is $Hg_2{}^{2+}$. It has a total charge of $+2$. Each bromide has a -1 charge. For a formula unit with zero charge, two bromide ions (total charge -2) are combined with one mercury(I) dimer (total charge $+2$). The formula of mercury(I) bromide is Hg_2Br_2.

Strontium peroxide: The strontium ion is Sr^{2+} and the peroxide ion is $O_2{}^{2-}$. For a neutral formula unit one strontium ion should be combined with one peroxide. The formula of strontium peroxide is SrO_2.

EXERCISE Write the formulas of tin(IV) chloride tetrahydrate, ammonium hydrogen carbonate, and barium fluoride.

ANSWER: $SnCl_4 \cdot 4H_2O$, NH_4HCO_3, BaF_2

KEY CONCEPT C Naming molecular compounds

At this stage we consider three categories of molecular compounds: (1) binary compounds that are nonacids, (2) binary compounds that are acids, and (3) oxoacids. A binary compound is one that contains only two elements; examples are HCl, H_2O, and PCl_5. Table 2.9 of the text shows some common and important binary compounds that are usually not referred to by systematic names. They have traditional, nonsystematic names that must be learned.

An acid is a compound that releases H^+ when it is placed in water. The formulas for inorganic acids usually have "H" as the first letter in the formula, as in the acids HBr and HNO_3. The formula H_2O is an exception because water is not considered an acid. CH_4 is not an acid even though it contains hydrogen; its hydrogens are not released as H^+ when it is dissolved in water.

Binary acids can exist as molecular gases when not dissolved in water. In this form, the first part of the name is "hydrogen" and the second part is the stem of the second element with the suffix *-ide*. When dissolved in water, the acid properties become apparent and the name changes. The first name is the prefix *hydro-* followed by the stem of the name of the second element in the formula, and then the suffix *-ide*. The second part of the name is "acid." Examples are shown in the table.

Molecular formula	Name of molecular gas	Name of dissolved acid
HF	hydrogen fluoride	hydrofluoric acid
HCl	hydrogen chloride	hydrochloric acid
H_2S	hydrogen sulfide	hydrosulfuric acid

Binary compounds that are not acids are named as follows. The first part of the name is the name of the first element in the formula with a Greek prefix to give the number of atoms of the element (except that the prefix *mono-* is dropped when one atom is present). The second part of the name is the stem of the name of the second element in the formula, with the suffix *-ide*. A Greek prefix is used to indicate the number of atoms of the element in the formula. Some examples are

PCl_3	phosphorus trichloride	N_2O	dinitrogen monoxide
NO	nitrogen monoxide	N_2O_5	dinitrogen pentoxide
ClO_2	chlorine dioxide	OF_2	oxygen difluoride

The oxoacids are all molecular compounds; their names are based on the name of the parent anion. An anion ending in *-ate* gives an acid ending in *-ic*, and an anion ending in *-ite* gives an acid ending in *-ous*. *Per* and *hypo* are used for the acid name if one is part of the anion name:

Molecular formula	Name of compound	Anion name and formula
H_2SO_4	sulfuric acid	sulfate ion (SO_4^{2-})
H_2SO_3	sulfurous acid	sulfite ion (SO_3^{2-})
H_3PO_4	phosphoric acid	phosphate ion (PO_4^{3-})
H_3PO_3	phosphorous acid	phosphite ion (HPO_3^{2-})
HNO_3	nitric acid	nitrate ion (NO_3^-)
HNO_2	nitrous acid	nitrite ion (NO_2^-)
HClO	hypochlorous acid	hypochlorite ion (ClO^-)

EXAMPLE Naming molecular compounds

Without looking at any references, write the names of the following compounds and state whether or not they are acids: HCl, N_2O_4, NH_3, and H_2CO_3.

SOLUTION HCl, hydrogen chloride (gas), hydrochloric acid (when dissolved in water), an acid; N_2O_4, dinitrogen tetroxide, nonacid; NH_3, ammonia, nonacid; H_2CO_3, carbonic acid, an acid.

EXERCISE Without looking at any references, name the following compounds and state whether or not they are acids: H_2O_2, HNO_3, PF_3, and HBr.

ANSWER: H_2O_2, hydrogen peroxide, nonacid; HNO_3, nitric acid, an acid; PF_3, phosphorus trichloride, nonacid; HBr, hydrogen bromide (gas), hydrobromic acid (when dissolved in water), an acid

PITFALL Sulfuric and hydrosulfuric acids

Be certain to distinguish sulfuric acid (H_2SO_4) from hydrosulfuric acid (H_2S).

DESCRIPTIVE CHEMISTRY TO REMEMBER

- The **alkali metals** are all soft, silvery metals with low melting points. They all react with water.
- The **noble gases** are all very unreactive.
- An **α particle** is a helium nucleus stripped of its electrons.
- **Heavy water** is water containing two atoms of deuterium instead of two atoms of hydrogen.
- **Avogadro's number** is 6.022×10^{23}.
- **NO_2 (nitrogen dioxide)** is a dark brown gas that contributes to the color of smog.
- Organic compounds contain primarily **carbon** and **hydrogen.**

SELF-TEST EXERCISES

Elements

For questions 1–4, associate the symbol of the element with the name.

1. Sodium
(a) So (b) S (c) nA (d) SO (e) Na

2. Phosphorus
(a) Ph (b) ph (c) P (d) Po (e) Pb

3. Kr
(a) kryptonite (b) kallium (c) krypton (d) korium (e) cobalt

4. Au
(a) gold (b) silver (c) actinium (d) arsenic (e) aluminum

5. Which is an alakli metal?
(a) Na (b) Ca (c) Al (d) Pb (e) Fe

6. Which element is a congener of Mg?
(a) Na (b) Ca (c) Li (d) Sc (e) Al

7. All of the following are typical properties of nonmetals except one. Which one?
(a) poor electrical conductor (b) dull appearance (c) brittle
(d) poor heat conductor (e) ductile

8. Which is a transition metal?
(a) Na (b) Ba (c) Fe (d) Ra (e) Pb

Atoms

9. The charge to mass ratio of the electron is -1.76×10^{8} coul/g and its mass is 9.11×10^{-28} g. What is the charge of the electron in coulombs?
(a) -6.23×10^{18} (b) -5.18×10^{-36} (c) -1.60×10^{-19}
(d) -1 (e) -1.93×10^{35}

10. The nuclear atom model states that atoms consist of a
(a) very large, heavy, positive nucleus surrounded by light, negative electrons.
(b) very large, light, negative nucleus surrounded by heavy, positive protons.
(c) very small, heavy, positive nucleus surrounded by light, negative electrons.
(d) jelly like positive substance made of protons in which negative electrons are embedded.
(e) small, light, negative nucleus surrounded by heavy, positive protons.

11. Use a periodic table to find the atomic number of potassium.
(a) 39 (b) 39.1 (c) 20 (d) 58 (e) 19

12. What element has an atomic number of 32?
(a) sulfur (b) oxygen (c) argon (d) rubidium (e) germanium

13. How many electrons are present in a zinc atom?
(a) 65 (b) 35 (c) 30 (d) 50 (e) 95

14. A 5.0-g sample of neon contains 1.49×10^{24} electrons. What percentage of the mass of the sample originates with the electrons? The mass of one electron is 9.11×10^{-28} g and 1 u = 1.6605×10^{-24} g.
(a) 0.014% (b) 0.027% (c) 0.88% (d) 0.038% (e) 0.68%

15. Which of the following represents an isotope of Ni ($p = 28$, $n = 58$) in which p is the number of protons and n the number of neutrons?
(a) $p = 29$, $n = 58$ (b) $p = 28$, $n = 59$ (c) $p = 29$, $n = 57$
(d) $p = 27$, $n = 60$ (e) $p = 59$, $n = 82$

16. What is the mass number of an atom with 25 electrons, 25 protons, and 30 neutrons?
(a) 25 (b) 50 (c) 55 (d) 80 (e) 35

17. How many neutrons are present in an atom of ^{56}Fe?
(a) 30 (b) 56 (c) 86 (d) 26 (e) 52

18. How many grams are present in an (average) oxygen atom? 1 u = 1.6605×10^{-24} g.
(a) 2.66×10^{-24} g (b) 1.04×10^{-25} g (c) 9.63×10^{24} g
(d) 2.66×10^{-23} g (e) 16.0 g

19. The following are the natural abundances and atomic weights of the naturally occurring isotopes of copper. Calculate the atomic weight of copper.

Element	Natural abundance (%)	Atomic mass (u)
Cu-63	69.17	62.9396
Cu-65	30.83	64.9278

(a) 63.93 u (b) 31.78 u (c) 63.55 u (d) 64.31 u (e) 64.00 u

20. Gallium's atomic weight is 69.72 u. What is the natural abundance of gallium-71? The atomic mass of the two naturally occurring isotopes of gallium are shown in the table.

Element	Atomic mass (u)
Ga-69	68.93
Ga-71	70.92

(a) 51.60% (b) 39.70% (c) 42.43% (d) 60.34% (e) 48.40%

21. How many calcium atoms are present in 12.5 g Ca? 1 u = 1.6605×10^{-24} g.
(a) 1.88×10^{23} (b) 5.26×10^{24} (c) 0.312 (d) 8.32×10^{-22} (e) 3.02×10^{26}

22. How many silicon atoms are present in 0.778 mol of silicon?
(a) 4.79×10^{23} (b) 7.74×10^{23} (c) 6.022×10^{23}
(d) 8.77×10^{24} (e) 2.22×10^{24}

23. A sample of uranium contains 4.59×10^{24} atoms. How many moles of uranium are present?
(a) 0.131 mol (b) 7.62 mol (c) 0.762 mol (d) 1.31 mol (e) 276 mol

24. How many moles of arsenic are present in 100 g of arsenic?
(a) 0.749 mol (b) 3.03 mol (c) 0.330 mol (d) 1.00 mol (e) 1.33 mol

25. What is the mass of 5.11×10^{23} argon atoms?
(a) 39.9 g (b) 33.9 g (c) 41.2 g (d) 47.1 g (e) 35.6 g

26. How many grams of strontium are there in 2.46 mol strontium atoms?
(a) 35.6 g (b) 0.0281 g (c) 4.62×10^{-3} (d) 216 g (e) 93.5 g

27. How many grams of nitrogen are present in 0.555 mol nitrogen atoms?
(a) 25.2 g (b) 7.78 g (c) 0.0396 g (d) 0.129 g (e) 3.89 g

Compounds

28. How many hydrogen atoms are there in one molecule of pentane (C_5H_{12})?
(a) 17 (b) 12 (c) 5 (d) 2 (e) 7

29. What is the total number of atoms in one molecule of cholesterol, $C_{27}H_{46}O$?
(a) 27 (b) 47 (c) 46 (d) 74 (e) 1242

30. What is the molar mass of cholesterol, shown in problem 29?
(a) 216 g/mol (b) 224 g/mol (c) 387 g/mol (d) 410 g/mol (e) 444 g/mol

31. What is the molar mass of the acid that gives vinegar its sour taste, acetic acid, $HC_2H_3O_2$?
(a) 32.0 g/mol (b) 44.2 g/mol (c) 60.1 g/mol (d) 102.4 g/mol (e) 98.1 g/mol

32. How many moles of $C_4H_8Br_2$ (dibromobutane) are present in 56.6 g of $C_4H_8Br_2$?
(a) 3.81 (b) 0.262 (c) 1.22×10^4 (d) 8.18×10^{-5} (e) 0.555

33. How many moles of $C_8H_{18}O$ (octanol) are present in 175 g of $C_8H_{18}O$?
(a) 1.34 (b) 1.43 (c) 0.744 (d) 0.697 (e) 22.7

34. How many grams of H_2SO_3 (sulfurous acid) are present in 5.22 mol H_2SO_3?
(a) 566 (b) 0.0636 (c) 15.7 (d) 2.34×10^{-3} (e) 428

35. How many grams of PF_5 (phosphorus pentafluoride) are there in 7.88×10^{-3} mol PF_5?
(a) 1.23 (b) 0.993 (c) 1.01 (d) 6.26×10^{-5} (e) 1.60×10^4

36. How many molecules of CH_4 (methane) are present in 50.0 g CH_4?
(a) 2.56×10^{23} (b) 1.88×10^{24} (c) 6.02×10^{23}
(d) 1.94×10^{23} (e) 5.12×10^{24}

37. What is the mass in grams of 5.57×10^{23} molecules of C_2H_6O (ethanol)?
(a) 45.5 g (b) 42.6 g (c) 50.1 g (d) 48.9 g (e) 38.9 g

38. How many electrons are there in the Ba^{2+} ion?
(a) 54 (b) 58 (c) 53 (d) 56 (e) 55

39. How many electrons are there in the S^{2-} ion?
(a) 16 (b) 17 (c) 15 (d) 18 (e) 14

40. How many electrons are there in the NO_2^- ion?
(a) 21 (b) 22 (c) 24 (d) 23 (e) 25

41. What is the formula of the compound formed by Ga^{3+} and O^{2-} ions?
(a) Ga_3O_2 (b) Ga_2O_3 (c) Ga_3O (d) GaO (e) GaO_2

42. What is the formula of the compound formed by Ba and S_8?
(a) BaS (b) Ba_8S_8 (c) Ba_2S (d) BaS_8 (e) BaS_2

43. Without looking at any references, name the ClO_3^- ion.
(a) perchlorate ion (b) chlorine trioxide ion (c) chlorate ion (d) hypochlorous ion

44. Which of the ions shown is the hydroxide ion?
(a) NH_2 (b) O^{2-} (c) OH^- (d) H^- (e) HS^-

45. What is the molar mass of $(NH_4)_2SO_4$?
(a) 114.1 g/mol (b) 146.2 g/mol (c) 210.2 g/mol
(d) 132.1 g/mol (e) 70.0 g/mol

46. How many moles of BaS are there in 2.58 kg of BaS?
(a) 4.38×10^5 (b) 0.0318 (c) 31.5 (d) 6.57×10^{-2} (e) 15.2

47. How many grams of $Cd(OH)_2$ are there in 1.4×10^5 mol $Cd(OH)_2$?
(a) 1.0×10^{-3} (b) 2.0×10^7 (c) 9.6×10^2 (d) 4.9×10^{-8} (e) 1.8×10^7

48. How many nitrogen atoms are present in 25.0 g $Ca(NO_3)_2$?
(a) 4.59×10^{22} (b) 1.83×10^{23} (c) 1.20×10^{24}
(d) 9.17×10^{22} (e) 6.02×10^{23}

49. What is the systematic name of $FeCO_3$?
(a) iron(III) carbonate (b) iron(II) carbonate
(c) iron carbonate (d) iron(VI) carbonate

50. What is the name of $Ba(OH)_2$?
(a) barium oxyhydride (b) barium(II) hydroxide (c) barium hydroxide
(d) barium dihydroxide (e) barium 2-hydroxide

51. What is the formula of potassium dichromate?
(a) $KCrO_4$ (b) K_2CrO_4 (c) KCr_2O_7 (d) $K_2Cr_2O_7$ (e) KCr_2

52. What is the formula of iridium(IV) chloride?
(a) Ir_4Cl (b) $IrCl_4$ (c) $IrCl_2$ (d) IrCl (e) Ir_2Cl_3

53. What is the formula of ammonia? (Do not refer to any references.)
(a) H_2O_2 (b) N_2H_4 (c) PH_3 (d) NH_3 (e) NH_4

54. What is the formula of dinitrogen pentoxide?
(a) N_2O_4 (b) NO (c) $(NO_5)_2$ (d) N_2O (e) N_2O_5

3 CHEMICAL REACTIONS

CHEMICAL EQUATIONS

KEY WORDS Define or explain each of the following terms in a written sentence or two.

chemical equation
combustion
decomposition
products
reactants
reagent
rearrangement
skeletal equation
stoichiometric coefficients
synthesis

3.1 SYMBOLIZING REACTIONS

KEY CONCEPT Symbolizing chemical reactions

During a chemical reaction, a specific substance (or substances) changes to a different substance (or substances). The substances we start with in the reaction are called **reactants,** and the new substances formed by the chemical reaction are called **products.** The chemical change from reactants to products is symbolized by a **chemical equation,** in which the formulas for the reactants are written to the left of an arrow and the products to the right of the arrow. The arrow symbolizes a chemical change. A **stoichiometric coefficient** in front of each substance is needed to insure that the number of atoms of each element is the same for the reactants as for the products. The physical state of the reactants and products is often indicated by placing a symbol next to each substance:

(*s*) solid (*l*) liquid (*g*) gas (*aq*) aqueous (substance dissolved in water)

EXAMPLE Writing a chemical equation

Write the chemical equation that symbolizes the reaction in which solid ammonium nitrate is heated to 200°C to form the gas dinitrogen oxide and water vapor.

SOLUTION Using techniques from Chapter 2, we get the following formulas for the substances involved: ammonium nitrate, NH_4NO_3; dinitrogen oxide, N_2O; and water, H_2O. We now assemble these into the required chemical equation. The statement in the problem makes clear that the reactant, or substance we start with, is NH_4NO_3 and that the products, or substances formed, are N_2O and H_2O. We indicate the physical state of each substance with the proper symbol and write 200°C over the arrow to indicate the temperature needed to make the reaction proceed:

$$NH_4NO_3(s) \xrightarrow{200^\circ C} N_2O(g) + 2H_2O(g)$$

The 2 is needed in front of H_2O to assure that the number of atoms of each element is the same to the left and right of the arrow.

EXERCISE When solid magnesium is heated to 600°C in the presence of oxygen, magnesium oxide is formed. Write the chemical equation for this reaction.

ANSWER: $2Mg(s) + O_2(g) \xrightarrow{600^\circ C} 2MgO(s)$

PITFALL Magnesium and manganese

Be certain to distinguish magnesium (Mg, atomic number 12) from manganese (Mn, atomic number 25).

3.2 BALANCING EQUATIONS

KEY CONCEPT Balancing chemical equations

All chemical equations must be balanced. We **balance a chemical equation** by adjusting the stoichiometric coefficients in the equation until the number of atoms of each element on the left-hand side of the arrow is equal to the number of atoms of the same element on the right-hand side. The simplest equations are balanced by inspection: in these only one or two coefficients have to be adjusted, and the equation is balanced in one or two steps by a commonsense approach. The reaction between H_2 and O_2 to form water is a good example:

$$H_2(g) + O_2(g) \longrightarrow H_2O(g) \qquad \triangle$$

Inspection of the unbalanced equation (symbolized by $\triangle$) reveals that there are two oxygen atoms on the left and only one on the right, so a 2 must be placed in front of $H_2O(g)$.

$$H_2(g) + O_2(g) \longrightarrow 2H_2O(g) \qquad \triangle$$

With this change, there are now four hydrogen atoms on the right, which are balanced by placing a 2 in front of $H_2(g)$.

$$2H_2(g) + O_2(g) \longrightarrow 2H_2O(g)$$

Atom	Left	Right
H	4	4
O	2	2

The equation is now balanced. We indicate the balance in the small table, which gives the number of each atom to the left and the right of the arrow. A second level of difficulty occurs when more than one or two steps are involved or more complicated compounds are involved. A few rules will suffice to balance most such equations. (The rules here are slightly different than the ones in the text: use the ones you are more comfortable with.)

Step 1. Balance any element that appears in only one compound on each side of the arrow.
Step 2. By inspection, balance any element that appears in more than one compound on either the right or left.
Step 3. Balance any element that appears in elemental form on the right or left.
Step 4. Clear any fractions present, if desired.
Hint. Balance polyatomic ions that appear on both sides of the equation as units rather than as separate atoms.

EXAMPLE Balancing chemical equations

The compound $C_2H_8N_2$ (dimethylhydrazine), related to hydrazine, is used as a rocket fuel in the following reaction. Balance this equation.

$$C_2H_8N_2(l) + N_2O_4(g) \longrightarrow N_2(g) + H_2O(g) + CO_2(g) \qquad \triangle$$

SOLUTION

Step 1. Balance any element that appears in only one compound on each side of the arrow. Carbon and hydrogen are balanced in this step. There are two C atoms on the left, so a 2 is placed in front of $CO_2(g)$. There are eight hydrogen atoms on the left, which require $4H_2O(g)$ molecules on the right:

$$C_2H_8N_2(l) + N_2O_4(g) \longrightarrow N_2(g) + 4H_2O(g) + 2CO_2(g) \qquad \triangle$$

Step 2. By inspection, balance any element that appears in more than one compound on either the right or the left. Only oxygen is balanced here, because nitrogen is balanced in step 3. There are four oxygen atoms on the left and eight on the right, so we place a 2 in front of $N_2O_4(g)$ to get eight oxygen atoms on the left:

$$C_2H_8N_2(l) + 2N_2O_4(g) \longrightarrow N_2(g) + 4H_2O(g) + 2CO_2(g) \qquad \triangle$$

Step 3. Balance any element that appears in elemental form on the right or left. Here, we balance the nitrogen atoms by placing a 3 in front of $N_2(g)$ on the right, so that there are six nitrogen atoms on each side of the arrow:

$$C_2H_8N_2(l) + 2N_2O_4(g) \longrightarrow 3N_2(g) + 4H_2O(g) + 2CO_2(g)$$

Atom	Left	Right
C	2	2
H	8	8
N	6	6
O	8	8

Step 4 is not used in this example.

EXERCISE Balance the following equation and include the table showing the number of atoms on the left and the right. Balance SO_4^{2-} as a unit, rather than as separate sulfur and oxygen atoms:

$$Fe_2(SO_4)_3(aq) + BaCl_2(aq) \longrightarrow BaSO_4(s) + FeCl_3(aq)$$

ANSWER: $Fe_2(SO_4)_3(aq) + 3BaCl_2(aq) \longrightarrow 3BaSO_4(s) + 2FeCl_3(aq)$

Atom or ion	Left	Right
Fe	2	2
SO_4^{2-}	3	3
Ba	3	3
Cl	6	6

PITFALL Changing formulas to balance an equation

Once the formulas of the products and reactants are written correctly, they *cannot* be changed to balance the equation. Changing the formulas would change the meaning of the equation.

PITFALL Incorrect formulas in an equation

Occasionally, with a complicated equation, you may find that balancing one atom unbalances another, which then has to be rebalanced, and that rebalancing this atom causes another (or the original atom) to go out of balance. If you find you are going in circles with this process, so that no matter what you do, one element always remains unbalanced, you probably have written a formula incorrectly or have forgotten to include a required formula. If balancing one element always unbalances another, with no end in sight, check your formulas!

PRECIPITATION REACTIONS

KEY WORDS Define or explain each of the following terms in a written sentence or two.

double replacement reaction
net ionic equation
net ionic reaction
precipitation reaction
solubility rules
spectator ions

3.3 NET IONIC EQUATIONS

KEY CONCEPT A Precipitation reactions

A **precipitation reaction** is one in which an insoluble solid forms as a product after two solutions are mixed. By definition, an insoluble solid does not dissolve. As it forms, it appears in the reaction mixture as a cloudy, opaque material. A precipitation may be represented as

Solution	+	Solution	$\longrightarrow$	Precipitate
Transparent, but possibly colored		Transparent, but possibly colored		Opaque, either white or colored

In the following safe home experiment you can see a precipitation reaction. Thoroughly dissolve $\frac{1}{2}$ teaspoon of epsom salts in $\frac{1}{2}$ cup of water; in a separate cup, throughly dissolve $\frac{1}{2}$ teaspoon of washing soda (not baking soda) in $\frac{1}{2}$ cup of water. You should now have two clear, colorless solutions.

Mix the two solutions in a clear glass: a white precipitate forms. The equation for this reaction is

$$\underset{\text{Epsom salt}}{MgSO_4(aq)} + \underset{\text{Washing soda}}{Na_2CO_3(aq)} \longrightarrow \underset{\text{White precipitate}}{MgCO_3(s)} + \underset{\text{Soluble salt}}{Na_2SO_4(aq)}$$

Both $MgSO_4$ and Na_2CO_3 are soluble and dissolve easily in water, giving colorless, transparent solutions. When they are mixed, a **double replacement** reaction occurs. The Mg^{2+} ion replaces the two Na^+ ions in Na_2CO_3 to form insoluble $MgCO_3$. The two Na^+ ions replace Mg^{2+} in $MgSO_4$ to form the soluble salt Na_2SO_4. The name *double replacement* refers to the two replacements that occur; a newer name for this type of reaction is **metathesis reaction.** A precipitation reaction is a type of metathesis reaction.

EXAMPLE Writing a precipitation equation

Write the equation for the precipitation reaction between NaF and $Ca(NO_3)_2$ to form products. Be certain to label all the substances appropriately with (*aq*) or (*s*). CaF_2 is insoluble.

SOLUTION We first deduce what the products are and which is a precipitate:

$$NaF + Ca(NO_3)_2 \longrightarrow ?$$

A precipitation reaction is a type of metathesis reaction. Thus, Na^+ replaces Ca^{2+} in $Ca(NO_3)_2$ and, in doing so, unites with NO_3^- to form $NaNO_3$; in addition, Ca^{2+} replaces Na^+ in NaF and, in doing so, unites with F^- to form CaF_2. So the unbalanced equation is

$$NaF + Ca(NO_3)_2 \longrightarrow NaNO_3 + CaF_2 \quad ⚠$$

Table 3.2 of the text states that salts of the Group I elements are soluble, so $NaNO_3$ does not form a precipitate. The problem states that CaF_2 is insoluble; therefore, it is the precipitate. (As expected, Table 3.2 also confirms that NaF and $Ca(NO_3)_2$ are soluble.) To balance the equation, we note that the left side has one F^- ion and that the right has two, so a 2 is placed in front of NaF to balance fluorine. In addition, there are two NO_3^- ions on the left but only one on the right, so a 2 is needed in front of $NaNO_3$. The balanced equation is

$$2NaF + Ca(NO_3)_2 \longrightarrow 2NaNO_3 + CaF_2$$

Ion or atom	Left	Right
Na	2	2
F	2	2
Ca	1	1
NO_3^-	2	2

Because CaF_2 is insoluble, it forms a solid precipitate and is labeled (*s*). The final equation is

$$2NaF(aq) + Ca(NO_3)_2(aq) \longrightarrow 2NaNO_3(aq) + CaF_2(s)$$

EXERCISE Write the equation for the precipitation reaction between $(NH_4)_2S$ and $AgNO_3$. Construct a table showing that the equation is balanced.

ANSWER: $(NH_4)_2S(aq) + 2AgNO_3(aq) \longrightarrow Ag_2S(s) + 2NH_4NO_3(aq)$

Ion or atom	Left	Right
NH_4^+	2	2
S	1	1
Ag	2	2
NO_3^-	2	2

PITFALL Writing the correct formulas in a metathesis reaction

When writing the products for a metathesis reaction, avoid the temptation to automatically carry forward the subscript of an ion in a reactant as the ion becomes part of the product. For example, in the metathesis reaction

$$FeCl_3(aq) + 3NaOH(aq) \longrightarrow Fe(OH)_3(s) + 3NaCl(aq) \quad \text{(correct)}$$

we take care not to automatically keep the subscript 3 on Cl^- and the (implied) 1 on OH^- and write

$$FeCl_3(aq) + 3NaOH(aq) \longrightarrow FeOH(s) + NaCl_3(aq) \quad \text{(incorrect)}$$

When writing the formulas of the products of a metathesis reaction, one must purposefully check the formulas of the products to write them correctly.

KEY CONCEPT B Net ionic equations

In the precipitation reactions that we have written, there are a total of four different ions in the reactants. Before we mix the reactant solutions, these ions exist as separated ions in aqueous solution. After they are mixed, two of the ions combine to form the solid precipitate, and the other two simply stay in aqueous solution. Chemically, nothing has happened to the two ions that stay in solution; they start out as aqueous ions and end up as aqueous ions. On the other hand, the ions that form the precipitate have undergone a chemical change, from separate aqueous ions to a solid bonded precipitate. We often choose to write a **net ionic equation** in which we show only the overall chemical change and ignore the ions that stay in solution. For example, consider the reaction between silver nitrate ($AgNO_3$) and potassium sulfate (K_2SO_4):

$$2AgNO_3(aq) + K_2SO_4(aq) \longrightarrow Ag_2SO_4(s) + 2KNO_3(aq)$$

At the start of the reaction, we have Ag^+, NO_3^-, K^+, and SO_4^{2-} ions in solution; the abbreviation (aq) tells us that the salts are dissolved. After the reaction, Ag^+ and SO_4^{2-} have reacted to form insoluble $Ag_2SO_4(s)$, whereas NO_3^- and K^+ stay in solution and do not react. The net ionic equation shows only the species that react and is, therefore,

$$2Ag^+(aq) + SO_4^{2-}(aq) \longrightarrow Ag_2SO_4(s)$$

K^+ and NO_3^- are called **spectator ions** because they do not participate in the reaction but stay in solution and "watch" as the reaction occurs.

EXAMPLE Writing a net ionic equation

Write the net ionic equation for the precipitation reaction between silver fluoride (AgF) and sodium sulfate (Na_2SO_4). Ag_2SO_4 is insoluble.

SOLUTION Because the way we write a net ionic equation depends on the products formed, the first step is to determine formulas of the products. In this case, Ag^+ replaces Na^+ in Na_2SO_4, resulting in Ag_2SO_4, and Na^+ replaces Ag^+ in AgF, resulting in NaF:

$$AgF(aq) + Na_2SO_4(aq) \longrightarrow Ag_2SO_4 + NaF \quad ⚠$$

Table 3.2 of the text informs us that NaF is soluble, so NaF exists as $Na^+(aq)$ ions and $F^-(aq)$ ions on both sides of the equation and are spectator ions. The net reaction is the reaction of $Ag^+(aq)$ and $SO_4^{2-}(aq)$ to form $Ag_2SO_4(s)$:

$$Ag^+(aq) + SO_4^{2-}(aq) \longrightarrow Ag_2SO_4(s) \quad ⚠$$

To balance the equation, two $Ag^+(aq)$ are needed:

$$2Ag^+(aq) + SO_4^{2-}(aq) \longrightarrow Ag_2SO_4(s)$$

Ion or atom	Left	Right
Ag^+	2	2
SO_4^{2-}	1	1

EXERCISE Write the net ionic equation for the precipitation reaction between $Pb(NO_3)_2$ and K_2CrO_4.

ANSWER: $Pb^{2+}(aq) + CrO_4^{2-}(aq) \longrightarrow PbCrO_4(s)$

PITFALL Spectator ions are present!

From looking at a net ionic equation, it is easy to imagine that the spectator ions are not present. This is not the case; the spectator ions are present. We just choose to ignore them because they are not important to the overall chemical process that occurs.

PITFALL Equations for nonexistent reactions

It is possible to write an apparent equation for a nonexistent process. For example, we can write the metathesis equation with $NaNO_3$ and K_2SO_4 as reactants:

$$2NaNO_3(aq) + K_2SO_4(aq) \longrightarrow Na_2SO_4(aq) + 2KNO_3(aq)$$

As Table 3.2 of the text confirms, the reactants exist as $Na^+(aq)$, $NO_3^-(aq)$, $K^+(aq)$ and $SO_4^{2-}(aq)$ in solution. However, the proposed products, Na_2SO_4 and KNO_3, are both soluble and also exist as ions in solution. Thus, no precipitate forms. We conclude from this observation that when $NaNO_3(aq)$ and $K_2SO_4(aq)$ are mixed, *no chemical process occurs.* The only process accomplished is the mixing of $NaNO_3(aq)$ with $K_2SO_4(aq)$. This situation is often represented by N.R., which stands for "no reaction," on the right side of the arrow.

$$NaNO_3(aq) + K_2SO_4(aq) \longrightarrow \text{N.R.}$$

ACID-BASE REACTIONS

KEY WORDS Define or explain each of the following terms in a written sentence or two.

Arrhenius acid
Arrhenius base
Brønsted acid
Brønsted base
neutralization
proton transfer
salt
weak electrolyte

3.5 ARRHENIUS ACIDS AND BASES

KEY CONCEPT Arrhenius acids and bases

Acids and bases are discussed together because acids react with bases in specific and characteristic ways. An **Arrhenius acid** is a substance that contains hydrogen and releases H^+ ions when dissolved in water. HCl, HNO_3, and H_2SO_4 are Arrhenius acids. The released H^+ immediately attaches to a water molecule; so when HCl is dissolved in water, the following process occurs:

$$HCl(aq) + H_2O(l) \longrightarrow H_3O^+(aq) + Cl^-(aq)$$

An **Arrhenius base** is a substance that produces OH^- ions in water. NaOH and $Ca(OH)_2$ are Arrhenius bases. CaO and NH_3 are also Arrhenius bases; they do not contain OH^- ions but produce them when dissolved in water:

$$NaOH(aq) \longrightarrow Na^+(aq) + OH^-(aq)$$
$$NH_3(aq) + H_2O(l) \longrightarrow NH_4^+(aq) + OH^-(aq)$$
$$CaO(s) + H_2O(l) \longrightarrow Ca^{2+}(aq) + 2OH^-(aq)$$

Chemists frequently (but not always) write the formulas of acids with the ionizable hydrogens (the ones that are released as H^+) as the first symbol in the formula. For instance, H_2SO_4 has two ionizable hydrogens, and $HC_2H_3O_2$ has one.

EXAMPLE Identifying and understanding Arrhenius acids and bases

Identify each of the following compounds as an Arrhenius acid, an Arrhenius base, or neither: H_2O, HBr, CH_3OH, $H_2C_2O_4$, KOH, NH_3, and C_3H_8. State the effect of aqueous solutions of each acid and base on litmus.

SOLUTION The formulas for Arrhenius acids frequently have hydrogen as the first letter in the formula. (One notable exception is water, which has H as the first letter it its formula but is not considered to be an acid or base.) Both HBr and $H_2C_2O_4$ have H as the first letter and both are acids (hydrobromic acid and oxalic acid, respectively). C_3H_8 (propane) contains eight hydrogen atoms, but they do not form hydrogen ions in water; so propane is not an acid. Because acids turn litmus red, aqueous solutions of HBr and $H_2C_2O_4$ will turn litmus red. Arrhenius bases produce OH^- when the base is dissolved in water, either because the base contains hydroxide ion or reacts with water to produce hydroxide. The bases that contain hydroxide are invariably ionic compounds, so KOH (potassium hydroxide) is recognized as a base. NH_3 (ammonia) is a base because it reacts with water to form hydroxide ion

$$NH_3(g) + H_2O(l) \longrightarrow NH_4^+(aq) + OH^-(aq)$$

CH_3OH (methanol) contains an OH group, but the group is not a hydroxide ion and does not ionize in water, so CH_3OH is not a base; it is an alcohol. Because bases turn litmus blue, an aqueous solution of KOH or NH_3 will turn litmus blue.

EXERCISE Identify each of the following as either an Arrhenius acid, an Arrhenius base, or neither: HI, C_2H_5OH, $Ca(OH)_2$, $HC_3H_5O_2$, and C_5H_{12}. State the effect of aqueous solutions of each acid and base on litmus.

ANSWER: Acids: HI, $HC_3H_5O_2$; aqueous solutions turn litmus red. Base: $Ca(OH)_2$; aqueous solution turns litmus blue.

3.6 NEUTRALIZATION

KEY CONCEPT Acid-base neutralization

When an acid reacts with a base, the acid and base are said to **neutralize** each other because the acid and base are both destroyed in an acid-base reaction. In neutralization reactions, a salt and water are often (but not always) formed. For example

$$\underset{\text{Acid}}{HBr(aq)} + \underset{\text{Base}}{KOH(aq)} \longrightarrow \underset{\text{Salt}}{KBr(aq)} + \underset{\text{Water}}{H_2O(l)}$$

Any ionic compound that does not contain OH^- ion or O^{2-} ion is called a **salt.** In an acid-base reaction, the anion of the salt formed originates with the acid and the cation originates with the base.

EXAMPLE 1 Writing a neutralization reaction

Write the neutralization reaction between nitric acid and calcium hydroxide. What is the name of the salt formed?

SOLUTION The reaction of an acid with a base results in the formation of a salt and water. The salt formed is calcium nitrate.

$$\underset{\text{Acid}}{2HNO_3(aq)} + \underset{\text{Base}}{Ca(OH)_2(aq)} \longrightarrow \underset{\text{Salt}}{Ca(NO_3)_2(aq)} + \underset{\text{Water}}{2H_2O(l)}$$

EXERCISE Write the neutralization reaction between rubidium hydroxide and sulfuric acid.

ANSWER: $2RbOH(aq) + H_2SO_4(aq) \longrightarrow Rb_2SO_4(aq) + 2H_2O(l)$

EXAMPLE 2 Using a neutralization reaction to synthesize a salt

Suggest a reaction for the synthesis of calcium sulfate through an acid-base neutralization.

SOLUTION The cation in a salt originates with a base and the anion with an acid. Thus, Ca^{2+} must come from $Ca(OH)_2$ and SO_4^{2-} from H_2SO_4; the reaction is

$$Ca(OH)_2(aq) + H_2SO_4(aq) \longrightarrow 2H_2O(l) + CaSO_4(aq)$$

EXERCISE Suggest a neutralization reaction that could be used to produce CsBr.

ANSWER: $CsOH(aq) + HBr(aq) \longrightarrow CsBr(aq) + H_2O(l)$

3.7 THE BRØNSTED DEFINITIONS

KEY CONCEPT Brønsted acids and bases

The Brønsted-Lowry concept of acids and bases adopts a broader view of acids and bases than does the Arrhenius theory; nevertheless, it is consistent with the Arrhenius concept. In the Brønsted approach a **Brønsted acid** is a proton (H^+) donor, and a **Brønsted base** is a proton acceptor. The reaction that occurs when HCl is dissolved in water is an example of a Brønsted acid-base reaction:

$$HCl(aq) + H_2O(l) \longrightarrow H_3O^+(aq) + Cl^-(aq)$$

In this reaction, HCl is a proton donor and is, therefore, a Brønsted acid. Water is a proton acceptor and is, therefore, a Brønsted base. This reaction is called the **ionization** of HCl. The **hydronium ion,** H_3O^+, forms whenever a Brønsted acid is dissolved in water. The reaction that occurs when ammonia (NH_3) is dissolved in water is also a Brønsted acid-base reaction:

$$NH_3(aq) + H_2O(l) \longrightarrow NH_4^+(aq) + OH^-(aq)$$

In this case, water is the proton donor and therefore the Brønsted acid; ammonia is the proton acceptor and the Brønsted base.

EXAMPLE 1 Identifying acids and bases

When carbonate ion is dissolved in water, the following reaction occurs:

$$CO_3^{2-}(aq) + H_2O(l) \longrightarrow HCO_3^-(aq) + OH^-(aq)$$

What are the Brønsted acid and the Brønsted base? Is the carbonate ion an Arrhenius acid, an Arrhenius base, or neither?

SOLUTION In this reaction, water donates a proton to the carbonate ion, so water is the Brønsted acid and carbonate ion is the Brønsted base. The dissolution of carbonate ion results in the formation of OH^- ion; therefore, carbonate ion is an Arrhenius base.

EXERCISE When ammonium ion is dissolved in water, the following reaction occurs:

$$NH_4^+(aq) + H_2O(l) \longrightarrow NH_3(aq) + H_3O^+(aq)$$

What are the Brønsted acid and the Brønsted base? Is the ammonium ion an Arrhenius acid, an Arrhenius base, or neither?

ANSWER: Brønsted acid, NH_4^+; Brønsted base, H_2O; ammonium, Arrhenius acid

EXAMPLE 2 Explaining an acid-base neutralization as a proton transfer reaction

When hydrochloric acid (HCl) is added to sodium sulfite (Na_2SO_3), a relatively large amount of sulfur dioxide gas (SO_2) is evolved. Account for this in terms of a proton-transfer reaction.

SOLUTION By analogy with the decomposition (given in the text) of carbonic acid (H_2CO_3) to carbon dioxide and water, we assume that SO_2 is produced by decomposition of H_2SO_3. The formation of H_2SO_3 occurs through a proton-transfer reaction in which the hydronium ion formed by HCl transfers a proton to a SO_3^{2-} ion (from Na_2SO_3). The net ionic equations for the overall process are

$$SO_3^{2-}(aq) + 2H_3O^+(aq) \longrightarrow 2H_2O(l) + H_2SO_3(aq)$$
$$H_2SO_3(aq) \longrightarrow H_2O(l) + SO_2(g)$$

EXERCISE Addition of a large amount of sodium hydroxide to ammonium chloride results in formation of ammonia gas. Show that a proton-transfer reaction is involved in this process.

ANSWER: $NH_4^+(aq) + OH^-(aq) \longrightarrow NH_3(g) + H_2O(l)$

REDOX REACTIONS

So far, we have discussed two classes of reactions: precipitation reactions and acid-base reactions. In this section we consider a third class: **oxidation-reduction** reactions. There are three different (but consistent) ways of looking at oxidation-reduction reactions: (1) as a loss or gain of electrons, (2) as a change in oxidation number, and (3) as a loss or gain of oxygen.

KEY WORDS Define or explain each of the following terms in a written sentence or two.

activity series	oxidation	redox reaction
disproportionation	oxidation number	reducing agent
electron transfer	oxidizing agent	reduction
half-reaction		

3.8 ELECTRON TRANSFER

KEY CONCEPT A Oxidation–reduction as electron transfer

Oxidation and reduction reactions involve the transfer of electrons from one substance to another. We start with the definitions that a substance is **oxidized** when it **loses electrons,** and is **reduced** when it **gains electrons.** (The mnemonic device "LEO the lion goes GER" can help you remember this: *L*ose *E*lectrons *O*xidation and *G*ain *E*lectrons *R*eduction). We should note that with this definition of oxidation and reduction, the electron transfer is assumed to be from one atom to another. Whenever oxidation occurs in a reaction, reduction *must* occur at the same time. It is impossible to have an oxidation without a reduction, or a reduction without an oxidation. A reaction in which oxidation and reduction occur is called a **redox** (reduction oxidation) reaction. The species that supplies electrons is the **reducing agent;** the one that accepts the electrons is the **oxidizing agent.**

EXAMPLE 1 Electron loss and gain in a redox reaction

Construct a table showing the changes that occur and identifying the oxidizing and reducing agents for the reaction

$$2AgNO_3(aq) + Cu(s) \longrightarrow Cu(NO_3)_2 + 2Ag(s)$$

SOLUTION We first analyze the charges of the reactants and products to see what electron transfers have occurred. For reactants, we have Ag^+ ion, NO_3^- ion, and a neutral Cu atom. The products contain the Cu^{2+} ion, the NO_3^- ion, and a neutral Ag atom. Thus, Ag^+ has gained an electron to form Ag and in doing so has been reduced (GER). Cu has lost two electrons to form Cu^{2+} and has therefore been oxidized (LEO). The NO_3^- ion is a spectator ion because it is in the same form on the right side of the equation as on the left. It has been neither oxidized nor reduced. The summary table is

Atom	Starting charge	Ending charge	Electron change (per atom)	Redox change	Agent
Cu	0	+2	loss of 2 electrons	oxidation	reducing agent
Ag	+1	0	gain of 1 electron	reduction	oxidizing agent

EXERCISE Construct a table showing the changes involved and the oxidizing and reducing agent for the reaction $Mg(s) + Cl_2(g) \rightarrow MgCl_2(s)$.

ANSWER:

Atom	Starting charge	Ending charge	Electron change (per atom)	Redox change	Agent
Mg	0	+2	loss of 2 electrons	oxidation	reducing agent
Cl	0	−1	gain of 1 electron	reduction	oxidizing agent

EXAMPLE 2 The substances oxidized and reduced; oxidizing and reducing agents

Identify the substance oxidized, the substance reduced, the oxidizing agent, and the reducing agent in the following equation:

$$Ca(s) + Cl_2(g) \longrightarrow CaCl_2(s)$$

SOLUTION The atom or substance that loses electrons is oxidized, and the atom or substance that gains electrons is reduced. In this reaction, the charge on calcium changes from zero to +2, so calcium loses electrons and is oxidized. The charge on chlorine changes from zero to −2, so chlorine gains electrons and is reduced. The electrons gained by chlorine come from calcium, so calcium is the reducing agent. The (same) electrons lost by calcium are accepted by chlorine, so chlorine is the oxidizing agent. As in all redox reactions, the oxidizing agent is reduced and the reducing agent is oxidized:

Reduced: Cl *Oxidized:* Ca
Oxidizing agent: Cl *Reducing agent:* Ca

EXERCISE Identify the substance oxidized, the substance reduced, the oxidizing agent, and the reducing agent in the following equation:

$$CuO(s) + H_2(g) \longrightarrow Cu(s) + H_2O(g)$$

ANSWER: Reduced, CuO (or Cu in CuO); oxidized, $H_2(g)$; oxidizing agent, CuO (or Cu in CuO); reducing agent, $H_2(g)$

KEY CONCEPT B Half-reactions in redox equations

It is profitable to think of a redox reaction as the simultaneous occurrence of two **half-reactions,** one involving oxidation, the other reduction. To illustrate, we consider the reaction between $Cu(s)$ and Ag^+ to give Cu^{2+} and $Ag(s)$. We have previously concluded that $Cu(s)$ is oxidized by loss of two electrons and that Ag^+ is reduced by gain of one electron in this reaction. To write the half-reactions, we write the oxidation and the reduction as separate reactions, with electrons as a reactant or product.

$$Cu(s) \longrightarrow Cu^{2+}(aq) + 2e^- \quad \text{oxidation half-reaction (loss of electrons)}$$

$$Ag^+(aq) + e^- \longrightarrow Ag(s) \quad \text{reduction half-reaction (gain of electrons)}$$

To add these half-reactions to get the overall reaction, we use the rule that an electron is never written as the reactant or product in a complete reaction, so the number of electrons lost through oxidization must be the same as the number gained through reduction. In our example, two Ag^+ ions are reduced by the electrons lost by one Cu(*s*):

$$\begin{array}{rll} Cu(s) & \longrightarrow Cu^{2+}(aq) + 2e^- & \text{oxidation (loss of 2 electrons)} \\ 2Ag^+(aq) + 2e^- & \longrightarrow 2Ag(s) & \text{reduction (gain of 2 electrons)} \\ \hline 2Ag^+(aq) + 2e^- + Cu(s) & \longrightarrow Cu^{2+}(aq) + 2e^- + 2Ag(s) & \end{array}$$

Canceling the two electrons that appear on both sides of the equation leaves us with the final net redox reaction:

$$2Ag^+(aq) + Cu(s) \longrightarrow Cu^{2+}(aq) + 2Ag(s)$$

▼ **EXAMPLE Writing redox equations**

Fe(*s*) reacts with $Cl_2(g)$ to give $FeCl_3(s)$. Write the half-reactions for this reaction and then add them to get the overall redox reaction. What is the oxidizing agent and what is the reducing agent?

SOLUTION We first analyze the changes in oxidation numbers that occur, in order to determine how electrons transfer. Fe(*s*) and $Cl_2(g)$ are both neutral elements with zero charge. The product contains the Fe^{3+} ion and Cl^- ion. Thus, Fe(*s*) loses three electrons to form Fe^{3+} (Fe is oxidized), and the Cl atoms in $Cl_2(g)$ each gain one electron to form Cl^- (Cl_2 is reduced):

$$\begin{array}{rll} Fe(s) & \longrightarrow Fe^{3+}(s) + 3e^- & \text{oxidation half-reaction} \\ Cl_2(g) + 2e^- & \longrightarrow 2Cl^-(s) & \text{reduction half-reaction} \end{array}$$

Because the Fe^{3+} ions and Cl^- ions are part of the solid $FeCl_3$, both are referred to with the symbol (*s*). To add the half-reactions to get the complete redox reaction, the number of electrons lost in the oxidation half-reaction must equal the number gained in the reduction half-reaction. To make the number of electrons equal, we multiply the oxidation half-reaction by 2 (to get $6e^-$) and the reduction half-reaction by 3 (to get $6e^-$). The result is

$$\begin{array}{rll} 2Fe(s) & \longrightarrow 2Fe^{3+}(s) + 6e^- & \text{oxidation half-reaction} \\ 3Cl_2(g) + 6e^- & \longrightarrow 6Cl^-(s) & \text{reduction half-reaction} \\ \hline 3Cl_2(g) + 6e^- + 2Fe(s) & \longrightarrow 2FeCl_3(s) + 6e^- & \end{array}$$

The Fe^{3+} ion and Cl^- ion do not exist as separate entities, but are bonded to each other in the ionic compound $FeCl_3$ and are written as such. The final complete reaction is

$$3Cl_2(g) + 2Fe(s) \longrightarrow 2FeCl_3(s)$$

Because Fe(*s*) is oxidized, it is the reducing agent; and because $Cl_2(g)$ is reduced, it is the oxidizing agent.

EXERCISE Write the half-reactions and add them to get the complete reaction in which the ore $MnO_2(s)$ reacts with Al(*s*) to form Mn(*s*) and $Al_2O_3(s)$. What are the oxidizing and reducing agents?

ANSWER:
$$\begin{array}{rl} 3Mn^{4+}(s) + 12e^- + 6O^{2-}(s) & \longrightarrow 3Mn(s) + 6O^{2-}(s) \\ 4Al(s) & \longrightarrow 4Al^{3+} + 12e^- \\ \hline 3MnO_2(s) + 4Al(s) & \longrightarrow 3Mn(s) + 2Al_2O_3(s) \end{array}$$

▲

(Note that, on the left side of this reaction, we break up the ionic compound MnO_2 into its separate ions to write the half-reaction and then reassemble the ions into MnO_2 to write the overall redox reaction. On the right side of the equation, the O^{2-} ion left over from the MnO_2 joins with the newly formed Al^{3+} ion to make the product Al_2O_3.)

3.9 THE ACTIVITY SERIES

KEY CONCEPT A Acids as oxidizing agents

Acids may act as oxidizing agents through an overall reaction in which the acid H^+ accepts electrons from a metal and forms H_2. Because the metal loses electrons, it is oxidized; and the acid, or more precisely the hydrogen in the acid, is the oxidizing agent. The metal, by donating electrons to H^+, reduces hydrogen and is, therefore, the reducing agent. Metals that lie above hydrogen in the activity series (such as Al, Zn, Fe, and Pb) are oxidized by acids in this manner whereas those that lie below hydrogen (such as Cu, Ag, and Hg) are not. Acids may also act as oxidizing agents through the behavior of the anion of the acid. For instance, the nitrate ion in nitric acid and the sulfate ion in sulfuric acid may, under the proper conditions, act at oxidizing agents.

EXAMPLE 1 Acids as oxidizing agents

Write the net ionic equation for the reaction between HCl and aluminum. State what is oxidized, what is reduced, and what the oxidizing and reducing agents are.

SOLUTION Acids act as oxidizing agents in two different ways: through the oxidizing ability of H^+ and through the oxidizing ability of the acid anion. In this case, because Al is above H_2 in the activity series, H^+ can oxidize aluminum; the reaction is

$$6H^+(aq) + 2Al(s) \longrightarrow 3H_2(g) + 2Al^{3+}(aq)$$

H^+ gains electrons and is reduced; it is the oxidizing agent. Al loses electrons and is oxidized; it is the reducing agent.

EXERCISE Write the net ionic equation for the reaction between iron and HCl. Assume conditions are such that iron forms Fe^{2+}. State what is oxidized, what is reduced, and what the oxidizing and reducing agents are.

ANSWER: $2H^+(aq) + Fe(s) \longrightarrow H_2(g) + Fe^{2+}(aq)$: Fe is oxidized and is the reducing agent; H^+ is reduced and is the oxidizing agent.

EXAMPLE 2 Acids as oxidizing agents

The net ionic equation for the reaction between copper and hot dilute nitric acid is

$$3Cu(s) + 8H^+(aq) + 2NO_3^-\ (aq,\ dilute) \longrightarrow 3Cu^{2+}(aq) + 2NO(g) + 4H_2O(l)$$

Name the substance oxidized, the substance reduced, the oxidizing agent, and the reducing agent.

SOLUTION Copper changes its charge from zero to +2, so it is oxidized and is the reducing agent. We have seen that when H^+ is reduced, it becomes H_2. Because no H_2 forms, we may conclude that H^+ is not the oxidizing agent and that NO_3^-, therefore, must be the oxidizing agent. It is reduced to NO. In summary:

Substance oxidized: Cu   *Species reduced:* NO_3^-
Reducing agent: Cu   *Oxidizing agent:* NO_3^-

EXERCISE The net ionic equation for the reaction between copper and hot concentrated nitric acid is

$$Cu(s) + 4H^+(aq) + 2NO_3^-\ (aq,\ conc.) \longrightarrow Cu^{2+}(aq) + 2NO_2(g) + 2H_2O(l)$$

Name the substance oxidized, the substance reduced, the oxidizing agent and the reducing agent.

ANSWER: Cu is oxidized and is the reducing agent; NO_3^- is reduced and is the oxidizing agent.

KEY CONCEPT B Oxidation numbers

An **oxidation number** (N_{ox}) is a number assigned to an atom according to the effective charge of the atom. The assignment is based on the set of rules given in Table 3.5 of the text. Work through the rules in the order given and stop when you have arrived at an oxidation number for an atom. Two inferences can be drawn from these rules (a) atoms in their elemental form (such as O_2, N_2, Na, and P_4) have zero oxidation number and (b) the oxidation state of a monatomic ion (such as Na^+, Cl^-, S^{2-}, and Al^{3+}) is the same as the charge of the ion. Rules higher in the table take precedence over rules that are lower. The rules will give the right answer for the oxidation numbers *most* of the time. (Some complicated compounds need a few additional rules, which we do not cover in this course.)

EXAMPLE Determining oxidation numbers

Determine the oxidation numbers of (a) Na in Na(*s*), (b) O in OF_2, and (c) Co in $Co_2O_4^{2-}$.

SOLUTION (a)Na in Na(*s*) is in the elemental form and, according to inference (a), must have oxidation number 0. (b) For OF_2, rule 1 tells us that the oxidation numbers of one O and two F must sum to zero. Rule 5 states that F must have oxidation number -1 in its compounds. The two F contribute -2 $(2 \times -1 = -2)$, so the O must have oxidation number $+2$. With this assignment, $N_{ox}(O) + 2 \times N_{ox}(F) = (+2) + (-2) = 0$; thus, the oxidation numbers sum to zero, as they should. The $+2$ oxidation number for O is very rare. (c) The sum of the oxidation numbers in $Co_2O_4^{2-}$ must sum to -2, the charge on the ion. Each O has a -2 oxidation number (rule 6), so the four O contribute -8 towards the -2 charge. This means that the two Co must contribute $+6$, since $(+6) + (-8) = -2$. If two Co contribute $+6$, each must contribute $+3$, and the oxidation number of Co is $+3$.

EXERCISE Determine the oxidation number of (a) Xe in XeO_6^{4-}, (b) H in H_2, and (c) Mg in $MgCl_2$.

ANSWER: (a) $+8$; (b) 0; (c) $+2$

PITFALL Oxidation numbers are not always actual charges

Even though the sum of all of the oxidation numbers of the atoms in a compound or molecular ion must equal the total charge on the compound or ion, it is not necessarily true that the individual oxidation number on an atom equals the actual charge on the atom. In some cases, the oxidation number equals the charge; in many cases, it doesn't.

KEY CONCEPT C Oxidation and reduction as a change in oxidation number

As we might expect from the term "oxidation number," the higher the oxidation number of an atom, the more oxidized it is. For example, SO_2 is oxidized when it reacts with O_2 to form SO_3. The oxidation number of S is $+4$ in SO_2 and $+6$ in SO_3, showing that the oxidation number is increased when oxidation occurs. From the point of view of electron transfer, sulfur must lose two electrons to go from a $+4$ oxidation number to a $+6$ oxidation number, confirming that oxidation, viewed as the loss of electrons, corresponds to oxidation, viewed as an increase in oxidation number. The following table summarizes the three points of view discussed in the text.

Process	Oxygen content	Electron transfer	Oxidation number
oxidation	increase in O content	loss of electrons	increase in oxidation number
reduction	decrease in O content	gain of electrons	decrease in oxidation number

EXAMPLE Using oxidation-number change to determine oxidation and reduction

Use the change in oxidation number to determine what is oxidized, what is reduced, the oxidizing agent, and the reducing agent in the reaction,

$$Cu(s) + 4HNO_3(aq,\ conc.) \longrightarrow Cu(NO_3)_2(aq) + 2NO_2(g) + 2H_2O(l)$$

SOLUTION When using the change in oxidation number to characterize a redox reaction, you must first determine the oxidation number of all the elements in the reaction. The table shows the changes.

	Left side of equation		Right side of equation		
Element	Species	Oxidation number	Species	Oxidation number	Process
Cu	Cu(*s*)	0	Cu^{2+}	+2	oxidation
H	H^+	+1	H_2O	+1	no change
N	NO_3^-	+5	NO_3^-	+5	no change
N	NO_3^-	+5	NO_2	+4	reduction
O	NO_3^-	−2	NO_3^-, NO_2, H_2O	−2	no change

Note that the oxidation number of Cu increases from 0 to +2, so Cu has been oxidized and is the reducing agent. The N in some of the NO_3^- has undergone a decrease in oxidation number, so it has been reduced. NO_3^- is the oxidizing agent, because the N that is reduced is part of the nitrate ion in HNO_3.

EXERCISE Use the change in oxidation number to determine what is oxidized, what is reduced, the oxidizing agent, and the reducing agent in the reaction that occurs in a lead storage battery:

$$2H_2SO_4(aq) + Pb(s) + PbO_2(s) \longrightarrow 2PbSO_4(s) + 2H_2O(l)$$

ANSWER: Pb(*s*) is oxidized and is the reducing agent; Pb in $PbO_2(s)$ is reduced; PbO_2 is the oxidizing agent.

3.10 BALANCING REACTIONS BY USING HALF-REACTIONS

KEY CONCEPT Balancing redox equations in acid solution

Some redox equations require a highly systematic approach for balancing. Because many of these reactions are run in aqueous solution with an acid or base present, we are allowed to add H_2O and H^+ (acid) or OH^- (base) in the balancing process. We outline the method for balancing with an example. Usually you will be provided with an unbalanced equation showing only the redox process that occurs:

$$As_2O_3(s) + HNO_3(aq) \longrightarrow H_3AsO_4(aq) + NO_2(g) \quad ⚠$$

Step 1. Write the net reaction for the unbalanced equation:

$$As_2O_3(s) + NO_3^-(aq) \longrightarrow AsO_4^{3-}(aq) + NO_2(g)$$

Step 2. Determine the oxidation numbers of all of the elements in the equation, and what has been oxidized and what reduced:

Atom	Left	Right	Process
As	+3	+5	oxidation
O	−2	−2	no change
N	+5	+4	reduction

Step 3. Write the half-reactions:

Oxidation half-reaction: $As_2O_3(s) \longrightarrow AsO_4^{3-}(aq)$
Reduction half-reaction: $NO_3^-(aq) \longrightarrow NO_2(g)$

Step 4. Mass balance the atom oxidized in the oxidation half-reaction and the atom reduced in the reduction half-reaction:

Oxidation half-reaction: $As_2O_3(s) \longrightarrow 2AsO_4^{3-}(aq)$
Reduction half-reaction: $NO_3^-(aq) \longrightarrow NO_2(g)$

Step 5. Balance O by adding the proper amount of H_2O to the right or left:

Oxidation half-reaction: $5H_2O + As_2O_3(s) \longrightarrow 2AsO_4^{3-}(aq)$
Reduction half-reaction: $NO_3^-(aq) \longrightarrow NO_2(g) + H_2O$

Step 6. Balance H by adding H^+ to the right or left:

Oxidation half-reaction: $5H_2O + As_2O_3(s) \longrightarrow 2AsO_4^{3-}(aq) + 10\,H^+$
Reduction half-reaction: $2H^+ + NO_3^-(aq) \longrightarrow NO_2(g) + H_2O$

Step 7. Balance the actual charge by adding electrons to the right or left. Charge is balanced when the total charge on the left equals the total charge on the right. The total charge may be positive, negative, or zero.

Oxidation half-reaction: $5H_2O + As_2O_3(s) \longrightarrow 2AsO_4^{3-}(aq) + 10H^+ + 4e^-$

Total charge = 0 Total charge = 0

Reduction half-reaction: $e^- + 2H^+ + NO_3^-(aq) \longrightarrow NO_2(g) + H_2O$

Total charge = 0 Total charge = 0

Notice that the electrons added are consistent with our notion that electron transfer can be used to describe redox processes. Four electrons are lost in the oxidation half-reaction as two arsenic atoms change from +2 to +4 oxidation number, and one electron is gained in the reduction half-reaction as a nitrogen atom changes from +5 to +4.

Step 8. Add the two half-reactions so that the electrons lost in the oxidation equal the electrons gained in the reduction. In this example, the reduction half-reaction must be multiplied by 4:

$$5H_2O(l) + As_2O_3(s) \longrightarrow 2AsO_4^{3-}(aq) + 10H^+(aq) + 4e^-$$
$$4e^- + 8H^+(aq) + 4NO_3^-(aq) \longrightarrow 4NO_2(g) + 4H_2O(l)$$
$$5H_2O(l) + As_2O_3(s) + 8H^+(aq) + 4NO_3^-(aq) \longrightarrow 2AsO_4^{3-}(aq) + 10H^+(aq) + 4NO_2(g) + 4H_2O(l)$$

Step 9. Cancel any species that appear on both sides of the equation. In our example, the four $H_2O(l)$ on the right cancel four of the five H_2O on the left and the eight H^+ on the left cancel eight of the ten on the right:

$$H_2O(l) + As_2O_3(s) + 4NO_3^-(aq) \longrightarrow 2AsO_4^{3-}(aq) + 2H^+(aq) + 4NO_2(g)$$

Step 10. Check mass balance and charge balance to make certain you have not made any mistakes:

Atom	**Left**	**Right**
H	2	2
O	16	16
As	2	2
N	4	4
charge	−4	−4

EXAMPLE Balancing a complex redox reaction in acid solution

Balance the following reaction in acid solution:

$$KIO_3(aq) + NaHSO_3(aq) \longrightarrow I_2(s) + K_2SO_4(aq)$$ ⚠

SOLUTION

Step 1. Write the net reaction for the unbalanced equation:

$$IO_3^-(aq) + HSO_3^-(aq) \longrightarrow I_2(s) + SO_4^{2-}(aq)$$

Step 2. Determine the oxidation numbers of all the elements in the equation, and find what has been oxidized and what reduced:

Atom	Left	Right	Process
I	+5	0	reduction
O	−2	−2	no change
S	+4	+6	oxidation
H	+1	+1	no change*

* We anticipate that H^+ will appear on the right after the equation is balanced.

Step 3. Write the half-reactions:

Oxidation half-reaction: $HSO_3^-(aq) \longrightarrow SO_4^{2-}(aq)$
Reduction half-reaction: $IO_3^-(aq) \longrightarrow I_2(s)$

Step 4. Mass balance the atom oxidized in the oxidation half-reaction and the atom reduced in the reduction half-reaction:

Oxidation half-reaction: $HSO_3^-(aq) \longrightarrow SO_4^{2-}(aq)$
Reduction half-reaction: $2IO_3^-(aq) \longrightarrow I_2(s)$

Step 5. Balance O by adding the proper amount of H_2O to the right or left:

Oxidation half-reaction: $H_2O(l) + HSO_3^-(aq) \longrightarrow SO_4^{2-}(aq)$
Reduction half-reaction: $2IO_3^-(aq) \longrightarrow I_2(s) + 6H_2O(l)$

Step 6. Balance H by adding H^+ to the right or left:

Oxidation half-reaction: $H_2O(l) + HSO_3^-(aq) \longrightarrow SO_4^{2-}(aq) + 3H^+$
Reduction half-reaction: $12H^+ + 2IO_3^-(aq) \longrightarrow I_2(s) + 6H_2O(l)$

Step 7. Balance the actual charge by adding electrons to the right or left. Charge is balanced when the total charge on the left equals the total charge on the right.

Oxidation half-reaction: $H_2O(l) + HSO_3^-(aq) \longrightarrow SO_4^{2-}(aq) + 3H^+ + 2e^-$

Total charge = −1   Total charge = −1

Reduction half-reaction: $10e^- + 12H^+ + 2IO_3^-(aq) \longrightarrow I_2(s) + 6H_2O(l)$

Total charge = 0   Total charge = 0

Step 8. Add the two half-reactions so that the electrons lost in the oxidation equal the electrons gained in the reduction:

$$5H_2O(l) + 5HSO_3^-(aq) \longrightarrow 5SO_4^{2-}(aq) + 15H^+ + 10e^-$$
$$10e^- + 12H^+ + 2IO_3^-(aq) \longrightarrow I_2(s) + 6H_2O(l)$$
$$5H_2O(l) + 5HSO_3^-(aq) + 12H^+ + 2IO_3^-(aq) \longrightarrow 5SO_4^{2-}(aq) + 15H^+ + I_2(s) + 6H_2O(l)$$

Step 9. Cancel any species that appear on both sides of the equation. In this example, the five $H_2O(l)$ on the left cancel five of the six on the right, and the twelve H^+ on the left cancel twelve of the fifteen on the right:

$$5HSO_3^-(aq) + 2IO_3^-(aq) \longrightarrow 5SO_4^{2-}(aq) + 3H^+(aq) + I_2(s) + H_2O(l)$$

Step 10. Check mass balance and charge balance to make certain you have not made any mistakes:

Atom	**Left**	**Right**
H	5	5
O	21	21
S	5	5
I	2	2
charge	−7	−7

EXERCISE Balance the equation $MnO_2(s) + 4HCl(aq) \rightarrow MnCl_2(aq) + Cl_2(g)$, in acid solution.

ANSWER: $MnO_2(s) + 4H^+(aq) + 2Cl^-(aq) \rightarrow Mn^{2+}(aq) + Cl_2(g) + 2H_2O(l)$

PITFALL Charge balancing an chemical equation

To be balanced, an equation must have the same total charge on the left as on the right. Balance does not mean that the charge on the right cancels the charge on the left but that charge on the left equals that on the right. For example, the following equation is mass balanced because the number of atoms of each element is the same on both sides of the equation. However, it is not charge balanced because the total charge on the left is +1 and the total charge on the right is +2:

$$Ag^+(aq) + Cu(s) \longrightarrow Cu^{2+}(aq) + Ag(s) \quad ⚠$$

Total charge +1 ≠ Total charge +2

To balance this equation, we multiply $Ag^+(aq)$ by 2 and $Ag(s)$ by 2:

$$2Ag^+(aq) + Cu(s) \longrightarrow Cu^{2+}(aq) + 2Ag(s)$$

Total charge +2 = Total charge +2

Atom	Left	Right
Ag	2	2
Cu	1	1
charge	+2	+2

DESCRIPTIVE CHEMISTRY TO REMEMBER

- A **synthesis reaction** is a reaction in which a substance is formed from simpler substances.
- A **decomposition reaction** is a reaction in which a substance is broken down into simpler substances.
- A **combustion reaction** is a reaction that occurs when a substance burns in oxygen.
- A **precipitation reaction** is a reaction in which a solid product is formed when two solutions are mixed.
- A **double replacement** (or **metathesis**) **reaction** is one in which two pairs of ions change partners during the reaction.
- **Ammonia** (NH_3) is an Arrhenius base.
- A **neutralization reaction** is a reaction in which an acid reacts with a base.
- A **disproportionation reaction** is a redox reaction in which a single substance is simultaneously oxidized and reduced.
- Sodium dichromate ($Na_2Cr_2O_7$), potassium dichromate ($K_2Cr_2O_7$) and potassium permanganate ($KMnO_4$) are common and useful **oxidizing agents.**
- A **photochemical reaction** is a reaction caused by light.

CHEMICAL EQUATIONS TO KNOW

- In a reaction typical of all carbonates, calcium carbonate ($CaCO_3$) decomposes to carbon dioxide and an oxide when heated.

$$CaCO_3(s) \xrightarrow{800^\circ} CaO(s) + CO_2(g)$$

- In a reaction typical of organic compounds, methane (CH_4) burns in a plentiful supply of air to form water and carbon dioxide. Butane (C_4H_{10}) undergoes the same reaction.

$$CH_4(g) + 2O_2(g) \longrightarrow CO_2(g) + 2H_2O(g)$$
$$2C_4H_{10}(g) + 13O_2(g) \longrightarrow 8CO_2(g) + 10H_2O(g)$$

- Bubbling nitrogen dioxide in water results in the formation of nitric acid and nitrogen monoxide.

$$3NO_2(g) + H_2O(l) \longrightarrow 2HNO_3(aq) + NO(g)$$

- The bright, yellow precipitate lead(II) chromate is formed when a solution of lead(II) nitrate is mixed with a solution of potassium chromate.

$$Pb(NO_3)_2(aq) + K_2CrO_4(aq) \longrightarrow 2KNO_3(aq) + PbCrO_4(s)$$

- In a reaction typical of metal oxides, calcium oxide (CaO) reacts with water to form calcium hydroxide:

$$CaO(s) + H_2O(l) \longrightarrow Ca(OH)_2(aq)$$

- In a redox reaction, zinc dissolves in hydrochloric acid to form hydrogen gas and aqueous zinc chloride:

$$Zn(s) + 2HCl(aq) \longrightarrow ZnCl_2(aq) + H_2(g)$$

- Copper reacts with hot dilute nitric acid to give copper(II) nitrate, nitric oxide (NO) and water:

$$3Cu(s) + 8HNO_3(aq) \longrightarrow 3Cu(NO_3)_2(aq) + 2NO(g) + 4H_2O(l)$$

- Copper reacts with hot concentrated nitric acid to give copper(II) nitrate, nitrogen dioxide (NO_2) and water

$$Cu(s) + 4HNO_3(aq) \longrightarrow Cu(NO_3)_2(aq) + 2NO_2(g) + 2H_2O(l)$$

- Chlorites have a tendency to disproportionate in aqueous solution, especially when heated.

$$3ClO_2^-(aq) \longrightarrow 2ClO_3^-(aq) + Cl^-(aq)$$

- The overall photochemical reaction that occurs when light hits photographic emulsions is

$$Ag^+(s) + Br^-(g) \xrightarrow{\text{light}} Ag(s) + Br(s)$$

SELF-TEST EXERCISES

Chemical equations

1. All of the equations given below, except one, are balanced. Which one is not?

(a) $6Li(s) + N_2(g) \longrightarrow 2Li_3N(s)$
(b) $2SeO_3(s) \longrightarrow 2SeO_2(s) + O_2(g)$
(c) $P_4O_{10}(s) + 5H_2O(l) \longrightarrow 4H_3PO_4(aq)$
(d) $2KNO_3(s) \longrightarrow 2KNO_2(s) + O_2(g)$
(e) $Cu(s) + 2H_2SO_4(aq) \longrightarrow CuSO_4(aq) + SO_2(g) + 2H_2O(l)$

2. Which of the following is a combustion reaction?
(a) $Zn(s) + 2HCl(aq) \longrightarrow ZnCl_2(aq) + H_2(g)$
(b) $SCl_4(aq) + 2H_2O(l) \longrightarrow SO_2(g) + 4HCl(aq)$
(c) $N_2(g) + 2H_2(g) \longrightarrow N_2H_4(l)$
(d) $2C_2H_6(g) + 7O_2(g) \longrightarrow 4CO_2(g) + 6H_2O(g)$
(e) $AgNO_3(aq) + NaCl(aq) \longrightarrow AgCl(s) + NaNO_3(aq)$

3. What coefficient appears in front of $O_2(g)$ when the following equation is balanced?

$$P_4(s) + O_2(g) \longrightarrow P_4O_{10}(s) \quad ⚠$$

(a) $\frac{5}{2}$ (b) 4 (c) 5 (d) 10 (e) 8

4. What coefficient appears in front of H_2O when the following equation is balanced?

$$CuSO_4 \cdot 5H_2O(s) \longrightarrow CuSO_4(s) + H_2O(g) \quad ⚠$$

(a) 5 (b) 1 (c) 10 (d) 3 (e) 4

5. What coefficient appears in front of H_2O when the following equation is balanced?

$$Ca(OH)_2(aq) + HCl(aq) \longrightarrow CaCl_2(aq) + H_2O(l) \quad ⚠$$

(a) 5 (b) 3 (c) 6 (d) 2 (e) 1

6. What coefficient appears in front of Se when the following equation is balanced?

$$Se_2Br_2(l) + H_2O(l) \longrightarrow HBr(g) + H_2SeO_3(aq) + Se(s) \quad ⚠$$

(a) 5 (b) 4 (c) 3 (d) 1 (e) 2

7. What coefficient appears in front of NaOH when the following equation is balanced?

$$MnO_2(s) + NaOH(aq) + O_2(g) \longrightarrow Na_2MnO_4(aq) + H_2O(l) \quad ⚠$$

(a) 2 (b) 4 (c) 1 (d) 6 (e) 3

8. What coefficient appears in front of O_2 when the following equation is balanced?

$$NH_3(g) + O_2(g) \longrightarrow NO(g) + H_2O(g) \quad ⚠$$

(a) 4 (b) 5 (c) 2 (d) 1 (e) 3

9. What coefficient appears in front of SiH_4 when the following equation is balanced?

$$HCl(g) + Mg_2Si(s) \longrightarrow MgCl_2(s) + SiH_4(g) \quad ⚠$$

(a) 2 (b) 1 (c) 4 (d) 3 (e) 5

Precipitation reactions

10. Which of the following is a precipitation reaction?
(a) $2Ca(s) + O_2(g) \longrightarrow 2CaO(s)$
(b) $2HNO_3(aq) + CaO(s) \longrightarrow Ca(NO_3)_2(aq) + H_2O(l)$
(c) $MgCO_3(s) \longrightarrow MgO(s) + CO_2(g)$
(d) $SiO_2(s) + 4HF(aq) \longrightarrow SiF_4(g) + 2H_2O(l)$
(e) $K_2SO_4(aq) + Ba(NO_3)_2(aq) \longrightarrow BaSO_4(s) + 2KNO_3(aq)$

11. What are the spectator ions in the following equation?

$$SrCl_2(aq) + K_2SO_4(aq) \longrightarrow SrSO_4(s) + 2KCl(aq)$$

(a) Sr^{2+} only (b) K^+ only (c) Sr^+, $SO_4{}^{2-}$ (d) Sr^{2+}, K^+ (e) K^+, Cl^-

12. What are the spectator ions in the following equation?

$$Na_2S(aq) + Ba(NO_3)_2(aq) \longrightarrow BaS(s) + 2NaNO_3(aq)$$

(a) Na^+, NO_3^- (b) Na^+ only (c) Na^+, Ba^{2+} (d) S^{2-}, NO_3^- (e) Ba^{2+}, S^{2-}

13. Which of the following is a metathesis reaction?
(a) $2H_2(g) + O_2(g) \longrightarrow 2H_2O(l)$
(b) $2AgNO_3(aq) + Cu(s) \longrightarrow 2Ag(s) + Cu(NO_3)_2(aq)$
(c) $H_2(g) + I_2(g) \longrightarrow 2HI(g)$
(d) $Fe_2O_3(s) + 3CO(g) \longrightarrow 2Fe(s) + 3CO_2(g)$
(e) $2KI(aq) + PbSO_4(aq) \longrightarrow PbI_2(s) + K_2SO_4(aq)$

14. What is the net ionic equation for the reaction of $Pb(NO_3)_2$ with H_2SO_4?
(a) $Pb(NO_3)_2(aq) + SO_4^{2-}(aq) \longrightarrow PbSO_4(s) + 2NO_3^-(aq)$
(b) $Pb(NO_3)_2(aq) + H_2SO_4(aq) \longrightarrow PbSO_4(s) + 2HNO_3(aq)$
(c) $Pb^{2+}(aq) + SO_4^{2-}(aq) \longrightarrow PbSO_4(s)$
(d) $Pb^{2+}(aq) + H_2SO_4(aq) \longrightarrow PbSO_4(s) + 2H^+(aq)$

Acid-base reactions

15. Which of the following is an Arrhenius acid?
(a) NaOH (b) CH_4 (c) H_2O (d) NH_3 (e) $HC_2H_3O_2$

16. All of the following but one are Arrhenius bases. Which one is not a base?
(a) NaOH (b) C_2H_5OH (c) NH_3 (d) CaO (e) $Mg(OH)_2$

17. In the following reaction, what acts as the Brønsted base?

$$CaO(aq) + H_2O(l) \longrightarrow Ca(OH)_2(aq)$$

(a) Ca^{2+} (b) O^{2-} (c) H^+ (d) H_2O (e) OH^-

18. An aqueous solution of $HCHO_2$ contains approximately 96 molecules of unionized $HCHO_2$ for every 4 ionized (into H^+ and CHO_2^-) molecules. How would you classify $HCHO_2$?
(a) weak acid (b) strong acid (c) strong base (d) weak base (e) salt

19. What is O^{2-} converted to when it is dissolves in water?
(a) H_2O (b) H_2O_2 (c) H_3O^+ (d) OH^- (e) O_2

20. Which of the following is a weak base?
(a) NaOH (b) CH_3NH_2 (c) CH_4 (d) $Ca(OH)_2$ (e) $HC_2H_3O_2$

21. Which of the reactions shown is a neutralization reaction?
(a) $CH_4(g) + 2O_2(g) \longrightarrow CO_2(g) + 2H_2O(l)$
(b) $Mg(s) + H_2O(g) \longrightarrow MgO(s) + H_2(g)$
(c) $KOH(aq) + H_3PO_4(aq) \longrightarrow KH_2PO_4(aq) + H_2O(l)$
(d) $MgCO_3(s) \longrightarrow MgO(s) + CO_2(g)$
(e) $CaO(s) + H_2O(l) \longrightarrow Ca(OH)_2(s)$

22. What salt is formed when HBr is neutralized with $Ca(OH)_2$?
(a) $CaBr_2$ (b) CaO (c) CaH_2 (d) $CaCl_2$ (e) $Ca(BrO_3)_2$

Redox reactions

23. What is the reducing agent in the reaction $MgO(s) + H_2(g) \rightarrow H_2O(l) + Mg(s)$?
(a) H_2 (b) MgO (c) H_2O (d) Mg

24. What is reduced in the reaction $C(s) + 2N_2O(g) \rightarrow CO_2(g) + 2N_2(g)$?
(a) C (b) N_2O (c) CO_2 (d) N_2

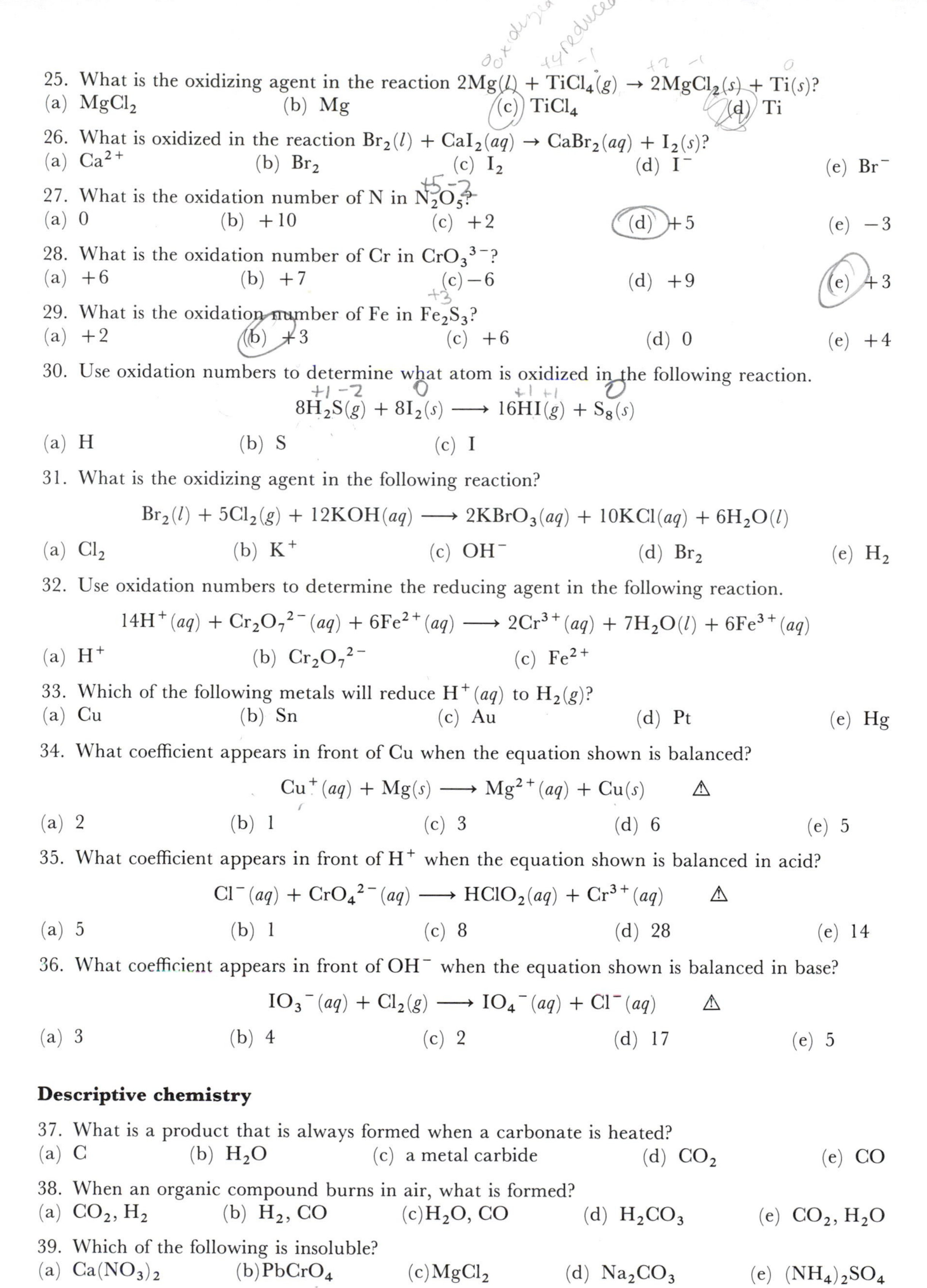

25. What is the oxidizing agent in the reaction $2Mg(l) + TiCl_4(g) \rightarrow 2MgCl_2(s) + Ti(s)$?
(a) $MgCl_2$ (b) Mg (c) $TiCl_4$ (d) Ti

26. What is oxidized in the reaction $Br_2(l) + CaI_2(aq) \rightarrow CaBr_2(aq) + I_2(s)$?
(a) Ca^{2+} (b) Br_2 (c) I_2 (d) I^- (e) Br^-

27. What is the oxidation number of N in N_2O_5?
(a) 0 (b) +10 (c) +2 (d) +5 (e) −3

28. What is the oxidation number of Cr in CrO_3^{3-}?
(a) +6 (b) +7 (c) −6 (d) +9 (e) +3

29. What is the oxidation number of Fe in Fe_2S_3?
(a) +2 (b) +3 (c) +6 (d) 0 (e) +4

30. Use oxidation numbers to determine what atom is oxidized in the following reaction.

$$8H_2S(g) + 8I_2(s) \longrightarrow 16HI(g) + S_8(s)$$

(a) H (b) S (c) I

31. What is the oxidizing agent in the following reaction?

$$Br_2(l) + 5Cl_2(g) + 12KOH(aq) \longrightarrow 2KBrO_3(aq) + 10KCl(aq) + 6H_2O(l)$$

(a) Cl_2 (b) K^+ (c) OH^- (d) Br_2 (e) H_2

32. Use oxidation numbers to determine the reducing agent in the following reaction.

$$14H^+(aq) + Cr_2O_7^{2-}(aq) + 6Fe^{2+}(aq) \longrightarrow 2Cr^{3+}(aq) + 7H_2O(l) + 6Fe^{3+}(aq)$$

(a) H^+ (b) $Cr_2O_7^{2-}$ (c) Fe^{2+}

33. Which of the following metals will reduce $H^+(aq)$ to $H_2(g)$?
(a) Cu (b) Sn (c) Au (d) Pt (e) Hg

34. What coefficient appears in front of Cu when the equation shown is balanced?

$$Cu^+(aq) + Mg(s) \longrightarrow Mg^{2+}(aq) + Cu(s)$$ ⚠

(a) 2 (b) 1 (c) 3 (d) 6 (e) 5

35. What coefficient appears in front of H^+ when the equation shown is balanced in acid?

$$Cl^-(aq) + CrO_4^{2-}(aq) \longrightarrow HClO_2(aq) + Cr^{3+}(aq)$$ ⚠

(a) 5 (b) 1 (c) 8 (d) 28 (e) 14

36. What coefficient appears in front of OH^- when the equation shown is balanced in base?

$$IO_3^-(aq) + Cl_2(g) \longrightarrow IO_4^-(aq) + Cl^-(aq)$$ ⚠

(a) 3 (b) 4 (c) 2 (d) 17 (e) 5

Descriptive chemistry

37. What is a product that is always formed when a carbonate is heated?
(a) C (b) H_2O (c) a metal carbide (d) CO_2 (e) CO

38. When an organic compound burns in air, what is formed?
(a) CO_2, H_2 (b) H_2, CO (c) H_2O, CO (d) H_2CO_3 (e) CO_2, H_2O

39. Which of the following is insoluble?
(a) $Ca(NO_3)_2$ (b) $PbCrO_4$ (c) $MgCl_2$ (d) Na_2CO_3 (e) $(NH_4)_2SO_4$

40. What ion is formed when NH_3 reacts with a Brønsted acid?
(a) NH_2^- (b) N^{3-} (c) H^+ (d) NH_4^+ (e) Cl^-

41. Which of the following tends to disproportionate when heated?
(a) Fe^{2+} (b) ClO_2^- (c) ClO_3^- (d) $CaCO_3$ (e) NH_3

42. What products form when nitrogen dioxide is bubbled in water?
(a) N_2O, H_2O_2 (b) HNO_3, NO (c) HNO_3, N_2O
(d) HNO_2, NO (e) HNO_2, N_2O

43. What photochemical reaction occurs when light hits photographic film?
(a) $Ag^+(s) + Br^-(s) \longrightarrow Ag(s) + Br(s)$ (b) $2Ag^+(s) + Cu(s) \longrightarrow Cu^{2+}(s) + 2Ag(s)$
(c) $Ag^+(s) + Br^-(s) \longrightarrow AgBr(s)$ (d) $2Ag(s) + S(s) \longrightarrow Ag_2S(s)$

4 REACTION STOICHIOMETRY

In this chapter we explore the quantitative relationships inherent in a balanced chemical equation. The law of conservation of mass tells us that the mass of the products in a chemical reaction must equal the mass of the reactants. Now we discover how the overall mass is divided among the various reactants and products.

INTERPRETING STOICHIOMETRIC COEFFICIENTS

KEY WORDS Define or explain each of the following terms in a written sentence or two.

empirical formula	limiting reactant	stoichiometric coefficients
gravimetric analysis	percentage yield	theoretical yield
law of constant composition		

4.1 MOLE CALCULATIONS

KEY CONCEPT A Stoichiometric coefficients as moles of a substance

The stoichiometric coefficients in a chemical equation can be interpreted as the number of moles of reactants and products that react:

$$\underset{1\text{ mol } N_2}{N_2(g)} + \underset{3\text{ mol } H_2}{3H_2(g)} \longrightarrow \underset{2\text{ mol } NH_3}{2NH_3(g)}$$

$$\underset{2\text{ mol } KClO_3}{2KClO_3(s)} \longrightarrow \underset{2\text{ mol } KCl}{2KCl(s)} + \underset{3\text{ mol } O_2}{3O_2(g)}$$

EXAMPLE Interpretating equations in terms of moles

Write an interpretation of the following equation in terms of the moles of reactants and products:

$$Mg_3N_2(s) + 6H_2O(l) \longrightarrow 3Mg(OH)_2(s) + 2NH_3(g)$$

SOLUTION The coefficient in front of a product or reactant represents the number of moles of each reactant participating in the reaction and the number of moles of each product that results. For this reaction, 1 mol Mg_3N_2 reacts with 6 mol H_2O to form 3 mol $Mg(OH)_2$ and 2 mol NH_3. More specifically,

$$\underset{1\text{ mol } Mg_3N_2}{Mg_3N_2(s)} + \underset{6\text{ mol } H_2O}{6H_2O(l)} \longrightarrow \underset{3\text{ mol } Mg(OH)_2}{3Mg(OH)_2(s)} + \underset{2\text{ mol } NH_3}{2NH_3(g)}$$

EXERCISE Write a sentence that interprets the following equation in terms of the moles of reactants and products:

$$C_2H_6(g) + \tfrac{7}{2}O_2(g) \longrightarrow 2CO_2(g) + 3H_2O(g)$$

ANSWER: 1 mol C_2H_6 reacts with 3.5 mol O_2 to form 2 mol CO_2 and 3 mol H_2O.

KEY CONCEPT B Mole calculations

The interpretation of stoichiometric coefficients as moles of reactants and products enables us to do calculations involving these quantities. We approach calculations by setting up the proper conversion factors, based on the coefficients, and using unit conversion. Let us consider the following equation,

$$P_4(s) + 5O_2(g) \longrightarrow P_4O_{10}(s)$$

and the meaning of the coefficients and the conversion factors that arise. The coefficients tell us that 1 mol P_4 reacts with 5 mol O_2 to yield 1 mol P_4O_{10}. This means that 1 mol P_4 requires 5 mol O_2 to completely react and that 5 mol O_2 requires 1 mol P_4 to completely react. In addition, when the 1 mol P_4 and 5 mol O_2 do react, 1 mol P_4O_{10} should result. From these coefficients, we get the following conversion factors:

Because 1 mol P_4 requires 5 mol O_2, we get $\frac{1 \text{ mol } P_4}{5 \text{ mol } O_2}$ and $\frac{5 \text{ mol } O_2}{1 \text{ mol } P_4}$

Because 1 mol P_4 can give 1 mol P_4O_{10}, we get $\frac{1 \text{ mol } P_4}{1 \text{ mol } P_4O_{10}}$ and $\frac{1 \text{ mol } P_4O_{10}}{1 \text{ mol } P_4}$

Because 5 mol O_2 can give 1 mol P_4O_{10}, we get $\frac{5 \text{ mol } O_2}{1 \text{ mol } P_4O_{10}}$ and $\frac{1 \text{ mol } P_4O_{10}}{5 \text{ mol } O_2}$

EXAMPLE Calculating moles of product from moles of reactant

How many moles of P_4O_{10} can be produced from 2.27 mol of O_2?

SOLUTION Here we seek a relationship between moles of P_4O_{10} and moles of O_2:

$$\frac{5 \text{ mol } O_2}{1 \text{ mol } P_4O_{10}} \quad \text{and} \quad \frac{1 \text{ mol } P_4O_{10}}{5 \text{ mol } O_2}$$

To set up the unit conversion, we start with the given amount of O_2 and multiply by the proper conversion factor to get the correct units for the answer. We note that the number of moles of P_4O_{10} formed is less than the number of moles of O_2, so we expect the answer to be less than 2.27 mol:

$$2.27 \cancel{\text{ mol } O_2} \times \frac{1 \text{ mol } P_4O_{10}}{5 \cancel{\text{ mol } O_2}} = 0.454 \text{ mol } P_4O_{10}$$

EXERCISE How many moles of P_4 are required to react with 2.27 mol O_2?

ANSWER: 0.454 mol

KEY CONCEPT C The masses taking part in reactions

We discovered that it is always possible to find the moles of a substance from the mass of the substance, or the mass from the moles, using the molar mass of the substance:

Moles $\xleftrightarrow{\text{molar mass}}$ Mass

Now, a chemical equation relates the moles of reactants and products in a reaction:

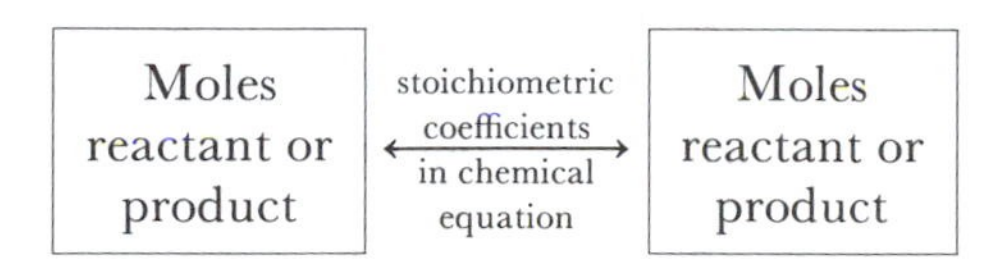

By combining these two relationships, we can in a three-step calculation relate the mass of a reactant or product in a chemical reaction to the mass of another reactant or product. In a typical problem, the mass of a reactant or product is given and we are asked to calculate the mass of another reactant or product (the unknown). The three steps are illustrated in the following example.

▼ **EXAMPLE 1 Calculating the mass of reactant needed to react with another reactant**

How many grams of H_2O are needed to react with 25.0 g of P_4O_{10} in the reaction given below?

$$P_4O_{10}(s) + 6H_2O(l) \longrightarrow 4H_3PO_4(aq)$$

SOLUTION We use a three-step calculation.

Step 1. Use the molar mass of the given substance to calculate the moles of the substance. In this case, we use the molar mass of P_4O_{10}, which is 283.89 g/mol:

$$25.0\ \cancel{g\ P_4O_{10}} \times \frac{1\ mol\ P_4O_{10}}{283.89\ \cancel{g\ P_4O_{10}}} = 0.0881\ mol\ P_4O_{10}$$

Step 2. Use the balanced chemical equation to determine the moles of the unknown reactant or product. The equation tells us that 6 mol of H_2O are required for every mole of P_4O_{10}:

$$0.0881\ \cancel{mol\ P_4O_{10}} \times \frac{6\ mol\ H_2O}{1\ \cancel{mol\ P_4O_{10}}} = 0.529\ mol\ H_2O$$

Step 3. Use the molar mass of the unknown to determine its mass. Here, we use the molar mass of H_2O, which is 18.02 g/mol:

$$0.529\ \cancel{mol\ H_2O} \times \frac{18.02\ g\ H_2O}{1\ \cancel{mol\ H_2O}} = 9.53\ g\ H_2O$$

We can also do this problem in a single step by using the three conversion factors, one after the other:

$$25.0\ \cancel{g\ P_4O_{10}} \times \frac{1\ \cancel{mol\ P_4O_{10}}}{283.89\ \cancel{g\ P_4O_{10}}} \times \frac{6\ \cancel{mol\ H_2O}}{1\ \cancel{mol\ P_4O_{10}}} \times \frac{18.02\ g\ H_2O}{1\ \cancel{mol\ H_2O}} = 9.52\ g\ H_2O$$

You may choose to do unit conversions with separate steps or with a single long step; use the method you are most comfortable with. The small difference in the two answers in step 3 is due to round-off error in the first method of calculation.

EXERCISE How many grams of H_3PO_4 would be expected to form from 100 g H_2O in the reaction just given?

▲ ANSWER: 363 g

▼ **EXAMPLE 2 Calculating the mass of product from the amount of reactant**

How many grams of hydrogen cyanide (HCN) can be produced from the complete reaction of 50.0 g O_2 from the reaction

$$2CH_4(g) + 3O_2(g) + 2NH_3(g) \longrightarrow 2HCN(g) + 6H_2O(g)$$

SOLUTION

Step 1. Use the molar mass of the given substance to calculatee the moles of the substance:

$$50.0\ \cancel{g\ O_2} \times \frac{1\ mol\ O_2}{32.0\ \cancel{g\ O_2}} = 1.56\ mol\ O_2$$

Step 2. Use the balanced chemical equation to determine the moles of the unknown reactant or product:

$$1.56\ \cancel{mol\ O_2} \times \frac{2\ mol\ HCN}{3\ \cancel{mol\ O_2}} = 1.04\ mol\ HCN$$

Step 3. Use the molar mass of the unknown to determine its mass:

$$1.04\ \cancel{mol\ HCN} \times \frac{27.06\ g}{1\ \cancel{mol\ HCN}} = 28.1\ g\ HCN$$

Again, we can do this calculation in one long step:

$$50.0\ \text{g } O_2 \times \frac{1\ \text{mol } O_2}{32.0\ \text{g}} \times \frac{2\ \text{mol HCN}}{3\ \text{mol } O_2} \times \frac{27.06\ \text{g}}{1\ \text{mol HCN}} = 28.2\ \text{g HCN}$$

EXERCISE How many grams of HNO_3 can be produced from the complete reaction of 65.0 g NO_2?

$$3NO_2(g) + H_2O(l) \longrightarrow 2HNO_3(aq) + NO(g)$$

ANSWER: 59.4 g

4.2 LIMITING REACTANTS

KEY CONCEPT A Limiting reactants

Part of the process of getting zinc metal from its ore involves converting zinc sulfide to its oxide:

$$2ZnS(s) + 3O_2(g) \longrightarrow 2ZnO(s) + 2SO_2(g)$$

In this reaction, the stoichiometric coefficients tell us that 3 mol O_2 are required to react with 2 mol ZnS. Now imagine a situation in which a chemist mixes 100.0 mol O_2 with 1.0 mol ZnS. Will all of the ZnS react? Will all of the O_2 react? (Answer these questions before you proceed.) It is evident that there is a great excess of oxygen. All of the ZnS will react, but most of the O_2 will not. In fact, we can calculate that 1.0 mol ZnS requires 1.5 mol O_2; thus, when all of the ZnS reacts, 98.5 mol O_2 remain unreacted (100.0 mol − 1.5 mol = 98.5 mol). In this situation, ZnS is called the **limiting reactant** because it limits the extent of reaction and the amount of product formed. To analyze a limiting reactant problem, we use the three-step procedure illustrated in the following example.

EXAMPLE Identifying the limiting reactant

48.0 g of O_2 is mixed with 85.0 g ZnS preparatory to converting the zinc sulfide to zinc oxide using the reaction $2ZnS(s) + 3O_2(g) \rightarrow 2ZnO(s) + 2SO_2(g)$. What is the limiting reactant?

SOLUTION We use a three-step procedure.

Step 1. Calculate the moles of each reactant present using the molar mass of each. The molar mass of O_2 is 32.0 g/mol and that of ZnS is 97.4 g/mol:

$$48.0\ \text{g } O_2 \times \frac{1\ \text{mol } O_2}{32.00\ \text{g } O_2} = 1.50\ \text{mol } O_2$$

$$85.0\ \text{g ZnS} \times \frac{1\ \text{mol ZnS}}{97.44\ \text{g ZnS}} = 0.872\ \text{mol ZnS}$$

Step 2. Calculate the moles of any one of the products that should form from the given amount of each reactant:

$$1.50\ \text{mol } O_2 \times \frac{2\ \text{mol ZnO}}{3\ \text{mol } O_2} = 1.00\ \text{mol ZnO}$$

$$0.872\ \text{mol ZnS} \times \frac{2\ \text{mol ZnO}}{2\ \text{mol ZnS}} = 0.872\ \text{mol ZnO}$$

Step 3. The reactant that results in the formation of the smallest amount of product is the limiting reactant. The amount of product that forms is determined by the limiting reactant. Because the smaller amount of product results from the amount of ZnS present, ZnS is the limiting reactant. (In addition, 0.872 mol of ZnO should form.)

EXERCISE 110 g CO is mixed with 355 g I_2O_5 for the following reaction; what is the limiting reactant?

$$5CO(g) + I_2O_5(s) \longrightarrow I_2(s) + 5CO_2(g)$$

ANSWER: CO

PITFALL The factors that determine the limiting reactant

The limiting reactant is not necessarily the reagent that is present with the smallest mass or the fewest moles. For example, in the problems just given, the following results were obtained:

Reagent	Grams present	Moles present	Limiting reagent?
ZnS	85.0	0.872	yes
O_2	48.0	1.50	no
CO	110	3.7	yes
I_2O_5	355	1.1	no

In the worked problem, there are more grams of ZnS than of O_2, but ZnS is still the limiting reactant. In the exercise problem, there are more moles of CO than of I_2O_5, but CO is still the limiting reactant. The identity of the limiting reactant depends on the mass of each reactant present, the molar mass of each, and the stoichiometric coefficients in the balanced chemical equation.

KEY CONCEPT B Theoretical and percentage yields

The **theoretical yield** is the mass of product we would expect if *all* of the limiting reactant converted to the expected product; it is, therefore, the maximum mass of product obtainable in a given reaction. At times, not all of a reactant converts to the desired product. For a variety of reasons, the actual amount of product obtained is often less than the expected theoretical yield. The **percentage yield** relates the theoretical yield to the amount of product obtained in an actual laboratory or industrial process:

$$\text{Percentage yield} = \frac{\text{actual mass of product obtained}}{\text{theoretical yield of product}} \times 100$$

EXAMPLE Calculating the percentage yield

When 2.3 kg CS_2 reacts with excess Cl_2, 3.6 kg CCl_4 is formed. What is the percentage yield of CCl_4?

$$CS_2(l) + 2Cl_2(g) \longrightarrow CCl_4(l) + 2S(s)$$

SOLUTION We must substitute into the equation

$$\text{Percentage yield} = \frac{\text{actual mass of product obtained}}{\text{theoretical yield of product}} \times 100$$

The actual mass of the product CCl_4 is given as 3.6 kg, so we have

$$\text{Percentage yield} = \frac{3.6\text{ kg}}{\text{theoretical yield of } CCl_4} \times 100$$

Now we need the theoretical yield, or the amount we expect, from the type of calculation illustrated earlier.

Step 1. Use the molar mass of the given substance to calculate the moles of the substance:

$$2.3\ \cancel{\text{kg } CS_2} \times \frac{10^3\ \cancel{\text{g } CS_2}}{1\ \cancel{\text{kg } CS_2}} \times \frac{1\text{ mol } CS_2}{76.13\ \cancel{\text{g } CS_2}} = 30\text{ mol } CS_2$$

Step 2. Use the balanced chemical equation to determine the moles of the unknown reactant or product:

$$30\ \cancel{\text{mol } CS_2} \times \frac{1\text{ mol } CCl_4}{1\ \cancel{\text{mol } CS_2}} = 30\text{ mol } CCl_4$$

Step 3. Use the molar mass of the unknown to determine its mass.

$$30\ \text{mol CCl}_4 \times \frac{154\ \text{g CCl}_4}{1\ \text{mol CCl}_4} \times \frac{1\ \text{kg}}{1000\ \text{g}} = 4.6\ \text{kg CCl}_4$$

Thus, 4.6 kg is the theoretical yield of CCl_4. It must be substituted in the defining equation for percentage yield to get the final answer:

$$\text{Percentage yield} = \frac{3.6\ \text{kg}}{4.6\ \text{kg}} \times 100 = 78\%$$

EXERCISE What is the percentage yield of Fe if the reaction of 233 g of Fe_2O_3 with excess C results in 128 g Fe?

$$3C(s) + 2Fe_2O_3(s) \longrightarrow 4Fe(s) + 3CO_2(g)$$

ANSWER: 78.5%

4.3 CHEMICAL COMPOSITION FROM MEASUREMENTS OF MASS

KEY CONCEPT Combustion analysis

Combustion analysis is typically done on organic compounds containing carbon, hydrogen, and perhaps some other element. The goal of the analysis is to determine either the mass percentage of each element in the compound or the empirical formula of the compound. The analysis is done by combusting the compound in a large excess of O_2, so that CO_2 and H_2O are the products. Three key facts underlie the experiment:

1. All the carbon atoms present in the original sample end up in the CO_2 formed by the combustion. In terms of conversion factors, there is 1 mol C in the original sample per mol CO_2 produced:

$$\frac{1\ \text{mol C (original sample)}}{1\ \text{mol CO}_2}$$

2. All of the hydrogen atoms in the original sample end up in the H_2O formed by the combustion. Because there are 2 mol H per mol H_2O (as indicated by the formula H_2O), 2 mol H in the original sample will produce 1 mol H_2O in the products. In terms of conversion factors:

$$\frac{2\ \text{mol H (original sample)}}{1\ \text{mol H}_2\text{O}}$$

3. The oxygen atoms in the CO_2 and H_2O produced do not originate with the sample.

EXAMPLE 1 Determining an empirical formula through combustion analysis

Analysis of 1.008 g of a compound containing only carbon and hydrogen results in the formation of 3.163 g CO_2 and 1.294 g H_2O in a combustion analysis experiment. What is the empirical formula of the compound?

SOLUTION For a compound containing only carbon and hydrogen, we use three steps.

Step 1. Use the molar mass of CO_2 and H_2O to calculate the moles of each produced:

$$3.163\ \text{g CO}_2 \times \frac{1\ \text{mol CO}_2}{44.01\ \text{g CO}_2} = 0.07187\ \text{mol CO}_2$$

$$1.294\ \text{g H}_2\text{O} \times \frac{1\ \text{mol H}_2\text{O}}{18.02\ \text{g H}_2\text{O}} = 0.07181\ \text{mol H}_2\text{O}$$

Step 2. Calculate the moles of C and moles of H in the original sample:

$$0.07187 \text{ mol } CO_2 \times \frac{1 \text{ mol C (original sample)}}{1 \text{ mol } CO_2} = 0.07187 \text{ mol C (in original sample)}$$

$$0.07181 \text{ mol } H_2O \times \frac{2 \text{ mol H (original sample)}}{1 \text{ mol } H_2O} = 0.1436 \text{ mol H (in original sample)}$$

Step 3. Find the simplest whole number ratio of moles of elements in the compound. This defines the empirical formula:

$$\frac{0.07187 \text{ mol C}}{0.07187} = 1 \text{ mol C}$$

$$\frac{0.1436 \text{ mol H}}{0.07187} = 1.998 \text{ mol H} = 2 \text{ mol H} \quad \text{(an integer)}$$

The empirical formula is CH_2.

EXERCISE Combustion of 1.333 g of a sample of a compound containing only carbon and hydrogen results in the formation of 4.334 g CO_2 and 1.332 g H_2O. What is the empirical formula of the compound?

ANSWER: C_2H_3

EXAMPLE 2 Determining an empirical formula through combustion analysis

Combustion of 1.426 g of an unknown sample containing carbon, hydrogen, and oxygen results in formation of 1.394 g CO_2 and 0.2855 g H_2O. What is the empirical formula of the compound?

SOLUTION Because the compound contains more than just hydrogen and carbon, some extra work (steps 2A and 2B) is required to calcuate the amount of oxygen present.

Step 1. Use the molar mass of CO_2 and H_2O to calculate the moles of each produced:

$$1.394 \text{ g } CO_2 \times \frac{1 \text{ mol } CO_2}{44.01 \text{ g } CO_2} = 0.03167 \text{ mol } CO_2$$

$$0.2855 \text{ g } H_2O \times \frac{1 \text{ mol } H_2O}{18.02 \text{ g } H_2O} = 0.01584 \text{ mol } H_2O$$

Step 2. Calculate the moles of C and moles of H in the original sample:

$$0.03167 \text{ mol } CO_2 \times \frac{1 \text{ mol C (original sample)}}{1 \text{ mol } CO_2} = 0.03167 \text{ mol C (in original sample)}$$

$$0.01584 \text{ mol } H_2O \times \frac{2 \text{ mol H (original sample)}}{1 \text{ mol } H_2O} = 0.03168 \text{ mol H (in original sample)}$$

Step 2A. Using the molar mass of C and H, calculate the grams of C and H in the original sample:

$$0.03167 \text{ mol C} \times \frac{12.01 \text{ g}}{1 \text{ mol C}} = 0.3804 \text{ g C}$$

$$0.03168 \text{ mol H} \times \frac{1.008 \text{ g}}{1 \text{ mol H}} = 0.03193 \text{ g H}$$

Step 2B. Calculate the mass of oxygen and moles of oxygen atoms in the original sample. Because the sample contains carbon, hydrogen, and oxygen only, the mass of oxygen in the sample equals the total mass of the sample minus the mass of carbon and mass of oxygen:

$$\text{mass O} = \text{mass sample} - (\text{mass carbon} + \text{mass hydrogen})$$
$$= 1.426 \text{ g} - (0.3804 \text{ g} + 0.03193 \text{ g}) = 1.014 \text{ g}$$

To calculate the moles of oxygen atoms, we use the molar mass of O, which is 16.00 g/mol:

$$1.014 \text{ g O} \times \frac{1 \text{ mol O}}{16.00 \text{ g O}} = 0.06338 \text{ mol O}$$

Step 3. Find the simplest whole number ratio of moles of elements in the compound. We calculated the moles of C and H in step 2 and the moles of O in step 2B. To find the simplest whole number of moles from the values already calculated, we divide each of the calculated values by the smallest one:

$$\frac{0.03168 \text{ mol H}}{0.03167} = 1.000 \text{ mol H} = 1 \text{ mol H} \quad \text{(an integer)}$$

$$\frac{0.03167 \text{ mol C}}{0.03167} = 1 \text{ mol C}$$

$$\frac{0.06338 \text{ mol O}}{0.03167} = 2.001 \text{ mol O} = 2 \text{ mol O} \quad \text{(an integer)}$$

The empirical formula is thus CHO_2.

EXERCISE What is the empirical formula of ethylene glycol if combustion of a 1.652-g sample results in formation of 2.343 g CO_2 and 1.438 g H_2O?

ANSWER: CH_3O

THE STOICHIOMETRY OF REACTIONS IN SOLUTION

KEY WORDS Define or explain each of the following terms in a written sentence or two.

acid-base titration
analyte
dilution
molar concentration
redox titration
stoichiometric point
titration
volumetric analysis

4.4 MOLAR CONCENTRATION

KEY CONCEPT A Molar concentration

Solutions have variable composition. To completely characterize a solution we must know its composition. The most widely used expression of solution composition is the **molarity,** which is the moles of solute per 1 liter of solution. To calculate the molarity, we determine the moles of solute in a solution and divide by the liters of solution:

$$\text{Molarity} = \frac{\text{moles of solute}}{\text{liters of solution}}$$

It is important to note that the factor in the denomintor is the volume of *solution,* not the volume of *solvent.* The volume of the solution includes the volume of solvent plus any changes made in the volume by the addition of the solute to the solvent.

EXAMPLE Determining the molarity of solute in a solution

Adding 53.5 g NaCl to 482 mL H_2O results in 500 mL of solution. What is the molarity of the NaCl solution?

SOLUTION We use the definition

$$\text{Molarity} = \frac{\text{moles of solute}}{\text{liters of solution}}$$

For NaCl:

$$\text{Molarity of NaCl} = \frac{\text{mol NaCl}}{\text{liters of solution}}$$

We calculate the moles of NaCl, using the molar mass of NaCl, which is 58.5 g/mol:

$$\text{Mol NaCl} = 53.5 \text{ g} \times \frac{1 \text{ mol NaCl}}{58.5 \text{ g}} = 0.915 \text{ mol}$$

The liters of solution are calculated from the 500 mL of solution given in the problem:

$$500 \text{ mL} \times \frac{1 \text{ L}}{1000 \text{ mL}} = 0.500 \text{ L}$$

Substituting these values into the defining equation gives

$$\begin{aligned} \text{Molarity of NaCl} &= \frac{\text{mol NaCl}}{\text{liters of solution}} \\ &= \frac{0.915 \text{ mol}}{0.500 \text{ L}} \\ &= 1.83 \text{ mol/L} \\ &= 1.83 \text{ M} \end{aligned}$$

Note that the amount of water (482 mL) is irrelevant to solving the problem. The volume of solvent is not important in calculating molarity; the volume of solution is.

EXERCISE Adding 50.0 g $MgSO_4$ to 200 mL of water results in 205 mL of solution. What is the molarity of the $MgSO_4$ in the solution?

ANSWER: 2.03 M

KEY CONCEPT B Moles of solute in a volume of solution

The molarity is widespread in use because it provides a convenient way to measure out a given number of moles of solute. The moles of solute in a given volume of solution is given by the equation

$$\text{Moles of solute} = \text{molarity} \times \text{volume (in L)}$$
$$\text{mol} = M \times V$$

where it is understood that the volume is in liters. Because chemists must frequently obtain a particular number of moles of a substance, solutions of known molarity are convenient, indeed. In addition, if a chemist is required to prepare a specific volume of a solution with a specific molarity, this equation specifies the number of moles (and, therefore, the number of grams) of solute to be used. We finally note that if the volume is expressed in milliliters, $M \times V$ gives the millimoles of solute. This is so because molarity equals millimole/milliliter (as well as mole/liter).

$$\text{mmol} = M \times V \quad \text{(volume in mL)}$$

EXAMPLE Calculating the milliliters of solution which contains a specific number of moles

A chemist is required to obtain 0.433 mol of Na_2SO_4, with a 2.00 M solution of Na_2SO_4. How many milliliters of solution are required?

SOLUTION For a solution of known molarity, the relationship mol = $M \times V$ relates the moles of solute to the liters of solution. We first substitute the stated molarity and moles of solute into this equation and then solve for the liters of solution:

$$0.433 \text{ mol } Na_2SO_4 = 2.00 \frac{\text{mol } Na_2SO_4}{\text{L}} \times V$$

$$V = \frac{0.433 \text{ mol } Na_2SO_4}{2.00 \text{ mol } Na_2SO_4\text{/L}} = 0.217 \text{ L}$$

Converting to milliliters gives

$$0.217 \text{ L} \times \frac{1000 \text{ mL}}{\text{L}} = 217 \text{ mL}$$

We could also, find the millimeters of solution needed by treating the molarity as a conversion factor. In this example the molarity of 2.00 M gives

$$2.00 \text{ mol } Na_2SO_4 = 1 \text{ L solution}$$

Because we started with 0.433 mol Na_2SO_4, the conversion problem becomes

$$0.433 \text{ mol } Na_2SO_4 \times \frac{1 \text{ L solution}}{2.00 \text{ mol } Na_2SO_4} = 0.217 \text{ L} = 217 \text{ mL}$$

As we expect, the answer is the same as we obtained earlier. The answer to a problem does not depend on which technique we use to solve it.

EXERCISE Write a recipe for preparing 250 mL of a 3.00 M solution of NaOH.

ANSWER: Add enough water to 30.0 g NaOH to make a final volume of 250 mL of solution.

KEY CONCEPT C Dilutions

To **dilute** a solution, we add solvent to the solution, which results in a new solution with a relatively smaller amount of solute per liter of solution. If we denote the molarity and volume of the original concentrated solution as M_{conc}, V_{conc}, and the molarity and concentration after dilution as M_{dil}, V_{dil}, then

$$M_{conc} \times V_{conc} = M_{dil} \times V_{dil}$$

EXAMPLE Making a dilute solution

A laboratory technician is asked to make up 2.00 L of a 0.200 M solution of NaBr from a bottle of 1.78 M NaBr. Calculate the volume of 1.78 M NaBr needed and write the steps to be taken to make the solution.

SOLUTION The equation $M_{conc} \times V_{conc} = M_{dil} \times V_{dil}$ is used in calculations involving the dilution of a solution when the composition of the solution is expressed as molarity. For this example we have

$$M_{conc} = 1.78 \text{ M} \qquad V_{conc} = ?$$
$$M_{dil} = 0.200 \text{ M} \qquad V_{dil} = 2.00 \text{ L}$$

Substituting into the equation gives

$$1.78 \text{ M} \times V_{conc} = 0.200 \text{ M} \times 2.00 \text{ L}$$
$$V_{conc} = \frac{0.200 \text{ M} \times 2.00 \text{ L}}{1.78 \text{ M}}$$
$$= 0.00225 \text{ L} = 2.25 \text{ mL}$$

To prepare the solution, the laboratory technician should measure 2.25 mL (0.00225 L) of the 1.78 M NaBr into a 2.00-L volumetric flask and add enough water (approximately 1998 mL) to make 2.00 L of solution.

EXERCISE Describe how you would make 200 mL of a 0.100 M solution of HCl from a 6.0 M stock solution.

ANSWER: Take 3.3 mL of the 6.0 M HCl and add enough water to make 200 mL of solution.

KEY CONCEPT D The pH of a solution

The **pH** is defined in terms of the concentration of hydrogen ions:

$$\text{pH} = -\log_{10}(\text{molar concentration of } H^+)$$

Because of the minus sign in the definition, the pH *decreases* as the hydrogen ion concentration *increases*. To calculate the molar concentration of hydrogen ion from the pH, we use the definition of the logarithm to get

$$\text{Molar concentration of } H^+ = 10^{-pH}$$

Both the log function and the 10^x function are available on scientific calculators.

EXAMPLE Using the pH

(a) What is the pH of a solution in which the molarity of hydrogen ion is 2.22×10^{-9} M? (b) Confirm your result by calculating the hydrogen ion molarity from your answer for the pH.

SOLUTION
(a) To obtain the pH, we use the defining equation for pH with the given hydrogen ion molar concentration of 2.22×10^{-9}:

$$\begin{aligned} pH &= -\log_{10}(\text{molar concentration of } H^+) \\ &= -\log_{10}(2.22 \times 10^{-9}) \end{aligned}$$

To calculate this value on your calculator, enter the number 2.22×10^{-9} and press the log key; the answer is

$$\begin{aligned} pH &= -(-8.65) \\ &= 8.65 \end{aligned}$$

(b) To obtain the molar concentration of H from the pH, we substitute the calculated value of the pH in the following equation:

$$\begin{aligned} \text{Molar concentration of } H^+ &= 10^{-pH} \\ &= 10^{-8.65} \\ &= 2.24 \times 10^{-9}\ M \end{aligned}$$

To calculate this value on your calculator, enter the number -8.65 and press the 10^x key; this key is a second-function key on many calculators, so you may have to depress the shift or second-function key first. If your calculator does not have a 10^x function, use the inverse-log function by pressing the inverse key first, and then the log key. The difference between the original 2.22×10^{-9} and the answer 2.24×10^{-9} is not important; it is due to round-off error in calculating $10^{-8.65}$.

EXERCISE What is the molar concentration of hydrogen ion in a solution with pH 5.78? Check your answer by calculating the pH from the molar concentration of hydrogen ion you get.

ANSWER: 1.66×10^{-6} M

4.5 THE VOLUME OF SOLUTION REQUIRED FOR REACTION

KEY CONCEPT The volume of solution required for reaction

We learned earlier that the quantitative relationships describing the moles of reactants and products that take part in a chemical reaction are given by the stoichiometric coefficients of the equation. In addition, we have just developed the relationship "molarity × volume = moles." By putting these two ideas together, we can calculate the volume of a solution that must be used to react with a certain amount of a reagent.

EXAMPLE Calculating the volume of solution required for reaction

How many milliliters of 0.112 M NaOH are required to neutralize 10.0 mL of a 0.352 M solution of H_2SO_4? Assume the reaction is

$$2NaOH(aq) + H_2SO_4(aq) \longrightarrow 2H_2O(l) + Na_2SO_4(aq)$$

SOLUTION Three steps are used to solve this problem.

Step 1. The number of moles of the reactant with known molarity and volume are calculated with the equation

$$\text{Moles} = \text{molarity} \times \text{volume}$$

We are concerned here with the H_2SO_4 because it is the reactant for which we have a known volume and known molarity. We note that 10.0 mL = 0.0100 L.

$$\text{Moles } H_2SO_4 = 0.352 \frac{\text{mol } H_2SO_4}{\cancel{\text{L}}} \times 0.0100 \cancel{\text{L}}$$
$$= 3.52 \times 10^{-3} \text{ mol } H_2SO_4$$

Step 2. The number of moles of the second reactant are calculated using the stoichiometric coefficients of the balanced equation. The equation tells us that 2 mol NaOH react with 1 mol H_2SO_4.

$$3.52 \times 10^{-3} \cancel{\text{mol } H_2SO_4} \times \frac{2 \text{ mol NaOH}}{1 \cancel{\text{mol } H_2SO_4}} = 7.04 \times 10^{-3} \text{ mol NaOH}$$

Step 3. The volume of solution of the second reactant is calculated by the equation

$$\text{Volume} = \frac{\text{moles}}{\text{molarity}}$$

The problem states that the molarity of the NaOH is 0.112 M.

$$\text{Volume NaOH} = \frac{7.04 \times 10^{-3} \cancel{\text{mol NaOH}}}{0.112 \cancel{\text{mol NaOH}}/\text{L}}$$
$$= 0.0629 \text{ L}$$
$$= 62.9 \text{ mL}$$

EXERCISE How many milliliters of 0.125 M $KMnO_4$ are required to react with 25.0 mL of 0.831 M $FeCl_2$? The reaction is

$$KMnO_4(aq) + 8HCl(aq) + 5FeCl_2(aq) \longrightarrow 5FeCl_3(aq) + MnCl_2(aq) + KCl(aq) + 4H_2O(l)$$

ANSWER: 33.2 mL

4.6 TITRATIONS

KEY CONCEPT A Acid-base titrations

A tritration is an experiment in which a chemical reaction is done with carefully measured amounts of reactants in order to determine the number of moles of one of the reactants. There are many ways to do titrations, but the following procedure illustrates the general approach taken.

1. The analyte (for instance, the acid in an acid-base titration) is placed into a clean flask. If it is a solid, it is weighed carefully on an analytical balance. If it is in solution, the volume of solution is determined accurately by obtaining it with a buret or a pipet.
2. A solution of the titrant is added to the analyte. The goal is to add the volume of titrant that will exactly react with the quantity of the analyte originally present. We know when we have added enough by using an indicator that changes color when the correct amount has been added; this point is called the endpoint. If too much or too little titrant is added, the titration must be redone.
3. The volume of the titrant needed to just react with the analyte is written down.
4. The data obtained are used to calculate the unknown moles of one of the reactants (either the analyte or titrant) from the known moles of the other. A titration is designed so that the moles of one of the reactants can be calculated from known factors, such as the molarity and volume of the reactant (for a solution) or its mass and molecular weight (for a solid). With this information and the stoichiometric coefficients of the balanced chemical equation the moles of the other reactant (the unknown) are calculated.

EXAMPLE Determining the molarity of an unknown base in an acid-base titration

A student prepares a solution of NaOH by placing 4 g of it into a flask and adding enough water to make 1 L of solution: an approximate 0.1 M solution. The exact molarity of the solution is determined by a titration, in which 10.00 mL of the solution requires 11.22 mL of a 0.1004 M solution of HCl to reach the endpoint. What is the exact molarity of the NaOH? The reaction is

$$\mathrm{NaOH}(aq) + \mathrm{HCl}(aq) \longrightarrow \mathrm{NaCl}(aq) + \mathrm{H_2O}(l)$$

SOLUTION The steps for this calculation are almost identical to the three steps in the previous example. You should compare the two calculations to see where they are similar and where they differ.

Step 1. The number of moles of the known reactant is determined by

$$\text{Moles} = \text{molarity} \times \text{volume}$$

The known reactant is the titrant HCl, because its molarity is given. We note that 11.22 mL = 0.01122 L.

$$\begin{aligned}\text{Moles HCl} &= 0.1004\,\frac{\text{mol HCl}}{\cancel{\text{L}}} \times 0.01122\,\cancel{\text{L}}\\ &= 0.001126 \text{ mol HCl}\\ &= 1.126 \times 10^{-3} \text{ mol HCl}\end{aligned}$$

Step 2. The stoichiometric coefficients of the balanced chemical equation are used to calculate the moles of the other reactant present:

$$1.126 \times 10^{-3}\,\cancel{\text{mol HCl}} \times \frac{1 \text{ mol NaOH}}{1\,\cancel{\text{mol HCl}}} = 1.126 \times 10^{-3} \text{ mol NaOH}$$

Step 3. The molarity of the unknown solution is calculated using the definition of molarity:

$$\text{Molarity} = \frac{\text{mole solute}}{\text{liters solution}}$$

1.126×10^{-3} mol NaOH is the number of moles that were originally present in the 10.00 mL of NaOH solution used in the titration. Because 10.00 mL = 0.01000 L,

$$\begin{aligned}\text{Molarity} &= \frac{1.126 \times 10^{-3} \text{ mol NaOH}}{0.0100 \text{ L}}\\ &= 0.1126 \text{ M}\end{aligned}$$

EXERCISE The student now uses the 0.1126 M standard NaOH solution to titrate 10.00 mL of an unknown H_2SO_4 solution. 23.65 mL of the NaOH were required. What is the molarity of the H_2SO_4? Assume the reaction is

$$2\mathrm{NaOH}(aq) + \mathrm{H_2SO_4}(aq) \longrightarrow \mathrm{Na_2SO_4}(aq) + 2\mathrm{H_2O}(l)$$

ANSWER: 0.1331 M

KEY CONCEPT B Redox titrations

A redox titration is conceptually no different from an acid-base titration. Only the details of the reaction differ. Instead of reacting an acid with a base to do the titration, we react an oxidizing agent with a suitable reducing agent. The following example illustrates a simplified version of the method used to determine the iron content of iron ore. Because a solid unknown is used and the desired information is not the molarity of a solution, the last step of the calculations differs from that just illustrated.

▼ **EXAMPLE Determining the percentage iron in a mixture of solids**

A 2.681-g mixture of solids, including salts with Fe^{2+} ion, is titrated using 25.71 mL of 0.1023 M $KMnO_4$. Assuming there is no other substance present that is oxidized by the $KMnO_4$ and that all of the iron is in the form Fe^{2+}, what is the percentage of iron in the solids? The net ionic equation is

$$5Fe^{2+}(aq) + MnO_4^-(aq) + 8H^+(aq) \longrightarrow 5Fe^{3+}(aq) + Mn^{2+}(aq) + 4H_2O(l)$$

SOLUTION

Step 1. We start the calculation in typical fashion, by calculating the moles of the known reactant. Because there is 1 mole of MnO_4^- ion per mole of $KMnO_4$, the concentration of MnO_4^- is the same as that of the $KMnO_4$. Also, 25.71 mL = 0.02571 L.

$$\begin{aligned} \text{Moles } MnO_4^- &= \text{molarity} \times \text{volume} \\ &= (0.1023 \text{ mol } MnO_4^-/\cancel{L}) \times 0.02571\ \cancel{L} \\ &= 0.002630 \text{ mol} \\ &= 2.630 \times 10^{-3} \text{ mol } MnO_4^- \end{aligned}$$

Step 2. The number of moles of the second reactant are calculated using the stoichiometric coefficients of the balanced equation:

$$2.630 \times 10^{-3}\ \cancel{\text{mol } MnO_4^-} \times \frac{5 \text{ mol } Fe^{2+}}{1\ \cancel{\text{mol } MnO_4^-}} = 1.315 \times 10^{-2} \text{ mol } Fe^{2+}$$

Step 3. The mass of the Fe present is essentially the same as the mass of Fe^{2+} present, and the atomic weight of both are essentially identical, so we calculate the total mass of Fe using the moles of Fe^{2+} and the molar mass of Fe, 55.85 g/mol:

$$\begin{aligned} 1.315 \times 10^{-2}\ \cancel{\text{mol } Fe^{2+}} \times \frac{55.85 \text{ g } Fe^{2+}}{1\ \cancel{\text{mol } Fe^{2+}}} &= 0.7344 \text{ g } Fe^{2+} \\ &= 0.7344 \text{ g Fe} \end{aligned}$$

The percentage Fe in the mixture is given by

$$\begin{aligned} \text{Percentage Fe} &= \frac{\text{mass Fe}}{\text{mass of sample}} \times 100 \\ &= \frac{0.7344\ \cancel{g}}{2.681\ \cancel{g}} \times 100 \\ &= 27.39\% \end{aligned}$$

EXERCISE A 1.555-g sample of solid containing Fe^{2+} requires 16.88 mL of 0.1122 M $Ce(SO_4)_2$ in a redox titration. What is the percentage Fe in the sample? The net ionic equation is

$$Ce^{4+}(aq) + Fe^{2+}(aq) \longrightarrow Ce^{3+}(aq) + Fe^{3+}(aq)$$

▲ ANSWER: 6.802%

DESCRIPTIVE CHEMISTRY TO REMEMBER

- **Acetylene** (C_2H_2) is a flammable gas.
- **Freon-12** is CCl_2F_2.
- **Combustion analysis** is a procedure in which the composition of an organic compound is determined by burning the compound in an unlimited supply of oxygen and measuring the masses of carbon dioxide and water produced.
- **Phosphorus pentoxide** (P_4O_{10}) acts as a drying agent because it strongly absorbs water.
- **Sodium hydroxide** (NaOH) strongly absorbs carbon dioxide and is used to determine the amount of carbon dioxide liberated in a combustion experiment.
- **Volumetric analysis** is the analysis of the composition of an unknown mixture or unknown substance by the measurement of volume.

CHEMICAL EQUATIONS TO KNOW

- In the Haber syntheses, hydrogen gas and nitrogen gas are reacted to form ammonia:

$$N_2(g) + 3H_2(g) \longrightarrow 2NH_3(g)$$

- Iron is produced from hematite ore (Fe_2O_3) through reduction with carbon monoxide:

$$Fe_2O_3(s) + 3CO(g) \longrightarrow 2Fe(s) + 3CO_2(g)$$

- Calcium carbide (CaC_2) reacts with water to form calcium hydroxide and acetylene (C_2H_2):

$$CaC_2(s) + 2H_2O(l) \longrightarrow Ca(OH)_2(s) + C_2H_2(g)$$

- The white precipitate silver chloride (AgCl) forms when aqueous solutions of silver nitrate and sodium chloride are mixed:

$$AgNO_3(aq) + NaCl(aq) \longrightarrow AgCl(s) + NaNO_3(aq)$$

Acid-base titration reactions

- Hydrochloric acid (HCl(aq)) reacts with sodium hydroxide (NaOH) on a one-to-one mole ratio to form sodium chloride and water:

$$HCl(aq) + NaOH(aq) \longrightarrow NaCl(aq) + H_2O(l)$$

- Potassium hydroxide reacts with phosphoric acid on a three-to-one mole basis to form sodium phosphate and water:

$$3KOH(aq) + H_3PO_4(aq) \longrightarrow K_3PO_4(aq) + H_2O(l)$$

- Acetic acid ($HC_2H_3O_2$) reacts with sodium hydroxide (NaOH) on a one-to-one mole ratio to form sodium acetate and water:

$$HC_2H_3O_2(aq) + NaOH(aq) \longrightarrow NaC_2H_3O_2(aq) + 3H_2O(l)$$

Redox titration reactions

- Cerium(IV) ion (Ce^{4+}, the oxidizing agent) reacts with iron(II) ion (Fe^{2+}, the reducing agent) on a one-to-one mole ratio to form cerium(III) ion and iron(III) ion:

$$Ce^{4+}(aq) + Fe^{2+}(aq) \longrightarrow Ce^{3+}(aq) + Fe^{3+}(aq)$$

- Permanganate ion (MnO_4^-, the oxidizing agent) reacts with iron(II) ion (Fe^{2+}, the reducing agent) on a one-to-five mole ratio in acid solution to form iron(III) ion, maganese(II) ion, and water:

$$5Fe^{2+}(aq) + MnO_4^-(aq) + 8H^+(aq) \longrightarrow 5Fe^{3+}(aq) + Mn^{2+}(aq) + 4H_2O(l)$$

MATHEMATICAL EQUATIONS TO KNOW AND UNDERSTAND

$\text{Percentage yield} = \dfrac{\text{actual yield}}{\text{theoretical yield}} \times 100$	definition of percentage yield
$\text{Molar concentration} = \dfrac{\text{moles solute}}{\text{liters solution}}$	definition of molarity
$M_{conc} \times V_{conc} = M_{dil} \times V_{dil}$	dilution of a solution
$pH = -\log(\text{molar } H^+ \text{ concentration})$	definition of pH
$\text{Molar } H^+ \text{ concentration} = 10^{-pH}$	molar H^+ concentration from pH

SELF-TEST EXERCISES

Interpreting stoichiometric coefficients

1. How many moles of $SiCl_4$ can be formed from 3.0 mol Cl_2?

$$Si(s) + 2Cl_2(g) \longrightarrow SiCl_4(g)$$

(a) 3.0 (b) 6.0 (c) 1.5 (d) 2.5 (e) 0.67

2. How many moles of HCl can be formed from 2.3 mol of H_2O?

$$PCl_5(s) + 4H_2O(l) \longrightarrow 5HCl(aq) + H_3PO_4(aq)$$

(a) 1.8 (b) 2.3 (c) 0.58 (d) 11.5 (e) 2.9

3. How many grams of Fe_2O_3 can be formed from 6.50 mol of Fe?

$$4Fe(s) + 3O_2(g) \longrightarrow 2Fe_2O_3(s)$$

(a) 519 (b) 1040 (c) 12.3 (d) 2078 (e) 653

4. How many kilograms of $NaHCO_3$ are required to produce 51.0 mol CO_2?

$$2NaHCO_3(s) \longrightarrow Na_2CO_3(s) + CO_2(g) + H_2O(g)$$

(a) 4.28 (b) 8.57 (c) 2.14 (d) 195 (e) 77.5

5. How many grams of PbO can be formed from 25.0 g PbS?

$$2PbS(s) + 3O_2(g) \longrightarrow 2PbO(s) + 2SO_2(g)$$

(a) 26.8 (b) 23.3 (c) 11.7 (d) 46.6 (e) 13.4

6. How many grams of HNO_2 are needed to produce 125 g of NO?

$$3HNO_2(aq) \longrightarrow HNO_3(aq) + 2NO(g) + H_2O(l)$$

(a) 8.82×10^3 (b) 196 (c) 294 (d) 588 (e) 131

7. How many kilograms of NH_4F can be prepared from 1.22 kg NH_3?

$$4NH_3(g) + 3F_2(g) \longrightarrow 3NH_4F(s) + NF_3(g)$$

(a) 1.99 (b) 1.56 (c) 2.66 (d) 3.54 (e) 2.87

8. What is the limiting reagent when 30.0 g CO and 10.0 g H_2 are mixed and reacted in the following reaction:

$$3H_2(g) + CO(g) \longrightarrow CH_4(g) + H_2O(g)$$

(a) CO (b) H_2 (c) neither

9. How much Fe can be produced if 100 g Fe_2O_3 and 100 g CO are mixed and reacted?

$$Fe_2O_3(s) + 3CO(g) \longrightarrow 2Fe(s) + 3CO_2(g)$$

(a) 97.9 g (b) 35.0 g (c) 147 g (d) 70.0 g (e) 200 g

10. When 50.0 g Ca are reacted with excess N_2, 41.8 g Ca_3N_2 are produced. What is the percentage yield of Ca_3N_2?

$$3Ca(s) + N_2(g) \longrightarrow Ca_3N_2(s)$$

(a) 67.9% (b) 83.6% (c) 22.6% (d) 100% (e) 33.4%

11. 19.9 g H_2O are formed when 45.0 g $Mg(OH)_2$ is reacted with excess HCl. What is the percentage yield of H_2O?

$$Mg(OH)_2(s) + 2HCl(aq) \longrightarrow 2H_2O(l) + MgCl_2(aq)$$

(a) 69.8% (b) 71.6% (c) 44.2% (d) 34.9% (e) 100%

12. A 1.532-g sample of a compound containing only carbon and hydrogen produces 4.807 g CO_2 and 1.968 g H_2O in a combustion analysis. What is the empirical formula of the compound?
(a) C_2H (b) CH (c) CH_2 (d) C_2H_3 (e) C_5H_2

13. A 0.8135-g sample of a compound containing carbon, hydrogen, and oxygen produces 2.104 g CO_2 and 0.4306 g H_2O in a combustion analysis. What is the empirical formula of the compound?
(a) CHO (b) C_2H_2O (c) CHO_2 (d) C_3HO_3 (e) C_4H_4O

14. What is the empirical formula of the compound deoxycholic acid, $C_{24}H_{40}O_4$?
(a) $C_{24}H_{40}O_4$ (b) $C_{18}H_{30}O_3$ (c) CHO (d) $C_{12}H_{20}O_2$ (e) $C_6H_{10}O$

15. What is the molecular formula of the compound with the empirical formula C_2H_3 and molar mass 135.2 g/mol?
(a) C_4H_6 (b) C_8H_{12} (c) $C_{10}H_{15}$ (d) C_2H_3 (e) CH

16. What is the molecular formula of a compound with the empirical formula C_5H_8O and molar mass 336.5 g/mol?
(a) C_5H_8O (b) $C_{20}H_{32}O_4$ (c) $C_{50}H_{80}O_5$ (d) $C_{21}H_{36}O_3$ (e) $C_{19}H_{28}O_5$

The stoichiometry of reactions in solution

17. A solution is prepared by adding 230 mL H_2O to 0.645 mol HNO_3, resulting in 250 mL of solution. What is the molar concentration of HNO_3?
(a) 0.161 M (b) 2.80 M (c) 2.58 M (d) 1.34 M (e) 0.645 M

18. 194 mL H_2O is added to 37.0 g $BaCl_2$, resulting in 200 mL of solution. What is the molar concentration of $BaCl_2$?
(a) 0.178 M (b) 0.466 M (c) 0.918 M (d) 0.883 M (e) 1.11 M

19. How many moles of NaCl are required to prepare 500 mL of a 0.250 M NaCl solution?
(a) 0.100 (b) 1.00 (c) 0.500 (d) 0.750 (e) 0.125

20. How many grams of $NaNO_3$ are required to prepare 125 mL of 3.00 M $NaNO_3$?
(a) 46.1 (b) 31.9 (c) 227 (d) 3.54 (e) 28.3

21. How many moles of $Zn(NO_3)_2$ are present in 115 mL of a 0.65 M $Zn(NO_3)_2$ solution?
(a) 0.25 (b) 13 (c) 0.075 (d) 5.7 (e) 0.18

22. How many grams of Na_2SO_4 are present in 250 mL of a 1.72 M solution of Na_2SO_4?
(a) 61.1 (b) 20.6 (c) 244 (d) 35.5 (c) 977

23. 25.0 mL of a 18 M H_2SO_4 solution are diluted to 500 mL with water. What is the molar concentration of the H_2SO_4?
(a) 0.45 M (b) 0.050 M (c) 0.23 M (d) 0.36 M (e) 0.90 M

24. 432 mL H_2O is added to 50.0 mL of a 3.00 M $NiCl_2$ solution, resulting in 500 mL of solution. What is the molar concentration of the $NiCl_2$?
(a) 0.347 M (b) 0.311 M (c) 0.300 M (d) 2.59 M (e) 3.33 M

25. What is the pH of a solution that has a hydrogen ion molar concentration of 4.65×10^{-5} M?
(a) 5.00 (b) 8.33 (c) 12.44 (d) 4.33 (e) 7.11

26. What is the pH of a solution that has a hydrogen ion molar concentration of 8.49×10^{-12} M?
(a) 5.88 (b) 12.00 (c) 12.92 (d) 2.93 (e) 11.07

27. What is the hydrogen ion molar concentration in a solution with pH 3.12?
(a) 1.38×10^{-6} M (b) 2.84×10^{-12} M (c) 7.99×10^{-3} M
(d) 7.59×10^{-4} M (e) 1.32×10^{3} M

28. What is the hydrogen ion molar concentration in a solution with pH 10.87?
(a) 1.35×10^{-11} M (b) 6.55×10^{-7} M (c) 7.41×10^{10} M
(d) 2.44×10^{-5} M (e) 8.71×10^{4} M

29. Which pH corresponds to the solution with the highest hydrogen ion molar concentration?
(a) 12.55 (b) 7.00 (c) 9.56 (d) 5.49 (e) 2.48

30. Which hydrogen ion molar concentration corresponds to the highest pH?
(a) 1.78×10^{-10} M (b) 8.95×10^{-13} M (c) 7.41×10^{-4} M
(d) 5.64×10^{-7} M (e) 6.66×10^{-9} M

31. How many milliliters of 3.00 M HCl are required to react with 20.0 g Zn?

$$Zn(s) + 2HCl(aq) \longrightarrow ZnCl_2(aq) + H_2(g)$$

(a) 102 (b) 612 (c) 306 (d) 204 (e) 408

32. How many milliliters of 0.652 M $AgNO_3$ are required to react with 25.0 milliliters of 0.468 M $BaCl_2$?

$$2AgNO_3(aq) + BaCl_2(aq) \longrightarrow 2AgCl(s) + Ba(NO_3)_2(aq)$$

(a) 17.9 (b) 71.8 (c) 34.8 (d) 17.4 (e) 35.9

33. A 25.00-mL sample of HCl requires 33.26 mL of 0.1026 M NaOH for titration. What is the molarity of the HCl?
(a) 0.2730 M (b) 0.06825 M (c) 0.7712 M (d) 0.1365 M (e) 0.3856 M

34. 5.00 mL of a citric acid ($H_3C_6H_5O_7$) solution require 27.35 mL of 0.1116 M NaOH for titration. What is the molarity of the citric acid solution?

$$3NaOH(aq) + H_3C_6H_5O_7(aq) \longrightarrow Na_3C_6H_5O_7(aq) + 3H_2O(l)$$

(a) 0.2035 M (b) 0.1256 M (c) 1.832 M (d) 0.6105 M (e) 0.06121 M

35. When a 0.7185-g sample of a solid acid is dissolved in about 10 mL of water and titrated with 0.1655 M NaOH, 21.26 mL of the NaOH are required. What is the molar mass of the acid? 1 mol NaOH is required to react with 1 mol of the acid.
(a) 43.4 g/mol (b) 351.8 g/mol (c) 36.46 g/mol
(d) 4.897 g/mol (e) 204.2 g/mol

36. A 25.00-mL sample of iron(II) sulfate solution requires 38.36 mL of 0.1234 M cerium(IV) sulfate for titration. What is the molarity of the iron(II) sulfate solution? The net ionic reaction is

$$Fe^{2+}(aq) + Ce^{4+}(aq) \longrightarrow Fe^{3+}(aq) + Ce^{3+}(aq)$$

(a) 0.08042 M (b) 0.1893 M (c) 0.1608 M (d) 0.09465 M (e) 0.1234 M

37. An 8.00-mL sample of iron(II) sulfate solution requires 12.63 mL of a 0.1250 M potasium permanganate solution for titration. What is the molarity of the iron(II) sulfate solution? The net ionic equation is

$$5Fe^{2+}(aq) + MnO_4^{-}(aq) + 8H^{+}(aq) \longrightarrow 5Fe^{3+}(aq) + Mn^{2+}(aq) + 4H_2O(l)$$

(a) 0.1973 M (b) 0.03875 M (c) 1.000 M (d) 0.9867 M (e) 0.3959 M

Descriptive chemistry

38. Combustion analysis is the determination of the composition of an unknown organic compound based on its combustion to
(a) $C + H_2$ (b) $CO + H_2$ (c) CH_4 (d) $CO_2 + H_2O$ (e) $CO_2 + H_2$

39. What is the reducing agent in the production of iron from hematite (Fe_2O_3)?
(a) C (b) H_2 (c) CO (d) O_2 (e) N_2

40. Acetylene (C_2H_2) is produced by the reaction of water with
(a) $CaCO_3$ (b) CaC_2 (c) C_2H_4 (d) NaF (e) $Mg(C_2H_3O_2)_2$

41. Which of the following is the equation for the Haber synthesis?
(a) $N_2(g) + 2H_2(g) \longrightarrow N_2H_4(g)$
(b) $Fe^{2+}(aq) + Ce^{4+}(aq) \longrightarrow Fe^{3+}(aq) + Ce^{3+}(aq)$
(c) $Fe_2O_3(s) + 3CO(g) \longrightarrow 2Fe(s) + 3CO_2(g)$
(d) $N_2(g) + 3H_2(g) \longrightarrow 2NH_3(g)$
(e) $NaOH(aq) + HCl(aq) \longrightarrow NaCl(aq) + H_2O(l)$

5 THE PROPERTIES OF GASES

THE GAS LAWS

KEY WORDS Define or explain each of the following terms in a written sentence or two.

Avogadro's principle
barometer
Boyle's law
Charles's law
equation of state
gas constant
ideal gas law
manometer
molar volume of a gas
pressure

5.1 PRESSURE

KEY CONCEPT A Pressure unit conversions

As a preliminary to many of the calculations in this chapter, we must consider how to convert pressures from one unit to another. This is done most directly with the unit conversion technique.

EXAMPLE Pressure unit conversions

It is thought that the pressure at the center of a tornado may be as low as 600 Torr. What is this pressure in atmospheres?

SOLUTION From Table 5.3 of the text, we know that 760 Torr = 1 atm exactly. With this information, the problem can be solved by the unit conversion method:

$$600 \not{\text{Torr}} \times \frac{1 \text{ atm}}{760 \not{\text{Torr}}} = 0.789 \text{ atm}$$

EXERCISE A typical high-pressure cooker can attain a pressure of 1.15 atm. What is this pressure in bars? (1 bar = 100 kPa, 1 atm = 101.325 kPa)

ANSWER: 1.17 bar

KEY CONCEPT B Reading a barometer

In reading a barometer, remember that the liquid (usually mercury or water) in the closed tube tries to "seek its own level," which means the level of the liquid in the open-to-the-air container. The pressure of the atmosphere prevents the liquid from finding its level and determines the actual level of the liquid. Thus, at zero atmospheric pressure the barometer looks like illustration (a), where the two liquid levels are the same. As the atmospheric pressure increases, it pushes the mercury column up the tube, as shown in (b) and (c). Because the pressure inside the closed barometer tube is (essentially) zero, the height of the mercury above the liquid level in the cup is a direct measure of atmospheric pressure.

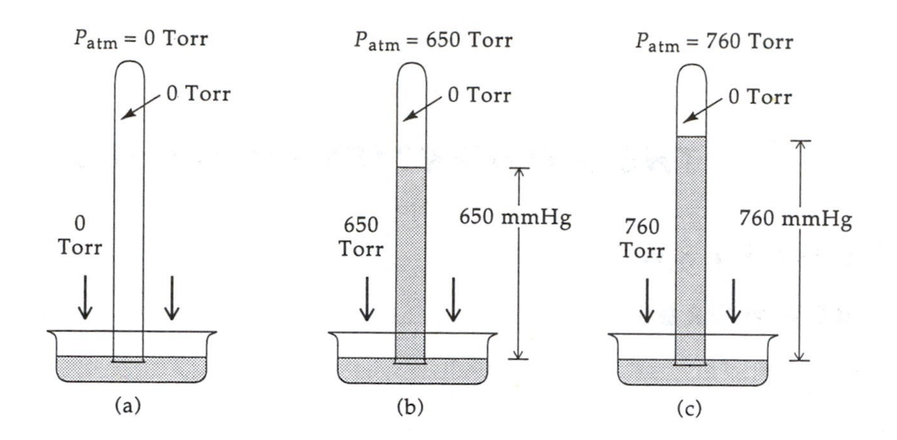

EXAMPLE Understanding a barometer

Would the barometer in illustration (b) function if there was a small leak at its top? Explain your answer.

SOLUTION No, it would not work. Air would leak into the barometer, and the pressure inside the barometer would eventually reach atmospheric pressure. Then, the inside pressure pushing the column downward would equal the outside pressure pushing it upward, and the two liquid levels would be the same.

EXERCISE Could a highly volatile liquid be used as a barometer liquid?

ANSWER: No. The liquid would evaporate into the empty space inside the barometer, thereby causing gas pressure to build up inside; this would result in inaccurate readings.

5.2 THE IDEAL GAS

KEY CONCEPT A Using the individual gas laws

The individual gas laws can be used to calculate changes in one of the variables P, V, T, or n when one or more of the other variables changes. There are a number of methods to do such calculations; in this study guide we use equations derived directly from the individual laws. The mathematical results are exactly the same as the examples in the text. To use the individual gas laws, we must determine which variables are to be held constant and which change, because this determines which law to use. The table that follows summarizes the derived equation for each law and which law to use under different circumstances. The subscript 1 indicates the value of a variable before some change occurs, and subscript 2 indicates the value after a change occurs.

Law	Equation	Constant variables	Variables that change	Derived equation
Boyle's law	$V \propto \frac{1}{P}$	n, T	P, V	$P_1V_1 = P_2V_2$
Charles's law	$V \propto T$	n, P	T, V	$\frac{V_1}{T_1} = \frac{V_2}{T_2}$
Avogadro's principle	$V \propto n$	P, T	n, V	$\frac{V_1}{n_1} = \frac{V_2}{n_2}$
combined gas law	$\frac{PV}{T}$ = constant	n	P, V, T	$\frac{P_1V_1}{T_1} = \frac{P_2V_2}{T_2}$

EXAMPLE 1 Boyle's law calculation

The pressure of a 5.2-L sample of gas at 700 Torr is increased to 780 Torr. What is the new volume of the gas? The temperature of the gas is held constant.

SOLUTION We must first ascertain that this is a Boyle's law problem. The problem states that a change in pressure results in a change in volume for a sample of gas at constant temperature. Thus, P and V change while n and T are held constant; thus, this is a Boyle's law problem with

$$P_1 = 700 \text{ Torr} \qquad P_2 = 780 \text{ Torr}$$
$$V_1 = 5.2 \text{ L} \qquad V_2 = ?$$

Substituting into Boyle's law gives

$$P_1V_1 = P_2V_2$$
$$(700 \text{ Torr})(5.2 \text{ L}) = (780 \text{ Torr})(V_2)$$

Dividing both sides of the equation by 780 Torr gives

$$V_2 = 5.2 \text{ L} \times \frac{700 \text{ Torr}}{780 \text{ Torr}}$$
$$= 4.7 \text{ L}$$

Notice that this calculation takes on the same form as those shown in the text, with the initial volume multiplied by the factor that makes the final volume smaller.

EXERCISE A 65-L sample of gas at 2.2 atm undergoes a change in pressure to 1.5 atm at constant temperature. What is the new volume of the gas?

ANSWER: 95 L

EXAMPLE 2 Charles's law calculation

A 200-mL sample of gas at 25°C undergoes a change in volume to 100 mL at constant pressure. What is the new temperature of the gas in kelvins?

SOLUTION Because the number of moles and the pressure of the gas are held constant while the volume and temperature are changed, this is a Charles's law problem. The temperature must be expressed in kelvins. Thus, using K = °C + 273, we have

$$V_1 = 200 \text{ mL} \qquad T_1 = 298 \text{ K}$$
$$V_2 = 100 \text{ mL} \qquad T_2 = ?$$

Substituting into Charles's law gives

$$\frac{V_1}{T_1} = \frac{V_2}{T_2}$$
$$\frac{200 \text{ mL}}{298 \text{ K}} = \frac{100 \text{ mL}}{T_2}$$

To solve this equation, we first multiply both sides of the equation by T_2 and by 298 K, and then divide both sides by 200 mL. This gives

$$T_2 = 298 \text{ K} \times \frac{100 \text{ mL}}{200 \text{ mL}}$$
$$= 149 \text{ K}$$

EXERCISE A 3.66-L sample of gas at 350°C undergoes a change in temperature to 283°C. What is the new volume of the gas?

ANSWER: 3.27 L

EXAMPLE 3 Avogadro's law calculation

A gas-phase reaction is run in which 1.25 L of gas-phase reactant converts to 3.75 L of gas-phase product. At the start of the reaction, 0.113 mol of reactant is present. How many moles of product are formed? The temperature and pressure of the reaction system are the same at the end of the reaction as at the start.

SOLUTION Because the number of moles and the volume of gas change at constant pressure and temperature, this is an Avogadro's law problem with

$$V_1 = 1.25 \text{ mL} \qquad n_1 = 0.113 \text{ mol}$$
$$V_2 = 3.75 \text{ mL} \qquad n_2 = ?$$

Substituting into the equation for Avogadro's principle gives

$$\frac{V_1}{n_1} = \frac{V_2}{n_2}$$

$$\frac{1.25 \text{ L}}{0.113 \text{ mol}} = \frac{3.75 \text{ L}}{n_2}$$

To solve this equation, we first multiply both sides of the equation by n_2 and 0.113 mol, and then divide both sides by 1.25 L. This gives

$$n_2 = 0.113 \text{ mol} \times \frac{3.75 \cancel{\text{L}}}{1.25 \cancel{\text{L}}}$$

$$= 0.339 \text{ mol}$$

EXERCISE During the course of a reaction, 0.25 mol of gas-phase reactant becomes 0.50 mol of gas-phase product. The initial volume of gaseous reactant is 125 mL. What is the volume of gaseous product?

ANSWER: 250 mL

PITFALL The correct temperature in gas law problems

In Charles's law and combined law calculations, be sure you express the temperature in kelvins.

PITFALL Is it V_1/V_2 or V_2/V_1?

For any of the calculations involving Boyle's and Charles's laws, Avogadro's principle, or the combined gas law, it is possible to inadvertently use a term such as V_1/V_2 when you really mcant to use the inverse, V_2/V_1.

If you look at the answer to all your calculations and ask yourself whether the answer seems reasonable, you can often catch and correct this type of error. For example, if a problem specifies that a sample undergoes a decrease in pressure at constant temperature, the volume must increase. If you solve the problem mathematically and end up with a smaller volume, you have probably made the kind of error mentioned above.

PITFALL There must be consistency of units in gas law problems

In gas law calculations, the units used for volume must be the same for V_1 and V_2. Similarly, the units for pressure must be the same for P_1 and P_2.

KEY CONCEPT B Using the combined gas law

If the number of moles of an ideal gas is kept constant and the temperature, pressure, and volume change, the change in conditions is described by the combined gas law equation:

$$\frac{P_1V_1}{T_1} = \frac{P_2V_2}{T_2}$$

▼ EXAMPLE Combined gas law calculation

If 3.2 L of a sample of an ideal gas at 25°C and 732 Torr undergoes a change to 50°C and 943 Torr, what is the new volume of the gas?

SOLUTION In this problem, changes in pressure and temperature cause a change in volume for a constant number of moles of an ideal gas. When n is constant and P, V, and T change, we use the combined gas law:

$$\frac{P_1V_1}{T_1} = \frac{P_2V_2}{T_2}$$

$P_1 = 732$ Torr $\qquad P_2 = 943$ Torr

$T_1 = 25 + 273.15 = 298$ K $\qquad T_2 = 50 + 273.15 = 323$ K

$V_1 = 3.2$ L $\qquad V_2 = ?$

Because V_2 is the unknown, we rearrange the original equation so that V_2 is on the left-hand side of the equation and everything else on the right:

$$V_2 = V_1 \times \frac{P_1}{P_2} \times \frac{T_2}{T_1}$$

Substituting the given values, we get

$$V_2 = 3.2 \text{ L} \times \frac{732 \cancel{\text{Torr}}}{943 \cancel{\text{Torr}}} \times \frac{323 \cancel{\text{K}}}{298 \cancel{\text{K}}}$$

$$= 2.7 \text{ L}$$

EXERCISE 5.5 L of an ideal gas at 0.955 atm and 12.2°C undergo a change in volume and pressure to 5.3 L and 1.22 atm. What is the new temperature of the gas?

ANSWER: 3.5×10^2 K

5.3 USING THE IDEAL GAS LAW

KEY CONCEPT A Direct substitution into the ideal gas law

The ideal gas law relates the pressure P, volume V, temperature T (in kelvins), and number of moles n of an ideal gas:

$$PV = nRT$$

In a typical problem using the ideal gas law, any three of the parameters P, V, T and n are given and the fourth must be calculated. The value of R, expressed in the proper units, is also needed. This can be looked up when needed and is usually supplied on examinations. When you use the ideal gas law equation, write the units with all of the values substituted in and cancel units correctly. If the units of the answer come out wrong, you most certainly have substituted into the equation incorrectly.

EXAMPLE 1 Ideal gas law calculation

A typical child's helium balloon has a volume of about 2 L. How many moles of helium (He) will the balloon hold if it is filled to a pressure of 770 Torr on a day when the temperature is 24°C? How many grams of helium does it hold?

SOLUTION The problem gives the volume, temperature, and pressure (V, T, P) of a gas and asks for the moles (n) of gas. The ideal gas law expresses the relationship among these parameters and is used to solve the problem. Because the unknown is the number of moles n, we divide both sides of the equation by RT to get the unknown on one side of the equation by itself:

$$PV = nRT$$

$$n = \frac{PV}{RT}$$

The values given for the known parameters are $P = 770$ Torr, $V = 2$ L, and $T = 24 + 273.15 = 297$ K. Because we use $R = 0.08206$ L atm/mol K, we first convert the pressure into units of atmospheres, using the fact that 1 atm = 760 Torr:

$$770 \,\cancel{\text{Torr}} \times \frac{1 \text{ atm}}{760 \,\cancel{\text{Torr}}} = 1.01 \text{ atm}$$

Substituting these values into the equation gives

$$\begin{aligned} n &= \frac{(1.01 \,\cancel{\text{atm}})(2 \,\cancel{\text{L}})}{(0.08206 \,\cancel{\text{L}}\, \cancel{\text{atm}}/\text{K mol})(297 \,\cancel{\text{K}})} \\ &= 0.0828422 \text{ mol He} \\ &= 0.08 \text{ mol He} \end{aligned}$$

The number of grams of He is calculated using the molar mass of He, which is 4.003 g/mol:

$$\begin{aligned} \text{Grams He} &= 0.08 \,\cancel{\text{mol He}} \times \frac{4.003 \text{ g He}}{1 \,\cancel{\text{mol He}}} \\ &= 0.3 \text{ g He} \end{aligned}$$

EXERCISE How many moles of an ideal gas occupy 125 mL at 35.2°C and 750 Torr?

ANSWER: 0.00487 mol

EXAMPLE 2 Ideal gas law calculation

What is the volume of 3.45 mol of an ideal gas at 35°C and 0.766 atm?

SOLUTION The temperature must be expressed in kelvins for an ideal gas law calculation: 35°C equals 308 K (273 + 35 = 308). Substituting into the ideal gas law gives

$$\begin{aligned} pV &= nRT \\ (0.766 \text{ atm})\,V &= (3.45 \text{ mol})(0.08206 \text{ L atm/mol K})(308 \text{ K}) \\ V &= \frac{(3.45 \,\cancel{\text{mol}})(0.08206 \text{ L} \,\cancel{\text{atm}/\text{mol K}})(308 \,\cancel{\text{K}})}{0.766 \,\cancel{\text{atm}}} \\ &= 114 \text{ L} \end{aligned}$$

EXERCISE What is the volume of 0.444 mol of an ideal gas at 0.95 atm and 77.3°C?

ANSWER: 13 L

EXAMPLE 3 Ideal gas law calculation

If 0.0640 mol of an ideal gas occupies 3.68 L at 742 Torr, what is the temperature of the gas?

SOLUTION Because we use R = 0.08206 L atm/mol K, the pressure must be converted to atmospheres before substituting into the ideal gas law:

$$742\ \text{Torr} \times \frac{1\ \text{atm}}{760\ \text{Torr}} = 0.976\ \text{atm}$$

$$PV = nRT$$

$$(0.976\ \text{atm})(3.68\ \text{L}) = (0.0640\ \text{mol})(0.08206\ \text{L atm/mol K})\,T$$

$$T = \frac{(0.976\ \text{atm})(3.68\ \text{L})}{(0.08206\ \text{L atm/mol K})(0.0640\ \text{mol})}$$

$$= 684\ \text{K}$$

EXERCISE If 2.33 mol of an ideal gas occupies 12,100 mL at 651 Torr, what is the temperature of the gas in kelvins?

ANSWER: 54.2 K

PITFALL The mass of a gas in the ideal gas law

At times an ideal gas calculation calls for the mass of a gas. Although the ideal gas law does not deal directly with the mass of a gas, remember that the mass can always be calculated from the moles and molar mass of the gas:

$$\text{Mass} = \text{moles} \times \text{molar mass}$$

KEY CONCEPT B The density of an ideal gas

By using the definition of density ($d = m/V$), the fact that the moles of a sample equal the mass of the sample divided by the molar mass ($n = m/\text{molar mass}$), and the ideal gas law, we can relate the density of an ideal gas to the temperature, pressure, and molar mass of the gas:

$$d = \frac{(\text{molar mass})P}{RT}$$

Substitution of the molar mass in grams per mole, pressure in atmospheres, temperature in kelvins, and R = 0.08206 L atm/mol K gives the density in grams per liter, the usual density unit for gases.

EXAMPLE 1 Calculating the density of an ideal gas

What is the density of carbon tetrafluoride (CF_4) at 1.0 atm and 298 K?

SOLUTION The molar mass of CF_4 is calculated first and then substituted into the preceding equation:

$$\begin{aligned} 1\ \text{C:}&\quad 1 \times 12.01 = 12.01 \\ 4\ \text{F:}&\quad 4 \times 19.00 = \underline{76.00} \\ &\qquad\qquad\quad 88.01\ \text{g/mol} \end{aligned}$$

$$d = \frac{(\text{molar mass})P}{RT}$$

$$= \frac{(88.01\ \text{g/mol})(1.0\ \text{atm})}{(0.08206\ \text{L atm/mol K})(298\ \text{K})}$$

$$= 3.6\ \text{g/L}$$

EXERCISE What is the density of acetylene (C_2H_2) at 50.0°C and 720 Torr?

ANSWER: 0.930 g/L

EXAMPLE 2 Calculating the molar mass of a gas from its density

The density of an ideal gas is 2.56 g/L at 100.0°C and 744 Torr. What is the molar mass of the gas?

SOLUTION For this question we have the following data: $d = 2.56$ g/L, $P = 744$ Torr $= 0.979$ atm, and $T = 100.0°C = 373.2$ K. Substituting into the relevant equation gives

$$d = \frac{(\text{molar mass})P}{RT}$$

$$2.56 \text{ g/L} = \frac{(\text{molar mass})(0.979 \text{ atm})}{(0.08206 \text{ L atm/mol K})(373.2 \text{ K})}$$

Multiplying both sides of the equation by (0.08206 L atm/mol K)(373.2 K) and dividing both sides by 0.979 atm gives the desired answer:

$$\text{Molar mass} = \frac{(2.56 \text{ g/}\cancel{\text{L}})(0.08206 \cancel{\text{L}}\ \cancel{\text{atm}}/\text{mol } \cancel{\text{K}})(373.2 \cancel{\text{K}})}{(0.979 \cancel{\text{atm}})}$$

$$= 80.0 \text{ g/mol}$$

EXERCISE The density of an ideal gas is 2.99 g/L at 120.0°C and 829 Torr. What is the molar mass of the gas?

ANSWER: 88.4 g/mol

THE STOICHIOMETRY OF REACTING GASES

KEY WORDS Define or explain each of the following terms in a written sentence or two.

Dalton's law of partial pressures
molar volume
partial pressure
standard temperature and pressure (STP)
vapor pressure

5.4 GAS VOLUMES AND REACTION STOICHIOMETRY

KEY CONCEPT A Molar volumes of gases

The **molar volume** V_m of a substance is the volume of 1 mol of the substance or, equivalently, the volume per mole of the substance. Because the volume of a gas depends on the temperature and pressure, chemists have defined a reference temperature and pressure for the discussion of gaseous volumes. The reference point is 0°C (273.15 K) and 1 atm (exactly) and is called **standard temperature and pressure** (STP). At STP, the volume of 1 mol of an ideal gas is 22.4 L:

$$V_m = 22.4 \text{ L/mol} \qquad \text{(at STP for an ideal gas)}$$

EXAMPLE Calculating the volume of a sample of gas at STP

What is the volume, at STP, of 11.0 g of CH_4 (methane; a gas)?

SOLUTION We first calculate the moles of CH_4 present using its molar mass (16.04 g/mol):

$$11.0 \cancel{\text{g CH}_4} \times \frac{1 \text{ mol CH}_4}{16.04 \cancel{\text{g}}} = 0.686 \text{ mol CH}_4$$

We know that 1 mol of an ideal gas occupies 22.4 L at STP, and we can use this fact to calculate the volume of any number of moles of a gas at STP:

$$0.686 \cancel{\text{mol CH}_4} \times \frac{22.4 \text{ L}}{1 \cancel{\text{mol}}} = 15.4 \text{ L CH}_4$$

EXERCISE What is the volume, at STP, of 11.0 g of the gas CF_4?

ANSWER: 2.80 L

KEY CONCEPT B The volume of gas taking part in a reaction

A balanced chemical equation gives the relative number of moles of reactant and product. Thus, in a typical stoichiometric problem, the moles of the substances in the reaction become the focus of the calculations. When a gaseous reactant or product is involved, we are often given the volume of the gas or asked to calculate the volume. For this type of problem, the ideal gas law or the molar volume of a gas at STP is used to determine the moles of the gas from the volume, or the volume from the moles, depending on the nature of the problem.

EXAMPLE 1 The volume of gaseous product from the mass of a second product

Magnesium reacts with cold dilute HCl to form the gas H_2:

$$Mg(s) + 2HCl(aq) \longrightarrow MgCl_2(aq) + H_2(g)$$

In a specific experiment, 12.2 g of $MgCl_2$ were formed. What volume of H_2 (at STP) was also formed?

SOLUTION We first calculate the moles of $MgCl_2$ formed, using the molar mass of $MgCl_2$, 95.2 g/mol:

$$12.2\ \cancel{g\ MgCl_2} \times \frac{1\ mol\ MgCl_2}{95.2\ \cancel{g}} = 0.128\ mol\ MgCl_2$$

The equation states that the number of moles of $MgCl_2$ formed is the same as the number of moles of H_2 formed. This information is used to calculate the moles of H_2:

$$0.128\ \cancel{mol\ MgCl_2} \times \frac{1\ mol\ H_2}{1\ \cancel{mol\ MgCl_2}} = 0.128\ mol\ H_2$$

Because the volume at STP is asked for, the easiest way to proceed is to use the molar volume of a gas at STP, 22.4 L/mol:

$$0.128\ \cancel{mol\ H_2} \times \frac{22.4\ L}{\cancel{mol}} = 2.87\ L\ H_2$$

EXERCISE How many liters of N_2 (at STP) are required to react with 100 g Li?

$$6Li(s) + N_2(g) \longrightarrow 2Li_3N(s)$$

ANSWER: 53.8 L

EXAMPLE 2 Calculating the volume of gas in a reaction

How many liters of $HCl(g)$, measured at STP, can be produced from 0.70 L of $Cl_2(g)$, also measured at STP?

$$H_2(g) + Cl_2(g) \longrightarrow 2HCl(g)$$

SOLUTION The equation tells us that 1 mol H_2 reacts with 1 mol Cl_2 to produce 2 mol HCl. By using Avogadro's law, we conclude that the volumes of reactants and products follow the same ratio as the number of moles of each. Thus, 1 L H_2 reacts with 1 L Cl_2 to produce 2 L HCl. Since 2 mol HCl are produced from 1 mol Cl_2, we get

$$0.70\ \cancel{L}\ Cl_2 \times \frac{2\ L\ HCl}{1\ \cancel{L}\ Cl_2} = 1.4\ L\ HCl$$

EXERCISE How many liters of O_2 can be produced from 0.22 mol O_3 in the reaction

$$2O_3(g) \longrightarrow 3O_2(g)$$

ANSWER: 0.33 L

PITFALL The volumes of gases that are not at STP

The molar volume of a gas is approximately 22.4 L only at STP. If the volume is requested at some other temperature or pressure, the ideal gas law must be used.

5.5 GASEOUS MIXTURES

KEY CONCEPT Dalton's law of partial pressures

Imagine we have a mixture of gases in a flask. The **partial pressure** of a gas in the mixture is the pressure the gas would have if it were alone in the flask. **Dalton's law of partial pressures** states that the total pressure of the mixture of gases is the sum of the partial pressures of the individual gases in the mixture. Dalton's law is a result of the fact that, for ideal gases, the presence of one gas has no effect on the other gases present.

EXAMPLE Using Dalton's law of partial pressures

A sample of nitrogen that exerts a pressure of 562 Torr in a 5.00-L flask and a sample of oxygen that exerts a pressure of 444 Torr in a 10.0-L flask are both placed in a 10.0-L flask. What is the total pressure of the mixture of gases in the 10.0-L flask? No reaction occurs.

SOLUTION According to Dalton's law of partial pressures, the total pressure of the two gases is the sum of the pressures each would exert if it were alone in the flask:

$$P_{\text{total}} = P_{O_2} + P_{N_2}$$

Because the O_2 exerts a pressure of 444 Torr when in a 10.0-L flask by itself, its partial pressure is 444 Torr. The N_2 exerts a pressure of 562 Torr when alone in a 5.00-L flask. When the N_2 is placed in a 10.0-L flask, Boyle's law tells us its pressure is halved, so the partial pressure of N_2 in the 10.0-L flask is 281 Torr ($\frac{1}{2}$ of 562 Torr). The total pressure in the flask is the sum of the partial pressures:

$$P_{O_2} = 444 \text{ Torr} \qquad P_{N_2} = 281 \text{ Torr}$$
$$P_{\text{total}} = 444 \text{ Torr} + 281 \text{ Torr}$$
$$= 725 \text{ Torr}$$

EXERCISE When 0.50 g of neon gas (Ne) and 0.50 g of argon gas (Ar) are placed in a 3.0-L flask at 27°C, what are the partial pressures of each gas and the total pressure of the mixture of gases, in atmospheres?

ANSWER: Ne, 0.21 atm; Ar, 0.10 atm; total, 0.31 atm

PITFALL Partial quantities in a mixture of gases

In a mixture of gases, each gas has its own partial presssure, but not its own partial temperature or partial volume. In the mixture, each gas occupies the same volume and has the same temperature. Of the three parameters P, V, and T, only the pressure of each gas has its own partial value, and only for the pressure is the total value the sum of the individual partial values.

THE KINETIC THEORY OF GASES

5.6 MOLECULAR SPEEDS

KEY CONCEPT A The kinetic theory

The kinetic theory has three main postulates:

1. A gas consists of a collection of molecules in continuous chaotic motion.
2. Molecules are infinitely small and move in straight lines until they collide.
3. Molecules do not affect each other except when they collide.

Postulate 2 indicates that a molecule of an ideal gas has approximately zero volume. If it were possible to take a snapshot of a flask containing an ideal gas, most of the volume of the flask would be "empty space." The fact that molecules travel in straight lines until the moment of collision means that the molecules do not attract or repel each other except at the instant of collision. In essence, the motion and behavior of each molecule are unaffected by the presence of the other molecules, unless molecules collide.

EXAMPLE Interpreting the postulates of the kinetic theory

Of the three gases listed below, which should most closely follow the ideal gas equation?

He: small size, weak intermolecular attractions
CH_4: medium size, intermediate intermolecular attractions
HCl: large size, strong intermolecular attractions

SOLUTION Based on the kinetic molecular theory, an ideal gas is one that is infinitely small in size and that is not affected by other molecules. A gas with small molecules and weak intermolecular forces is most likely to act ideally. On the other hand, when intermolecular attractions are large, molecules have a substantial effect on the motion of nearby molecules. By both criteria (size and effect on nearby molecules), helium should follow the ideal gas law most closely.

EXERCISE As the volume of a sample of gas is decreased, the gas molecules are forced nearer each other. Will a gas act more like an ideal gas when it is enclosed in a large volume or a small volume?

ANSWER: Large volume

KEY CONCEPT B Molecular speeds

The average speed of a sample of molecules is usually measured using the root-mean-square (rms) speed rather than the arithmetical average. For a typical sample of gas, the rms speed is close to but a bit larger than the average speed:

$$\text{rms speed} = v = \sqrt{\frac{\text{sum of squares of speeds of individual molecules}}{\text{number of molecules}}}$$

The root-mean-square speed of a sample of gas molecules is related to both the temperature and the molar mass of the molecules:

$$v = \sqrt{\frac{3RT}{\text{molar mass (in kg/mol)}}}$$

In this equation, R should be expressed in units that give the speed in units of meters per second; thus, $R = 8.314$ J/mol K should be used. If the molar mass is expressed in grams per mole, the equation becomes

$$v = \sqrt{\frac{3000\ RT}{\text{molar mass (in g/mol)}}}$$

EXAMPLE Calculating the rms speed of a sample of molecules

What is the rms speed of carbon monoxide (CO, an extremely toxic gas found in automobile exhaust fumes) at 27°C?

SOLUTION If the temperature and molar mass of a gas are known, the rms speed can be calculated using the preceding equation. The molar mass of carbon monoxide is 28.01 g/mol and the absolute temperature is 300 K (273 + 27 = 300 K). Thus,

$$v = \sqrt{\frac{3000 \times 8.314\ \text{J/}\cancel{\text{mol}}\ \cancel{\text{K}} \times 300\ \cancel{\text{K}}}{28.01\ \text{g/}\cancel{\text{mol}}}} = \sqrt{2.67 \times 10^5\ \frac{\text{m}^2}{\text{s}^2}} = 517\ \text{m/s}$$

EXERCISE What is the root-mean-square speed of carbon monoxide at 100°C? Is the speed greater or less than that at 27°C?

ANSWER: 576 m/s; greater

PITFALL The units of the equation for rms speed

Let's be certain that the units in the equation for the rms speed are clear. We first need the basic units of the joule:

$$1\,\mathrm{J} = 1\,\frac{\mathrm{kg\,m^2}}{\mathrm{s^2}}$$

Then, by using the mathematical relationship

$$\frac{1}{a/b} = \frac{b}{a}$$

we conclude that, for *units only,*

$$\frac{1}{\text{molar mass (in kg/mol)}} = \frac{1}{\text{kg/mol}} = \frac{\text{mol}}{\text{kg}}$$

If we consider the *units only* in the equation for v and remember that the factor of 1000 converts the molar mass from grams per mole to kilograms per mole, we get

$$\begin{aligned} v &= \sqrt{\frac{\mathrm{J}}{\cancel{\mathrm{K}}\,\cancel{\mathrm{mol}}} \times \cancel{\mathrm{K}} \times \frac{\cancel{\mathrm{mol}}}{\mathrm{kg}}} \\ &= \sqrt{\mathrm{J} \times \frac{1}{\mathrm{kg}}} \\ &= \sqrt{\frac{\cancel{\mathrm{kg}}\,\mathrm{m^2}}{\mathrm{s^2}} \times \frac{1}{\cancel{\mathrm{kg}}}} \\ &= \sqrt{\frac{\mathrm{m^2}}{\mathrm{s^2}}} \\ &= \mathrm{m/s} \end{aligned}$$

KEY CONCEPT C Diffusion and effusion

The rate of diffusion and rate of effusion of a gas are each proportional to the rms speed of the gas. The time it takes for effusion or diffusion to occur is proportional to 1/rate. These facts lead directly to **Graham's law of effusion,** which relates the relative times it takes for an equal number of moles of two gases at the same temperature and pressure to effuse:

$$\frac{t_A}{t_B} = \sqrt{\frac{\text{molar mass of gas A}}{\text{molar mass of gas B}}}$$

In this equation, t_A is the time it takes for gas A to effuse and t_B is the time for gas *B*. This equation also approximately describes the relative times of diffusion for two gases at the same temperature and pressure.

EXAMPLE Calculating the time needed for effusion

At a given temperature and pressure, a certain amount of propane (C_3H_8) requires 235 s to effuse through a porous plug. How long will it take an equivalent number of moles of carbon monoxide (CO) to diffuse at the same conditions?

SOLUTION The relationship that expresses the relative times needed for effusion of an equal number of moles of two gases at the same experimental conditions is

$$\frac{t_A}{t_B} = \sqrt{\frac{\text{molar mass of gas A}}{\text{molar mass of gas B}}}$$

The molar mass of propane is 44.09 g/mol and that of carbon monoxide is 28.01 g/mol. Substituting into the equation gives

$$\frac{t_{CO}}{t_{C_3H_8}} = \sqrt{\frac{28.01 \text{ g/mol}}{44.09 \text{ g/mol}}} = 0.797$$

Since $t_{C_3H_8} = 235$ s,

$$t_{CO} = (0.797)(235 \text{ s}) = 187 \text{ s}$$

As we expect, the lighter carbon monoxide diffuses in less time than the heavier propane.

EXERCISE At a given temperature and pressure, a certain amount of argon (Ar) requires 277s to diffuse through a porous plug. How long will it take an equivalent number of moles nitrogen (N_2) to diffuse at the same conditions?

ANSWER: 232 s

PITFALL Square roots on a calculator

Be certain you know how to get a square root and a square on your calculator before trying the calculations in this section. Many calculators use the same key for these two functions, with one or the other activated by a second function or inv (inverse) key. Try a few simple examples such as the square and square root of 4 and 9 to be certain you understand how your calculator handles these functions.

5.7 REAL GASES

KEY CONCEPT The van der Waals equation

A comparison between an ideal gas and a **real gas** is presented in the following table.

Ideal gas	**Real gas**
molecules are infinitely small	molecules have a small but finite volume
molecules do not affect each other except during collisions	molecules experience attractive forces at relatively long distances

The van der Waals equation contains the parameters a and b, which correct for the effects of intermolecular attractions (a) and the finite volume of the gas molecules (b).

EXAMPLE Using the van der Waals equation

What is the pressure exerted by 100 mol of carbon dioxide (CO_2) in a 22.4-L vessel at 273 K, as calculated from the van der Waals equation?

SOLUTION The van der Waals equation is

$$P = \frac{nRT}{V - nb} - \frac{an^2}{V^2}$$

From the statement of the problem in this example and Table 5.5 of the text, we have the following:

$$n = 100 \text{ mol} \qquad R = 0.08206 \text{ L atm/mol K}$$
$$T = 273 \text{ K} \qquad V = 22.4 \text{ L}$$
$$a = 3.59 \text{ L}^2 \text{ atm/mol}^2 \qquad b = 0.043 \text{ L/mol}$$

We evaluate each of the two terms in the equation separately:

$$\frac{nRT}{V - nb} = \frac{100 \cancel{\text{mol}} \times 0.08206 \text{ L atm}/\cancel{\text{mol}}\,\cancel{\text{K}} \times 273 \cancel{\text{K}}}{22.4 \text{ L} - (100 \cancel{\text{mol}} \times 0.043 \text{ L}/\cancel{\text{mol}})}$$
$$= \frac{2241 \cancel{\text{L}} \text{ atm}}{18.1 \cancel{\text{L}}}$$
$$= 124 \text{ atm}$$

$$\frac{an^2}{V^2} = \frac{(3.59 \text{ L}^2 \text{ atm}/\cancel{\text{mol}^2}) \times (100 \cancel{\text{mol}})^2}{(22.4 \text{ L})^2}$$
$$= \frac{3.59 \times 10^4 \cancel{\text{L}^2} \text{ atm}}{502 \cancel{\text{L}^2}}$$
$$= 71.5 \text{ atm}$$

According to the equation, the pressure is the difference between these two terms:

$$P = 124 \text{ atm} - 71.5 \text{ atm}$$
$$= 53 \text{ atm}$$

EXERCISE Use the van der Waals equation to predict the pressure exerted by 120 mol of argon (Ar) enclosed in a 10.0-L flask at 100°C. Compare your result with the pressure predicted by the ideal gas law.

ANSWER: 402 atm (van der Waals): 367 atm (ideal gas law)

DESCRIPTIVE CHEMISTRY TO REMEMBER

- 1 Torr = 1 mmHg
- 1 atm = 760 Torr exactly
- The escape of a substance through a small hole is called **effusion.**
- The spreading of one substance through another is called **diffusion.**
- **Uranium** is enriched in U-235 by producing the volatile solid UF_6 and allowing the vapor to diffuse through a series of porous barriers. The molecules containing the lighter U-235 diffuse slightly faster than those containing U-238, so an enrichment is accomplished.
- Solid carbon dioxide is called **dry ice.**
- The cooling of a gas as it expands is called the **Joule–Thomson effect.**
- The volume of 1 mol of an ideal gas at standard temperature and pressure (STP, 273.15 K, 1 atm) is **22.4 L.**

CHEMICAL EQUATIONS TO KNOW

- Sulfur dioxide is a product when solid sulfur is burned in oxygen.

$$S_8(s) + 8O_2(g) \longrightarrow 8SO_2(g)$$

- Potassium superoxide reacts with carbon dioxide to form potassium carbonate and oxygen.

$$4KO_2(s) + 2CO_2(g) \longrightarrow 2K_2CO_3(s) + 3O_2(g)$$

- An explosion of nitroglycerin ($C_3H_5N_3O_9$) generates a large amount of gaseous products, 29 mol of gas for every 4 mol of nitroglycerin.

$$4C_3H_5N_3O_9(l) \longrightarrow 6N_2(g) + O_2(g) + 12CO_2(g) + 10H_2O(g)$$

- Lead azide [$Pb(N_3)_2$] explodes to release three moles of gas per mole of lead azide.

$$Pb(N_3)_2(s) \longrightarrow Pb(s) + 3N_2(g)$$

- Potassium chlorate decomposes into potassium chloride and oxygen when heated in the presence of the catalyst manganese dioxide (MnO_2).

$$2KClO_3(s) \xrightarrow{MnO_2} 2KCl(s) + 3O_2(g)$$

MATHEMATICAL EQUATIONS TO KNOW AND UNDERSTAND

$P = \frac{\text{force}}{\text{area}}$	definition of pressure
$P_1V_1 = P_2V_2$	Boyle's law for a sample of gas at constant temperature
$\frac{V_1}{T_1} = \frac{V_2}{T_2}$	Charles's law, for a sample of gas at constant pressure
$\frac{V_1}{n_1} = \frac{V_2}{n_2}$	Avogadro's principle
$\frac{P_1V_1}{T_1} = \frac{P_2V_2}{T_2}$	combined gas law
$PV = nRT$	ideal gas law
$d = \frac{P \times \text{molar mass}}{RT}$	density of an ideal gas
$v = \sqrt{\frac{3RT}{\text{molar mass (in kg/mol)}}} = \sqrt{\frac{3000RT}{\text{molar mass (in g/mol)}}}$	root-mean-square speed of an ideal gas
$\frac{t_A}{t_B} = \sqrt{\frac{\text{molar mass of gas A}}{\text{molar mass of gas B}}}$	Graham's law of effusion
$\left[P + \frac{an^2}{V^2}\right](V - nb) = nRT \qquad P = \frac{nRT}{V - nb} - \frac{an^2}{V^2}$	two forms of the van der Waals equation

SELF-TEST EXERCISES

The gas laws

1. Pressure is defined as
(a) force (b) mass × acceleration (c) change in speed/time (d) force/area

2. A pressure of 683 Torr (760 Torr = 1 atm) is equivalent to
(a) 0.899 atm (b) 1.00 atm (c) 5.19×10^5 atm (d) 1.93×10^{-6} atm (e) 1.11 atm

3. What is atmospheric pressure for the mercury barometer pictured?

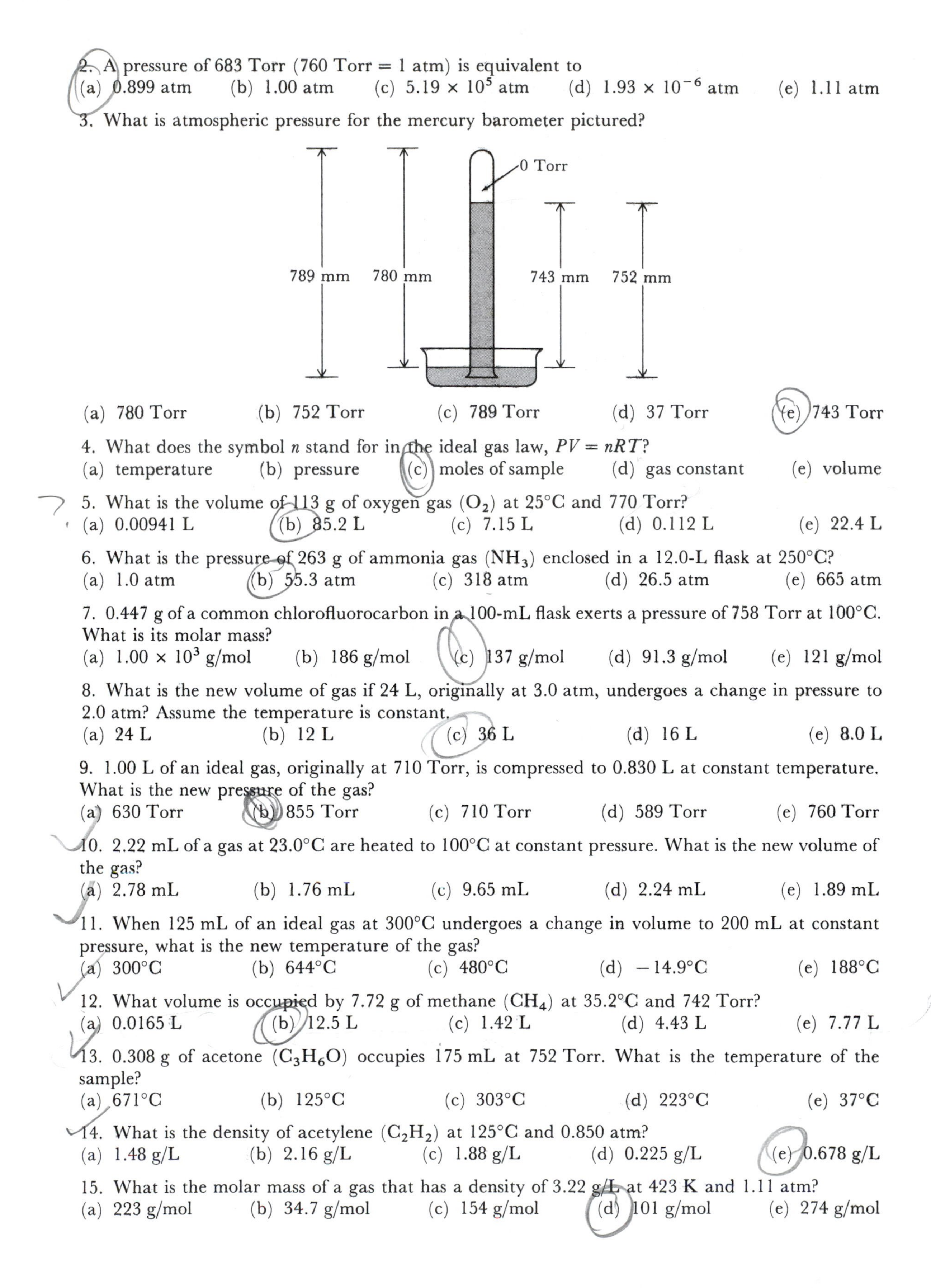

(a) 780 Torr (b) 752 Torr (c) 789 Torr (d) 37 Torr (e) 743 Torr

4. What does the symbol n stand for in the ideal gas law, $PV = nRT$?
(a) temperature (b) pressure (c) moles of sample (d) gas constant (e) volume

5. What is the volume of 113 g of oxygen gas (O_2) at 25°C and 770 Torr?
(a) 0.00941 L (b) 85.2 L (c) 7.15 L (d) 0.112 L (e) 22.4 L

6. What is the pressure of 263 g of ammonia gas (NH_3) enclosed in a 12.0-L flask at 250°C?
(a) 1.0 atm (b) 55.3 atm (c) 318 atm (d) 26.5 atm (e) 665 atm

7. 0.447 g of a common chlorofluorocarbon in a 100-mL flask exerts a pressure of 758 Torr at 100°C. What is its molar mass?
(a) 1.00×10^3 g/mol (b) 186 g/mol (c) 137 g/mol (d) 91.3 g/mol (e) 121 g/mol

8. What is the new volume of gas if 24 L, originally at 3.0 atm, undergoes a change in pressure to 2.0 atm? Assume the temperature is constant.
(a) 24 L (b) 12 L (c) 36 L (d) 16 L (e) 8.0 L

9. 1.00 L of an ideal gas, originally at 710 Torr, is compressed to 0.830 L at constant temperature. What is the new pressure of the gas?
(a) 630 Torr (b) 855 Torr (c) 710 Torr (d) 589 Torr (e) 760 Torr

10. 2.22 mL of a gas at 23.0°C are heated to 100°C at constant pressure. What is the new volume of the gas?
(a) 2.78 mL (b) 1.76 mL (c) 9.65 mL (d) 2.24 mL (e) 1.89 mL

11. When 125 mL of an ideal gas at 300°C undergoes a change in volume to 200 mL at constant pressure, what is the new temperature of the gas?
(a) 300°C (b) 644°C (c) 480°C (d) −14.9°C (e) 188°C

12. What volume is occupied by 7.72 g of methane (CH_4) at 35.2°C and 742 Torr?
(a) 0.0165 L (b) 12.5 L (c) 1.42 L (d) 4.43 L (e) 7.77 L

13. 0.308 g of acetone (C_3H_6O) occupies 175 mL at 752 Torr. What is the temperature of the sample?
(a) 671°C (b) 125°C (c) 303°C (d) 223°C (e) 37°C

14. What is the density of acetylene (C_2H_2) at 125°C and 0.850 atm?
(a) 1.48 g/L (b) 2.16 g/L (c) 1.88 g/L (d) 0.225 g/L (e) 0.678 g/L

15. What is the molar mass of a gas that has a density of 3.22 g/L at 423 K and 1.11 atm?
(a) 223 g/mol (b) 34.7 g/mol (c) 154 g/mol (d) 101 g/mol (e) 274 g/mol

16. 6.00 L of an ideal gas at 1.00 atm pressure and 273 K undergo a change in temperature and pressure to 1.63 atm and 255 K. What is the new volume of the gas?
(a) 3.94 L (b) 9.14 L (c) 10.5 L (d) 3.44 L (e) 3.59 L

17. A sample of gas at 735 Torr occupies a volume of 2.55 L at 25°C. The pressure is suddenly increased to 800 Torr and the volume decreased to 2.45 L. What is the new temperature of the gas?
(a) −10°C (b) 12°C (c) 26°C (d) 65°C (e) 39°C

18. 12.0 L of an ideal gas at 60.0°C and 2.25 atm are expanded to 16.0 L and increased in temperature to 120°C. What is the new pressure of the gas?
(a) 1.99 atm (b) 3.38 atm (c) 1.43 atm (d) 2.54 atm (e) 3.54 atm

19. 2.00 mol of O_2 occupy 40.0 L at a certain temperature and pressure. What volume do 2.00 mol of CH_4 occupy at the same conditions?
(a) 80.0 L (b) 20.0 L (c) 40.0 L (d) 10.0 L (e) 160 L

The stoichiometry of reacting gases

20. Use the fact that the molar volume of an ideal gas is 22.4 L to calculate the volume occupied by 60.0 g of hydrogen sulfide at STP.
(a) 39.4 L (b) 23.9 L (c) 22.4 L (d) 31.6 L (e) 18.0 L

21. A 0.643-g sample of a gas occupies 100.0 mL at STP. What is the molar mass of the gas?
(a) 64.3 g/mol (b) 144 g/mol (c) 162 g/mol (d) 2.87 g/mol (e) 187 g/mol

22. 0.650 g of carbon is burned in oxygen to produce CO_2. What volume of CO_2 (measured at STP) is formed?

$$C(s) + O_2(g) \longrightarrow CO_2(g)$$

(a) 14.6 L (b) 1.21 L (c) 8.88 L (d) 22.4 L (e) 15.3 L

23. How many liters of gaseous H_2O are formed by the reaction of 0.36 L O_2? All volumes are measured at STP.

$$2H_2(g) + O_2(g) \longrightarrow 2H_2O(g)$$

(a) 0.36 L (b) 0.72 L (c) 0.18 L (d) 0.090 L (e) 1.44 L

24. How many liters of N_2 (measured at STP) are produced by the reaction of 1.22 kg lead azide, $Pb(N_3)_2$?

$$Pb(N_3)_2 \longrightarrow Pb(s) + 3N_2(g)$$

(a) 93.8 L (b) 281 L (c) 8.23 L (d) 31.3 L (e) 162 L

25. A 25.0-L flask contains 20.0 g of carbon monoxide and 20.0 g of carbon dioxide at 298 K. What is the partial pressure of the carbon monoxide?
(a) 622 Torr (b) 380 Torr (c) 442 Torr (d) 245 Torr (e) 530 Torr

26. A flask containing two gases, A and B, is at a temperature such that the pressure of A, if it were in the flask alone, would be 655 Torr, and the pressure of B, if it were in the flask alone, would be 22 Torr. What is the pressure of the mixture of the two gases in the flask?
(a) 1.3×10^5 Torr (b) 633 Torr (c) 677 Torr (d) 33 Torr (e) 760 Torr

27. 115 g each of fluorine and chlorine are enclosed in a 250-mL flask at −30°C. What is the partial pressure of each of the gases? Answers are given as "partial pressure F_2, partial pressure Cl_2." Assume ideal gas behavior.
(a) 242 atm, 129 atm (b) 302 atm, 161 atm (c) 242 atm, 161 atm
(d) 302 atm, 129 atm (e) 185 atm, 185 atm

28. For the mixture described in problem 27, what is the total pressure?
(a) 113 atm (b) 141 atm (c) 371 atm (d) 463 atm (e) 760 atm

29. A sample of oxygen gas is collected in a bottle, over water, at 25°C at a total pressure of 748.2 Torr. What is the actual pressure of the oxygen molecules in the bottle? The vapor pressure of water at 25°C is 23.8 Torr.
(a) 760.0 Torr (b) 748.2 Torr (c) 736.2 Torr (d) 724.4 Torr (e) 772.0 Torr

30. 285 mL of oxygen gas were collected in a gas bottle, over water, at 16.0°C at a total pressure of 761.1 Torr. How many millimoles of oxygen were collected? The vapor pressure of water at 16.0°C is 13.6 Torr.
(a) 12.2 (b) 11.8 (c) 221 (d) 12.0 (e) 213

The kinetic theory of gases

31. What is the rms speed of gaseous carbon tetrachloride (CCl_4) at 85°C?
(a) 241 m/s (b) 23.9 m/s (c) 3.37×10^9 m/s (d) 3.28×10^5 m/s (e) 7.62 m/s

32. What is the rms speed of four cars with the following speeds: 32.6 mph, 35.5 mph, 38.6 mph, and 45.2 mph?
(a) 38.0 mph (b) 38.2 mph (c) 38.9 mph (d) 37.5 mph (e) 38.5 mph

33. A sample of an unknown gas takes 434 s to diffuse through a porous plug. An equal amount of N_2 takes 176 s to diffuse through the same plug at the same conditions. What is the molar mass of the unknown?
(a) 6.30 g/mol (b) 115 g/mol (c) 161 g/mol (d) 170 g/mol (e) 69.0 g/mol

34. By what factor does the rms speed of any ideal gas change when the temperature is increased from 100°C to 200°C?
(a) 1.00 (b) 2.00 (c) 1.13 (d) 1.28 (e) 4.00

35. What is the rms speed of He atoms at the temperature at which the rms speed of Xe is 165 m/s?
(a) 28.8 m/s (b) 5.42×10^3 m/s (c) 5.33 m/s (d) 945 m/s (e) 165 m/s

36. At low temperatures most molecules have speeds ________ to/from the rms speed; at high temperatures, a high proportion of molecules have speeds that are ________ to/from the rms speed.
(a) very different, close (b) close, close (c) identical, identical
(d) very different, very different (e) close, very different

37. Use the van der Waals equation to calculate the pressure exerted by 500 mol of N_2 enclosed in a 40.0-L vessel at 298 K.
(a) 186 atm (b) 221 atm (c) 379 atm (d) 226 atm (e) 100 atm

Descriptive chemistry

38. What products are formed when potassium chlorate ($KClO_3$) is heated in the presence of manganese dioxide (MnO_2)?
(a) Mn, KCl, O_2 (b) $MnCl_2$, KO_2 (c) $KMnO_4$, KCl
(d) KCl, O_2 (e) MnO_2, KCl

39. What is the formula of lead(II) azide?
(a) PbN (b) PbN_3 (c) $Pb_2(N_3)_2$ (d) Pb_2N_3 (e) $Pb(N_3)_2$

40. What low temperature can be reached by adding solid chips of carbon dioxide to a low freezing liquid?
(a) −77°C (b) −35°C (c) 0°C (d) −196°C (e) −105°C

41. The Joule–Thompson effect is the name given to the fact that many gases ________ as they expand.
(a) warm (b) cool (c) do not change temperature (d) solidify (e) react

ENERGY, HEAT, AND THERMOCHEMISTRY

Virtually every chemical process is accompanied by either a release or an absorption of heat. For a variety of reasons, chemists would like to be able to predict the heat changes that accompany a reaction. **Thermochemistry,** the study of the heat effects that accompany chemical reactions, is a principal topic of this chapter.

ENTHALPY AND CALORIMETRY

KEY WORDS Define or explain each of the following terms in a written sentence or two.

endothermic
enthalpy
exothermic
heat
heat capacity
joule
specific heat capacity
state property
surroundings
system

6.1 ENERGY AND HEAT

KEY CONCEPT A The energy of chemical and physical processes

In a chemical reaction, if the energy of the reactants is greater than the energy of the products, energy is released to the surroundings as the reaction occurs. That is, because the reactants must lose energy to become products, energy is lost to the surroundings. On the other hand, if the products have a higher energy than the reactants, the reactants must gain energy from the surroundings for the reaction to occur. The energy transfer in a reaction often occurs as heat, but other types of energy may be involved. A reaction that releases energy to the surroundings is **exothermic,** and a reaction that takes in energy from the surroundings is **endothermic;** these terms are also used to describe the energy transfer that accompanies physical changes (such as melting, boiling, and condensation).

EXAMPLE Recognizing the direction of energy transfer

A refrigerator does its job by taking advantage of how energy is transferred when a refrigerant changes its physical state. Use the term *exothermic* or *endothermic* to describe the change in the physical state of the refrigerant that occurs in the coils inside a refrigerator.

SOLUTION The purpose of the refrigerant is to remove energy from the inside of the refrigerator and release it (the energy) to the outside. An *endothermic* process (usually a vaporization) occurs in the coils inside the refrigerator; energy (as heat) is absorbed from the contents of the refrigerator as the vaporization occurs, so the refrigerator cools. This energy is later released to the outside of the refrigerator by an exothermic process (usually a condensation); you can sense this release by touching the coils on the back of your refrigerator.

EXERCISE When coming out of the water after a swim, you normally feel cool. This is due to the evaporation of the water on your skin. Is evaporation an endothermic or exothermic process?

ANSWER: Endothermic

KEY CONCEPT B The units of energy

Two common units of energy are the **joule** and the **calorie.** The relation between these units is

$$1 \text{ cal} = 4.184 \text{ J} \quad \text{(exactly)}$$

The kilojoule (1 kJ = 1000 J) and kilocalorie (1 kcal = 1000 cal) are also commonly used. Because 4.184 J is the heat required to raise the temperature of 1 g of water by 1°C, 1 cal is the heat required to raise the temperature of 1 g of water by 1°C.

EXAMPLE Calculating an energy transfer

When 50 mL of an aqueous solution of sodium hydroxide are mixed with 50 mL of an aqueous solution of hydrochloric acid, a reaction occurs and the temperature of the resulting 100 mL of solution increases by 3.8°C. How much heat (in joules) was released by the reaction. Is the reaction exothermic or endothermic? Assume the solution is very dilute, so it behaves like water.

SOLUTION We know that when 1 g of water absorbs 4.184 J of heat, the temperature is increased by 1°C. This fact translates into the conversion factor:

$$\frac{4.184\ \text{J}}{1\ \text{g}\ °\text{C}}$$

Thus, if we know the mass of water and the temperature change, we can calculate the heat absorbed. The volume of water is 100 mL, and we assume it has a density of 1.00 g/mL; so the mass of water is 100 g. Thus, the heat absorbed is

$$100\ \text{g} \times 3.8°\text{C} \times \frac{4.184\ \text{J}}{1\ \text{g}\ °\text{C}} = 1.6 \times 10^3\ \text{J}$$

$$= 1.6\ \text{kJ}$$

Because the temperature of the water increases, energy (as heat) is released, and the reaction is exothermic.

EXERCISE When a beaker containing 750.0 g of water is placed inside a refrigerator, the temperature of the water changes from 23.2°C to 14.8°C. How much energy did the water lose?

ANSWER: 26 kJ

KEY CONCEPT C Heat capacity and specific heat capacity

The **heat capacity** is an extensive property (that is, it depends on the size of the sample) that relates the heat transfer of a specific sample to the temperature change the sample undergoes:

$$\underset{\text{joule}}{\text{Heat transfer}} = \underset{\frac{\text{joule}}{\text{K}}}{\text{heat capacity}} \times \underset{\text{K}}{\text{change in temperature}}$$

The **specific heat capacity** is an intensive property (that is, it does not depend on the size of the sample) that relates the heat transfer of a sample of a substance of given mass to the temperature change the sample undergoes:

$$\underset{\text{joule}}{\text{Heat transfer}} = \underset{\text{K}}{\text{change in temperature}} \times \underset{\text{g}}{\text{mass of sample}} \times \underset{\frac{\text{joule}}{\text{g}\cdot\text{K}}}{\text{specific heat capacity}}$$

Because the celsius degree and the kelvin are the same size, the change in °C is the same as the change in absolute temperature, and either scale may be used to express the change in temperature.

EXAMPLE 1 Using the specific heat capacity

How much heat must be supplied to a 870-g piece of copper in order to warm it from 25°C to 432°C? The specific heat capacity of copper is 0.38 J/(g·K).

SOLUTION The relationship that relates the heat gain to the temperature change of a sample is

Heat = change in temperature × mass of sample × specific heat capacity

In this problem, mass = 870 g, specific heat capacity = 0.38 J/(K·g), and change in temperature (ΔT) = 432°C − 25°C = 407°C. Since a ΔT of 407°C corresponds to a ΔT of 407 K,

$$\text{Heat} = 407\ \text{K} \times 870\ \text{g} \times 0.38\ \frac{\text{J}}{\text{K}\cdot\text{g}}$$

$$= 1.3 \times 10^5\ \text{J}$$

$$= 1.3 \times 10^2\ \text{kJ}$$

EXERCISE How much heat is required to raise the temperature of 25.0 g of ethanol from 0.0°C to 100.0°C? The specific heat capacity of ethanol is 2.42 J/(K·g).

ANSWER: 6.05 kJ

EXAMPLE 2 Using the specific heat capacity

A 225-g sample of benzene at 55.3°C is warmed by the addition of 432 J of heat. What is the new temperature of the sample? The specific heat of benzene is 1.05 J/(K·g).

SOLUTION The relationship between the heat withdrawn from or added to a sample and the change in temperature of the sample is

Heat transfer = change in temperature × mass of sample × specific heat capacity

In this problem, mass = 225 g, specific heat capacity = 1.05 J/(K·g), and heat transfer = 432 J; the change in temperature is the unknown in the equation:

$$432\ \text{J} = \text{change in temperature} \times 225\ \text{g} \times 1.05\ \frac{\text{J}}{\text{K}\cdot\text{g}}$$

$$\text{Change in temperature} = \frac{432\ \text{J}}{1.05\ \text{J}/(\text{K}\cdot\text{g}) \times 225\ \text{g}}$$

$$= 1.89\ \text{K}$$

EXERCISE A 166-g sample of water at 88.8°C is warmed by the addition of 752 J of heat. What is the change in temperature and the new temperature of the sample?

ANSWER: ΔT = 1.08°C; new temperature = 89.9°C

PITFALL The units used for the change in temperature ΔT

The kelvin scale and celsius scale use different numbers for the temperature of a sample, but since the size of the celsius degree is the same as the size of the kelvin, the change in temperature in celsius degrees is the same as the change in kelvins:

$$T(\text{celsius}) \neq T(\text{kelvins}) \qquad \text{but} \qquad \Delta T\ (\text{celsius}) = \Delta T\ (\text{kelvins})$$

6.2 ENTHALPY

KEY CONCEPT A Heat transfer: exothermic versus endothermic reactions

It is common practice to run reactions in a vessel open to the atmosphere so that the pressure on the reaction system stays constant at atmospheric pressure. Under conditions of constant pressure, the heat transfer that occurs is attributed to a change in **enthalpy** (ΔH) of the chemical system.

$$\text{Change in enthalpy of system} = \text{heat transfer that accompanies reaction}$$
$$\Delta H \qquad\qquad\qquad q_p$$

It is important to realize that the enthalpy is a type of energy and, therefore, has units of energy. The change in enthalpy is defined as

$$\Delta H = \text{final value of enthalpy} - \text{initial value of enthalpy}$$

When the initial enthalpy is greater than the final enthalpy, the preceding equation tells us that ΔH is negative; in this case, heat is lost to the surroundings. When the initial enthalpy is smaller than the final enthalpy, ΔH is positive and the surroundings transfer heat to the chemical system. Thus, a negative change in enthalpy corresponds to an exothermic process and a positive change in enthalpy to an endothermic process.

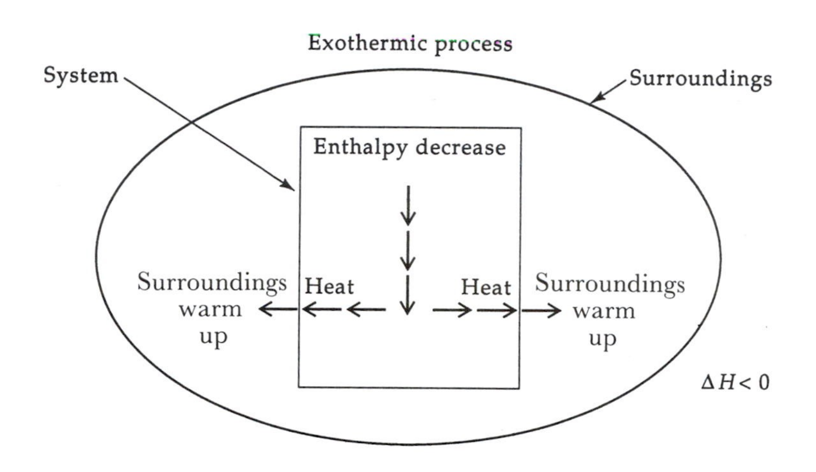

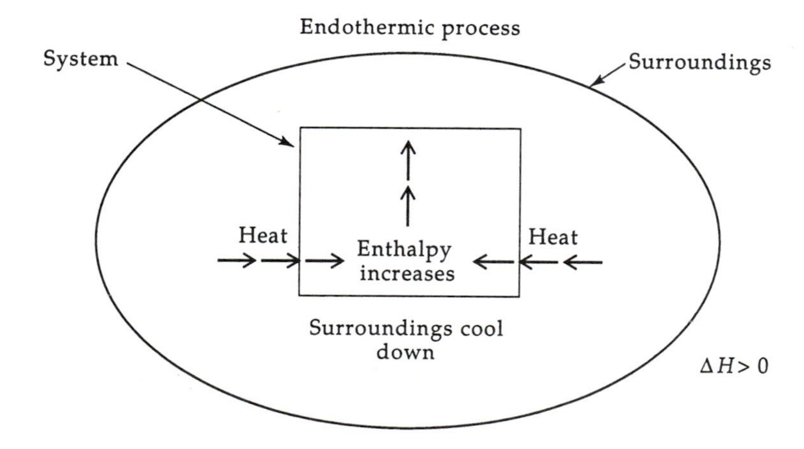

EXAMPLE 1 Using the definition of enthalpy

A chemical reaction run in an open beaker releases 1.23 kJ of heat. The same reaction run in a closed stainless steel cylinder releases 1.25 kJ. What is ΔH for the reaction?

SOLUTION The change in enthalpy for a chemical reaction is the heat released or absorbed when the reaction is run at constant pressure. When a reaction is run in an open container, the pressure is constant at atmospheric pressure, and the heat released or absorbed is ΔH. Because in this example heat is released, a negative sign is associated with the change. Thus, $\Delta H = -1.23$ kJ.

EXERCISE A reaction run at constant volume absorbs 3.55 kJ of heat from the surroundings. When run in an open container at atmospheric pressure, the same reaction absorbs 3.47 kJ of heat. What is ΔH for the reaction?

ANSWER: +3.47 kJ

EXAMPLE 2 Classifying a reaction as exothermic or endothermic

A chemical reaction occurs at constant pressure in which 11 kJ of energy (as heat) leaves the system and enters the surroundings. What is ΔH for the reaction? Classify the reaction as endothermic or exothermic and discuss what happens to the temperature of the surroundings.

SOLUTION Because the reaction is run at constant pressure, the heat gained or lost is equal to the change in the enthalpy of the system, ΔH. A loss of energy from the system as heat corresponds to a decrease in the enthalpy and requires a negative ΔH, so $\Delta H = -11$ kJ. Because energy is transferred to the surroundings, the reaction is exothermic; the surroundings warm up.

EXERCISE A reaction run a constant pressure causes 7.8 kJ of energy as heat to leave the surroundings and enter the system. Is the reaction exothermic or endothermic? What happens to the temperature of the surroundings? What is ΔH?

ANSWER: Endothermic; temperature of surroundings decreases; $\Delta H = +7.8$ kJ

KEY CONCEPT B State properties

A **state property** of a system is a property that depends only on the present condition of the system, not on how it got there. The enthalpy, volume, temperature, and pressure of a system are state properties. As an example, let's consider starting with a gas at a pressure of 700 Torr, and changing the pressure to 750 Torr in two different ways, by path (A + B) and by path C:

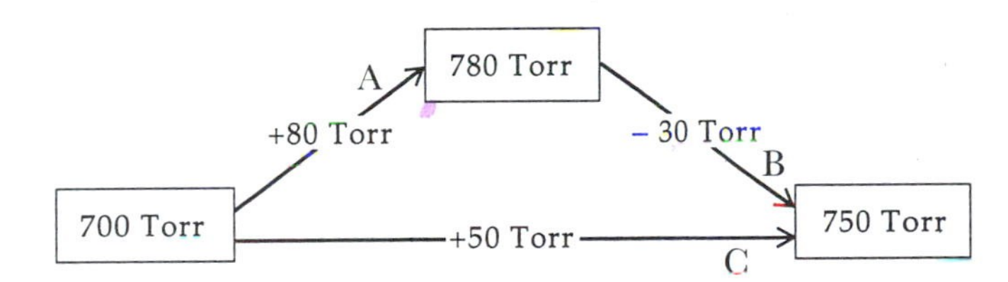

By path (A + B), we change the pressure in two steps. First, we increase the pressure by 80 Torr to get 780 Torr; second, we decrease it by 30 Torr to arrive at the final pressure of 750 Torr. By path C, we simply increase the pressure by 50 Torr to get to the final pressure of 750 Torr. The final pressure is 750 Torr in both cases and does not depend on the path taken. We express the change in a state property by subtracting the value of the property before the change (the initial value) from the value of the property after the change (the final value). A Greek upper case delta (Δ) is used to symbolize a change, so ΔP means change in pressure:

$$\Delta P = P_{\text{final}} - P_{\text{initial}}$$

Note that, with this definition, a change from a higher to a lower value for a property results in a negative change, and that a change from a lower to a higher value gives a positive change.

Change	Initial value	Final value	Δ value
change T from 350 K to 325 K	350 K	325 K	−25 K
change P from 700 Torr to 750 Torr	700 Torr	750 Torr	+50 Torr

EXAMPLE State properties

A sample of copper at a temperature of 250 K undergoes the following changes in temperature, in the order given: (1) an increase of 23 K, to 273 K; (2) a decrease of 7 K, to 266 K; (3) a decrease of 12 K, to 254 K; (4) an increase of 9 K, to 263 K; (5) and an increase of 16 K, to 279 K. What is ΔT for the copper?

SOLUTION Because temperature is a state property, if we know the initial and final temperatures, we need not be concerned with the individual changes that got the copper to its final temperature. We may use the formula

$$\Delta T = T_{\text{final}} - T_{\text{initial}}$$

From the data given, $T_{\text{final}} = 279$ K and $T_{\text{initial}} = 250$ K; thus,

$$\begin{aligned}\Delta T &= 279 \text{ K} - 250 \text{ K}\\ &= +29 \text{ K}\end{aligned}$$

EXERCISE A system undergoes two changes in enthalpy, one that increases enthalpy by 52 J and one that decreases enthalpy by 78 J. What is ΔH for the overall change?

ANSWER: −26 J

KEY CONCEPT C Calorimeter calibration

A calorimeter operates by absorbing or losing heat. Part of doing an experiment in a calorimeter is the precise and accurate measurement of its heat capacity, so that the heat lost or gained by a reaction can be correctly calculated. This is called **calibrating** the calorimeter. In one method of calibration, the temperature rise in the calorimeter is measured as a known electric current is passed through a heating coil in the calorimeter for a known amount of time. The energy the heater supplies is calculated using the equation

$$q \text{ (in joules)} = \text{current in amperes} \times \text{volts} \times \text{time in seconds}$$

This heat and the change in temperature can then be used to construct a conversion factor that expresses the heat transfer for a 1-K change in absolute temperature:

$$\text{Heat transfer per 1-K change} = \frac{q}{\Delta T}$$

Another way to supply a known amount of heat for calibration is to run a reaction with a known ΔH in the calorimeter.

EXAMPLE Using a calorimeter calibration

A 0.3250-A current passing through a heating coil in a calorimeter for 212.6 s causes a temperature increase of 2.224°C in the calorimeter when the voltage drop across the coil is 10.28 V; the calorimeter is open to atmospheric pressure. When a reaction is then run in the same calorimeter, the temperature increases by 3.446°C. What is the enthalpy change of the reaction?

SOLUTION We anticipate that we must first use the information from the heating coil to construct a conversion factor that relates the temperature change in the calorimeter to the heat transfer to the calorimeter. For calculation of the heat produced by the coil, current = 0.3250 A, voltage = 10.28 V, and time = 212.6 s.

$$\begin{aligned}\text{Heat produced by coil} &= 0.3250 \text{ A} \times 10.28 \text{ V} \times 212.6 \text{ s}\\ &= 710.3 \text{ J}\end{aligned}$$

Because the temperature change in the calorimeter was 2.224°C, $\Delta T = 2.224$ K:

$$\begin{aligned}\text{Heat transfer per 1-kelvin change} &= \frac{q}{\Delta T}\\ &= \frac{710.3 \text{ J}}{2.224 \text{ K}}\end{aligned}$$

With this conversion factor, we can calculate the heat output of the chemical reaction:

$$\text{Heat output} = 3.446\ \text{K} \times \frac{710.3\ \text{J}}{2.224\ \text{K}}$$
$$= 1101\ \text{J}$$

A heat output corresponds to a negative change in enthalpy for the reaction system:

$$\Delta H = -1101\ \text{J}$$

EXERCISE A temperature increase of 2.558°C is observed in a calorimeter when a 0.446-A current is passed through a heating coil inside the calorimeter for 313.5 s; the voltage drop across the coil is 10.57 V. When a reaction is run inside the calorimeter, the temperature drops by 1.492°C. What is ΔH for the reaction?

ANSWER: $\Delta H = +862.1$ J

PITFALL Endothermic reactions in calorimeters

As a practical matter, most of the reactions run in calorimeters are exothermic, and most of the examples in the text and study guide reflect that fact. However, there is no difference, in principle, in the methodology and calculations if the reaction under investigation is endothermic.

KEY CONCEPT D Enthalpies of physical change

Although it is impossible to measure the enthalpy of a substance (only enthalpy *changes* can be measured), it is safe to say that for any substance at a given temperature

$$H_{\text{vapor}} > H_{\text{liquid}} > H_{\text{solid}}$$

Thus, a phase change is accompanied by a change in enthalpy, which is manifested by the absorption or release of heat. The change in enthalpy for a phase change is given by

$$\Delta H = H_2 - H_1$$

The following table summarizes the enthalpies of phase change.

Phase change	Enthalpy change	Exo- or endothermic	Value for water (kJ/mol)
vaporization	$\Delta H_{\text{vap}} = H_{\text{vapor}} - H_{\text{liquid}}$	endothermic	+40.7
condensation*	$\Delta H = H_{\text{liquid}} - H_{\text{vapor}}$	exothermic	−40.7
melting	$\Delta H_{\text{melt}} = H_{\text{liquid}} - H_{\text{solid}}$	endothermic	+6.01
freezing	$\Delta H = H_{\text{solid}} - H_{\text{liquid}}$	exothermic	−6.01
sublimation	$\Delta H_{\text{sub}} = H_{\text{vapor}} - H_{\text{solid}}$	endothermic	+46.7
condensation*	$\Delta H = H_{\text{solid}} - H_{\text{vapor}}$	exothermic	−46.7

* The term *condensation* refers to the change from vapor to liquid and to the change from vapor to solid.

EXAMPLE Measuring the molar enthalpy of melting

It is found that it takes 15.0 kJ of heat to melt 45.0 g of ice at its melting temperature of 0°C. What is the enthalpy of the melting of water at 0°C in kilojoules per mole?

SOLUTION The enthalpy of melting is the heat needed to melt 1 mol of H_2O at its melting point. The problem does not give us the heat for 1 mol, but for 45.0 g. To get the enthalpy for 1 mol, we must first find the number of moles of H_2O in 45.0 g H_2O. For this we use the molar mass of water, 18.02 g/mol:

$$45.0\ \text{g} \times \frac{1\ \text{mol H}_2\text{O}}{18.02\ \text{g}} = 2.50\ \text{mol H}_2\text{O}$$

To calculate the heat per mole, we divide the heat required for the 2.50 mol (45.0 g) by 2.50 mol:

$$\Delta H_{melt} = \frac{15.0 \text{ kJ}}{2.50 \text{ mol}}$$
$$= 6.00 \text{ kJ/mol}$$

EXERCISE It takes 3.4 kJ of heat to melt 56.3 g of iodine (I_2) at its melting point of 387°C. Calculate the enthalpy of melting of iodine in kilojoules per mole.

ANSWER: 15 kJ/mol

PITFALL Enthalpies of physical change and temperatures of physical change

Values for enthalpies of physical change, such as those given in Table 6.2 of the text, are correct only when the phase change occurs at the normal temperature for the change, such as 0°C for the melting or freezing of water. If the phase change occurs at a different temperature, the value for the enthalpy change is different. Many problems involving the change in enthalpy for a phase change will state the temperature to let us know that it is permissible to use the type of data given in Table 6.2. However, the temperature itself is not used in the calculation.

KEY CONCEPT E The enthalpy for a reverse process

Because the enthalpy is a state property, the enthalpy change of the reverse of a process is the negative of the enthalpy change for the forward process. For instance, the enthalpy of the melting of water is +6.01 kJ/mol; therefore, the enthalpy of freezing is −6.01 kJ/mol.

$$H_2O(s) \longrightarrow H_2O(l) \qquad \Delta H = +6.01 \text{ kJ}$$
$$H_2O(l) \longrightarrow H_2O(s) \qquad \Delta H = -6.01 \text{ kJ}$$

This change of sign for the enthalpy change for a reverse process works for chemical reactions as well as phase changes. In fact, it works for any process.

EXAMPLE The enthalpy change for a reverse process

Table 6.2 of the text states that it takes 8.2 kJ to vaporize 1 mol of methane (CH_4) at its boiling temperature. How much heat is released when 1 mol of methane condenses from the vapor to the liquid at the same temperature?

SOLUTION Condensation from the vapor to the liquid is the reverse of vaporization. Thus, the change in enthalpy for one process will be the negative of the change in enthalpy for the other, and 8.2 kJ of heat is released when 1 mol of methane condenses. This is shown by the following equations:

$$CH_4(l) \longrightarrow CH_4(g) \qquad \Delta H = +8.2 \text{ kJ}$$
$$CH_4(g) \longrightarrow CH_4(l) \qquad \Delta H = -8.2 \text{ kJ}$$

EXERCISE The decomposition of HF(*g*) requires 542.2 kJ/mol of heat:

$$2HF(g) \longrightarrow H_2(g) + F_2(g) \qquad \Delta H = +542.2 \text{ kJ}$$

What is ΔH for the reaction

$$H_2(g) + F_2(g) \longrightarrow 2HF(g) \qquad \Delta H = ?$$

ANSWER: −542.2 J

THE ENTHALPY OF CHEMICAL CHANGE

KEY WORDS Define or explain each of the following terms in a written sentence or two.

carbohydrate	Hess's law	standard enthalpy of combustion
enthalpy density	protein	standard enthalpy of formation
fat	reaction enthalpy	standard reaction enthalpy
fossil fuel	specific enthalpy	standard state

6.3 REACTION ENTHALPIES

KEY CONCEPT A Enthalpy change for a chemical equation

In general, a chemical reaction is accompanied by a change in enthalpy. The enthalpy change is called the **reaction enthalpy.** The reaction enthalpy is usually written next to the equation for the reaction:

$$C_2H_4(g) + 3O_2(g) \longrightarrow 2CO_2(g) + 2H_2O(g) \qquad \Delta H = -1411 \text{ kJ}$$

This way of indicating the enthalpy has a very specific meaning: it reflects the enthalpy change for the exact reaction that is written. For the preceding reaction, the interpretation is as follows: when 1 mol of C_2H_4 (ethylene) in the gaseous state reacts with 3 mol of O_2 in the gaseous state, resulting in the production of 2 mol of CO_2 in the gaseous state and 2 mol of H_2O in the gaseous state, the enthalpy of the reaction system decreases by 1411 kJ; and, therefore, 1411 kJ of heat is released.

EXAMPLE Using reaction enthalpies

How much heat is released when 25.0 g of propane (C_3H_8) are completely burned?

$$C_3H_8(g) + 5O_2(g) \longrightarrow 3CO_2(g) + 4H_2O(l) \qquad \Delta H = -2220 \text{ kJ}$$

SOLUTION The chemical equation tells us that when 1 mol of propane is burned, 2220 kJ of heat are released:

$$1 \text{ mol } C_3H_8 = 2220 \text{ kJ}$$

This can be used as a conversion factor to calculate the heat produced by any amount of propane. The moles of propane are calculated using its molar mass, 44.09 g/mol:

$$25.0 \text{ g } C_3H_8 \times \frac{1 \text{ mol } C_3H_8}{44.09 \text{ g}} = 0.567 \text{ mol } C_3H_8$$

$$0.567 \text{ mol } C_3H_8 \times \frac{2220 \text{ kJ}}{1 \text{ mol } C_3H_8} = 1260 \text{ kJ}$$

$$= 1.26 \text{ kJ}$$

EXERCISE How much heat is released when 75.0 g of methanol (CH_3OH) are burned completely in air?

$$2CH_3OH(l) + 3O_2(g) \longrightarrow 2CO_2(g) + 4H_2O(l) \qquad \Delta H = -726 \text{ kJ}$$

ANSWER: 850 kJ

KEY CONCEPT B Standard states and standard reaction enthalpies

The reaction enthalpy depends on the temperature and pressure of the reactants and products, as well as on their physical states. It is easy to imagine that chemists might have a great deal of difficulty comparing results from different laboratories if each laboratory chose a different set of conditions for reporting the reaction enthalpy. In response to this concern, **standard states** have been defined for all substances, and reaction enthalpies are frequently reported with all of the reactants and products in their standard states and at 298.15 K. The standard state of a substance is its pure form at 1 atm pressure. The reaction enthalpy for a reaction with all of the reactants and products in their respective standard states is called the **standard reaction enthalpy** and is symbolized by $\Delta H°$ rather than ΔH. Thus,

$$\text{Reactants} \longrightarrow \text{products} \qquad \text{reaction enthalpy} = \Delta H$$

$$\text{Reactants in standard states} \longrightarrow \text{products in standard states} \qquad \text{reaction enthalpy} = \Delta H°$$

EXAMPLE 1 Understanding the standard reaction enthalpy

When methane is burned in air in a bunsen burner, is the observed enthalpy change the standard reaction enthalpy?

$$CH_4(g) + 2O_2(g) \longrightarrow CO_2(g) + 2H_2O(g) \qquad \Delta H \stackrel{?}{=} \Delta H°$$

SOLUTION For the reaction enthalpy to be the standard reaction enthalpy, all of the reactants and products must be in their standard states. This means that gases must be present at a partial pressure of 1 atm. The oxygen in the air has a partial pressure of 0.2 atm and is not in its standard state; thus, the reaction enthalpy cannot be the standard reaction enthalpy and $\Delta H \neq \Delta H°$. In addition, some reflection should indicate that it is unlikely that any of the other gaseous products exists at exactly 1 atm partial pressure, bolstering the fact that the observed reaction enthalpy is not the standard reaction enthalpy. A set of well-established, albeit complex, calculations is required to convert the observed reaction enthalpy to the standard reaction enthalpy.

EXERCISE A sample of water is boiled at an atmospheric pressure of 1 atm, and the water vapor at 1 atm pressure is allowed to come to equilbrium with the liquid water. Is the observed enthalpy the standard enthalpy of boiling?

ANSWER: Yes, because the water is pure and at 1 atm pressure, and the gaseous water has a pressure of 1 atm

EXAMPLE 2 Understanding standard reaction enthalpies

For which of the following reactions is the reaction enthalpy the standard reaction enthalpy? All the reactions are at 25°C, and all gases are at 1 atm pressure.

(a) $C(g) + O_2(g) \longrightarrow CO_2(g) \qquad \Delta H = -1110 \text{ kJ}$

(b) $CH_4(g) + 2O_2(g) \longrightarrow CO_2(g) + 2H_2O(l) \qquad \Delta H = -890 \text{ kJ}$

SOLUTION For a reaction enthalpy to be a standard reaction enthalpy, all products and reactants must be present in their standard states. In (a), carbon is not in its standard state, so the reaction enthalpy of -1110 kJ is not the standard reaction enthalpy. In (b), all the reactants and products are in their standard state, so the reaction enthalpy is the standard reaction enthalpy.

EXERCISE For which of the following reactions is the reaction enthalpy the standard reaction enthalpy? All the reactions are at 25°C, and all gases are at 1 atm pressure.

(a) $C_2H_4(g) + 3O_2(g) \longrightarrow 2CO_2(g) + 2H_2O(l) \qquad \Delta H = -1411 \text{ kJ}$

(b) $C(\textit{diamond}) + O_2(g) \longrightarrow CO_2(g) \qquad \Delta H = -395 \text{ kJ}$

ANSWER: (a) Yes; (b) no

KEY CONCEPT C Hess's Law and combining reaction enthalpies

Enthalpy is a state property; therefore, the change in enthalpy for any process is independent of the steps taken to accomplish the process. Hess's law restates this proposition in slightly different terms. It states that the overall reaction enthalpy is the sum of the reaction enthalpies of any sequence of reactions that accomplishes the same overall reaction. Hess's law makes it possible to determine the reaction enthalpy for any reaction, using the following three steps:

Step 1. Write any number of intermediate reactions that go from the original reactants to the desired products; even imaginary reactions are all right to use.

Step 2. Write the reaction enthalpies for each of the intermediate reactions.

Step 3. Sum all of the intermediate reaction enthalpies to get the reaction enthalpy for the overall reaction.

EXAMPLE Using Hess's law

The standard enthalpy of combustion of $S_8(s)$ to $SO_3(g)$ is -3166 kJ/mol S_8; for combustion of $S_8(s)$ to $SO_2(g)$, the standard enthalpy of combustion is -2374 kJ/mol S_8. Use this information to calculate the standard reaction enthalpy for

$$2SO_2(g) + O_2(g) \longrightarrow 2SO_3(g) \qquad \Delta H^\circ = ?$$

SOLUTION Because there are so many details to keep track of in this kind of calculation, it is advisable to write down all the information we have to see how to proceed. From the problem, we have these two reactions:

$$S_8(s) + 12O_2(g) \longrightarrow 8SO_3(g) \qquad \Delta H^\circ = -3166 \text{ kJ} \qquad (1)$$

$$S_8(s) + 8O_2(g) \longrightarrow 8SO_2(g) \qquad \Delta H^\circ = -2374 \text{ kJ} \qquad (2)$$

We now envision a two-step process that will accomplish the same overall reaction as that given in the problem. First, we use reaction (2) to write a reaction with SO_2 as a reactant, because SO_2 is one of the reactants in the original problem. The result, reaction (3) (which follows), is the reverse of reaction (2); so $\Delta H = +2374$ kJ. Second, we note that the products of reaction (3) are S_8 and O_2; so we write a reaction that has S_8 and O_2 as reactants and SO_3 (the desired product in the original problem) as a product. This is reaction (4), which is the same as reaction (1). We arrive at the desired reaction, reaction (5), by summing reactions (3) and (4) and canceling the substances that appear on both sides (S_8 and $8O_2$):

$$8SO_2(g) \longrightarrow S_8(s) + 8O_2(g) \qquad \Delta H^\circ = +2374 \text{ kJ} \qquad (3)$$

$$S_8(s) + 12O_2(g) \longrightarrow 8SO_3(g) \qquad \Delta H^\circ = -3166 \text{ kJ} \qquad (4)$$

$$8SO_2(g) + 4O_2(g) \longrightarrow 8SO_3(g) \qquad \Delta H^\circ = -792 \text{ kJ} \qquad (5)$$

However, this is not quite the the result we seek. If we divide all of the coefficients in reaction (5) by 4, and also divide the reaction enthalpy by 4, we arrive at the desired answer:

$$2SO_2(g) + O_2(g) \longrightarrow 2SO_3(g) \qquad \Delta H^\circ = -198 \text{ kJ}$$

EXERCISE Use the given reaction enthalpies to calculate the unknown reaction enthalpy. (The unknown reaction is a hypothetical one.)

$$2N_2 + O_2 \longrightarrow 2N_2O \qquad \Delta H^\circ = 164.1 \text{ kJ}$$

$$N_2 + O_2 \longrightarrow 2NO \qquad \Delta H^\circ = 180.5 \text{ kJ}$$

$$N_2 + 2O_2 \longrightarrow N_2O_4 \qquad \Delta H^\circ = 9.2 \text{ kJ}$$

$$2NO + 2O_2 + 2N_2O \longrightarrow 2N_2O_4 + N_2 \qquad \Delta H^\circ = ?$$

ANSWER: -326.2 kJ

KEY CONCEPT D Standard enthalpies of combustion

The **standard enthalpy of combustion** of a substance is the reaction enthalpy when 1 mol of the substance undergoes complete combustion under standard conditions. When you write the combustion reaction, CO_2, H_2O, N_2, and SO_2 should be used as the products for carbon, hydrogen, nitrogen, and sulfur, respectively. Two combustion reactions that follow this stipulation are

$$2C_2H_6S(g) + 9O_2(g) \longrightarrow 4CO_2(g) + 6H_2O(l) + 2SO_2(g)$$

$$2C_6H_8N(l) + 16O_2(g) \longrightarrow 12CO_2(g) + 8H_2O(l) + N_2(g)$$

EXAMPLE Using standard enthalpies of combustion

Use the data in Table 6.3 of the text to calculate the reaction enthalpy for the hypothetical reaction $C_6H_6(l) + CH_4(g) \longrightarrow C_7H_8(l) + H_2(g)$. The standard enthalpy of combustion of toluene, $C_7H_8(l)$, is -3910 kJ/mol.

SOLUTION We anticipate that we shall need the combustion reactions for each of the reactants and products in the given equation, so we first write the correct equations with their enthalpies:

$$2C_6H_6(l) + 15O_2(g) \longrightarrow 12CO_2(g) + 6H_2O(l) \qquad \Delta H^\circ = -6536 \text{ kJ} \qquad (1)$$
$$CH_4(g) + 2O_2(g) \longrightarrow CO_2(g) + 2H_2O(l) \qquad \Delta H^\circ = -890 \text{ kJ} \qquad (2)$$
$$2C_7H_8(l) + 18O_2(g) \longrightarrow 14CO_2(g) + 8H_2O(l) \qquad \Delta H^\circ = -7820 \text{ kJ} \qquad (3)$$
$$2H_2(g) + O_2(g) \longrightarrow 2H_2O(l) \qquad \Delta H^\circ = -572 \text{ kJ} \qquad (4)$$

Notice that for all but equation (2) the enthalpy is twice that given in Table 6.3 because 2 mol of each reactant are required for a balanced equation. We now use Hess's law to add these equations in order to generate the desired equation. Because we want the CO_2 and H_2O to completely cancel, we must reverse equations (3) and (4) and multiply equation (2) by 2. This also results in the proper ratios of $C_6H_6(l)$, $CH_4(g)$, $C_7H_8(l)$, and $H_2(g)$ in the final equation:

$$2C_6H_6(l) + 15O_2(g) \longrightarrow 12CO_2(g) + 6H_2O(l) \qquad \Delta H^\circ = 1 \times (-6536 \text{ kJ}) = -6536 \text{ kJ}$$
$$2CH_4(g) + 4O_2(g) \longrightarrow 2CO_2(g) + 4H_2O(l) \qquad \Delta H^\circ = 2 \times (-890 \text{ kJ}) = -1.78 \times 10^3 \text{ kJ}$$
$$14CO_2(g) + 8H_2O(l) \longrightarrow 2C_7H_8(l) + 18O_2(g) \qquad \Delta H^\circ = 1 \times (+7820 \text{ kJ}) = +7820 \text{ kJ}$$
$$2H_2O(l) \longrightarrow 2H_2(g) + O_2(g) \qquad \Delta H^\circ = 1 \times (+572 \text{ kJ}) = +572 \text{ kJ}$$

$$2C_6H_6(l) + 2CH_4(g) + 19O_2(g) + 10H_2O(l) + 14CO_2(g) \longrightarrow C_7H_8(l) + 2H_2(g) + 19O_2(g) + 10H_2O(l) + 14CO_2(g) \qquad +76 \text{ kJ}$$

$$2C_6H_6(l) + 2CH_4(g) \longrightarrow 2C_7H_8(l) + 2H_2(g) \qquad \Delta H^\circ = +76 \text{ kJ}$$

This equation is the one asked for in the problem, multiplied by two. We arrive at the desired result by dividing the equation (and its enthalpy) by 2.

$$C_6H_6(l) + CH_4(g) \longrightarrow C_7H_8(l) + H_2(g) \qquad \Delta H^\circ = +32 \text{ kJ}$$

EXERCISE Use the data in Table 6.3 of the text to calculate the standard reaction enthalpy for the reaction of 3 mol of acetylene (C_2H_2) to form 1 mol of benzene (C_6H_6).

ANSWER: -632 kJ

6.4 ENTHALPIES OF FORMATION

KEY CONCEPT Standard enthalpies of formation

The **standard enthalpy of formation** ΔH_f° of a substance is the standard reaction enthalpy, per mole of the substance, for formation of the substance from its elements in their most stable form. This definition assumes that the standard enthalpy of formation of an element is zero (if the element is in its most stable form). We can calculate the standard reaction enthalpy for any reaction for which we know the standard enthalpies of formation of all of the reactants and products. To do this, we substitute into the formula

$$\Delta H^\circ(\text{for a reaction}) = (\text{sum of } \Delta H_f^\circ \text{ of all products}) - (\text{sum of all } \Delta H_f^\circ \text{ of all reactants})$$

Values of ΔH_f° are tabulated in Table 6.4 and Appendix 2A of the text.

EXAMPLE Using enthalpies of formation

Use enthalpies of formation to calculate the standard enthalpy of combustion of pentane $C_5H_{12}(g)$. Assume water formed is in the liquid state.

SOLUTION Before doing any calculations, we write the chemical equation involved:

$$C_5H_{12}(g) + 8O_2(g) \longrightarrow 5CO_2(g) + 6H_2O(l) \qquad \Delta H^\circ = ?$$

Next, we look up the standard enthalpy of formation for each of the reactants and products and write it above each of the substances in the equation:

$$\begin{matrix} -146.44\ \text{kJ/mol} & 0\ \text{kJ/mol} & & -393.51\ \text{kJ/mol} & -285.83\ \text{kJ/mol} \\ C_5H_{12}(g) & +\ 8O_2(g) & \longrightarrow & 5CO_2(g) & +\ 6H_2O(l) \end{matrix}$$

Next, we insert the standard enthalpies of formation into the equation for calculating $\Delta H°$, being certain to multiply each standard enthalpy of formation by the stoichiometric coefficient in front of each respective substance in the equation:

$$\begin{aligned} \Delta H°(\text{for a reaction}) &= [\text{sum of } \Delta H_f° \text{ of all products}] - [\text{sum of } \Delta H_f° \text{ of all reactants}] \\ &= [5\ \text{mol} \times \Delta H_f° \{CO_2(g)\} + 6\ \text{mol} \times \Delta H_f° \{H_2O(l)\}] \\ &\qquad - [1\ \text{mol} \times \Delta H_f° \{C_5H_{12}(g)\} + 8\ \text{mol} \times \Delta H_f° \{O_2(g)\}] \\ &= [(5\ \cancel{\text{mol}} \times -393.51\ \text{kJ}/\cancel{\text{mol}}) + (6\ \cancel{\text{mol}} \times -285.83\ \text{kJ}/\cancel{\text{mol}})] \\ &\qquad - [(1\ \cancel{\text{mol}} \times -146.44\ \text{kJ}/\cancel{\text{mol}}) + (8\ \text{mol} \times 0\ \text{kJ/mol})] \end{aligned}$$

Next, perform all of the calculations inside the parentheses:

$$\Delta H° = [-3682.53\ \text{kJ}] - [-146.44\ \text{kJ}]$$

Finally, perform the subtraction indicated. Do not forget that minus × minus = plus.

$$\Delta H° = -3536.09\ \text{kJ}$$

This reaction corresponds to the combustion of 1 mol of pentane, as can be seen from the equation

$$C_5H_{12}(g) + 8O_2(g) \longrightarrow 5CO_2(g) + 6H_2O(l) \qquad \Delta H° = -3536.09\ \text{kJ}$$

Thus, the enthalpy of combustion of pentane is -3536.09 kJ/mol C_5H_{12}

EXERCISE Repeat the calculation of the enthalpy of combustion of pentane, but assume the water formed is in the gaseous state.

ANSWER: $\Delta H = -3272.03$ kJ/mol

PITFALL Reaction enthalpies and the physical state of substances

We have seen many examples in the text and study guide that indicate that the reaction enthalpy depends on the physical state (solid, liquid, or gas) of the reactants and products. You should take great care when using a table like Table 6.4, not only to choose the correct compound but also to make sure the compound is in the correct physical state.

PITFALL Negative signs in calculating ΔH with $\Delta H_f°$

When using $\Delta H°$(for a reaction) = [sum of $\Delta H_f°$ of all products] − [sum of $\Delta H_f°$ of all reactants], be careful to (1) use the correct sign for the enthalpies of formation and (2) properly take into account the negative sign in the equation itself. By using the calculational technique illustrated in the preceding example—substituting in each individual $\Delta H_f°$ with its own sign and then performing the calculation *inside* the brackets first—you can avoid errors.

6.5 THE ENTHALPIES OF FUELS

KEY CONCEPT Enthalpy as a resource, specific enthalpy and enthalpy density

We use the word enthalpy here instead of the more common term *energy* because most of the chemical reactions used to extract heat from fuels are run at constant pressure. The source of the enthalpy is the solar energy absorbed and stored by green plants. Green plants absorb and store approximately

10^{19} kJ of solar energy annually. The energy is stored through the production of about 6×10^{14} kg of glucose ($C_6H_{12}O_6$) yearly through a complex series of reactions called photosynthesis, in which the glucose is formed from CO_2 and H_2O. The overall reaction is

$$6CO_2 + 6H_2O \xrightarrow{\text{sunlight}} C_6H_{12}O_6 + 6O_2$$

Much of the glucose produced is chemically changed to more complex carbohydrates such as starch and cellulose. The rate at which energy is stored in a useful form is a good deal less than the overall 10^{19} kJ absorbed each year. Losses occur through oxidation; in addition, much of the storage does not result in the high concentrations of energy-rich substances that are needed for economical harvesting of the energy. The result is that only about 10^{11} kJ is stored in a useful form, considerably less than the 3×10^{17} kJ of energy a year that we use worldwide.

Whenever humans use fuel (either as food or as fuel), a chemical oxidation reaction occurs. In most cases, fuels are oxidized through combustion reactions. A fuel with a high specific enthalpy is used when a lot of heat is needed from a small mass of fuel: a high **specific enthalpy** for a fuel means that a relatively small mass of the fuel releases a relatively large amount of heat when it is combusted. A fuel with a high energy density is used when a lot of heat is needed from a small volume of fuel; a high **energy density** for a fuel means that a relatively small volume of the fuel releases a relatively large amount of heat when it is combusted.

EXAMPLE Calculating the energy stored by plants

The text states that the photosynthesis of 1 g of glucose needs 16 kJ of energy. Show that the production of 6×10^{14} kg of glucose results in the storage of about 10^{19} kJ of energy, as stated in the text.

SOLUTION We first use the fact that it takes 16 kJ of energy to produce 1 g of glucose to calculate the energy required to produce 1 kg of glucose:

$$\frac{16\ \text{kJ}}{1\ \cancel{\text{g glucose}}} \times \frac{1000\ \cancel{\text{g}}}{1\ \text{kg}} = 16 \times 10^3 \frac{\text{kJ}}{\text{kg glucose}}$$

We can then proceed, using unit analysis, to calculate the energy required to produce 6×10^{14} kg of glucose:

$$6 \times 10^{14}\ \cancel{\text{kg glucose}} \times \frac{16 \times 10^3\ \text{kJ}}{1\ \cancel{\text{kg glucose}}} = 9.6 \times 10^{18}\ \text{kJ}$$

We note that 9.6×10^{18} kJ is about 10^{19} kJ.

EXERCISE Give an example, other than those in the text, of a situation in which a fuel with both a high specific enthalpy and a high energy density would be desirable.

ANSWER: A hiker, who must carry as much equipment as possible with the lowest possible weight and volume, would appreciate such a fuel.

6.6 THE ENTHALPIES OF FOODS

KEY CONCEPT Foods as fuels

Foods are the fuels that all living cells use to survive, grow, and reproduce. The energy from foods is obtained through biochemical oxidation reactions rather than through combustion. Just as with other fuels, foods are evaluated on the basis of their specific enthalpy or enthalpy density. The three main classes of foods and their approximate specific enthalpies are digestible **carbohydrates,** 17 kJ/g; **fats,** 84 kJ/g; and **proteins,** 17 kJ/g.

EXAMPLE Specific enthalpies of foods

Why is the specific enthalpy of fats larger than the specific enthalpy of carbohydrates?

SOLUTION Carbohydrates have a larger proportion of oxygen in their molecules than do fats. For example, glucose ($C_6H_{12}O_6$, a model carbohydrate) contains 45% oxygen, whereas tristearin ($C_{57}H_{110}O_6$, a typical fat) contains only 11% oxygen. The reaction that releases the enthalpy of a food as heat or work is an oxidation reaction. Compounds that contain relatively more oxygen are already more oxidized than compounds that contain less oxygen, and because they are more oxidized, there will be a smaller enthalpy change for their complete oxidation to carbon dioxide and water.

EXERCISE Is all the enthalpy in foods released as heat with biochemical oxidation?

ANSWER: No; many types of work in living cells (for instance, mass transport, synthesis of compounds, and electrical propagation of nerve signals) are powered by the oxidation of foods.

DESCRIPTIVE CHEMISTRY TO REMEMBER

- The salt ammonium nitrate (NH_4NO_3) has a large **endothermic heat of solution,** and produces a pronounced cooling effect when it is dissolved in water.
- **Graphite** is the most stable form of carbon at 25°C.
- Green plants use the energy captured from the sun in the process of **photosynthesis** to build carbohydrates $[C_n(H_2O)_m]$ from carbon dioxide and water.
- One of the simplest carbohydrates is **glucose** ($C_6H_{12}O_6$), the main energy source of the human body; cellulose and starch are carbohydrates that consist of linked glucose molecules.
- Most of the molecules in gasoline are **hydrocarbons** (molecules containing only carbon and hydrogen) that possess about eight carbons.
- **Hydrogen** condenses to a liquid only at a very low temperature, and the liquid form has a very low density (0.089 g/mL).
- Compounds with approximate formulas such as $\mathbf{FeTiH_2}$ release hydrogen gas when heated or treated with acid. They are being studied as possible fuels that would release hydrogen when needed, but they suffer from the problem of having a low specific enthalpy.
- **Foods** supply energy that is obtained by oxidation of their components, especially by oxidation of carbohydrates to give carbon dioxide and water; digestible carbohydrates are broken down into glucose, which is the actual molecule used by cells to obtain energy.
- The three main components of food, other than water, and their specific enthalpies are **carbohydrates** (17 kJ/g), **fats** (84 kJ/g), and **proteins** (17 kJ/g).

CHEMICAL EQUATIONS TO KNOW

- The reaction of the white crystalline solid, barium hydroxide octahydrate $[Ba(OH)_2 \cdot 8H_2O]$ with the white crystalline solid, ammonium thiocyanate (NH_4SCN) is very endothermic.

$$Ba(OH)_2 \cdot 8H_2O(s) + 2NH_4SCN(s) \longrightarrow Ba(SCN)_2(aq) + 2NH_3(g) + 10H_2O(l)$$

- Hydrocarbon fuels release a great deal of heat when they are combusted. Complete combustion results in the formation of carbon dioxide and water. The water formed may be in the gaseous or liquid state, so its state is not specified in the following equations.

methane: $CH_4(g) + 2O_2(g) \longrightarrow CO_2(g) + 2H_2O$

propane: $C_3H_8(g) + 5O_2(g) \longrightarrow 3CO_2(g) + 4H_2O$

butane: $2C_4H_{10}(g) + 13O_2(g) \longrightarrow 8CO_2(g) + 10H_2O$

octane: $2C_8H_{18}(l) + 25O_2(g) \longrightarrow 16CO_2(g) + 18H_2O$

benzene: $2C_6H_6(l) + 15O_2(g) \longrightarrow 12CO_2(g) + 6H_2O$

- In a thermite reaction, aluminum metal reacts with a metal oxide such as iron(III) oxide in a fiery exothermic reaction.

$$2Al(s) + Fe_2O_3(s) \longrightarrow Al_2O_3(s) + 2Fe(s)$$

- Phosphorus trichloride can be synthesized by the direct reaction of phosphorus with chlorine.

$$P_4(s) + 6Cl_2(g) \longrightarrow 4PCl_3(l)$$

- Ammonium nitrate, when heated, decomposes (sometimes explosively) into dinitrogen monoxide (nitrous oxide) and water.

$$NH_4NO_3(s) \longrightarrow N_2O(g) + 2H_2O(g)$$

- The photosynthesis of glucose ($C_6H_{12}O_6$) from carbon dioxide and water is driven by energy from the sun.

$$6CO_2(g) + 6H_2O(l) \xrightarrow{\text{sunlight}} C_6H_{12}O_6(s) + 6O_2(g)$$

When the human body uses the glucose as an energy source, the reverse reaction occurs.

- Three reactions that occur in the atmosphere are

$$2O_3(g) \longrightarrow 3O_2(g)$$
$$O_2(g) \longrightarrow 2O(g)$$
$$NO(g) + O_3(g) \longrightarrow NO_2(g) + O_2(g)$$

MATHEMATICAL EQUATIONS TO KNOW AND UNDERSTAND

$\Delta X = X_{\text{final}} - X_{\text{initial}}$	change or difference in any property X
heat = heat capacity × ΔT	heat absorbed or release
heat = ΔT × mass × specific heat capacity	heat absorbed or released
$\Delta H = q_p$	definition of enthalpy change
heat = time × power = current in amperes × voltage × time in seconds	heat supplied by electrical heater
ΔH° = (sum of all ΔH_f° of all products) − (sum of all ΔH_f° of all reactants)	use of ΔH_f° to calculate ΔH°

SELF-TEST EXERCISES

Enthalpy and calorimetry

1. The energy of a system is a measure of its
(a) capacity to do work or supply heat
(b) temperature
(c) speed of movement
(d) height above the ground
(e) tendency to emit electromagnetic radiation

2. Which of the following is an endothermic process?
(a) freezing of water
(b) vaporization of ethanol
(c) oxidation of natural gas (as in a Bunsen burner)
(d) burning of gasoline
(e) condensation of carbon tetrachloride vapor

3. How many joules of heat are required to increase the temperature of 25.0 g of water by 3.40°C?
(a) 105 (b) 85.0 (c) 356 (d) 30.8 (e) 20.3

4. A reaction run in a calorimeter containing 500 mL of water releases 8.62 kJ of heat; during the reaction the temperature of the calorimeter and contents increases by 1.821°C. What is the heat capacity of the calorimeter, in kJ/K?
(a) 15.7 (b) 2.09 (c) 11.6 (d) 4.73 (e) 9.21

5. How many joules are there in 123 cal? 4.184 J = 1 cal.
(a) 127 (b) 515 (c) 119 (d) 29.4 (e) 632

6. How much heat is lost from a 25.0-g block of copper when it cools by 7.1°C? The specific heat capacity of copper is 0.38 J/(K·g).
(a) 4.7 kJ (b) 9.5 J (c) 1.3 J (d) 11 J (e) 67 J

7. How much heat is required to raise the temperature of 2.00 kg of water from 22.0°C to 26.3°C?
(a) 8.6 kJ (b) 8.2 kJ (c) 1.2×10^3 kJ (d) 2.2×10^2 kJ (e) 36 kJ

8. What is the final temperature of a 40-g sample of ethanol (C_2H_6O) at 25.0°C that absorbs 355 J of heat? The specific heat capacity of ethanol is 2.42 J/(K·g).
(a) 46.5°C (b) 59.4°C (c) 28.7°C (d) 25.5°C (e) 21.3°C

9. How much heat is required to increase the temperature of 2.0 mol of benzene (C_6H_6) from 10.0°C to 25.0°C? The specific heat capacity of benzene is 1.05 J/(K·g).
(a) 41 J (b) 2.5 kJ (c) 32 J (d) 4.1 kJ (e) 9.8 kJ

10 A 34.2-g block of aluminum increases in temperature from 22.3°C to 27.7°C when 166 J of heat is added to it. What is the specific heat capacity of aluminum, in J/(K·g)?
(a) 26 (b) 1.0 (c) 0.90 (d) 31 (e) 0.18

11 What is the molar heat capacity of ethanol (C_2H_6O), which has a specific heat capacity of 2.42 J/(K·g)? All answers are given in J/(K·mol).
(a) 111 (b) 5.25×10^{-2} (c) 19.0 (d) 9.01×10^{-3} (e) 72.6

12 What is the heat capacity of 500 g of copper, in J/K? The specific heat capacity of copper is 0.38 J/(K·g).
(a) 5.3×10^{-3} (b) 1.3×10^3 (c) 1.9×10^2 (d) 1.2×10^4 (e) 3.0

13. 65.5 kJ of heat are required to vaporize a sample of carbon tetrachloride inside a closed vessel. Because the vessel is closed, the pressure increases from 1.0 atm to 6.4 atm. What is ΔH for this process?
(a) +65.5 kJ (b) −65.5 kJ (c) +354 kJ
(d) −354 kJ (e) need more information to tell

14. A reaction run at constant pressure releases 512 kJ of heat. At constant volume, the same reaction releases 518 kJ of heat. What is ΔH for the reaction?
(a) +518 kJ (b) −6 kJ (c) −512 kJ (d) −506 kJ (e) +6 kJ

15. A chemical system releases heat as it undergoes a reaction at constant pressure. What happens to the enthalpy of the system as the reaction occurs?
(a) decreases (b) increases (c) stays the same

16. Which of the following values for ΔH corresponds to an endothermic reaction?
(a) +612 kJ (b) −667 kJ (c) −12 kJ (d) 0 kJ (e) −344 kJ

17. You drive from your home to Washington, D.C. Which of the following is a "state property" for your trip?
(a) The time it takes to make the trip.
(b) The actual distance, in miles, between your home and Washington, D.C.
(c) The number of miles you actually drive to get to Washington.
(d) The gallons of gasoline required to make the trip.
(e) The average speed you traveled, including time off the road.

18. The pressure of a system is changed from 750 Torr to 700 Torr. What is ΔP?
(a) 725 Torr (b) 700 Torr (c) 750 Torr (d) −50 Torr (e) 50 Torr

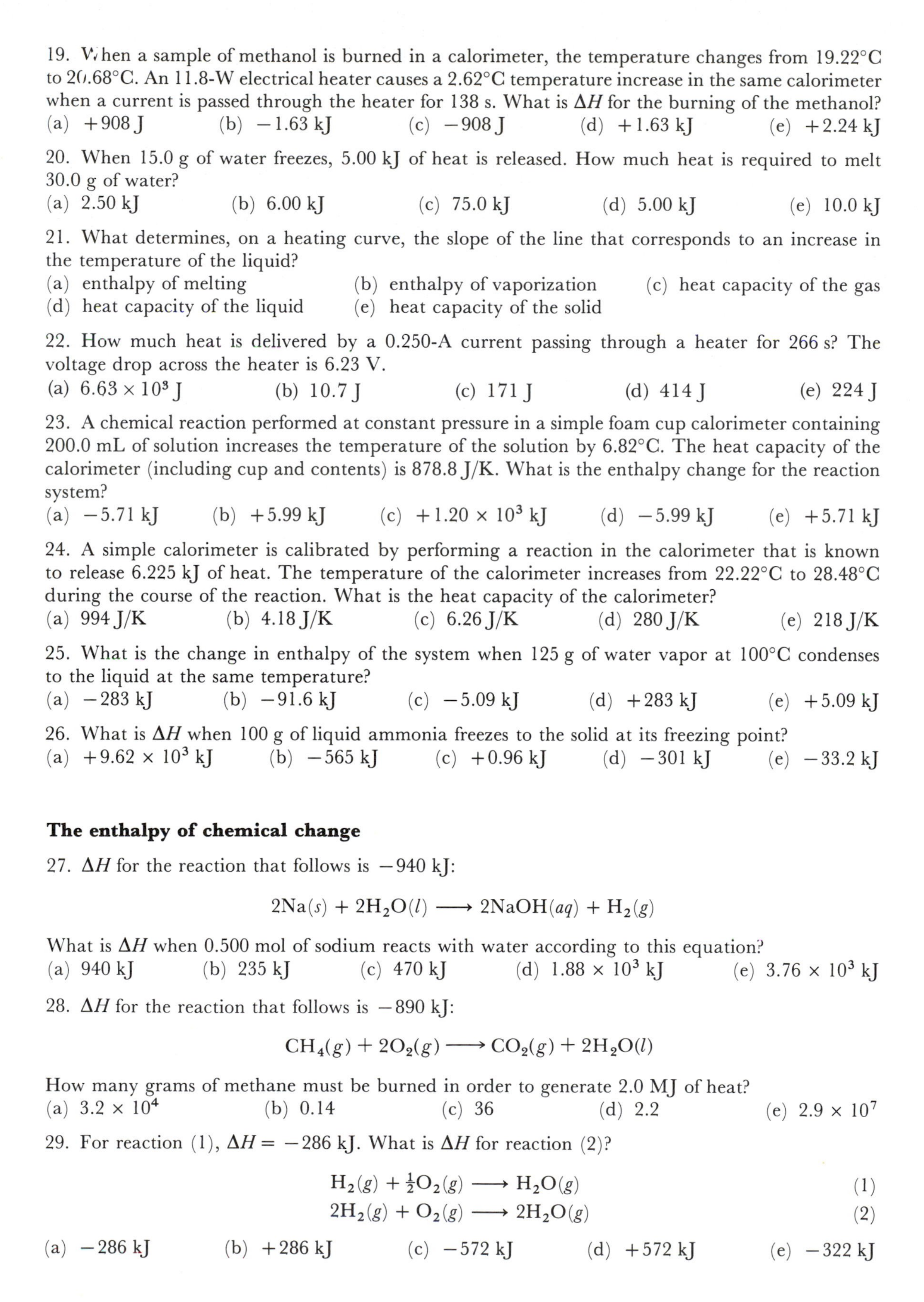

19. When a sample of methanol is burned in a calorimeter, the temperature changes from 19.22°C to 20.68°C. An 11.8-W electrical heater causes a 2.62°C temperature increase in the same calorimeter when a current is passed through the heater for 138 s. What is ΔH for the burning of the methanol?
(a) +908 J (b) −1.63 kJ (c) −908 J (d) +1.63 kJ (e) +2.24 kJ

20. When 15.0 g of water freezes, 5.00 kJ of heat is released. How much heat is required to melt 30.0 g of water?
(a) 2.50 kJ (b) 6.00 kJ (c) 75.0 kJ (d) 5.00 kJ (e) 10.0 kJ

21. What determines, on a heating curve, the slope of the line that corresponds to an increase in the temperature of the liquid?
(a) enthalpy of melting (b) enthalpy of vaporization (c) heat capacity of the gas
(d) heat capacity of the liquid (e) heat capacity of the solid

22. How much heat is delivered by a 0.250-A current passing through a heater for 266 s? The voltage drop across the heater is 6.23 V.
(a) 6.63×10^3 J (b) 10.7 J (c) 171 J (d) 414 J (e) 224 J

23. A chemical reaction performed at constant pressure in a simple foam cup calorimeter containing 200.0 mL of solution increases the temperature of the solution by 6.82°C. The heat capacity of the calorimeter (including cup and contents) is 878.8 J/K. What is the enthalpy change for the reaction system?
(a) −5.71 kJ (b) +5.99 kJ (c) $+1.20 \times 10^3$ kJ (d) −5.99 kJ (e) +5.71 kJ

24. A simple calorimeter is calibrated by performing a reaction in the calorimeter that is known to release 6.225 kJ of heat. The temperature of the calorimeter increases from 22.22°C to 28.48°C during the course of the reaction. What is the heat capacity of the calorimeter?
(a) 994 J/K (b) 4.18 J/K (c) 6.26 J/K (d) 280 J/K (e) 218 J/K

25. What is the change in enthalpy of the system when 125 g of water vapor at 100°C condenses to the liquid at the same temperature?
(a) −283 kJ (b) −91.6 kJ (c) −5.09 kJ (d) +283 kJ (e) +5.09 kJ

26. What is ΔH when 100 g of liquid ammonia freezes to the solid at its freezing point?
(a) $+9.62 \times 10^3$ kJ (b) −565 kJ (c) +0.96 kJ (d) −301 kJ (e) −33.2 kJ

The enthalpy of chemical change

27. ΔH for the reaction that follows is −940 kJ:

$$2Na(s) + 2H_2O(l) \longrightarrow 2NaOH(aq) + H_2(g)$$

What is ΔH when 0.500 mol of sodium reacts with water according to this equation?
(a) 940 kJ (b) 235 kJ (c) 470 kJ (d) 1.88×10^3 kJ (e) 3.76×10^3 kJ

28. ΔH for the reaction that follows is −890 kJ:

$$CH_4(g) + 2O_2(g) \longrightarrow CO_2(g) + 2H_2O(l)$$

How many grams of methane must be burned in order to generate 2.0 MJ of heat?
(a) 3.2×10^4 (b) 0.14 (c) 36 (d) 2.2 (e) 2.9×10^7

29. For reaction (1), $\Delta H = -286$ kJ. What is ΔH for reaction (2)?

$$H_2(g) + \tfrac{1}{2}O_2(g) \longrightarrow H_2O(g) \quad (1)$$
$$2H_2(g) + O_2(g) \longrightarrow 2H_2O(g) \quad (2)$$

(a) −286 kJ (b) +286 kJ (c) −572 kJ (d) +572 kJ (e) −322 kJ

30. For which of the following reactions is ΔH the standard reaction enthalpy? Assume all reactions are run at 25°C and all gases are at 1 atm pressure.
(a) $CH_4(g) + 2O_2(g) \longrightarrow CO_2(g) + 2H_2O(g)$
(b) $C(graphite) + 2Cl_2(g) \longrightarrow CCl_4(l)$
(c) $C(diamond) + O_2(g) \longrightarrow CO_2(g)$
(d) $C(g) + 2H_2(g) \longrightarrow CH_4(g)$

31. How much heat is produced by burning 25 kg of liquid methanol at constant pressure? Assume all reactants and products are at 25°C.

$$2CH_4O(l) + 3O_2(g) \longrightarrow 2CO_2(g) + 4H_2O(l)$$

(a) 0.93 MJ (b) 5.8×10^2 MJ (c) 0.57 MJ (d) 9.1 kJ (e) 1.8 MJ

32. Which of the following indicates a substances in its standard state at 25°C?
(a) $NaCl(aq)$
(b) $H_2O(g, 20\ \text{Torr})$
(c) $H_2O(l, 1\ \text{atm})$
(d) $NaCl(g, 0.05\ \text{Torr})$
(e) $O_2(g, 0.95\ \text{atm})$

33. The density of methanol is 0.791 g/mL. Calculate the heat released when 10.0 mL of methanol burns under standard conditions at 25°C.
(a) 179 kJ (b) 196 kJ (c) 227 kJ (d) 184 kJ (e) 5.56×10^4 kJ

34. Given the reaction enthalpies

$$2P(s) + 3Cl_2(g) \longrightarrow 2PCl_3(g) \qquad \Delta H = -574\ \text{kJ}$$
$$2P(s) + 5Cl_2(g) \longrightarrow 2PCl_5(l) \qquad \Delta H = -887\ \text{kJ}$$

what is the reaction enthalpy for the reaction

$$PCl_3(g) + Cl_2(g) \longrightarrow PCl_5(l) \qquad \Delta H = ?$$

(a) −157 kJ (b) −313 kJ (c) −1461 kJ (d) +1461 kJ (e) +222 kJ

35. The dissolution of sodium chloride is endothermic:

$$NaCl(s) \longrightarrow Na^+(aq) + Cl^-(aq) \qquad \Delta H = +3.9\ \text{kJ}$$

What is the reaction enthalpy for the crystallization of sodium chloride?

$$Na^+(aq) + Cl^-(aq) \longrightarrow NaCl(s) \qquad \Delta H = ?$$

(a) 0 kJ (b) +3.9 kJ (c) −3.9 kJ (d) +2.6 kJ (e) −2.6 kJ

36. Given the standard reaction enthalpy at 25°C for the following reaction,

$$N_2(g) + 3H_2(g) \longrightarrow 2NH_3(g) \qquad \Delta H = -92.22\ \text{kJ}$$

what is the standard enthalpy of formation at 25°C for $NH_3(g)$, in kJ/mol?
(a) −92.22 (b) −46.11 (c) +92.22 (d) +30.74 (e) +46.11

37. Use standard enthalpies of formation to calculate ΔH for the reaction

$$2H_2O(l) + 2F_2(g) \longrightarrow 4HF(aq) + O_2(g)$$

(a) −1902 kJ (b) −758.9 kJ (c) −1814 kJ (d) −846.9 kJ (e) −46.8 kJ

38. What is the standard reaction enthalpy at 25°C for the decomposition of calcium carbonate (calcite) to calcium oxide and carbon dioxide?

$$CaCO_3(s) \longrightarrow CaO(s) + CO_2(g) \qquad \Delta H = ?$$

(a) −2235.5 kJ (b) −965.3 (c) +1206.9 kJ (c) +178.3 kJ (e) +654.8 kJ

39. What is the standard enthalpy of combustion at 25°C for cyclopropane [$C_3H_6(g)$]? The standard enthalpy of formation of cyclopropane is +53.30 kJ/mol at 25°C.
(a) −161 kJ/mol
(b) −108 kJ/mol
(c) −2091 kJ/mol
(d) −3325 kJ/mol
(e) −733 kJ/mol

40. The text states that the energy of the solar radiation absorbed by vegetation on earth is enough to build about 6×10^{14} kg of glucose a year. Given that the photosynthesis of 1 g of glucose requires 16 kJ of energy, calculate the energy absorbed by the vegetation per year.
(a) 1×10^{19} kJ (b) 1×10^{16} kJ (c) 4×10^{16} kJ (d) 4×10^{19} kJ

41. The standard enthalpy of combustion of ethylene [$C_2H_4(g)$] is -1411 kJ/mol. What is the specific enthalpy of ethylene?
(a) 8.88 kJ/g (b) 19.1 kJ/g (c) 50.3 kJ/g (d) 39.6 kJ/g (e) 26.2 kJ/g

42. The density of heptane [$C_7H_{16}(l)$] is 0.684 g/mL and its enthalpy of combustion is 4.85×10^3 kJ/mol. What is the enthalpy density of heptane, in MJ/L?
(a) 48.4 (b) 26.5 (c) 19.2 (d) 81.1 (e) 33.1

Descriptive chemistry

43. Which of the following is not a fossil fuel?
(a) coal (b) hydrogen gas (c) oil (d) natural gas

44. Which of the following is a carbohydrate?
(a) C_6H_6 (b) CO_2 (c) $C_6H_{12}O_6$ (d) C (*graphite*) (e) NH_4NO_3

45. All but one of the following is a major component of our foods. Which one is not?
(a) proteins (b) hydrocarbons (c) fats (d) carbohydrates (e) water

46. Which two products are always obtained when a hydrocarbon is burned in an excess of oxygen?
(a) $CO_2 + H_2O$ (b) $CO + H_2O$ (c) $C + H_2$ (d) $CO_2 + H_2$ (e) $CO + H_2$

47. The major component of natural gas is
(a) H_2 (b) CO_2 (c) CH_4 (d) C_2H_4 (e) CO

48. Photosynthesis in plants results in the formation of
(a) H_2CO_3 (b) CO_2 (c) C_2H_6 (d) H_2O (e) $C_6H_{12}O_6$

7 ATOMIC STRUCTURE AND THE PERIODIC TABLE

The periodic table is a powerful tool that chemists use to help organize the properties and reactivities of the elements. As we have already seen, the arrangement of the periodic table reflects the observed properties of the elements. However, an underlying factor that is responsible for the periodicity of the elements is their **atomic structure,** which is the arrangement of the electrons around the nucleus. This chapter explores the atomic structure of the elements.

LIGHT AND SPECTROSCOPY

KEY WORDS Define or explain each of the following terms in a written sentence or two.

boundary surface
orbital
photoelectric effect
photon
quantization
quantum number
spectrum
Stern-Gerlach experiment
uncertainty principle
wave-particle duality

7.1 THE CHARACTERISTICS OF LIGHT

KEY CONCEPT Electromagnetic radiation as waves

Electromagnetic radiation can be considered to be a wavelike disturbance in space (not in air), much like a water wave is a disturbance in water. Like any wave, it is described by its **wavelength** (λ), **frequency** (ν), and **speed of propagation** (c). The wavelength has units of length, such as meters; the frequency has units of 1/s (1/second). The unit 1/s is also called the **hertz** (Hz = 1/s). The wavelength of a wave times its frequency is the speed of propagation of the wave. The symbol c is used for the speed of electromagnetic radiation (or speed of light), 3.00×10^8 m/s.

$$\lambda \times \nu = c$$

Because the speed of light is a constant, the wavelength and frequency are not independent of each other. The value of one determines the other. Finally, we note that for visible light, the frequency (or wavelength) determines the color of the light.

EXAMPLE Calculating the frequency of electromagnetic radiation

The wavelength of the light to which the eye is most sensitive is 556 nanometers (nm). What is the frequency of this light? What color is it?

SOLUTION The relationship between the wavelength and frequency of light is $\lambda \times \nu = c$. $\lambda = 556$ nm and $c = 3.00 \times 10^8$ m/s. Substitution of these values into the equation will allow us to solve for the frequency ν. Because c is given in m/s, we must first convert the wavelength from nanometers to meters:

$$556\ \text{nm} \times \frac{1\ \text{m}}{10^9\ \text{nm}} = 5.56 \times 10^{-7}\ \text{m}$$

Now, we substitute into the equation:

$$\lambda \times \nu = c$$
$$(5.56 \times 10^{-7}\ \text{m}) \times \nu = 3.00 \times 10^8\ \text{m/s}$$
$$\nu = 5.40 \times 10^{14}\ \text{1/s} = 5.40 \times 10^{14}\ \text{Hz}$$

It is possible to determine the color by noting that 556 nm is between the 580 nm of yellow and 530 nm of green. We conclude that 556 nm is a yellow-green (Table 7.1 of the text); this is the color of many modern emergency vehicles.

EXERCISE Electromagnetic radiation with a frequency of 4.0×10^{14} Hz is in the infrared region of the electromagnetic spectrum. What is the wavelength of this electromagnetic radiation?

ANSWER: 7.5×10^{-7} m

7.2 QUANTIZATION AND PHOTONS

KEY CONCEPT A Photons

Electromagnetic radiation often behaves like a wave. Rainbows are a spectacular reminder of the wave nature of light because only waves can be refracted and dispersed. However, in many experiments light acts like a stream of massless particles, called **photons.** Each particle has a frequency and wavelength. The brightness of the light is determined by the number of photons; a larger number of photons results in a more intense beam of light. By feeling the warmth of sunlight on a cool spring day, we can observe that electromagnetic radiation carries energy. The energy of one photon is given by

$$E = h \times \nu$$

where ν is the frequency in Hz (1/s) and h is **Planck's constant,** 6.63×10^{-34} J/Hz.

EXAMPLE Calculating the energy of a photon

What is the energy of a photon of blue light of wavelength 455 nm?

SOLUTION The energy of a photon is given by the equation $E = h \times \nu$, with $h = 6.63 \times 10^{34}$ J/Hz, so we need the frequency of the light to calculate the energy of the photon. To get the frequency from the wavelength, we substitute into the equation

$$\nu \times \lambda = c$$

$$\nu = \frac{c}{\lambda} = \frac{3.00 \times 10^{8}\ \text{m/s}}{4.55 \times 10^{-7}\ \text{m}}$$

$$= 6.59 \times 10^{14} (1/\text{s})$$

$$= 6.59 \times 10^{14}\ \text{Hz} \qquad [\text{recall that } (1/\text{s}) = \text{Hz}]$$

Finally, we use the value of the frequency in the equation for the energy of a photon:

$$E = h \times \nu$$

$$= 6.63 \times 10^{-34} \frac{\text{J}}{\text{Hz}} \times 6.59 \times 10^{14}\ \text{Hz}$$

$$= 4.37 \times 10^{19}\ \text{J}$$

EXERCISE What is the wavelength of a photon that has an energy equal to 5.33×10^{-19} J?

ANSWER: 373 nm (3.73×10^{-7} m)

KEY CONCEPT B The photoelectric effect

Some metals emit electrons when bombarded with electromagnetic radiation above a certain frequency; the frequency depends on the metal. In this process, a light photon collides with an electron at the surface of the metal and transfers all of its energy to the electron. Some of the energy is used to kick the electron out of the metal. Any left-over energy goes into the kinetic energy of the ejected electron:

$$\text{Energy of photon} = \begin{pmatrix}\text{energy needed}\\ \text{to eject electron}\\ \text{from metal}\end{pmatrix} + \begin{pmatrix}\text{kinetic energy}\\ \text{of electron}\end{pmatrix}$$

If the photons impinging on the metal do not have the minimum energy required to cause the electron to be ejected from the metal, no electrons will be ejected because only one photon at a time can collide with an electron. In such a situation, the electron is excited to a higher energy state momentarily, but stays bound in the metal.

EXAMPLE The kinetic energy of an electron in the photoelectric effect

The energy requir-d to eject an electron from cesium is 3.43×10^{-19} J. What is the kinetic energy of the electron ejected when light with a wavelength of 483 nm impinges on cesium?

SOLUTION The equation that relates the kinetic energy of an electron ejected by the photoelectric effect to the energy of the light impinging on the metal is

$$\text{Energy of photon} = \begin{pmatrix}\text{energy needed}\\ \text{to eject electron}\\ \text{from metal}\end{pmatrix} + \begin{pmatrix}\text{kinetic energy}\\ \text{of electron}\end{pmatrix}$$

Using the symbol E_k for the kinetic energy of the electron and the fact that the energy of a photon equals $h\nu$, we get

$$h\nu = 3.43 \times 10^{-19}\,\text{J} + E_k$$

We use the fact that 483 nm $= 4.83 \times 10^{-7}$ mm and the equation $\lambda \times \nu = c$ to calculate the photon frequency:

$$\nu = \frac{c}{\lambda} = \frac{3.00 \times 10^{8}\,\text{m/s}}{4.83 \times 10^{-7}\,\text{m}}$$
$$= 6.21 \times 10^{14}\,\text{Hz} \qquad \text{(recall that (1/s) = Hz)}$$

Substituting this result into the original equation along with Planck's constant, $h = 6.63 \times 10^{-34}$ J/Hz, gives

$$(6.63 \times 10^{-34}\,\text{J/Hz}) \times (6.21 \times 10^{14}\,\text{Hz}) = 3.43 \times 10^{-19}\,\text{J} + E_k$$
$$4.11 \times 10^{-19}\,\text{J} = 3.43 \times 10^{-19}\,\text{J} + E_k$$
$$E_k = 4.11 \times 10^{-19}\,\text{J} - 3.43 \times 10^{-19}\,\text{J}$$
$$= 0.68 \times 10^{-19}\,\text{J}$$
$$= 6.8 \times 10^{-20}\,\text{J}$$

EXERCISE When lead is bombarded with electromagnetic radiation of wavelength 182 nm, electrons with a kinetic energy of 4.09×10^{-19} J are ejected. What is the minimum energy required to eject an electron from lead?

ANSWER: 6.84×10^{-19} J

THE STRUCTURE OF THE HYDROGEN ATOM

7.3 THE SPECTRUM OF ATOMIC HYDROGEN

KEY CONCEPT A The hydrogen atom spectrum

When hydrogen gas is heated or placed in an intense electric field, it glows with a purplish color due to light emitted from hydrogen atoms. If all of the electromagnetic radiation emitted in such an experiment (not only the visible light) is analyzed, it is found that only certain specific frequencies are present. An analysis of all of the frequencies of the emitted electromagnetic radiation discloses the surprising result that all can be calculated from only one equation:

$$\nu = \mathcal{R} \times \left(\frac{1}{n_l^2} - \frac{1}{n_u^2}\right)$$

In using this equation, n_l can be any nonzero integer and n_u an integer greater than n_l. $\mathcal{R}$ is the **Rydberg constant** and has the value 3.29×10^{15} Hz. If we substitute any values of n_l and n_u such that $n_l < n_u$ into the equation, the value of ν matches one of the frequencies of light emitted.

EXAMPLE Calculating the frequency of light emitted from a heated hydrogen gas

What frequency of light emitted from a heated hydrogen gas is associated with $n_l = 1$ and $n_u = 3$?

SOLUTION To solve this problem, we substitute the given values of n_l and n_u into the equation:

$$\nu = \mathcal{R} \times \left(\frac{1}{n_l^2} - \frac{1}{n_u^2}\right)$$

The value of the Rydberg constant $\mathcal{R} = 3.29 \times 10^{15}$ Hz must also be used.

$$\begin{aligned} \nu &= \mathcal{R} \times \left(\frac{1}{1^2} - \frac{1}{3^2}\right) \\ &= \mathcal{R} \times \left(\frac{1}{1} - \frac{1}{9}\right) \\ &= (3.29 \times 10^{15}\ \text{Hz})\left(\frac{8}{9}\right) \\ &= 2.92 \times 10^{15}\ \text{Hz} \end{aligned}$$

It should be noted that 1 and 3 are exact numbers, so the number of significant figures in the answer is determined by the three significant figures used for $\mathcal{R}$.

EXERCISE What frequency and color of light from heated hydrogen is associated with $n_l = 2$ and $n_u = 4$?

ANSWER: 6.17×10^{14} Hz; blue-green

PITFALL Evaluating terms such as $\left(\frac{1}{n_l^2} - \frac{1}{n_u^2}\right)$

When evaluating such an expression, the lowest common denominator must be found and the subtraction then completed. For example, if $n_l = 2$ and $n_u = 3$, the lowest common denominator is 36 (4×9). The preceding expression is

$$\frac{1}{2^2} - \frac{1}{3^2} = \frac{1}{4} - \frac{1}{9} = \frac{9}{36} - \frac{4}{36} = \frac{5}{36}$$

KEY CONCEPT B The frequency of light emitted by a high-energy atom

Heating an atom or placing it in an electric field increases its energy. As a natural process, the atom spontaneously loses energy by emitting a photon of electromagnetic radiation; it does not like to stay in the higher energy state. The energy loss of the atom must equal the energy of the emitted photon $h\nu$ in order to conserve energy:

$$\Delta E = h\nu$$

ΔE should be expressed as a positive energy because it represents the actual decrease in energy.

EXAMPLE Using the equation $\Delta E = h\nu$

A hydrogen atom undergoes a change in energy from -8.72×10^{-20} J to -5.45×10^{-19} J. What is the energy of the emitted photon? What is the frequency of the photon?

SOLUTION First we calculate ΔE for the atom. We must be careful to keep track of the negative signs associated with the energies:

$$\begin{aligned}\Delta E &= E_i - E_f \\ &= (-5.45 \times 10^{-19}\text{ J}) - (-8.72 \times 10^{-20}\text{ J}) \\ &= -4.58 \times 10^{-19}\text{ J}\end{aligned}$$

This energy (expressed as a positive number because photons always have positive energy) is equal to the energy of the emitted photon,

$$E_{\text{photon}} = 4.58 \times 10^{-19}\text{ J}$$

and the energy of the photon is equal to $h\nu$, which allows us to solve for the frequency ν:

$$\begin{aligned}h\nu &= 4.58 \times 10^{-19}\text{ J} \\ \nu &= \frac{4.58 \times 10^{-19}\text{ J}}{6.63 \times 10^{-34}\text{ J/Hz}} \\ &= 6.91 \times 10^{14}\text{ Hz}\end{aligned}$$

EXERCISE An atom with energy equal to -1.35×10^{-19} J loses energy by emission of a photon with a frequency of 3.08×10^{15} Hz. What is the energy of the atom after emission of the photon?

ANSWER: -2.18×10^{-18} J

KEY CONCEPT C Quantization of energy

One of the remarkable features of the loss of energy by atoms is that only a few of the seemingly infinite possible photon energies occur, resulting in the observed line spectra. The reason that only a few specific photon energies are observed is that the atom can exist in only a few specific energy states; and when it loses energy, it changes from one specific energy to another specific energy, releasing a photon with an energy that is the difference between the two specific energy states.

EXAMPLE Quantized energy states

A hypothetical atom can exist in only the energy states shown below. List all of the energies of all of the photons that can be emitted from this atom when it is heated.

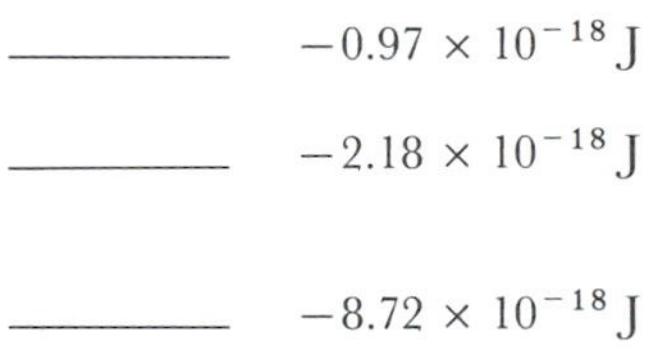

-0.97×10^{-18} J

-2.18×10^{-18} J

-8.72×10^{-18} J

SOLUTION An atom releases energy as a photon when it goes from a higher energy state to a lower energy state. The energy of the photon equals the difference in energies between the higher energy state and the lower one. The hypothetical atom pictured in the problem can go from a higher to lower state in three possible ways, indicated by A, B, and C in the following diagram:

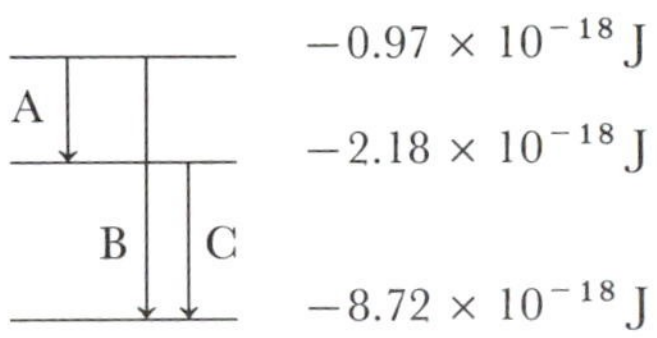

-0.97×10^{-18} J

A

-2.18×10^{-18} J

B C

-8.72×10^{-18} J

The energy of the photon for each transition is the difference in the energies of the two states involved.

Transition	Difference in energies of two states	= Energy of emitted photon
A	$(-0.97 \times 10^{-18}\,J) - (-2.18 \times 10^{-18}\,J)$	$1.21 \times 10^{-18}\,J$
B	$(-0.97 \times 10^{-18}\,J) - (-8.72 \times 10^{-18}\,J)$	$7.75 \times 10^{-18}\,J$
C	$(-2.18 \times 10^{-18}\,J) - (-8.72 \times 10^{-18}\,J)$	$6.54 \times 10^{-18}\,J$

The last column lists the energies asked for in the problem.

EXERCISE A hypothetical atom can exist in only the energy states shown. List all of the energies of all of the photons that will be released from this atom when it is heated.

———— $-0.36 \times 10^{-18}\,J$

———— $-0.81 \times 10^{-18}\,J$

———— $-3.22 \times 10^{-18}\,J$

ANSWER: $0.45 \times 10^{-18}\,J$; $2.86 \times 10^{-18}\,J$; $2.41 \times 10^{-18}\,J$

KEY CONCEPT D The Bohr model of the atom

The Bohr model of the atom explains the origin of the quantization of the energy of the atom. In this model, the electron is allowed to exist in only certain specific orbits, with each orbit corresponding to a specific energy given by

$$E_n = -h \times \frac{\mathcal{R}}{n^2}$$

To find the energies of all of the orbits, we set n equal to 1 for the first orbit, then 2 for the second, 3 for the third; and so on; n is called the **quantum number** of the orbit. The energy loss that occurs when an electron drops from a higher energy orbit to a lower energy orbit is given by

$$\text{Energy lost} = \mathcal{R} \times h \times \left(\frac{1}{n_l^2} - \frac{1}{n_u^2}\right)$$

The energy lost is the same as the energy of the emitted photon.

EXAMPLE Calculating the energy of a transition in the Bohr atom

What is the energy of the photon emitted when an electron in hydrogen drops from the $n = 5$ orbit to the $n = 2$ orbit?

SOLUTION The energy of the emitted photon is the same as the energy lost by the electron. We substitute $n_u = 5$ and $n_l = 2$ and the universal constants $\mathcal{R} = 3.29 \times 10^{15}$ Hz and $h = 6.63 \times 10^{-34}$ J/Hz into the equation just given:

$$\begin{aligned}\text{Energy of emitted photon} &= (6.63 \times 10^{-34}\,\text{J/Hz}) \times (3.29 \times 10^{15}\,\text{Hz})\left(\frac{1}{2^2} - \frac{1}{5^2}\right)\\ &= (2.18 \times 10^{-18}\,\text{J})\left(\frac{1}{4} - \frac{1}{25}\right)\\ &= (2.18 \times 10^{-18}\,\text{J})\,\frac{21}{100}\\ &= 4.58 \times 10^{-19}\,\text{J}\end{aligned}$$

EXERCISE What is the energy of the photon emitted when an electron in hydrogen drops from the $n = 6$ state to the $n = 3$ state?

ANSWER: 1.82×10^{-19} J

7.4 PARTICLES AND WAVES

KEY CONCEPT A The de Broglie relation

The de Broglie relation quantitatively expresses the observation that all matter exhibits both a particle character and a wave character. Whether the matter exhibits its wave character or particle character depends on the experiment we use to observe the matter. Certain experiments "force" the particle character to appear, whereas others "force" the wave character to appear. The quantitative expression of this wave-particle duality is

$$\text{Wave length of particle} = \frac{h}{\text{mass of particle} \times \text{velocity of particle}}$$

$$\lambda = \frac{h}{\text{mass} \times \text{velocity}}$$

EXAMPLE Using the de Broglie relation

What is the de Broglie wavelength of a 4.0-oz (114-g) rock moving at 73 m/s?

SOLUTION If we know the mass and velocity of an object, we can calculate the de Broglie wavelength using the equation just given. In this problem, mass = 114 g, velocity = 73 m/s, $h = 6.63 \times 10^{-34}$ J s.

$$\lambda = \frac{h}{\text{mass} \times \text{velocity}}$$

$$= \frac{6.63 \times 10^{-34}\ \text{J s}}{114\ \text{g} \times 73\ \text{m/s}}$$

$$= \frac{6.63 \times 10^{-34}\ \text{kg}\frac{\text{m}^2}{\text{s}^2}\ \text{s} \times \frac{1000\ \text{g}}{\text{kg}}}{8.32 \times 10^3\ \text{g}\frac{\text{m}}{\text{s}}}$$

Because the units in the numerator contain kilograms while those in the denominator contain grams, we have included a conversion from kilograms to grams. Thus,

$$\lambda = 7.96 \times 10^{-35}\ \text{m}$$

EXERCISE What is the mass, in grams, of a particle with a de Broglie wavelength of 450 picometers (pm) and a velocity of 2500 km/s?

ANSWER: 5.89×10^{-31} g

KEY CONCEPT B Shells, subshells, and orbitals

The quantum mechanical picture of the atom results in a probabilistic picture of an electron in the atom, in which it is impossible to know, with absolute certainty, where the electron is. The best that we can do in locating an electron is to state the probability of finding it at some position. An **orbital** is a volume in space in which there is a high probability of finding an electron. Every orbital is described by the three **quantum numbers**—n, l, and m_l—which (with the spin quantum number described below) label the state of an electron in the orbital and specify the value of every property associated with the electron. Only certain quantum numbers are allowed with each other, so a specific pattern of orbitals arises. The orbitals for the three lowest values of n are shown in the following figure. Each orbital is represented by a square.

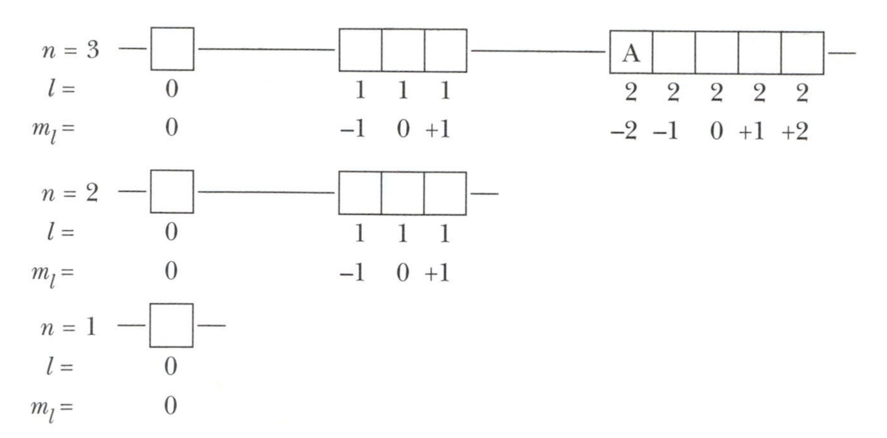

The values of the quantum numbers for the orbital labeled A are $n = 3$, $l = 2$, and $m_l = -2$. All of the orbitals with the same value of n are said to constitute a **shell.** All of the orbitals with the same value of n and the same value of l constitute a **subshell.** For example, in the $n = 3$ shell, there are three subshells, one with $l = 0$ (1 orbital), one with $l = 1$ (3 orbitals), and one with $l = 2$ (5 orbitals). Subshells are often labeled according the value of l: an *s* subshell has $l = 0$, a *p* subshell has $l = 1$, a *d* subshell has $l = 2$, and an *f* subshell has $l = 3$. Orbitals are designated with the same symbols; the orbital labeled A in the preceding figure is called a 3*d* orbital.

EXAMPLE The quantum numbers of orbitals

What quantum numbers are possible for a 3*p* orbital?

SOLUTION By definition, a 3*p* orbital has $n = 3$. The *p* in the label tells us that $l = 1$ for the orbital. Finally, the preceding illustration indicates that for $l = 1$, m_l may be -1 or 0 or $+1$. Thus, there are three possible sets of quantum numbers for the orbital. These are

$$n = 3, \quad l = 1, \quad m_l = -1$$
$$n = 3, \quad l = 1, \quad m_l = 0$$
$$n = 3, \quad l = 1, \quad m_l = +1$$

EXERCISE Which two orbitals in the first three shells have the same three quantum numbers?

ANSWER: No two orbitals have the same three quantum numbers.

KEY CONCEPT C Electron spin

Experimental evidence indicates that electrons have what might be considered an intrinsic spin. The spin is labeled with a quantum number called the **spin magnetic quantum number** (m_s). Because experiments indicate there are only two possible spin states for any electron, there are only two possible spin magnetic quantum numbers: $m_s = +\frac{1}{2}$ and $m_s = -\frac{1}{2}$. An electron with $m_s = +\frac{1}{2}$ is represented by an up arrow (↑) and one with $m_s = -\frac{1}{2}$ by a down arrow (↓). Every electron must have one of these two spins. For a very large number of electrons in any sample of matter, about one-half have $m_s = +\frac{1}{2}$ and about one-half have $m_s = -\frac{1}{2}$.

▼ **EXAMPLE Understanding quantum numbers**

What four quantum numbers are consistent with an electron represented by an up arrow in a $2p$ orbital?

$$\boxed{\uparrow}$$

$2p$

SOLUTION The quantum numbers for the orbital are associated with an electron in the orbital, so we must first deduce which quantum numbers are permitted for the orbital. By definition, $n = 2$ for a $2p$ orbital and for a p orbital, $l = 1$. The three allowed values of m_l for $l = 1$ are $m_l = -1$, $m_l = 0$, and $m_l = +1$. Finally, the up arrow indicates that $m_s = +\frac{1}{2}$ for this electron. Any one of the following sets of four quantum numbers is allowed for this $2p$ electron:

$$n = 2, \quad l = 1, \quad m_l = -1, \quad m_s = +\frac{1}{2}$$

$$n = 2, \quad l = 1, \quad m_l = 0, \quad m_s = +\frac{1}{2}$$

$$n = 2, \quad l = 1, \quad m_l = +1, \quad m_s = +\frac{1}{2}$$

EXERCISE Which four quantum numbers are consistent with an electron represented by a down arrow in a $3d$ orbital?

$$\boxed{\downarrow}$$

$3d$

ANSWER: Five sets of quantum numbers are consistent. These are

$n = 3, \quad l = 2, \quad m_l = -2, \quad m_s = -\frac{1}{2}$
$n = 3, \quad l = 2, \quad m_l = -1, \quad m_s = -\frac{1}{2}$
$n = 3, \quad l = 2, \quad m_l = 0, \quad m_s = -\frac{1}{2}$
$n = 3, \quad l = 2, \quad m_l = +1, \quad m_s = -\frac{1}{2}$
$n = 3, \quad l = 2, \quad m_l = +2, \quad m_s = -\frac{1}{2}$

▲

THE STRUCTURES OF MANY-ELECTRON ATOMS

KEY WORDS Define or explain each of the following terms in a written sentence or two.

building-up process	electron configuration	Hund's rule
diamagnetic	exclusion principle	paramagnetic
effective nuclear charge	ground state configuration	valence shell

7.5 ORBITAL ENERGIES

KEY CONCEPT A Effective nuclear charge in a many-electron atom

When there is more than one electron in an atom, electron-electron repulsions are present. Any electron in an atom is attracted toward the positive nucleus, but electrons closer to the nucleus repel outer electrons away from the center of the atom, in effect canceling some of the effect of the positive nuclear charge. This effect is called **shielding** because the closer-in electron essentially shields the farther-out electron from some of the nuclear charge. Orbital shapes determine, to some extent, the shielding ability of an electron. For instance, a $4s$ electron, on average, is closer to the nucleus than a $4p$ electron. It is said to **penetrate** closer to the nucleus and, therefore, shields other electrons from the nucleus more effectively than a $4p$ electron. In addition, because the $4s$ electron is closer in to the nucleus and less shielded than a $4p$ electron, it feels a higher **effective nuclear charge** than the $4p$ electron. When an electron feels a higher nuclear charge, it is lower in energy; so a $4s$ electron has a lower energy than a $4p$ electron. For any shell, penetration and shielding effects result in the following order of subshell energies:

Lowest energy $ns < np < nd < nf$ highest energy

EXAMPLE Subshell energies

Without looking at any references, predict which of the two subshells, 2*s* or 2*p*, has the lower energy.

SOLUTION For any shell, *s* orbitals penetrate closer to the nucleus and shield other electrons more than *p* orbitals. Therefore, an electron in a 2*s* orbital feels a higher effective nuclear charge and is lower in energy than an electron in a 2*p* orbital. This is equivalent to saying that the 2*s* subshell is lower in energy than the 2*p* subshell.

EXERCISE Which subshell is higher in energy, the 3*d* or the 3*p*?

ANSWER: 3*d*

PITFALL Penetration and shielding for electrons in different shells

The discussions in the text and study guide concern the relative energies of orbitals and subshells in the same shell. The same concepts can be used to understand the relative energies of subshells in different shells, but the details are quite different. At this point, we have not discussed the relative penetration and shielding of subshells in different shells, so we could not easily decide, for example, whether the 3*d* subshell is higher or lower in energy than the 4*s* subshell.

KEY CONCEPT B The exclusion principle

The exclusion principle consists of two related observations that have been verified experimentally countless times:

1. An orbital can hold a maximum of two electrons.
2. When there are two electron's in an orbital, they must have opposite spins, that is, the spins must be paired.

There are no exceptions to these rules. When two electrons in an orbital have opposite spins, the spins are said to be **paired.** Paired spins are represented by an up arrow next to a down arrow, ↑↓. Another, more common, way of stating these two rules is

- No two electrons in an atom can have the same four quantum numbers.

EXAMPLE Using the exclusion principle

Assume two electrons must be placed in a 2*p* orbital. Which of the following represents a permissible way of placing the electrons in the orbital?

(a)	(b)	(c)
—[↑↓]—	—[↑↑]—	—[↓↓]—

SOLUTION According to the exclusion principle, two electrons in any orbital must have paired spins, that is, they must have opposite spins. In (b) and (c), the two electrons have the same spin; thus, the spins are not paired and the configuration is not allowed. In (a), the spins are paired, so (a) represents the only way to place two electrons in an orbital.

EXERCISE Is it possible for all of the electrons in a lithium atom to be in the 1 *s* orbital?

ANSWER: No, because lithium has three electrons and an orbital can only hold a maximum of two.

KEY CONCEPT C The electron configurations of hydrogen, helium, and lithium

We now turn our attention to the **electron configurations** of atoms, in which the number of electrons in each orbital or subshell of an atom is listed. We are interested in atoms that are in their lowest energy state, which is called the **ground state.** In the ground state, electrons are always in the lowest energy orbitals possible, subject to the exclusion principle. As an example, for hydrogen in its ground state, the single electron must be placed in the lowest energy orbital available, which is the 1s orbital. We symbolize this by $1s^1$, indicating that one electron is in the $1s$ orbital. For lithium, with atomic number 3, the first two electrons go into the $1s$ orbital. The third electron cannot fit into the $1s$ orbital because this would violate the exclusion principle; it goes in the next lowest energy orbital available, the $2s$ orbital, resulting in a $1s^22s^1$ configuration. When a shell in an atom contains the maximum number of electrons it can hold, it is called a **closed shell.** Similarly, a filled subshell is called a **closed subshell.** Helium and lithium each have a closed $1s$ shell, because the $1s$ shell is filled when it contains 2 electrons.

EXAMPLE The electron configuration of lithium

Is it possible for lithium in the ground state to have the electron configuration $1s^12s^2$?

SOLUTION No. For an atom to have its ground-state (lowest energy) configuration, the lower energy orbitals must fill with electrons (two electrons per orbital) before electrons go into higher energy orbitals. The $1s^12s^2$ configuration for lithium is an excited (high-energy) state configuration.

EXERCISE How many closed shells are there in ground-state lithium?

ANSWER: One

7.6 THE BUILDING-UP PRINCIPLE

KEY CONCEPT A Hund's Rule

As we build up the electron configurations of larger atoms, we are occasionally faced with the problem of where to place electrons when orbitals of equal energy are available. For instance, how do we place two electrons into a p subshell, in which there are three equal-energy orbitals available? **Hund's rule** supplies the answer to this question by stating that electrons in a subshell will occupy different orbitals in the subshell before filling any orbital, and that the electrons in the singly occupied orbitals all have the same spin. This rule is illustrated in the following figure for the problem of placing two electrons in three p orbitals.

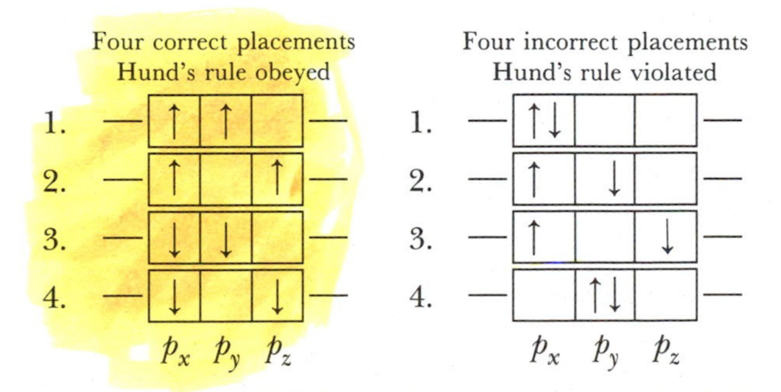

When electrons have their spins in the same direction, as in all of the correct configurations just shown, the spins are referred to as **parallel spins.**

EXAMPLE Using Hund's rule

Draw a correct electron configuration, different from those just shown for two electrons in a p subshell.

SOLUTION To obey Hund's rule, we must place the two electrons in separate orbitals with their spins parallel (in the same direction). We can choose any two orbitals and can use either up spins (↑) or down spins (↓). The following configuration follows these restrictions and is different:

↓	↓	
p_x	p_y	p_z

EXERCISE When four electrons are placed into a p subshell, how many of the electrons will have their spins unpaired?

ANSWER: Two

KEY CONCEPT B The building-up (Aufbau) principle

The **building-up principle** gives us a straightforward way to predict the electron configuration of many of the atoms in the periodic table. To get an electron configuration, we first determine the number of electrons in the atom from its atomic number. Then, with care to follow the exclusion principle and Hund's rule, we place the electrons into orbitals starting with the lowest energy orbital and moving to higher energy orbitals until all of the electrons have been placed. The first ten subshells in a large number of neutral many-electron atoms are ordered as follows. (It should be noted that there are some slight changes in this ordering in many ions.)

$$\text{Lowest energy} \quad 1s < 2s < 2p < 3s < 3p < 4s < 3d < 4p < 5s < 4d \quad \text{highest energy}$$

Thus, the order of filling of orbitals using the building-up principle for atoms of up to 38 electrons is usually,

$$\text{Fill first} \quad 1s, 2s, 2p, 3s, 3p, 4s, 3d, 4p, 5s, 4d \quad \text{fill last}$$

Outer shell electrons are known as **valence electrons.** To save space, inner electrons that mimic a noble-gas configurattion may be symbolized by the noble-gas elemental symbol enclosed in brackets [ ]. The electron configurations of five elements are shown in the following table, with valence electrons indicated in boldface.

Element	**1s**	**2s**	**$2p_x$**	**$2p_y$**	**$2p_z$**	**3s**	**Configuration**
lithium, $Z = 3$	2	**1**					$1s^2 2s^1$ or $[\text{He}]2s^1$
carbon, $Z = 6$	2	**2**	**1**	**1**			$1s^2 2s^2 2p^2$ or $[\text{He}]2s^2 2p^2$
oxygen, $Z = 8$	2	**2**	**2**	**1**	**1**		$1s^2 2s^2 2p^4$ or $[\text{He}]2s^2 2p^4$
fluorine, $Z = 9$	2	**2**	**2**	**2**	**1**		$1s^2 2s^2 2p^5$ or $[\text{He}]2s^2 2p^5$
sodium, $Z = 11$	2	2	2	2	2	**1**	$1s^2 2s^2 2p^6 3s^1$ or $[\text{Ne}]3s^1$

EXAMPLE Electron configurations

Predict the electron configuration of silicon.

SOLUTION Silicon has atomic number 14, so 14 electrons must be placed into the many-electron atomic orbitals. The order of filling is

$$\text{Fill first} \quad 1s, 2s, 2p, 3s, 3p, 4s, 3d, 4p, 5s, 4d \quad \text{fill last}$$

Care must be taken to obey the exclusion principle and Hund's rule as the 14 electrons are placed into the orbitals. The filling proceeds as follows:

1. The first and second electrons go into the 1*s* orbital with opposite spins, filling the 1*s* orbital.
2. The third and fourth electrons go into the 2*s* orbital with opposite spins, filling the 2*s* orbital.
3. The fifth–seventh electrons go, one each, into the three 2*p* orbitals with parallel spins.
4. The eighth–tenth electrons complete the 2*p* subshell.
5. The eleventh and twelfth electrons go into the 3*s* orbital with their spins paired; this fills the 3*s* orbital.
6. The last two electrons, the thirteenth and fourteenth, go into two different 3*p* orbitals with parallel spins.

The final electron configuration is $1s^22s^22p^63s^23p^2$ or $[Ne]3s^23p^2$.

EXERCISE Predict the electron configuration of Cl.

ANSWER: $1s^22s^22p^63s^23p^5$ or, equivalently, $[Ne]3s^23p^5$

KEY CONCEPT C The filling of *d*-orbitals

The third period of the periodic table ends with argon, which has the configuration $1s^22s^22p^63s^23p^6$. Following the order of filling 1*s*, 2*s*, 2*p*, 3*s*, 3*p*, 4*s*, 3*d*, 4*p*, 5*s*, 4*d*, the 4*s* orbital fills next with potassium (K) and calcium (Ca). The 3*d* subshell then fills; ten electrons are required to fill this subshell, corresponding to the ten transition elements scandium (Sc) to zinc (Zn). Scandium, titanium, and vanadium have the configurations predicted by the building-up principle. Vanadium, for example, has the configuration $1s^22s^22p^63s^23p^63d^34s^2$ or $[Ar]3d^34s^2$. Chromium, however, has an unexpected configuration due to the relatively low energy of half-filled subshells; its configuration is $1s^22s^22p^63s^23p^63d^54s^1$ or $[Ar]3d^54s^1$. As we continue across the period, cobalt, for example, has the configuration $1s^22s^22p^63s^23p^63d^74s^2$ or $[Ar]3d^74s^2$. Copper, like chromium, has an unexpected configuration. In this case, the relatively low energy of a completed subshell is responsible for its configuration, which is $1s^22s^22p^63s^23p^63d^{10}4s^1$ or $[Ar]3d^{10}4s^1$. One more electron completes the fourth-period transition elements with zinc, which has the configuration $1s^22s^22p^63s^23p^63d^{10}4s^2$ or $[Ar]3d^{10}4s^2$. Even though the 3*d* subshell fills after the 4*s* subshell, it is quite common to write electron configurations with the 4*s* electrons written after the 3*d*, because the 4*s* electrons are lost first when the transition elements ionize.

EXAMPLE Predicting the electron configuration of a transition element

Predict the electron configuration of the fifth-period transition element niobium (Nb) from the building-up principle. Is this the correct configuration for Nb? How many unpaired electrons does Nb have?

SOLUTION The fifth period starts with filling the 5*s* orbital in rubidium (Rb) and strontium (Sr). The transition elements that follow involve filling the 4*d* subshell. Nb is the third transition element in this period and is, therefore, predicted to have three 4*d* electrons and the configuration $[Kr]4d^35s^2$. Reference to Appendix 2C of the text indicates the predicted configuration is not the actual one. Nb has the actual configuration $[Kr]4d^45s^1$. It turns out that there are many such irregular configurations in the fifth-period transition series. Also, Nb has five unpaired electrons because Hund's rule must be obeyed in placing electrons in the 4*d* subshell. This is shown in the following figure.

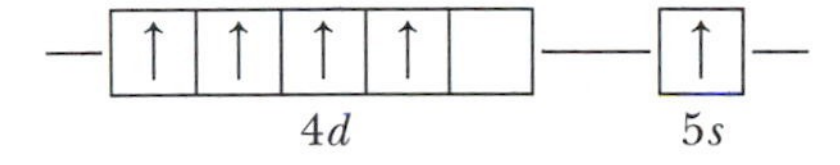

EXERCISE What is the predicted and the actual electron configuration of technetium (Tc)? How many unpaired electrons does it have?

ANSWER: Predicted and actual (from text), $[Kr]4d^55s^2$; five unpaired electrons.

KEY CONCEPT D The configurations of positive monatomic ions

Metals lose their most loosely bound electrons to form ions. The electron configuration of an atom determines how tightly electrons are bound and, therefore, affects the electron configuration of the ions formed by the atom. For the transition elements, many ions are formed by loss of one or two loosely held outer *s* electrons. As an example, four ions formed in the first transition series by loss of the outer 4*s* electrons are shown in the following table.

Element	Atomic configuration	Electron change	Ion	Ion configuration
Ti	$[Ar]3d^2 4s^2$	lose two 4*s* electrons	Ti^{2+}	$[Ar]3d^2$
Fe	$[Ar]3d^6 4s^2$	lose two 4*s* electrons	Fe^{2+}	$[Ar]3d^6$
Ni	$[Ar]3d^8 4s^2$	lose two 4*s* electrons	Ni^{2+}	$[Ar]3d^8$
Cu	$[Ar]3d^{10} 4s^1$	lose one 4*s* electron	Cu^+	$[Ar]3d^{10}$

EXAMPLE Predicting the electron configuration of an ion

State the charge and configuration of a common ion of silver (Ag).

SOLUTION Ag has the electron configuration $[Kr]4d^{10}5s^1$. Because all of the elements in the first transion series (except scandium) can form an ion by loss of their outer *s* electrons, we expect Ag to behave similarly and predict it will form an Ag^+ ion by loss of its 5*s* electron. This results in the configuration $[Kr]4d^{10}$.

EXERCISE State the charge and configuration of a common ion formed by Pb.

ANSWER: Pb^{4+}; $[Xe]4f^{14}5d^{10}$

PITFALL The oxidation states of the transition elements

More than 50 different simple ions have been observed for the ten elements Sc to Zn. We have mentioned only a few of these in the text and study guide up to now. More will be described later in the course.

KEY CONCEPT E Paramagnetism and diamagnetism

Experimentally, electrons act like little magnets. An electron with an up spin (↑) behaves like an bar magnet in one orientation and an electron with a down spin (↓) like a bar magnet in exactly the opposite orientation. Thus, the magnetic effects cancel for two paired electrons; however, magnetic effects are easily observed experimentally if any unpaired electrons are present in an atom or molecule. Atoms with closed shells and subshells have all their electrons paired and do not show any dramatic response to a magnetic field; they are called **diamagnetic** atoms. Atoms that have one or more unpaired electrons respond markedly to external magnetic fields; these are called **paramagnetic** atoms.

EXAMPLE Predicting whether an atom is paramagnetic or diamagnetic

Which of the following atoms is paramagnetic and which diamagnetic? Ba, Mn, S.

SOLUTION An atom is paramagnetic if it has one or more unpaired electrons and diamagnetic if it has no unpaired electrons. To answer the question, we must look at the electron configuration of each atom and determine which has unpaired electrons.

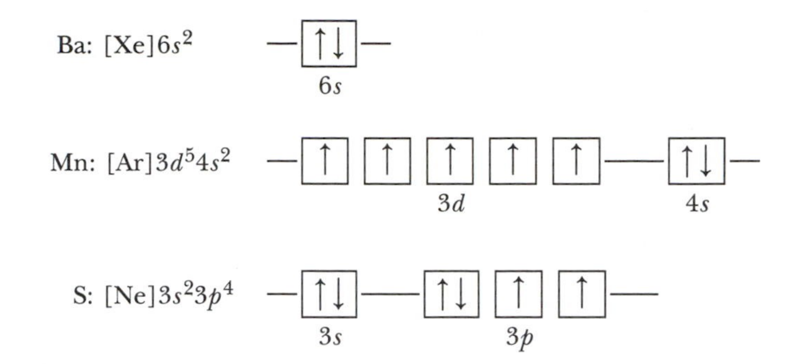

We first note that a noble-gas core consists entirely of paired electrons, so only the electrons outside the core must be considered; these are shown in the orbital diagrams next to each atom. For barium, two electrons in a $6s$ orbital must be paired; barium has no unpaired electrons and is diamagnetic. Manganese has five electrons in a d subshell. According to Hund's rule, this leads to five unpaired electrons; manganese is paramagnetic. The two electrons in the $3s$ orbital of sulfur are, of course, paired. However, when there are four electrons in a p subshell, two of them are unpaired, because Hund's rule must be obeyed. Thus, sulfur becomes an interesting case. Even though it has an even number of electrons, it is paramagnetic.

EXERCISE Which of the following atoms is paramagnetic and which diamagnetic? Cd, Br, Ru.

ANSWER: Cd, diamagnetic; Br, Ru, paramagnetic.

A SURVERY OF PERIODIC PROPERTIES

KEY WORDS Define or explain each of the following terms in a written sentence or two.

atomic radius
diagonal relationship
electron affinity
electronegativity
electron gain enthalpy
ionic radius
ionization energy
main group elements
s, *p*, and *d* block
transition elements

7.7 BLOCKS, PERIODS, AND GROUPS

KEY CONCEPT Categorizing the elements in the periodic table

Chemists have adopted a variety of ways of categorizing the periodic table. The different ways complement each other and allow us to better understand the various relationships in the periodic table. It would be wise to refer to a periodic table as you read the following list of categorizations.

Blocks: The elements are divided according to the subshell that received the last electron in the building-up process. The periodic table has an *s block,* a *p block,* a *d block,* and an *f block.* Refer to Figure 7.26 of the text to see where each is.

Main group elements and others: The elements are also divided into three broad classes called the **main group elements,** the **transition elements,** and the **inner transition elements.** The *s*-block elements and *p*-block elements grouped together are collectively called the main group elements. In this classification scheme, the *d*-block elements are called the transition elements and the *f*-block elements are the inner transition elements.

Periods: Each row of elements in the periodic table is called a **period.** The period number is the same as the shell number of the valence shell. For example, selenium is a fourth-period element with electron configuration $[Ar]3d^{10}4s^24p^4$ and its $n = 4$ shell is its valence shell.

Groups: Columns of elements in the periodic table are called **groups.** There is currently no universally agreed on way to number the groups. The periodic table we use labels the *s*-block elements

Groups I and II and the p-block groups III–VIII. With this particular numbering system, the number of each group corresponds to the sum of the number of s and p electrons in the valence shell. Some of the groups also have tranditional, but still commonly used, names. These are Group I, alkali metals; Group II, alkaline earth metals; Group VII, halogens; and Group VIII, noble gases.

EXAMPLE Categorizing the periodic table

What s-block element has two electrons in its valence shell and is in the fifth period?

SOLUTION The s-block elements are the elements of Group I and Group II. Because the element we are seeking has two electrons in its valence shell, it must be a Group II element. The fifth period starts with Rb and ends with Xe. The only element that is in Group II and in the fifth period is Sr.

EXERCISE Name the main group element which has only two 6p electrons.

ANSWER: Pb (lead)

PITFALL Helium and the p block

Helium is located in the p block even though the last subshell filled is an s subshell.

7.8 PERIODICITY OF PHYSICAL PROPERTIES

KEY CONCEPT A Atomic and ionic radii

If we think of atoms and ions as small spheres, then the radius r of the atom or ion is a measure of its size. Atomic and ionic radii are determined by experiment. We should note the following facts and trends regarding atomic and ionic radii.

1. As we go from left to right in a period in the periodic table, atomic radii tend to decrease. As we go from top to bottom of a group, atomic radii tend to increase. This is illustrated by the following examples. (See also Figure 7.27 of the text.)

Period	Group I	Group VII
third period	Na (r = 180 pm)	Cl (r = 100 pm)
fifth period	Rb (r = 235 pm)	I (r = 140 pm)

The same trend holds for ionic radii. (See Figure 7.29 of the text and be certain to compare cations with other cations and anions with other anions.)

2. Cations are smaller than the parent atom from which they are formed. For example,

K	r = 220 pm	Ca	r = 180 pm
K^+	r = 133 pm	Ca^{2+}	r = 99 pm

3. Anions are larger than the parent atom from which they are formed. For example,

Cl	r = 100 pm	S	r = 115 pm
Cl^-	r = 181 pm	S^{2-}	r = 198 pm

EXAMPLE Periodic trends in atomic and ionic radii

Use the periodic table to decide whether Se or Cl has the largest atomic radius.

SOLUTION The main tools we have to solve this problem are the periodic trends of atomic radii. As we go down a group, atomic radii tend to increase. Even though Se and Cl are not in the same group, Se is below Cl, so we suspect that it may have a larger radius than Cl. The second periodic trend is that as we go from left to right in a period, atomic radii tend to decrease. Again, Se and Cl are not in the same period, but Se is to the left of Cl, so it probably has a larger radius than Cl. Both trends indicate that Se should be larger than Cl. Se has the larger radius.

EXERCISE Which has the larger radius Rb or Ca?

ANSWER: Rb

KEY CONCEPT B Ionization energies

It is possible to knock the outermost electron off an atom. This is called **ionization;** the minimum energy needed to do this is called the **ionization energy** of the atom. Electrons can also be knocked off molecules and ions, so ionization energies can also be measured for ions and molecules. Using the data in Figure 7.31 of the text, we can express the first two ionizations of nitrogen, as

$$N \longrightarrow N^+ + e^- \qquad I_1 = 1400 \text{ kJ}$$
$$N^+ \longrightarrow N^{2+} + e^- \qquad I_2 = 2860 \text{ kJ}$$

I_1 is called the first ionization energy and is the minimum energy required to remove an electron from the nitrogen atom. I_2 is the second ionization energy and is the minimum energy required to remove an electron from the N^+ ion. For atoms, the ionization energy is largely (but not exclusively; see example 7.11 of the text) determined by the size of the atom because it is the outer electron that is removed. Ionization energies follow the periodic trend expected from atomic size. As we go across the periodic table from left to right in a period, ionization energy tends to increase. As we go from top to bottom down a group , ionization energy tends to decrease. Thus, it is quite difficult to remove an electron from O, F, or N. On the other hand, atoms on the bottom of the periodic table and to the left have low ionization energies; for example, it is *relatively* easy to remove an electron from Cs, Fr, or Ra. Metals, which form positive ions in their ionic compounds, have low ionization energies. Nonmetals, which form negative ions in their ionic compounds, have high ionization energies; they do not readily form positive ions.

EXAMPLE Periodic trends in ionization energy

Three elements and their atomic radii are Ca(180 pm), Fe(140 pm), Br(115 pm). Without consulting any references, list these in order of increasing ionization energy. Does the order properly reflect expected periodic trends?

SOLUTION When an atom undergoes ionization, the outer electron is removed. When this electron is far away from the positively charged nucleus, it is generally easier to remove than when it is closer to the nucleus. Thus, all other things being equal, a smaller atom should have a higher ionization energy than a larger atom. The order of ionization energies should be

$$\text{Highest ionization energy} \qquad \text{Br} > \text{Fe} > \text{Ca} \qquad \text{lowest ionization energy}$$

This does follow the expected trend in the periodic table. All three elements are in the fourth period, with calcium furthest to the left and bromine furthest to the right. Ionization energies tend to increase in going from left to right in the periodic table.

EXERCISE Without consulting any references except a periodic table, list the three atoms Li, K, and Cs in order of decreasing ionization energy.

ANSWER: Largest Li > K > Cs smallest

KEY CONCEPT C Electron-gain enthalpy and electron affinity

It is possible to conceive of a process in which an electron becomes attached to a neutral atom. In some cases it is necessary to force the atom to accept the electron; that is, it takes energy to place the electron on the atom. In other cases, energy is released when the electron is attached to the neutral atom, and the electron "wants" to attach to the atom. The **electron affinity** is the energy *released* when an electron becomes attached to a gaseous atom or other chemical species. Thus, the electron affinity is positive when energy is released and negative when it is taken in (this is opposite to the signs associated with enthalpy changes). The **electron-gain enthalpy** is defined as the enthalpy change when an electron becomes attached to an atom or chemical species. Thus, if we assume that the electron affinity is measured at constant pressure, we would have as examples

$$P(g) + e^-(g) \longrightarrow P^-(g) \qquad \Delta H_{gain} = -72\ kJ \qquad E_{ea} = +72\ kJ$$
$$Ne(g) + e^-(g) \longrightarrow Ne^-(g) \qquad \Delta H_{gain} = 29\ kJ \qquad E_{ea} = -29\ kJ$$

Electron-gain enthalpies do not follow strong periodic trends, but elements in the upper right of the periodic table (nonmetals) have the most negative electron-gain enthalpies; these elements, therefore, form anions more readily than the other elements in the periodic table.

PITFALL The sign of electron affinity and electron-gain enthalpy

An element with a negative electron-gain enthalpy (and, therefore, a positive electron affinity) is one for which gaining an electron is a *favorable* process; whereas an element with a positive electron-gain enthalpy (and, therefore, a negative electron affinity) is one for which gaining an electron is an *unfavorable* process.

EXAMPLE Using electron-gain enthalpies

What energy change occurs when sulfur accepts an electron to form S^- ion? Refer to the text for information.

SOLUTION When a neutral atom accepts an electron, the energy change involved is given by the electron-gain enthalpy. Figure 7.35 of the text gives the electron-gain enthalpy of S as −200 kJ/mol. Thus, when 1 mol of sulfur atoms accepts 1 mol of electrons to form 1 mol of S^- ions, 200 kJ of energy is released.

EXERCISE What energy change occurs when S^- accepts an electron to form S^{2-}.

ANSWER: For 1 mol of S^- to become 1 mol of S^{2-} 532 kJ must be supplied.

PITFALL The physical state in electron gain

Ionization energies and electron affinities refer to processes that occur in the gas phase. They do not apply directly to similar processes that occur in aqueous solutions.

KEY CONCEPT D Electronegativity

The **electronegativity** χ of an atom is a measure of its ability to attract electrons in a compound toward itself when the atom is part of the compound. Atoms with a high electronegativity, such as F, N, O, and Cl, strongly attract electrons and tend to form anions. Atoms with a low electronegativity, such as K, Ca, Zn, and Al, do not attract electrons very strongly. They often lose all or part of an electron to an atom with higher electronegativity, so they tend to form cations.

Electronegativities follow a periodic trend. As we go to the right in a period, the electronegativity tends to increase. As we go down a group, the electronegativity tends to decrease. That is, a high electronegativity is a nonmetal trait. Fluorine, one of the two most nonmetallic elements, has the highest electronegativity (4.0); francium, the most metallic main group element, has the lowest electronegativity, (0.7).

EXAMPLE Periodic trends in electronegativity

Let's assume that a new element with atomic number 117 is synthesized; thus, it fits under astatine (At) in the periodic table. Estimate its electronegativity and decide whether it is a nonmetal, metal, or metalloid.

SOLUTION From Figure 7.36 of the text, we note the following decreases in electronegativity as we go down Group VII from chlorine to astatine (we ignore fluorine because the first element in a group is often unusual in its properties).

Cl to Br	$\Delta\chi = 3.0 - 2.8 = 0.2$
Br to I	$\Delta\chi = 2.8 - 2.5 = 0.3$
I to At	$\Delta\chi = 2.5 - 2.2 = 0.3$

From the changes observed, we would expect the next element below astatine to have an electronegativity about 0.3 less than astatine. For the new element $\chi \approx 1.9$. Other elements with electronegativities near this value (As, Ge, Sb, Bi, Po) are either metals or metalloids. We, therefore, expect the new element to also be a metal or metalloid—a rather surprising prediction for a Group VII element!

EXERCISE A new element is discovered and its electronegativity is measured to be 2.3. Would you classify the element as a metal, a nonmetal, or a metalloid?

ANSWER: nonmetal

PITFALL Are electron-gain enthalpy and electronegativity the same?

The electron-gain enthalpy and electronegativity are related to each other, but they are not the same. Both are concerned with the attraction of an electron to an atom. However, the electron-gain enthalpy applies to an isolated atom, that is, one that is in the gaseous state. The electronegativity refers to an atom that is bound in an ionic compound or molecule.

7.9 TRENDS IN CHEMICAL PROPERTIES

KEY CONCEPT A The relationship among properties, electronic structure, and position in the periodic table

The position of an element in the periodic table determines its properties because the properties are intimately related to the underlying electronic structure. There is a three-way connection among periodic properties, position in the periodic table, and electronic structure, as shown in the figure.

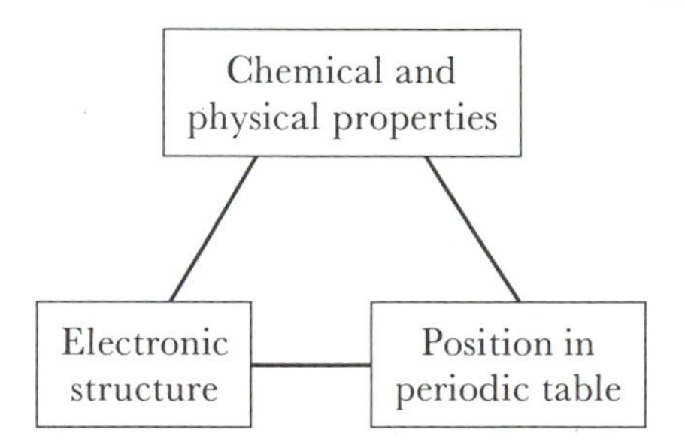

This relationship makes it possible to discuss the properties of the elements based on whether the element is an *s*-block, *p*-block, or *d*-block element.

EXAMPLE Predicting properties of the elements

Which of the elements Se, P, As, or Br is most similar to S in its properties?

SOLUTION All of the elements in the question are *p*-block elements, so we expect to find some similarities among them. However, both S and Se have similar ns^2np^4 electronic configurations, and we would correctly suspect that Se is the closest to S in many of its properties. For instance, both form ions with a charge of -2 by gaining two electrons to form a noble-gas configuration (Ar configuration for S; Kr configuration for Se) and both are nonconducting solids. Bromine, on the other hand, is a liquid and arsenic is a semiconductor (it conducts electricity, but not well). Phosphorus forms an ion with a -3 charge and bromine an ion with a -1 charge.

EXERCISE Of the elements Kr, I, Te and C, which is most similar to Cl. Why?

ANSWER: I; both I and Cl have a ns^2np^5 configuration.

KEY CONCEPT B Gradual changes in properties in the periodic table

The position of an element in the periodic table is a clue to the properties of the element; in addition as we move from one part of the table to another, we see gradual changes in properties, rather than abrupt changes. This has two important ramifications:

1. There are elements with properties intermediate between metals and nonmetals. As we move from right to left in the periodic table, properties gradually change from nonmetallic to metallic; the dividing line between metals and nonmetals is the diagonal staircase starting to the left of boron and ending between polonium and astatine. The elements in this "gray" area possess some of the properties of metals and some of the porperties of nonmetals.
2. *d*-block elements near the *p* block tend to be similar to nearby *p*-block elements in their properties, whereas *d*-block elements near the *s* block tend to be similar to nearby *s*-block elements in their properties.

EXAMPLE Gradual changes in the periodic table

Of the elements Ni, Ge, and Br, all fourth-period elements, which possesses properties intermediate between those of a metal and nonmetal?

SOLUTION This is not a memorization question. If we look at a periodic table, we note that Ni is a *d*-block element, well to the left of the dividing line between metals and nonmetals, and is, therefore, a metal; all of the *d*-block elements are metals. Br is a *p*-block element, well to the right of the dividing line between metals and nonmetals; it is a nonmetal. Ge (germanium) is on the dividing line between metals and nonmetals and is, therefore, expected to have properties intermediate between a metal and nonmetal.

EXERCISE Of the elements Sc (scandium), Ru (ruthenium), and Ag (silver), which is expected to be most like Pb (lead) in its chemical reactivity?

ANSWER: Ag

PITFALL How reliable are trends in the periodic table?

We have discussed a variety of similarities and trends based on position in the periodic table, but care must be taken regarding how general such relationships are. Depending on the elements involved, similarities and trends are observed for some properties but not for others. As an example, even though they are in the same group, Li and Na do not react similarly when burned in air. Sodium forms an oxide and peroxide, whereas lithium forms some nitride and some oxide.

DESCRIPTIVE CHEMISTRY TO REMEMBER

- **Visible light** has wavelengths from 700 nm (red light) to 400 nm (violet light).
- The **Bohr radius,** the radius of the Bohr orbit with $n = 1$, is 53 pm.
- The members of the *s* and *p* blocks are called **main group elements.**
- **Cesium** is so reactive that it reduces water explosively.
- In, Tl, Sn, Sb, and Bi have an **inert pair of electrons** and, therefore, lose electrons in stages; this results in ions of two different charges for each element.
- **Group I** elements form cations with +1 charge, and **Group II** elements form cations with +2 charge.
- The following pairs of elements share a **diagonal relationship:** Li and Mg, Be and Al, B and Si.
- The ***s*-block metals** are all very reactive.
- Many ***p*-block metals** are not nearly as reactive as the *s*-block metals.
- Steel cans are **tin plated** to protect them against corrosion.
- All of the ***d*-block elements** (transition elements) are metals.
- **Transition elements** near the left of the *d*-block resemble *s*-block elements and are, therefore, quite reactive. Examples are Sc, Ti, Zr, and Ta.
- **Transition elements** near the right of the *d*-block resemble *p*-block metals and are, therefore, less chemically reactive than those near the left of the *d* block. Examples are Cu, Ag, Au, Hg, and Pt (but Zn and Cd are quite reactive).

CHEMICAL EQUATIONS TO KNOW

- All of the Group I metals reduce water to hydrogen. For example,

$$2K(s) + 2H_2O(l) \longrightarrow 2KOH(aq) + H_2(g)$$

- All of the Group II metals (except Be) reduce water to form hydrogen. For example,

$$Ca(s) + 2H_2O(l) \longrightarrow Ca(OH)_2(aq) + H_2(g)$$

- All of the *s*-block elements have basic oxides that react with water to form hydroxides. For instance,

$$CaO(s) + H_2O(l) \longrightarrow Ca(OH)_2(aq)$$

- Lithium and magnesium both form a nitride when burned in air. (Magnesium also forms some oxide when it is burned in air.)

$$6Li(s) + N_2(g) \longrightarrow 2Li_3N(s)$$
$$3Mg(s) + N_2(g) \longrightarrow Mg_3N_2(s)$$

- Beryllium and aluminum both reduce acids to form hydrogen:

$$Be(s) + 2H^+(aq) \longrightarrow Be^{2+}(aq) + H_2(g)$$
$$2Al(s) + 6H^+(aq) \longrightarrow 2Al^{3+}(aq) + 3H_2(g)$$

- Beryllium and aluminum both react with aqueous bases to form a complex ion and hydrogen:

$$Be(s) + 2OH^-(aq) + 2H_2O(l) \longrightarrow [Be(OH)_4]^{2-}(aq) + H_2(g)$$
$$2Al(s) + 2OH^-(aq) + 6H_2O(l) \longrightarrow 2[Al(OH)_4]^-(aq) + 3H_2(g)$$

- Lead and tin both form amphoteric oxides. For example, SnO reacts both with acids and bases:

$$SnO(s) + 2H^+(aq) \longrightarrow Sn^{2+}(aq) + H_2O(l)$$
$$SnO(s) + OH^-(aq) + H_2O(l) \longrightarrow [Sn(OH)_3]^-$$

- When tin is heated in air, it forms tin(IV) oxide. Lead, on the other hand, shows a stronger inert pair effect and forms lead(II) oxide rather than lead(IV) oxide when heated in air:

$$Sn(s) + O_2(g) \longrightarrow SnO_2(g)$$
$$2Pb(s) + O_2(g) \longrightarrow 2PbO(s)$$

MATHEMATICAL EQUATIONS TO KNOW AND UNDERSTAND

$\lambda \times \nu = c$	electromagnetic radiation
$E = h \times \nu$	energy of a photon
$\nu = \mathscr{R} \times \left(\frac{1}{n_l^2} - \frac{1}{n_u^2}\right)$	Rydberg equation
$\Delta E = h \times \nu$	frequency of light generated by an excited atom
$E = \frac{-h \times \mathscr{R}}{n^2}$	energy of electron in the hydrogen atom
Energy lost $= \mathscr{R} \times h \times \left(\frac{1}{n_l^2} - \frac{1}{n_u^2}\right)$	energy lost by electron in hydrogen atom
$\lambda = \frac{h}{\text{mass} \times \text{velocity}}$	de Broglie wavelength

SELF-TEST EXERCISES

Light and Spectroscopy

1. What is the wavelength of electromagnetic radiation with a frequency of 4.4×10^{12} Hz?
(a) 1.5×10^4 m (b) 3.0×10^5 m (c) 6.8×10^{-5} m
(d) 1.3×10^{21} m (e) 4.4×10^{10} m

2. What is the frequency of light with $\lambda = 561$ nm?
(a) 1.87×10^{-15} Hz (b) 5.35×10^5 Hz (c) 5.35×10^{14} Hz
(d) 1.87×10^{-6} Hz (e) 168 Hz

3. What is the energy of a photon with a frequency of 6.5×10^9 Hz?
(a) 1.0×10^{-43} J (b) 9.8×10^{42} J (c) 5.2×10^{-19} J
(d) 4.3×10^{-24} J (e) 2.3×10^{23} J

4. What is the energy of a mole of photons with wavelength 3.3 cm?
(a) 6.6×10^{-25} J (b) 3.6 J (c) 1.3×10^{-10} J (d) 5.4×10^{31} J (e) 21 J?

5. Approximately 1.0% of the photons emitted from a 150-W 3200-K photoflood lamp have wavelength 450 nm. How many photons per second are emitted at 450 nm?
(a) 3.4×10^{18} (b) 8.2×10^{24} (c) 7.4×10^{21} (d) 4.4×10^{20} (e) 3.40×10^9

6. A photon with energy 5.6×10^{-19} J impinges on a metal and causes ejection of an electron with a kinetic energy of 7.2×10^{-20} J. What is the minimum energy required to remove an electron from the metal?
(a) 4.9×10^{-19} J (b) 5.6×10^{-19} J (c) 7.2×10^{-20} J (d) 7.8 J (e) 0.12 J

7. A photon impinges on a metal for which 4.30×10^{-19} J are required to remove an electron. When the photon hits the metal, an electron with a kinetic energy of 2.20×10^{-20} J is ejected from the metal. What is the wavelength of the photon?

(a) 463 nm (b) 904 nm (c) 315 nm (d) 442 nm (e) 2.26 nm

8. What is the frequency of the light emitted when a hydrogen atom changes energy states from the $n = 7$ state to the $n = 5$ state?
(a) 1.37×10^{14} Hz (b) 2.65×10^{14} Hz (c) 1.88×10^{14} Hz
(d) 1.65×10^{15} Hz (e) 6.45×10^{13} Hz

9. When the electron in a hydrogen atom drops to the $n = 1$ level, light with frequency 3.223×10^{15} Hz is emitted. What is the value of n for the energy level the electron originated in?
(a) 2 (b) 3 (c) 7 (d) 4 (e) 5

10. An atom with energy -3.23×10^{-19} J changes its energy to -7.27×10^{-19} J by emitting a photon. What is the energy of the photon?
(a) 10.5×10^{-19} J (b) 0 J (c) 4.04×10^{-19} J
(d) 3.23×10^{-19} J (e) 7.27×10^{-19} J

11. A potassium atom emits a photon with energy 2.59×10^{-21} J in changing energy states. What is the frequency of the photon?
(a) 8.63×10^{-30} Hz (b) 3.91×10^{12} Hz (c) 1.71×10^{-54} Hz
(d) 2.56×10^{-13} Hz (e) 5.44×10^{15} Hz

12. Calculate the energy of the $n = 3$ level of the hydrogen atom.
(a) -2.18×10^{-18} J (b) -7.27×10^{-19} J (c) -3.54×10^{-18} J
(d) -5.89×10^{-17} J (e) -2.42×10^{-19} J

13. What is the change in energy of a hydrogen atom when an electron falls from the $n = 6$ state to the $n = 1$ state?
(a) -2.74×10^{15} J (b) -1.82×10^{-18} J (c) -3.20×10^{15} J
(d) -2.18×10^{-18} J (e) -2.12×10^{-18} J

14. A mole of hydrogen atoms changes energy from the $n = 3$ state to the $n = 2$ state. How much energy is released?
(a) 4.36×10^{-19} J (b) 3.03×10^{-19} J (c) 262 kJ (d) 182 kJ (e) 109 kJ

15. Calculate the deBroglie wavelength of a proton moving at 350 km/s.
(a) 1.13×10^{-12} m (b) 1.39×10^{5} nm (c) 0.139 nm
(d) 1.39×10^{-4} nm (e) 1.13 nm

16. The deBroglie wavelength of an electron is 410 nm. How fast is the electron moving? The mass of an electron is 9.109×10^{-31} kg; $h = 6.63 \times 10^{-34}$ J/Hz.
(a) 2.54×10^{3} m/s (b) 1.78×10^{3} ms (c) 550 m/s
(d) 6.55×10^{3} m/s (e) 2.11×10^{4} m/s

17. The wave function ψ of an electron in a hydrogen is 0.22 (i.e., $\psi = 0.22$) at a point just outside the nucleus. What is the probability of finding the electron at this point?
(a) 0.048 (b) 0.22 (c) 0.47 (d) 0.78 (e) 0.028

18. How many subshells are present in the $n = 3$ shell?
(a) 3 (b) 4 (c) 9 (d) 18 (e) 2

19. How many orbitals are present in the $n = 2$ shell?
(a) 4 (b) 2 (c) 8 (d) 16 (e) 1

20. What three quantum numbers are permissible for a $3p$ orbital? The answers are expressed as (n, l, m_l)
(a) (3, 0, 0) (b) (3, 2, −2) (c) (3, 2, −1) (d) (2, 1, 0) (e) (3, 1, −1)

21. How many orbitals are present in the $7d$ subshell?
(a) 10 (b) 5 (c) 4 (d) 7 (e) 14

22. What four quantum numbers are permissible for a $3d$ electron? The answers are expressed as (n, l, m_l, m_s).

(a) $(3, 0, 0, -\frac{1}{2})$ (b) $(3, 2, 2, -\frac{1}{2})$ (c) $(3, 2, 2, 1)$
(d) $(2, 1, 0, \pm\frac{1}{2})$ (e) $(3, 1, -1, -\frac{1}{2})$

23. An electron with the quantum numbers $n = 4$, $l = 2$, $m_l = 0$, and $m_s = -\frac{1}{2}$ is designated as what kind of electron?

(a) 6s (b) $2s$ (c) $2p$ (d) $4d$ (e) $4s$

The structures of many-electron atoms

24. Which of the following orbitals penetrates most?

(a) $4s$ (b) $4p$ (c) $4d$ (d) $4f$

25. What is the maximum number of electrons permitted in a $4f$ orbital?

(a) 7 (b) 14 (c) 1 (d) 3 (e) 2

26. What is the electron configuration of N?

(a) $1s^2 2s^2 2p^2 3s^1$ (b) $1s^2 2s^2 2p^6 3s^2 3p^2$ (c) $1s^2 2s^2 2p^3$ (d) $1s^6 2s^1$ (e) $1s^2 2s^2 3s^2 4f^1$

27. What is the electron configuration of Mn?

(a) $[\mathrm{Ar}]3d^7$ (b) $[\mathrm{Ar}]3d^6 4s^1$ (c) $[\mathrm{Ar}]3d^5 4s^2$ (d) $[\mathrm{Ar}]4s^2 4p^5$ (e) $[\mathrm{Ar}]3d^3 4s^2 4p^2$

28. How many unpaired electrons are present in an oxygen atom?

(a) 4 (b) 8 (c) 1 (d) 6 (e) 2

29. What is the electron configuration of lithium in the ground state?

(a) $1s^3$ (b) $1s^1 2s^2$ (c) $2s^3$ (d) $1s^1 2s^1 2p^1$ (e) $1s^2 2s^1$

30. Which of the following configurations is permitted for four electrons in a p subshell?

(a) [↑ | ↑ | ↑↓] (b) [↑↑ | ↑ | ↑] (c) [↑↓ | ↑↓ |]

(d) [↑↑↑ | ↓ |] (e) [↑↓ | ↓ | ↑]

31. How many valence shell electrons are present in Cl?

(a) 7 (b) 5 (c) 10 (d) 17 (e) 2

32. Which of the following atoms possesses only closed shells in the ground state?

(a) K (b) F (c) Ar (d) Se (e) Zn

33. What is the electron configuration of Pb?

(a) $[\mathrm{Xe}]6s^2 6p^2$ (b) $[\mathrm{Xe}]4f^{14} 5d^{10} 6s^2 6p^2$ (c) $[\mathrm{Xe}]5d^{10} 6s^2 6p^2$
(d) $[\mathrm{Xe}]6s^2 6p^6 6d^{10} 6f^{10}$ (e) $[\mathrm{Xe}]6s^2 6p^2 6d^{10} 6f^{14}$

A survey of periodic properties

34. Which of the following is a p-block element?

(a) Ca (b) Pd (c) U (d) Br (e) V

35. Which of the following has the largest atomic radius?

(a) Rb (b) K (c) Ca (d) I (e) S

36. Which Group I element has the highest ionization energy?

(a) Li (b) Na (c) K (d) Rb (e) Cs

37. Which of the following has the highest ionization energy?

(a) Al^{2+} (b) Mg^+ (c) Mg^{2+} (d) Na (e) Mg

38. A hypothetical element Z which possesses an inert pair forms an ion with a +1 charge. What charge would be expected for a second ion of Z, based on the fact that it posseses an inert pair?
(a) +2 (b) +7 (c) +4 (d) +3 (e) no other ions

39. The electron-gain enthalpy is the enthalpy of which of the following processes? (E = element)
(a) $E(aq) + e^- \longrightarrow E^-(aq)$ (b) $E(g) \longrightarrow E^+(g) + e^-$ (c) $2E(g) \longrightarrow E_2(g)$
(d) $E(s) \longrightarrow E(g)$ (e) $E(g) + e^- \longrightarrow E^-(g)$

Descriptive chemistry

40. What element shares a diagonal relationship with Mg?
(a) Sc (b) Li (c) B (d) Ca (e) K

41. Which of the following elements reacts explosively with water?
(a) Cs (b) Fe (c) Al (d) He (e) I

42. On the basis of its position in the periodic table, which metal would be expected to be the least reactive?
(a) Na (b) Ca (c) V (d) Sb (e) U

43. When lead is heated in air, it forms
(a) Pb_3N_4 (b) Pb_3N_2 (c) PbO (d) PbO_2 (e) no reaction occurs

44. Which element is expected to have properties intermediate between a metal and nonmetal?
(a) Mn (b) Si (c) Xe (d) O (e) Ag

45. A characteristic feature found in the *d*-block elements is
(a) high reactivity.
(b) a mixture of metallic and nonmetallic properties.
(c) a tendency for elements to gain electrons to form a closed-shell configuration.
(d) an ability for elements to form ions with different charges.
(e) the occurrence of liquid state at room temperature.

46. Only one of the alkali metals forms both a nitride and an oxide when burned in air. Which one?
(a) Li (b) Na (c) K (d) Rb (e) Cs

47. Which of the following wavelengths corresponds to visible light?
(a) 340 nm (b) 530 nm (c) 150 m (d) 750 nm (e) 3 cm

8 THE CHEMICAL BOND

Atoms may join together to form compounds. A **chemical bond** is the interaction that holds atoms together in a compound or molecular element. In this chapter, we explore how chemical bonds are formed and some of the ramifications of bond formation.

IONIC BONDS

KEY WORDS Define or explain each of the following terms in a written sentence or two.

Born-Haber cycle
ionic bond
lattice
lattice enthalpy
pseudonoble-gas configuration

8.1 THE ENERGETICS OF IONIC BOND FORMATION

KEY CONCEPT A Energy lowering as a criterion for bond formation

An important thread of thought that runs through this whole chapter is that a bond can form if the energy of the bonded atoms is less than the energy of the separate atoms. The converse is also true; if the energy of the bonded atoms is more than the energy of the separate atoms, no bond will form. Although it is not immediately obvious, it is also true that if two different bonding situations result in different decreases of energy, the path with the greatest decrease of energy will be the favored one. Thus, given the three processes pictured below, path III will be the most likely for the bonding of atoms A and B to each other because it results in the greatest decrease of energy.

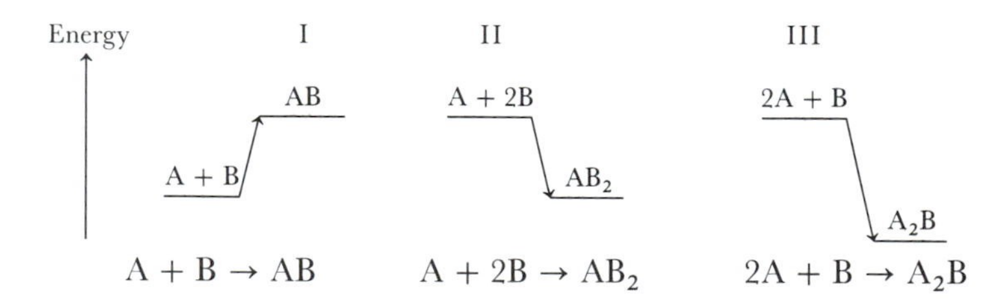

$A + B \rightarrow AB$ $\quad$ $A + 2B \rightarrow AB_2$ $\quad$ $2A + B \rightarrow A_2B$

EXAMPLE Predicting which path a reaction will take

If calcium and chlorine react, two possible products that can be envisioned are CaCl and $CaCl_2$. Formation of CaCl results in a net decrease of energy of approximately 180 kJ, whereas formation of $CaCl_2$ results in a net decrease of energy of approximately 800 kJ. Which product should form?

SOLUTION The path that results in the greatest decrease of energy is the favored one; thus, $CaCl_2$ should form.

EXERCISE Which has higher energy, the separated elements sodium and chlorine, or the elements bonded together as sodium chloride? Explain your answer.

ANSWER: We know that when atoms bond, a lower energy results, so the bonded atoms must have lower energy than the separated elements.

KEY CONCEPT B The three contributions to the formation of an ion pair from atoms

Let us consider the energetics of the process in which a gas-phase potassium atom and gas-phase chlorine atom react to form a pair of gas-phase ions separated by distance *d*:

$$K(g) + Cl(g) \longrightarrow K^+(g) + Cl^-(g) \qquad \text{(distance between ions} = d)$$

There are three energy contributions to this process. First, the potassium atom ionizes to form potassium ion, requiring the ionization energy +425.02 kJ/mol. Second, an electron is added to the chlorine atom, releasing an energy equal to the enthalpy of electron attachment for chlorine, −354.81 kJ/mol. Finally, the ions are formed some distance *d* from each other, and the third contribution to the overall energy is the potential energy of interaction of the two ions at that distance *d*. The only variable in this process is the distance between the ions; the overall energy of the process depends on this distance. This is shown in the following table, in which we compare the results of forming the ions at three different distances from each other.

Energy to form K^+, kJ	Energy to form Cl^-, kJ	Distance between ions formed, pm	Potential energy of interaction*, kJ	Overall energy of process, kJ
+425	−355	5000	−28	+42
+425	−355	2000	−70	+0
+425	−355	314	−443	−484

* Calculated by $E = (1.39 \times 10^5 \text{ kJ/mol}) \times (Z_A Z_B)/d$, where Z_A and Z_B are the charges on the ions.

EXAMPLE The contributions to formation of a pair of ions

A $Na^+(g)$ ion and a $F^-(g)$ ion are formed, from their respective atoms, 500 pm away from each other. What is the overall energy of this process? Is this an energetically favorable process?

$$Na(g) + F(g) \longrightarrow Na^+(g) + F^-(g) \qquad \text{(500 pm apart)}$$

SOLUTION There are three energy contributions to this process: the ionization energy of sodium, the enthalpy of electron attachment of fluorine, and the potential energy of interaction of the two ions (which are 500 pm apart). We first calculate the energy of interaction:

$$E = (1.39 \times 10^5 \text{ kJ/mol}) \times \frac{Z_A Z_B}{d}$$

$$= (1.39 \times 10^5 \text{ kJ/mol}) \times \frac{(+1)(-1)}{500}$$

$$= -278 \text{ kJ/mol}$$

The ionization energy of sodium is +494 kJ/mol and the electron-gain enthalpy of fluorine −328 kJ/mol. The overall energy of the process is the sum of these three energies. Thus, for one mole of ions, the total energy is −278 kJ + 494 kJ − 328 kJ = −112 kJ. The fact the energy change for the process is negative means that it is energetically favorable.

EXERCISE What would be the energy of the process in the example just given if one mole of the ions are formed 1000 pm apart? Is the process still energetically favorable?

ANSWER: +27 kJ; no

KEY CONCEPT C Lattice enthalpies

We now consider a process in which gas-phase positive ions and negative ions come together to form a solid rather than a pair of gas-phase ions. For NaF, this would be

$$Na^{+}(g) + F^{-}(g) \longrightarrow NaF(s) \qquad \Delta H = -929 \text{ kJ}$$

We assume here that the ions on the left side of the equation are very far apart, so their energy of interaction is zero. Also, it should be recognized that every positive ion in the solid interacts with all of the other ions in the solid (both positive and negative), and every negative ion also interacts with every other ion in the solid; the enthalpy of reaction results from this myriad of interactions, not from the interaction of just one positive ion with one negative ion. Because the dominant interaction is the favorable one between cations and anions, forming the solid from the gas-phase ions is always highly exothermic. The enthalpy of the reverse reaction is called the **lattice enthalpy** of the ionic compound involved. For NaF(*s*), for instance, the lattice enthalpy is +929 kJ.

EXAMPLE Understanding the magnitude of lattice enthalpies

According to Table 8.1 of the text, the lattice enthalpy of LiF is 1046 kJ/mol whereas that of KI is 645 kJ/mol. Suggest a reason for this difference.

SOLUTION The size of the lattice enthalpy depends predominantly on two factors: the size of the ions and the charge of the ions. Smaller ions can get closer together than larger ions. Thus, all other things the same, smaller ions experience stronger interactions and compounds with small ions have a larger lattice enthalpy than compounds with large ions. With all other things the same, ions of higher charge experience stronger interactions than ions of lower charge; this results in a higher lattice enthalpy for compounds with ions of higher charge. For our problem, the charge on all of the ions (Li^{+}, F^{-}, K^{+}, I^{-}) is +1 or −1, so differences in charge cannot account for the different lattice enthalpies. However, K^{+} is significantly larger than Li^{+}, and I^{-} is significantly larger than F^{-}. The larger ions in KI, relative to those in LiF, are responsible for the higher lattice enthalpy of LiF.

EXERCISE Which has the larger lattice enthalpy, BaS or KCl?

ANSWER: BaS (due to +2 and −2 charges on ions)

KEY CONCEPT D The Born–Haber cycle

When dealing with ionic solids, we are frequently interested in the value of the standard enthalpy of formation of the solid from the elements. For NaF, this is the enthalpy of reaction for the reaction:

$$Na(s) + \frac{1}{2}F_2(g) \longrightarrow NaF(s) \qquad \Delta H = \text{standard enthalpy of formation}$$

The **Born–Haber cycle** breaks down the overall reaction into five steps, as illustrated below for NaF. We start with Na(*s*) and $F_2(g)$ and use the various enthalpies from the the text.

Step 1. Sublime 1 mol Na	$Na(s) \longrightarrow Na(g)$	$\Delta H^0 = +108.4$ kJ
Step 2. Ionize 1 mol Na	$Na(g) \longrightarrow Na^{+}(g) + e^{-}$	$\Delta H^0 = +494$ kJ
Step 3. Dissociate $\frac{1}{2}$ mol $F_2(g)$	$\frac{1}{2}F_2(g) \longrightarrow F(g)$	$\Delta H^0 = +78.99$ kJ
Step 4. Attach electron to 1 mol F(*g*)	$F(g) + e^{-} \longrightarrow F^{-}(g)$	$\Delta H^0 = -328$ kJ
Step 5. Form solid lattice from 1 mol $Na^{+}(g)$ and 1 mol $F^{-}(g)$	$Na^{+}(g) + F^{-}(g) \longrightarrow NaF(s)$	$\Delta H^0 = -929$ kJ

These reactions all sum to the reaction for formation of solid NaF from the elements, and the sum of the enthalpies is the standard enthalpy of formation of NaF:

$$Na(s) + \frac{1}{2}F_2(g) \longrightarrow NaF(s) \qquad \Delta H^0 = -576 \text{ kJ}$$

The large exothermicity of this reaction tells us that the formation of NaF from its elements is energetically a very favorable process.

EXAMPLE Using the Born–Haber cycle

Would we expect $NaF_2(s)$ to form from its elements? Assume that the lattice enthalpy of $NaF_2(s)$ is about the same as that of $MgF_2(s)$, +2913 kJ. Comment on the result. (Experimentally, $NaF_2(s)$ does not exist.)

SOLUTION If the enthalpy of formation of $NaF_2(s)$ is negative, we would predict that it would form from its elements. On the other hand, a positive value for the enthalpy of formation indicates it would not form. The Born–Haber cycle can be used to calculate the standard energy of formation of this (hypothetical) ionic solid:

$$Na(s) + F_2(g) \longrightarrow NaF_2(s) \qquad \Delta H^0 = ?$$

From the Born–Haber cycle, we get these steps:

Step 1. Sublime metal	$Na(s) \longrightarrow Na(g)$	$\Delta H^0 = +108.4$ kJ
Step 2. Ionize metal to +1 ion	$Na(g) \longrightarrow Na^+(g) + e^-$	$\Delta H^0 = +494$ kJ
Step 3. Ionize +1 metal ion to +2 ion	$Na^+(g) \longrightarrow Na^{2+}(g) + e^-$	$\Delta H^0 = +4562$ kJ
Step 4. Dissociate 1 mol $F_2(g)$	$F_2(g) \longrightarrow 2F(g)$	$\Delta H^0 = +157.98$ kJ
Step 5. Attach electrons to 2 mol $F(g)$	$2F(g) + 2e^- \longrightarrow 2F^-(g)$	$\Delta H^0 = -656$ kJ
Step 6. Form solid lattice from $Na^{2+}(g)$ and $F^-(g)$	$Na^+(g) + 2F^-(g) \longrightarrow NaF_2(s)$	$\Delta H^0 = -2913$ kJ

The sum of the enthalpies of these reactions is +1753 kJ:

$$Na(s) + F_2(g) \longrightarrow NaF_2(s) \qquad \Delta H^0 = +1753$$

The positive enthalpy of formation indicates that $NaF_2(s)$ will not form when $Na(s)$ and $F_2(g)$ react. The main reason for this result is the enormous amount of energy required to form the Na^{2+} ion in step 3.

EXERCISE Use the Born–Haber cycle to calculate the enthalpy of formation of $AgCl(s)$.

ANSWER: −127 kJ/mol

8.2 IONIC BONDS AND THE PERIODIC TABLE

KEY CONCEPT A The formation of ionic compounds

We can consider that the formation of ionic compounds from elements occurs through two hypothetical steps:

$$\text{Elements} \longrightarrow \text{ions in gaseous form}$$
$$\text{Ions in gaseous form} \longrightarrow \text{ionic solid}$$

The first step requires energy whereas the second releases energy; an ionic compound can form when the energy that is released in the second step (the lattice enthalpy) is large enough to compensate for the energy required in the first step. This occurs when small ions of high charge are formed. Some generalizations for formation of such ions are (a) *s*-block elements lose all their valence electrons, forming ions with charge +1 for Group I and +2 for Group II; (b) nonmetals in the *p* block gain electrons to complete a valence shell and form ions with charge equal to group number − 8; (c) breaking into a closed shell to form an ion is not favored because this takes too much energy; (d) elements (metals) in the lower left of the *p* block form ions with charge equal to group number or group number − 2 (which ion forms depends on the compound formed and the conditions of reaction). A general rule is that an ionic compound is (usually) formed when a metal reacts with a nonmetal.

EXAMPLE Predicting the formula of a compound

What is the formula of the compound formed by the combination of sodium and sulfur.

SOLUTION Because sodium is a metal and sulfur a nonmetal, the compound formed is likely to be ionic. Sodium is a Group I element, so it will form an ion with a +1 charge (Na^+); sulfur is a Group VI element, so it will form an ion with charge −2 (S^{2-}). To form a compound with zero charge, two sodium ions must combine with one sulfide ion; the formula of the compound is Na_2S.

EXERCISE Predict the formula of the compound that forms between aluminum and oxygen. (This is an ionic compound.)

ANSWER: Al_2O_3

PITFALL Does combining a metal with a nonmetal always result in an ionic compound?

In almost all cases, the combination of a metal with a nonmetal results in an ionic compound. However, there are exceptions, especially with beryllium and occasionally with aluminum. Such exceptions will be discussed in more detail later in the course.

KEY CONCEPT B Pseudonoble-gas configurations

It is energetically unfavorable for elements near the right of the *d* block to lose enough electrons to form an ion with a noble-gas configuration. Other configurations are observed for ions in this part of the periodic table. For example:

1. Cu, Zn, Ag, and Cd may lose their outer *s* electrons when they form ions, resulting in a **pseudonoble-gas configuration,** a noble-gas core plus a filled *d* subshell.
2. Cu may, in addition, lose one *d* electron from the pseudonoble-gas core, resulting in a different ion from that with the pseudonoble-gas configuration. Thus, Cu displays **variable valence,** that is, the possibility of forming two different ions. (When a choice of ions is possible, which one is formed depends on the conditions used in the chemical reaction to make the compound containing the ion.)

Element	Ion with pseudonoble-gas configuration	Second ion formed by loss of additional electron
Cu $[Ar]3d^{10}4s^1$	Cu^+ $[Ar]3d^{10}$	Cu^{2+} $[Ar]3d^9$

3. Au may lose its outer *s* electrons to form a configuration consisting of a noble-gas core plus a filled *d* subshell and a filled *f* subshell. Au also displays variable valence.

Element	Ion with noble-gas core plus filled *d* and *f* subshells	Second ion formed by loss of two more electrons
Au $[Xe]4f^{14}5d^{10}6s^1$	Au^+ $[Xe]4f^{14}5d^{10}$	Au^{3+} $[Xe]4f^{14}5d^8$

4. Varied configurations are observed in other areas of the *d* block. Loss of the outer *s* electrons is quite common, as are configurations with half-filled subshells. Variable valence occurs for many transition elements. Fe, for instance, forms a +2 ion by loss of its two outer *s* electrons; loss of one additional electron results in a +3 ion with a half filled *d* subshell.

Element	One ion	Second ion
Fe $[Ar]3d^64s^2$	Fe^{2+} $[Ar]3d^6$	Fe^{3+} $[Ar]3d^5$

EXAMPLE Predicting an ion electron configuration

Lutetium (Lu) has the configuration $[Xe]4f^{14}5d^{1}6s^{2}$. Based on the trends observed for *d*-block elements, predict the charge on the only common ion formed by Lu.

SOLUTION We have observed that an electron configuration of a noble-gas core plus filled subshells is stable and common. We, therefore, predict that Lu tends to lose its outer three electrons to form a +3 ion, Lu^{3+}, with the configuration $[Xe]4f^{14}$.

EXERCISE Predict the charge and configuration of the only ion formed by Zn.

ANSWER: Zn^{2+} $[Ar]3d^{10}$

PITFALL Unexpected electron configurations for transition metal ions

Some types of electron configurations have relatively low energy and are found in many ions. The ones we have seen so far are the noble-gas configuration; the pseudonoble-gas configuration; the configuration consisting of a noble-gas core plus filled subshells (including inert pair *s* electrons); and configurations containing half-filled subshells. However, ion formation is a complex process, and there are many ions, especially in the *d*-block and *f*-block elements, that do not have such predictable configurations. Some ions that have "unexpected" configurations are Cr^{3+}, $[Ar]3d^{4}$; Co^{2+}, $[Ar]3d^{7}$; Pt^{4+}, $[Xe]4f^{14}5d^{6}$; and Pd^{2+}, $[Kr]4d^{8}$.

KEY CONCEPT C Electron dot diagrams

Ionic compounds form through the loss of electrons by a metal and the gain of electrons by a nonmetal. A conventional method of keeping track of the electrons in a compound starts with the **electron dot diagrams** of atoms, in which each of the valence electrons of an atom is indicated by a dot next to the chemical symbol of the element. In drawing electron dot diagrams, it is helpful to recall that the number of valence electrons for any main-group element is the same as the group number. Some examples, keyed to their groups, are

I	II	III	IV	V	VI	VII	VIII
Na·	Mg:	·Al:	·Si:	·P:	:S:	:Cl:	:Ar:

If two electrons are in the same orbital, the dots are placed next to each other, as for the two 3*s* electrons in magnesium. Electrons in separate orbitals are placed away from each other; thus, the three *p* electrons in phosphorus, which are each in separate orbitals, are represented by three single dots. Because all the atoms in a group have similar electron configurations, all the elements in a group have similar electron dot diagrams. For instance, in Group II, we have

Element	Be	Mg	Ca	Sr	Ba	Ra
Configuration	$[He]2s^{2}$	$[Ne]3s^{2}$	$[Ar]4s^{2}$	$[Kr]5s^{2}$	$[Xe]6s^{2}$	$[Rn]7s^{2}$
Electron dot diagram	Be:	Mg:	Ca:	Sr:	Ba:	Ra:

The orientation of dots around the symbol is immaterial, as long as the number of dots and their presence in the same or different orbitals are correctly shown. For example, all of the following electron dot diagrams are correct for oxygen:

:O· :O: :O· ·O· ·O: ·O:

EXAMPLE 1 Drawing electron dot diagrams

Draw the electron dot diagram of Te.

SOLUTION It is not necessary to write the entire electron configuration of an element to draw its element dot diagram. Tellurium is a Group VI element, so it has six valence electrons. In addition, because all Group

VI elements have a ns^2np^4 configuration, four of the electrons will be paired in two orbitals (an *s* orbital and one of the three *p* orbitals) and the fifth and sixth electrons are each alone in a *p* orbital. Thus, any one of the electron dot diagrams shown here is correct:

$$:\ddot{\text{Te}}\cdot \quad :\dot{\text{Te}}: \quad :\dot{\text{Te}}\cdot \quad \cdot\ddot{\text{Te}}\cdot \quad \cdot\ddot{\text{Te}}: \quad \cdot\dot{\text{Te}}:$$

EXERCISE Draw the electron dot diagram of Ga.

ANSWER: $\ddot{\text{Ga}}\cdot$

EXAMPLE 2 Drawing electron dot diagrams of ionic compounds

Use electron dot diagrams to describe the formula of magnesium arsenide.

SOLUTION We first note that magnesium is a Group II element and will, therefore, form an ion with charge +2; similarly, arsenic is a Group V element and will form an ion with charge −3 (group number − 8 = 5 − 8 = −3). Each magnesium attains an octet of electrons by losing two electrons and, in doing so, gets a +2 charge. Each nitrogen attains its octet by gaining three electrons and, in doing so, gets a −3 charge. Electron balance is maintained by having three magnesiums each lose two electrons, for a total of six electrons; these electrons are then gained by two nitrogens, which each gain three:

$$3\text{Mg}: + 2\,:\dot{\text{N}}\cdot \longrightarrow 3\text{Mg}^{2+} + 2\,:\ddot{\text{N}}:^{3-} \longrightarrow \text{Mg}_3\text{N}_2$$

EXERCISE Use electron dot diagrams to interpret the formula of sodium sulfide.

ANSWER: $2\text{Na}\cdot + :\ddot{\text{S}}\cdot \longrightarrow 2\text{Na}^+ + :\ddot{\text{S}}:^{2-} \longrightarrow \text{Na}_2\text{S}$

COVALENT BONDS

KEY WORDS Define or explain each of the following terms in a written sentence or two.

complex
covalent bond
double bond
expanded octet
incomplete octet
Lewis acid
Lewis base
Lewis structure
lone pair
octet rule
radical
resonance
triple bond

8.3 THE ELECTRON-PAIR BOND

KEY CONCEPT A The electron-pair bond

When a nonmetal atom bonds to another nonmetal atom, formation of an ionic bond is energetically unfavorable. Instead, the atoms bond to each other by sharing one or more pairs of electrons, to form what is called a **covalent bond.** Generally, the energetically most favorable bonding is achieved when each atom attains a noble-gas configuration of eight valence electrons (two valence electrons for hydrogen). For the purpose of counting valence electrons, shared electrons are counted as being in the valence shell of each atom sharing them and are therefore counted twice. The tendency for atoms to attain a noble-gas configuration is called the **octet rule** (eight valence electrons) or **duet rule** (two valence electrons for hydrogen.)

EXAMPLE Counting valence electrons

Water can be represented by the picture $\text{H}:\ddot{\text{O}}:\text{H}$. (a) For this picture, draw circles around the electrons that have formed covalent bonds by being shared by two atoms. (b) How many covalent bonds does water possess? (c) Show that all of the atoms in water obey the octet rule.

SOLUTION (a) The electrons that are shared are those that are drawn between atoms. These are circled:

H(:)Ö(:)H

(b) There are two covalent bonds because each shared pair of electrons is called a bond. The two electrons that make up a bond are often indicated by a line (—) rather than by a pair of dots (:), so water may also be written as H—Ö—H. (c) To count valence electrons for purposes of checking the octet rule, shared electrons are counted with both atoms that share them. The circles indicate the electrons counted with each atom:

(H—)Ö—H
2 valence electrons for hydrogen

H(—Ö—)H
8 valence electrons for hydrogen

H—Ö(—H)
2 valence electrons for hydrogen

Each hydrogen has achieved the He noble-gas configuration by having two electrons in its valence shell, and oxygen has achieved the neon noble-gas configuration by having eight valence electrons. The octet (duet for hydrogen) rule has been satisfied.

EXERCISE (a) How many valence electrons does the nitrogen atom in ammonia have?

```
H—N̈—H
  |
  H
```

(b) How many covalent bonds are there in this molecule? (c) Do all of the atoms have a noble-gas configuration?

ANSWER: (a) 8; (b) 3; (c) yes

PITFALL What atoms form covalent bonds?

The formation of covalent bonds always occurs when one nonmetal bonds to another nonmetal. However, a variety of metals, including Li, Be, Mg, Al, and Sn, form covalent bonds with some nonmetals, as in $BeCl_2$ and $AlCl_3$.

KEY CONCEPT B Lone pairs in Lewis structures

A pair of electrons that is shared by two atoms is called a covalent bond. Many compounds have pairs of electrons that are not shared but that belong exclusively to one atom. Such a pair is called a **lone pair.** Lone pairs of electrons have profound effects on the chemistry and structure of compounds, so it is essential to recognize them in the Lewis structure of a compound. When you count electrons to see if an atom obeys the octet rule, lone pairs are counted only with the single atom to which they belong.

EXAMPLE Counting lone pairs

How many lone pairs of electrons does water possess?

SOLUTION Electrons that are not shared to form a covalent bond are lone pairs. Water has two lone pairs (four electrons), which are circled:

H—(:)O(:)—H

Two lone pairs

EXERCISE How many lone pairs of electrons are present in ammonia, NH_3? Which atom do they belong to?

ANSWER: One lone pair; nitrogen

KEY CONCEPT C Multiple bonds

At times more than one pair of electrons is shared by two atoms. When two pairs of electrons are shared by two atoms, the resulting bond is called a **double bond.** The Lewis structure for sulfur dioxide has a double bond:

$$:\ddot{O}=\ddot{S}-\ddot{O}:$$

Double bond

A triple bond consists of three pairs of electrons shared by two atoms. Acetylene has a triple bond:

$$H-C\equiv C-H$$

Triple bond

Double and triple bonds are counted in the usual way when you check to see if an atom obeys the octet rule. You should confirm that all of the atoms in SO_2, including the double-bonded atoms, have eight electrons in their valence shells. Similarly, both carbon atoms in acetylene are surrounded by eight electrons. It is useful to remember that C, O, N, and S often form multiple bonds.

EXAMPLE Double bonds and the octet rule

How many double bonds are present in carbon dioxide, $:\ddot{O}=C=\ddot{O}:$? Do all of the atoms in carbon dioxide obey the octet rule?

SOLUTION Carbon dioxide has two double bonds:

$$:\ddot{O}=C=\ddot{O}:$$

Double bond Double bond

All of the atoms in carbon dioxide have eight valence electrons in their outer shell and, therefore, obey the octet rule:

$$:\ddot{O}=C=\ddot{O}: \quad :\ddot{O}=C=\ddot{O}: \quad :\ddot{O}=C=\ddot{O}:$$

Oxygen has eight valence electrons. Carbon has eight valence electrons. Oxygen has eight valence electrons.

EXERCISE Carbon dioxide (CO_2) has three atoms, each of which obeys the octet rule. Does this mean that CO_2 has 24 (3×8) valence electrons?

ANSWER: No, because shared electrons are counted twice. CO_2 has 16 valence electrons.

8.4 LEWIS STRUCTURES OF POLYATOMIC MOLECULES

KEY CONCEPT A Writing Lewis structures I: Drawing atoms in the proper arrangement

We use three steps to draw the correct Lewis structure of a compound or ion:

Step 1. Draw the structure of the compound (or ion) with the correct arrangement of atoms, that is, so that each atom ends up bonded to the correct atom(s).

Step 2. Determine the number of valence electrons in the compound (or ion).
Step 3. Draw in bonds and, possibly, lone pairs following the octet rule where possible. Multiple bonds may be required.

Here we focus on the first (and most difficult) step, drawing the atoms in their correct positions. It is helpful to note that symmetrical arrangements, especially those with a less electronegative atom surrounded by more electronegative atoms, are quite common. For instance:

For CO_2 O C O   is correct, not   O O C

For SO_3

```
   O
   S
O     O
```

is correct, not S O O O *or* O O S O *or*

```
   S
   O
O     O
```

Also, certain elements tend to form a specific number of bonds. For example, H, one bond; C, four bonds; N, three bonds; O, two bonds; F, Cl, Br, and I, one bond. Even with these aids to drawing structures, getting the correct arrangement of atoms can be tricky. Practice and experience will make you proficient in this task.

EXAMPLE Arranging atoms for a Lewis structure

What is the correct arrangement of atoms in the carbonate ion, CO_3^{2-}.

SOLUTION CO_3^{2-} contains three of the same atom (O) and one other atom (C). In addition, oxygen has a higher electronegativity than carbon. A symmetrical structure, with the less electronegative atom at the center and the more electronegative atoms surrounding it, gives the desired arrangement:

```
   O
   C
O     O
```

EXERCISE What is the arrangement of atoms in CCl_4?

ANSWER:

```
    Cl
Cl  C  Cl
    Cl
```

PITFALL The number of bonds formed by an element

The listing of the number of bonds formed by certain elements reflects a tendency, not a certainty. Bonding is a complex matter, and exceptions to these numbers occur.

KEY CONCEPT B Writing Lewis structures II: Counting valence electrons

Here we focus on the second step for drawing Lewis structures, counting the number of valence electrons that belong in the structure. The total number of electrons in a Lewis structure is the sum of the valence electrons of the atoms in the structure, plus any adjustments in that sum resulting from the charge on ionic structures. As an example, let's work out the number of valence electrons in nitrogen trifluoride, NF_3. Nitrogen is a Group V element, so it possesses five valence electrons. Fluorine is in group VII, so each fluorine possesses seven valence electrons, resulting in a total of 21 for the three fluorines. The sum of valence electrons from the atoms is 26:

$$\begin{array}{rl} \text{N: } 1 \times 5 = & 5 \text{ valence electrons} \\ \text{3F: } 3 \times 7 = & \underline{21} \text{ valence electrons} \\ \text{Total: } & 26 \text{ valence electrons} \end{array}$$

Because there is no charge on NF_3, there is no need to go any further. NF_3 possesses 26 valence electrons. We recall that for a charged species, a negative charge results from the gain of electrons and a positive charge from the loss of electrons. The following table presents some examples of how to take into consideration the charge on a species.

Species	**Electrons gained or lost by original atoms**	**Total number of valence electrons**
neutral molecule	0	sum of valence electrons of atoms
−1 ion	1 gained	(sum of valence electrons) + 1
−2 ion	2 gained	(sum of valence electrons) + 2
+1 ion	1 lost	(sum of valence electrons) − 1
+2 ion	2 lost	(sum of valence electrons) − 2

EXAMPLE Counting valence electrons

How many valence electrons are present in the carbonate ion CO_3^{2-}?

SOLUTION We first determine the number of valence electrons that originate with the single carbon atom and three oxygen atoms, using the fact that carbon is a Group IV element and oxygen a Group VI element. Then we take into account the −2 charge on the ion, which occurs because the carbonate ion has gained two electrons above and beyond those supplied by the original atoms:

$$\begin{array}{rl} \text{1 C: } 1 \times 4 = & 4 \text{ valence electrons} \\ \text{3 O: } 3 \times 6 = & 18 \text{ valence electrons} \\ -2 \text{ charge: } & \underline{+2} \text{ valence electrons} \\ \text{Total: } & 24 \text{ valence electrons} \end{array}$$

CO_3^{2-} has 24 valence electrons.

EXERCISE Now many valence electrons are present in the methylammonium ion, $CH_3NH_3^+$?

ANSWER: 14

PITFALL Count valence electrons, not all the electrons

Be certain you count only valence electrons for a Lewis structure. Do not count the total number of electrons. Only for hydrogen is the number of valence electrons the same as the total number of electrons for an atom.

KEY CONCEPT C Drawing Lewis structures III: Drawing in bonds and lone pairs

Here we focus on the last step used in drawing Lewis structures, drawing in bonds, and possibly lone pairs, following the octet rule where possible. Multiple bonds may be required. In this step we first draw a single bond (—) between any two bonded atoms. Atoms surrounding a central atom are all bonded to the central atom, but not to each other. If any atom, after this step, does not satisfy the octet rule, form multiple bonds, if possible. Do not use a multiple bond if it results in too many electrons for an atom; remember, the *usual* numbers of bonds formed are H (one bond), C (four bonds), N (three bonds), O (two bonds), and halogens (one bond). Finally, add the remaining valence electrons as lone pairs. We should note that N, O, and the halogens usually have one or more lone pairs, whereas hydrogen never has any and carbon rarely does. Some of the electron pairs that were originally drawn in as multiple bonds may have to be changed into lone pairs to use the correct number of valence electrons. (Hint: Use a pencil so that you can move electrons easily on paper.) Finally, check the structure by (1) counting the valence electrons to make sure you have the correct number and (2) checking that those atoms that tend to follow the octet rule (C, O, N, F) or duet rule (H) have the correct number of valence electrons.

EXAMPLE Drawing in bonds and lone pairs

We have ascertained the arrangement of atoms and number of valence electrons (24) in the carbonate ion, CO_3^{2-}. Complete the Lewis structure by drawing in bonds and lone pairs.

SOLUTION The arrangement of atoms is

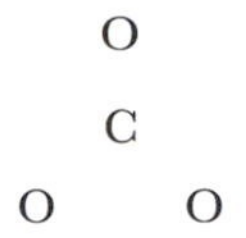

```
    O
    C
  O   O
```

We first draw in single bonds:

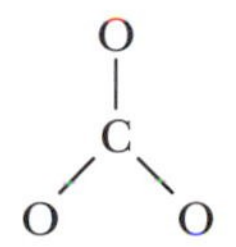

```
    O
    |
    C        (6 electrons used)
   / \
  O   O
```

Now, carbon "always" follows the octet rule, as does oxygen. Making one double bond gives carbon its eight valence shell electrons:

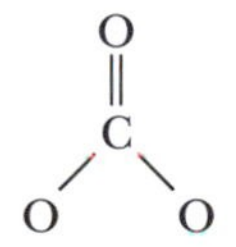

```
    O
    ‖
    C        (8 electrons used)
   / \
  O   O
```

No more double bonds can be made because doing so would put ten electrons around carbon. Therefore, we add the remaining electrons as lone pairs on oxygen. (Carbon already has eight electrons and normally does not have lone pairs.)

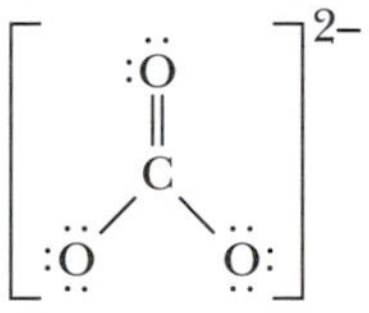

```
 ┌    ..      ┐2-
 │   :O       │
 │    ‖       │
 │    C       │   (24 electrons used)
 │ .. / \ ..  │
 │ :O     O:  │
 └  ..    ..  ┘
```

No further adjustment is needed because all of the electrons have been used and the octet rule has been satisfied. The charge is always indicated with a Lewis structure, generally by drawing large brackets around any ionic structure and writing the charge as a superscript.

EXERCISE Draw the Lewis structure of XeO_3.

```
                 ..
                :O:
          ..     |     ..
ANSWER:  :O — Xe — O:
          ..    ..     ..
```

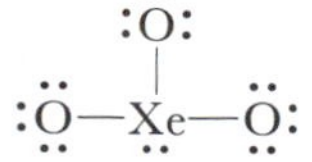

8.5 LEWIS ACIDS AND BASES

KEY CONCEPT The definition of Lewis acids and bases

From the Lewis acid/base point of view, a base is a chemical species with an unshared pair of electrons and an acid is a species that can form a covalent bond by accepting an unshared pair of electrons. An acid-base reaction, therefore, results in formation of a covalent bond:

$$H_3N\!: + Ag^+ \longrightarrow Ag{-}NH_3{}^+$$

Complex (label pointing to $Ag{-}NH_3{}^+$)

Species with unshared pair of electrons (base) (label pointing to $H_3N\!:$)

Species that can accept unshared pair to form a covalent bond (acid) (label pointing to Ag^+)

With this reaction in mind we can make the following definitions: A **Lewis base** is an electron pair donor; a **Lewis acid** is an electron pair acceptor; and a **complex** is the species formed when a Lewis acid reacts with a Lewis base.

EXAMPLE 1 Recognizing a Lewis base from its structure

Which of the following can function as a Lewis base? (a) OH^-, (b) CO, (c) $NH_4{}^+$

SOLUTION Because a Lewis base is characterized by possession of at least one unshared pair of electrons, we must investigate the electron-dot structure of each species to see if unshared pairs of electrons are present. The Lewis structure of each is

(a) $[:\ddot{\underset{\cdot\cdot}{O}}{-}H]^-$ (b) $:C{\equiv}O:$ (c) $\left[\begin{array}{c} H \\ | \\ H{-}N{-}H \\ | \\ H \end{array}\right]^+$

As we can see, both the hydroxide ion and carbon monoxide have unshared pairs of electrons, so both can function as Lewis bases. The ammonium ion does not have any unshared pairs of electrons, so it cannot function as a Lewis base.

EXERCISE Which of the following can function as a Lewis base? (a) H_2O, (b) CN^-, (c) BF_3

ANSWER: H_2O, CN^-

PITFALL Does a Lewis base lose its electrons?

It should be noted that when a Lewis base "donates" its electrons, it does not lose them (as when a Brønsted acid donates a proton). The electrons stay with the Lewis base as they are donated, and a covalent bond between the Lewis base and Lewis acid results.

EXAMPLE 2 Lewis acid-base reactions

Adding acid to a solution of sodium cyanide can be very dangerous because the reaction of H^+ with CN^- results in formation of the very toxic gas HCN. Explain this reaction as a Lewis acid-base reaction.

SOLUTION We recognize that the cyanide ion, with two unshared pairs of electrons can function as a Lewis base; in fact, it is a strong Lewis base. The hydrogen ion has an empty 1*s* orbital that can accept an electron pair from cyanide. Thus, the reaction proceeds by the donation of a pair of electrons by cyanide to hydrogen ion:

$$\underset{\text{Lewis acid}}{H^+} + \underset{\text{Lewis base}}{:C\equiv N:^-} \longrightarrow \underset{\text{Complex}}{H-C\equiv N:}$$

EXERCISE Heating copper(II) sulfate ($CuSO_4$) to a temperature in excess of 600°C results in decomposition to copper(II) oxide and sulfur trioxide. Explain this reaction in terms of Lewis acids and bases.

ANSWER: $\underset{\text{Complex}}{CuSO_4} \longrightarrow \underset{\text{Acid}}{SO_3} + \underset{\text{Base}}{CuO}$

8.6 RESONANCE

KEY CONCEPT Resonance structures

Two problems with some Lewis structures are simultaneously resolved by the concept of resonance. Let's look at the problems first.

1. For some species, more than one Lewis structure with the same energy can be drawn and there is no way to decide which structure is correct. For example, the gas sulfur trioxide (SO_3) has three equal-energy structures, which are called **equivalent structures:**

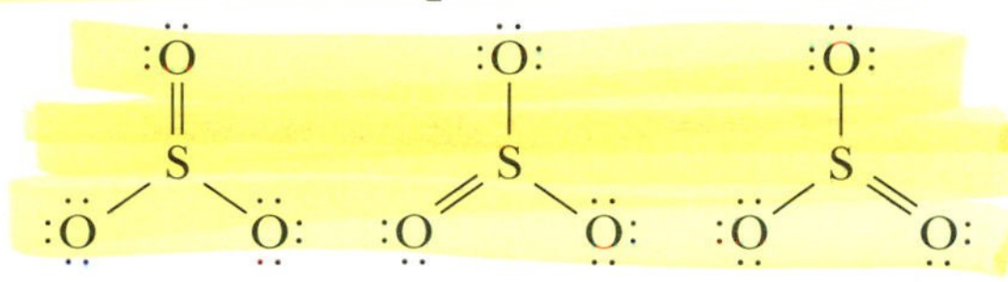

2. When more than one equivalent Lewis structure exists, the actual molecular properties do not agree with those of any one of the structures. For example, from any one of the Lewis structures shown, we would predict that SO_3 has two single bonds and one double bond. However, measurements show that all three bonds are equivalent, with properties intermediate between those of a single bond and those of a double bond.

We solve both of these problems by assuming that the actual SO_3 molecule found in nature is not represented by any one of the equivalent Lewis structures, but by a blending of all three. Because the actual SO_3 is a composite of the three equivalent Lewis structures and no single Lewis structure by itself is correct, we do not have to choose one of the Lewis structures as the correct one; all are used. This solves the first problem. Our second problem is solved because a blending of the three Lewis structures leads to the prediction that each S—O bond is intermediate between an S—O single bond and S=O double bond. This blending of structures is called **resonance,** and the composite structure that results is called a **resonance hybrid** of the individual Lewis structures that contribute to the composite. Thus, SO_3 is a resonance hybrid of the three structures shown.

EXAMPLE Drawing contributing Lewis structures for a resonance hybrid molecule

Draw the two principal contributing Lewis structures for nitromethane, H_3C-NO_2.

SOLUTION Nitromethane has a total of 24 valence electrons. Its arrangement of atoms is suggested by the way it is written in the problem, with carbon bonded to nitrogen and the hydrogens and oxygens as terminal atoms. We arrive at two equivalent contributing Lewis structures:

```
   H  :Ö              H  :Ö:
   |   ||             |   |
H—C—N—Ö:          H—C—N=Ö:
   |                  |
   H                  H
```

In these structures, only the placement of electrons is different, as is required for Lewis structures that make major contributions to a resonance hybrid.

EXERCISE For which of the following species is a single Lewis structure a good representation of the actual species: CO_2, O_3, SO_2, CO_3^{2-}.

ANSWER: CO_2, because it does not undergo resonance

8.7 EXCEPTIONS TO THE LEWIS OCTET RULE

KEY CONCEPT A Expanded octets

Because of the large size of the atom and the presence of low-energy empty *d* orbitals, *p*-block elements in the third, fourth, and fifth periods can accommodate more than eight electrons in their valence shells. Therefore, these elements may violate the octet rule. An atom in a Lewis structure with more than eight valence electrons in its valence shell is said to possess an **expanded octet.** This behavior is especially prominent with P, S, Se, Te, Br, I, Kr, and Xe. How can we tell when an expanded octet is formed? In most cases, the only way to fit all of the valence electrons into the structure will be by using an expanded octet; the expanded octet is forced on us by the structure itself. Occasionally, accurate quantum mechanical calculations indicate that an expanded octet should be used even where one does not seem required, as in the sulfate ion SO_4^{2-}, in which a structure with an expanded octet on sulfur has a lower energy than one in which sulfur obeys the octet rule.

EXAMPLE Drawing a Lewis structure with an expanded octet

Draw the Lewis structure of XeF_4.

SOLUTION This molecule has the expected symmetric structure, with the four more-electronegative fluorines surrounding the (presumably) less electronegative xenon. There are a total of 36 valence electrons in the structure, eight from xenon (Group VIII) and seven each from four fluorines, for a total of 28 from the fluorines. Any structure in which all of the atoms obey the octet rule does not have the correct number of valence electrons; for instance,

:F:
|
:F—Xe—F: (incorrect, only 32 valence electrons)
|
:F:

To use 36 valence electrons, we put two lone pairs on xenon, giving xenon an expanded octet:

:F:
|
:F—:Xe:—F:
|
:F:

We do not form double bonds with these two additional pairs of electrons, because fluorine does not violate the octet rule or form double bonds.

EXERCISE Draw the Lewis structure of SF_6.

ANSWER:

:F:
:F | F:
   S
:F | F:
:F:

KEY CONCEPT B Incomplete octets

Boron and beryllium frequently have less than eight electrons in their valence shells in their Lewis structures. This situation is generally "forced" onto the structure by the lack of enough valence electrons to fill the shells. Any atom that has less than a noble-gas configuration of electrons in its valence shell in a Lewis structure is said to have an **incomplete octet.**

EXAMPLE Drawing a Lewis structure containing an incomplete octet

Draw the Lewis structure of $BeCl_2$ (which exists only in the gas phase at high temperature).

SOLUTION Beryllium (Group II) has two valence electrons and each chlorine (Group VII) has seven valence electrons, for a total of 16 electrons in the Lewis structure. The expected arrangement of atoms has the less electronegative Be surrounded by the more electronegative chlorines:

Cl Be Cl

Chlorine does not form double bonds and usually has an octet of eight valence electrons in a Lewis structure, so we arrive at the following structure for $BeCl_2$:

$$:\ddot{\underset{..}{Cl}}-Be-\ddot{\underset{..}{Cl}}:$$

Note that chlorine obeys the octet rule in this structure but that beryllium has only four electrons in its valence shell, so beryllium has an incomplete octet.

EXERCISE Draw the Lewis structure of BCl_3. Which atom in this structure has an incomplete octet?

ANSWER: $:\ddot{\underset{..}{Cl}}-\underset{}{\overset{:\ddot{Cl}:}{\overset{|}{B}}}-\ddot{\underset{..}{Cl}}:$; boron has an incomplete octet.

KEY CONCEPT C Odd-electron species

An interesting and important class of chemical entities have an odd number of valence electrons. Any such entity must have an unpaired electron and must therefore have at least one atom that violates the octet rule; an odd number of electrons cannot be all paired. When drawing the Lewis structure of such a compound, there is usually a choice of where to place the unpaired electron. In general, it belongs on the less electronegative of the possible atoms. So, for ClO_2,

$$:\ddot{\underset{..}{O}}-\dot{\underset{..}{Cl}}-\ddot{\underset{..}{O}}: \quad \text{is preferred over} \quad \cdot\ddot{\underset{..}{O}}-\ddot{\underset{..}{Cl}}-\ddot{\underset{..}{O}}:$$

Odd-electron species may arise when a bond is broken in a reaction in such a way that the electrons originally in the bond become unpaired. When this happens, pieces of molecules with unpaired electrons are formed; this type of highly reactive molecular fragment is called a **radical.** The methyl radical $CH_3\cdot$ and hydroxyl radical $\cdot OH$ are two important and commonly encountered radicals. (When writing the symbol of a radical, the unpaired electron is explicitly drawn as a dot.)

EXAMPLE Drawing the Lewis structure of an odd-electron species

Draw the Lewis structure of NO_2.

SOLUTION One nitrogen and two oxygens possess 17 valence electrons. The less electronegative nitrogen is in the center of the molecule, and the two oxygens surround it. We place the odd electron on the less electronegative nitrogen to get

$$:\ddot{\underset{..}{O}}-\dot{N}=\ddot{O}:$$

EXERCISE Draw the Lewis structure of the $\cdot CF_3$ radical.

ANSWER: 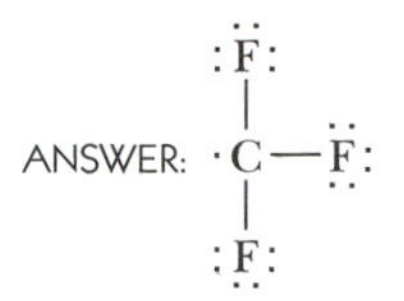

THE SHAPES OF MOLECULES

KEY WORDS Define or explain each of the following terms in a written sentence or two.

angular
hybrid orbital
linear
octahedral
polar molecule
tetrahedral
trigonal bipyramidal
trigonal planar
trigonal pyramidal
VSEPR theory

8.8 ELECTRON PAIR REPULSIONS

KEY CONCEPT A The arrangement of electron pairs in VSEPR theory

Probably the most important concept in **valence-shell electron-pair repulsion (VSEPR) theory** is that molecules take on the shapes they do because of electron-electron repulsions. We start out by concerning ourselves with how these repulsions cause electron pairs to arrange themselves around a central atom. It makes no difference at this point whether an electron pair is a lone pair or a bonding pair. We just imagine that a certain number of electron pairs are attached to a central atom and that they arrange themselves so as to be as far apart as possible to minimize electron–electron repulsions. The arrangements that result are given in Table 8.2 of the text and are repeated here.

Number of electron pairs	Arrangement around central atom
2	linear
3	trigonal planar
4	tetrahedral
5	trigonal bipyramidal
6	octahedral
7	pentagonal bipyramidal

It is imperative that you understand the three-dimensional arrangements to which each of these correspond. A good way of doing this is to build models of the shapes, using toothpicks to represent electron pairs and a styrofoam ball (or gumdrop) to represent the central atom. In your models, you should note that these shapes do result in electron pairs being as far apart as possible.

EXAMPLE Understanding molecular shapes

Use a three-dimensional model of the trigonal planar arrangement to determine the angle between electron pairs in this arrangement.

SOLUTION In the trigonal planar arrrangement of electron pairs, the three electron pairs are all in the same plane. A representation of this is shown in the following figure, with M as the central atom and lines as the electron pairs.

The angle between electron pairs is 120°.

EXERCISE In the trigonal bipyramidal arrangement, three of the electron pairs are in a plane and 120° apart; lets label these pairs a, b, and c. If we label the fourth and fifth pairs d and e, what is the angle between pair d and pair a? What is the angle between pair d and pair e? (Hint: Use a model.)

ANSWER: d and a, 90°; d and e, 180°

PITFALL Octa means "eight," doesn't it?

Don't be mislead by the term *octahedral,* which refers to an eight-sided solid. A molecule with an octahedral shape has only six atoms on its periphery, not eight. Similarly, an octahedral arrangement of electron pairs refers to six electron pairs around the central atom. The *eight* in octahedral refers to the number of faces in the regular solid that has six vertices.

KEY CONCEPT B Lewis structures with single bonds

The shape of molecules and molecular ions that have Lewis structures with single bonds follows directly from the arrangement of electron pairs around the atom. The electron pairs arrange themselves as we have already noted, but now we recognize that the electron pairs can be either bonding pairs or lone pairs. In structures with no lone pairs, all the electron pairs are bonding pairs; and because each electron pair has an atom attached to it, the shape of the molecule is the same as the arrangement of electron pairs around the central atom. To determine the shape, we need only draw the Lewis structure of the molecule or molecular ion, count the electron pairs around the central atom, and then deduce the shape from Table 8.2 of the text. The following table gives some examples.

Molecule	Number of electron pairs	Arrangement of electrons	Number of lone pairs	Shape of molecule
$BeCl_2$	2	linear	0	linear
BF_3	3	trigonal planar	0	trigonal planar
CH_4	4	tetrahedral	0	tetrahedral
PCl_5	5	trigonal bipyramidal	0	trigonal bipyramidal
SF_6	6	octahedral	0	octahedral

EXAMPLE Using VSEPR theory

Predict the shape of silane, SiH_4.

SOLUTION The first step in using VSEPR theory to predict the structure of a compound is to draw the Lewis structure of the compound. For SiH_4 this is

```
      H
      |
  H—Si—H
      |
      H
```

The number of electron pairs around the central atom is four. That is, each Si—H bond counts as one pair of electrons. We use the preceeding table to determine the arrangement of the electron pairs and the shape of the molecule. Four electron pairs adopt a tetrahedral arrangement, so we conclude that SiH_4 is tetrahedral.

EXERCISE Predict the shape of IF_6^+.

ANSWER: Octahedral

KEY CONCEPT C VSEPR and the effect of lone pairs

Lone pairs are treated like any other pair of electrons when counting electron pairs around the central atom. However, because no atom is bonded to a lone pair, the shape of a molecule containing a lone pair is not the same as the arrangement of electron pairs around the central atom. This is because the shape of a molecule is based on the location of the atoms in the molecule. Let's consider the electron-deficient compound (discussed later in the chapter) GeF_2 as an example. Its Lewis structure is

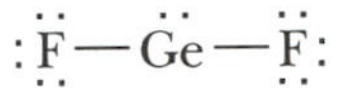

$$:\ddot{\underset{..}{F}}-\ddot{Ge}-\ddot{\underset{..}{F}}:$$

There are three electron pairs around Ge, the two bonding pairs and the lone pair. These take on a trigonal planar shape around the Ge, as shown in the following figure.

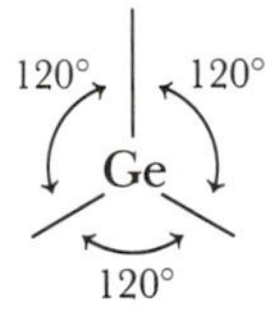

120° 120° Ge 120°

But only two of the three pairs have atoms attached. So the three atoms in the molecule form a bent structure, not a trigonal planar structure. At this point, we would predict that the bond angle in GeF_2 is 120°, the normal trigonal planar bond angle. However, lone pairs have a predictable effect on the structure. A lone pair repels bonding pairs very strongly and pushes them away; in GeF_2 this repulsion results in a bond angle smaller than 120°.

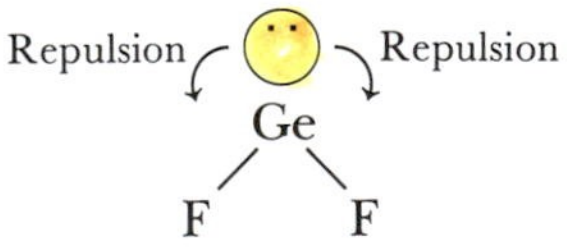

Repulsion Repulsion Ge F F

EXAMPLE Using VSEPR theory

Predict the shape of $AsCl_3$.

SOLUTION We first draw the Lewis structure of the compound.

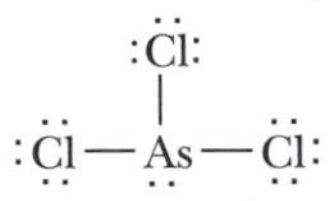

$$:\ddot{\underset{..}{Cl}}-\underset{..}{As}(-\ddot{Cl}:)-\ddot{\underset{..}{Cl}}:$$

The four electrons take on a tetrahedral arrangement around the central As atom. Only three of the electron pairs have atoms attached, resulting in a trigonal pyramidal structure. (A trigonal pyramid is a pyramid with a three-sided base.) The lone pair–bonding pair repulsions cause the bond angles to be smaller than the normal 109° tetrahedral bond angle.

EXERCISE Predict the shape of SF_2.

ANSWER: Bent, with a bond angle less than 109°

KEY CONCEPT D VSEPR and lone pairs on larger molecules

When we determine the shape of molecules with two, three, or four electron pairs, we do not have to decide where to spatially locate unshared pairs and bonding pairs; each position around the central atom is equally likely to have an unshared pair. However, when five electron pairs are present, and the trigonal bipyramidal arrangement of electrons is obtained, we do have a choice to make. For this arrangement of electrons, lone pairs are found at equatorial positions and not at axial positions. (The equatorial positions are the three positions that point at the vertices of an equilateral triangle and are, therefore, 120° apart. The axial positions are the two positions that protrude from the top and bottom of the plane of the equilateral triangle and are 180° apart.) For the octahedral arrangement, we again have no decision to make when one lone pair is present; all six positions are equivalent. However, two lone pairs will always be positioned 180° apart from each other, that is, across from each other with the central atom in between. The names given to the various shapes that arise when lone pairs are present on both small and large structures are summarized in the following table for the most common shapes found.

Number of electron pairs	Arrangement of electron pairs	Number of lone pairs	Shape of molecule
3	trigonal planar	1	angular
		2	linear
4	tetrahedral	1	trigonal pyramidal
		2	angular
5	trigonal bipyramidal	1	seesaw
		2	T-shaped
		3	linear
6	octahedral	1	square pyramidal
		2	square planar

EXAMPLE Using VSEPR theory

Predict the shape of XeF_4.

SOLUTION We start with the Lewis structure:

```
        :F:
         |
  :F — Xe — F:
         |
        :F:
```

The molecule has six electron pairs around xenon, so there is an octahedral arrangement of electrons. The two lone pairs are positioned across from each other, 180° apart. The four fluorine atoms are positioned at the remaining four positions, at the corners of a square. Thus, the molecule has a square planar shape.

EXERCISE Predict the shape of IF_2^-.

ANSWER: Linear

8.9 MOLECULES WITH MULTIPLE BONDS

KEY CONCEPT In VSEPR theory, multiple bonds count as one pair of electrons

Although it may seem paradoxical, double and triple bonds are counted as only *one* electron pair for the purposes of VSEPR theory. Because both bonds in a double bond follow the molecular framework, if we count a multiple bond as a single electron pair for the purposes of VSEPR theory, we do predict the correct shapes of molecules. This is illustrated in the following example.

EXAMPLE Using VSEPR theory

Predict the shape of SO_2.

SOLUTION The Lewis structure of SO_2 is

$$:\ddot{\underset{..}{O}}=\ddot{S}-\ddot{\underset{..}{O}}:$$

If we count the double bond as one electron pair (as we must for VSEPR theory), there are three electron pairs around sulfur. That is, one lone pair + one single-bond pair + one pair for the double bond, which equals three pairs of electrons. Thus, the electrons take on a trigonal planar arrangement around the central sulfur atom. Only two of the electron pairs have atoms attached, so SO_2 is a bent molecule. The presence of the lone pair results in a bond angle smaller than the usual trigonal planar angle of 120°.

EXERCISE Predict the shape of HCN.

ANSWER: Linear

DESCRIPTIVE CHEMISTRY TO REMEMBER

- **Ionic compounds** typically form between a metal and a nonmetal.
- **Variable valence** is frequently observed in the *d* block.
- **Interhalogen** compounds are those in which two (or more) different Group VII elements combine. Examples are ClF and ClF_4.
- Calcium oxide (CaO) is called *quick lime;* when water is added to calcium oxide, then calcium hydroxide [$Ca(OH)_2$], called *slaked lime,* is formed.
- The oxide ion (O^{2-}) is a strong **Lewis base.**
- Solid **phosphorus pentoxide** (PCl_5) is actually an ionic compound consisting of $[PCl_4]^+$ cations and $[PCl_6]^-$ anions. In the gas phase, however, it exists as PCl_5 molecules.
- BF_3 and BCl_3 are common **Lewis acids.**
- **Benzene** (C_6H_6) is much less reactive than expected owing to resonance.
- **Aluminum chloride** is a gas of Al_2Cl_6 between 180°C and 200°C; above 200°C, it becomes a gas of $AlCl_3$.
- **Free radicals** such as methyl radical ($\cdot CH_3$) and hydroxyl radical ($HO\cdot$) are highly reactive.

CHEMICAL EQUATIONS TO KNOW

- Gold(I) readily disproportionates in water.

$$3Au^+(aq) \longrightarrow 2Au(s) + Au^{3+}(aq)$$

- In a Lewis acid–base reaction, hydrogen ion (Lewis acid) and ammonia (Lewis base) react to form ammonium ion (complex).

$$H^+(aq) + NH_3(aq) \longrightarrow NH_4^+(aq)$$

- In a Lewis acid–base reaction, aluminum ion (Lewis acid) and water (Lewis base) react to form a water-aluminum complex ion.

$$Al^{3+}(aq) + 6H_2O(l) \longrightarrow [Al(H_2O)_6]^{3+}(aq)$$

- The decomposition of carbonate ion (a complex) into oxide ion (Lewis base) and carbon dioxide (Lewis acid) can be interpreted using Lewis acid–base theory.

$$CO_3^{2-} \longrightarrow O^{2-} + CO_2$$

- In a Lewis acid–base reaction, hydrogen ion (Lewis acid) and water (Lewis base) react to form hydronium ion (complex).

$$H^+(aq) + H_2O(l) \longrightarrow H_3O^+(aq)$$

- In a Lewis acid–base reaction, oxide ion (Lewis acid) and water (Lewis base) react to form hydroxide ion (complex).

$$O^{2-}(aq) + H_2O(aq) \longrightarrow 2OH^-(aq)$$

- In a Lewis acid–base reaction, hydrogen ion (Lewis acid) from ammonia and water (Lewis base) react to form ammonium ion (complex) and hydroxide ion.

$$H_2O(l) + NH_3(aq) \longrightarrow OH^-(aq) + NH_4^+(aq)$$

- In a limited supply of chlorine, phosphorus reacts to form phosphorus trichloride.

$$P_4(g) + 6Cl_2(g) \longrightarrow 4PCl_3(l)$$

- In a plentiful supply of chlorine, phosphorus reacts to form phosphorus pentachloride.

$$P_4(g) + 10Cl_2(g) \longrightarrow 4PCl_5(s)$$

- Boron trifluoride is produced in large amounts by heating boron oxide and calcium fluoride together.

$$B_2O_3(s) + 3CaF_2(s) \longrightarrow 2BF_3(s) + 3CaO(s)$$

- In a Lewis acid–base reaction, boron trifluoride (Lewis acid) and fluoride ion (Lewis base) react to form tetrafluoroborate ion (complex).

$$F^- + BF_3 \longrightarrow BF_4^-$$

- In a Lewis acid–base reaction, boron trifluoride (Lewis acid) and ammonia (Lewis base) react to form a complex of the two molecules.

$$NH_3 + BF_3 \longrightarrow NH_3BF_3$$

- Nitric oxide is produced by the reaction of ammonia with oxygen. It is also produced in the hot gases of automobile exhausts and jet engines.

$$4NH_3(g) + 5O_2(g) \xrightarrow{1000°C,\ Pt} 6H_2O(g) + 4NO(g)$$
$$N_2(g) + O_2(g) \longrightarrow 2NO(g)$$

SELF-TEST EXERCISES

Ionic bonds

1. What is the potential energy of an Ag^+ ion and F^- ion separated by 150 pm?
(a) +927 kJ/mol (b) −209 kJ/mol (c) +209 kJ/mol
(d) −618 kJ/mol (e) −927 kJ/mol

2. For which equation is the enthalpy of reaction equal to the lattice enthalpy of NaCl?
(a) $NaCl(s) \longrightarrow Na^+(g) + Cl^-(g)$ (b) $2NaCl(s) \longrightarrow 2Na(s) + Cl_2(g)$
(c) $NaCl(g) \longrightarrow Na^+(g) + Cl^-(g)$ (d) $NaCl(g) \longrightarrow Na(s) + Cl(g)$
(e) $NaCl(s) \longrightarrow Na^+(aq) + Cl^-(aq)$

3. What factor is most important in determining the type of bond formed between two atoms?
(a) the size of the atoms (b) the energy changes that accompany bond formation
(c) the physical state of the elements (d) the Lewis dot structure of the atoms
(e) the charge on the nuclei of the atoms

4. Without consulting any references, predict which of the following has the highest lattice enthalpy.
(a) KF (b) NaCl (c) MgO (d) BaTe (e) MgS

5. Use the Born–Haber cycle to calculate the lattice enthalpy of NaBr. Use 108 kJ for the formation of 1 mol of Na(*g*) from Na(*s*) and 112 kJ for the formation of 1 mol of Br(*g*) from $Br_2(l)$.
(a) 678 kJ (b) 1400 kJ (c) 28 kJ (d) 511 kJ (e) 750 kJ

6. Use the Born–Haber cycle to calculate the lattice enthalpy of $MgCl_2$. Use 146 kJ for the formation of 1 mol of Mg(*g*) from Mg(*s*), 242 kJ for the formation of 2 mol of Cl(*g*) from $Cl_2(g)$, and −641 kJ for the enthalpy of formation of $MgCl_2$.
(a) 2517 kJ (b) 1067 kJ (c) 811 kJ (d) 2168 kJ (e) 1621 kJ

7. Use the Born–Haber cycle to calculate the enthalpy of formation of BaO. Use 143 kJ/mol for the enthalpy of sublimation of Ba(*s*).
(a) −211 kJ/mol (b) −1303 kJ/mol (c) −1162 kJ/mol
(d) −553 kJ/mol (e) −1425 kJ/mol

8. Which of the following is the electron dot structure of nitrogen?
(a) $:\overset{\cdot\cdot}{N}\cdot$ (b) $:\underset{\cdot\cdot}{\overset{\cdot\cdot}{N}}\cdot$ (c) $\vdots N:$ (d) $:\underset{\cdot}{\overset{\cdot}{N}}\cdot$ (e) $\vdots\overset{\cdot}{N}\vdots$

9. Which of the following is the electron dot structure of the chloride ion, Cl^-?
(a) $:\underset{\cdot\cdot}{\overset{\cdot\cdot}{Cl}}\cdot^-$ (b) $:\underset{\cdot}{\overset{\cdot}{Cl}}\cdot^-$ (c) $:\underset{\cdot\cdot}{\overset{\cdot\cdot}{Cl}}:^-$ (d) $\cdot\underset{\cdot}{\overset{\cdot}{Cl}}\cdot^-$ (e) $:Cl^-$

10. Which of the following ions has a noble-gas configuration?
(a) Fe^{2+} (b) Pb^{2+} (c) Tl^+ (d) Sn^{2+} (e) Ca^{2+}

11. All but one of the ions shown have a pseudonoble-gas configuration. Which one does not?
(a) In^{3+} (b) Pb^{2+} (c) Cd^{2+} (d) Tl^{3+} (e) Ag^+

12. What is the formula of the compound formed by strontium and nitrogen?
(a) SrN (b) Sr_2N_3 (c) SrN_2 (d) Sr_3N (e) Sr_3N_2

Covalent bonds

13. Which of the following is the correct Lewis structure for Br_2?
(a) $:\underset{\cdot\cdot}{\overset{\cdot\cdot}{Br}}{-}Br:$ (b) $:Br{-}Br:$ (c) $:\overset{\cdot\cdot}{Br}{=}\overset{\cdot\cdot}{Br}:$ (d) $:\underset{\cdot\cdot}{\overset{\cdot\cdot}{Br}}{-}\underset{\cdot\cdot}{\overset{\cdot\cdot}{Br}}:$ (e) $:Br{\equiv}Br:$

14. How many lone pairs of electrons are present in hydrazine, N_2H_4?

H   H
 N—N
H ·· ·· H

(a) 0 (b) 1 (c) 2 (d) 7 (e) 8

15. How many electrons are there in a double bond?
(a) 0 (b) 6 (c) 1 (d) 2 (e) 4

16. What is the correct arrangement of atoms in the Lewis structure of phosphorus trichloride, PCl_3?
(a) Cl P Cl Cl
(b)
P
Cl
Cl Cl
(c)
Cl
P
Cl Cl
(d) P Cl Cl Cl

17. What is the correct arrangement of atoms in the Lewis structure of NO_2Cl?
(a)
N
O
Cl O
(b)
Cl
N
O O
(c) N O O Cl
(d) Cl N O O

18. How many valence electrons are there in the Lewis structure of difluoromethane, CH_2F_2?
(a) 20 (b) 32 (c) 8 (d) 26 (e) 18

19. How many valence electrons are there in the Lewis structure of the perchlorate ion, ClO_4^-?
(a) 32 (b) 31 (c) 30 (d) 17 (e) 16

20. Which of the following is the correct Lewis structure for the sulfite ion, SO_3^{2-}?

```
(a) [    :Ö:     ]2-   (b) [    :S̈:     ]2-   (c) [    :Ö:     ]2-   (d) [    :Ö:     ]2-
    [     ||     ]         [     ||     ]         [     |      ]         [     |      ]
    [     S      ]         [     O      ]         [     S      ]         [     S̈      ]
    [   //  \\   ]         [   /   \    ]         [   /   \    ]         [   /   \\   ]
    [ :O̤     Ö:  ]         [ :Ö̤     Ö̤:  ]         [ :Ö̤     Ö̤:  ]         [ :Ö̤     Ö:  ]
```

21. Which of the following is the correct Lewis structure for ammonia, NH_3?

```
(a) H—N=H—H      (b)   :N̈:        (c)    H         (d)    H
                        |                ||                |
                        H                N                 N
                       / \              / \               /··\
                      H   H            H   H             H    H
```

22. Each of the elements shown except one is capable of achieving an expanded octet. Which one is not?
(a) I (b) S (c) O (d) Xe (e) P

23. Which of the following compounds has an atom with an expanded octet?
(a) SF_5 (b) H_2O (c) PCl_3 (d) HBr (e) NH_3

24. Which element displays variable valence?
(a) C (b) O (c) S (d) F (e) Al

25. Which element often has an incomplete octet in its compounds?
(a) F (b) O (c) B (d) Xe (e) P

26. A complex is formed when a molecule or ion
(a) decomposes. (b) coordinates to another species. (c) reacts.
(d) loses energy. (e) bonds to a gas.

27. Which of the following is an odd-electron molecule?
(a) N_2O (b) N_2O_4 (c) NH_3 (d) NO_3 (e) N_2O_5

28. Which of the following is a radical?
(a) NO_2 (b) CH_4 (c) NO_3^- (d) OH^- (e) CH_3

29. Which of the following can act as a Lewis base?
(a) Al^{3+} (b) CH_4 (c) H^+ (d) H_2 (e) OH^-

30. Which of the following can act as a Lewis acid?
(a) H^+ (b) CF_4 (c) H_2O (d) Cl_2 (e) Cl^-

31. A species with a ______ charge and ______ radius is likely to be a strong Lewis base.
(a) high, large (b) low, small (c) low, large (d) high, small

32. All but one of the following species exhibit the phenomenon of resonance. Which one does not?
(a) NO_3^- (b) SO_2 (c) CO_2 (d) SO_4^{2-} (e) C_6H_6

33. All but one of the statements regarding resonance is correct. Which one is incorrect?
(a) A species that undergoes resonance changes rapidly from one form to another.
(b) Resonance lowers the energy of a species.
(c) One Lewis structure will not properly describe a molecule that undergoes resonance.
(d) Resonance involves changing the positions of double bonds in Lewis structures.
(e) Resonance averages out certain bond lengths in a Lewis structure.

34. Which of the following structures represent resonance structures of a single species?

$$:\ddot{X}=\ddot{A}-\underset{\cdot\cdot}{\ddot{X}}: \qquad :\underset{\cdot\cdot}{\ddot{X}}-\ddot{A}=\ddot{X}: \qquad :\underset{\cdot\cdot}{\ddot{X}}-\ddot{X}=\ddot{A}:$$

I II III

(a) I and III (b) I and II (c) II and III (d) I, II, and III (e) none of these

35. Which molecule has a central atom with an expanded octet?
(a) CO_2 (b) XeF_4 (c) CF_4 (d) SO_3 (e) NCl_3

36. Which molecule has a central atom with an incomplete octet?
(a) CH_4 (b) BeF_2 (c) N_2 (d) SO_2 (e) PBr_3

37. Which of the following species has a single unpaired electron on nitrogen?
(a) NO_3^- (b) N_2F_4 (c) NO_2^- (d) N_2O (e) NO_2

The shapes of molecules

38. Within the framework of VSEPR theory, how many electron pairs surround the central atom in carbon tetrabromide, CBr_4?
(a) 0 (b) 4 (c) 2 (d) 3 (e) 32

39. Within the framework of VSEPR theory, how many electron pairs surround the central atom in sulfur tetrafluoride, SF_4?
(a) 5 (b) 4 (c) 34 (d) 6 (e) 3

40. Within the framework of VSEPR theory, how many electron pairs surround the central atom in carbonate ion, CO_3^{2-}?
(a) 4 (b) 6 (c) 5 (d) 3 (e) 4

41. Within the framework of VSEPR theory, how many electron pairs surround the central atom in SO_2?
(a) 2 (b) 5 (c) 4 (d) 6 (e) 3

42. What is the arrangement of four electron pairs (in the VSEPR sense) around a central atom?
(a) linear (b) octahedral (c) tetrahedral (d) trigonal planar (e) square planar

43. What is the arrangement of six electron pairs (in the VSEPR sense) around a central atom?
(a) octahedral (b) hexagonal (c) trigonal planar
(d) trigonal bipyramidal (e) tetrahedral

44. What is the effect of lone pairs of electrons on bond angles?
(a) make angles smaller (b) make angles larger (c) have no effect

45. What is the shape of an XY_4E_2 molecule?
(a) tetrahedral (b) octahedral (c) linear (d) trigonal planar (e) square planar

46. What is the shape of an XY_3E_2 molecule?
(a) trigonal bipyramidal (b) trigonal planar (c) octahedral
(d) trigonal pyramidal (e) T-shaped

47. What is the shape of $GaCl_3$?
(a) trigonal bipyramidal (b) tetrahedral (c) trigonal planar
(d) linear (e) T-shaped

48. What is the shape of $TeCl_4$?
(a) square planar (b) seesaw (c) trigonal pyramidal
(d) tetrahedral (e) square pyramidal

Descriptive chemistry

49. Which of the following is the symbol for a methyl group?
(a) CO_3^{2-} (b) Mn (c) CO_2H (d) CH_3 (e) OH

50. Which ion readily disproportionates in aqueous solution?
(a) Na^+ (b) O^{2-} (c) Mg^{2+} (d) F^- (e) Au^+

51. In a limited supply of Cl_2, phosphorus reacts to form
(a) PCl_3 (b) PCl_5 (c) PCl (d) P_2Cl (e) P_4Cl_2

52 In the industrial production of BF_3, calcium fluoride is one of the reactants. What is the other?
(a) B_2O_3 (b) B (c) B_2H_6

THE STRUCTURES OF MOLECULES

In this chapter we start by exploring the quantitative ideas that are used to describe chemical bonds; bond length, bond strength, and the charge distribution that can occur in different bonding situations are considered. We end by exploring the two preeminent theories of chemical bonding, the valence bond theory and molecular orbital theory.

BOND PARAMETERS

KEY WORDS Define or explain each of the following terms in a written sentence or two.

average bond enthalpy   bond enthalpy   bond length   covalent radius

9.1 BOND STRENGTH

KEY CONCEPT A The factors affecting bond strength

There are four major factors that affect the strength of a bond between two atoms:

1. Multiple bonds. For any two atoms A and B, triple bonds are typically stronger than double bonds and double bonds stronger than single bonds. (A and B may be the same.)

$$\text{Strongest} \quad A{\equiv}B > A{=}B > A{-}B \quad \text{weakest}$$

$$C{\equiv}C > C{=}C > C{-}C$$

$$C{\equiv}O > C{=}O > C{-}O$$

2. Lone pairs. Lone pairs repel each other, so the presence of lone pairs on the bonding atoms makes bonds weaker than they would otherwise be, especially if atoms are small.
3. Size of atoms. All other things being equal, bonds between large atoms are weaker than similar bonds between small atoms.
4. Resonance. A bond that is part of a resonance structure has a different energy than would be predicted from a single Lewis structure. (A⋯B represents a resonance effected bond, as in the text).

$$A \overset{\cdots}{-} B \quad \text{weaker than} \quad A{=}B$$

$$A \overset{\cdots}{-} B \quad \text{stronger than} \quad A{-}B$$

Factors 1–3 are rules of thumb and cannot be relied on in all circumstances; there are many subtleties that affect bond strength.

EXAMPLE Estimating relative bond strengths

Nitrogen and fluorine atoms are similar in size, but the N_2 molecule is far more stable than the F_2 molecule. Explain why.

SOLUTION We first draw the Lewis structures of each molecule in order to see what bonds and unshared pairs are present:

$$:\ddot{\underset{..}{F}}{-}\ddot{\underset{..}{F}}: \qquad :N{\equiv}N:$$

Each atom in F_2 has three unshared pairs; with small atoms such as fluorine, the unshared pairs on one atom repel those on the other atom quite strongly and tend to drive the atoms apart. In addition, F_2 has a single bond. Both factors result in a relatively weak bond in F_2, and the molecule is, therefore, relatively unstable. On the other hand, each nitrogen atom in N_2 has only one unshared pair, and the N_2 bond is a triple bond. Both these factors make the N_2 very strong, and the N_2 is, therefore, very stable. In fact, N_2 is one of the most stable diatomic species known.

EXERCISE Based on average bond enthalpies of 210 kJ/mol for an N—O bond and 630 kJ/mol for an N=O bond, and the Lewis structure of NO_3^-, we would predict that 1050 kJ (2×210 kJ $+ 1 \times 630$ kJ) are required to completely break all of the bonds in NO_3^-. Is this prediction correct?

ANSWER: No; NO_3^- undergoes resonance and, therefore, is more stable than predicted by any one Lewis structure. It will take more than 1050 kJ to break all the bonds in NO_3^-.

KEY CONCEPT B Using bond enthalpies to predict enthalpies of reaction

As you might imagine, it takes energy to break a chemical bond. The enthalpy change that accompanies the breaking of a covalent bond in such a way that the two electrons in the bond become unpaired is called the **bond enthalpy.** Average bond enthalpies (Table 9.2 of the text) and the measured bond enthalpies of diatomic molecules (Table 9.1 of the text) are valuable for estimating the enthalpy of a reaction. For such a calculation, we must remember that breaking a bond requires energy (positive contribution to ΔH) and that forming a bond releases energy (negative contribution to ΔH). The reaction enthalpy is estimated by adding the bond enthalpies of all of the bonds broken in the reaction and subtracting from this result the sum of the bond enthalpies for all of the bonds formed. The result is only an estimate, not a precise calculation of the enthalpy reaction; it works best when the reactants and products are in the gas phase.

EXAMPLE Estimating the enthalpy of reaction using bond enthalpies

Use bond enthalpies to estimate the enthalpy of reaction for the following reaction. The C—Cl bond enthalpy is 330 kJ/mol.

$$CH_4(g) + Cl_2(g) \longrightarrow CH_3Cl(g) + HCl(g)$$

SOLUTION We must determine which bonds are broken and which are formed in order to use bond enthalpies to estimate the enthalpy of reaction. Writing the Lewis structures of the reactants and products makes this task a straightforward matter:

```
     H                               H
     |           ..   ..             |     ..           ..
H—C✕H  +  :Cl✕Cl:  ⟶  H—C⊖Cl:  +  H⊖Cl:
     |           ..   ..             |     ..           ..
     H                               H
```

One C—H bond and one Cl—Cl bond are broken and one C—Cl and one H—Cl bond are formed. (The bonds broken are crossed and the bonds made are circled.) Using the fact that breaking bonds takes energy and making bonds release energy, we arrive at the following:

Break 1 mol C—H bonds:	+412 kJ
Break 1 mol of Cl—Cl bonds:	+242 kJ
Make 1 mol of C—Cl bonds:	−330 kJ
Make 1 mol of H—Cl bonds:	−431 kJ
Estimated reaction enthalpy:	−107 kJ

EXERCISE Estimate the enthalpy of reaction for the oxidation of ethane to enthanol in the vapor phase. (The Lewis structure of ethanol is shown in Example 9.2 of the text).

$$C_2H_6(g) + \frac{1}{2}O_2(g) \longrightarrow C_2H_5OH(g)$$

ANSWER: −163 kJ

9.2 BOND LENGTHS

KEY CONCEPT The factors affecting bond length

The bond length of a bond is the distance between the nuclei of the two atoms joined by the bond. For instance, in the F_2 molecule the bond length is 142 picometers (pm).

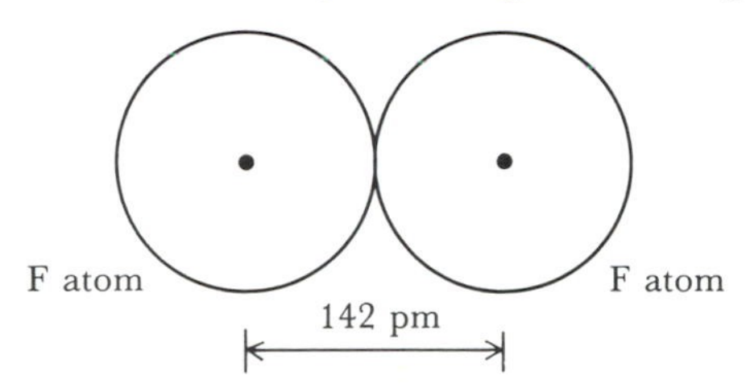

A variety of factors affect bond length:

1. Bonds between heavier atoms tend to be longer than bonds between lighter atoms.
2. Multiple bonds are shorter than single bonds between the same atoms, for instance,

$$\text{Shortest } A{\equiv}B < A{=}B < A{-}B \text{ longest}$$
$$C{\equiv}C < C{=}C < C{-}C$$
$$C{\equiv}O < C{=}O < C{-}O$$

3. Resonance averages the lengths of the bonds participating in the resonance.

Bond lengths can be estimated by summing the covalent radii (Table 9.4 of the text) of the atoms in the bond.

EXAMPLE Estimating bond lengths

Use the covalent radii in Table 9.4 of the text to show that a typical carbon–oxygen double bond is shorter than a typical carbon–oxygen single bond.

SOLUTION The pertinent covalent radii from Table 9.4 are shown in the following table, along with the sums that give the desired approximate bond length:

	Carbon radius, pm	Oxygen radius, pm	Carbon + oxygen, pm
C—O	77	74	151
C=O	67	60	127

The double bond length (127 pm) is predicted to be less than the single bond length (151 pm), as expected. It should be noted that the values predicted from this calculation are not the same as the average values given in Table 9.3; this emphasizes that summing covalent radii to get a bond length is only an approximation.

EXERCISE Without consulting any references, predict which bond is longer: S—Cl or S—F. Confirm your prediction by estimating the bond lengths using covalent radii.

ANSWER: S—Cl is predicted to be longer; estimated lengths are S—Cl, 200 pm, and S—F, 174 pm.

CHARGE DISTRIBUTIONS IN COMPOUNDS

KEY WORDS Define or explain each of the following terms in a written sentence or two.

formal charge
oxidation number
polar bond
polarizable
polarizing power

9.3 IONIC VERSUS COVALENT BONDING

KEY CONCEPT A The variation of ionic character with electronegativity

Almost every bond is partially ionic and partially covalent. The covalent character results in the presence of electron density between the two atoms in the bond. The ionic character is a manifestation of the movement of electron density away from one atom and towards the other; this results in the development of a positive charge on one atom (the one that electron density moves away from) and development of a negative charge on the other atom (the one that electron density moves toward). The extent of the development of ionic character and its attendant positive and negative charges depends on the difference in electronegativities of the two bonded atoms. A large difference in electronegativity leads to a high ionic character; a small difference in electronegativity leads to a low ionic character. If the difference in electronegativities is above 2, the bond is considered ionic; if it is below 1, the bond is considered covalent. When the difference in electronegativities is between 1 and 2, the bond is neither clearly ionic nor clearly covalent.

Bond	Difference in electronegativities	Ionic character, %*	Covalent character, %	Partial positive charge on	Partial negative charge on	Type of bond
Cs—F	3.3	95	5	Cs	F	ionic
Si—Cl	2.2	75	25	Si	Cl	ionic
O—H	1.4	40	60	H	O	unclear
P—Cl	0.9	20	80	P	Cl	covalent
C—H	0.4	5	95	H	C	covalent
F—F	0	0	100	—	—	covalent

* From Figure 9.3 of the text. These values are approximate.

EXAMPLE Determining the ionic character in a bond

What is the percentage ionic character in a S—Cl bond? On which atoms are the positive and negative charges? Is the bond ionic, covalent, or ambiguous in character?

SOLUTION The electronegativity of sulfur is 2.5 and that of chlorine is 3.0. Because chlorine has the higher electronegativity, it will carry a partial negative charge; sulfur will, therefore, have a partial positive charge. The difference in electronegativities is 0.5; from Figure 9.3 of the text, the ionic character is about 8%. A difference of 0.5 in electronegativity implies a covalent bond, because 0.5 is less than 1.

EXERCISE What is the percentage ionic character in a Ca—Cl bond? On which atoms are the positive and negative charge? Is the bond ionic, covalent, or ambiguous in character?

ANSWER: About 65%; Ca has partial positive charge, Cl has partial negative charge; bond is (probably) ionic.

KEY CONCEPT B Polarizability and polarizing power

Imagine that we have a compound in which a small, compact cation is next to a large, diffuse anion. The small size and overall positive charge of the cation keep its electrons very tightly held in place. On the other hand, the large size and overall negative charge of the anion allow its electrons more freedom to spread out and move away from the ion itself. In this situation, the cation attracts some of the electron charge cloud from the anion, resulting in the movement of some electron density into the space between the ions. The positioning of electrons between two atoms is precisely what constitutes a covalent bond, so the movement of electrons has added some covalent character to what was formerly an ionic bond. Atoms and ions that have charge clouds that can be easily distorted are said to be **polarizable.** Ions that can cause this electron charge cloud distortion have a high **polarizing power.** When an ion with high polarizing power is bonded to an easily polarizable ion, there will be substantial covalent character in the bond. Small highly charged cations have high polarizing power, and large highly charged anions are very polarizable.

Some cations with high polarizing power	Some highly polarizable anions
Li^{+}, Be^{2+}, Mg^{2+}, Al^{3+}, B^{3+}	I^{-}, Se^{2-}, Te^{2-}, As^{3-}, P^{3-}

EXAMPLE The polarization of compounds

One of the compounds, KCl, BCl_3 and $BaCl_2$, is covalent. Which one?

SOLUTION The anion for each of these compounds is the same, so we must decide which of the cations has the highest polarizing power. Of the three cations, K^{+}, Ba^{2+}, and B^{3+}, the B^{3+} is the smallest because it combines being "high" and "to the right" in the periodic table. It is also the most highly charged. These factors make it the cation with the highest polarizing power, so we conclude that BCl_3 is the covalent compound. Even though Cl^{-} is not listed as a highly polarizable anion, it can be polarized by a strongly polarizing cation.

EXERCISE Explain why $SnCl_4$ is a covalent compound.

ANSWER: The Sn^{4+} ion is highly polarizing. (It is so highly polarizing that it never really exists as a simple ion.)

KEY CONCEPT C Dipole moments and polar bonds

The electronegativity of an atom is a measure of the atom's ability to attract electrons in a bond. As a result, if two bonded atoms have different electronegativities, they will indulge in a tug-of-war for the electrons in the bond. The atom with the higher electronegativity tends to get an excess of electron density; the one with lower electronegativity loses some of its electron density. As the negatively charged electron-charge cloud undergoes this shift, the more electronegative atom acquires a partial negative charge and the less electronegative atom a partial positive charge. The ultimate outcome, a positive charge next to a negative charge of the same magnitude, is called an **electric dipole.** The Debye is the unit of the electric dipole moment; the dipole moment on a bond is approximately equal to the difference in the electronegativities of the two atoms in the bond ($\mu = \chi_A - \chi_B$).

When an electric dipole is formed on bonded atoms, the bond is called a **polar bond.** The magnitude of the polarity of the bond depends on the difference in electronegativities of the two atoms in the bond. A large difference in electronegativities results in a highly polar bond. If the difference in electronegativities is zero, the bond is usually completely nonpolar, which means that no electric dipole exists. (However, consult the text for remarks on ozone, O_3.) The examples that follow illustrate these ideas. Electronegativities are given in Figure 7.36 of the text.

Bond	Difference in electronegativities	Polarity of bond	Partial + charge on	Partial − charge on
Si—Cl	2.2	highly polar	Si	Cl
O—H	1.4	very polar	H	O
P—Cl	0.9	polar	P	Cl
C—H	0.4	slightly polar	H	C
F—F	0	nonpolar	—	—

EXAMPLE Predicting the polarity of bonds

Based on electronegativities, which of the following bonds would you predict to be most polar, S—Cl, N—F, or P—O?

SOLUTION The difference in electronegativities of the bonding atoms is usually a good measure of the polarity of a bond. For the bonds listed, we calculate the following differences, using the electronegativities in Figure 7.36 of the text: S—Cl, 0.5; N—F, 1.0; P—O, 1.4. The largest difference occurs for the P—O bond, so this bond is predicted to be the most polar.

EXERCISE Which of the following bonds is most polar, Sn—Cl, C—F, or Cl—Cl?

ANSWER: C—F

KEY CONCEPT D The polarity of molecules

We learned earlier that a polar bond is a bond that possesses an electric dipole. A **polar molecule** is a molecule that as a whole posseses an electric dipole. The sources of an electric dipole on a molecule are electric dipoles on the bonds in the molecule. However, there is an important qualifier; the presence of bond dipoles in a molecule does not guarantee that a molecular dipole will exist. It is possible for the shape of the molecule to position bonds so that the bond dipoles all cancel to give a net zero electric dipole. Thus, it is possible for a molecule to have polar bonds but still be a nonpolar molecule. The canceling of individual bond dipoles occurs when a number of the same atoms are bonded to a central atom in a highly symmetrical way. A summary of some common molecular shapes and their polarity is given in Table 9.6 of the text.

EXAMPLE Predicting the polarity of a molecule

Is ClF_5 polar?

SOLUTION At a first glance, we might imagine that because this is an AB_5 molecule (Table 9.6 of the text), it is nonpolar. But beware! It is not only the formula but also the shape of a molecule that determines its polarity. The Lewis structure of ClF_5 is

```
        F
        |
   F — Cl — F
       / \
      F   F
```

It possesses a lone pair and, therefore, has an octahedral arrangement of electrons and a square pyramidal shape. The square pyramidal shape results in a polar molecule.

EXERCISE Is CS_2 polar?

ANSWER: No

9.4 ASSESSING THE CHARGE DISTRIBUTION

KEY CONCEPT A Formal charge

The **formal charge** on an atom is the charge the atom would have if we arbitrarily assign valence electrons to it in the following way:

1. All of the lone pairs on the atom are assigned to the atom.
2. Half of the electrons in bonds on the atom are assigned to the atom.

The formal charge on the atom equals the number of valence electrons on the neutral atom (from the group number) minus the number of electrons assigned to the atom:

$$\text{Formal charge} = \begin{pmatrix}\text{number of valence electrons} \\ \text{on the neutral atom}\end{pmatrix} - \begin{pmatrix}\text{number assigned} \\ \text{to the atom}\end{pmatrix}$$

The sum of the formal charges on all of the atoms in a Lewis structure must equal the overall charge on the Lewis structure; if not, the formal charges have been assigned incorrectly.

EXAMPLE Determining formal charge

Assign formal charge to all of the atoms in the two inequivalent Lewis structures shown for dinitrogen oxide (nitrous oxide, N_2O):

$$:\text{N}{\equiv}\text{N}{-}\ddot{\underset{..}{\text{O}}}: \qquad :\ddot{\underset{..}{\text{N}}}{-}\text{N}{\equiv}\text{O}:$$

Structure I   Structure II

SOLUTION For each atom, we determine the number of valence electrons for the neutral atom from the group number. Then we assign to each atom in the structure half the electrons in every bond to that atom and all the electrons in unshared pairs.

Structure I

Atom	A, electrons in bonds	B, electrons in unshared pairs	C, electrons in neutral atom	D, electrons assigned $= \frac{1}{2}A + B$	Formal charge $= C - D$
N (terminal)	6	2	5	5	0
N (central)	8	0	5	4	+1
O	2	6	6	7	−1

Structure II

Atom	A, electrons in bonds	B, electrons in unshared pairs	C, electrons in neutral atom	D, electrons assigned $= \frac{1}{2}A + B$	Formal charge $= C - D$
N (terminal)	2	6	5	7	−2
N (central)	8	0	5	4	+1
O	6	2	6	5	+1

The answer is in the last column of each table.

EXERCISE Assign formal charge to all of the atoms in the two inequivalent Lewis structures shown for xenon difluoride:

:F̤=X̤e—F̤:   :F̤—Ẍ̤e—F̤:

I   II

ANSWER: I: F(double bonded), −1; Xe, +1; F(single bonded), 0
II: F, O; Xe, 0

KEY CONCEPT B Formal charge and plausible structure

Formal charges can be valuable in judging the relative energies of different Lewis structures of the same molecule. A Lewis structure in which the atoms do not have much formal charge is usually lower in energy than a structure that has atoms with a lot of formal charge. If one plausible structure has about the same amount of formal charge as another, the structure that puts negative charge on the more electronegative atoms is likely to have the lower energy.

EXAMPLE Choosing between Lewis structures

Of the two structures given for N_2O in the previous example, which is preferable?

SOLUTION Structure I has one atom with formal charge +1 and one with formal charge −1. Structure II has one atom with formal charge −2 and two with formal charge +1. Because structure I has less formal charge, it is the lower energy structure and, therefore, the preferred one.

EXERCISE Of the two structures given for XeF_2 in the previous exercise, which is preferred?

ANSWER: II

KEY CONCEPT C Oxidation number

We learned earlier how to assign oxidation numbers according to a set of rules. In some complicated compounds these rules lead to incorrect oxidation numbers. Also, there are some compounds that the rules simply do not cover. For these situations it is possible to assign oxidation numbers in the following way:

1. Assign all the electrons in all the bonds to the more electronegative atom involved in the bond.
2. Assign all lone pairs to the atom to which they belong.
3. Count the electrons around each atom in the compound.
4. The oxidation number of each atom is equal to the normal number of valence electrons for the atom (from the group number) minus the number of electrons assigned according to the above rules.

EXAMPLE Determining oxidation numbers

What is the oxidation number of S in $SOCl_2$?

SOLUTION The rules in Table 3.5 of the text do not provide any guidance in assigning oxidation numbers for Cl or S. Thus, we draw the Lewis structure of the compound and assign electrons to each atom based on the electronegativities of the atoms. The circles indicate the electron assignment, and the following table gives the assignment of oxidation numbers. As the following shows, sulfur has a +4 oxidation number.

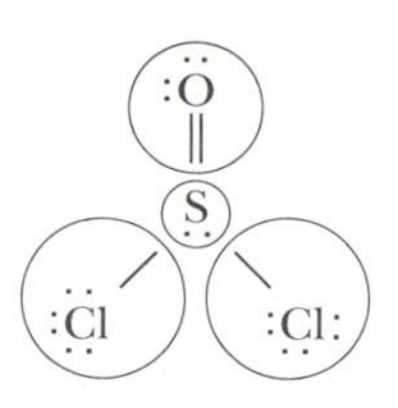

Atom	Number of electrons assigned	Number of valence electrons from group number	Oxidation number
O	8	6	$6 - 8 = -2$
Cl	8	7	$7 - 8 = -1$
S	2	6	$6 - 2 = +4$

EXERCISE What is the oxidation number of Br in the $BrCl_2^-$ ion?

ANSWER: +1

PITFALL Formal charges versus oxidation number: I

The formal charge on an atom and the oxidation number of an atom are different entities and should not be confused.

PITFALL Formal charges versus oxidation number: II

The method for assigning electrons to determine the formal charge on an atom is different from the method used for determining oxidation states. In formal charge calculations, each of the two electrons in a bond is assigned to one of the bonded atoms; for the purpose of computing oxidation states, both electrons in a bond are assigned to the most electronegative atom sharing the bond.

KEY CONCEPT D Chemical consequences of oxidation number

We saw earlier that oxidation numbers can be used to identify the oxidizing and reducing agents in a redox reaction. Here, we discover that oxidation numbers are valuable in describing some aspects of the chemical behavior of the elements.

1. Compounds in which an element has an oxidation number low in its range are often good reducing agents.
2. Compounds in which an element has an oxidation number in the middle of its range tend to undergo disproportionation.
3. Compounds in which an element has an oxidation number high in its range are often good oxidizing agents.

These ideas are rather intuitive. An element with a relatively low oxidation number will act as a reducing agent as it goes to a higher oxidation number. Similarly, an element with a relatively high oxidation number will act as a oxidizing agent as it goes to a lower oxidation number. Finally, an element in the midrange of its oxidation numbers can simultaneously go to higher and to lower oxidation numbers through disproportionation.

EXAMPLE Using oxidation numbers

Permanganate ion, MnO_4^-, frequently participates in redox reactions. Is it a good reducing agent, a good oxidizing agent, or does it tend to disproportionate? (Manganese exists with oxidation states from -3 to $+7$).

SOLUTION Manganese in permanganate has oxidation state $+7$. This is the highest possible oxidation state for manganese, so we predict it will act as an oxidizing agent. In fact, potassium permanganate is a common laboratory oxidizing agent.

EXERCISE One of these species, UF_5, $FeCl_2$, or CrO_4^{2-}, disproportionates readily. Which one? The stable oxidation states of the elements involved are: uranium $+3$ to $+6$; iron $+2$ and $+3$; and chromium, $+3$, $+4$, and $+6$.

ANSWER: UF_5

THE VALENCE-BOND MODEL OF BONDING

KEY WORDS Define or explain each of the following terms in a written sentence or two.

double bond	single bond	π bond
hybrid orbital	torsionally rigid	σ bond
orbital overlap	triple bond	

9.5 BONDING IN DIATOMIC MOLECULES

KEY CONCEPT A σ bonds in diatomic molecules

Three things occur when a σ bond forms:

1. An unpaired electron on one atom pairs with an unpaired electron on another atom (for most σ bonds).
2. An atomic orbital on one atom overlaps and merges with an atomic orbital on the other atom to form a new, larger orbital that encompasses both bonding atoms. For orbitals such as atomic *p* orbitals, which have a long axis, the merging occurs so that the long axis of one orbital is in line with the long axis of the other, much like placing two hot dogs end-to-end.
3. Each bond results in the sharing of two electrons by the two bonding atoms.

A σ bond concentrates a lot of electron density directly between the two nuclei of the bonded atoms; also, if you slice the orbital perpendicular to its long axis, the cross section is roughly circular (see Figures 9.8, 9.9, and 9.10 of the text). The sigma bonds in diatomic molecules can form from the overlap of various types of orbitals:

Molecule	Orbitals that overlap
H_2	H(1*s*) + H(1*s*)
HCl	H(1*s*) + Cl(2*p*)
Cl_2	Cl(2*p*) + Cl(2*p*)

EXAMPLE Understanding σ bonds

Describe how a σ bond forms in HBr. Use orbital diagrams.

SOLUTION The valence electron configurations and orbital diagrams of H and Br are as follows:

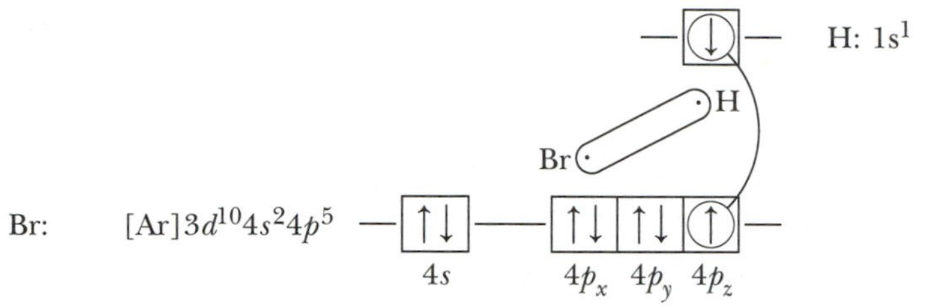

H: $1s^1$

Br: $[Ar]3d^{10}4s^24p^5$

4s $4p_x$ $4p_y$ $4p_z$

Hydrogen has an unpaired electron in an 1*s* orbital, and bromine has an unpaired electron in a 2*p* orbital. Head-to-head overlap of the H(1s) and Br(3*p*) orbitals allows the electrons to pair; a new larger orbital that includes both the hydrogen atom and the bromine atom forms. Formation of this single new orbital with two electrons in it holds the two atoms together; thus, a σ bond is formed.

EXERCISE Describe the bonding in F_2.

ANSWER: A 2*p* orbital in one fluorine overlaps a 2*p* orbital in the other to form a σ bond:

F : $1s^22s^22p^5$

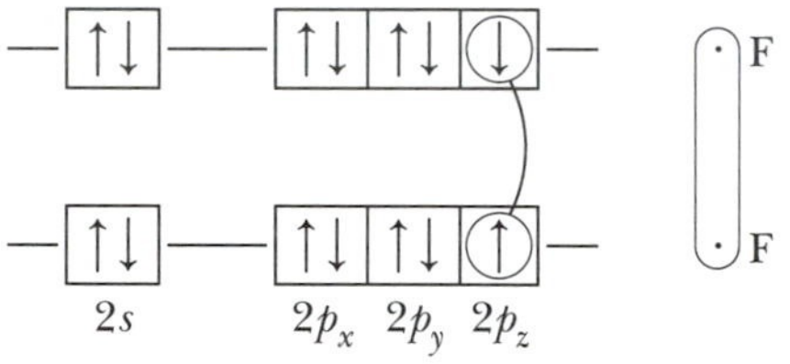

2s $2p_x$ $2p_y$ $2p_z$

KEY CONCEPT B π bonds

In certain situations atomic orbitals with unpaired electrons cannot overlap in a head-to-head fashion to form a bond, but they can overlap in a side-by-side fashion. This situation generally occurs when two atoms, each having a *p* orbital containing a single electron, are brought close to each other through σ bonding. The side-by-side overlap results in formation of a π bond; the two electrons (one from each atom) pair and are shared by both atoms, resulting in a covalent bond. A π bond forms only after a σ bond is formed. When a single π bond exists with the σ bond, a double bond results; when two π bonds exist with the σ bond, a triple bond results. A π bond concentrates electron density above and below the internuclear axis rather than in between the two bonded atoms (see Figure 9.11 of the text); hence, it is weaker than the σ bond. A single π bond has two lobes, one above the internuclear axis and one below the internuclear axis.

EXAMPLE Understanding π bonds

Why are there no π bonds in the Cl_2 molecule?

SOLUTION Chlorine has the electron configuration $1s^22s^22p^63s^23p^5$ and the orbital diagram (for valence electrons) that follows:

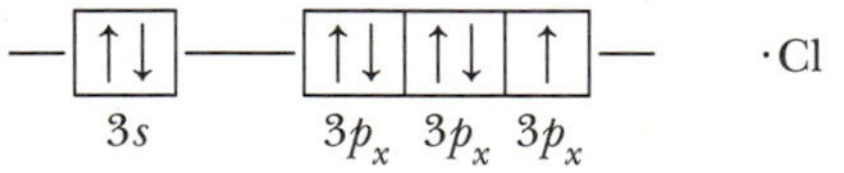

3s $3p_x$ $3p_x$ $3p_x$ ·Cl

A σ bond is formed by the overlap of the 3*p* orbital on one chlorine with the 3*p* orbital on the other, accompanied by the pairing and sharing of the electrons in the orbitals. At this point, all of the electrons on both chlorines are paired and no new bonds can form. Hence, no π bond forms.

EXERCISE Account for the location of all of the valence electrons in the cyanide ion, CN^-.

ANSWER: CN^- has 10 valence electrons, 2 in a σ bond, 4 in two π bonds, and 4 in two unshared pairs.

9.6 HYBRIDIZATION

KEY CONCEPT A Hybridization of atomic orbitals

Our bonding theory at this point has two problems. The first is that, in many cases, the shapes of molecules are inconsistent with the location of atomic orbitals that form bonds; the second is that the number of bonds to an atom frequently does not match the number of unpaired electrons in the unbonded atom. For example, according to our (so far, deficient) ideas, the smallest compound between hydrogen and carbon should be an angular CH_2 molecule, whereas, as we know, it is, a tetrahedral CH_4 molecule. We can patch up both problems by invoking orbital **hybridization,** in which new largely *equivalent* atomic orbitals form from old ones:

$$\text{Old atomic orbitals} \xrightarrow{\text{hybridization}} \text{new atomic orbitals}$$

For the carbon in methane, for instance,

$$2s + 2p_x + 2p_y + 2p_z \xrightarrow{\text{hybridization}} sp^3 + sp^3 + sp^3 + sp^3$$

(A way to imagine what occurs during hybridization is that the four electron clouds of the old atomic orbitals combine into a single cloud, which is then broken up into four new equivalent electron clouds.) A mathematical analysis indicates that the four sp^3 orbitals take on a tetrahedral arrangement, which agrees with the observed shape of the methane molecule. The four new sp^3 hybrid orbitals all have the same energy, and (following Hund's rule) each has one of the four valence electrons in it; thus, carbon can form four σ bonds, again in agreement with the actual methane molecule. To arrive at other molecular shapes, different atomic orbitals are hybridized; the common types of hybridization and the arrangement of electrons associated with each are given in Table 9.7 of the text.

EXAMPLE Predicting hybridization

What is the hybridization of Xe in XeF_2?

SOLUTION We must determine the number of electron pairs around Xe in order to ascertain its hybridization. To do this we first draw the Lewis structure of XeF_2:

$$:\ddot{\underset{\cdot\cdot}{F}}-\ddot{\underset{\cdot\cdot}{Xe}}\cdot-\ddot{\underset{\cdot\cdot}{F}}:$$

We then count the number of electron pairs around Xe, five in this case (two bonding pairs and three lone pairs). Table 9.7 of the text indicates that five electron pairs result in sp^3d hybridization.

EXERCISE What is the hybridization of As in $AsCl_3$?

ANSWER: sp^3

KEY CONCEPT B Formation of σ bonds with hybridized orbitals

Hybrid atomic orbitals are very effective in forming σ bonds because they can achieve significant orbital overlap. The steps involved in hybridization and bonding are as follows.

Step 1. Electrons are promoted from low-energy orbitals to higher energy orbitals to create enough unpaired electrons to form the required number of σ bonds.

Step 2. Half-filled low-energy orbitals are hybridized to create enough hybrid orbitals to form the required number of σ bonds. Remember that the number of hybridized orbitals that result must equal the number of orbitals that are used to create them.

$$\text{Number of hybridized orbitals} = \text{number of orbitals used to create the hybridized orbitals}$$

(Low-energy orbitals with pairs of electrons may also have to be hybridized because unshared pairs often exist in hybridized orbitals.)

Step 3. Orbitals then overlap to form σ bonds. Once the orbitals overlap, two electrons pair and become shared between two atoms; hence, a familiar covalent bond results.

Many of the σ bonds in the familiar species we have already encountered involve overlap of hybridized orbitals, as the following table indicates.

σ bond	**Atomic orbital on first atom**	**Atomic orbital on second atom**
C—H in CH_4	C(sp^3)	H(1s)
B—F in BF_3	B(sp^2)	F(sp^3)
S—F in SF_6	S(sp^3d^2)	F(sp^3)
C—C in C_2H_4	C(sp^2)	C(sp^2)

EXAMPLE Using promotion and hybridization for bond formation

Describe how promotion and hybridization may be used to explain the bonding in AsF_5.

SOLUTION The Lewis structure of AsF_5 is

The central arsenic must form five σ bonds to fluorine atoms. The electron configuration for arsenic is [Ar]$3d^{10}4s^24p^3$. Its valence electron orbital diagram is

To form five bonds, we need five unpaired electrons; promotion of a 4s electron to a 4d orbital gives us the required number:

We finally hybridize the singly occupied 4s, 4p, and 4d orbitals to generate five sp^3d orbitals. This results in the correct shape for AsF_5 as well as maximal overlap of atomic orbitals as bonds form

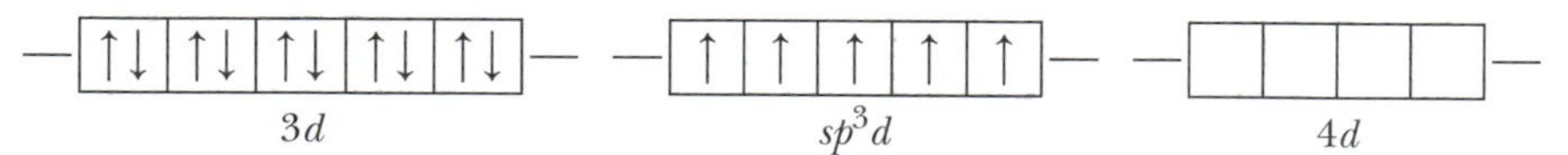

To form the compound, we overlap a singly occupied 2p orbital on each of five fluorine atoms with each of the singly occupied sp^3d orbitals on arsenic. The electrons pair, and each pair is shared by a fluorine and the arsenic to form a covalent bond. As the bond is formed, it is likely that the valence orbitals on each fluorine hybridize to form four sp^3 orbitals; this gives maximal orbital overlap. Each As—F bond is a (As sp^3d, F sp^3) bond.

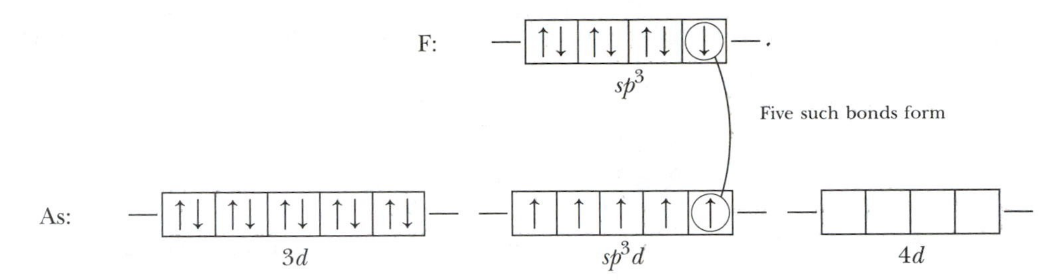

EXERCISE Describe how promotion and hybridization may be used to explain the bonding in $BeCl_2$.

ANSWER: Two Be—Cl σ (Be sp, Cl sp^3) bonds form.

KEY CONCEPT C Formation of π bonds with hybridized orbitals

A π bond is part of the make up of any double or triple bond. We will use the double bond in ethylene to illustrate formation of a π bond with hybridized orbitals:

$$H_2C{=}CH_2$$

From the point of view of VSEPR theory, each carbon atom has three electron pairs around it, and each, therefore, has a trigonal planar structure. Thus, each carbon must also be sp^2 hybridized. The sp^2 hybridization uses a $2s$ orbital and two $2p$ orbitals on carbon, but leaves one $2p$ orbital unaffected. This $2p$ orbital protrudes above and below the carbon, perpendicular to the trigonal planar sp^2 hybrid orbitals, as shown in the next figure.

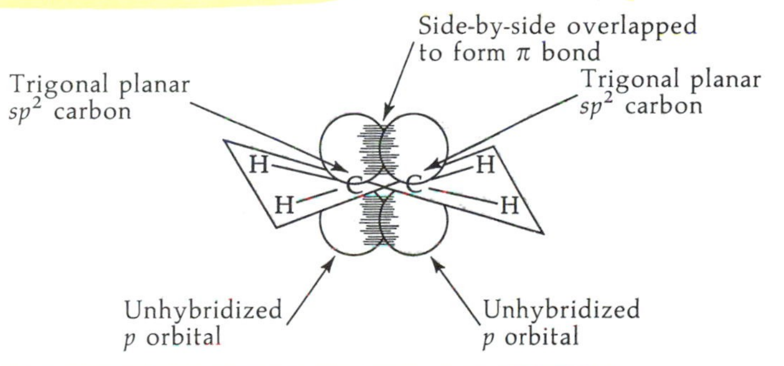

The two $2p$ orbitals now overlap with each other in a side-by-side manner (rather than head-on, as in a σ bond) to form a **π bond.** The two electrons (one from each $2p$ orbital) are paired and shared by the two carbon atoms, so a covalent bond forms. Hybridization does not occur in π bond formation, because hydridization would result in less side-by-side overlap and, therefore, less effective bond formation.

EXAMPLE Describe how π bonds may form in CO_2

SOLUTION VSEPR theory predicts that CO_2 is linear; and, therefore, the central carbon is sp hybridized.

C: —[↑↓]— ($2s$) —[↑|↑|]— ($2p$) $\xrightarrow{\text{hybridize}}$ —[↑|↑]— (sp) —[↑|↑]— ($2p$ (unhybridized))

The electrons in the sp hybrid orbitals are used for the two σ bonds, one to each oxygen atom. The electrons in the unhybridized $2p$ orbitals are used to form π bonds, one to each oxygen atom.

EXERCISE Based on promotion and hybridization, how many π bonds are there in XeO_4?

ANSWER: None

MOLECULAR ORBITAL THEORY

KEY WORDS Define or explain each of the following terms in a written sentence or two.

antibonding orbital
bond order
bonding orbital
electron-deficient compound
molecular orbital
molecular potential energy curve
π molecular orbital
σ molecular orbital

9.7 MOLECULAR ORBITALS

KEY CONCEPT A Failures of the valence-bond theory

The valence-bond theory of bonding works for many compounds, but in a few cases it does not provide an accurate model for bonding. One of the problems is its failure to explain the bonding in **electron-deficient** compounds, which are compounds that do not have enough valence electrons for a Lewis structure to be written. A second problem with the valence-bond theory is its inability to describe the bonding in the familiar O_2 molecule. According to the valence-bond theory, the structure of O_2 is :Ö=Ö:, a structure that, like any proper Lewis structure, has all of its electrons paired. However, experiments unequivocally show that O_2 is **paramagnetic;** paramagnetism is a type of magnetism that is associated with unpaired electrons. The valence-bond theory, again, disagrees with experiment.

Compound	Valence-bond theory	Observed
B_2H_6	14 valence electrons	12 valence electrons
O_2	all electrons paired: nonmagnetic	2 electrons unpaired; paramagnetic

Despite these failures, and a few others we have yet to encounter, we must remember that the valence-bond concept of bonding works well for a vast number of compounds.

EXAMPLE Electron-deficient compounds

Show that the valence-bond theory predicts that B_2H_6 has 14 valence electrons; explain how this prediction is a failure of the valence-bond theory.

SOLUTION If we want to understand the valence-bond model of a compound, we should first draw the predicted structure of the compound. Recall that in the context of the valence-bond theory, hydrogen can form only one single bond. The only plausible Lewis structure for B_2H_6 is

```
     H   H
     |   |
 H — B — B — H
     |   |
     H   H
```

Incorrect structure; 14 valence electrons required

Because valence-bond theory states that a chemical bond contains 2 valence electrons, the seven bonds in this structure should require a total of 14 valence electrons. However, in actuality, each boron atom has 3 valence electrons and each hydrogen atom has 1 valence electron; so in B_2H_6 there are 12 valence electrons, or 2 less than demanded by the Lewis structure. The valence-bond concept simply does not work. Because the actual compound has fewer electrons than predicted from the valence-bond theory, it is called an electron-deficient compound.

EXERCISE Which compound, Be_2Cl_4, C_2H_6, or PCl_3, is electron deficient?

ANSWER: Be_2Cl_4

KEY CONCEPT B Molecular orbitals and molecular orbital theory

The most successful description of how atoms bond to form molecules is done with **molecular orbital theory.** In this approach, bonding occurs through the formation of a **molecular orbital** rather than through sharing a pair of electrons. A molecular orbital describes the position and energy of an electron in a molecule, just as an atomic orbital describes the position and energy of an electron in an atom. Molecular orbitals are similar to atomic orbitals in many ways but are different in one important aspect, as can be seen from the following comparison.

Atomic orbital	Molecular orbital
wave function localized on atom	wave function spread throughout molecule
holds maximum of two electrons	holds maximum of two electrons
electrons obey Pauli exclusion principle and Hund's rule	electrons obey Pauli exclusion principle and Hund's rule
characterizes position and energy of electron	characterizes position and energy of electron

EXAMPLE Molecular orbitals

Assume two electrons are to be placed into two molecular orbitals of equal energy. Draw the orbital energy diagram that results.

SOLUTION We first draw a representation of the two molecular orbitals, which we simply call MO 1 (molecular orbital number 1) and MO 2 (molecular orbital number 2), with the energy of the orbital indicated by its position on an energy *y* axis.

Energy ↑ —□— —□—
MO 1 MO 2

When two electrons are placed in molecular orbitals of equal energy, they must, according to Hund's rule, spread out in the orbitals with parallel spins before filling. Thus, the two possible correct configurations are

Energy ↑ —[↑]— —[↑]—
MO 1 MO 2

Energy ↑ —[↓]— —[↓]—
MO 1 MO 2

Some impermissible configurations are

—[↑]— —[↓]— Violates Hund's rule

—[↑↓]— —□— Violates Hund's rule

—[↑↑]— —□— Violates Pauli exclusion principle

EXERCISE Draw an allowed orbital electron digram for placing four electrons in the three molecular orbitals shown. (Assume a ground-state configuration, so that the system is in the lowest possible energy state.)

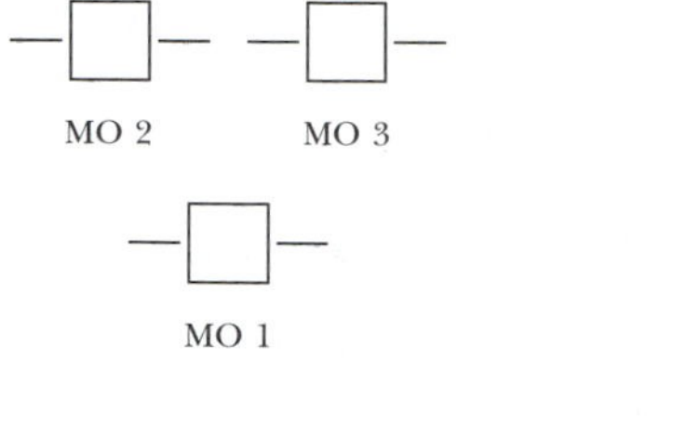

ANSWER : —[↑]— —[↑]—
MO 1 MO 2

—[↑↓]—
MO 3

PITFALL The configurations we consider are for ground-state molecules

The configurations we consider that follow Hund's rule are for the ground state (lowest energy state) of a molecule. Excited states can violate Hund's rule.

KEY CONCEPT C σ molecular orbitals

We can picture the formation of a pair of molecular orbitals as occurring through a three-step process, illustrated next for the two σ orbitals in H_2.

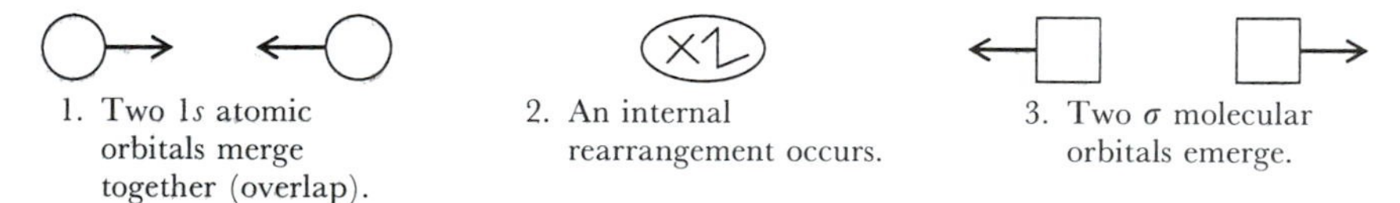

We should note that the number of molecular orbitals formed is always the same as the number of atomic orbitals that originally overlap; for instance, when two atomic 1s orbitals merge, two σ molecular orbitals result. Electrons in the **bonding orbital** (σ) tend to encourage formation of a molecule by lowering the energy of the molecule. Electrons in the **antibonding orbital** (σ*) tend to discourage formation of a molecule by raising the energy of the molecule. We can better illustrate the relative energies of the bonding and antibonding molecular orbitals by showing their formation with an energy *y* axis present.

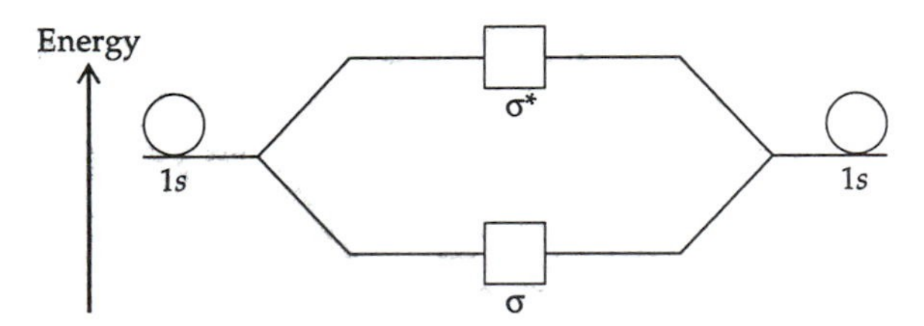

This **molecular orbital energy diagram** can be used to describe the bonding (or lack of bonding) in any homonuclear diatomic chemical species with four or fewer electrons, such as H_2, H_2^+, H_2^-, He_2, He_2^+, and He_2^{2+}, because the two molecular orbitals shown can accommodate a maximum of four electrons.

EXAMPLE Bonding in small homonuclear diatomic species

Which is more stable, H_2 or H_2^-?

SOLUTION Both H_2 and H_2^- are described by the molecular orbital picture given previously because both have fewer than four electrons. To judge their relative stabilities, we must place the requisite number of electrons into the available molecular orbitals and determine the effect of the bonding electrons and anti-bonding electrons present. H_2 has two electrons (one from each hydrogen atom), which, when placed in the molecular orbital energy diagram, result in the following arrangement:

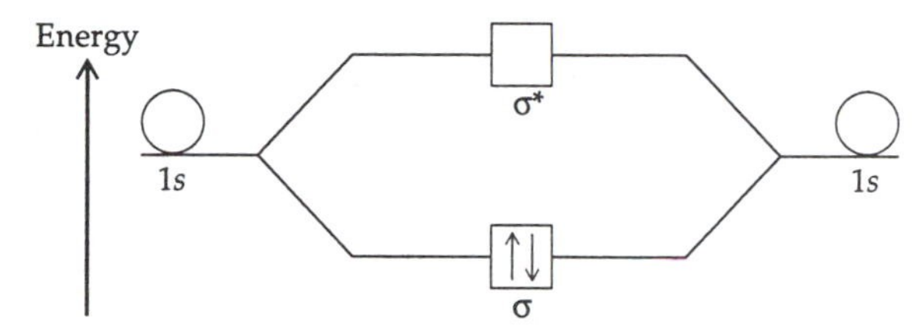

H_2^- possesses three electrons (2 electrons from two hydrogen atoms + 1 electron to form the negative ion = 3 electrons), resulting in the molecular-orbital energy diagram in the next figure:

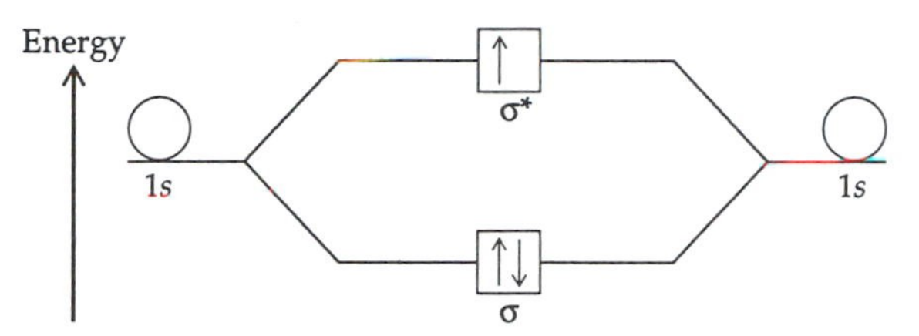

H_2^- contains two bonding electrons and one antibonding electron. The one antibonding electron increases the energy of the molecule and, in effect, cancels the energy effect of one of the bonding electrons; so the net bonding in H_2^- results from one bonding electron. H_2 has two bonding electrons, so its bonding results from two bonding electrons. Thus, the bonding in H_2 is stronger than that in H_2^- because two bonding electrons lower the energy of a species more than one bonding electron.

EXERCISE Is the H_2^{2-} ion stable?

ANSWER: No; two antibonding electrons cancel the effect of two bonding electrons.

9.8 BONDING IN PERIOD 2 DIATOMIC MOLECULES

KEY CONCEPT A Molecular orbitals for period 2 homonuclear diatomic molecules

We can carry foward the treatment used for small molecules to larger ones. The fundamental ideas remain the same although some details change. The molecules we want to consider are the homonuclear diatomics of the second period: Li_2, Be_2, B_2, C_2, N_2, O_2, F_2, and Ne_2. The molecular orbitals are constructed from atomic orbitals just as they are in smaller molecules. However, we encounter a new situation here with the merging of the 2*p* atomic orbitals to form molecular orbitals. When two 2*p* orbitals merge by mixing in a head-to-head fashion, two σ molecular orbitals are formed, a bonding σ orbital and an antibonding σ^* orbital:

$$2p + 2p \longrightarrow 2p\sigma + 2p\sigma^*$$

When two $2p$ atomic orbitals mix in a side-by-side fashion, two π molecular orbitals are formed, a bonding π orbital and an antibonding π^* orbital:

$$2p + 2p \longrightarrow 2p\pi + 2p\pi^*$$

We use the prefix $2p$ in symbols such as $2p\sigma$ or $2p\pi$ to indicate that a molecular orbital forms by the mixing of $2p$ atomic orbitals. The mixing of three p orbitals on one atom with three p orbitals on a second atom results in the formation of six molecular orbitals. The perpendicular arrangement of the three p orbitals on an atom mandates that two of the molecular orbitals formed will be σ molecular orbitals (one σ and one σ^*) and four will be π molecular orbitals (two π and two π^*). The formation of the 10 molecular orbitals in a period 2 diatomic molecule is depicted in the next equation. (The energy ordering of the molecular orbitals is shown in Figure 9.25 of the text.) The valence-shell atomic orbitals and the molecular orbitals they lead to are shown in boldface.

$$\underset{\text{Atom 1}}{(1s, \mathbf{2s}, \mathbf{2p_x}, \mathbf{2p_y}, \mathbf{2p_z})} + \underset{\text{Atom 2}}{(1s, \mathbf{2s}, \mathbf{2p_x}, \mathbf{2p_y}, \mathbf{2p_z})} \longrightarrow \underset{\text{Diatomic molecule}}{(1s\sigma, 1s\sigma^*, \mathbf{2s\sigma}, \mathbf{2s\sigma^*}, \mathbf{2p\sigma}, \mathbf{2p\pi}, \mathbf{2p\pi}, \mathbf{2p\pi^*}, \mathbf{2p\pi^*}, \mathbf{2p\sigma^*})}$$

▼ EXAMPLE Predicting the bonding in a period 2 homonuclear diatomic molecule

Is C_2 paramagnetic?

SOLUTION A paramagnetic species is one that contains one or more unpaired electrons. To determine if C_2 has any unpaired electrons, we must construct the pertinent molecular orbital energy diagram (showing valence MOs only) for C_2 and place eight valence electrons (four per carbon atom) into the molecular orbitals indicated. The orbital energy diagram in Figure 9.25 of the text is the one that should be used. Placing eight electrons into it following the building-up rules gives

$2p\sigma^*$	[ ]
$2p\pi^*$	[ ][ ]
$2p\pi$	[↑][↑]
$2p\sigma$	[↑↓]
$2s\sigma^*$	[↑↓]
$2s\sigma$	[↑↓]

We can see that C_2 has two unpaired electrons, so it is paramagnetic.

EXERCISE Is O_2^{2+} paramagnetic?

▲ ANSWER: No

KEY CONCEPT B Bond order

Electrons in bonding molecular orbitals are responsible for the formation of bonds, but, as we have seen, the favorable effect of such electrons on bond formation is canceled by the presence of electrons in antibonding molecular orbitals. We can put this idea on a more formal footing by definition of the **bond order,** which is calculated as follows

$$\text{Bond order} = \frac{\text{no. of electrons in bonding MOs} - \text{no. of electrons in antibonding MOs}}{2}$$

The effect of antibonding electrons is taken into account by subtracting their number from the number of bonding electrons. Thus, in effect, the bond order tells us how strong a bond is; the higher the bond order, the stronger the bond. Most importantly, if the bond order equals 0, there is no bond. If a Lewis structure can be drawn for a compound, the bond order calculated from molecular orbital theory correlates with the Lewis structure in the following way.

Lewis structure	Bond order
single bond	1
double bond	2
triple bond	3

Noninteger bond orders occur. For example, both H_2^+ and H_2^- have bond order $\frac{1}{2}$.

EXAMPLE Calculating bond order

What is the bond order of O_2^+?

SOLUTION To calculate the bond order, we must determine the number of electrons in bonding orbitals and the number in antibonding orbitals and apply the formula given. Each oxygen contributes 6 valence electrons, and 1 electron is lost to form the positive ion with +1 charge, so the number of valence electrons is $(2 \times 6) - 1 = 11$. Placing 11 electrons into the valence MO diagram gives

$2p\sigma^*$	[ ]	
$2p\pi^*$	[↑][ ]	1 antibonding electron
$2p\pi$	[↑↓][↑↓]	4 bonding electrons
$2p\sigma$	[↑↓]	2 bonding electrons
$2s\sigma^*$	[↑↓]	2 antibonding electrons
$2s\sigma$	[↑↓]	2 bonding electrons

The total number of bonding electrons is 8, and the total number of antibonding electrons is 3. Inserting into the formula for calculating bond order gives,

$$\text{Bond order} = \frac{\text{no. of electrons in bonding MOs} - \text{no. of electrons in antibonding MOs}}{2}$$

$$= \frac{8-3}{2} = \frac{5}{2}$$

EXERCISE What is the bond order of O_2^-?

ANSWER: $\frac{3}{2}$

9.9 ORBITALS IN POLYATOMIC MOLECULES

In large molecules, molecular orbitals encompass the whole molecule, resulting in **delocalized orbitals.** In benzene (C_6H_6), for instance, the six carbon atoms are sp^2-hybridized; after the sigma bonds in the benzene form, the six unhybridized p orbitals all merge to form large π molecular orbitals that encompass all the carbon atoms in the molecule. Because six atomic orbitals are used, six molecular orbitals are formed. The energies and orbital shapes of the π molecular orbitals of benzene are shown in Figure 9.26 of the text. Six electrons (one from each carbon) must be placed into these orbitals to derive the benzene orbital occupancies.

▼ EXAMPLE Delocalized orbitals

Which is more stable, benzene or benzene negative ion ($C_6H_6^-$)?

SOLUTION The π molecular orbital occupancies of both species are indicated in the following diagrams.

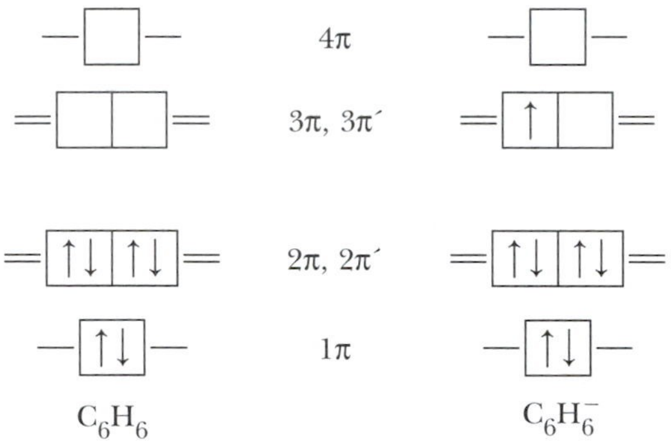

C_6H_6		$C_6H_6^-$
[ ]	4π	[ ]
[ ] [ ]	3π, $3\pi'$	[↑] [ ]
[↑↓] [↑↓]	2π, $2\pi'$	[↑↓] [↑↓]
[↑↓]	1π	[↑↓]

The six electrons in benzene are all in bonding orbitals whereas the benzene negative ion has seven electrons, six in bonding orbitals and one in an antibonding orbital. The extra electron in an antibonding orbital destabilizes benzene negative ion relative to benzene itself.

EXERCISE How many π molecular orbitals are there in the nitrite ion, NO_2^-?

▲ ANSWER: 3

DESCRIPTIVE CHEMISTRY TO REMEMBER

- **Ethylene** is $CH_2{=}CH_2$; **acetylene** is $CH{\equiv}CH$.
- Stannane, SnH_4, decomposes into tin and hydrogen at room temperature; plumbane, PbH_4, is completely unstable and has been prepared in only trace amounts.
- $BeCl_2$ consists of individual $BeCl_2$ molecules in the vapor and of long chains of covalently bonded $BeCl_2$ units in the solid.
- The covalent character of the silver halides increases from AgF to AgI. Thus, AgF is largely ionic and soluble in water, and solubility in water decreases from AgF to AgI.
- Vision depends on the molecule retinal, which contains a double bond. When light hits the molecular, the π bond in the double bond is broken, allowing the molecule to rotate around the remaining σ bond. The π bond then reforms, imparting a new shape to the molecule. The new shape imparts a signal along the optic nerve.
- Period 3 atoms are too large to allow side-by-side p orbital overlap; hence, no significant multiple bonding occurs with these atoms.
- Diborane (B_2H_6) is a colorless gas that bursts into flame upon contact with air.
- Molecular oxygen, O_2, is paramagnetic with bond order 2.

CHEMICAL EQUATIONS TO KNOW

- Beryllium chloride is prepared by the action of carbon tetrachloride on beryllium oxide.

$$2BeO(s) + CCl_4(g) \xrightarrow{800°C} 2BeCl_2(g) + CO_2(g)$$

- The recovery of sulfur from the hydrogen sulfide found in natural gas and petroleum uses the Claus process.

$$2H_2S(g) + SO_2(g) \longrightarrow 3S(s) + 2H_2O(l)$$

- Copper(I) ions disproportionate in solution.

$$2Cu^+(aq) \longrightarrow Cu^{2+}(aq) + Cu(s)$$

- CrO_4^{2-} is stable in basic solutions but in acid is converted to dichromate without change of oxidation number.

$$2CrO_4^{2-}(aq) + 2H^+(aq) \longrightarrow Cr_2O_7^{2-}(aq) + H_2O(l)$$

- Diborane (B_2H_6) is produced by the action of boron trifluoride on lithium tetrahydroborate ($LiBH_4$).

$$3LiBH_4 + 4BF_3 \longrightarrow 2B_2H_6 + 3LiBF_4$$

MATHEMATICAL EQUATIONS TO KNOW AND UNDERSTAND

$\mu(D) \approx \chi_A - \chi_B$ electric dipole moment

$$\text{Bond order} = \frac{1}{2} \times (\text{no. of electrons in bonding orbitals} - \text{no. of electrons in antibonding orbitals})$$

SELF-TEST EXERCISES

Bond parameters

1. Which of the following bonds is shortest?
(a) C≡C (b) C═C (c) C—C

2. Which bond is weakest?
(a) HO—H (b) HS—H (c) HSe—H (d) HTe—H

3. Use bond enthalpies to estimate the reation enthalpy for the reaction of ethanol with oxygen to form acetic acid. (You will have to draw the Lewis structures of the reactants and products to answer this question.)

$$CH_3CH_2OH(g) + O_2(g) \longrightarrow HC_2H_3O_2(g) + H_2O(g)$$

(a) −612 kJ (b) −688 kJ (c) −298 kJ (d) +350 kJ (e) −349 kJ

4. Which has no effect on bond energies?
(a) existence of multiple bonds (b) existence of unshared pairs
(c) number of neutrons in nuclei (d) size of atoms
(e) none of these; all effect bond energies

5. Use covalent radii to predict the approximate length of a C—S single bond.
(a) 90 pm (b) 72 pm (c) 102 pm (d) 179 pm (e) 204 pm

6. Which molecule has the largest bond length?
(a) O_2 (b) Te_2 (c) Se_2 (d) S_2

Charge distribution in compounds

7. Which bond is predominantly ionic? (Consult Figure 7.36 of the text.)
(a) Ge—F (b) Sn—Cl (c) As—Br (d) N—O (e) Si—O

8. Which ion has the highest polarizing power?
(a) Pb^{2+} (b) Al^{3+} (c) Se^{2-} (d) Ca^{2+} (e) Br^-

9. Which ion is most polarizable?
(a) Na^+ (b) O^{2-} (c) Cl^- (d) Mg^{2+} (e) Te^{2-}

10. What is the approximate value of the dipole moment on the SO molecule? (Consult Figure 7.36 of the text.)
(a) 0 D (b) 6.0 D (c) 1.0 D (d) 4.5 D (e) 3.0 D

11. Which of the following symbolizes a dipole moment? Each sign refers to a charge on a species.
(a) ⊕ ⓪ (b) ⊕ ⊖ (c) ⊖ ⊖ (d) ⊕ ⊕ (e) ⊕ ⊖ ⊕

12. Which of the following is polar?
(a) CO_2 (b) NH_3 (c) CF_4 (d) BF_3 (e) SeO_3

13. A molecule consists of a central atom surrounded by some number of atoms of the same element. What shape molecule of this type is likely to be polar?
(a) linear (b) octahedral (c) square planar (d) T-shaped (e) trigonal planar

14. Which of the following is nonpolar?
(a) SeO_2 (b) CH_3F (c) H_2O (d) ClF_3 (e) XeF_4

15. What is the oxidizing agent in the following reaction?

$$8H^+ + MnO_4^- + 5Fe^{2+} \longrightarrow 5Fe^{3+} + Mn^{2+} + 4H_2O$$

(a) H^+ (b) MnO_4^- (c) Fe^{2+} (d) H_2O (e) Fe^{3+}

16. Each hypothetical element is listed with a particular oxidation number and its range of oxidation numbers. Which one is most likely to act as an oxidizing agent?
(a) Z^{2-}(range: -2 to $+2$) (b) X^{2+}(range: $+2$ to $+7$)
(c) Y^{6+}(range: $+3$ to $+6$) (d) M^{4+}(range: $+3$ to $+6$)

The valence bond model of bonding

17. The hypothetical element Q has the valence-shell configuration shown. According to valence-bond theory, what is the Lewis structure of an Q_2 molecule?

—[↑↓]— —[↑↓|↑|↑]—

(a) :Q̤̈—Q̤̈: (b) :Q̈=Q̈: (c) :Q≡Q:

18. One of the following does not apply to a covalent bond. Which one?
(a) electrons pair (b) electrons are shared
(c) orbitals overlap (d) may be polar
(e) electrons localized on one atom

19. What hybridization is associated with a trigonal planar arrangement of electron pairs?
(a) sp^3d (b) sp^2 (c) sp^3d^2 (d) sp (e) sp^3

20. What hybridization is associated with an octahedral arrangement of electron pairs?
(a) sp^3 (b) sp (c) sp^3d^2 (d) sp^3d (e) sp^2

21. What arrangement of electron pairs occurs with sp^3 hybridization?
(a) trigonal planar (b) T-shaped (c) octahedral
(d) tetrahedral (e) trigonal bipyramidal

22. An octahedral arrangement of electron pairs can lead to all of the molecular shapes given except one. Which molecular shape cannot arise from an octahedral arrangement of electron pairs?
(a) trigonal planar (b) octahedral (c) square planar
(d) square pyramidal (e) linear

23. An AB_4 molecule has two lone pairs. What is the hybridization of the central atom?
(a) sp (b) sp^3d (c) sp^3 (d) sp^2 (e) sp^3d^2

24. Head-to-head overlap of an sp^3 hybrid orbital with an sp^2 hybrid orbital creates a
(a) double bond. (b) σ bond. (c) π bond. (d) triple bond.

25. Side-by-side overlap of a p orbital with another p orbital creates a
(a) single bond. (b) σ bond. (c) π bond. (d) triple bond.

26. A triple bond consists of
(a) three σ bonds. (b) one σ bond and two π bonds.
(c) two σ bonds and one π bond. (d) three π bonds.

27. An atom with the valence electron configuration shown is expected to form four σ bonds. What hybridization is expected?

[↑↓] [↑][↑][ ]
s p

(a) sp (b) sp^3 (c) sp^3d (d) sp^2 (e) sp^3d^2

28. The orbital diagram that follows shows the valence-shell configuration of an atom after promotion of electrons but before hybridization. What hybridization is most likely?

[↑] [↑][↑][↑] [↑][↑][ ][ ][ ]
s p d

(a) sp (b) sp^3 (c) sp^3d (d) sp^2 (e) sp^3d^2

Molecular orbital theory

29. Which of the following is electron deficient?
(a) CH_4 (b) B_2H_6 (c) SO_3 (d) O_2 (e) IF_5

30. What is responsible for paramagnetism in compounds?
(a) Fe atoms (b) s orbitals (c) π bonds
(d) unpaired electrons (e) electrons in hybridized orbitals

31. Which molecule is actually paramagnetic but is predicted to be nonmagnetic according to its Lewis structure?
(a) H_2 (b) N_2 (c) Cl_2 (d) F_2 (e) O_2

32. Electrons in an antibonding orbital
(a) tend to force atoms to move apart. (b) lower the energy of a molecule.
(c) have no effect on the stability of a molecule. (d) are always unpaired.

33. Which of the following comments about molecular orbitals is untrue?
(a) They are localized on one atom.
(b) They hold two electrons.
(c) The low-energy molecular orbitals fill first.
(d) Two electrons in one molecular orbital must be paired.
(e) They follow Hund's rule when filling.

34. According to molecular orbital theory, one of the following is unstable and won't form? Which one?
(a) Ne_2^+ (b) Cl_2^- (c) C_2 (d) O_2^+ (e) Be_2

35. According to molecular orbital theory, one of the following is stable. Which one?
(a) He_2 (b) C_2^{2+} (c) H_2^{2-} (d) Li_2^{2+} (e) F_2^{2-}

36. A molecule with 10 electrons in bonding molecular orbitals and 7 electrons in antibonding molecular orbitals has a bond order of
(a) 5 (b) $\frac{3}{2}$ (c) $\frac{7}{2}$ (d) 3 (e) 1

37. What is the bond order of C_2?
(a) 1 (b) $\frac{1}{2}$ (c) $\frac{3}{2}$ (d) $\frac{5}{2}$ (e) 2

38. What is the bond order of O_2^{2+}?
(a) 2 (b) $\frac{3}{2}$ (c) $\frac{5}{2}$ (d) 3 (e) 1

39. What is the bond order of Ne_2?
(a) 0 (b) 1 (c) -1 (d) 2 (e) $\frac{1}{2}$

40. Which of the following is paramagnetic?
(a) Li_2 (b) C_2 (c) B_2 (d) H_2 (e) F_2

Descriptive chemistry

41. Which of the following describes molecular oxygen, O_2?
(a) Nonmagnetic, single bonded (b) Paramagnetic, single bonded
(c) Nonmagnetic, double bonded (d) Paramagnetic, double bonded

42. The formula for diborane is
(a) B_2 (b) B_2H_6 (c) BH_3 (d) B_6H_6 (e) BF_3

43. Which two compounds produce diborane (B_2H_6) when reacted?
(a) $Li[BH_4]$, $AlCl_3$ (b) $LiCl$, BF_3 (c) BF_3, H_2 (d) $Li[BH_4]$, BF_3 (e) $Li[BF_4]$, H_2

44. Which of the silver halides is the least soluble in water?
(a) AgF (b) AgI (c) AgCl (d) AgBr

45. Which Group IV hydride is so unstable that its existence is in doubt?
(a) CH_4 (b) SiH_4 (c) PbH_4 (d) SnH_4

46. Which of the following is ethylene?
(a) CH_4 (b) $CH_2{=}CH_2$ (c) $CH{\equiv}CH$ (d) $CH_3{-}CH_3$ (e) NH_3

47. $BeCl_2$ is prepared by the action of which reagent on BeO?
(a) NaCl (b) $CaCl_2$ (c) CCl_4 (d) HCl (e) Cl_2

48. Period 3 atoms rarely form double bonds because
(a) they are too large.
(b) their atomic number is too great.
(c) they don't always follow the octet rule.
(d) they don't contain *p* orbitals.
(e) their electronegativity is too low.

10 LIQUIDS AND SOLIDS

Everyday experience tells us that matter commonly exists in three physical states: solid, liquid, and gas. In this chapter, we are concerned with the solid and liquid states. We first explore the nature of the forces that hold solids and liquids together. Then we consider the structure and some properties of liquids and solids. Each physical state of a substance (liquid, solid, gas, allotropic form) is called a **phase.** The structure and properties of a substance depends on its phase.

FORCES BETWEEN ATOMS, IONS AND MOLECULES

KEY WORDS Define or explain each of the following terms in a written sentence or two.

dipole–dipole interaction
hydrogen bond
intermolecular force
ion–dipole interaction
ion–ion interaction
ionic solid
London force
molecular solid
network solid
phase

10.1 ION AND DIPOLE FORCES

KEY CONCEPT A The four types of solids

The four types of solids differ in the unit the solid is built of and/or the forces that hold it together. These differences, in turn, lead to vastly different properties for the different types of solids. A summary of some of the features of the different solids is given in Table 10.7 of the text.

EXAMPLE The properties of a solid

A sample of a solid is extremely hard, is completely insoluble in water, has a very high melting point, and does not conduct electricity. What type of solid is it?

SOLUTION Table 10.7 of the text lists properties of solids. The properties of this sample match those of a network solid.

EXERCISE Solid CO_2 sublimes quite rapidly at a very low temperature. What type of solid is likely to do this?

ANSWER: Molecular (the forces holding molecules together are very weak)

KEY CONCEPT B Ion–ion interactions

The energy of interaction between two ions is proportional to the charge on the ions and inversely proportional to the distance between the ions. This relationship implies that

1. The energy of interaction increases rapidly as the charge on the ions increases.
2. The energy of interaction decreases moderately slowly as the distance between the ions increases.

These facts lead, in turn, to the conclusions that small ions (that can get close to each other) with high charge have the highest energy of interaction. In addition, the slow fall off of the interaction with distance means that an ion in a solid crystal interacts not only with its nearby neighbors in the crystal but also with ions that are further away.

EXAMPLE The effect of the size and charge of ions on physical properties

Without consulting any references, predict which has the higher melting point, NaCl or KCl.

SOLUTION The difference in these two compounds lies with the cations, Na^+ and K^+. Because both cations have the same charge, we look to their size to explain the different melting points. Because Na^+ is smaller than K^+, the two ions can get closer together in NaCl than in KCl; thus, the ion–ion attractions are stronger in NaCl than in KCl. We predict, therefore, that NaCl will have the higher melting point. [This prediction is correct: KCl has melting point 776°C and NaCl 801°C].

EXERCISE Salt A has a melting point of 650°C and salt B a melting point of 735°C. Which salt would you predict has the higher lattice enthalpy? Explain.

ANSWER: Salt B should have the higher lattice enthalpy because the relative melting points indicate that salt B has stronger ion–ion attractions.

PITFALL Unusually high polarization effects

We normally expect that a compound with highly polarizing cations and easily polarizable anions will experience strong ion–ion attractions and, therefore, will have high melting and boiling temperatures. However, when polarization effects are exceptionally large, an appreciable amount of covalent bonding can be introduced into a crystal; this lowers the melting and boiling temperatures of a compound. For instance, LiCl has a lower melting temperature than expected, as is evidenced by the following melting temperatures: CsCl, 646°C; RbCl, 715°C; KCl, 776°C; NaCl, 801°C; and LiCl, 614°C.

KEY CONCEPT C Ion–dipole interactions

This interaction, as the name implies, occurs between an ion and a molecule with a permanent dipole, that is, a polar molecule. When an ionic substance is dissolved in water, strong ion–dipole interactions assist the salt in dissolving. Both the cation and the anion experience an ion–dipole interaction on dissolving; the cation interacts with the negative end of the water dipole and the anion with the positive end. Waters of hydration in salts such as $MgSO_4 \cdot 10H_2O$ and $CaCl_2 \cdot 6H_2O$ owe their presence in a crystal to the ion–dipole interaction of the polar water molecule with the ions in the crystal. The ion–dipole interaction decreases more rapidly than the ion–ion interaction with distance; it is proportional to $1/d^2$ rather than $1/d$. This means that at small distances, the ion–dipole interaction can be very important, but that at long distances it is quite unimportant. Because the charges on a polar molecule are smaller than on an ion, the ion–dipole interaction is less strong than the ion–ion interaction.

EXAMPLE Hydration and the ion–dipole attraction

Of the three salts NaBr, KBr, and RbBr, only one exists in a hydrated form. Which one and why?

SOLUTION The waters of hydration in a crystal are trapped in the crystal partially as the result of ion–dipole attractions. This attraction is strongest when the ion is small because this permits the water molecule to get closer to the ion and maximize this favorable interaction. The anion is the same in the three salts given, so it could not account for any differences in the three. Of the three cations Na^+, K^+, and Rb^+, the $Na^+$is the smallest, so we expect it to have the strongest ion–dipole attraction for a water molecule. We predict that NaBr exists in a hydrated form.

EXERCISE Of the two salts $BaCl_2$ and KCl, only one exists in a hydrated form. Which one and why?

ANSWER: $BaCl_2$, because of the higher charge on Ba^{2+} relative to K^+; Ba^{2+} and K^+ are about the same size.

KEY CONCEPT D Dipole–dipole interactions

Two polar molecules, that is two molecules with permanent dipole moments, experience an attractive interaction when the negative end of the dipole on one molecule approaches the positive end of the dipole on the other. The attractive interaction is called a **dipole–dipole interaction.** As with the other interactions discussed in this chapter, the attraction is due to the fundamental attraction of a positive charge for a negative charge. The form of the dipole–dipole interaction depends on whether the molecules are rotating or rigid. For molecules in a solid (rigid molecules), the attraction between molecules is proportional to $1/d^3$, whereas for molecules in a gas or liquid (which rotate), the interaction is proportional to $1/d^6$. As expected, the larger the charge on the electric dipole—that is, the more polar the molecule—the stronger the interaction is.

The $1/d^3$ and $1/d^6$ dependence of the energy of interaction on distance implies that the energy is very sensitive to distance and drops off rapidly as the distance between the molecules increases, so this interaction has a very short range. The negative energies for the interaction reflect the fact that the interaction is a favorable one. The dipole–dipole interaction is weaker than the ion–ion interaction and the ion–dipole interaction because of the smaller charges on the dipole of a polar molecule relative to the charge on an ion.

EXAMPLE The polarity of molecules and the dipole–dipole interaction

Which of the following molecules in the solid phase, HCl or HBr, is likely to experience a stronger dipole–dipole interaction?

SOLUTION The strength of the dipole–dipole interaction increases as the electric dipole on the molecule gets larger and as the molecules get closer to each other. Because Cl has a larger electronegativity (3.0) than Br (2.8), we expect, if all other things are the same, that HCl is more polar than HBr; this means it has a larger electric dipole. In addition, Cl is smaller than Br, so we expect that HCl molecules in solid HCl can get closer to each other than HBr molecules in solid HBr. Both the higher electric dipole on HCl and its smaller size indicate that HCl experiences a stronger dipole–dipole interaction than HBr in the solid phase. (The bond length of HCl is smaller than that of HBr; this has a slight effect on the dipole moment.)

EXERCISE Which will have the higher boiling point, *cis*-1,2-dichloroethene or *trans*-1,2-dichloroethene?

```
H       Cl          H       Cl
 \     /             \     /
  C=C                 C=C
 /     \             /     \
Cl      H           Cl      H
```

trans-1,2-dichloroethene   *cis*-1,2-dichloroethene

ANSWER: *cis*-1,2-dichloroethene, 60°C; *trans*-1,2-dichloroethene, 48°C

KEY CONCEPT E The London force

The London force arises because, for an instant, the electrons in a molecule can be slightly "out of position" relative to the nuclei of the atoms in the molecule. When such an event occurs, a slight positive charge appears in one part of the molecule (the part with slightly less electron charge cloud than it should have), and a slight negative charge appears elsewhere in the molecule (where there is slightly more electron charge cloud than there should be). Thus, a small transitory electric dipole appears in the molecule. This transitory electric dipole can encourage formation of a similar transitory electric dipole in a neighboring molecule so that the negative end of one of the transitory electric dipoles is next to the positive end of a neighboring one. Thus, a slight attractive interaction, called the **London force,** results. The London force increases as the electron polarizability increases because highly polarizable electrons move out of position most easily. Large molecules experience the greatest London forces; they have a large number of electrons and, therefore, a substantial electron charge cloud density far removed from the atomic nuclei. The London force is a very short range interaction. The shape of a molecule also determines the strength of the London force. For two different molecules with the same number of electrons, the London force will be higher for the least spherical molecule. All molecules experience the London force. The large number of London force interactions make it a potent intermolecular attractive force, despite its weakness.

EXAMPLE Predicting the relative effect of the London force

Which element, Cl_2 or F_2, is predicted to have the highest boiling point? Why?

SOLUTION Both Cl_2 and F_2 are nonpolar, so the major intermolecular attraction that holds the molecules of each in the liquid state is the London force. Because both molecules are diatomic molecules, both have the same shape; so it is safe to assume that the number of electrons in the molecule alone will determine the strength of the London force. Cl_2 with its 34 electrons is more polarizable than F_2 with its 18 electrons; therefore, Cl_2 experiences stronger London forces than F_2. The Cl_2 molecules are held more strongly together, and Cl_2 has the higher boiling point.

EXERCISE Which of the following has the highest boiling point: Ne, Ar, or Kr?

ANSWER: Kr

PITFALL All molecules experience the London force

In polar molecules, the London force usually is more important than dipole–dipole interactions.

10.2 HYDROGEN BONDING

KEY CONCEPT Hydrogen bonding

Hydrogen bonding is an especially strong intermolecular attractive force that occurs when a hydrogen atom that is covalently bonded to F, O, or N moves into position close to a second F, O, or N atom. For example, water undergoes extensive hydrogen bonding, as illustrated in the following figure.

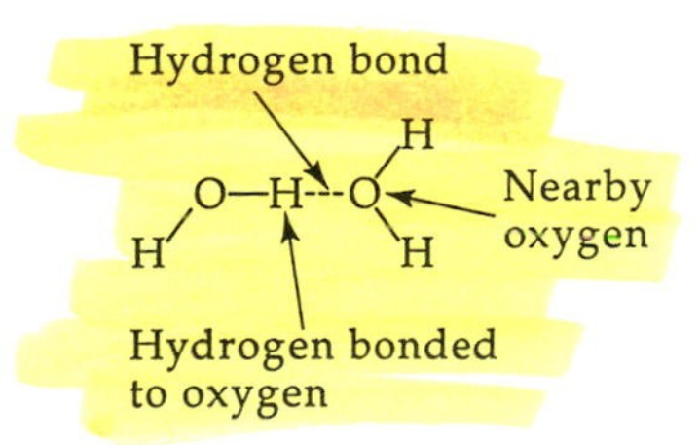

Three effects are at play here that together make hydrogen bonding particularly strong. First, the electric dipoles on the molecules involved are very large because F, O, and N are among the most electronegative atoms in the periodic table. Thus, the partial positive charge on hydrogen is exceptionally high and the partial negative charge on F, O, or N (whichever is involved) is also very high; this leads to a very strong attractive interaction. Second, the atoms involved (H, F, O, and N) are all quite small, so there is an excellent opportunity for the H atom to get very close to the F, O, or N; this close approach also increases the strength of the attractive interaction. Third, F, O, and N usually have lone pairs of electrons that can be distorted, and therefore attracted, to the positively charged hydrogen, further increasing the attractive interaction.

EXAMPLE Hydrogen bonding

Explain why hydrogen bonding is important for HF and not for HCl.

SOLUTION To answer this question we will evaluate HF and HCl for the three factors that lead to hydrogen bonds.

1. Both HF and HCl are highly polar molecules with substantial electric dipoles. F, with electronegativity 4.0, and Cl, with electronegativity 3.0, both cause a large electric dipole to exist. However, this difference alone cannot account for the difference in hydrogen bonding because N, with electronegativity 3.0 (the same as Cl), does participate in hydrogen bonding.
2. F in HF and Cl in HCl each have three lone pairs, so the presence of lone pairs cannot account for the difference.
3. F has a covalent radius of 72 pm and Cl, 99 pm. The smaller size of the F allows the closer approach of a hydrogen and is the dominant factor that accounts for the different behavior of F and Cl with regard to hydrogen bonding.

EXERCISE Three of the following undergo hydrogen bonding. Which one doesn't?

$$H_2O,\ H_3C{-}NH_2,\ CH_3F,\ \text{and}\ CH_3OH$$

ANSWER: CH_3F

THE PROPERTIES OF LIQUIDS

KEY WORDS Define or explain each of the following terms in a written sentence or two.

capillary action	meniscus	Trouton's Rule
critical temperature	phase diagram	vapor pressure
dynamic equilibrium	triple point	viscosity

10.3 SURFACE TENSION

KEY CONCEPT A Viscosity

Viscosity can be thought of as a measure of the resistance of a substance to flow. Materials with a high resistance to flow, such as honey or toothpaste, have a high viscosity. Water, which flows much more easily, has a lower viscosity than they do. When intermolecular attractions between molecules are very strong, the molecules cannot move past each other very well, and the liquid does not flow easily. When the intermolecular attractions are weak, molecules can move and jostle past each other with little difficulty, and the liquid flows easily. Large, chainlike molecules that can intertwine and tangle together also cannot move past each other very easily. Flow is hindered for such molecules; they have high viscosities.

EXAMPLE Factors affecting viscosity

Large biological molecules such as those found in egg whites are often long and chainlike. Account for the fact that egg whites undergo an increase in viscosity when they are cooked.

SOLUTION We expect, from knowledge of the factors that affect viscosity, that cooking the egg increases intermolecular attractions and/or increases the chainlike character of the molecules. In fact, both occur in this case. An uncooked molecule forms a roughly spherical structure by intertwining with itself, much like a piece of string rolled into a ball. Cooking the egg causes the molecule to "unwind." It can then intertwine with other molecules, and resistance to flow is increased. In addition, the unwound molecule can undergo many more intermolecular attractions than a self-intertwined molecule; these attractions also contribute to the increase in viscosity that occurs when an egg is cooked. (Some chemical cross-linking also occurs in the cooking process.)

EXERCISE Of the three compounds acetone, propyl alcohol and ethyl methyl ether, which has the highest viscosity? These compounds are all approximately the same size.

```
    H   O   H            H   H   H                H       H   H
    |   ||  |            |   |   |                |   ..  |   |
H — C — C — C — H    H — C — C — C — ÖH       H — C — O — C — C — H
    |       |            |   |   |                |   ..  |   |
    H       H            H   H   H                H       H   H

     Acetone              Propyl alcohol           Ethyl methyl ether
```

ANSWER: Propyl alcohol (due to hydrogen bonding)

KEY CONCEPT B Surface tension

A liquid has a surface tension because a molecule finds it slightly more favorable, energetically, to be buried in the body of a liquid sample, where it is surrounded on all sides by other molecules, than to be on the surface of the liquid, where it has fewer neighbors. It takes energy to force a molecule onto the surface of a liquid. This energy (which might better be called the "surface energy") is called the **surface tension.** The surface tension has units of joules/square meter (J/m^2); it is the energy needed to create a square meter of surface area. In any liquid sample, the surface area formed is the minimum possible under the circumstances; the creation of additional surface area requires energy and is an unfavorable process. The surface tension phenomenon results from intermolecular attractions. Substances that have the highest intermolecular attractions have the highest surface tension, because they require the most energy to move a molecule from the body of the liquid to the surface.

EXAMPLE Surface tension effects

A water-skipper insect can walk on the surface of a pond due to surface tension effects. Explain how.

SOLUTION We will idealize the insect's foot as a disk. A sketch of the foot on the surface of the pond (a) and how it would look if it sank (b) follows.

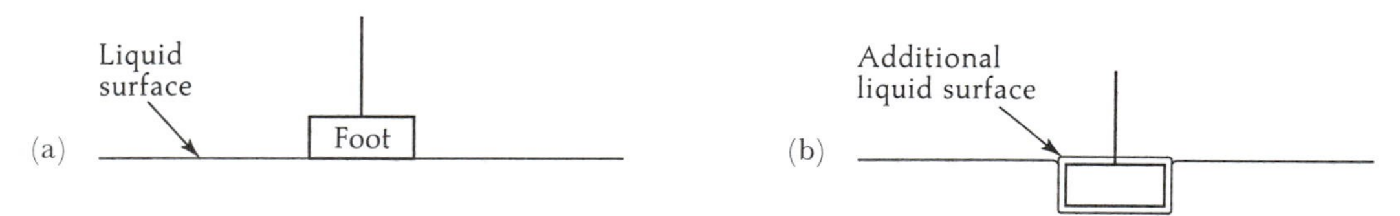

As the foot sinks, liquid surface area is created around the foot. Because it takes energy to create surface area, this is an unfavorable process. If the insect is light enough and has a large enough foot, the gravitational energy supplied (due to gravity on the insect's mass) will not be enough to create the new surface area; so the insect will not sink. (A fat water skipper with small feet would sink, however.)

EXERCISE Considering surface tension effects only, would 0.05 mL of water exist as one large drop or as 5 drops of 0.01 mL each?

ANSWER: As one larger drop

10.4 VAPOR PRESSURE

KEY CONCEPT A Vapor pressure

Imagine that we set up an experiment in which a beaker of water is placed on a laboratory bench and covered with a glass bell jar. The bell jar has a pressure gauge attached (which is set to 0 Torr at the start of the experiment) and is sealed with an airtight seal. Now, we observe the level of water in the beaker and the reading of the pressure gauge over a period of a few days.

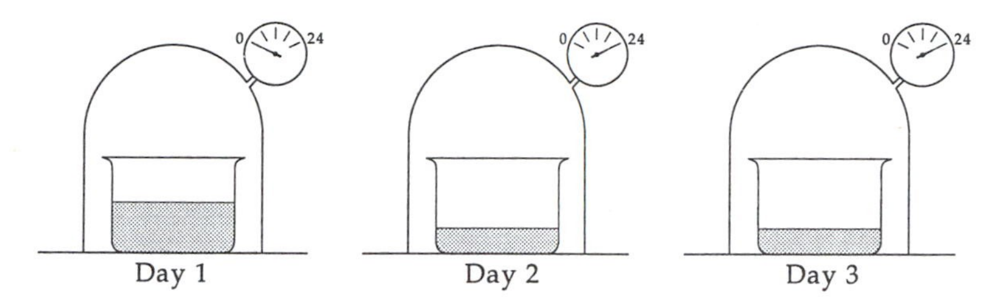

Because water molecules leave the liquid and enter the gas phase as the water in the beaker evaporates, the level of the water in the beaker on day 2 is lower than on day 1. Also, on day 2 additional gas-phase molecules are trapped in the volume enclosed by the bell jar; so the pressure inside the bell jar increases. On day 3, however, we observe that the level of water and the pressure are the same as on day 2. At this point, the number of gas-phase water molecules that collide with the surface of the liquid and rejoin the liquid equals the number that evaporate from the liquid; so the level of the liquid stays the same. In addition, because the number entering the gas phase equals the number leaving the gas phase, the number of gas-phase molecules stays the same, and the pressure does not change. At this point, a **dynamic equilibrium** exists. *Equilibrium* exists because there is no observable change in time for the system under observation; it is *dynamic* because there is an underlying motion or movement that is responsible for maintaining the equilibrium—that is, the continuous evaporation and condensation of the water. A good way of thinking about dynamic equilibria is "Whatever is done is undone." The pressure recorded by the pressure gauge is called the **vapor pressure** of the liquid; it increases as the temperature of the liquid increases.

EXAMPLE Understanding the vapor-pressure experiment

In the experiment just described, the level of water in the beaker stayed the same after day 2 and would, presumably, stay the same from then on. Given this fact, why does a puddle of water completely evaporate to dryness on a summer day?

SOLUTION The difference between the experiment and a puddle lies in the presence of the bell jar. The bell jar traps the gas-phase water molecules in a small volume and does not let them escape from the vicinity of the liquid. The random motion of these molecules then permits a large number of them to collide with the surface of the liquid and recondense into the liquid. The molecules that evaporate from a puddle are under no such constraints. They diffuse far away from the puddle after evaporation and do not recondense into the liquid. Eventually, all the liquid evaporates because the liquid phase is not being replenished to make up for the molecules lost by evaporation.

EXERCISE How will the vapor pressure of a liquid change if the surface area of the liquid is doubled?

ANSWER: It will not change at all.

KEY CONCEPT B Vapor pressure and molecular structure

A liquid that evaporates easily has a higher vapor pressure than one that does not evaporate easily. The vapor pressure of a liquid is determined by the net number of molecules that evaporate into the enclosed volume above the liquid. If evaporation occurs easily, a large number of molecules will evaporate before the rate of condensation catches up to and equals the rate of evaporation. The large number of vapor-phase molecules then results in a high vapor pressure. The ease of evaporation itself depends on the intermolecular attractive forces present in the liquid. If the attractive forces are large, molecules are held back in the liquid phase and evaporation does not occur easily; the vapor pressure of the liquid is low. If the attractive forces are weak, molecules can leave the liquid with ease; evaporation occurs easily, and the vapor pressure is high.

EXAMPLE Predicting the relative vapor pressures of different substances

Methane (CH_4) and carbon tetrafluoride (CF_4) are both liquids at 125 K. Predict which has the higher vapor pressure at this temperature.

SOLUTION Both CH_4 and CF_4 are tetrahedral and are therefore nonpolar. The intermolecular attractions in both are due to London forces. Because both molecules have the same shape, the strength of the London forces depends largely on the number of electrons in each compound. The compound with fewer electrons is less polarizable, will experience the weaker London forces, and will have the higher vapor pressure. CH_4 has less electrons than CF_4, so it has the higher vapor pressure.

EXERCISE Methanol (CH_3OH) and fluoromethane (CH_3F) are both liquids at 0°C. Both have approximately the same molecular weight. Which has the higher vapor pressure and why?

ANSWER: CH_3F; because it does not hydrogen bond as does CH_3OH

KEY CONCEPT C Vapor pressure and boiling point

Boiling is characterized by formation of large bubbles in a liquid. These bubbles are not air bubbles; they are filled with molecules from the liquid that have formed a small pocket of gas-phase molecules. This occurs because the vapor pressure of the liquid is equal to the external pressure on the liquid, and molecules entering the gas phase can successfully push against the applied external pressure to form the bubble. The bubbles in a boiling liquid are small "balloons" filled with molecules from the liquid. The key idea is that a liquid boils when its vapor pressure equals the external pressure (usually atmospheric pressure) on the liquid. Because the boiling point (actually the boiling temperature) of a liquid depends on the external pressure, it is convenient to define some type of standard boiling point; the **normal boiling point** of a liquid is such a standard and is defined as the temperature at which the vapor pressure of the liquid equals exactly 1 atm. Remember, when you boil water in the laboratory or at home, the applied atmospheric pressure is usually *not* exactly 1 atm, and the temperature of the boiling water is not exactly 100.00°C.

EXAMPLE Predicting the boiling point of water at a given pressure

Assume we double the pressure on a sample of water from 0.4 atm to 0.8 atm. Does the boiling point (in kelvins) also double?

SOLUTION To answer this question, we must determine the actual boiling point of water at the two given pressures. The actual boiling point is the temperature at which the vapor pressure equals the external pressure. From Figure 10.15 of the text, the vapor pressure of water is 0.4 atm at 76°C, or 349 K. Thus, the boiling point of water at 0.4 atm is 76°C (349 K). Similarly, at 0.8 atm, the boiling point is 94°C (367 K). 367 K is not twice 349 K, so the boiling temperature does not double when the pressure doubles.

EXERCISE Instructions for canning many nonacid foods call for cooking the food in a pressure cooker filled with water at a total pressure of 1.7 atm. The reason for the high pressure is to increase the boiling temperature of the water so that it is high enough to assure killing the bacteria that cause botulism. What is the boiling point of water at 1.7 atm? (Use Figure 10.15 of the text.)

ANSWER: 112°C

KEY CONCEPT D Trouton's rule

Based on the observed normal boiling points in kelvins (T_b) and the observed enthalpy of vaporizations (ΔH°_{vap}) of a large number of nonhydrogen-bonded liquids, the following approximate relationship, known as **Trouton's rule,** has been found:

$$\frac{\Delta H^\circ_{vap}}{T_b} = 85 \text{ J/K·mol}$$

Trouton's rule implies that as ΔH°_{vap} gets larger, T_b must also get larger, or else their ratio would not stay approximately constant at 85 J/K·mol. This is to be expected because liquids with strong intermolecular attractions should have a high enthalpy of vaporization (it takes a relatively large amount of energy to pull a molecule out of the liquid at the boiling point) and a relatively high boiling point (the temperature of the liquid must be high to get the vapor pressure to 1 atm).

EXAMPLE Using Trouton's rule

Use Trouton's rule to estimate the boiling point of acetone (C_3H_6O), which has an enthalpy of vaporization of 3.20×10^4 J/mol at its boiling point.

SOLUTION If we take Trouton's rule and (1) divide both sides by 85 J/K·mol and (2) multiply both sides by T_b, we get

$$T_b = \frac{\Delta H^\circ_{vap}}{85\ \text{J/K}\cdot\text{mol}}$$

Substituting in 3.20×10^4 J/mol for ΔH°_{vap} gives

$$T_b = \frac{3.20 \times 10^4\ \cancel{\text{J}}/\cancel{\text{mol}}}{85\ \cancel{\text{J}}/\text{K}\cdot\cancel{\text{mol}}} = 380\ \text{K}$$

The actual boiling point of acetone is 329 K. Although the difference in the two results may seem large, the estimate is only off by 16%.

EXERCISE Does ethanol ($\Delta H^\circ_{vap} = 40.5$ kJ/mol, $T_b = 78.5$°C) follow Trouton's rule? If not, why?

ANSWER: No, because $\Delta H^\circ_{vap}/T_b = 115$ J/mol·K; it doesn't follow Trouton's rule because it hydrogen bonds.

PITFALL Proper units in Trouton's rule

Remember to express the enthalpy of vaporization with units of joules/mole (not kilojoules/mole) when using Trouton's rule.

KEY CONCEPT E The critical temperature

The temperature of a substance can be increased to the point where it is impossible to liquefy the substance, no matter how high the applied pressure. The temperature above which it is impossible to liquefy a substance is called the **critical temperature** of the substance and is denoted by the symbol T_c.

EXAMPLE Interpreting the critical temperature

The critical temperature of methane is 191 K. What pressure must be applied to methane at 298 K in order to liquefy it?

SOLUTION The critical temperature of a substance is the maximum temperature at which it can exist in the liquid form. Above the critical temperature it is impossible for the substance be liquefied. Because 298 K is higher than methane's critical temperature of 191 K, it is impossible to liquefy the methane at 198 K, no matter what pressure is applied.

EXERCISE Solid carbon dioxide is called "dry ice" and, as you may have observed, does not form a liquid at 25°C and 1 atm. The critical temperature of carbon dioxide is 31°C. Is the failure of dry ice to form a liquid at 25°C and 1 atm due to the relative values of its actual temperature and critical temperature?

ANSWER: No

10.5 SOLIDIFICATION

KEY CONCEPT A Solidification

For most substances the molecules in the solid are closer together than in the liquid. An increase in pressure pushes the molecules together and makes it easier for the solid to form; the liquid freezes at a higher temperature when the pressure on it is increased. The effect is small but measureable. Water is unusual in this respect. Water molecules are farther apart in the solid than in the liquid, so applying pressure to water tends to favor the liquid state. That is, increasing the pressure makes it easier to form the liquid, and liquid water must be cooled to a lower temperature than expected to get it to freeze when pressure is applied; the freezing temperature of water drops as pressure is applied.

EXAMPLE Predicting the change in freezing temperature as a function of pressure

A science fiction story about the life of an android living on the surface of Jupiter (where the atmospheric pressure is enormous) contains a scene in which the android makes an ax by first melting some water over a fire and then letting it freeze in an ax-shaped mold. Even though the writer assumed the temperature was very cold, could the high pressure make freezing impossible?

SOLUTION Yes. Increasing the pressure on water lowers its freezing point. At the high pressure described in the story, the freezing point of water might be so low that it would not freeze, even at the temperature used by the writer.

EXERCISE The text, in Section 10.5, describes how the weight of a glacier causes a film of water to form under the glacier; the glacier then slides downhill on the lubricating film of water. Could the same mechanism work if the glacier was made of ethyl alcohol instead of water? Explain your answer.

ANSWER: No; water is the only common substance for which the melting point decreases as pressure increases. The melting point of ethyl alcohol increases as pressure increases, so it would not melt under the pressure of a glacier.

KEY CONCEPT B Phase diagrams

A **phase diagram** is a graph that shows the physical state of a substance as a function of the pressure and temperature of the substance. A sketch of a phase diagram is shown in the following figure.

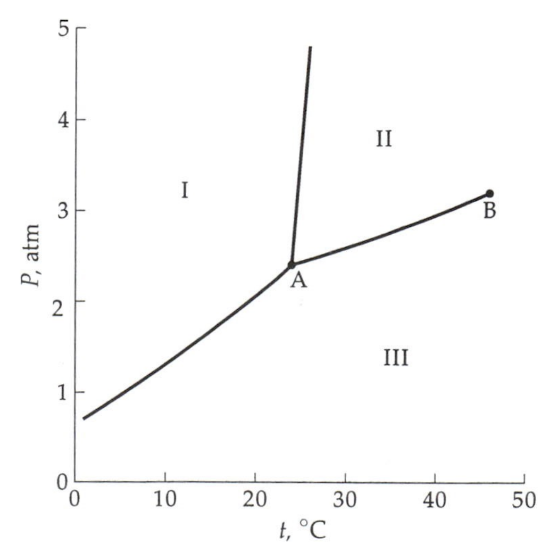

The x axis is a temperature axis and refers to the actual temperature of the sample. The y axis is a pressure axis. When a vapor in equilibrium with some other phase or a pure vapor is present, the y axis refers to the pressure of the vapor; when no vapor is present, it refers to the applied pressure on the sample. The three separate areas on the graph show where the substance exists as a solid (I), as a liquid (II), and as a vapor (III). As we expect, the solid is present at high pressures and low temperatures, and the vapor at high temperatures and low pressures. Each of the three lines represents the temperatures and pressures at which two phases are at equilibrium. The word *equilibrium* is important here; two (or more) phases may be present at almost any temperature and pressure, but the only temperatures and pressures at which two phases are at equilibrium fall on a line. The point at which the three lines meet is called the **triple point** (A). It is unique point, representing the only temperature and pressure at which the solid, liquid, and vapor can coexist at equilibrium. The triple point for the diagram above is at $T = 24°C$ and $P = 2.4$ atm.

EXAMPLE Interpreting a phase diagram

For the previous phase diagram, what is the lowest pressure at which the substance can exist as a liquid?

SOLUTION The liquid region of the curve defines all of the temperatures and pressures at which the substance can exist as a liquid. The lowest pressure for existence of the liquid is at the triple point; at lower pressures, the graph indicates that only the solid and vapor may exist. The pressure at the triple point, 2.4 atm, is the lowest pressure at which the liquid can exist.

EXERCISE What phase(s) of the substance are present at $T = 46°C$ and $P = 3.2$ atm, point B?

ANSWER: Liquid and vapor are present at equilibrium.

KEY CONCEPT C Cooling and heating curves

A cooling curve is a graph that shows the temperature of a substance (on the y axis) as heat is withdrawn (the amount of heat withdrawn is on the x axis); a heating curve shows the temperature as heat is added to a substance. Important points to notice regarding such a curve are the following:

1. Some parts of the curve correspond to changes in temperature. During these changes, no phase change occurs; added (or withdrawn) heat causes a change in molecular motion and, therefore, a change in temperature. To calculate the temperature change that accompanies the withdrawal (or addition) of a certain amount of heat, you must use the specific heat of the substance (for the correct physical state).
2. Other parts of the curve (the flat parts) correspond to a constant temperature. At these points, a phase change occurs; added (or withdrawn) heat, instead of changing molecular motion, causes the phase change to occur. You should consider this carefully; heat can be added or withdrawn with no accompanying temperature change. To calculate the amount of heat required to accomplish a particular phase change, you must use the enthalpy of fusion (for freezing or melting) or the enthalpy of vaporization (for boiling or condensation).
3. The flat parts of the curve correspond to the melting temperature and boiling temperature of the substance.

EXAMPLE Calculating a portion of a heating curve

We add 1000 J to a 15.0-g sample of liquid carbon tetrachloride at 55.0°C. Describe what happens.

Substance	Freezing point, °C	Enthalpy of fusion, J/g	Specific heat capacity (liquid), J/g·K	Boiling point, °C	Heat of vaporization, J/g	Specific heat capacity (vapor), J/g·K
CCl_4	−23.0	21.3	0.862	76.7	194	0.540

SOLUTION We note that at 55.0°C carbon tetrachloride is a liquid. Adding heat increases the temperature of the liquid up to the boiling temperature; at this point a phase change occurs. We first calculate the heat required to warm the liquid to its boiling point, which involves the specific heat capacity of the liquid:

$$\text{Heat} = \text{mass} \times \text{specific heat capacity} \times \text{change in temperature}$$

$$= 15.0\ \text{g} \times 0.862\ \frac{\text{J}}{\text{g K}} \times (76.7°\text{C} - 55.0°\text{C})$$

$$= 15.0\ \cancel{\text{g}} \times 0.862\ \frac{\text{J}}{\cancel{\text{g}}\ \cancel{\text{K}}} \times (21.7\ \cancel{\text{K}}) \qquad [\text{remember } \Delta T\ (°\text{C}) = \Delta T\ (\text{K})]$$

$$= 281\ \text{J}$$

At this point, the liquid is at its boiling temperature with 719 J (1000 J − 281 J) of heat left to add. As this heat is added, the liquid vaporizes. The heat of vaporization is 194 J/g, so it is clear that not all of the carbon tetrachloride can be vaporized; we must calculate what mass of carbon tetrachloride can be vaporized with the remaining 719 J.

$$\text{Heat} = \text{mass} \times \text{heat of vaporization (in J/g)}$$

$$\text{Mass} = \frac{\text{heat}}{\text{heat of vaporization}}$$

$$= \frac{719\ \cancel{\text{J}}}{194\ \cancel{\text{J}}/\text{g}}$$

$$= 3.71\ \text{g}$$

Thus, we are left with 11.3 g of liquid carbon tetrachloride at its boiling temperature of 76.7°C and 3.71 g of gaseous carbon tetrachloride at 76.7°C.

EXERCISE How much heat is required vaporize the remaining carbon tetrachloride and warm the whole sample to 85.0°C?

ANSWER: 2.26 kJ

THE STRUCTURES AND PROPERTIES OF SOLIDS

KEY WORDS Define or explain each of the following terms in a sentence or two.

allotrope
alloy
amorphous solid
close-packed structure
conduction band
coordination number
liquid crystal
network solid
semiconductor
superconductor
unit cell

10.6 METALS AND SEMICONDUCTORS

KEY CONCEPT A Close packing in metals

A **close-packed structure** is one in which the atoms packed together occupy the smallest possible volume. In the following figure, (a) shows a structure that is not close-packed and (b) shows a close-packed structure.

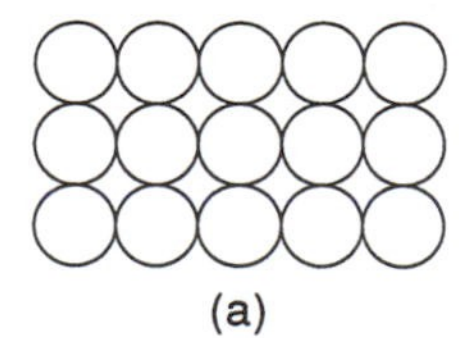
(a)

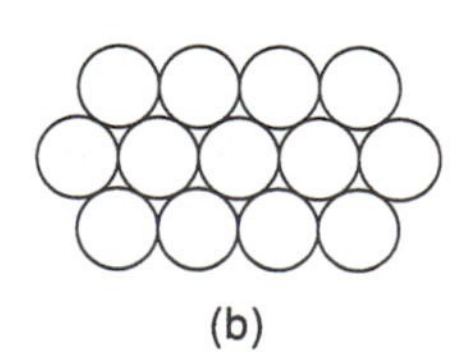
(b)

Two different close-packed structures are possible for packing spheres in three dimensions. To understand them, we imagine we have two layers of spheres, with the second layer fitting in the depressions made by the first. We now add a third layer. In the structure called **hexagonal close packing (hcp),** each sphere in the third layer is placed so that it lies directly above a sphere in the first layer; the first and third layers duplicate each other. In **cubic close packing (ccp),** each sphere in the third layer is placed in a depression that does not lie directly above a sphere in the first layer; the first and third layers are now not duplicates of each other. These two ways for making the third layer are the only ones possible for close packing, and each results in a complete third layer. Both close-packing schemes are equally efficient in packing spheres into the smallest possible volume. (Building these structures in three dimensions with styrofoam balls should be a great aid in understanding packing.) Every packing arrangement of spheres has an associated **coordination number,** which is the number of nearest neighbors an atom has in a particular structure. The coordination numbers for a few common arrangements are given in the following table.

Structure	**Coordination number**
hexagonal close packing (hcp)	12
cubic close packing (ccp); generates face-centered cubic (fcc) unit cell	12
body-centered cubic (bcc) unit cell	8

▼ EXAMPLE Counting coordination numbers

What is the coordination number of atom A in the two-dimensional structure shown below?

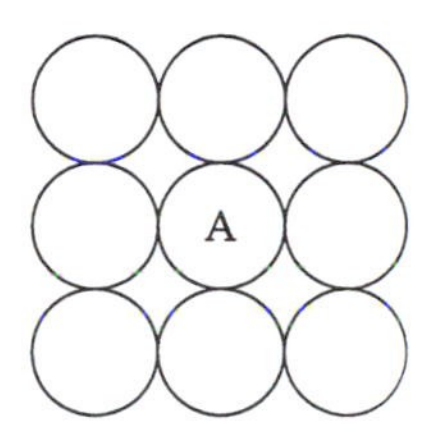

SOLUTION The coordination number of an atom is the number of "nearest neighbors." To find the nearest neighbors, we look for all atoms that are closer to A than any other atoms. In the next figure, the atoms marked NN (nearest neighbor) are closer to A than the other atoms and are its nearest neighbors. Atom A has four nearest neighbors and a coordination number of four:

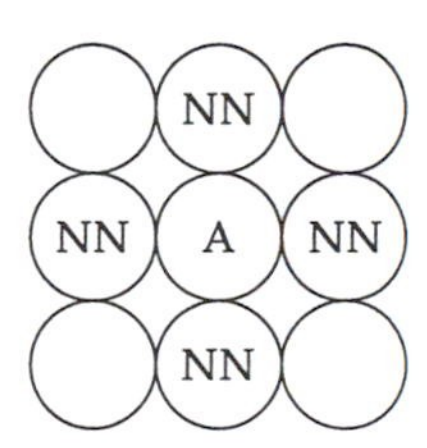

EXERCISE Use about 20 equally sized styrofoam spheres to build a model of a two-layer close-packed structure (because a third layer is not present, we cannot call the model hcp or ccp). By looking at your model, confirm that there are "holes" (air spaces) inside the structure. Looking from the top, you will see two different kinds of holes, one that lies directly above an atom in the lower layer and one that allows you to see through the two-layered structure to the table top. Now, take a very small sphere (a small marble will do) and place it in one of the holes that allows you to see to the table top. What is the coordination of the marble, as determined by the number of styrofoam-sphere nearest neighbors?

▲ ANSWER: 6

KEY CONCEPT B Unit cells

A crystal is characterized by a repetitive, orderly arrangement of molecules, atoms, or ions. The smallest group of molecules, atoms, or ions that repeats in space in three dimensions to make up the crystal is called the **unit cell** for that crystal. A simple analogy is to imagine a rubber stamp that would be used over and over again to stamp out the pattern of a crystal; the information on the rubber stamp would be the unit cell. Consider the following two-dimensional tile pattern, and try to imagine what the unit cell, or repeat unit, for this pattern is

The repeat unit or unit cell is the pattern that when repeated in both dimensions reproduces the whole tile pattern. In this case, the repeat unit is

In crystals, unit cells often contain parts of atoms at corners or in faces, because this makes it convenient to describe three-dimensional structures with highly symmetric unit cells.

▼ EXAMPLE Counting the atoms in a unit cell

What is the total number of atoms contained in a face-centered cubic (fcc) unit cell?

SOLUTION Some square wood blocks, each representing a cubic unit cell, might prove helpful in thinking about this problem. But first, we note that a fcc unit cell contains *parts* of 14 atoms; an atom is located at each of the 8 corners of the unit cell, and an atom is located in each of the 6 faces. Let's consider what part of an atom in a face is in the unit cell. Putting two blocks together, face to face, should convince you that an atom in a face is shared by two unit cells; thus, each unit cell contains $\frac{1}{2}$ of an atom in the face. Now place eight of the blocks together in a cube-shaped structure. Buried inside the structure is a point at which the corners of all eight blocks come together; you can't see it, but it is there. The implication is that an atom in a corner is shared by eight unit cells; so each unit cell contains $\frac{1}{8}$ of an atom at the corner of the unit cell. The total number of atoms in the unit cell is calculated as

$$\begin{aligned}\text{Total atoms in fcc unit cell} &= 6 \text{ faces} \times \frac{1}{2} \text{ atom per face} + 8 \text{ corners} \times \frac{1}{8} \text{ atom per corner} \\ &= 3 + 1 \\ &= 4\end{aligned}$$

EXERCISE How many atoms are contained in a body-centered cubic unit cell?

▲ ANSWER: 2

KEY CONCEPT C Metallic conductors and insulators

Molecular orbitals form when the atoms of a metal bond together in a sample of metal. Some of the molecular orbitals extend over the whole sample of metal; when a huge number of molecular orbitals with similar energies form, they merge together to form a *band.* If such a set of molecular orbitals is partially filled with electrons, the electrons are free to move from place to place in the sample and electrical conduction is possible. Such a set of molecular orbitals is called a **conduction band.** Heating a metal increases its resistance to electrical conduction; the increase in temperature causes the metal atoms to vibrate more vigorously and to collide with electrons in the conduction band, making conduction less efficient.

An **insulator** is a substance that does not conduct electricity. In many insulators, the band with electrons is completely filled, so electrons cannot move from place to place; such a band is called a **valence band.** Some insulators may have empty high-energy bands that extend over the whole sample, but the electrons in lower energy orbitals cannot move into the band; so conduction is not possible. The difference in energy between the potential conduction band and the orbitals that contain electrons is called the **band gap.** Insulators have a large band gap; metals have a zero band gap.

EXAMPLE Band gaps

Imagine that by putting a high mechanical pressure on an insulator, we can change the positions of the atoms so that the band gap is decreased to zero. Will the insulator become a conductor?

SOLUTION Yes. Once the band gap is zero, electrons from lower energy orbitals can easily move into the conduction band. Once they are in the conduction band, they are free to move from place to place in the solid; so the solid is a conductor.

EXERCISE Why are liquids such as water insulators? (Pure water is not a good conductor of electricity.)

ANSWER: The rapid motion of molecules prevents establishment of a permanent conduction band.

KEY CONCEPT D Semiconductors

Semiconductors usually have a modest ability to conduct electricity; they also have the suprising property that an increase in temperature will increase their ability to conduct electricity. The change in conductivity with temperature is explained by the fact that semiconductors have a small but nonzero band gap. As the temperature increases, electrons are easily excited into the normally empty conduction band, and so conduction increases. The higher temperature also causes more atom–electron collisions (as in a metal), but this effect is completely overwhelmed by the increase in the number of conduction electrons, so the overall outcome is an increase in conductivity.

Most commercial semiconductors are made from exceptionally pure silicon, into which a small amount of an impurity has been added. The addition of the impurity is called **doping.** When silicon is doped with a Group III element, such as indium, there are fewer electrons present than if the whole device were made of silicon because indium has one less valence electron than silicon. When an electrical current runs through the semiconductor, the semiconductor acts as though the lesser number of valence electrons has created positive charge (in actuality, the semiconductor remains electrically neutral). Thus, the device is called a **p-type semiconductor** (*p* for positive). If a group V element, such as arsenic, is used as the doping agent, excess electrons are present (relative to pure silicon), and electrical conduction appears to involve negative charge, resulting in a **n-type semiconductor** (*n* for negative).

EXAMPLE Doping silicon to form a semiconductor

What type of semiconductor is formed when silicon is doped with antimony?

SOLUTION Antimony (Sb) is a Group V element, whereas silicon is a Group IV element. The antimony atoms have five valence electrons and the silicon atoms four, so the addition of antimony results in the presence of excess electrons relative to silicon. This means there is excess negative charge, so the semiconductor is an *n*-type semiconductor.

EXERCISE Why do semiconductors undergo an increase in conductivity as the temperature is increased?

ANSWER: Increasing the temperature excites electrons across the band gap into the conduction band.

10.7 IONIC SOLIDS

KEY CONCEPT The rock salt and cesium chloride structures in ionic crystals

Unlike metals, ionic crystals are often made up of different-sized "spheres." That is, the anion in the crystal might be very much different from the cation in size. The problem of how the ions pack together is much more complicated than in a metal. It turns out that the relative sizes of the anion and cation are very important in determining the actual packing arrangement. In the **rock-salt structure,** the larger anions pack in face-centered cubic unit cells. The smaller cations fit into holes between the anions in such a way that the cations also form a face-centered cubic unit cell; unlike the anions, however, the cations are not right next to each other. In this structure each anion has six cations as nearest neighbors and each cation six anions, so the structure has (6,6)-coordination. In the **cesium chloride structure** (which is the structure of cesium chloride as well as many other salts), the very large cations occupy the corners of a cube and a smaller anion is at the center of the cube. The nearest neighbors of the centrally located anion are the eight cations at the corners of the cube, so the anion has a coordination number of 8. From another point of view, the anions in the cesium chloride structure are also located at the corners of a cube, with the larger cation at the center of the cube. The cation therefore also has coordination number 8, so the cesium chloride structure has (8,8)-coordination. The coordination number is expressed as two numbers: cation coordination number and anion coordination number; the two numbers are not always the same.

▼ **EXAMPLE The effect of size on the structure of an ionic solid**

Is CaS more likely to crystallize in the rock-salt structure or the cesium chloride structure?

SOLUTION The relative size of the ions in a crystal largely (but not solely!) determines the structure found in the crystal. Sodium chloride crystallizes in the rock-salt structure and cesium chloride in the cesium chloride structure, so we compare the size of the ions in CaS to the size of the ions in NaCl and CsCl.

	Size of ion, pm		
Ion	**NaCl**	**CsCl**	**CaS**
anion	181	181	184
cation	102	170	100

As we can see, the Cl^- ion is about the same size as the S^{2-} ion; but the Ca^{2+} ion with its 100-pm radius is much closer in size to the Na^+ ion with its 102-pm radius than to the Cs^+ ion with a 170-pm radius. Because the ions in CaS are much closer in actual size to those in NaCl than to those in CsCl, we conclude that CaS crystallizes in the NaCl structure. We should be careful to note that it is the size of the ions relative to each other that determines the structure, not their absolute size. LiF, for example, is composed of ions that are much smaller than those in NaCl, but it also crystallizes in the rock-salt structure. This is because the size of F^- relative to Li^+ is the same as that of Cl^- relative to Na^+.

EXERCISE Is it possible for a salt with the general formula MX to crystallize in a structure in which the unit cell has anions at the corners of a cube and cations in the faces?

▲ ANSWER: No, because the unit cell would not have zero charge

10.8 OTHER TYPES OF SOLIDS

KEY CONCEPT A Network solids

A network solid is characterized by an extensive network of covalent bonds that links all of the atoms in the solid. If we imagine that the covalent bonds are roads, it would be possible to drive from any atom to every other atom in the solid. Network solids are often very hard because the extensive network of bonds holds the atoms in place very well. However, if some of the bonds in the solid are weak, as in graphite and $AlCl_3$, the solid is not hard. The bonds in a network solid are usually highly localized, so they do not possess low-energy molecular orbitals that extend throughout the solid in a continuous fashion. Thus, they are usually insulators. (Graphite is an exception.) Graphite and diamond are both network solids made of linked carbon atoms; graphite and diamond are also **allotropes,** or different forms of an element in which the atoms are bonded differently.

EXAMPLE **Bonding in network solids**

Many texts claim that a diamond is actually one large molecule. In what way is this true?

SOLUTION A molecule is a group of atoms bonded together as a unit. A diamond consists of a very large number of carbon atoms bonded to each other, so in some sense it can be thought of as a molecule. However, most chemists would agree that a molecule should follow the law of constant composition; that is, a molecule always has the same number of atoms of each kind in it. Diamond of different sizes have different numbers of carbon atoms. So in some ways a diamond is a huge molecule; in some ways it is not.

EXERCISE Is the fact that the network solid graphite is soft consistent with being a conductor?

ANSWER: Yes, because both properties indicate that some of its bonds are not highly localized

KEY CONCEPT B Molecular solids

In a molecular solid, individual molecules are held together in the solid phase by one or more of the weak intermolecular forces we discussed earlier (dipole–dipole attractions, London forces, and hydrogen bonds). The weakness of the force holding the molecules together means it is easy to pull the molecules away from each other. Therefore, molecular solids are generally low melting and relatively soft. **Liquid crystals** are a special type of molecular solid intermediate in properties between a liquid and solid. This physical state, which shows a liquid-crystalline behavior is called a **mesophase.** Three different mesophases, which differ primarily in the amount of order in the material, are the **smectic phase** (most ordered, almost crystalline), **cholesteric phase** (ordered, but in a noncrystalline manner), and the **nematic phase** (least ordered). Not all liquid crystals can adopt each of these phases; the possibilities are dictated by the structure of the molecules involved.

EXAMPLE **The properties of solids**

Should a molecular solid be a good electrical conductor?

SOLUTION Electrical conduction occurs when an extensive molecular orbital extends throughout the solid. In a molecular solid, there are molecular orbitals within each molecule, but because the molecules do not covalently bond to each other, no molecular orbital extends throughout the solid. The lack of an extended molecular orbital leads us to predict that molecular solids are insulators.

EXERCISE For a material that can adopt both a smectic phase and a nematic phase, which mesophase should be present at relatively high temperatures?

ANSWER: Nematic

DESCRIPTIVE CHEMISTRY TO REMEMBER

- Lithium and sodium frequently form **hydrated salts,** but their larger cogeners in group I (potassium, rubidium, and cesium) rarely do.
- **Barium** and **lanthanum** salts are often hydrated because of the small size and high charge on the metal ion.
- **Hydrocarbons** with up to 4 carbon atoms are gases, those with 5 through 17 carbons are liquids, and those containing 18 or more carbons are solids.

- **Water** is unusual because it expands when it freezes.
- Magnesium and zinc crystallize in a **hcp** structure; aluminum, copper, silver, and gold in a ccp structure; iron, sodium, and potassium in a **bcc** structure.
- **Titanium chloride,** $TiCl_4$, is a liquid that boils at 136°C and freezes at −25°C to a molecular solid.
- **Rhombic sulfur,** a molecular solid, of S_8 rings, melts at 113°C to form a mobile straw-colored liquid.
- **Mercury** forms a meniscus that bulges upward; water forms a meniscus that bulges downward.
- **Boron nitride,** BN, is a white, fluffy, slippery powder; it is an insulator. Under pressure at 1650°C, it changes to a very hard structure known as borazon, which is almost as hard as diamond but more resistant to oxidation.
- Common glass is an **amorphous** solid.
- Compressing graphite at over 80,000 atm and 1500°C in the presence of a trace amount of a metal such as chromium or iron results in the formation of natural **diamond.**
- **Graphite** is a black, lustrous, electrically conducting, slippery solid which sublimes at 3700°C.
- **Boron carbide** ($B_{12}C_3$) is made up of B_{12} units and C_3 units bonded to each other. It is a very hard network solid.
- **Ice** is an open network of water molecules held together by hydrogen bonds.

CHEMICAL EQUATIONS TO KNOW

- Boron nitride (BN) is formed when boron trichloride (BCl_3) and ammonia (NH_3) are heated strongly together.

$$BCl_3(g) + NH_3(g) \longrightarrow BN(s) + 3HCl(g)$$

MATHEMATICAL EQUATIONS TO KNOW AND UNDERSTAND

(All mathematical equations in this chapter are optional; check with your instructor.)

$$h = \frac{2\gamma}{dgr}$$ height of liquid

$$\ln P' = \ln P + \frac{\Delta H_{\text{vap}}}{R}\left[\frac{1}{T} - \frac{1}{T'}\right]$$ Clausius–Clapeyron equation

$$I = \frac{V}{R}$$ Ohm's law

$$\Delta H^{\circ}_{\text{vap}} = T_{\text{b}} \times 85\ \text{J/(K·mol)}$$ Trouton's rule

$$2d \sin \theta = \lambda$$ Bragg equation

SELF-TEST EXERCISES

Forces between atoms, ions, and molecules

1. Which type of solid is best described as "cations held together by a sea of electrons"?
(a) ionic (b) molecular (c) network (d) metal

2. Which type of force is least affected by changes in distance?
(a) ion–ion (b) ion–dipole (c) dipole–dipole (d) London force

3. What is the ratio of the energies of interaction, E_1/E_2, of two ions if the distance between them triples from $d_1 = 100$ pm to $d_2 = 300$ pm?
(a) 9/1 (b) 1/9 (c) 3/1 (d) 1/1 (e) 1/3

4. Which of the following is predicted to have the highest lattice enthalpy?
(a) $CaCl_2$ (b) MgO (c) LiF (d) BaO (e) KBr

5. The hydration of cations in solution is due largely to which interaction?
(a) London forces (b) ion–ion (c) dipole–dipole (d) ion–dipole

6. Which cation is likely to be the most strongly hydrated? The radius of the cation is shown for each answer.
(a) Sr^{2+} (113 pm) (b) K^+ (133 pm) (c) Ba^{2+} (135 pm)
(d) NH_4^+ (143 pm) (e) La^{3+} (115 pm)

7. The energy of interaction of the dipole–dipole interaction is proportional to which of the following for a nonrotating dipole?
(a) $1/d$ (b) $1/d^6$ (c) $1/d^3$ (d) $1/d^2$ (e) $1/d^{12}$

8. Which molecule experiences London forces only?
(a) HCl (b) H_2Se (c) H_2O (d) P_4 (e) HF

9. Which molecule experiences London forces only?
(a) NH_3 (b) H_2O (c) CO_2 (d) CH_3OH (e) SO_2

10. Which of the atoms shown experiences the largest London forces?
(a) Kr (b) He (c) Xe (d) Ne (e) Ar

11. Predict which of the following has the highest boiling point.
(a) C_4H_{10} (*n*-butane, cylindrical) (b) C_4Cl_{10} (perchoro-*n*-butane, cylindrical)
(c) C_4H_{10} (isobutane, roughly spherical) (d) C_4Cl_{10} (perchloro-isobutane, roughly spherical)

12. Which of the following undergoes hydrogen bonding?
(a) HCl (b) H_2 (c) CH_4 (d) CFH_3 (e) CH_3OH

13. A hydrogen bond is
(a) the covalent bond between two hydrogen atoms in H_2.
(b) an especially strong intermolecular attractive force.
(c) a covalent bond between hydrogen and oxygen.
(d) an ionic bond between H^- and a Group I metal cation.

14. All but one of the following contribute to the hydrogen bonding. Which one doesn't?
(a) Small size of hydrogen.
(b) Polarizability of hydrogen.
(c) Small size of the more electronegative atom (O, F, N).
(d) High electronegativity of the more electronegative atom (O, F, N).
(e) Low electronegativity of hydrogen.

The properties of liquids

15. Which has the highest viscosity?
(a) water (b) gasoline (c) rubbing alcohol (d) honey

16. For most liquids the viscosity ______ as temperature ______.
(a) decreases, increases (b) increases, increases

17. On the basis of its structure, which of the following would be predicted to have the highest viscosity?

```
        H   H   H               OH  OH  OH              H   H   H
        |   |   |               |   |   |               |   |   |
(a) H — C — C — C — OH   (b) H — C — C — C — H   (c) H — C — C — C — H
        |   |   |               |   |   |               |   |   |
        H   H   H               H   H   H               H   H   H
```

18. A molecule at the surface of a liquid has ______ energy than a molecule in the bulk of the liquid.
(a) higher (b) lower (c) the same

19. If a liquid in a glass tube forms a meniscus that is essentially flat (it bulges neither upward nor downward), the forces between liquid molecules are ______ the forces between the liquid molecules and the glass.
(a) greater than (b) equal to (c) less than

20. Capillary action refers to a phenomenon in which a liquid
(a) flows through a pipe because of a pressure differential.
(b) is pushed up a tube by a pressure differential.
(c) forms a sphere because of cohesive forces.
(d) forms a meniscus.
(e) rises up a narrow tube because of adhesive forces.

21. Which liquid, on the basis of its structure, has the lowest vapor pressure?

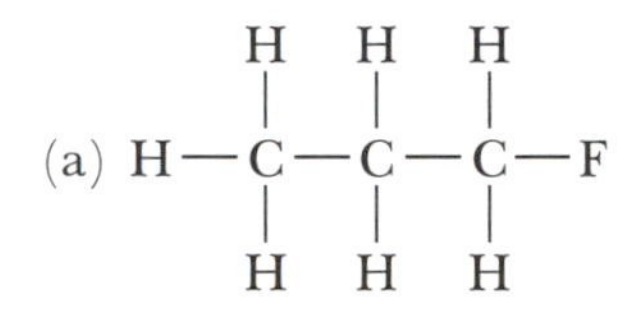

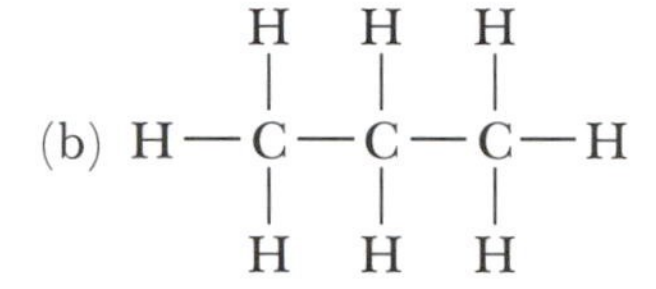

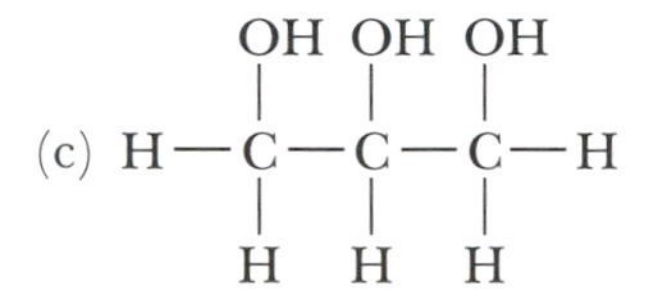

22. What is the normal boiling point of the liquid that has the vapor pressure curve shown in the following figure?

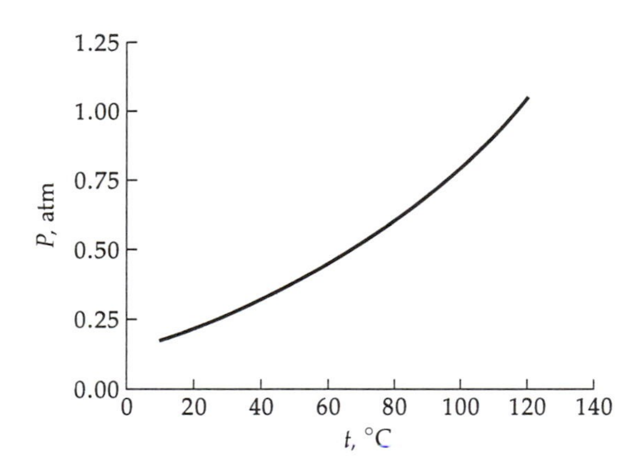

(a) 18°C (b) 126°C (c) 83°C (d) 100°C (e) 118°C

23. Which of the liquids in question 21 has the lowest boiling point?

24. Based on Trouton's rule, what is the enthalpy of vaporization of carbon tetrachloride? Its boiling point is 76.7°C.
(a) 4.1 J/mol (b) 30 kJ/mol (c) 243 kJ/mol (d) 6.6 kJ/mol (e) 111 J/mol

25. The critical temperature of oxygen is −118°C. What pressure is required to liquefy oxygen at room temperature, about 25°C?
(a) 1 atm (b) 10 atm (c) 25 atm
(d) 1000 atm (e) It is impossible to liquefy oxygen at 25°C.

26. In which state of matter is it very difficult for a molecule to move from one place to another?
(a) gas (b) solid (c) liquid

Use the following phase diagram for questions 27–29.

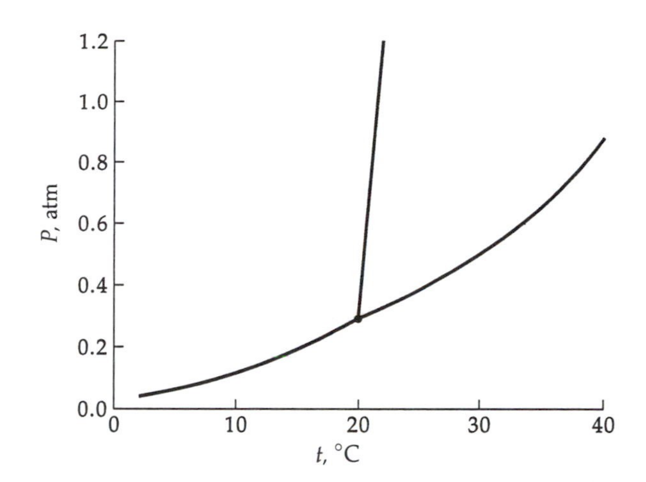

27. What is the physical state of the substance at 10°C and 1.0 atm?
(a) solid (b) liquid (c) vapor
(d) vapor in equilibrium with liquid (e) solid in equilibrium with liquid

28. What is the minimum pressure that must be applied to liquefy the substance?
(a) 1.2 atm (b) 0.2 atm (c) 0.1 atm (d) 0.3 atm (e) 0.5 atm

29. At what temperature and pressure are only the solid and vapor in equilibrium?
(a) 10°C, 0.2 atm (b) 20°C, 0.5 atm (c) 18°C, 0.3 atm
(d) 30°C, 0.5 atm (e) 15°C, 0.2 atm

30. How much heat must be withdrawn from 12.0 g of liquid carbon tetrachloride (CCl_4) at 25.0°C to completely solidify it?

Substance	Freezing point, °C	Enthalpy of fusion, J/g	Specific heat capacity (liquid), J/g·K	Boiling point, °C	Heat of vaporization, J/g	Specific heat capacity (vapor), J/g·K
CCl_4	−23.0	21.3	0.862	76.7	194	0.540

(a) 863 J (b) 256 J (c) 752 J (d) 497 J (e) 514 J

The structures and properties of solids

31. Four of the following phrases apply to a crystalline solid. Which one does not?
(a) near random arrangement of molecules (b) contains unit cell (c) repetitive pattern
(d) orderly array (e) well-defined forces

32. What is the distance between layers of atoms if x-rays of wavelength 110 pm constructively interfere when the beam of x-rays is at an angle of 25° to the layers?
(a) 144 pm (b) 130 pm (c) 260 pm (d) 110 pm (e) 93.0 pm

33. What is the coordination number of a metal atom in a hexagonal close-packed structure?
(a) 6 (b) 12 (c) 8 (d) 10 (e) 4

34. What percentage of the volume of a face-centered cubic (fcc) unit cell is occupied by atoms in the cell. (The fcc unit cell is a cubic close-packed arrangement; the correct answer corresponds to the most efficient way to pack equal-sized spheres.)
(a) 66.7% (b) 44.4% (c) 81.5% (d) 100% (e) 74.0%

35. A representation of a brick wall is shown below.

What is the number of bricks in the unit cell for this two-dimensional pattern?
(a) 2 (b) 1 (c) 3 (d) $1\frac{1}{2}$ (e) 4

36. Which mixture is an alloy?
(a) $NaCl + H_2O$ (dissolved) (b) $NaCl + KNO_3$ (ground together)
(c) $H_2O + CH_3OH$ (dissolved together) (d) Au + Sn (dissolved together)

37. Which is a semiconductor?
(a) Cu (b) Co doped in Au (c) NaCl (d) graphite (e) As doped in Si

38. What happens to the resistance of a conductor when its temperature is increased?
(a) decreases (b) increases (c) stays the same

39. What is a typical band gap in metals?
(a) 2 kJ (b) 0 kJ (c) 30 kJ

40. What is a typical resistance for a superconductor?
(a) zero (b) very high (c) very low (d) depends on the superconductor

41. Doping silicon with which element results in an *n*-type semiconductor?
(a) In (b) Ge (c) As (d) Ga (e) B

42. What type of coordination is present in the rock-salt structure?
(a) (8, 8) (b) (6, 6) (c) (12, 12) (d) (8, 4) (e) (6, 3)

43. Which describes a typical network solid?
(a) atoms all connected to each other (b) very small band gap
(c) soft, and low melting (d) contains a vast array of ions
(e) held together by intermolecular attractive forces

44. Which of the following is a pair of allotropes?
(a) ^{12}C, ^{14}C (b) H_2O, H_2O_2 (c) ice, water (d) S_8, S_2 (e) Na, K

45. Which of the following forms a molecular solid?
(a) C (graphite) (b) H_2O (c) NaCl (d) C (diamond) (e) $AlCl_3$

46. A mesophase
(a) is highly crystalline. (b) has regular faces.
(c) has both liquid and solid properties. (d) has a very low viscosity.
(e) contains a regular array of ions.

Descriptive chemistry

47. Which substance is noted for the fact that it expands when it freezes?
(a) copper (b) silicon (c) iron(III) oxide
(d) water (e) all substances expand when they freeze

48. Which salt can be found in hydrated form?
(a) Na_2SO_4 (b) KCl (c) $BaSO_4$ (d) NH_4NO_3 (e) $Sr(NO_3)_2$

49. Boron nitride can be formed by heating together
(a) $B + N_2$ (b) $B + NO_2$ (c) $BCl_3 + N_2$ (d) $B + NH_3$ (e) $BCl_3 + NH_3$

50. Rhombic sulfur (the normal room-temperature form of sulfur) exists as
(a) S_2 diatomic molecules. (b) S_8 rings. (c) S_4 tetrahedra.
(d) S atoms. (e) S_3 rings.

11 THE PROPERTIES OF SOLUTIONS

MEASURES OF CONCENTRATION

KEY WORDS Define or explain each of the following terms in a written sentence or two.

electrolyte	molar concentration (molarity, M)	nonelectrolyte
mass concentration	molar fraction (x)	parts per million (ppm)
molality (m)		

11.1 EMPHASIZING THE AMOUNT OF SOLUTE IN SOLUTION

KEY CONCEPT A Molarity review

The molarity (molar concentration) gives the amount of solute present in a specific volume of solution:

$$\text{Molarity M} = \frac{\text{moles solute}}{\text{liters solution}}$$

This way of specifying concentration is convenient because a chemist can measure out a desired number of moles of solute by measuring a specific volume of solution. This becomes evident if we rearrange the preceding formula:

$$\text{Moles solute} = (\text{molarity}) \times (\text{liters solution})$$

Thus, by using the correct volume of solution of known molarity, we can obtain a desired number of moles of solute. It is sometimes convenient to deal with millimoles (1 mmol = 1×10^{-3} mol), in which case the following formulas apply:

$$\text{Molarity} = \frac{\text{millimoles solute}}{\text{milliliters solution}}$$

$$\text{Millimoles solute} = (\text{molarity}) \times (\text{milliliters solution})$$

EXAMPLE Using the molar concentration

How many milliliters of 0.222 M NaOH are required to deliver 3.00 mmol NaOH?

SOLUTION The problem gives us the molarity of a solution and the number of millimoles of solute required. We start with the formula for millimole solute and divide both sides by the molarity:

$$\text{Millimole solute} = (\text{molarity}) \times (\text{milliliter solution})$$

$$\text{Milliliter solution} = \frac{\text{millimole solute}}{\text{molarity}}$$

Substituting the values given in the problem gives

$$\text{Milliliter solution} = \frac{3.00 \text{ mmol NaOH}}{0.222 \text{ M}}$$

$$= \frac{3.00 \cancel{\text{mmol NaOH}}}{0.222 \cancel{\text{mmol NaOH}}/\text{mL}}$$

$$= 13.5 \text{ mL}$$

EXERCISE A student pours 25.0 mL of 0.866 M HCl into a beaker. How many moles of HCl are in this volume of solution?

ANSWER: 0.0217 mol HCl

KEY CONCEPT B Mass concentration

The mass concentration, like the molarity, gives the amount of solute present in a specific volume of solution:

$$\text{Mass concentration of solute B} = \frac{\text{grams of B}}{\text{liters of solution}}$$

The units for mass concentration are grams per liter (g/L). Algebraic manipulation of the defining formula for mass concentration gives

$$\text{Grams solute} = \text{mass concentration} \times \text{liters solution}$$

EXAMPLE Using mass concentration

How many grams of $FeCl_2$ are there in 15.0 mL of a solution with a mass concentration $FeCl_2$ = 25.0 g/L?

SOLUTION We note that 15.0 mL equals 0.0150 L and substitute the given values of mass concentration and liters solution into the preceding formula:

$$\text{Grams solute} = \text{mass concentration} \times \text{liter solution}$$
$$\text{Grams } FeCl_2 = (25.0 \text{ g } FeCl_2/\cancel{L}) \times 0.0150 \cancel{L}$$
$$= 0.375 \text{ g } FeCl_2$$

EXERCISE 1.25 g $AgNO_3$ is dissolved in enough water to make 25.0 mL of solution. What is the mass concentration of the solution?

ANSWER: 50.0 g/L

11.2 EMPHASIZING RELATIVE AMOUNTS OF SOLUTE AND SOLVENT MOLECULES

KEY CONCEPT A Mole fraction

We now discuss a slightly different kind of concentration unit, one that gives the number of moles of one substance relative to the total number of moles of all of the substances in the solution. The volume of solution or solvent is not involved in the mole fraction, only the moles of substance present. Even though a solute and solvent are present, the mole fraction concentration unit treats all substances equally. So, if we have a solute mixed with a solvent,

$$\text{Mole fraction solute} = \frac{\text{mole solute}}{\text{mole solute} + \text{mole solvent}}$$

$$\text{Mole fraction solvent} = \frac{\text{mole solvent}}{\text{mole solute} + \text{mole solvent}}$$

Using the symbols

n_A = mol solvent n_B = mol solute

x_A = mole fraction solvent x_B = mole fraction solute

we get

$$x_A = \frac{n_A}{n_A + n_B} \qquad x_B = \frac{n_B}{n_A + n_B}$$

It is always true that the sum of the mole fractions of all of the substances in a solution equals 1. For the solution just discussed, $x_A + x_B = 1$. The mole fraction is dimensionless.

EXAMPLE Calculating mole fractions

If we add 100 g C_2H_5OH to 100 g H_2O, what is the mole fraction of each component of the solution? Do the mole fractions sum to 1?

SOLUTION The mole fraction of each component is given by

$$x_{C_2H_5OH} = \frac{n_{C_2H_5OH}}{n_{C_2H_5OH} + n_{H_2O}} \qquad x_{H_2O} = \frac{n_{H_2O}}{n_{C_2H_5OH} + n_{H_2O}}$$

where $n_{C_2H_5OH}$ = mol C_2H_5OH and n_{H_2O} = mol H_2O. We calculate the moles of each component using their respective molar mass:

$$n_{C_2H_5OH} = 100 \text{ g} \times \frac{1 \text{ mol } C_2H_5OH}{46.1 \text{ g}} = 2.17 \text{ mol } C_2H_5OH \qquad n_{H_2O} = 100 \text{ g} \times \frac{1 \text{ mol } H_2O}{18.0 \text{ g}} = 5.56 \text{ mol } H_2O$$

Substituting these values into the defining equations for mole fraction gives

$$x_{C_2H_5OH} = \frac{n_{C_2H_5OH}}{n_{C_2H_5OH} + n_{H_2O}} = \frac{2.17 \cancel{\text{mol}}}{2.17 \cancel{\text{mol}} + 5.56 \cancel{\text{mol}}} = 0.281 \qquad x_{H_2O} = \frac{n_{H_2O}}{n_{C_2H_5OH} + n_{H_2O}} = \frac{5.56 \cancel{\text{mol}}}{2.17 \cancel{\text{mol}} + 5.56 \cancel{\text{mol}}} = 0.719$$

For the second part of the question, the mole fractions sum to 1, as they should:

$$x_{C_2H_5OH} + x_{H_2O} = 0.281 + 0.719 = 1.000$$

EXERCISE What is the mole fraction of each component of the solution made by dissolving 25.0 g CCl_4 in 50.0 g C_5H_{12}. Do the mole fractions sum to 1?

ANSWER: $x_{CCl_4} = 0.190$, $x_{C_5H_{12}} = 0.810$; yes

KEY CONCEPT B Molality

The molality *m* of a solution is obtained by dividing the number of moles of solute by the kilograms of solvent. For this unit of concentration, the volume of solution is irrelevant.

$$\text{Molality} = \frac{\text{moles of solute}}{\text{kilograms of solvent}}$$

This unit is convenient because the concentration expressed in molality does not change as the temperature of the solution changes. The units of molality are moles per kilogram (mol/kg.)

EXAMPLE Preparing a solution of known molality

What is the molality of a solution made by adding 125 g Na_2SO_4 to 500 g H_2O?

SOLUTION The molality is defined as

$$\text{Molality} = \frac{\text{moles of solute}}{\text{kilograms of solvent}}$$

For this problem, we have

$$\text{Molality } Na_2SO_4 = \frac{\text{mol } Na_2SO_4}{\text{kg } H_2O}$$

The moles of Na_2SO_4 are calculated with the molar mass of Na_2SO_4, which is 142.1 g/mol:

$$125 \text{ g} \times \frac{1 \text{ mol } Na_2SO_4}{142.1 \text{ g}} = 0.880 \text{ mol } Na_2SO_4$$

We note that 500 g = 0.500 kg and calculate the molality by substituting into the equation for molality:

$$\text{Molality} = \frac{0.880 \text{ mol } Na_2SO_4}{0.500 \text{ kg } H_2O} = 1.76\ m$$

EXERCISE What is the molality of the solution made by dissolving 10.0 g ethylene glycol ($C_2H_6O_2$) in 50.0 g water?

ANSWER: 3.22 *m*

PITFALL Distinguishing molality from molarity

Care must be taken to distinguish the concentration unit molality from the concentration unit molarity. Molarity is defined as moles solute per liter solution and is symbolized by an upper case Roman M. Molality is defined as moles solute per kilogram solvent and is symbolized by a lower case italic *m*.

$$\text{Molarity (M)} = \frac{\text{moles solute}}{\text{liter solution}} \qquad \text{Molality } (m) = \frac{\text{moles solute}}{\text{kilogram solvent}}$$

Because the density of water is close to 1 kg/L, the molarity and molality of a dilute aqueous solution are very close to the same number. For a nonaqueous solution in which the density of solvent is not 1 or for a concentrated aqueous solution, the molarity and molality can be very different.

SOLUBILITY

KEY WORDS Define or explain each of the following terms in a written sentence or two.

disorder of matter
enthalpy of solution
Henry's law
hydrated
hydrophilic group
hydrophobic group
Le Chatelier's principle
saturated solution
solubility
solvated

11.3 SATURATION AND SOLUBILITY

KEY CONCEPT A Solubility and saturated solutions

For most solutes, there is a limit to the amount of solute that can be dissolved in a given amount of a particular solvent. For instance, we can imagine what would happen if we tried to dissolve a 5-lb bag of sugar in a cup of water. At some point, the water would contain as much sugar as it can possibly hold and no more would dissolve; any more added sugar would just sink to the bottom of the cup. At this point, the solution is called a **saturated solution.** The amount of sugar that we dissolved, which is the maximum amount the water can hold, is called the **solubility** of the sugar. The solubility of any liquid or solid depends on the solvent and the temperature. When a mass of undissolved solid is in contact with a saturated solution, like our undissolved sugar on the bottom of a cup, a dynamic equilibrium is set up between the undissolved solute and the solute in the solution:

$$\text{Sugar}(s) \rightleftharpoons \text{sugar}(aq)$$

The mass of dissolved solute and of undissolved solute stays the same, but there is a constant interchange of dissolved solute molecules for undissolved solute molecules and vice versa. For every solute molecule that dissolves from the solid mass of solute, a solute molecule rejoins the solid. "Whatever is done is undone."

EXAMPLE Determining if a solution is saturated

If we add 75.0 g of solid $CaCl_2$ to 100 g of water in a beaker at 0°C, does a saturated solution result? If, so what is the mass of the undissolved $CaCl_2$ at the bottom of the beaker? Does a dynamic equilibrium exist? The solubility of $CaCl_2$ in water at 0°C is 59.5 g/100 g H_2O.

SOLUTION The solubility of 59.5 g/100 g H_2O means that the maximum amount of $CaCl_2$ that can be dissolved in 100 g of water at 0°C is 59.5 g. If we add 75.0 g $CaCl_2$ to 100 g water at 0°C, 59.5 g dissolves and 15.5 g remains undissolved. The solution would now contain as much $CaCl_2$ as it could hold, so it is saturated. In this situation, the undissolved solid becomes involved in a dynamic equilibrium with solute in the solution.

EXERCISE What is the molarity of a saturated aqueous solution of CaF_2 at 0°C? Because CaF_2 is only sparingly soluble, assume that dissolving CaF_2 in 1000 mL of water results in 1000 mL of solution; that is, the addition of a small amount of CaF_2 to 1000 mL of water results in no appreciable change in the volume of the water. The solubility of CaF_2 is 1.7×10^{-3} g/100 g H_2O.

ANSWER: 2.2×10^{-4} M

KEY CONCEPT B The dependence of solubility on the solvent

A rough rule of thumb used to predict whether a solute is soluble in a specific solvent is "likes dissolve likes." This translates to the following:

1. Nonpolar solutes are relatively soluble in nonpolar solvents but are relatively insoluble in polar and hydrogen-bonded solvents.
2. Ionic, polar, and hydrogen-bonded solutes are relatively soluble in polar and hydrogen-bonded solvents but are relatively insoluble in nonpolar solutes.

These rules are only approximations. In exacting work, where solubilities must be accurately known, they should be looked up.

EXAMPLE Predicting solubilities

Which of the following is least soluble in H_2O: benzene (C_6H_6), hydrogen peroxide (H_2O_2), or dinitrogen trioxide (N_2O_3)?

SOLUTION Water is a hydrogen-bonded solvent, so we expect that polar, ionic, and hydrogen-bonded substances will be relatively soluble in it and that nonpolar substances will be relatively insoluble. H_2O_2 forms hydrogen bonds; it is extremely soluble in water. N_2O_3 is polar; it is soluble in water. C_6H_6 is nonpolar; it is relatively insoluble in water and is the least soluble of the three compounds given.

EXERCISE Is CCl_4 more soluble in benzene or in water?

ANSWER: Benzene

KEY CONCEPT C The consequences of different solubilities

The solubilities of various substances affect our biology, agriculture, economy and hygiene. In this section, we consider an example of each.

Biology. Our bones are composed largely of calcium phosphate [$Ca_3(PO_4)_2$], which is insoluble in water. Because we have so much water in our bodies, any body structural material must be insoluble.

Agriculture. The phosphate ion (PO_4^{3-}) is essential to plant life (as well as animal life), and many fertilizers contain a form of phosphate. Nutrients are absorbed by plant roots as aqueous solutions; commercial fertilizers often contain phosphorus in the form of the soluble salt calcium hydrogen phosphate, $CaHPO_4$.

Economy. The formation of boiler scale ($CaCO_3$) inside hot water pipes exacts an enormous economic cost in the United States. Hard water contains, among other things, the water-soluble calcium hydrogen carbonate. When the water is heated, however, the hydrogen carbonate ion decomposes:

$$2HCO_3^-(aq) \longrightarrow CO_3^{2-}(aq) + H_2O(l) + CO_2(g)$$

Dissolved calcium ion now reacts with the newly formed carbonate ion to form insoluble calcium carbonate (boiler scale) on the inside of the pipes:

$$Ca^{2+}(aq) + CO_3^{2-}(aq) \longrightarrow CaCO_3(s)$$

The difference in solubilities of $Ca(HCO_3)_2$ and $CaCO_3$ is responsible for this costly problem.

Hygiene. Much of the dirt and grime we want to wash from our clothes and bodies is nonpolar in nature, so it does not dissolve in water very well. A soap or detergent contains long molecules that have a polar end and a nonpolar end. In formation of a **micelle**, the nonpolar end dissolves and traps the nonpolar dirt molecules; the polar end allows the micelle to remain dissolved in the polar water. The overall effect is that water can be used to dissolve a nonpolar material. A nonpolar species is **hydrophobic** (water fearing) and a polar or hydrogen-bonded species is **hydrophilic** (water loving), so the type of species found in soap is said to have a hydrophobic end and a hydrophilic end.

EXAMPLE Understanding solubilities

Suggest a method, based on solubilities, for removing boiler scale (calcium carbonate) from the inside of a hot water pipe.

SOLUTION Calcium hydrogen carbonate is much more soluble than calcium carbonate. Running a dilute HCl solution through the pipe should dissolve the scale through the reaction

$$CaCO_3(s) + H_3O^+(aq) \longrightarrow Ca^{2+}(aq) + HCO_3^-(aq) + H_2O(l)$$

The pipe must be made of a material that does not dissolve in the acid.

EXERCISE Is it possible, in principle, for the molecules in soap to cause a polar substance to be dissolved in a nonpolar solvent?

ANSWER: Yes

11.4 THE EFFECT OF PRESSURE ON GAS SOLUBILITY

KEY CONCEPT A The effect of pressure on gas solubility and Henry's law

Henry's law states that the solubility of a gas increases linearly with the partial pressure of the gas:

$$\text{Solubility of gas} = \text{constant} \times \text{partial pressure of gas}$$

$$S = k_H \times P_{gas}$$

The Henry's law constant k_H depends on the identity of the gas, the solvent, and the temperature.

EXAMPLE Using Henry's law

How many grams of CO_2 are trapped in a 500-mL bottle of soda water under a pressure of 5.0 atm at 25°C? (Assume that the partial pressure of CO_2 is 3.0 atm.)

SOLUTION Because the soda is mostly water, we can use the Henry's law constants given in the text (Table 11.3). For CO_2, $k_H = 23$ mM/atm. Henry's law gives, for the solubility of CO_2,

$$S = k_H \times P_{gas}$$

$$S(CO_2) = \frac{23 \text{ mM}}{\cancel{\text{atm}}} \times 3.0 \cancel{\text{ atm}}$$

$$= 69 \text{ mM}$$

Thus, 69 mmol of CO_2 dissolve per liter of soda. This is equivalent to 35 mmol (69/2) per 500 mL (500 mL = 0.500 L), and

$$35 \cancel{\text{ mmol}} \times \frac{1 \cancel{\text{ mol}}}{1000 \cancel{\text{ mmol}}} \times \frac{44.0 \text{ g}}{1 \cancel{\text{ mol } CO_2}} = 1.5 \text{ g } CO_2$$

EXERCISE What mass of N_2 (in mg) is dissolved in 1 L of pond water at 25°C at a total atmospheric pressure of 1.0 atm? Assume the atmosphere is 79% N_2.

ANSWER: 2×10 mg

KEY CONCEPT B Le Chatelier's principle

Imagine that we have a system in dynamic equilibrium and that we disturb the equilibrium state in some manner. For example, let's assume we have a dissolved gas in equilibrium with undissolved gas above the liquid, and we suddenly increase the partial pressure of the gas above the liquid by introducing some of the gas into the experimental apparatus. According to Henry's law, increasing the partial pressure makes the gas more soluble; so some of the introduced gas dissolves in the liquid and the partial pressure of the gas decreases, as the following figure illustrates.

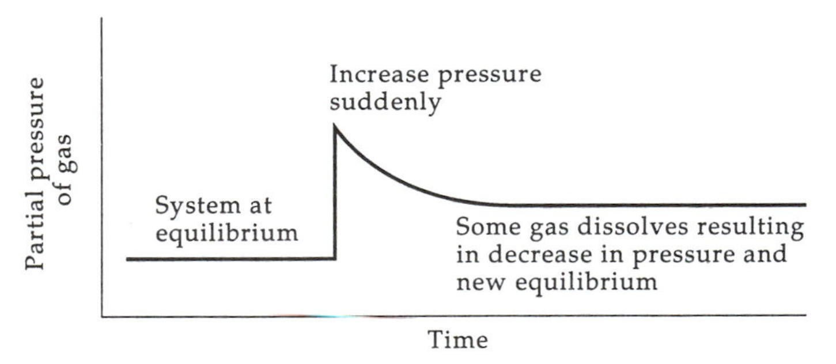

Le Chatelier's principle states that when a system at equilibrium is disturbed, it adjusts so as to partially cancel the effect of the disturbance. For the system shown, the disturbance is the sudden increase in pressure, and we can see that the way the system responds results in a partial canceling of the increase and a new state of equilibrium. This is a manifestation of Le Chatelier's principle.

EXAMPLE Using Le Chatelier's principle

Assume we have a system in a dynamic equilibrium and that the molarity of a compound is determined by the equilibrium. The equilibrium is disturbed by adding some of the compound, increasing the molarity from the equilibrium value of 1.0 M to 2.0 M. Use Le Chatelier's principle to predict how the system responds to the sudden increase in the molarity.

SOLUTION Le Chatelier's principle predicts that a system at equilibrium responds to a disturbance in the equilibrium by canceling some of the effect of the disturbance. It does so by establishing a new equilibrium state. In this case, we predict that, by establishment of a new equilibrium state, the molarity of the compound will fall below the 2.0 M achieved by the disturbance but above the starting molarity of 1.0 M.

EXERCISE Assume a dynamic equilibrium is responsible for determining the volume of a system. The system has volume V_1; we suddenly increase the volume to V_2. How does the system respond to the increase?

ANSWER: It establishes a new equilibrium with volume larger than V_1 but smaller than V_2.

11.5 THE EFFECT OF TEMPERATURE ON SOLUBILITY

KEY CONCEPT A Using Le Chatelier's principle to predict the effect of temperature on solubility

The **solubility** of a substance is generally dependent on temperature. We can use Le Chatelier's principle to predict the effect of changes in temperature on solubility, whether a solute liberates heat when it dissolves (exothermic process) or absorbs heat when it dissolves (endothermic process). A summary of the predicted effects is shown in the following table.

Temperature change	Substance that dissolves exothermically	Substance that dissolves endothermically
increase	becomes less soluble	becomes more soluble
decrease	becomes more soluble	becomes less soluble

EXAMPLE Predicting the solubility change of an ionic compound as the temperature changes

Which of the potassium salts listed in Table 11.4 of the text should increase in solubility as the temperature decreases?

SOLUTION According to Le Chatelier's principle, a compound that releases heat as it dissolves (one that dissolves exothermically) becomes more soluble as the temperature decreases. Such a compound has a negative enthalpy of solution. The potassium compounds in Table 11.4 with negative enthalpies of solution are KF, KOH, and K_2CO_3; these are the potassium compounds that become more soluble at lower temperatures.

EXERCISE Which of the lithium compounds listed in Table 11.4 of the text should decrease in solubility as temperature increases.

ANSWER: All except LiF

KEY CONCEPT B Contributions to the enthalpy of solution

We would like to know what factors contribute to the enthalpy of solution, what makes it exothermic or endothermic. We can do this by envisioning that a two-step process occurs when an ionic compound dissolves. First, the ionic compound breaks up into gas-phase ions; second, the ions dissolve in the solvent. Breaking the solid into the gas-phase ions is always endothermic and requires an amount of energy that we earlier called the lattice enthalpy. When the ions dissolve, the solvent molecules usually arrange themselves around the ions in a low-energy configuration (a process called **solvation**), so energy is released during this process. In water, the process is called **hydration** and the energy released called the **enthalpy of hydration.** For NaCl dissolving in water, the two step process is as follows:

Solid breaks up into gas phase ions:

$$NaCl(s) \longrightarrow Na^+(g) + Cl^-(g) \qquad \Delta H^\circ_L = \text{lattice enthalpy}$$

Water molecules hydrate ions as they dissolve:

$$Na^+(g) + Cl^-(g) \longrightarrow Na^+(aq) + Cl^-(aq) \qquad \Delta H^\circ_H = \text{hydration enthalpy}$$

The (aq) after an ion indicates it is hydrated. Summing these two reactions gives the reaction for the dissolving of the solid. According to Hess's law, the enthalpy of solution equals the sum of the two reactions, that is, the sum of the lattice enthalpy plus the hydration enthalpy:

$$NaCl(s) \longrightarrow Na^+(aq) + Cl^-(aq) \qquad \Delta H^\circ_{sol} = \Delta H^\circ_L + \Delta H^\circ_H$$

EXAMPLE Estimating the hydration enthalpy

The lattice enthalpy of AgCl is +905 kJ/mol, and its heat of solution in water is +65.5 kJ/mol. Calculate the hydration enthalpy of AgCl.

SOLUTION The sum of the lattice enthalpy and hydration energy equals the enthalpy of solution:

$$\Delta H^\circ_{sol} = \Delta H^\circ_L + \Delta H^\circ_H$$

We are given $\Delta H^\circ_{sol} = +65.5$ kJ/mol and $\Delta H^\circ_L = +905$ kJ/mol. Substitution of these into the given equation permits calculation of ΔH°_H:

$$\begin{aligned} 65.5 \text{ kJ/mol} &= 905 \text{ kJ/mol} + \Delta H^\circ_H \\ \Delta H^\circ_H &= 65.5 \text{ kJ/mol} - 905 \text{ kJ/mol} \\ &= -840 \text{ kJ/mol} \end{aligned}$$

EXERCISE Calculate the enthalpy of hydration of AgF. Considering the difference between the Cl^- ion and F^- ion, does this result seem reasonable, compared with the result just obtained for AgCl?

ANSWER: −993 kJ/mol; the difference is reasonable because the smaller F^- should hydrate more strongly than Cl^- and result in a more negative ΔH°_H for AgF than for AgCl.

KEY CONCEPT C Individual ion hydration energies

An individual ion hydration enthalpy (ΔH°_H) is the enthalpy change that occurs when the gas-phase ion dissolves and becomes hydrated. For instance, for sodium ion, ΔH°_H is −444 kJ; thus,

$$Na^+(g) \longrightarrow Na^+(aq) \qquad \Delta H^\circ_H = -444 \text{ kJ}$$

A large number of individual ion enthalpies have been determined; some are shown in Table 11.6 of the text. From this table, we should note that a small highly charged ion hydrates very strongly, resulting in a large negative enthalpy change for the hydration process. Conversely, a large ion with low charge does not hydrate very strongly; its hydration energy is not as large, and hydration of such an ion is not as energetically favorable a process as it is for a smaller more highly charged ion.

EXAMPLE Using and interpreting individual ion enthalpies

Use individual ions enthalpies to calculate the enthalpy of hydration of $MgBr_2$.

SOLUTION We want the enthalpy change for the overall process

$$Mg^{2+}(g) + 2Br^-(g) \longrightarrow Mg^{2+}(aq) + 2Br^-(aq) \qquad \Delta H^\circ_H = ?$$

Using Hess's law we can determine this value by summing the equations that represent hydration of the individual ions. The individual ion enthalpies are read from Table 11.6 of the text.

$$\begin{array}{lll} Mg^{2+}(g) \longrightarrow Mg^{2+}(aq) & \Delta H^\circ_H = 1 \cancel{\text{mol}} \times (-2003 \text{ kJ}/\cancel{\text{mol}}) = & -2003 \text{ kJ} \\ 2Br^-(g) \longrightarrow 2Br^-(aq) & \Delta H^\circ_H = 2 \cancel{\text{mol}} \times (-309 \text{ kJ}/\cancel{\text{mol}}) = & -618 \text{ kJ} \\ \hline Mg^{2+}(g) + 2Br^-(g) \longrightarrow Mg^{2+}(aq) + 2Br^-(aq) & \Delta H^\circ_H = & -2621 \text{ kJ} \end{array}$$

EXERCISE Use individual ions enthalpies to calculate the enthalpy of hydration of SrI_2. Comment on why the value differs from that for $MgBr_2$, calculated in the example.

ANSWER: −2116 kJ; it is more positive because Sr^{2+} is larger than Mg^{2+} and I^- is larger than Br^-.

KEY CONCEPT D Enthalpy, solubility, and disorder

Although it is not always obvious, nature favors disorder and randomness over order. When given a chance, molecules tend to spread throughout as large a volume as possible. In a similar fashion, the available energy in a system tends to spread out over as many quantum levels as possible, as chaotic thermal motion. The spread-out state is a more disordered state. When a solid solute dissolves, the molecules change from a highly ordered, confined state (the compact solid) to a disordered, spread-out state (the molecules spread throughout the solution); so the dissolving process is a favorable process insofar as the disorder of molecules is concerned. What happens energetically depends on whether the enthalpy of solution is exothermic or endothermic. If the solute dissolves exothermically, energy that was confined to the relatively few molecules of the dissolving system spreads to all the molecules of the surroundings. As far as the energy is concerned, this is a change from a more ordered to a more disordered state, and the process is favorable. However, if the the solute dissolves endothermically, energy that was spread among all the molecules of the surroundings becomes confined to the relatively few molecules of the system. Energetically, the change is from a more disordered to a more ordered state and is unfavorable.

	Exothermic dissolving	**Endothermic dissolving**
molecules	ordered → disordered	ordered → disordered
energy	ordered → disordered	disordered → ordered

Exothermic dissolving is always a favorable process because the changes in the disorder of molecules and energy are both favorable. Endothermic dissolving occurs only if the favorable ordered → disordered change for the molecules overwhelms the unfavorable disordered → ordered change for the energy.

EXAMPLE Disorder and order in the solution process

When a nonpolar solute dissolves in a nonpolar solvent, the enthalpy of solution is often close to zero. Since the energy change is unimportant, why does the solute dissolve?

SOLUTION The solute dissolves because the solute molecules change from a more ordered to a more disordered state during the dissolving process, and such a change is favorable.

EXERCISE When a helium balloon pops in a room, why do the helium molecules spread throughout the room rather than stay in place where they were before the balloon broke?

ANSWER: If we assume the helium and air to be ideal gases, a zero enthalpy change will occur when the helium molecules disperse throughout the room (ideal gas molecules have zero energy of interaction). The spreading out of the molecules is a favorable process because of the increase in disorder and because there are no opposing unfavorable energetic processes.

COLLIGATIVE PROPERTIES

KEY WORDS Define or explain each of the following terms in a written sentence or two.

boiling point elevation
freezing point lowering
ideal solution
osmometry
osmosis
Raoult's law
semipermeable membrane
van't Hoff factor
vapor pressure lowering
zone refining

11.6 CHANGES IN VAPOR PRESSURE, BOILING POINTS, AND FREEZING POINTS

KEY CONCEPT A A survey of colligative properties

Addition of a nonvolatile solute to a solvent affects the solvent in such a way that many of the physical properties of the resulting solution are different from those of the pure solvent. The properties thus affected are called *colligative properties*. The properties we will consider and the difference between the pure solvent and a solution are summarized in the table.

Colligative property	Difference between solution and pure solvent
vapor pressure	solution vapor pressure lower than pure solvent
boiling point	solution boiling point higher than pure solvent
freezing point	solution freezing point lower than pure solvent
osmotic pressure	exists only when solution is in contact with pure solvent with semipermeable membrane present

Colligative properties depend only on the relative number of solute particles (molecules or ions) present and not on the specific identity of the solute. For instance, NaCl ionizes in water to produce two ions per dissolved formula unit, whereas glucose, which does not ionize, produces one molecule per dissolved molecule. For equal concentrations of glucose and NaCl, the NaCl produces twice as many solute particles and has approximately twice the effect on the four colligative properties just listed. For example, if a certain concentration of glucose raises the boiling point of water by 0.5°C, the same concentration of NaCl raises it by approximately twice as much, or 1.0°C.

EXAMPLE Effect of the number of solute particles on colligative properties

A 0.50-*m* aqueous glucose solution boils at 100.3°C; that is, the boiling point of the solution is 0.3°C above that of pure water. What is the boiling point of a 0.50-*m* aqueous $Ca(NO_3)_2$ solution?

SOLUTION Glucose does not ionize in solution whereas $Ca(NO_3)_2$ ionizes into three ions ($1Ca^{2+} + 2NO_3^- = 3$ ions) when it is dissolved. There are three times as many solute particles present in a 0.50 *m* $Ca(NO_3)_2$ solution as there are in a 0.50 *m* glucose solution, so we expect the increase in the boiling point to be three times larger in 0.50 *m* $Ca(NO_3)_2$ than in 0.50 *m* glucose. Because $3 \times 0.3°C = 0.9°C$, the boiling point of the $Ca(NO_3)_2$ is predicted to be 100.9°C ($100.0°C + 0.9°C = 100.9°C$).

EXERCISE Addition of 1.5 mol NaCl to 1.0 kg H_2O lowers the freezing point of the solution to −5.6°C. What is the freezing point of the solution made by adding 1.5 mol Na_2SO_4 to 1.0 kg water?

ANSWER: −8.4°C

PITFALL Nonvolatile solutes only!

In this chapter we consider nonvolatile solutes only. These are solutes that do not evaporate easily and, for the purposes of our discussion, are considered to have zero vapor pressure. Ionic compounds and sugars (such as glucose, fructose, and sucrose) are examples of nonvolatile solutes; in contrast, the alcohols that easily dissolve in water are generally quite volatile.

KEY CONCEPT B The lowering of vapor pressure and Raoult's law

Addition of a nonvolatile solute to a solvent lowers the vapor pressure, so the vapor pressure of the resulting solution is less than that of the pure solvent. The effect is quantitatively described by

Raoult's law:

$$\text{Vapor pressure of solution} = \text{mole fraction solvent} \times \text{vapor pressure of pure solvent}$$
$$P_A \quad = \quad x_A \quad \times \quad P_A^*$$

where P_A = vapor pressure of solution, x_A = mole fraction of solvent, and P_A^* = vapor pressure of pure solvent.

EXAMPLE Calculation of the lowering of the vapor pressure

What is the vapor pressure, at 25°C, of the solution that results from adding 60.0 g glucose ($C_6H_{12}O_6$) to 500 g water? The vapor pressure of water at 25°C is 23.8 Torr. Assume ideal solution behavior.

SOLUTION Because glucose is a nonvolatile solute, Raoult's law can be used to calculate the vapor pressure of the solution:

$$P_A = x_A \times P_A^*$$

For our problem, in which water is the solvent, this translates to

$$P_{\text{solution}} = x_{H_2O} \times P_{H_2O}^*$$

The problem states that $P_{H_2O}^* = 23.8$ Torr. The mole fraction of water in the solution x_{H_2O} is given by

$$x_{H_2O} = \frac{n_{H_2O}}{n_{H_2O} + n_{C_6H_{12}O_6}}$$

We use the molar mass of water (18.0 g/mol) and glucose (180.2 g/mol) to calculate the moles of each in the solution:

$$n_{H_2O} = 500 \cancel{g} \times \frac{1 \text{ mol } H_2O}{18.0 \cancel{g}} \qquad n_{C_6H_{12}O_6} = 60.0 \cancel{g} \times \frac{1 \text{ mol } C_6H_{12}O_6}{180.2 \cancel{g}}$$
$$= 27.8 \text{ mol} \qquad\qquad = 0.333 \text{ mol}$$

Substituting these values into the equation for the mole fraction of water gives

$$x_{H_2O} = \frac{27.8}{27.8 + 0.333} = \frac{27.8}{28.1} = 0.988$$

Substituting this value and the vapor pressure of pure water into Raoult's law gives the desired answer:

$$P_{\text{solution}} = 0.988 \times 23.8 \text{ Torr} = 23.5 \text{ Torr}$$

As expected, the vapor pressure of the solution is lower than that of the pure solvent.

EXERCISE What is the vapor pressure, at 23°C, of the solution that results from adding 75.0 g I_2 to 500 g carbon tetrachloride, CCl_4. The vapor pressure of CCl_4 at 23°C is 100 Torr. Assume I_2 is nonvolatile, and assume ideal solution behavior.

ANSWER: 91.5 Torr

KEY CONCEPT C Boiling point elevation

Addition of a nonvolatile solute to a solvent increases the boiling point of the resulting solution above that of the pure solvent. Quantitatively, the boiling point elevation (elevation above the pure solvent boiling point) is given by the equation

$$\text{Boiling point elevation} = i \times k_b \times \text{molality of solute}$$

The constant k_b (the **boiling point constant**) depends on the identity of the solvent. The factor i is the van't Hoff factor and is equal to the number of moles of solute particles that result from 1 mol of solute. This is illustrated in the table.

Compound	Solute particles formed upon dissolving	Number of moles of solute particles per mole solute	i
$C_6H_{12}O_6$	$C_6H_{12}O_6$	1	1
NaCl	Na^+, Cl^-	2	2
Na_2SO_4	$2Na^+$, SO_4^{2-}	3	3
$Ca(NO_3)_2$	Ca^{2+}, $2NO_3^-$	3	3

The van't Hoff factor i is dimensionless; the dimensions of k_b are kelvins/molality (K/m). Molecular solutes such as glucose do not ionize, and $i = 1$. With such solutes we get

$$\text{Boiling point elevation} = k_b \times \text{molality of solute}$$

EXAMPLE Calculating the boiling point elevation of a solution

What is the boiling point of the solution made by mixing 65.0 g KCl with 500 g water?

SOLUTION The equation for calculating the boiling point elevation of a solution is

$$\text{Boiling point elevation} = i \times k_b \times \text{molality of solute}$$

Because KCl dissociates into two ions when it dissolves in water, $i = 2$. k_b depends on the identity of the solvent, and for water $k_b = 0.51$ K/m. To calculate the molality of the solute, we use the molar mass of KCl, 74.6 g/mol, to first calculate the moles of KCl:

$$\text{Mol KCl} = 65.0\ \cancel{g} \times \frac{1\ \text{mol KCl}}{74.6\ \cancel{g}} = 0.871\ \text{mol KCl}$$

We use the fact that 500 g H_2O = 0.500 kg H_2O and substitute into the defining equation for molality:

$$\begin{aligned}\text{Molality} &= \frac{\text{moles solute}}{\text{kg solvent}} \\ &= \frac{0.871\ \text{mol KCl}}{0.500\ \text{kg}} \\ &= 1.74\ m\end{aligned}$$

Finally, we substitute all of the required values into the equation for the boiling point elevation:

$$\begin{aligned}\text{Boiling point elevation} &= i \times k_b \times \text{molality of solute} \\ &= 2 \times 0.51\ \text{K}/\cancel{m} \times 1.74\ \cancel{m} \\ &= 1.8\ \text{K}\end{aligned}$$

A change in temperature of 1.8 K is the same as a change of 1.8°C. Because the 1.8°C is a boiling point elevation, the boiling point of the solution is higher than that of the pure solvent. The boiling point of the solution is 101.8°C (100.0°C + 1.8°C = 101.8°C).

EXERCISE What is the boiling point of a 0.75 m solution of naphthalene (a molecular solute) in acetone? Treat naphthalene as a nonvolatile solute; the boiling point of acetone is 56.5°C.

ANSWER: 57.8°C

PITFALL The van't Hoff constant

The ideal values of i we have given are valid only for very dilute solutions. At concentrations at which ions or molecules begin to interact with each other, experimental values of i must be used.

KEY CONCEPT D The freezing point depression

Addition of a nonvolatile solute to a solvent results in a solution with a lower freezing point than that of the pure solvent. The solution must be cooled to a lower temperature than the pure solvent in order to get it to freeze. Quantitatively, the effect can be calculated with the relation

$$\text{Freezing point depression} = i \times k_f \times \text{molality of solute}$$

Here i has the same meaning as previously discussed and k_f (the **freezing point constant**) is a constant that depends on the identity of the solvent. (In this context, the term depression means "lowering"). For molecular solutes, $i = 1$, and we get

$$\text{Freezing point depression} = k_f \times \text{molality of solute}$$

EXAMPLE 1 Calculating a freezing point depression

What is the freezing point of a mixture consisting of 0.300 g S_8 dissolved in 20.0 g camphor? The freezing point of pure camphor is 179.8°C; S_8 is a nonvolatile molecular solute.

SOLUTION The equation used to calculate the lowering of the freezing point for a solution is

$$\text{Freezing point depression} = i \times k_f \times \text{molality of solute}$$

For a molecular solute $i = 1$; k_f for camphor is 39.7 K/m (Table 11.7 of the text). The molar mass of S_8 is 256 g/mol, so its molality is calculated as follows.

$$\text{Molality of } S_8 = \frac{\text{mol } S_8}{\text{kg camphor}}$$

$$\text{Mol } S_8 = 0.300 \text{ g} \times \frac{1 \text{ mol } S_8}{256 \text{ g}} = 1.17 \times 10^{-3} \text{ mol } S_8$$

$$\text{kg camphor} = 20.0 \text{ g} \times \frac{1 \text{ kg}}{1000 \text{ g}} = 0.0200 \text{ kg}$$

$$\text{Molality of } S_8 = \frac{1.17 \times 10^{-3} \text{ mol } S_8}{0.0200 \text{ kg}} = 0.0585\ m$$

Substituting into the equation for the freezing point depression gives

$$\begin{aligned}\text{Freezing point depression} &= i \times k_f \times \text{molality of solute}\\ &= 1 \times 39.7 \text{ K}/m \times 0.0585\ m\\ &= 2.3 \text{ K} = 2.3°\text{C}\end{aligned}$$

Because the freezing point of pure camphor is 179.8°C and the freezing point depression (lowering) is 2.3°C, the freezing point of the solution is 177.5°C.

EXERCISE What is the freezing point (in °C) of the solution made by adding 5.0 g $Ca(NO_3)_2$ to 400 g water?

ANSWER: −0.43°C

EXAMPLE 2 Using a freezing point depression to measure a molar mass

The addition of 0.226 g of a nonionizing unknown compound to 100 g camphor lowers the freezing point of the camphor by 0.42 K. What is the molar mass of the unknown? $k_f = 39.7\ \text{K}/m$

SOLUTION The equation relating molality to freezing point depression can be rearranged to

$$\text{Molality} = \frac{1}{k_f} \times \text{freezing point depression}$$

We have $k_f = 39.7\ \text{K}/m$ and freezing point depression = 0.42 K. Substituting into the equation gives

$$\begin{aligned}\text{Molality} &= \frac{1}{39.7\ \cancel{\text{K}}/m} \times 0.42\ \cancel{\text{K}} \\ &= 0.011\ m\end{aligned}$$

This molality means that there is 0.0106 mol unknown dissolved in 1000 g (1 kg) camphor. We can now calculate the moles of unknown in 100 g camphor.

$$100\ \cancel{\text{g camphor}} \times \frac{0.011\ \text{mol unknown}}{1000\ \cancel{\text{g camphor}}} = 0.0011\ \text{mol}$$

The problem states there is 0.226 g unknown in 100 g camphor; therefore, 0.226 g unknown corresponds to 0.0011 mol unknown and

$$\begin{aligned}\text{Molar mass} &= \frac{0.226\ \text{g}}{0.0011\ \text{mol}} \\ &= 2.1 \times 10^2\ \text{g/mol}\end{aligned}$$

EXERCISE The addition of 0.243 g of a nonionizing unknown to 250 g carbon tetrachloride lowers its freezing point by 0.12 K. What is the molar mass of the unknown? $k_f = 29.8\ \text{K}/m$ for carbon tetrachloride.

ANSWER: 2.4×10^2 g/mol

PITFALL Choosing the correct boiling point and freezing point constant

It is important to remember that the boiling point constant k_b and freezing point constant k_f used in calculations of boiling point elevations and freezing point depressions depend on the identity of the solvent and not the solute.

PITFALL Is the change in temperature subtracted or added?

Boiling points are elevated (increased) by the presence of a solute. After calculating the boiling point elevation, add the elevation to the boiling point of the pure solvent so that the boiling point of the solution is higher than that of the solvent. Freezing points are depressed (lowered) by the presence of a solute. After calculating the freezing point depression, subtract the depression from the freezing point of the pure solvent so that the freezing point of the solution is less than the freezing point of the pure solvent. A good way to remember these points is the statement "Boiling points start higher and go higher; freezing points start lower and go lower."

11.7 OSMOSIS

KEY CONCEPT Osmosis and osmotic pressure

The osmosis process depends on the existence of a **semipermeable membrane,** which lets some species pass through but is a barrier to the passage of other species. Many important semipermeable membranes allow water molecules to pass through the membrane but do not allow passage of larger biological molecules or ions. Osmosis occurs when a pure solvent is placed on one side of a semipermeable membrane and a solution on the other; the solvent molecules pass through the membrane but the solute molecules do not. The solvent molecules have a natural tendency to flow from where they are more concentrated (the pure solvent, which is 100% solvent) to where they are less concentrated (the solution, which is less than 100% solvent).

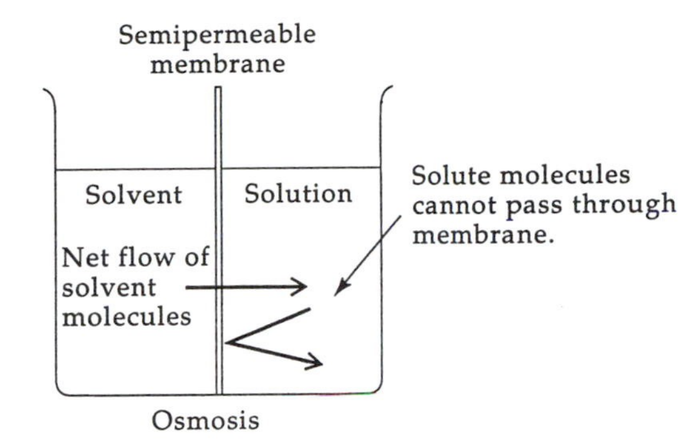

As osmosis occurs, solvent moves to the solution side of the membrane, so the level of the solution in the container increases and that of the solvent decreases. The pressure that must be applied to the top of the solution side to stop its level from rising is called the **osmotic pressure (Π)** and is given by

$$\Pi = i \times RT \times \text{molarity of solute}$$

In this equation, i = van't Hoff factor, R = ideal gas constant, and T = temperature in kelvins.

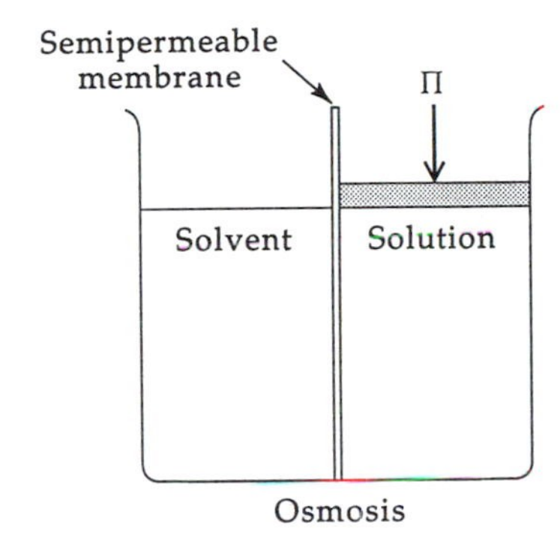

EXAMPLE Calculating the osmotic pressure

We dissolve 3.00 g of a nonionizing biomolecule with molar mass 5.0×10^3 g/mol in enough water to make 250 mL of solution. What is the osmotic pressure of the solution at 25.0°C?

SOLUTION The osmotic pressure of a solution is given by

$$\Pi = i \times RT \times \text{molarity}$$

The molarity of the solute is given by

$$\text{Molarity} = \frac{\text{mol solute}}{\text{liter solution}}$$

$$\text{Mol solute} = 3.00\ \cancel{g} \times \frac{1\ \text{mol}}{5.0 \times 10^3\ \cancel{g}} = 6.0 \times 10^{-4}\ \text{mol}$$

$$\text{Liter solution} = 250\ \cancel{\text{mL}} \times \frac{1\ \text{L}}{1000\ \cancel{\text{mL}}} = 0.250\ \text{L}$$

$$\text{Molarity} = \frac{6.0 \times 10^{-4}\ \text{mol}}{0.250\ \text{L}} = 2.4 \times 10^{-3}\ \text{M}$$

$i = 1$ because the solute is nonionizing; $T = 25 + 273 = 298$ K, and $R = 0.0821$ L atm/mol K. Substituting these values into the original equation gives

$$\Pi = i \times RT \times \text{molarity}$$
$$= 1 \times \left(0.0821 \frac{\text{L atm}}{\text{mol K}}\right) \times (298\ \text{K}) \times \left(2.4 \times 10^{-3} \frac{\text{mol}}{\text{L}}\right)$$
$$= 0.059\ \text{atm}$$

EXERCISE What is the osmotic pressure at 50°C of the solution made by dissolving 1.55 g sucrose ($C_{12}H_{22}O_{11}$) in enough water to make 100 mL of solution?

ANSWER: 1.20 atm

MIXTURES OF LIQUIDS

11.8 RAOULT'S LAW FOR MIXTURES OF LIQUIDS

KEY CONCEPT The vapor pressure of a mixture of two volatile liquids

When two volatile liquids (liquids that evaporate easily) are mixed, the vapor pressure of the mixture originates with both liquids. If the two liquids are similar to each other, each individually follows Raoult's law; so the vapor pressure above the mixture is the sum of the individual vapor pressures of the two liquids according to Raoult's law. Thus, for a mixture of two liquids A and B,

$$\text{Total vapor pressure} = \text{vapor pressure of liquid A} + \text{vapor pressure of liquid B}$$
$$= x_A P_A^* + x_B P_B^*$$

In this equation, x_A = mole fraction A, P_A^* = vapor pressure of pure A, x_B = mole fraction B, and P_B^* = vapor pressure of pure B.

EXAMPLE Calculating the vapor pressure of a binary mixture of volatile liquids

50.0 g of benzene (C_6H_6) and 50.0 g of toluene (C_7H_8) are mixed at 25°C. What is the vapor pressure of the mixture? The vapor pressures of the pure liquids are benzene, 94.6 Torr; toluene, 29.1 Torr.

SOLUTION Both liquids have a finite vapor pressure, so the vapor pressure of the mixture is caused by both. We will assume that Raoult's law applies and that the vapor pressure is the sum of the individual vapor pressures:

$$\text{Total vapor pressure} = x_{\text{benzene}} P^*_{\text{benzene}} + x_{\text{toluene}} P^*_{\text{toluene}}$$

We first must determine the number of moles of each component in order to calculate the mole fraction of each:

$$\text{Mol benzene} = 50.0\ \text{g} \times \frac{1\ \text{mol}}{78.1\ \text{g}} = 0.640\ \text{mol}$$

$$\text{mol toluene} = 50.0\ \text{g} \times \frac{1\ \text{mol}}{92.1\ \text{g}} = 0.543\ \text{mol}$$

$$x_{\text{benzene}} = \frac{0.640\ \text{mol}}{0.640\ \text{mol} + 0.543\ \text{mol}} = 0.541$$

$$x_{\text{benzene}} = \frac{0.543\ \text{mol}}{0.640\ \text{mol} + 0.543\ \text{mol}} = 0.459$$

We now substitute these values into the equation for the total vapor pressure

$$\begin{aligned}\text{Total vapor pressure} &= x_{\text{benzene}}P^*_{\text{benzene}} + x_{\text{toluene}}P^*_{\text{toluene}} \\ &= (0.541 \times 94.6\ \text{Torr}) + (0.459 \times 29.1\ \text{Torr}) \\ &= 64.5\ \text{Torr}\end{aligned}$$

EXERCISE 250 g of water and 125 g of methanol (CH_3OH) are mixed at 25°C. What is the vapor pressure of the mixture? The vapor pressure of water is 23.8 Torr and that of methanol 120 Torr.

ANSWER: 44.8 Torr

DESCRIPTIVE CHEMISTRY TO REMEMBER

- The high solubility of **nitrates** (ionic compounds containing NO_3^-) in water results in their rarely being found in mineral deposits. Rainwater dissolves and washes away any potential nitrate mineral deposits.
- Bone is largely **calcium phosphate** [$Ca_3(PO_4)_2$]. Like most phosphates it is insoluble, a desirable feature for the skeletons of creatures that are mostly water.
- **Calcium hydrogen phosphate** [$Ca(HPO_4)_2$] is more soluble than calcium phosphate and is included in many commercial fertilizers.
- Hard water contains dissolved **magnesium** and **calcium** salts.
- Soaps react with the **calcium ions** in hard water to form scum. Pretreatment of the water with washing soda (Na_2CO_3) precipitates the calcium as $CaCO_3$, so the formation of scum is avoided.
- Polyphosphate ions, which are negatively charged large chains and rings, can be added to hard water to avoid the formation of scum. The large, negatively charged structures surround and **sequester** the positive Ca^{2+} ions and prevent the large soap molecules from gaining access to the ion.
- Most **ionic** and **molecular solids** are more soluble in hot water than in cold water. All **gases** are less soluble in hot water than in cold water.
- **Thermal pollution** is the damage caused to the environment by waste heat. One consequence of thermal pollution is that O_2 is driven out of solution. This results in the death of the bacteria that degrade dead organic matter.
- Organic chemists often check the purity of a compound by checking its **melting point.** An impure compound melts at a lower temperature than the pure compound because of the freezing point depression of the compound caused by the impurity.
- Biological cell walls act like **semipermeable membranes.** They allow water and small ions to pass but not the large biomolecules synthesized and required inside the cell.
- **Salting** preserves meat because the flow of water from bacteria to the concentrated salt solution formed on the meat by salting dehydrates the bacteria and kills them.
- A form of **reverse osmosis** is the forcing of water molecules across a semipermeable membrane from a solution (seawater) to a pure-water collector by the application of pressure to the solution in contact with the membrane.

CHEMICAL EQUATIONS TO KNOW

- Treating phosphate rock with sulfuric acid results in the formation of calcium sulfate.

$$Ca_5(PO_4)_3OH(s) + 5H_2SO_4(aq) \longrightarrow 3H_3PO_4(aq) + 5CaSO_4(s) + H_2O(l)$$

- Carbon dioxide reacts with water to form carbonic acid.

$$CO_2(g) + H_2O(l) \longrightarrow H_2CO_3(aq)$$

- Calcium carbonate ($CaCO_3$) reacts with dissolved carbonic acid (H_2CO_3) to form the more soluble calcium hydrogen carbonate [$Ca(HCO_3)_2$].

$$CaCO_3(s) + H_2CO_3(aq) \longrightarrow Ca(HCO_3)_2(aq)$$

- When aqueous calcium hydrogen carbonate [$Ca(HCO_3)_2$] is heated in hot water pipes and boilers, carbon dioxide is driven off and calcium carbonate (boiler scale) precipitates.

$$Ca(HCO_3)_2(aq) \longrightarrow CaCO_3(s) + H_2O(l) + CO_2(g)$$

- Polyphosphate ions, formed when phosphates are heated, consist of chains and rings of PO_4 groups.

$$HO-\overset{\overset{\displaystyle O}{\|}}{\underset{\underset{\displaystyle OH}{|}}{P}}-OH + HO-\overset{\overset{\displaystyle O}{\|}}{\underset{\underset{\displaystyle OH}{|}}{P}}-OH \longrightarrow HO-\overset{\overset{\displaystyle O}{\|}}{\underset{\underset{\displaystyle OH}{|}}{P}}-O-\overset{\overset{\displaystyle O}{\|}}{\underset{\underset{\displaystyle OH}{|}}{P}}-OH + H_2O$$

MATHEMATICAL EQUATIONS TO KNOW AND UNDERSTAND

$\text{Mass percentage solute} = \dfrac{\text{mass of solute}}{\text{mass of solution}} \times 100$	definition of mass percentage
Molar concentration = moles of solute/liters of solution	definition of molarity
Mass concentration = mass of solute/liters of solution	definition of mass concentration
$x_A = \dfrac{n_A}{n_A + n_B}$	definition of mole fraction
Molality = moles of solute/kilograms of solvent	definition of molality
Solubility of gas = k_H × partial pressure of gas	Henry's law
$P_A = x_A \times P_A^*$	Raoult's law
Boiling point elevation = $i \times k_b$ × molality of solute	boiling point elevation
Freezing point depression = $i \times k_f$ × molality of solute	freezing point depression
$\Pi = i \times RT$ × molarity of solute	osmotic pressure

SELF-TEST EXERCISES

Measures of concentration

1. What is the molarity of the solution made by adding enough water to 85.0 g $HC_2H_3O_2$ to make 600 mL solution?
(a) 0.236 M (b) 2.36 M (c) 1.11 M (d) 0.142 M (e) 0.813 M

2. How many grams of Na_2S_2 are present in 10.0 mL of a 0.20 M solution?
(a) 110 g (b) 2.0×10^{-3} g (c) 0.22 g (d) 220 g (e) 2.2×10^{-3} g

3. How many milliliters of 3.0 M HCl are required to obtain 50 mmol of HCl?
(a) 60 (b) 15 (c) 170 (d) 150 (e) 17

4. What is the mass concentration of a solution made by adding 65.0 g glucose to enough water to make 811 mL of solution?
(a) 12.5 g/L (b) 80.1 g/L (c) 47.2 g/L (d) 52.7 g/L (e) 31.9 g/L

5. The mass concentration of a Na_2SO_4 solution is 30 g/L. How many grams of Na_2SO_4 are present in 125 mL of this solution?
(a) 3.8 (b) 4.2 (c) 0.26 (d) 120 (e) 3.8×10^3

6. 200 g glycerol ($C_3H_8O_3$) is added to 100 g water. What is the mole fraction of glycerol?
(a) 1.00 (b) 0.521 (c) 0.718 (d) 3.56 (e) 0.281

7. What is the mole fraction iodine (I_2) in a solution of I_2 in CCl_4 in which the mole fraction CCl_4 is 0.28.
(a) 1.28 (b) 0.28 (c) 0.56 (d) 0.72 (e) 1.72

8. What is the molality of the solution made by adding 30.0 g $Ca(NO_3)_2$ to 250 g water?
(a) 0.732 *m* (b) 0.183 *m* (c) 0.0458 *m* (d) 0.120 *m* (e) 0.618 *m*

9. What is the molality of the solution made by dissolving 125 g Br_2 in 750 g $CHCl_3$?
(a) 0.0938 *m* (b) 0.167 *m* (c) 0.666 *m* (d) 0.782 *m* (e) 1.04 *m*

10. How many grams of KBr should be added to 100 g water in order to prepare a 0.250 *m* solution?
(a) 4.00 (b) 29.8 (c) 0.336 (d) 2.98 (e) 25.0

11. What is the mass concentration of a 0.40 *m* $AgNO_3$ aqueous solution? The density of the solution is 1.05 g/mL.
(a) 67.9 g/L (b) 61.3 g/L (c) 66.8 g/L (d) 63.6 g/L (e) 71.3 g/L

12. What is the molality of a solution of cysteamine (C_2H_7NS) in propanol (C_3H_8O) if the mole fraction cysteamine is 0.110?
(a) 1.10 *m* (b) 3.21 *m* (c) 2.57 *m* (d) 0.0840 *m* (e) 2.06 *m*

13. What is the molarity of a 2.00 *m* aqueous KOH solution? The density of the solution is 1.18 g/mL. Use 1.00 g/mL for the density of water.
(a) 1.91 M (b) 1.52 M (c) 1.63 M (d) 2.00 M (e) 2.12 M

Solubility

14. Which of the following is an electrolyte when dissolved in water?
(a) $NaNO_3$ (b) C_2H_5OH (ethanol) (c) sugar (d) N_2 (e) Ar

15. Imagine an experiment in which, at the start, a sample of solid sugar composed entirely of radioactive sugar molecules is in contact with a saturated aqueous solution of sugar. The sugar molecules in the saturated solution are all nonradioactive. After a week or two, which of the following is true?
(a) All the molecules in the solid are radioactive.
(b) All the molecules in the solution are radioactive.
(c) All the molecules in the solid are nonradioactive.
(d) All the molecules in the solution are nonradioactive.
(e) Some of the molecules in the solution and some in the solid are radioactive.

16. Which of the following is most likely to be soluble in CCl_4?
(a) P_4 (b) KOH (c) sucrose (d) HCl (e) H_2O

17. All of the following statements about soap except one are true. Which one is untrue?
(a) The hydrophobic end of a soap molecule is soluble in water.
(b) Soap molecules entrap dirt in a micelle.
(c) A soap molecule is a long chainlike molecule.
(d) One end of a soap molecule is soluble in nonpolar substances.
(e) Soap molecules react with Ca^{2+} to form a solid precipitate.

18. The solubility of a gas in water at 25°C is 6.0 mM. What is the solubility if the partial pressure of the gas above the water is doubled?
(a) 36 mM (b) 3.0 mM (c) 12 mM (d) 6.0 mM (e) 2.4 mM

19. What mass of argon gas is dissolved at 20°C in 100 mL of solution if the partial pressure of argon above the solution is 1.0×10^{-3} atm?
(a) 4.0×10^{-3} g (b) 0.060 g (c) 6.0 g (d) 6.0×10^{-6} g (e) 0.27 g

20. The solubility of NH_4NO_3 in water at 30°C is 242 g/100 mL. Its enthalpy of solution is +35.0 kJ/mol. One of the three answers is the solubility of NH_4NO_3 at 40°C. Which one?
(a) 192 g/100 mL (b) 297 g/100 mL (c) 242 g/100 mL

21. What is the enthalpy of solution of $MgCl_2$(s) as calculated from the lattice enthalpy (2493 kJ/mol) and individual ion hydration enthalpies (Mg^{2+}, −1926 kJ/mol; Cl^-, −378 kJ/mol).
(a) −5175 kJ/mol (b) +189 kJ/mol (c) +1575 kJ/mol (d) −189 kJ/mol
(e) +567 kJ/mol

22. Calculate the individual ion hydration enthalpy of Li^+ from the lattice enthalpy of LiCl (846 kJ/mol), the ethalpy of solution of LiCl (−37.2 kJ/mol), and the individual ion hydration of Cl^- (−378 kJ/mol).
(a) −1261 kJ/mol (b) −883 kJ/mol (c) −1187 kJ/mol (d) −505 kJ/mol
(e) −431 kJ/mol

23. Which ion is expected to have the highest individual ion hydration enthalpy?
(a) Na^+ (b) Al^{3+} (c) Se^{2-} (d) Br^- (e) Au^+

24. An ionic solid with a ________ lattice energy and ________ hydration energy is likely to have an exothermic enthalpy of solution.
(a) high, high (b) low, low (c) high, low (d) low, high

25. Which of the following describes a process that is certain to be a spontaneous process?
(a) Energy becomes disordered; molecules become disordered.
(b) Energy becomes disordered; molecules become ordered.
(c) Energy becomes ordered; molecules become disordered.
(d) Energy becomes ordered; molecules become ordered.

26. When a solid solute dissolves in a solvent, the molecules become more
(a) ordered. (b) disordered.

Colligative properties

27. When 2.0 mol of ethylene glycol, a molecular solvent, is dissolved in a certain mass of water, the freezing point of the resulting solution is −0.77°C. What is the freezing point of the solution made by adding 2.0 mol $Mg(NO_3)_2$ to the same mass of water?
(a) 0.0°C (b) −0.77°C (c) −0.51°C (d) −1.5°C (e) −2.3°C

28. What is the vapor pressure, at 30°C, of an aqueous solution of hexose in water in which $x_{hexose} = 0.16$? The vapor pressure of pure water at 30°C is 31.8 Torr.
(a) 5.1 Torr (b) 17 Torr (c) 38 Torr (d) 32 Torr (e) 27 Torr

29. What is the vapor pressure, at 25°C, of the solution made by adding 100 g P_4 to 500 g $CHCl_3$. The vapor pressure of pure $CHCl_3$ at 25°C is 182 Torr. Assume ideal solution behavior.
(a) 153 Torr (b) 29.4 Torr (c) 146 Torr (d) 36.4 Torr (e) 182 Torr

30. What is the boiling point of a 0.40 *m* solution of KF in H_2O?
(a) 100.2° (b) 99.8°C (c) 100.4°C (d) 100.0°C (e) 99.6°C

31. What is the boiling point of the solution made by adding 75.0 g S_8 to 500 g carbon tetrachloride (CCl_4)? The normal boiling point of CCl_4 is 76.5°C; $k_b = 4.95$ K/*m*.
(a) 77.9°C (b) 73.6°C (c) 75.1°C (d) 79.4°C (e) 76.5°C

32. What is the freezing point of a 0.050 *m* solution of LiCl in H_2O?
(a) −0.19°C (b) −0.093°C (c) +0.093°C (d) +0.19°C (e) 0.00°C

33. What is the freezing point of a 5.00 g sample of camphor ($C_{10}H_{16}O$) that has been contaminated with 0.025 g naphthalene ($C_{10}H_8$)? Pure camphor freezes at 179.8°C; $k_f = 39.7$ K/m for camphor.
(a) 178.3°C (b) 179.8°C (c) 181.4°C (d) 180.8°C (e) 178.8°C

34. What is the osmotic pressure, at 25°C, of a 4.4×10^{-3} M solution of sodium stearate in water? Sodium stearate ionizes into sodium ion and stearate ion when it dissolves.
(a) 0.018 atm (b) 0.0090 atm (c) 0.22 atm (d) 0.11 atm (e) 0.16 atm

35. What is the osmotic pressure at 37°C of the solution prepared by dissolving 310 mg of a large nonionizing biomolecule (molar mass = 4.0×10^5 g/mol) in enough water to make 5.0 mL of solution?
(a) 0.36 Torr (b) 5.6 Torr (c) 3.0 Torr (d) 2.8 Torr (e) 6.0 Torr

36. An aqueous solution of a nonionizing solute with molar mass 161 g/mol has a vapor pressure of 22.6 Torr at 25°C. What is the freezing point of the solution? The vapor pressure of pure water at 25°C is 23.8 Torr.
(a) +5.4°C (b) −5.4°C (c) +10.8°C (d) −10.8°C (e) 0.0°C

37. When 50.0 mg of an unknown compound is dissolved in 12.0 g of melted camphor, the camphor freezes at 177.5°C. What is the molar mass of the unknown? The freezing point of pure camphor is 179.8°C; $k_f = 39.7$ K/m.
(a) 60 g/mol (b) 5.5×10^2 g/mol (c) 1.7×10^2 g/mol (d) 1.5×10^2 g/mol (e) 72 g/mol

38. When 21.2 mg of a nonionizing unknown is dissolved in enough water to make 10.0 mL of solution at 25°C, the solution develops an osmotic pressure of 8.20 Torr. What is the molar mass of the unknown?
(a) 4.03×10^3 g/mol (b) 813 g/mol (c) 4.80×10^3 g/mol (d) 3.61×10^3 g/mol
(e) 1.55×10^3 g/mol

39. 150 g of two nonionizing compounds, one with molar mass 161 g/mol and one with molar mass 1052 g/mol are each separately dissolved in enough water to make 1.0 L of solution. Which one has the largest effect on colligative properties, that is, causes the greatest vapor pressure lowering, boiling point elevation, freezing point depression and osmotic pressure?
(a) Compound with molar mass = 161 g/mol (b) Compound with molar mass = 1052 g/mol
(c) Neither, because equal masses of each are used

40. 75.0 g of chloromethane (CH_3Cl) and 215.0 g of bromomethane (CH_3Br) are mixed at −38.5°C. What is the vapor pressure of the mixture? The vapor pressures of the pure liquids at this temperature are chloromethane, 394 Torr; bromomethane, 110 Torr.
(a) 190 Torr (b) 835 Torr (c) 252 Torr (d) 222 Torr (e) 282 Torr

Descriptive chemistry

41. Which of the following is the form in which phosphorus is included is most commercial fertilizers?
(a) $Ca_3(PO_4)_2$ (b) P_4 (c) P_2O_5 (d) Na_3PO_4 (e) $Ca(HPO_4)_2$

42. The presence of which two ions makes water "hard"?
(a) Na^+, K^+ (b) Ca^{2+}, Mg^{2+} (c) Cl^-, HCO_3^- (d) F^-, Cl^- (e) Na^+, Cl^-

43. What is the chemical composition of boiler scale?
(a) $Ca(HCO_3)_2$ (b) Na_2CO_3 (c) $CaCO_3$ (d) NaCl (e) $Na(HCO_3)$

44. What is chemical sequestering?
(a) reacting anything with water
(b) surrounding an ion with a large structure of opposite charge
(c) removing boiler scale with acid
(d) making certain one procedure follows another in correct sequential order

12 THE RATES OF REACTIONS

THE DESCRIPTION OF REACTION RATES

KEY WORDS Define or explain each of the following terms in a written sentence or two.

first-order reaction
half-life
initial rate
instantaneous rate
integrated rate law
pseudo first-order reaction
overall order
rate constant
rate law
second-order reaction
zeroth-order reaction

12.1 REACTION RATES

KEY CONCEPT A The reaction rate

When we talk about a rate, we usually are thinking about how long it takes for something to happen. The same is true of reaction rates. A reaction in which a lot of product is formed and a lot of reactant consumed in a short amount of time is a fast reaction. Similarly, a slow reaction is one in which very little product is formed and very little reactant consumed over a long time. The quantitative definition of rate follows; the rate has units of concentration divided by time.

$$\text{Rate of reaction} = \frac{\text{change in concentration of reactant or product}}{\text{time it takes for the change to occur}}$$

For instance, for the following reaction, the rate could be defined in any of three ways.

$$2NOBr(g) \longrightarrow 2NO(g) + Br_2(g)$$

$$\text{Rate of reaction} = \frac{\text{increase in concentration of } Br_2}{\text{time for the increase to occur}} = \frac{\Delta[Br_2]}{\Delta t}$$

$$\text{Rate of reaction} = \frac{\text{increase in concentration of NO}}{\text{time for increase to occur}} = \frac{\Delta[NO]}{\Delta t}$$

$$\text{Rate of reaction} = \frac{\text{decrease in concentration of NOBr}}{\text{time for decrease to occur}} = \frac{\Delta[NOBr]}{\Delta t}$$

EXAMPLE Calculating the rate of a reaction from changes in concentration

When the reaction shown is run in a 1.0 L flask, 8.0×10^{-4} mol of Br_2 is formed during the first 5.0 s of reaction. What is the rate of reaction as defined by the change in concentration of Br_2?

$$2NOBr(g) \longrightarrow 2NO(g) + Br_2(g)$$

SOLUTION The concentration of Br_2 at the start of the reaction is zero, that is, $[Br_2] = 0$ M. After formation of 8.0×10^{-4} mol in a 1.0 L flask, $[Br_2] = 8.0 \times 10^{-4}$ mol/1.0 L or 8.0×10^{-4} M. Thus,

$$\text{Change in concentration of } Br_2 = 8.0 \times 10^{-4}\text{ M} - 0\text{ M}$$
$$= 8.0 \times 10^{-4}\text{ M}$$

This change occurs in a 5.0 s time interval, so $\Delta t = 5.0$ s. We substitute these values into the equation for rate that involves Br_2:

$$\text{Rate of reaction} = \frac{\text{increase in concentration of } Br_2(g)}{\text{time for the increase to occur}}$$
$$= \frac{8.0 \times 10^{-4}\ \text{M}}{5.0\ \text{s}}$$
$$= 1.6 \times 10^{-4}\ \text{M/s}$$

EXERCISE In the same reaction, what is the rate of reaction, as measured by the loss of NOBr?

ANSWER: 3.2×10^{-4} M/s

PITFALL The consistency of reaction rates

How fast a reaction proceeds does not change as we change the method of measurement. It is the same no matter which product or reactant concentration we measure. The number we obtain for the rate of reaction, however, may depend on what product or reactant concentration is measured.

PITFALL The sign of the rate of reaction

The rate of reaction is always a positive quantity. If we remember that the rate is determined by the *decrease* in concentration of a reactant over time, or the *increase* in concentration of a product over time, we always end up with a positive rate.

KEY CONCEPT B The instantaneous reaction rate

The instantaneous rate of reaction is the rate at some instant in time. At the beginning of a reaction, the reaction rate is relatively fast. As the reaction proceeds, the speed of the reaction slows down, that is, the rate decreases. Finally, the reaction stops. At every instant in time, the reaction rate is different than at the previous instant in time. To determine the instantaneous rate, we use the equation

$$\text{Rate} = \frac{\Delta[\text{reactant or product}]}{\Delta t}$$

and make both the change in concentration and the time interval as small as possible. One way to accomplish this graphically (as explained in Appendix 1D of the text) is to plot the concentration of reactant (or product) as a function of time and draw a tangent line on the curve at the point where you want to know the instantaneous rate. The slope of the tangent is the instantaneous rate. The plot must have concentration on the *y* axis and time on the *x* axis in order to get the correct answer for the instantaneous rate.

EXAMPLE Determine the instantaneous rate of reaction

When the reaction $2N_2O_5 \longrightarrow 4NO_2 + O_2$ is run in liquid bromine as a solvent at 55°C, the following results are obtained for the concentration of N_2O_5 as a function of time.

Time, s	0	100	200	300	400	500	600	700	800
$[N_2O_5]$	0.100	0.081	0.066	0.054	0.044	0.035	0.029	0.023	0.019

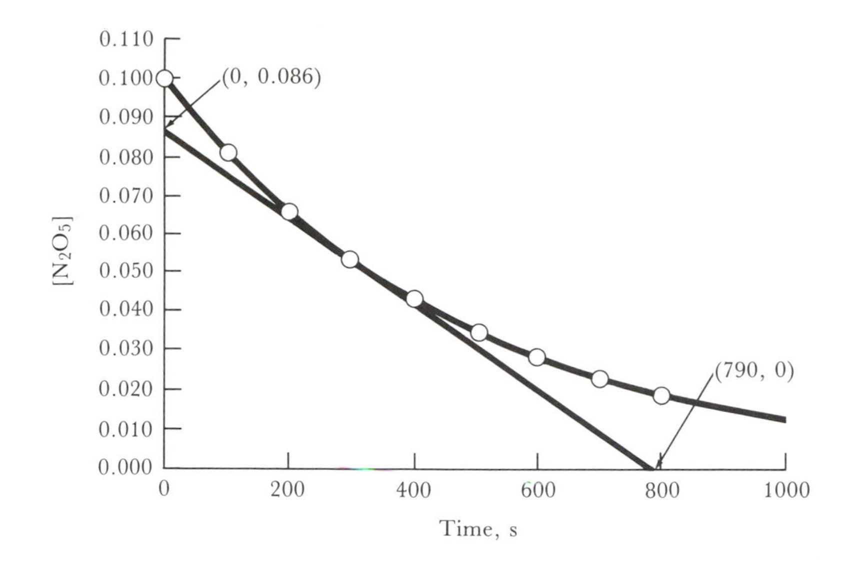

Plot the data, with time as the x axis and the molarity of N_2O_5 as the y axis. From the plot, determine the instantaneous rate of reaction at $t = 300$ s by drawing a tangent at $t = 300$ s and determining the slope of the tangent.

SOLUTION The figure shows the required plot, with a tangent line drawn at $t = 300$ s.

The point at which the tangent crosses the y axis is (0 s, 0.086 M); it crosses the x axis at (790 s, 0 M). We can use these values to determine the slope of the line:

$$\text{Slope} = \frac{\text{rise}}{\text{run}} = \frac{\Delta y}{\Delta x}$$
$$= \frac{0.086\ \text{M} - 0\ \text{M}}{0\ \text{s} - 790\ \text{s}}$$
$$= -1.1 \times 10^{-4}\ \text{M/s}$$

The negative sign reflects the fact that we are dealing with the loss of reactant. The rate of loss of N_2O_5, which must be positive, is the magnitude of the slope:

$$\text{Rate of loss of } N_2O_5 = 1.1 \times 10^{-4}\ \text{M/s}$$

EXERCISE Assume the reaction A $\longrightarrow$ (products) gives the following results for the molarity of A as a function of time.

Time, s	0	10	20	30	40	50	60	70
[A]	0.20	0.15	0.11	0.078	0.057	0.041	0.030	0.022

Determine the instantaneous rate of reaction at $t = 30$ s by plotting the values, drawing a tangent at $t = 30$ s and using the slope of the tangent line to get the instantaneous rate. Your answer may be a *little* different from the one given because drawing a tangent line is not an exact procedure.

ANSWER: 2.6×10^{-3} M/s

KEY CONCEPT C Initial rates and the concentration dependence of initial rates

The initial rate of a reaction is the instantaneous rate of reaction right at the start of the reaction, before any significant amount of product has formed. It has been found that the initial rate of reaction almost always depends on the concentration of reactant at the start of the experiment. The quantitative dependence is

$$\text{Initial rate of reaction} = \text{constant} \times (\text{concentration of reactant})^n$$

where n is usually a smaller integer. The value of the constant (the rate of constant, k) depends on which reaction occurs and the temperature; the value of n depends on which reaction occurs. For example:

Reaction	Rate	Value of k	Temperature, °C
$2N_2O_5(g) \longrightarrow 4NO_2(g) + O_2(g)$	initial rate $= k[N_2O_5]_0$	4.9×10^{-3}/s 3.1×10^{-5}/s	65 25°
$2NO_2(g) \longrightarrow 2NO(g) + O_2(g)$	initial rate $= k[NO_2]_0^2$	0.54 L/mol·s	300°

The equation for the initial rate can be used to calculate the initial rate for any starting concentration of reactant.

EXAMPLE Calculating the initial rate of reaction

Calculate the initial rate of decomposition of $NO_2(g)$ at 300°C with the data given in the previous table when the initial concentration of NO_2 is 0.012 M.

SOLUTION The initial rate of reaction for the decomposition of $NO_2(g)$ is

$$\text{Initial rate} = k[NO_2]_0^2$$

At 300°C, $k = 0.54$ L/mol·s, and we are given $[NO_2]_0 = 0.012$ M; therefore,

$$\begin{aligned}\text{Initial rate} &= (0.54\ \text{L/mol·s})(0.012\ \text{mol/L})^2 \\ &= (0.54\ \cancel{\text{L}}/\cancel{\text{mol}}\text{·s})(0.000144\ \text{mol}^{\cancel{2}}/\text{L}^{\cancel{2}}) \\ &= 7.8 \times 10^{-5}\ \text{mol/L·s} \\ &= 7.8 \times 10^{-5}\ \text{M/s}\end{aligned}$$

EXERCISE Calculate the initial rate of reaction for the decomposition of N_2O_5 at 25°C when the initial concentration of N_2O_5 is 0.045 M.

ANSWER: 1.4×10^{-6} M/s

12.2 RATE LAWS

KEY CONCEPT A Rate laws and reaction order

If the products formed in a reaction do not affect the rate of the reaction, the equation that describes the initial rate applies for the rate of reaction throughout the course of the reaction. As an example, for the decomposition of NO_2,

$$\text{Instantaneous rate at any time} = k[NO_2]^2$$

Here the symbol $[NO_2]$ represents the concentration at the time that the rate is calculated. This equation, called the **rate law** for the reaction, is usually shortened to the form

$$\text{Rate} = k[NO_2]^2$$

The constant k is called the **rate constant** and depends on both the temperature and the identity of the reactants, but not on their concentrations. The power to which the concentration is raised in the rate law defines the **order** of the reaction. Some examples of reaction orders and rate laws are given in the following table.

Reaction	Rate law	Order
$SO_2Cl_2(g) \longrightarrow SO_2(g) + Cl_2(g)$	rate = $k[SO_2Cl_2]$	first
$2NH_3(g) \longrightarrow N_2(g) + 3H_2(g)$	rate = k	zero
$2N_2O(g) \longrightarrow 2N_2(g) + O_2(g)$	rate = $k[N_2O]$	first
$2NO_2(g) \longrightarrow NO(g) + NO_3(g)$	rate = $k[NO_2]^2$	second

Many reactions involve more than one reactant in the rate law. For example, consider the following reaction and rate law.

$$2NO(g) + H_2(g) \longrightarrow N_2O(g) + H_2O(g) \qquad \text{Rate} = k[NO]^2[H_2]$$

The rate law is said to be "second order in NO" because the concentration of NO is raised to the power 2, and "first order in H_2" because the concentration of H_2 is raised to the power 1. The **overall order** of the reaction is the sum of all of the powers of all of the reactant concentrations in the rate law. For our example, 2 + 1 = 3, so the overall order of the reaction is 3; it is third-order overall.

EXAMPLE Recognizing the order of reaction

A reaction and the rate equation associated with the reaction are

$$2NO_2(g) + F_2(g) \longrightarrow 2NO_2F(g) \qquad \text{Rate} = k[NO_2][F_2]$$

What is the order of reaction for each reactant and the overall rate of reaction?

SOLUTION In the rate equation given, the concentration of NO_2 is raised to the power 1 and that of F_2 to the power 1. This means that the order in NO_2 is 1 and the order in F_2 is 1. The sum of the orders is 2 (1 + 1 = 2), so the overall order is 2. The reaction is first order in NO_2, first order in F_2, and second order overall.

EXERCISE A reaction and the rate law for the reaction are

$$2NO(g) + Br_2(g) \longrightarrow 2NOBr(g) \qquad \text{Rate} = k[NO]^2[Br_2]$$

What is the order for each reactant and the overall order?

ANSWER: Second order in NO; first order in Br_2; third order overall

PITFALL Do the coefficients in a chemical equation tell us the orders of the rate?

In general the coefficients in the balanced chemical equation *do not* determine the orders expressed in the rate equation. For most chemical reactions, the *only* way to determine the order for each reactant is through an experiment. There is frequently a coincidental correspondence between the orders and the coefficients in the balanced chemical equation, but because there is no way of knowing in advance when such a coincidence will occur, an experiment must still be done to determine reaction orders. An important exception to this general idea will be encountered later in the chapter.

KEY CONCEPT B How order of reaction affects rate of reaction

One of the interesting aspects of the order of a reactant is that it tells us how the rate changes as the concentration of the reactant changes. For example, consider the rate law

$$\text{Rate} = k[N_2O]$$

Let us assume we have an experiment situation in which $[N_2O] = 0.20$ M. The rate is then

$$\text{Rate} = 0.20\ k$$

Now, if we double the concentration of N_2O, so that $[N_2O] = 0.40$ M, we get for the new rate

$$\text{Rate} = 0.40\ k$$

which is twice the original rate. When the concentration of reactant is doubled, the rate doubles. Now consider the rate equation

$$\text{Rate} = k[\text{NOCl}]^2$$

Let us assume that we have an experiment in which the concentration of reactant is the same as in the previous example, that is, $[\text{NOCl}] = 0.20$ M. The rate is then

$$\text{Rate} = (0.20)^2 k = 0.040\ k$$

If we now double the concentration of NOCl, so that $[\text{NOCl}] = 0.40$ M, we get

$$\text{Rate} = (0.40)^2 k = 0.16\ k$$

The rate has now quadrupled; that is, the new rate is four times the old one when the concentration is doubled ($0.16\ k/0.04\ k = 4$). The point of these two calculations is that the change in rate when the concentration is changed depends on the order of reaction. We can use this fact to determine the order of a reaction. By changing the concentration of the reactant, or of one reactant at a time for a more complex reaction, we can often determine the order of reaction.

EXAMPLE 1 Determining the order of a reaction from rate data

The data in the table show the rate of reaction for different concentrations of reactant for the following reaction. Write the rate law for the reaction and state the order with respect to each reactant and the overall order.

$$2\text{NO}(g) + \text{Cl}_2(g) \longrightarrow 2\text{NOCl}(g)$$

Experiment	**[NO]**	**[Cl_2]**	**Rate, M/h**
1	0.30	0.30	0.85
2	0.30	0.60	1.70
3	0.60	0.30	3.40

SOLUTION The change in the rate when the concentration of reactant changes tells us the order of reaction for each reactant. In experiments 1 and 2, the concentration of NO doesn't change, so all of the change in rate is attributable to the change in concentration of Cl_2. For these experiments, the Cl_2 concentration doubles; at the same time the rate doubles, so the reaction must be first order in Cl_2. In experiments 1 and 3, the Cl_2 concentration is held constant as the NO concentration is doubled. In these two experiments, the rate quadruples ($4 \times 0.85 = 3.40$). Because the rate quadruples as the NO concentration alone doubles, the reaction must be second order in NO. The rate law is

$$\text{Rate} = k[\text{NO}]^2[\text{Cl}_2]$$

The reaction is overall third order ($2 + 1 = 3$).

EXERCISE The data shown give the rate of reaction for different concentrations of reactant for the following reaction. Write the rate equation and state the order with respect to each reactant. Also state the overall order of the reaction.

$$2\text{NO}(g) + \text{H}_2(g) \longrightarrow \text{N}_2\text{O}(g) + \text{H}_2\text{O}(g)$$

Experiment	[NO]	[H_2]	Rate, M/s
1	0.10	0.10	1.6
2	0.10	0.20	3.2
3	0.20	0.10	6.4

ANSWER: Rate = $k[NO]^2[H_2]$; second order in NO; first order in H_2; third order overall

EXAMPLE 2 Determining the order of a reaction from rate data

The data shown give the rate of reaction for different concentrations of reactant for the following reaction. What is the rate equation for the reaction? What is the overall order of the reaction?

$$A + B \longrightarrow C + D$$

Experiment	[A]	[B]	Rate, M/s
1	0.20	0.20	2.60
2	0.20	0.25	4.06
3	0.25	0.20	3.25

SOLUTION In this case, the concentrations of reactants are not doubled or tripled, so the relationship of rate to the order for each reactant is not apparent. It is necessary to substitute the data into the rate equation, rate = $k[A]^m[B]^n$, with the two orders m and n as unknowns. We start with the first two experiments.

$$2.60 \text{ M/s} = k(0.20)^m(0.20)^n$$
$$4.06 \text{ M/s} = k(0.20)^m(0.25)^n$$

We now divide the second equation by the first.

$$\frac{4.06 \cancel{\text{M/s}}}{2.60 \cancel{\text{M/s}}} = \frac{\cancel{k}(0.20)^m(0.25)^n}{\cancel{k}(0.20)^m(0.20)^n}$$

$$1.56 = \frac{(0.25)^n}{(0.20)^n}$$
$$= (0.25/0.20)^n$$
$$= 1.25^n$$

To solve for n, we take the log of both sides and use the fact that $\log x^b = \text{b} \log x$.

$$\log 1.56 = \log 1.25^n$$
$$= n \log 1.25$$
$$0.193 = n\ (0.0969)$$
$$n = (0.193/0.0969)$$
$$= 1.99$$
$$= 2$$

To solve for m, we do the same manipulations using experiments 1 and 3. This gives us $m = 1$. Thus, the rate equation is

$$\text{Rate} = k[A]^2[B]$$

Because 2 + 1 = 3, the overall order is 3.

EXERCISE The data show the rate of reaction for different concentrations of reactant for the reaction

$$A + B \longrightarrow C + D$$

Experiment	[A]	[B]	Rate, M/s
1	3.0	3.0	0.55
2	2.2	3.0	0.40
3	3.0	2.3	0.32

What is the rate equation for the reaction? What is the overall order of the reaction?

ANSWER: Rate $= k[A][B]^2$; overall order = 3

KEY CONCEPT C The integrated rate law

For a first-order reaction, that is, one with the rate law

$$\text{Rate} = k[A]$$

where A is the reactant, it can be shown with the methods of calculus that

$$\ln\left(\frac{[A]_0}{[A]}\right) = kt$$

In this equation, t is the time at which the concentration of reactant A is measured, [A] is the concentration of A at that time, $[A]_0$ is the concentration at the start of the reaction, when $t = 0$, and k is the rate constant; ln symbolizes the natural log function. This equation, called the **integrated rate law,** tells us how the concentration of reactant [A] varies with time t for a first-order reaction. The form of the equation lends itself to a useful graphic technique for determining if a reaction of unknown order is first order or not. To make this determination, the concentration of A is measured experimentally as a function of time and a plot is made with ln [A] on the y axis and time on the x axis. If the plot is a straight line, the reaction is first order and the slope of the line equals $-k$. If the resulting plot is not a straight line, the reaction is not first order.

EXAMPLE 1 Using the integrated rate equation

The reaction given below is first order in sucrose, with a rate constant equal to 3.65×10^{-3} min^{-1} at 23°C. Water does not appear in the rate equation because it is the solvent. Assume an experiment is run with a starting concentration of sucrose equal to 0.310 M. What is the concentration of sucrose after 120 min?

$$\text{Sucrose} + \text{water} \longrightarrow \text{glucose} + \text{fructose}$$

SOLUTION We are given $k = 3.65 \times 10^{-3}$ min^{-1} and $t = 120$ min, so the integrated rate law becomes

$$\ln\left(\frac{[A]_0}{[A]}\right) = kt = (3.65 \times 10^{-3}\,\cancel{\text{min}}^{-1})(120\,\cancel{\text{min}})$$
$$= 0.438$$

To get rid of the ln function, we must take the antilog of both sides of the equation. (The antilog of a number x corresponds to e^x on most calculators.) Antilog(ln x) = x for any x, and antilog(0.438) = $e^{0.438} = 1.550$.

$$\text{Antilog}\left\{\ln\left(\frac{[A]_0}{[A]}\right)\right\} = \text{antilog } 0.438$$
$$\frac{[A]_0}{[A]} = e^{0.438}$$
$$= 1.550$$

The problem gives $[A]_0 = 0.310$ M; multiplying both sides of the equation by [A] and dividing both sides by 1.550 gives

$$[A] = \frac{0.310\ \text{M}}{1.550}$$
$$= 0.200\ \text{M}$$

EXERCISE The decomposition of H_2O_2 in aqueous solution at 20°C is first order, with $k = 1.06 \times 10^{-3}\ \text{min}^{-1}$. Assume that a starting concentration of 0.0600 M is used in the reaction. What is the concentration after 400 min?

ANSWER: 0.0393 M

EXAMPLE 2 Using the integrated rate equation

Assume the reaction

$$X \longrightarrow \text{products}$$

is first order, with rate constant $0.0444\ \text{s}^{-1}$, and a reaction starts with $[X] = 0.227$ M. How many seconds does it take for 75.0% of X to react?

SOLUTION The dependence of concentration on time for a first-order reaction is given by the integrated rate equation for a first-order reaction:

$$\ln\left(\frac{[A]_0}{[A]}\right) = kt$$

We do not actually need the initial concentration of X because the problem states that 75.0% of X reacts. This means that 25.0% of X remains:

$$\frac{[A]}{[A]_0} = 0.250$$

$$\frac{[A]_0}{[A]} = 4.00$$

Substituting this value and $k = 0.0444\ \text{s}^{-1}$ into the integrated rate equation gives

$$\ln(4.00) = 0.0444\ \text{s}^{-1}(t)$$
$$1.39 = 0.0444\ \text{s}^{-1}\ (t)$$

Dividing both sides of the equation by $0.0444\ \text{s}^{-1}$ gives

$$t = \frac{1.39}{0.0444\ \text{s}^{-1}}$$
$$= 31.2\ \text{s}$$

EXERCISE For the reaction given in the example, how long does it take for 50.0% of X to react?

ANSWER: 15.6 s

EXAMPLE 3 Using the integrated rate equation to determine *k*

The decomposition of di-*t*-butyl peroxide, $(CH_3)_3COOC(CH_3)_3$, gives the following time versus concentration data at 160°C. The symbol X is used for the peroxide.

$$X \longrightarrow \text{products}$$

Time, min	0.00	1.00	2.00	3.00	4.00	5.00
[X]	0.244	0.239	0.234	0.229	0.224	0.219

Plot a graph of ln [A] versus time and determine (a) if the reaction is first order and (b) the value of the rate constant if it is first order.

SOLUTION Using the data, we take the natural log of each concentration to construct a plot of ln [A] versus time.

Time, min	0.00	1.00	2.00	3.00	4.00	5.00
[X]	0.244	0.239	0.234	0.229	0.224	0.219
ln [X]	−1.411	−1.431	−1.452	−1.474	−1.496	−1.519

The resulting plot is

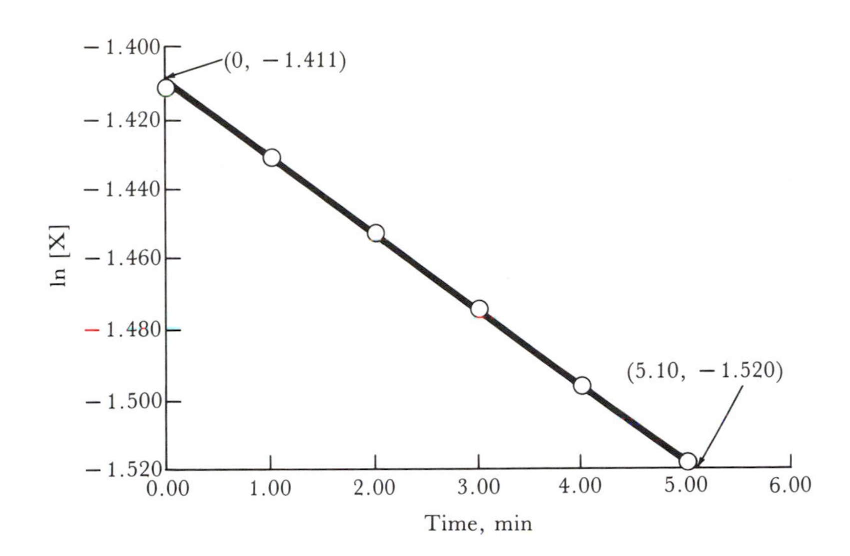

Because the points lie on a straight line, the reaction is first order. The value of k is equal to the negative of the slope of the line. The slope can be calculated with any two points on the line; we choose, for convenience, the points on the x and y axis (0 min, −1.411) and (510 min, −1.520).

$$\text{Slope} = \frac{\Delta x}{\Delta y}$$

$$= \frac{-1.411 - (-1.520)}{0 \text{ min} - 5.10 \text{ min}}$$

$$= \frac{0.109}{-5.10 \text{ min}}$$

$$= -0.0214 \text{ min}^{-1}$$

Because k is equal to the negative of the slope, we obtain

$$k = -\text{slope} = 0.0214 \text{ min}^{-1}$$

EXERCISE The time versus concentration data for the given reaction are shown in the table.

$$\text{Y} \longrightarrow \text{products}$$

Time, s	0	10	20	30	40	50
[Y]	0.155	0.137	0.125	0.109	0.0969	0.0859

Use a plot to determine (a) if the reaction is first order and (b) the value of k if it is first order.

ANSWER: (a) Reaction is first order; (b) $k = 0.012 \text{ s}^{-1}$

KEY CONCEPT D Half-life

The **half-life** ($t_{1/2}$) of a reaction is the time it takes for half of the reactants to react. For a first-order reaction the half-life is the same throughout the course of the reaction. For a second-order reaction, the half-life depends on the concentration of reactant and changes as the reaction proceeds. Using the symbol $t_{1/2}$ for the half-life, we have

First-order reaction: $t_{1/2} = \frac{0.693}{k}$ where k = first-order rate constant

Second-order reaction: $t_{1/2} = \frac{1}{k[A]_0}$ where k = second-order rate constant and $[A]_0$ = the concentration of reactant at the start of the half-life measurement

EXAMPLE Calculating and using the half-life of a chemical reaction

The decomposition of HI(g) is second order, with a rate constant of 0.060 L/mol·s at 630 K. If a reaction starts with a concentration of HI equal to 0.20 M, how long will it take for the concentration to fall to 0.050 M?

SOLUTION We want to calculate the time it takes for the concentration to drop to one fourth its original value (0.050 M/0.20 M = $\frac{1}{4}$). Because the concentration drops by a factor of $\frac{1}{2}$ in one half-life, it will take two half-lives to get to one fourth the original concentration, and we must calculate the length of two half-lives. For a second-order reaction, the half-life is given by

$$t_{1/2} = \frac{1}{k[A]_0}$$

Substituting k = 0.060 L/mol·s and $[A]_0$ = 0.20 M into this equation gives

$$t_{1/2} = \frac{1}{(0.060 \text{ L/}\cancel{\text{mol}}\cdot\text{s})(0.20 \text{ }\cancel{\text{mol}}\text{/L})} = 83 \text{ s}$$

This is the time it takes for the original concentration to drop to half its original value, from 0.20 to 0.10 M. For the second half-life, we must use the concentration at the start of the half-life, in this case 0.10 M. Substituting into the equation for the half-life gives

$$t_{1/2} = \frac{1}{(0.060 \text{ L/mol}\cdot\text{s})(0.10 \text{ M})} = 167 \text{ s}$$

This is the time it takes for the concentration to drop from 0.10 M to 0.050 M. The total time required for the concentration to drop from 0.20 M to 0.050 M is the sum of the two half-lives, or 250 s (83 s + 167 s = 250 s).

EXERCISE The decomposition of N_2O_5 is a first-order reaction with $k = 5.0 \times 10^{-4}\ s^{-1}$ at 318 K. Assume we start a reaction with $[N_2O_5]$ = 0.24 M. How long will it take for the concentraction of N_2O_5 to reach 0.030 M?

ANSWER: 4.2×10^3 s

CONTROLLING RATES OF REACTIONS

KEY WORDS Define or explain each of the following terms in a written sentence or two.

activated complex
activation energy
active site
Arrhenius parameters
catalyst
enzyme
frequency factor
heterogeneous catalyst
homogeneous catalyst
substrate

12.3 THE TEMPERATURE DEPENDENCE OF REACTION RATES

KEY CONCEPT A Arrhenius behavior

The relationship between the rate constant k and temperature is often given by the Arrhenius equation,

$$\ln k = \ln A - \frac{E_a}{RT}$$

In the equation, A is the **frequency factor,** E_a is the **activation energy,** and R is the gas constant, 8.314 J/K mol. A and E_a are called the **Arrhenius parameters.** Reactions for which this relationship is obeyed are said to display **Arrhenius behavior.** There are three important ways to use this relationship.

1. A plot of ln k versus $1/T$ gives a straight line with slope equal to $-E_a/R$. Thus, by measuring the value of k at a series of different temperatures, E_a can be determined.
2. If the rate constant at one temperature, k, and the activation energy are known, the rate constant at a second temperature, k', can be calculated by a form of the relationship

$$\ln \frac{k'}{k} = \frac{E_a}{R}\left[\frac{1}{T} - \frac{1}{T'}\right]$$

3. If E_a and A are known, the rate constant can be calculated.

EXAMPLE 1 Measuring an activation energy

The rate constant for the following second-order reaction in methanol solution depends on the temperature as shown. What is the activation energy for the reaction?

$$CH_3I + Br^- \longrightarrow CH_3Br + I^-$$

t, °C	3	13	23	35
k, L/mol·s	5.0×10^{-6}	1.7×10^{-5}	6.1×10^{-5}	2.2×10^{-4}

SOLUTION We assume that the reaction displays Arrhenius behavior; therefore, a plot of ln k versus $1/T$ should give a straight line with slope = $-E_a/R$. We first set up a table showing the values of ln k and $1/T$.

t, °C	**T, K**	**$1/T$, K^{-1}**	**k, (L/mol·s)**	**ln k**
3	276	3.62×10^{-3}	5.0×10^{-6}	−12.21
13	286	3.50×10^{-3}	1.7×10^{-5}	−10.98
23	296	3.38×10^{-3}	6.1×10^{-6}	−9.70
33	306	3.27×10^{-3}	2.0×10^{-4}	−8.50

The required plot is shown in the following figure.

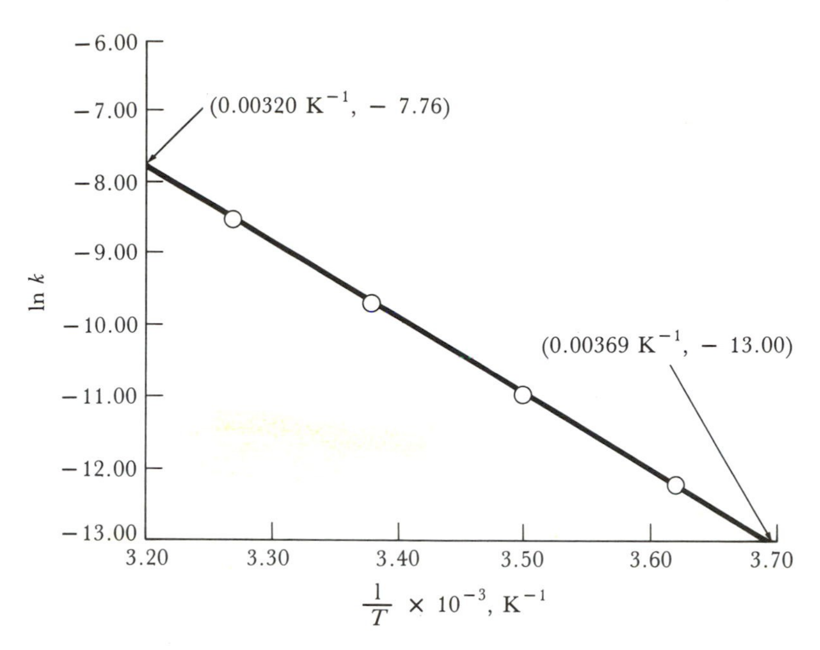

The slope of the line can be calculated using any two points on the line; we choose the two intercepts ($0.00320\ K^{-1}$, -7.76) and ($0.00369\ K^{-1}$, -13.00) for convenience.

$$\begin{aligned}\text{Slope} &= \frac{\Delta y}{\Delta x}\\ &= \frac{-7.76-(-13.00)}{(0.00320\ K^{-1}-0.00369\ K^{-1})}\\ &= \frac{5.24}{-4.9\times 10^{-4}\ K^{-1}}\\ &= -1.14\times 10^{4}\ K\\ &= -1.1\times 10^{4}\ K\end{aligned}$$

Because the slope $= -E_a/R$, we obtain

$$\begin{aligned}E_a &= -\text{slope}\times R\\ &= -(-1.1\times 10^{4}\ \cancel{K})\times 8.314\ \frac{kJ}{mol\cdot \cancel{K}}\\ &= 91\ kJ/mol\end{aligned}$$

EXERCISE The temperature dependence of k for a first-order reaction is given in the table

t, °C	39	52	81	98
k, s^{-1}	3.4×10^{-8}	2.4×10^{-7}	1.0×10^{-5}	7.3×10^{-5}

What is the activation energy for this reaction?

ANSWER: 125 kJ/mol

EXAMPLE 2 Calculating k at one temperature when k at a second temperature and E_a are known

For the second-order gas phase reaction between H_2 and I_2, $k = 1.32 \times 10^{-4}$ L/mol·s at 302°C. What is k at 348°C? $E_a = 163$ kJ/mol.

SOLUTION The equation that relates two values of k at different temperatures is

$$\ln\frac{k'}{k} = \frac{E_a}{R}\left[\frac{1}{T} - \frac{1}{T'}\right]$$

The parameters for the equation are (using k' for the unknown k)

$$k = 1.32 \times 10^{-4} \text{ L/mol·s} \qquad T = 302 + 273 = 575 \text{ K}$$
$$k' = ? \qquad T = 348 + 273 = 621 \text{ K}$$

We first calculate the temperature factor in brackets, using the lowest common denominator TT' for the subtraction:

$$\begin{aligned}\frac{1}{T} - \frac{1}{T'} &= \frac{T' - T}{TT'} \\ &= \frac{(621 - 575)\text{ K}}{621 \text{ K} \times 575 \text{ K}} \\ &= \frac{46 \text{ K}}{3.57 \times 10^5 \text{ K}^2} \\ &= 1.3 \times 10^{-4} \text{ K}^{-1}\end{aligned}$$

Substituting this result, $E_a = 163$ kJ/mol $= 1.63 \times 10^5$ J/mol, and $R = 8.314$ J/mol·K into the original equation gives

$$\begin{aligned}\ln\frac{k'}{k} &= \frac{1.63 \times 10^5 \text{ J/mol} \times 1.3 \times 10^{-4}\text{K}^{-1}}{8.314 \text{ J/mol·K}} \\ &= 2.55 \\ \frac{k'}{k} &= e^{2.55} \\ &= 12.9 \\ k' &= 12.9k\end{aligned}$$

Finally, the value of k, 1.32×10^{-4} L/mol·s is substituted into this last equation to get k'

$$\begin{aligned}k' &= 12.9 \times 1.32 \times 10^{-4} \text{ L/mol·s} \\ &= 1.70 \times 10^{-3} \text{ L/mol·s}\end{aligned}$$

The value of the rate constant at the higher temperature is larger, as expected.

EXERCISE The rate constant for a first-order reaction at 12°C is 5.32×10^{-5} min^{-1}. What is the rate constant at 41°C? The activation energy for the reaction is 96.3 kJ/mol.

ANSWER: 2.2×10^{-3} min^{-1}

KEY CONCEPT B Collision theory

The collision theory of reaction rates states that the rate of reaction depends on at least two factors: (1) the rate of collisions and (2) the fraction f of collisions that successfully lead to a reaction.

$$\begin{aligned}\text{Rate of reaction} &= \text{fraction of collisions that result in a reaction} \times \text{rate of collisions} \\ &= f \times \text{rate of collisions}\end{aligned}$$

The value of f is given by

$$\ln f = \frac{E_{min}}{RT}$$

In this relationship, E_{min} is the minimum total kinetic energy required for a reaction to occur. If two molecules collide with total kinetic energy less than E_{min}, no reaction occurs and the molecules simply rebound; if they collide with total kinetic energy greater than E_{min}, a reaction can occur. This is the meaning we earlier attached to the activation energy; so E_a and E_{min} are the same and

$$\ln f = \frac{E_a}{RT}$$

A third factor may enter into the reaction rate. In many cases, a collision must occur with the molecules in the correct orientation for a reaction to occur. Even if the molecules have the requisite energy, if they meet at the wrong orientation, they will rebound without a reaction occurring.

EXAMPLE Calculating the rate of collisions

The rate of the gase-phase reaction between H_2 and I_2 is 2.5×10^{-3} M/s at 630 K; the activation energy for the reaction is 163 kJ/mol. Calculate the rate of collisions between H_2 and I_2.

SOLUTION Because the rate of reaction is given by

$$\text{Rate} = f \times \text{rate of collisions}$$

we can calculate the rate of collisions as

$$\text{Rate of collisions} = \frac{\text{rate of reaction}}{f}$$

We calculate f using

$$\begin{aligned}\ln f &= \frac{-E_a}{RT}\\ &= \frac{-1.63 \times 10^5 \text{ J/mol}}{(8.314 \text{ J/mol}\cdot\text{K})(630 \text{ K})}\\ &= -31.1\\ f &= e^{-31.1}\\ &= 3.1 \times 10^{-14}\end{aligned}$$

We can now substitute rate of reaction $= 2.5 \times 10^{-3}$ M/s and $f = 3.1 \times 10^{-14}$ into the original equation.

$$\begin{aligned}\text{Rate of collisions} &= \frac{\text{rate of reaction}}{f}\\ &= \frac{2.5 \times 10^{-3}\text{ M/s}}{3.1 \times 10^{-14}}\\ &= 8.1 \times 10^{10}\text{ M/s}\end{aligned}$$

EXERCISE Repeat the calculation shown above at 780 K, at which temperature the rate of reaction is 1.3 M/s.

ANSWER: 1.1×10^{11} M/s

KEY CONCEPT C Activation barriers

According to **activated complex theory,** when two molecules collide with energy greater than the activation energy, they "stick together" momentarily to form an **activated complex.** Bond making and bond breaking may occur in the activated complex to form products, which then separate and move away from each other. If bond making and breaking do not occur, the activated complex breaks apart into the original reactant molecules. The activated complex has a relatively high potential energy, gathered from the kinetic energy of the colliding molecules. The following graph shows the change of the potential energy of the reaction system as the reaction progresses. The activated complex exists at the highest point of the curve.

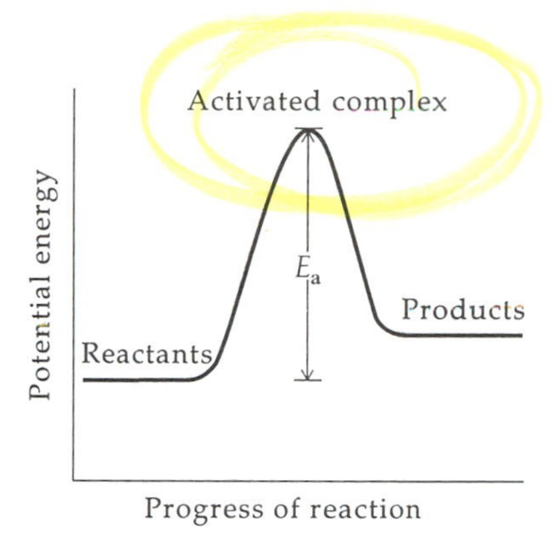

The energy hill between reactants and products is called the **activation barrier.** Molecules that collide with less than the required energy "roll back down" the activation energy hill to reactants. Thus, the energy difference between the potential energy of the reactants and the activated complex is E_a, the activation energy for the reaction; the higher the activation barrier, the slower is the reaction. A graph such as the preceeding one, which shows the potential energy changes for the complete reaction, is called a **reaction profile.**

EXAMPLE Interpreting a reaction profile

Three reaction profiles for three single-step reactions follow. Which reaction is fastest?

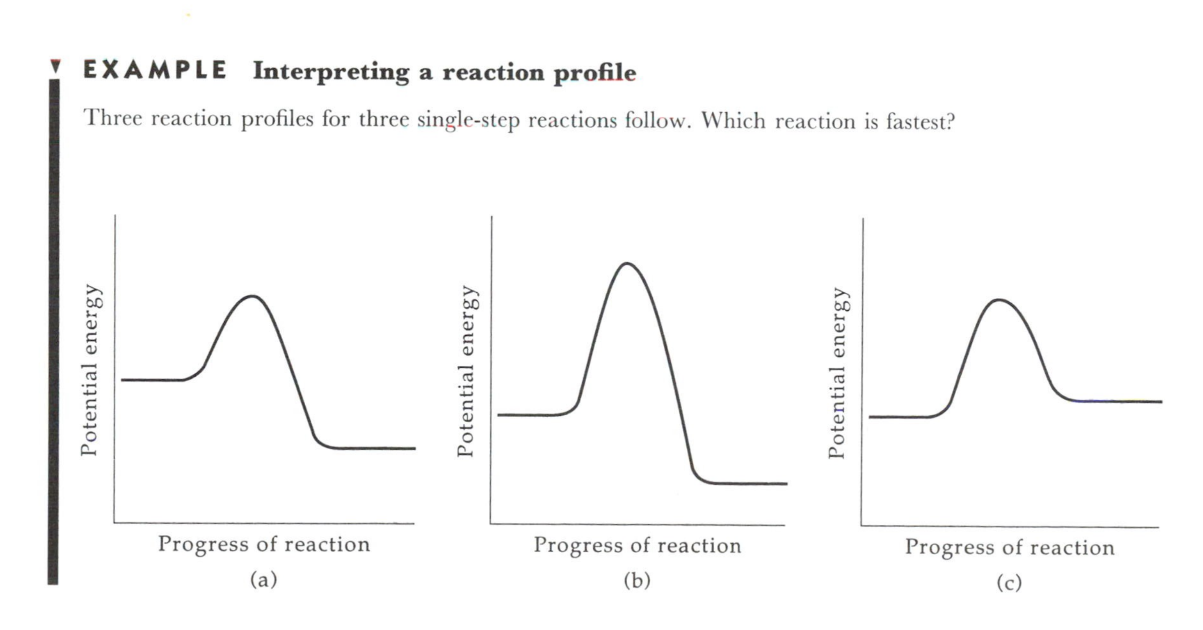

SOLUTION Profile (a) corresponds to the fastest reaction because it has the lowest activation energy.

EXERCISE Sketch the reaction profile for a two-step reaction with a fast first step and slow second step. Be certain to label the x and y axes.

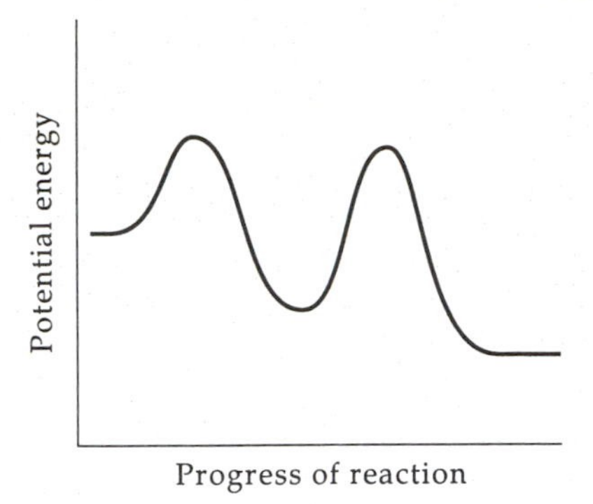

ANSWER: This profile is one of many possible correct answers. Any profile with two activation energy hills, with the second barrier higher than the first (measured from the reactants for that step to the top of the activation energy hill), is a correct answer.

12.4 CATALYSIS

KEY CONCEPT Different types of catalysts

A **catalyst** is a substance that speeds up the rate of a reaction without itself being consumed in the reaction. A **homogeneous catalyst** is one that is present in the same phase as the reactants. A **heterogeneous catalyst** is one that is present in a different phase from that of the reactants. A common type of heterogeneous catalyst is a solid that has a very large surface area that gas-phase or liquid-phase molecules can attach to in a process called **adsorption** (not absorption). When the molecules adsorb onto the surface, they may be brought into the correct orientation for reaction and/or a bond that must be broken in the reaction is weakened. A substance that irreversibly adsorbs onto a catalytic surface prevents reactant molecules from adsorbing and **poisons** the catalyst. An **enzyme** is a biological catalyst. A reactant molecule (called the **substrate**) has the proper shape to fit into the enzymatic **active site,** where the reaction occurs. After reaction, the newly formed product molecule leaves the active site and another reactant molecule can enter the site for reaction. The way in which enzymes work is not well understood, but it is known that they speed up biological reactions considerably.

EXAMPLE 1 Recognizing a catalyst

The two steps for a reaction that occurs in aqueous solution follow. What is the catalyst in the reaction? Is it homogeneous or heterogeneous? What is the overall chemical reaction?

$$V^{3+}(aq) + Cu^{2+}(aq) \longrightarrow V^{4+}(aq) + Cu^{+}(aq)$$
$$Cu^{+}(aq) + Fe^{3+}(aq) \longrightarrow Cu^{2+}(aq) + Fe^{2+}(aq)$$

SOLUTION A catalyst is a substance that speeds up a reaction without itself being consumed. In the reaction shown, Cu^{2+} is used in the first step but is regenerated in the second step; it is not consumed, so it is the catalyst. Because the Cu^{2+} catalyst is a dissolved solute and the reactants are also dissolved solutes, it is a homogeneous catalyst. The overall reaction can be derived by summing the two given reactions

$$V^{3+}(aq) + Fe^{3+}(aq) \longrightarrow V^{4+}(aq) + Fe^{2+}(aq)$$

EXERCISE The two steps for a reaction that occurs in the gas phase follow. What is the catalyst in the reaction? Is it homogeneous or heterogeneous? What is the overall chemical reaction?

$$2NO(g) + Br_2(g) \longrightarrow 2NOBr(g)$$
$$2NOBr(g) + Cl_2(g) \longrightarrow 2NOCl(g) + Br_2(g)$$

ANSWER: Br_2 is the homogeneous catalyst; overall reaction is $2NO(g) + Cl_2(g) \longrightarrow 2NOCl(g)$.

EXAMPLE 2 Heterogeneous catalysis

The decomposition of NH_3 is catalyzed by the presence of a tungsten surface.

$$2NH_3(g) \longrightarrow N_2(g) + 3H_2(g)$$

Addition of H_2 to the reaction vessel slows down the reaction whereas addition of N_2 does not. Explain this observation.

SOLUTION Because addition of N_2 has no effect on the rate of reaction, we can conclude that it is not simply the addition of a product molecule that is important. The observation is explained if H_2 poisons the tungsten catalyst by adsorbing onto the tungsten more strongly than the reactant NH_3 molecules.

EXERCISE What happens to the activation energy of the preceding reaction when H_2 is added to the reaction vessel?

ANSWER: Because the reaction slows down, the activation energy has likely increased.

REACTION MECHANISMS

KEY WORDS Define or explain each of the following terms in a written sentence or two.

bimolecular reaction
chain reaction
elementary reaction
initiation
molecularity
propagation
rate-determining step
reaction intermediate
reaction mechanism
termination
unimolecular reaction

12.5 ELEMENTARY REACTIONS

KEY CONCEPT A Elementary reactions

Most reactions occur through a series of two or more simple steps. For example, the reaction

$$2NO(g) + H_2(g) \longrightarrow N_2O(g) + H_2O(g)$$

does not occur through the collision of an H_2 molecule with NO molecules to result in products. Instead it occurs through two steps:

$$2NO \longrightarrow N_2O_2$$
$$N_2O_2 + H_2 \longrightarrow N_2O + H_2O_2$$

In the first step, two NO molecules collide to produce the intermediate N_2O_2. In the second step, the N_2O_2 collides with an H_2 molecule to produce the products N_2O and H_2O. These steps are called **elementary reactions** because they show a single simple process that occurs for the species portrayed in the step. In an elementary reaction, when two species react, the product is formed because of a direct collision between the reactants. In this respect, an elementary reaction differs from an overall chemical reaction. The *elementary reaction* shows what actually happens in a single step; an *overall reaction* shows the result of a series of steps. A series of elementary steps that describes how the reactants become products for a reaction is called a **mechanism** for the reaction.

▼ EXAMPLE Deciding if a reaction is an elementary process

In the reaction of a deuterium atom with a hydrogen molecule a deuterium atom collides with a hydrogen molecule. At the instant of collision, an H—D bond is formed as an H—H bond is broken, resulting in the direct formation of products. Is this an elementary reaction?

$$D + H_2 \longrightarrow HD + H$$

SOLUTION This is an elementary reaction. One type of elementary reaction portrays a single simple step in which the products are formed as a direct result of a collision. The reaction in the problem is this kind of elementary reaction.

EXERCISE The reaction between nitrogen dioxide and carbon monoxide is

$$NO_2(g) + CO(g) \longrightarrow CO_2(g) + NO(g)$$

The reaction works in the following way. Two NO_2 molecules collide to form NO_3 and NO (NO is one of the products). The NO_3 thus formed collides with CO to form NO_2 and CO_2 (CO_2 is the other product).

$$2NO_2 \longrightarrow NO_3 + NO$$
$$NO_3 + CO \longrightarrow NO_2 + CO_2$$

The overall result is the formation of one NO molecule and one CO_2 molecule from one NO_2 molecule and one CO molecule, as portrayed in the chemical equation shown. Is the reaction $NO_2(g) + CO(g) \rightarrow CO_2(g) + NO(g)$ an elementary reaction?

ANSWER: No; the products CO_2 and NO are not formed in a single step as the result of a collision between the reactants NO_2 and CO. ▲

PITFALL There is more than one type of elementary reaction

The preceding example and exercise focus on one type of elementary reaction, in which two reactants collide to form product(s). We should remember that another type is considered in the text, one in which a single molecule breaks apart into fragments. There are still other types of elementary reactions, but they will not be discussed in this course.

KEY CONCEPT B Unimolecular and bimolecular reactions

A **unimolecular reaction** is an *elementary* reaction in which the reactant is a single molecule. The rate law for a unimolecular reaction is always a first order law.

$$\text{Unimolecular reaction: } A \longrightarrow \text{products} \qquad \text{Rate} = k[A]$$

A **bimolecular reaction** is an *elementary* reaction in which two reactants collide to form products. The rate of reaction is always first order in each reactant.

$$\text{Bimolecular reaction: } A + B \longrightarrow \text{products} \qquad \text{Rate} = k[A][B]$$

▼ EXAMPLE Writing the rate equation for an elementary reaction

The decomposition of acetone (C_3H_6O) starts with the reaction

$$C_3H_6O \longrightarrow CH_3 + C_2H_3O$$

In this reaction, acetone is heated to a very high temperature so that it starts vibrating violently. Eventually, the molecule falls apart into two fragments. What is the rate law for this reaction?

SOLUTION From the description of the reaction, we recognize it as a unimolecular elementary reaction. Such a reaction is always described by a first-order rate law.

$$\text{Rate} = k[C_3H_6O]$$

EXERCISE One of the steps in the decomposition of nitrous acid (HNO_2) is

$$NO_2 + NO_2 \longrightarrow N_2O_4$$

In this reaction, two NO_2 molecules collide and bond together to form a N_2O_4 molecule. What is the rate law for this reaction?

ANSWER: Rate $= k[NO_2]^2$

PITFALL When is it possible to deduce a rate law from a balanced equation?

We must be careful when attempting to write a rate law from a balanced equation. This can be done reliably *only* for an elementary reaction. For reactions that are not elementary, the rate law must be determined by an experiment. It cannot be deduced from the balanced equation.

KEY CONCEPT C The overall reaction

One of the goals of studying the rates of reaction is to deduce the mechanism for the reaction, that is, the series of elementary processes that occur when the reactants change to products. There are two basic requirements for any mechanism:

1. The sequence of elementary reactions in the mechanism must account for the overall chemical change from reactants to products. Practically, this means that the elementary steps in a mechanism must add up to the overall reaction.
2. The mechanism must be consistent with the experimental rate law that has been determined for the overall reaction.

Many mechanisms involve formation of one or more **intermediates.** An intermediate is a chemical species that is formed as a product in an elementary reaction and then becomes a reactant in another elementary reaction, so it is used up and does not appear as a product in the overall chemical equation. Intermediates are generally highly reactive.

EXAMPLE Analysis of a reaction mechanism

A proposed mechanism for the reaction of nitrogen dioxide with fluorine is shown. Does the proposed mechanism account for the overall chemical change from reactants to products? Are there any intermediates in the mechanism?

$$\text{Overall reaction: } 2NO_2(g) + F_2(g) \longrightarrow 2NO_2F(g)$$

$$\text{Proposed mechanism: } NO_2 + F_2 \longrightarrow NO_2F + F$$

$$F + NO_2 \longrightarrow NO_2F$$

SOLUTION We check to see if the mechanism accounts for the overall chemical changes in the chemical equation by adding the equations representing the elementary reactions in the mechanism:

$$NO_2 + F_2 \longrightarrow NO_2F + F$$

$$F + NO_2 \longrightarrow NO_2F$$

$$NO_2 + F_2 + F + NO_2 \longrightarrow NO_2F + F + NO_2F$$

F appears on both sides of the chemical equation, so it is canceled out of the overall reaction. There are two NO_2 molecules on the left-hand side of the equation and two NO_2F on the right,

$$2NO_2 + F_2 \longrightarrow 2NO_2F$$

which is the overall chemical equation. The proposed mechanism accounts for the overall chemical change from reactants to products. We note that F is formed as a product in the first step of the mechanism but is then consumed in the second step and does not appear in the overall equation. It is an intermediate.

EXERCISE Is the following mechanism consistent with the overall reaction given? What intermediates (if any) are present?

$$\text{Overall reaction: } 2NO(g) + O_2(g) \longrightarrow 2NO_2(g)$$
$$\text{Proposed mechanism: } NO + O_2 \longrightarrow NO_2 + O$$
$$O + NO \longrightarrow NO_2$$

ANSWER: Mechanism is consistent with overall equation; O is an intermediate.

PITFALL Adding the elementary reactions in a mechanism

The steps in a mechanism show what happens to different molecules and atoms in the mechanism but don't always explicitly show the number of times each step occurs as reactants become products. For instance, the following mechanism is generally accepted as the one that accounts for the reaction of oxygen to form ozone in the upper atmosphere.

$$\text{Overall reaction: } 3O_2(g) \longrightarrow 2O_3(g)$$
$$\text{Accepted mechanism: } O_2(g) \longrightarrow 2O(g)$$
$$O(g) + O_2(g) \longrightarrow O_3(g)$$

If the two steps portrayed in the mechanism are added, the result is not the overall equation. In actuality, the second step occurs twice each time the reaction occurs. Taking this into consideration gives the correct result when the reactions in the mechanism are added.

$$O_2(g) \longrightarrow 2O(g)$$
$$O(g) + O_2(g) \longrightarrow O_3(g) \qquad \text{(once)}$$
$$O(g) + O_2(g) \longrightarrow O_3(g) \qquad \text{(twice)}$$

These three reactions sum to the overall reaction. We *may not* write

$$2O(g) + 2O_2(g) \longrightarrow 2O_3(g) \qquad \text{(incorrect!)}$$

to express the step that occurs twice because this would imply that two O atoms collide with two O_2 molecules in a single step to form two O_3 molecules. Such a four-particle collision is improbable.

KEY CONCEPT D Rate-determining steps

Many reaction mechanisms contain one step that is much slower than the rest. When this occurs, the slow step acts like a bottleneck and determines how fast the whole reaction can occur. The slow step is called the **rate-determining step** in the mechanism.

EXAMPLE Writing a rate law from a mechanism

An overall reaction and mechanism are given. What rate law is predicted for the overall reaction?

$$\text{Overall reaction: } (CH_3)_3CBr(aq) + OH^-(aq) \longrightarrow (CH_3)_3COH(aq) + Br^-(aq)$$
$$\text{Accepted mechanism: } (CH_3)_3CBr \longrightarrow (CH_3)_3C^+ + Br^- \qquad \text{(slow)}$$
$$(CH_3)_3C^+ + OH^- \longrightarrow (CH_3)_3COH \qquad \text{(fast)}$$

SOLUTION In this mechanism, there is a slow step followed by a fast one. The slow step determines how fast the reaction can go, that is it is the rate-determining step. The rate of the overall reaction will be the same as the rate of the slow step. Thus, the rate law for the overall reaction is

$$\text{Rate} = k[(CH_3)_3CBr]$$

EXERCISE What is the predicted rate law for the overall reaction shown if it proceeds by the following mechanism.

$$\text{Overall reaction: } 2NO(g) + Cl_2(g) \longrightarrow 2NOCl(g)$$

$$\text{Accepted mechanism: } NO + Cl_2 \longrightarrow NOCl_2 \quad \text{(slow)}$$

$$NOCl_2 + NO \longrightarrow 2NOCl \quad \text{(fast)}$$

ANSWER: Rate = $k[NO][Cl_2]$

KEY CONCEPT E Elementary reactions that involve a dynamic equilibrium

In many mechanisms, a dynamic equilibrium occurs for one of the elementary reactions in the mechanism. As an example, consider the following elementary reaction (the "forward" reaction):

$$A \longrightarrow B \qquad \text{Rate} = k[A]$$

Let's assume that after a certain amount of product B forms, it reacts to reform the reactant A,

$$B \longrightarrow A \qquad \text{Rate} = k'[B]$$

where the rate constant k' refers to the reverse reaction. A dynamic equilibrium exists when the rate of the reverse reaction equals the rate of the forward reaction, so that for every A molecule that becomes a B molecule, a B molecule reverts back to an A molecule. ("Whatever is done is undone.") The concentration of A and B do not change. When dynamic equilibrium is achieved, we have

$$\text{Rate of forward reaction} = \text{rate of reverse reaction}$$

$$k[A] = k'[B]$$

It is now possible, by dividing both sides of this equation by k', to express the concentration of B in terms of k, k', and the concentration of A:

$$[B] = \frac{k}{k'}[A]$$

EXAMPLE Predicting the rate law for a mechanism with a dynamic equilibrium

What is the predicted rate law for the overall reaction if it proceeds by the following mechanism. (These mechanistic steps have only recently been proposed to be important in the high temperature reaction of H_2 with I_2).

$$\text{Overall reaction: } H_2(g) + I_2(g) \longrightarrow 2HI(g)$$

$$\text{Proposed mechanism: } I_2 \rightleftharpoons 2I \quad \text{(rapid dynamic equilibrium)}$$

$$H_2 + 2I \longrightarrow 2HI \quad \text{(slow)}$$

SOLUTION The rate of the overall reaction is determined by the slow step in the mechanism

$$\text{Rate} = k_2[H_2][I]^2$$

The rate constant is called k_2 to distinguish it from k_1, the rate constant for the first step in the mechanism. The rate law is correct but not acceptable as a rate law for the overall reaction because it contains the concentration of I, and I is not one of the reactants in the overall reaction. It is an intermediate and should not appear in the rate law for the overall reaction. To eliminate [I], we recognize that the rapid dynamic equilibrium in the first step gives

$$\text{Rate of forward reaction} = k_1[I_2] \qquad \text{Rate of reverse reaction} = k'_1[I]^2$$

$$k_1[I_2] = k'_1[I]^2$$

$$[I]^2 = \frac{k_1}{k'_1}[I_2]$$

This value of $[I]^2$ is now substituted in the original rate equation for the slow step to give

$$\text{Rate} = \frac{k_2k_1}{k'_1}[H_2][I_2]$$

Recognizing that the factor k_2k_1/k'_1 equals a constant, we define an overall rate constant k with the relation $k = k_2k_1/k'_1$ and end up, finally, with the rate law for the overall reaction:

$$\text{Rate} = k[H_2][I_2]$$

EXERCISE What is the rate law for the following overall reaction if it proceeds by the given mechanism?

Overall reaction: $2NO(g) + H_2(g) \longrightarrow N_2O(g) + H_2O(g)$

Proposed mechanism:

$2NO \rightleftharpoons N_2O_2$ (rapid dynamic equilibrium)

$N_2O_2 + H_2 \longrightarrow N_2O + H_2O$ (slow)

ANSWER: Rate $= k[H_2][NO]^2$

12.6 CHAIN REACTIONS

KEY CONCEPT Radical chain reactions

A chain reaction is a reaction in which an intermediate reacts to form another intermediate, which then proceeds to form some of the first intermediate. This cyclic process occurs over and over again, forming some product in each cycle. A distinctive feature of a chain reaction is related to how the reaction is sustained. In a nonchain reaction, the reaction proceeds by having the mechanism occur, from the first step to the last, over and over again. A chain reaction works somewhat differently. After the first step in the mechanism (**initiation**), two or more product-producing steps cycle through many times in a self-sustaining way (**propagation**). This occurs because an intermediate produced as a product in one of the cycling steps becomes a reactant for another cycling step. The cycling continues until a random reaction consumes one of the intermediates required for the cycling process (**termination**). In a nonchain reaction, if the first step in the mechanism occurs 1000 times, 1000 sets of product molecules are formed. In a chain reaction, 1000 initiations result in many more than 1000 sets of product molecules, because each initiation is followed by many cycles of the product-producing steps. **Retardation** occurs when an intermediate reacts to form something other than product, but also forms a reactive chain carrier. **Inhibition** occurs when an intermediate reacts with a foreign substance to produce nonreactive substances. In a **chain-branching** step, a single radical intermediate reacts to form two or more radical intermediates. Chain branching causes a rapid increase in the overall rate of reaction because the new radicals formed start new chains, more branching occurs in the new chains, more new radicals start more new chains, and so on.

EXAMPLE Identifying and understanding the steps in a chain reaction

The overall reaction and mechanism for the chlorination of methane—a radical chain reaction—is

Overall reaction: $CH_4(g) + Cl_2(g) \longrightarrow CH_3Cl(g) + HCl(g)$

Accepted mechanism:

$Cl_2 \longrightarrow 2Cl\cdot$

$Cl\cdot + CH_4 \longrightarrow CH_3\cdot + HCl$

$CH_3\cdot + Cl_2 \longrightarrow CH_3Cl + Cl\cdot$

$Cl\cdot + CH_3\cdot \longrightarrow CH_3Cl$

Identify and briefly discuss the initiation, propagation, and termination steps. Be certain to carefully explain how propagation works.

SOLUTION The first step, in which a chlorine molecule breaks up into two chlorine atoms, is the initiation step. The chlorine atom is a radical intermediate. The next two steps are the propagation steps:

$$Cl\cdot + CH_4 \longrightarrow CH_3\cdot + HCl$$
$$CH_3\cdot + Cl_2 \longrightarrow CH_3Cl + Cl\cdot$$

In the first of these steps, a chlorine atom formed in initiation removes a hydrogen atom from methane to form HCl and a methyl radical, $CH_3\cdot$. The methyl radical, which is an intermediate, reacts with a chlorine molecule in the next step to form a product molecule, CH_3Cl, and a chlorine atom. Now, the chlorine atom can start the propagation sequence over again by finding another methane molecule to react with. The propagation steps can cycle, over and over, as long as reactants are available and termination doesn't occur. The last step is a termination step because it involves the combination of two radicals to produce a nonradical product.

EXERCISE Suggest a possible retardation step for the mechanism given in the example.

ANSWER: $CH_3\cdot + HCl \longrightarrow CH_4 + Cl\cdot$ and
$Cl\cdot + CH_3Cl \longrightarrow CH_3\cdot + Cl_2$

DESCRIPTIVE CHEMISTRY TO REMEMBER

- **Ozone** molecules (O_3) in the upper atmosphere absorb much of the intense ultraviolet radiation from the sun.
- The **antioxidants** added to food, plastics, and rubber inhibit oxidation by removing chain-propagating radicals.
- Finely divided particles, such as iron dust (or other dusts) burn very rapidly in air because of the **large surface area** exposed to the air.
- Catalytic converters in automobiles contain solid **heterogeneous catalysts** that oxidize carbon monoxide to carbon dioxide and reduce nitrogen oxides to nitrogen (N_2).
- **Sulfur** in automobile fuels is a problem because the catalysts in catalytic converters convert SO_2 to SO_3, which can then react with water to form sulfuric acid.
- The action of many biological poisons is thought to stem from their ability to block the action of **enzymes.**

CHEMICAL EQUATIONS TO KNOW

Gas-phase reactions

- $2HI(g) \longrightarrow H_2(g) + I_2(g)$
- $2N_2O_5(g) \longrightarrow 4NO_2(g) + O_2(g)$
- $2NO_2(g) \longrightarrow 2NO(g) + O_2(g)$
- $2O_3(g) \longrightarrow 3O_2(g)$

$H_2(g) + Br_2(g) \longrightarrow 2HBr(g)$
$2ICl(g) \longrightarrow I_2(g) + Cl_2(g)$
$C_3H_6(g) \longrightarrow CH_3{-}CH{=}CH_2(g)$
$NO_2(g) + CO(g) \longrightarrow CO_2(g) + NO(g)$

Solution reactions

- $S_2O_8{}^{2-}(aq) + 3I^-(aq) \longrightarrow 2SO_4{}^{2-}(aq) + I_3{}^-(aq)$
- $BrO_3{}^-(aq) + 5Br^-(aq) + 6H^+(aq) \longrightarrow 3Br_2(aq) + 3H_2O(l)$
- $CH_3Br(aq) + OH^-(aq) \longrightarrow CH_3OH(aq) + Br^-(aq)$
- $C_2H_5Br(aq) + OH^-(aq) \longrightarrow C_2H_5OH(aq) + Br^-(aq)$

Reactions that work best when catalyzed (heterogeneous catalyst)

- $2NH_3(g) \xrightarrow{\Delta,\ Pt} N_2(g) + 3H_2(g)$
- $2KClO_3(s) \xrightarrow{\Delta,\ MnO_2} 2KCl(s) + 3O_2(g)$
- $N_2(g) + 3H_2(g) \xrightarrow{\Delta,\ Fe} 2NH_3(g)$
- $2SO_2(g) + O_2(g) \xrightarrow{V_2O_5} 2SO_3(g)$
- $4NH_3(g) + 5O_2(g) \xrightarrow{\Delta,\ Pt} 4NO(g) + 6H_2O(g)$

Reactions that work best when catalyzed (homogeneous catalyst)

- $2H_2O_2(aq) \xrightarrow{Br_2} 2H_2O(l) + O_2(g)$

MATHEMATICAL EQUATIONS TO KNOW AND UNDERSTAND

$\text{Rate} = \dfrac{\Delta[A]}{\Delta t}$	definition of rate
$\ln\left(\dfrac{[A]_0}{[A]}\right) = kt$	first-order reaction integrated rate law
$\ln\dfrac{k'}{k} = \dfrac{E_a}{R}\left[\dfrac{1}{T} - \dfrac{1}{T'}\right]$	dependence of k on temperature
$\ln k = \ln A - E_a/RT$	Arrhenius equation for k
$t_{1/2} = 0.693/k$	half-life for first order reaction
$\dfrac{1}{[A]} - \dfrac{1}{[A]_0} = kt$	second-order reaction integrated rate law
$t_{1/2} = \dfrac{1}{k[A]_0}$	half-life for second order reaction
$\ln f = -E_a/RT$	definition of f from collision theory

SELF-TEST EXERCISES

The description of reaction rates

1. The concentration of Cl_2 as a function of time is shown for the following reaction. What is the rate of reaction during the time interval $t = 5.0$ s to $t = 10.0$ s as measured by the decrease in $[Cl_2]$?

$$2NO(g) + Cl_2(g) \longrightarrow 2NOCl(g)$$

Time, s	0.0	5.0	10.0	15.0	20.0
$[Cl_2]$	0.121	0.108	0.100	0.094	0.091

(a) 1.6×10^{-3} M/s (b) 1.0×10^{-2} M/s (c) 2.2×10^{-2} M/s
(d) 3.2×10^{-2} M/s (e) 2.1×10^{-3} M/s

2. For the reaction shown, what is the rate of disappearance of H_2 when the rate of appearance of N_2 is 5.2 mM/s?

$$2NO(g) + 2H_2(g) \longrightarrow 2H_2O(g) + N_2(g)$$

(a) 1.3 mM/s (b) 2.6 mM/s (c) 5.2 mM/s (d) 21 mM/s (e) 10 mM/s

The following data for the reaction "A → products" are for Exercises 3–5.

Time, min	0.0	1.0	2.0	3.0	4.0	5.0
[A]	1.30	1.01	0.788	0.614	0.478	0.372

3. From a plot of [A] versus t, determine the rate of reaction at $t = 2.5$ min. (Because plotting involves a graphic technique, your answer may be somewhat different from the author's. Choose the closest answer.)
(a) 0.26 M/min (b) 1.2 M/min (c) 0.17 M/min
(d) 5.7 M/min (e) 3.8 M/min

4. From the plot used in exercise 3, determine the initial rate. (Because this involves a graphic technique, your answer may be somewhat different from the author's. Choose the closest answer.)
(a) 5.6 M/min (b) 0.32 M/min (c) 0.44 M/min
(d) 2.3 M/min (e) 1.7 M/min

5. Use the initial rate determined in exercise 4 and the initial concentration to determine the rate constant. The rate is given by (initial rate) = k[A].
(a) 4.0 min^{-1} (b) 0.25 min^{-1} (c) 0.11 min^{-1}
(d) 3.2×10^{-3} min^{-1} (e) 5.44×10^{-3} min^{-1}

6. Which of the following is a correct unit for a second-order rate constant?
(a) M/min (b) hr^{-1} (c) mol/s (d) L/mol·s (e) s·L/mol

For Exercises 7 and 8, use the reaction $2NO(g) + O_2(g) \rightarrow 2NO_2(g)$, which follows the rate law: rate of reaction = $k[NO]^2[O_2]$.

7. What is the reaction order with respect to NO?
(a) first (b) second (c) third (d) zeroth

8. What is the overall order of reaction?
(a) second (b) first (c) third (d) zeroth

9. The following data show how the initial rate of reaction changes with initial concentration of reactants for the reaction 2A + 2B → products. What is the rate law for the reaction?

$[A]_0$	$[B]_0$	Initial rate, M/s
0.22	0.22	0.816
0.44	0.22	3.26
0.22	0.44	1.63

(a) rate = $k[A][B]$ (b) rate = $k[A]^2[B]$ (c) rate = $k[B]^2[A]^2$ (d) rate = $k[A][B]^2$

10. The following data show how the initial rate of reaction changes with initial concentration of reactants for the reaction 2A + B → products. What is the rate law for the reaction?

$[A]_0$	$[B]_0$	Initial rate, M/s
0.52	0.46	2.24
0.52	0.70	2.24
0.68	0.73	3.83

(a) rate = $k[A][B]$ (b) rate = $k[A]^2[B]$ (c) rate = $k[B]$
(d) rate = $k[A]^2[B]^2$ (e) rate = $k[A]^2$

11. At high temperatures the decomposition of N_2O is first order with $k = 3.4 \times 10^{-5}\ s^{-1}$. What percentage of N_2O is left 1.0 h after the start of a reaction?
(a) 88.4% (b) 71.6% (c) 23.8% (d) 41.1% (e) 2.65%

12. A first-order reaction with $[A]_0 = 0.84$ M and $k = 0.112\ min^{-1}$ is run until $[A] = 0.62$ M. For how long was the reaction run?
(a) 2.7 min (b) 12 min (c) 1.4 min (d) 0.15 min (e) 6.6 min

13. What is $t_{1/2}$ for a first-order reaction with $k = 1.22\ h^{-1}$ and $[A]_0 = 0.44$ M?
(a) 0.820 h (b) 1.22 h (c) 0.845 h (d) 0.568 h (e) 1.76 h

14. A reaction with $t_{1/2} = 10.0$ min starts with $[A]_0 = 3.2$ M. What is [A] after 40.0 min of reaction?
(a) 0.20 M (b) 0.80 M (c) 1.6 M (d) 0.32 M (e) 0.16 M

15. What is the half-life of a second-order reaction with $k = 8.2 \times 10^{-2}$ L/mol·h and $[A]_0 = 0.45$ M?
(a) 8.5 h (b) 12 h (c) 5.5 h (d) 5.7×10^{-2} h (e) 27 h

16. The reaction "A + B → products" is pseudo first order, with rate = k'[A] and $k' = 6.29\ s^{-1}$ when [B] = 6.0 M. What is k for the rate law, rate = k[A][B]?
(a) 12.3 L/mol·s (b) 6.29 L/mol·s (c) 38 L/mol·s
(d) 16.3 L/mol·s (e) 1.0 L/mol·s

Controlling the rates of reaction

17. $E_a = 132$ kJ/mol and $A = 4.9 \times 10^8$ L/mol·s for the second-order reaction $NO_2(g) + CO(g) \longrightarrow NO(g) + CO_2(g)$. What is k at 300°C?
(a) 3.6×10^5 L/mol·s (b) 2.1×10^{-9} L/mol·s (c) 5.1×10^{-15} L/mol·s
(d) 4.8×10^8 L/mol·s (e) 4.5×10^{-4} L/mol·s

18. The rate constant for a second-order reaction shows the following dependence on temperature. What is the activation energy for the reaction? (Choose the closest answer.)

Temperature, °C	0	5	10	15	20
k, L/mol·s	5.60×10^{-5}	9.50×10^{-5}	19.8×10^{-5}	37.0×10^{-5}	73.3×10^{-5}

(a) 151 kJ/mol (b) 86.5 kJ/mol (c) 15.2 kJ/mol
(d) 66.3 kJ/mol (e) 108 kJ/mol

19. What is k at 59°C for a reaction with $k = 7.3 \times 10^{-2}\ s^{-1}$ at 25°C and $E_a = 82.6$ kJ/mol?
(a) $5.5 \times 10^{-2}\ s^{-1}$ (b) $0.15\ s^{-1}$ (c) $3.2 \times 10^{-2}\ s^{-1}$ (d) $2.2\ s^{-1}$ (e) $0.17\ s^{-1}$

20. What fraction of collisions between molecules has energy greater than 3.5 kJ at room temperature (25°C)?
(a) 0 (b) 0.76 (c) 0.24 (d) 0.99 (e) 0.50

21. What activation energy corresponds to the slowest reaction?
(a) 89 kJ/mol (b) 101 kJ/mol (c) 34 kJ/mol

22. Which reaction profile corresponds to a reaction with a fast first step followed by a slow second step?

Potential energy
Progress of reaction
(a)

Potential energy
Progress of reaction
(b)

Potential energy
Progress of reaction
(c)

23. Addition of a catalyst to a reaction ______ the activation energy of the reaction profile.
(a) raises (b) lowers (c) has no effect on

24. What is the catalyst in the following mechanism?

$$Ce^{4+}(aq) + Ag^{+}(aq) \longrightarrow Ce^{3+}(aq) + Ag^{2+}(aq)$$
$$Ag^{2+}(aq) + Tl^{+}(aq) \longrightarrow Tl^{2+}(aq) + Ag^{+}(aq)$$
$$Tl^{2+}(aq) + Ce^{4+}(aq) \longrightarrow Tl^{3+}(aq) + Ce^{3+}(aq)$$

(a) Ce^{4+} (b) Ag^{+} (c) Ag^{2+} (d) Tl^{+} (e) Tl^{2+}

25. An enzyme is
(a) a reactant molecule in a biological reaction. (b) the active site on a biological catalyst.
(c) any molecule found in a living cell. (d) a biological catalyst.

Reaction mechanisms

26. What is the molecularity of the following elementary reaction?

$$CH_3 + NO \longrightarrow CH_3NO$$

(a) unimolecular (b) bimolecular (c) termolecular

27. What is the rate law for the elementary reaction shown in question 26?
(a) rate = $k[CH_3]$ (b) rate = $k[CH_3][NO]^2$ (c) rate = $k[NO]$
(d) rate = $k[CH_3][NO]$ (e) need more information to tell

The following mechanism is for Exercises 28 and 29.

$$Cl_2 \longrightarrow 2Cl$$
$$Cl + CO \longrightarrow COCl$$
$$COCl + Cl \longrightarrow COCl_2$$

28. What is the overall reaction for the mechanism?
(a) $Cl_2 + 2CO \longrightarrow 2COCl$ (b) $2Cl + CO \longrightarrow COCl_2$ (c) $CO + Cl_2 \longrightarrow COCl_2$

29. What intermediates are present?
(a) $COCl_2$, CO (b) Cl, Cl_2 (c) Cl, COCl (d) CO, Cl_2 (e) CO, Cl

30. The following mechanism has been proposed for the reaction between NO and Br_2 to form NOBr.

$$Br_2 + NO \rightleftharpoons NOBr_2 \quad \text{fast equilibrium}$$
$$NOBr_2 + NO \longrightarrow 2NOBr \quad \text{slow}$$

What is the rate law for this mechanism?
(a) rate = $k[NO]^2[Br_2]$ (b) rate = $k[NOBr_2][NO]$ (c) rate = $k[NO][Br_2]$
(d) rate = $k[Br_2][NOBr_2]$ (e) rate = k

Use the following mechanism for Exercises 31–33:

Step 1:	$Br_2 \longrightarrow 2Br$
Step 2:	$Br + H_2 \longrightarrow HBr + H$
Step 3:	$H + Br_2 \longrightarrow HBr + Br$
Step 4:	$H + HBr \longrightarrow H_2 + Br$
Step 5:	$Br + Br \longrightarrow Br_2$
Overall reaction:	$H_2(g) + Br_2(g) \longrightarrow 2HBr(g)$

31. What steps are chain-propagating steps?
(a) 2, 3 (b) 1 (c) 1, 2 (d) 2, 4 (e) 4

32. What step is a chain-retardation step?
(a) 1 (b) 2 (c) 3 (d) 4 (e) None exists.

33. What step is a chain-termination step?
(a) 3 (b) 2 (c) 5 (d) 1 (e) 4

34. With a chain reaction, the occurrence of an explosion often is due to
(a) chain inhibition (b) chain termination (c) chain retardation
(d) chain branching (e) chain initiation

Descriptive chemistry

35. Which compound in the upper atmosphere absorbs intense ultraviolet light from the sun?
(a) CO_2 (b) O_2 (c) O_3 (d) N_2 (e) H_2O

36. The antioxidants found in foodstuffs, rubber, and plastics are involved in what chain-reaction mechanistic step?
(a) initiation (b) termination (c) branching (d) propagation (e) inhibition

37. One of the conversions (which are not shown as balanced reactions) is not an important one for catalytic converters. Which one?
(a) $NO_x \longrightarrow N_2$ (b) $CO \longrightarrow CO_2$ (c) $SO_2 \longrightarrow SO_3$ (d) $H_2O \longrightarrow H_2O_2$

38. What products are formed by the decomposition of N_2O_5 in the gas phase?
(a) NO, O_2 (b) N_2, O_2 (c) NO_2, N_2O_4 (d) NO_2, O_2 (e) N_2O, NO

39. What catalyst is used in the following reaction?

$$2KClO_3(s) \longrightarrow 2KCl(s) + 3O_2(g)$$

(a) $Pt(s)$ (b) $MnO_2(s)$ (c) $Fe(s)$ (d) $V_2O_5(s)$ (e) $Br_2(g)$

13 CHEMICAL EQUILIBRIUM

THE DESCRIPTION OF CHEMICAL EQUILIBRIUM

13.1 REACTIONS AT EQUILIBRIUM

KEY WORDS Define or explain each of the following terms in a written sentence or two.

chemical equilibrium
decomposition vapor pressure
equilibrium constant
heterogeneous equilibrium
homogeneous equilibrium
law of mass action
reaction quotient

KEY CONCEPT **Reactions at equilibrium**

Imagine a chemical reaction in which the reactants are placed in a reaction vessel and allowed to form products:

$$\text{Reactants} \longrightarrow \text{products}$$

For many reactions, the products can also back-react to form reactants:

$$\text{Products} \longrightarrow \text{reactants}$$

At some point, the rate of formation of products equals the rate of formation of reactants, and a **dynamic equilibrium** is attained. At this point there is no *net* formation of reactants or products, so the concentration of each is constant. Such an equilibrium is represented by double arrows $\rightleftharpoons$.

$$\text{Reactants} \rightleftharpoons \text{products}$$

EXAMPLE **Understanding equilibrium**

Equilibrium exists at a certain time for the following reaction, with $[SO_2] = 0.50$ M, $[O_2] = 0.60$ M, and $[SO_3] = 4.11 \times 10^{-2}$ M. If no changes are made in the experimental conditions, how do the concentrations of each substance change with time?

$$2SO_2(g) + O_2(g) \rightleftharpoons 2SO_3(g)$$

SOLUTION The concentrations do not change with time once equilibrium has been established.

EXERCISE In the preceding equilibrium, assume that we set up an experiment in which, initially, there is only radioactive oxygen in the O_2 molecules. After a relatively long time, will all the radioactive oxygen still be only in O_2 molecules?

ANSWER: No. The equilibrium is dynamic, with a constant exchange of atoms among the substances in the equilibrium; so the radioactive oxygen atoms will be distributed among all three oxygen-containing species, SO_2, O_2, and SO_3.

13.2 THE EQUILIBRIUM CONSTANT

KEY CONCEPT A The equilibrium constant

Experimentation has shown that, at equilibrium, the ratio of products to reactants is a constant at any temperature. More specifically, for the equilibrium

$$aA + bB \rightleftharpoons cC + dD$$

the concentrations of reactants and products obey the relationship

$$K_c = \frac{[C]^c[D]^d}{[A]^a[B]^b}$$

K_c is a constant at any temperature and is called the **equilibrium constant.** The concentrations used in the calculation of K_c must be the equilibrium concentrations.

EXAMPLE 1 Calculating an equilibrium constant

For the reaction of an alcohol with an acid at 50°C, the concentration of each substance at equilibrium is [acid] = 2.2 M, [alcohol] = 3.6 M, [ester] = 5.2 M, and $[H_2O]$ = 3.8 M. What is K_c for this equilibrium?

$$\text{Acid} + \text{alcohol} \rightleftharpoons \text{ester} + H_2O$$

SOLUTION The equilibrium constant for this equilibrium is

$$K_c = \frac{[\text{ester}]^1[H_2O]^1}{[\text{acid}]^1[\text{alcohol}]^1}$$

We have been careful to remember that the concentration of each substance in the equilibrium must be raised to the power given by its coefficient in the chemical equation. In this case, each coefficient is 1. (In the future, powers of 1 will not be written.) Substituting each of the equilibrium values into the equation gives

$$K_c = \frac{5.2\ M \times 3.8\ M}{2.2\ M \times 3.6\ M} = 2.5$$

EXERCISE Calculate K_c for the equilibrium shown if the equilibrium concentrations of gases are $[CO_2]$ = 0.40 M, $[H_2]$ = 0.18 M, [CO] = 0.30 M, and $[H_2O]$ = 0.42 M.

$$CO_2(g) + H_2(g) \rightleftharpoons CO(g) + H_2O(g)$$

ANSWER: 1.8

EXAMPLE 2 Calculating an equilibrium constant

What is K_c for the equilibrium shown if the equilibrium concentrations of each substance are $[SO_2]$ = 0.68 M, $[O_2]$ = 0.88 M, and $[SO_3]$ = 2.3 M.

$$2SO_2(g) + O_2(g) \rightleftharpoons 2SO_3(g)$$

SOLUTION For this equilibrium,

$$K_c = \frac{[SO_3]^2}{[SO_2]^2[O_2]}$$

Substituting in the equilibrium concentrations gives

$$K_c = \frac{(2.3\ M)^2}{(0.68\ M)^2(0.88\ M)} = 13\ M^{-1}$$

EXERCISE What is K_c for the equilibrium shown if the equilibrium concentrations of each of the substances in the equilibrium are $[NO_2] = 0.016$ M, $[O_2] = 0.13$ M, and $[NO] = 0.26$ M.

$$2NO_2(g) \rightleftharpoons O_2(g) + 2NO(g)$$

ANSWER: 34 M

KEY CONCEPT B The magnitude of K_c

The size of K_c tells us whether an equilibrium favors reactants, or products, or neither. Because

$$K_c = \frac{\text{concentration of products}}{\text{concentration of reactants}}$$

a large K_c indicates that the concentration of products is larger than the concentration of reactants at equilibrium. On the other hand, a small K_c indicates that the concentration of reactants is larger than that of products at equilibrium. If K_c is close to one, the concentrations of products and reactants are similar at equilibrium.

$K_c > 1000$	concentration of products > concentration of reactants
$1000 > K_c > 0.001$	concentration of products ≈ concentration of reactants
$K_c < 0.001$	concentration of products < concentration of reactants

EXAMPLE Qualitative interpretation of an equilibrium constant

$K_c = 0.10$ for the equilibrium $2ICl(g) \rightleftharpoons I_2(g) + Cl_2(g)$. Without doing any calculations, decide which set of equilibrium values are correct for the equilibrium:

Reagent	Set 1	Set 2	Set 3
ICl	0.14 M	0.88 M	6.3×10^{-5} M
Cl_2	6.3×10^{-5} M	0.23 M	0.24 M
I_2	7.3×10^{-5} M	0.34 M	0.44 M

SOLUTION Because the value of K_c is less than 1000 and more than 0.001, we predict that the concentration of products and reactants will be similar to each other at equilibrium. Set 2 is the set of concentrations that most closely meets this criterion.

EXERCISE $K_c = 4.7 \times 10^4$ for the equilibrium $CO(g) + H_2O(g) \rightleftharpoons CO_2(g) + H_2(g)$. Would you expect the concentration of CO or CO_2 to be higher at equilibrium?

ANSWER: CO_2

PITFALL Using K_c to estimate the relative amounts of products and reactants

Estimates of the relative amounts of product and reactant based on the size of K_c assume that if more than one product or more than one reactant is present, the concentrations of all the products are close to each other, and/or the concentrations of all the reactants are similar. If this is not true, such estimates can be wrong.

KEY CONCEPT C The equilibrium constant and the chemical equation

The value of K_c for an equilibrium depends on how we choose to write the chemical equation for the equilibrium. If we represent an equilibrium as $aA \rightleftharpoons bB$, we get the following:

Form of equilibrium used	Equilibrium constant
$aA \rightleftharpoons bB$	equilibrium constant $= K_c$
$bB \rightleftharpoons aA$	equilibrium constant $= \dfrac{1}{K_c}$
$naA \rightleftharpoons nbB$	equilibrium constant $= (K_c)^n$

The last equilibrium in the table represents the original equilibrium multiplied by a number n. This number may be a fraction such as $\frac{1}{2}$. In using the relationship, equilibrium constant $= (K_c)^n$, with fractional values of n, we should remember that

$$(K_c)^{1/z} = \sqrt[z]{K_c}$$

EXAMPLE Values of the equilibrium constant for different forms of a chemical equation

An equilibrium equation and the associated equilibrium constant are

$$2ICl(g) \rightleftharpoons I_2(g) + Cl_2(g) \qquad K_c = 0.10$$

What is the equilibrium constant for the reverse equilibrium,

$$I_2(g) + Cl_2(g) \rightleftharpoons 2ICl(g) \qquad K_c = ?$$

SOLUTION The relationship between the equilibrium constant for a forward reaction, K_c(forward), and a reverse reaction, K_c(reverse), is

$$K_c(\text{reverse}) = \frac{1}{K_c(\text{forward})}$$

Thus, for our problem,

$$K_c(\text{forward}) = \frac{1}{0.10} = 10$$

EXERCISE What is the equilibrium constant for the equation

$$\frac{1}{2}I_2(g) + \frac{1}{2}Cl_2(g) \rightleftharpoons ICl(g)$$

ANSWER: 3.2

KEY CONCEPT D The direction of reaction and the reaction quotient

Imagine we have a flask containing NH_3, N_2, and H_2, in which equilibrium has not yet been attained:

$$2NH_3(g) \rightleftharpoons N_2(g) + 3H_2(g) \qquad K_c = 7.4 \times 10^{-4}\ M^2$$

For example, assume $[NH_3] = 0.51$ M, $[N_2] = 0.040$ M, and $[H_2] = 0.12$ M. It is possible to calculate a quantity Q_c that symbolically resembles the equilibrium constant but is not numerically equal to the equilibrium constant because equilibrium concentrations are not present:

$$Q_c = \frac{[N_2][H_2]^3}{[NH_3]^2}$$

$$= \frac{(0.040\ M)(0.12\ M)^3}{(0.51\ M)^2}$$

$$= 2.7 \times 10^{-4}\ M^2$$

Note that Q_c ($2.7 \times 10^{-4}\ M^2$) is not equal to K_c ($7.4 \times 10^{-4}\ M^2$), a confirmation that equilibrium concentrations are not present. Q_c is called the **reaction quotient.** It tells us in which direction the reaction must occur in order to reach equilibrium.

When $Q_c > K_c$, reactants must form from products in order to reach equilibrium.
When $Q_c = K_c$, the reaction is already at equilibrium.
When $Q_c < K_c$, products must form from reactants to reach equilibrium.

The value of the reaction quotient tells us only what must occur for equilibrium to be reached. It does not give information regarding how fast the equilibrium will be attained, but it does tell us whether reactants or products have a tendency to form. For our example, $Q_c < K_c$, so some NH_3 must decompose into N_2 and H_2 in order to attain equilibrium.

EXAMPLE Predicting the direction of reaction

$K_c = 160$ at 500 K for the following equilibrium. If 20 mM HI, 10 mM H_2, and 10 mM I_2 are introduced into a flask and the temperature increased to 500 K, in what direction must the reaction proceed in order to reach equilibrium?

$$H_2(g) + I_2(g) \rightleftharpoons 2HI(g)$$

SOLUTION We must calculate Q_c in order to answer this problem. Because Q_c is dimensionless we do not have to convert millimolar concentrations to molar concentrations.

$$Q_c = \frac{[HI]^2}{[H_2][I_2]}$$

$$= \frac{(20\ mM)^2}{(10\ mM)(10\ mM)}$$

$$= 4.0$$

Because $Q_c < K_c$, some H_2 and I_2 must react to form HI in order to attain equilibrium.

EXERCISE $K_c = 0.050$ at 2200°C for the following equilibrium; 0.50 M of each gas in the equilibrium is introduced into a flask and the temperature increased to 2200°C. Calculate Q_c and state what must occur for equilibrium to be attained.

$$N_2(g) + O_2(g) \rightleftharpoons 2NO(g)$$

ANSWER: $Q_c = 1$, and some NO must decompose into N_2 and O_2 in order to attain equilibrium.

KEY CONCEPT E Rate constants and equilibrium constants

Imagine an equilibrium in which both the reverse and the forward reactions are elementary bimolecular reactions. In such a situation, the equilibrium constants and rate constants for the forward and reverse reactions are

$$aA + bB \rightleftharpoons cC + dD \qquad K_c = \frac{[C]^c[D]^d}{[A]^a[B]^b}$$

$$aA + bB \longrightarrow cC + dD \qquad \text{rate} = k[A]^a[B]^b$$

$$cC + dD \longrightarrow aA + bB \qquad \text{rate} = k'[C]^c[D]^d$$

Because the rate of forward reaction equals the rate of reverse reaction at equilibrium,

$$K_c = \frac{k}{k'}$$

Similar expressions can be derived for more complicated mechanisms. In almost all cases, it is possible to express the equilibrium constant in terms of the forward and reverse rate constants for the steps in the mechanism.

K_c = a function of rate constants for forward and reverse reactions of steps in mechanism

EXAMPLE Relating rate constants to the equilibrium constant

$K_c = 6.3 \times 10^{14}$ at 1000 K for the following equilibrium. Both the forward and reverse reactions in the equilibrium are elementary bimolecular reactions. $k = 24$ L/mol·s for the forward reaction. What is k', the rate constant for the reverse reaction?

$$NO(g) + O_3(g) \rightleftharpoons NO(g) + O_2(g)$$

SOLUTION Because the equilibrium is accomplished through forward and reverse bimolecular reactions,

$$K_c = \frac{k}{k'}$$

$$6.3 \times 10^{14} = \frac{24\ \text{L/mol}\cdot\text{s}}{k'}$$

$$k' = \frac{24\ \text{L/mol}\cdot\text{s}}{6.3 \times 10^{14}}$$

$$= 3.9 \times 10^{-14}\ \text{L/mol}\cdot\text{s}$$

EXERCISE Assume the equilibrium shown is accomplished through forward and reverse elementary bimolecular reactions with $k = 1.3 \times 10^2$ L/mol·s and $k' = 6.1 \times 10^7$ L/mol·s. What is K_c?

$$A + B \rightleftharpoons C + D$$

ANSWER: 2.1×10^{-6}

KEY CONCEPT F The equilibrium constant in terms of partial pressures

For equilibria that involve gases but not dissolved solutes, it is possible to write an equilibrium constant K_P that uses the partial pressures of the gases rather than their concentrations. For instance, for the equilibrium

$$2NO_2(g) \rightleftharpoons 2NO(g) + O_2(g)$$

$$K_P = \frac{(P_{NO})^2 P_{O_2}}{(P_{NO_2})^2}$$

The reaction quotient Q_P is defined similarly. The equilibrium constant K_c for this equilibrium is

$$K_c = \frac{[NO]^2[O_2]}{[NO_2]^2}$$

It is important to understand that K_P and K_c describe the same equilibrium and that the amounts of reactants and products present do not depend on which equilibrium constant we choose to use. The two equilibrium constants are related to each other by the relation

$$K_P = K_c(RT)^{\Delta n}$$

Δn = moles gaseous product in the equation − moles gaseous reactant in the equation

In our example, there are 3 mol of products and 2 mol of reactants, so $\Delta n = +1$. Δn may be negative. If the moles of product are the same as the moles of reactant, then $\Delta n = 0$ and $K_c = K_P$. When we use $R = 0.0821$ L atm/mol·K, the partial pressures must be expressed in atmospheres.

EXAMPLE 1 Relating K_P to K_c

For the equilibrium shown, at 650 K, $K_c = 4.0 \times 10^5$ M. What is K_P?

$$2NO_2(g) \rightleftharpoons 2NO(g) + O_2(g)$$

SOLUTION $\Delta n = 1 (3 - 2 = 1)$ for this equilibrium. Substituting this result, $R = 0.0821$ L atm/mol·K, and $T = 500$ K into the relationship between K_P and K_c gives

$$\begin{aligned} K_P &= K_c(\mathrm{RT})^{\Delta n} \\ &= 4.0 \times 10^5\ \mathrm{M} \times (0.0821\ \mathrm{L\ atm/mol \cdot K} \times 650\ \mathrm{K})^1 \\ &= 2.1 \times 10^7\ \mathrm{atm} \end{aligned}$$

EXERCISE At 250°C, $K_c = 3.2 \times 10^3\ \mathrm{M}^{-1}$ for the equilibrium shown in the example. What is K_P?

ANSWER: $1.4 \times 10^5\ \mathrm{atm}^{-1}$

EXAMPLE 2 Relating K_P to K_c

At 500 K, $K_P = 2.5 \times 10^{10}$/atm for the equilibrium shown. What is K_c at this temperature?

$$2SO_2(g) + O_2(g) \rightleftharpoons 2SO_3(g)$$

SOLUTION $\Delta n = -1\ (2 - 3 = -1)$ for this equilibrium. Substituting this result, $R = 0.0821$ L atm/mol·K, and $T = 500$ K into the relationship between K_P and K_c gives

$$\begin{aligned} K_P &= K_c(RT)^{\Delta n} \\ 2.5 \times 10^{10}\ \mathrm{atm}^{-1} &= K_c\left(0.0821\ \frac{\mathrm{L\ atm}}{\mathrm{mol\ K}} \times 500\ \mathrm{K}\right)^{-1} \\ &= \frac{K_c}{\left(0.0821\ \frac{\mathrm{L\ atm}}{\mathrm{mol}\ \cancel{\mathrm{K}}} \times 500\ \cancel{\mathrm{K}}\right)} \\ &= \frac{K_c}{41.1\ \frac{\mathrm{L\ atm}}{\mathrm{mol}}} \end{aligned}$$

$$\begin{aligned} K_c &= 2.5 \times 10^{10}\ \cancel{\mathrm{atm}^{-1}} \times 41.1\ \frac{\mathrm{L}\ \cancel{\mathrm{atm}}}{\mathrm{mol}} \\ &= 1.0 \times 10^{12}\ \mathrm{M}^{-1} \end{aligned}$$

EXERCISE At 700 K, $K_P = 54$ for the equilibrium shown. What is K_c?

$$H_2(g) + I_2(g) \rightleftharpoons 2HI(g)$$

ANSWER: 54

13.3 HETEROGENEOUS EQUILIBRIA

KEY CONCEPT Heterogeneous equilibria

Because the concentration of a pure solid or pure liquid is constant, the equilibrium expression for an equilibrium involving a pure liquid or pure solid does not explicitly contain the concentration of either. The constant concentration instead is collapsed into the equilibrium constant. Thus, the equilibrium constants reported in tables and books are based on the assumption that the concentrations of pure liquids and pure solids do not appear in the equilibrium expression. For instance, the equilibrium expression for the following equilibrium is

$$CuCl_2 \cdot 5H_2O(s) \rightleftharpoons CuCl_2(s) + 5H_2O(g)$$
$$K_c = [H_2O]^5 \quad \text{or} \quad K_P = (P_{H_2O})^5$$

EXAMPLE Writing the equilibrium constant for a heterogeneous equilibrium

Write the equilibrium expression for K_c for the equilibrium

$$S(s) + O_2(g) \rightleftharpoons SO_2(g)$$

SOLUTION This is a heterogeneous equilibrium because both a solid and gases are present. The concentration of the sulfur, which is a pure solid, does not appear in the equilibrium expression. Thus,

$$K_c = \frac{[SO_2]}{[O_2]}$$

EXERCISE Write the equilibrium expression for K_c for the equilibrium

$$Al_2O_3(s) + 3C(s) + 3Cl_2(g) \rightleftharpoons 2AlCl_3(s) + 3CO(g)$$

ANSWER: $K_c = \frac{[CO]^3}{[Cl_2]^3}$

EQUILIBRIUM CALCULATIONS

13.4 SPECIFIC INITIAL CONCENTRATIONS

KEY CONCEPT A Calculating equilibrium constants from concentrations

When the concentrations of products and reactants at equilibrium are known or can be calculated in a straightforward way, the equilibrium constant can be calculated. In the two examples that follow, we determine the equilibrium concentrations in two ways:

1. The equilibrium concentrations of all substances are measured directly.
2. The equilibrium concentrations are calculated from intial known concentrations and the measured equilibrium concentration of one substance.

EXAMPLE 1 Calculating the equilibrium constant, method 1

The concentrations of the substances in the following reaction are measured at equilibrium and found to be $[Br_2] = 0.17$ M, $[Cl_2] = 0.17$ M, and $[BrCl] = 0.030$ M.

$$2BrCl(g) \rightleftharpoons Cl_2(g) + Br_2(g)$$

What is K_c for the equilibrium?

SOLUTION We first write the expression for the equilibrium constant to determine what information is required.

$$K_c = \frac{[Cl_2][Br_2]}{[BrCl]^2}$$

Because we have the equilibrium concentration of each substance in the equilibrium expression, we can substitute the given values into it to calculate K_c:

$$K_c = \frac{(0.017\ \text{M})(0.17\ \text{M})}{(0.030\ \text{M})^2} = 0.032$$

EXERCISE The concentrations of the substances in the following reaction are measured at equilibrium and found to be $[SO_2] = 0.025$ M, $[Cl_2] = 0.025$ M, and $[SO_2Cl_2] = 4.1 \times 10^{-4}$ M. What is the value of K_c?

$$SO_2(g) + Cl_2(g) \rightleftharpoons SO_2Cl_2(g)$$

ANSWER: $0.66\ \text{M}^{-1}$

EXAMPLE 2 Calculating the equilibrium constant, method 2

A sample of $HI(g)$ is placed in a flask at a pressure of 0.10 atm. After the system comes to equilibrium, the partial pressure of $HI(g)$ is 0.050 atm. What is K_P for the equilibrium?

$$2HI(g) \rightleftharpoons H_2(g) + I_2(g) \qquad K_P = ?$$

SOLUTION We solve this by means of a table showing the partial pressures of each substance before and after equilibrium. Also included in the table is the change in partial pressure of each substance expressed as the variable x.

	$2HI(g)$ $\rightleftharpoons$	$H_2(g)$ +	$I_2(g)$
before equilibrium	0.10 atm	0 atm	0 atm
change to reach equilibrium	$-2x$	$+x$	$+x$
equilibrium values	$0.10 - 2x$	x	x

Note that the values of x used in the row "change to reach equilibrium" must agree with the stoichiometry of the equilibrium; that is, the 2 in front of HI requires the change in the partial pressure of HI to be $2x$. In this problem, we know the equilibrium partial pressure of HI is 0.050 atm. This allows us to solve for x.

$$0.10\ \text{atm} - 2x = 0.050\ \text{atm}$$
$$-2x = 0.050\ \text{atm} - 0.10\ \text{atm}$$
$$x = 0.025\ \text{atm}$$

Because x is the equilibrium partial pressure of H_2 and I_2, we can calculate K_P. That is, $P_{H_2} = P_{I_2} = x = 0.025$ atm, and from the problem, $P_{HI} = 0.050$ atm; thus,

$$K_P = \frac{P_{H_2} \times P_{I_2}}{(P_{HI})^2} = \frac{0.025\ \text{atm} \times 0.025\ \text{atm}}{(0.050\ \text{atm})^2} = 0.25$$

EXERCISE A sample of $CO(g)$ is placed in a flask at a pressure of 0.200 atm. After the system comes to equilibrium, the partial pressure of $CO(g)$ is 0.190 atm. What is K_P for the equilibrium?

$$2CO(g) \rightleftharpoons C(s) + CO_2(g)$$

ANSWER: $0.139\ \text{atm}^{-1}$

KEY CONCEPT B Calculations with one unknown concentration

For some calculations, the equilibrium constant and all but one of the concentrations are known. Substitution of the data into the equilibrium expression leads to one equation with one unknown, so we can solve for the unknown concentration in a straightforward manner.

EXAMPLE 1 Solving for an unknown concentration or partial pressure

$K_P = 160$ for the equilibrium shown. What is the partial pressure of H_2 if the partial pressure of I_2 is 0.35 atm and that of HI is 6.3 atm?

$$H_2(g) + I_2(g) \rightleftharpoons 2HI(g)$$

SOLUTION The equilibrium constant expression for this equilibrium is

$$K_P = \frac{(P_{HI})^2}{(P_{H_2})(P_{I_2})}$$

Substituting the values $P_{HI} = 6.3$ atm, $P_{I_2} = 0.35$ atm, and $K_P = 160$ gives

$$160 = \frac{(6.3 \text{ atm})^2}{P_{H_2}(0.35 \text{ atm})}$$

$$P_{H_2} = \frac{39.7 \text{ atm}^{\cancel{2}\,1}}{(0.35 \cancel{\text{atm}})(160)}$$

$$= 0.71 \text{ atm}$$

It is prudent to always check your results by substituting the equilibrium values obtained into the equilibrium constant expression to see if you get the correct value of the equilibrium constant:

$$\frac{(6.3 \cancel{\text{atm}})^{\cancel{2}}}{(0.71 \cancel{\text{atm}})(0.35 \cancel{\text{atm}})} = 160$$

EXERCISE $K_c = 0.10$ for the equilibrium shown. What is the concentration of I_2 when $[ICl] = 0.50$ M and $[Cl_2] = 0.063$ M?

$$2ICl(g) \rightleftharpoons Cl_2(g) + I_2(g)$$

ANSWER: 0.40 M

EXAMPLE 2 Solving for an unknown concentration

$K_c = 4.4 \times 10^4 \text{ M}^{-2}$ for the equilibrium given. What is the molarity of H_2 at equilibrium when the molarity of N_2 is 0.11 M and that of NH_3 is 1.5 M?

$$N_2(g) + 3H_2(g) \rightleftharpoons 2NH_3(g)$$

SOLUTION The equilibrium expression for the equilibrium is

$$K_c = \frac{[NH_3]^2}{[N_2][H_2]^3}$$

Substitution of the given quantities $K_c = 4.4 \times 10^4 \text{ M}^{-2}$, $[N_2] = 0.11$ M, and $[NH_3] = 1.5$ M gives

$$4.4 \times 10^4 \text{ M}^{-2} = \frac{(1.5 \text{ M})^2}{(0.11 \text{ M})[H_2]^3}$$

$$[H_2]^3 = \frac{2.3 \text{ M}^2}{(4.4 \times 10^4 \text{ M}^{-2})(0.11 \text{ M})}$$

$$= 4.8 \times 10^{-4} \text{ M}^3$$

$$[H_2] = (4.8 \times 10^{-4} \text{ M}^3)^{1/3}$$

Remember that raising a quantity to the $\frac{1}{3}$ power is the same as taking the cube root of the quantity. To accomplish this on a calculator that uses algebraic notation:

Enter 4.8×10^{-4}

Press Y^x (this may be X^y on some calculators)

Enter 0.33333333

Press =

Display shows 7.8297 − 02

This is your calculator's way of indicating 7.8297×10^{-2}. The correct answer to the problem is 7.8×10^{-2} M. (Recall that $\{M^3\}^{1/3} = M$.) Substitution of these values into the equlibrium expression gives the correct value for the equilibrium constant:

$$\frac{(1.5\ \text{M})^2}{(0.11\ \text{M})(7.8 \times 10^{-2}\ \text{M})^3} = 4.3 \times 10^4\ \text{M}^{-2}$$

EXERCISE For the equilibrium in the example, what is $[NH_3]$ at equilibrium when $[H_2] = 6.1 \times 10^{-3}$ M and $[N_2] = 0.15$ M?

ANSWER: 0.039 M

KEY CONCEPT C Equilibrium calculation for the decomposition of a single substance

When a single substance decomposes, the concentrations (or partial pressures) of each of the substances in the equilibrium can be determined from the stoichiometry of the chemical equation. The concentrations are expressed in terms of an unknown x, and x is solved for by using the equilibrium expression.

EXAMPLE 1 Equilibrium concentrations when a single substance decomposes

$K_c = 0.10$ for the equilibrium shown. What are the equilibrium concentrations of each of the substances in the equilibrium when ICl, initially at 0.75 M, is allowed to come to equilibrium?

$$2ICl(g) \rightleftharpoons I_2(g) + Cl_2(g)$$

SOLUTION The first task is to set up a table showing the initial concentrations, the changes necessary to attain equilibrium, and the equilibrium concentrations. The changes and the equilibrium concentrations are expressed in terms of an unknown x, with x defined so that $2x$ is the amount of ICl that reacts to reach equilibrium.

	2ICl	$\rightleftharpoons$	**I_2**	+	**Cl_2**
initial concentration	0.75 M		0 M		0 M
change to reach equilibrium	$-2x$		$+x$		$+x$
equilibrium concentration	0.75 M − $2x$		x		x

Note that the changes follow the stoichiometry of the balanced equation. That is, where $2x$ M ICl decomposes, x M I_2 and x M Cl_2 form. We next substitute the equilibrium concentrations into the equilibrium expression:

$$K_c = \frac{[I_2][Cl_2]}{[ICl]^2}$$

$$0.10 = \frac{(x)(x)}{(0.75\ \text{M} - 2x)^2}$$

$$= \frac{x^2}{(0.75\ \text{M} - 2x)^2}$$

We now take the square root of both sides of the equation:

$$0.32 = \frac{x}{0.75\text{ M} - 2x}$$

$$0.32(0.75\text{ M} - 2x) = x$$

$$x = 0.24\text{ M} - 0.64x$$

$$1.64x = 0.24\text{ M}$$

$$x = 0.15\text{ M}$$

From the value of x, we can calculate the equilibrium concentrations of each of the substances in the equilibrium:

$$[ICl] = 0.75\text{ M} - 2x = 0.45\text{ M}$$

$$[Cl_2] = [I_2] = x = 0.15\text{ M}$$

Substitution of these values into the equilibrium expression gives the correct value for the equilibrium constant.

$$\frac{(0.15\text{ M})(0.15\text{ M})}{(0.45\text{ M})^2} = 0.11$$

EXERCISE $K_c = 1.9 \times 10^{-2}$ for the equilibrium shown. What are the equilibrium concentrations of each of the substances in the equilibrium if HI, initially at 35 mM, is allowed to come to equilibrium?

$$2HI(g) \rightleftharpoons H_2(g) + I_2(g)$$

ANSWER: $[HI] = 27$ mM; $[H_2] = [I_2] = 3.8$ mM

EXAMPLE 2 Equilibrium concentrations when a single substance decomposes

$K_P = 0.050$ atm at 900 K for the equilibrium shown. What is the equilibrium concentration of C_2H_6 when 2.0 atm of C_2H_6 is placed in a flask and allowed to come to equilibrium?

$$C_2H_6(g) \rightleftharpoons C_2H_4(g) + H_2(g)$$

SOLUTION We first set up a table showing the partial pressures of all of the substances involved before and after equilibrium.

	$C_2H_6(g)$ $\rightleftharpoons$	$C_2H_4(g)$ +	$H_2(g)$
before equilibrium	2.0 atm	0 atm	0 atm
change to reach equilibrium	$-x$	$+x$	$+x$
equilibrium values	$2.0 - x$	x	x

We next set up the equilibrium expression and substitute the equilibrium partial pressures, written in terms of the unknown x, into the expression.

$$K_P = \frac{P_{C_2H_4} \times P_{H_2}}{P_{C_2H_6}}$$

$$= \frac{x^2}{2.0 - x}$$

We now solve for x using the quadratic equation. We first must get the equation into the form $ax^2 + bx + c = 0$.

$$\frac{x^2}{2.0 - x} = 0.050\text{ atm}$$

$$x^2 = (0.050\text{ atm})(2.0 - x)$$

$$x^2 + 0.050x - 0.10 = 0$$

Thus, $a = 1$, $b = 0.050$, and $c = -0.10$. Substituting these into the quadratic formula gives

$$x = \frac{-b \pm \sqrt{b^2 - 4ac}}{2a}$$

$$= \frac{-0.050 \pm \sqrt{0.0025 - 4(1)(-0.10)}}{2}$$

$$= \frac{-0.050 \pm \sqrt{0.4025}}{2}$$

$$= \frac{-0.050 \pm 0.634}{2}$$

$$x = \frac{-0.050 + 0.634}{2} = 0.29 \text{ atm} \qquad x = \frac{-0.050 - 0.634}{2} = -0.34 \text{ atm}$$

x represents the partial pressure of two gases; and because a partial pressure cannot be negative, we reject the negative answer as physically meaningless. We next substitute the equilibrium partial pressures of each component into the equilibrium expression to check our answer.

$$\frac{x^2}{2.0 - x} = \frac{(0.29 \text{ atm})^2}{2.0 - x}$$

$$= \frac{0.084 \text{ atm}^2}{1.71 \text{ atm}}$$

$$= 0.049 \text{ atm}$$

This result (0.049 atm) is the value of K_P (if we realize that some round-off error occurs) and shows that we have calculated the correct equilibrium concentrations. To answer the question: $P_{C_2H_6} = 1.71$ atm.

EXERCISE $K_P = 25$ atm at 500 K for the equilibrium shown. What is the equilibrium partial pressure of PCl_5 if 6.0 atm of PCl_5 are placed in a flask and allowed to come to equilibrium?

$$PCl_5(g) \rightleftharpoons PCl_3(g) + Cl_2(g)$$

ANSWER: 1.0 atm

13.5 ARBITRARY INITIAL CONCENTRATIONS

KEY CONCEPT Equilibrium calculations with arbitrary initial conditions

Solving an equilibrium problem often involves four distinct steps:

Step 1. Write the equilibrium equation and expression for K_c or K_P. If necessary, calculate Q_c or Q_P to decide in which direction the reaction must shift to reach equilibrium.
Step 2. Construct a table, showing the concentrations of components before and after equilibrium. Express the change in concentrations with a single variable x.
Step 3. Use the required mathematical tools to solve for x.
Step 4. Calculate the equilibrium amounts of each component with the value of x obtained.

Step 3 is often the most time consuming because the equation to solve often contains x^2, x^3, or x to a higher power. Thus, more than one root exists for the equation. In most cases the reasonableness of the answer allows us to choose which root to use; for instance, the value of x used cannot result in a negative concentration. In some cases, the different values of x that you find must be checked to see which one results in equilibrium concentrations that give the correct equilibrium constant when substituted in the equilibrium expression.

▼ EXAMPLE 1 Equilibrium calculations using the quadratic equation

$K_c = 6.5 \times 10^{-3}$ M for the equilibrium given. What are the equilibrium concentrations of the substances in the equilibrium if N_2O_4, initially at 0.40 M, is allowed to come to equilibrium?

$$N_2O_4(g) \rightleftharpoons 2NO_2(g)$$

SOLUTION The first step, as usual, is to set up a table showing the initial concentrations, changes to attain equilibrium, and equilibrium concentrations.

	N_2O_4	$\rightleftharpoons$ $2NO_2$
initial concentration	0.40 M	0 M
change to attain equilibrium	$-x$	$+2x$
equilibrium concentration	0.40 M $- x$	$2x$

These values are substituted into the equilibrium expression to solve for x.

$$K_c = \frac{[NO_2]^2}{[N_2O_4]}$$

$$6.5 \times 10^{-3}\ \text{M} = \frac{(2x)^2}{0.40\ \text{M} - x}$$

$$= \frac{4x^2}{0.40\ \text{M} - x}$$

To solve for x, we rearrange the equation into the form $ax^2 + bx + c = 0$ and use the quadratic equation.

$$(6.5 \times 10^{-3})(0.40 - x) = 4x^2$$

$$2.6 \times 10^{-3} - 6.5 \times 10^{-3}x = 4x^2$$

$$4x^2 + 6.5 \times 10^{-3}x - 2.6 \times 10^{-3} = 0$$

This equation shows the desired form, with $a = 4$, $b = 6.5 \times 10^{-3}$, and $c = -2.6 \times 10^{-3}$. We can now use the quadratic equation to solve for x:

$$x = \frac{-b \pm \sqrt{b^2 - 4ac}}{2a}$$

$$= \frac{-6.5 \times 10^{-3} \pm \sqrt{(6.5 \times 10^{-3})^2 - 4(4)(-2.6 \times 10^{-3})}}{2(4)}$$

$$= \frac{-6.5 \times 10^{-3} \pm \sqrt{4.3 \times 10^{-5} + 4.16 \times 10^{-2}}}{8}$$

$$= \frac{-6.5 \times 10^{-3} \pm \sqrt{4.2 \times 10^{-2}}}{8}$$

$$= \frac{-6.5 \times 10^{-3} \pm 0.20}{8}$$

Because $2x$ is the concentration of NO_2, x cannot be a negative number; so we ignore the negative root of the equation.

$$x = \frac{-6.5 \times 10^{-3} + 0.20}{8}$$

$$= 0.025\ \text{M}$$

With this result, it is possible to calculate the equilibrium concentrations of NO_2 and N_2O_4.

$$[NO_2] = 2x = 0.050\ \text{M}$$

$$[N_2O_4] = 0.40 - x = 0.38\ \text{M}$$

Substitution of these values into the equilibrium expression gives the correct value for the equilibrium constant:

$$\frac{(0.050\ M)^2}{(0.38\ M)} = 6.6 \times 10^{-3}\ M$$

EXERCISE $K_c = 6.8 \times 10^{-2}$ M for the equilibrium given. What are the equilibrium concentrations of I_2 and I when I_2, initially at 3.5 M, is allowed to come to equilibrium?

$$I_2(g) \rightleftharpoons 2I(g)$$

ANSWER: $[I_2] = 3.3$ M; $[I] = 0.47$ M

EXAMPLE 2 Equilibrium calculations using approximations

$K_c = 7.7 \times 10^{-11}$ for the equilibrium shown. What are the equilibrium concentrations of all the substances in the equilibrium when HBr, initially at 0.20 M, and H_2, initially at 0.15 M, are allowed to come to equilibrium?

$$2HBr(g) \rightleftharpoons H_2(g) + Br_2(g)$$

SOLUTION The first step is to set up a table showing the initial concentrations, the changes required to attain equilibrium and the equilibrium concentrations.

	2HBr	$\rightleftharpoons$	**H_2**	+ **Br_2**
initial concentration	0.20 M		0.15 M	0 M
change to attain equilibrium	$-2x$		$+x$	$+x$
equilibrium concentration	0.20 M − $2x$		0.15 M + x	x

We now substitute these values and the value of K_c into the equilibrium expression to solve for x.

$$K_c = \frac{[H_2][Br_2]}{[HBr]^2}$$

$$7.7 \times 10^{-11} = \frac{(0.15\ M + x)(x)}{(0.20\ M - x)^2}$$

This equation could be solved by putting it into the form $ax^2 + bx + c = 0$ and using the quadratic formula, but a simpler approach is possible. Because of the small size of K_c, we make the assumption that x is very small—so small, in fact, that adding x to 0.15 results in 0.15 and subtracting x from 0.20 results in 0.20.

$$0.15\ M + x \approx 0.15\ M$$
$$0.20\ M - x \approx 0.20\ M$$

With this assumption, the equation to solve now becomes

$$7.7 \times 10^{-11} = \frac{(0.15\ M)x}{(0.20\ M)^2}$$

$$x = \frac{(7.7 \times 10^{-11})(0.20\ M)^{\not{2}\,1}}{(0.15\ \not{M})}$$
$$= 2.1 \times 10^{-11}\ M$$

Before proceeding, we check to see if the assumption regarding the smallness of x is appropriate. A general rule of thumb is that x should be less than 5% of the quantity relative to which it is assumed to be small. In our example,

$$\frac{x}{0.15} \times 100 = \frac{2.1 \times 10^{-11}}{0.15} \times 100 = 1.4 \times 10^{-8}\%$$

x is even a smaller percentage of 0.20. Thus, x is less than 5% of both quantities relative to which it is assumed to be small, and the assumption that x is small is justified. The desired equilibrium concentrations can now be calculated.

$$[Br_2] = x = 2.1 \times 10^{-11}\ M$$
$$[H_2] = 0.15 + x = 0.15\ M$$
$$[HBr] = 0.20 - x = 0.20\ M$$

Substitution of these values into the equilibrium expression gives the correct value for the equilibrium constant:

$$\frac{(2.1 \times 10^{-11}\ \cancel{M})(0.15\ \cancel{M})}{(0.20\ \cancel{M})^2} = 7.9 \times 10^{-11}$$

EXERCISE $K_c = 1.6 \times 10^{-9}$ M for the equilibrium shown. What are the equilibrium concentrations of the substances in the equilibrium when SO_3, initially at 0.65 M, and O_2, initially at 0.010 M, are allowed to come to equilibrium?

$$2SO_3(g) \rightleftharpoons 2SO_2(g) + O_2(g)$$

ANSWER: $[O_2] = 0.010$ M; $[SO_2] = 2.6 \times 10^{-4}$ M; $[SO_3] = 0.65$ M

THE RESPONSE OF EQUILIBRIA TO THE REACTION CONDITIONS

13.6 THE EFFECT OF ADDED REAGENTS

KEY CONCEPT The effect of adding or removing a reagent from an equilibrium mixture

When a reagent is added or removed from a system at equilibrium, the equilibrium is disturbed and adjusts to attain a new equilibrium state. As an example, consider the equilbrium shown, with $[HI] = 0.21$ M and $[H_2] = [I_2] = 0.017$ M.

$$2HI(g) \rightleftharpoons H_2(g) + I_2(g) \qquad K_c = 6.3 \times 10^{-3}$$

Imagine that HI is suddenly added to the equilibrium mixture to make its concentration 0.46 M. There is now too much HI for equilibrium to exist; some of it has to decompose into H_2 and I_2 to establish a new equilibrium. The following graph portrays the concentrations of the substances in the equilibrium before and after the addition of the extra HI.

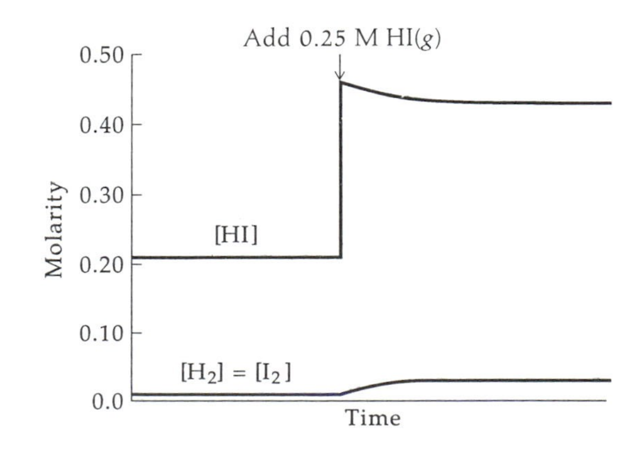

The new equilibrium concentrations after addition of the HI are [HI] = 0.43 M and $[H_2] = [I_2] =$ 0.034 M. As can be seen from the graph, addition of HI results in formation of additional products. This result can be generalized:

- Addition of reactants or removal or products in an equilibrium results in loss of reactant and formation of additional products.
- Addition of products or removal of reactants in an equilibrium results in loss of product and formation of additional reactants.

These effects are in accord with Le Chatelier's principle.

EXAMPLE 1 Shifting an equilibrium by adding reagents

Describe the effect on the equilibrium shown of (a) addition of SO_2, (b) addition of SO_3, (c) removal of O_2, and (d) removal of SO_3.

$$2SO_2(g) + O_2(g) \rightleftharpoons 2SO_3(g)$$

SOLUTION (a) Addition of a reactant to an equilibrium causes formation of additional product, so addition of SO_2 results in the reaction of some SO_2 with O_2 to form SO_3. (b) Addition of a product to an equilibrium causes formation of additional reactant, so addition of SO_3 results in the decomposition of some SO_3 into SO_2 and O_2. (c) Removal of reactant results in formation of additional reactants, so removal of O_2 results in the decomposition of some SO_3 into SO_2 and O_2. (d) Removal of product results in formation of additional product, so removal of SO_3 results in the reaction of some SO_2 with some O_2 to form SO_3.

EXERCISE Describe the effect on the equilibrium shown of (a) addition of HI, (b) addition of H_2, (c) removal of I_2, and (d) removal of HI.

$$H_2(g) + I_2(g) \rightleftharpoons 2HI(g)$$

ANSWER: (a) Reactants form, (b) products form, (c) reactants form, and (d) products form.

EXAMPLE 2 Addition of a pure liquid or solid to an equilibrium

Describe the effect of the addition of C(s) to the equilibrium shown.

$$C(s) + O_2(g) \rightleftharpoons CO_2(g)$$

SOLUTION Addition of a pure liquid or pure solid to an equilibrium has no effect on the equilibrium. This is why the concentration of a pure liquid or pure solid does not appear in the equilibrium expression for an equilibrium.

EXERCISE Describe the effect of removal of exactly half of the $CuSO_4 \cdot 5H_2O$ from the equilibrium shown.

$$CuSO_4 \cdot 5H_2O(s) \rightleftharpoons CuSO_4(s) + 5H_2O(g)$$

ANSWER: No effect

13.7 THE EFFECT OF PRESSURE

KEY CONCEPT How changes in pressure change equilibria

A change in pressure has a marked effect on the equilibrium concentrations of any gas-phase equilibrium. One way to think about this problem is to imagine what happens to the equilibrium when the pressure is increased by decreasing the volume. Le Chatelier's principle predicts that the gas molecules respond to a decrease in volume by reacting to form fewer molecules, thereby "fitting" into the smaller volume better. For the equilibrium

$$2A(g) \rightleftharpoons B(g)$$

the A molecules would react to form B molecules, resulting in fewer gas molecules overall. On the other hand, a decrease in pressure that is accomplished by an increase in volume causes the gas to react in such a way as to produce more gas molecules, thereby filling the additional volume created. For the preceding equilibrium, some B molecules would decompose into A molecules, resulting in more gas-phase molecules overall. The following graphs show the results of an increase in pressure for different types of equilibria. The increase in pressure occurs at the arrow.

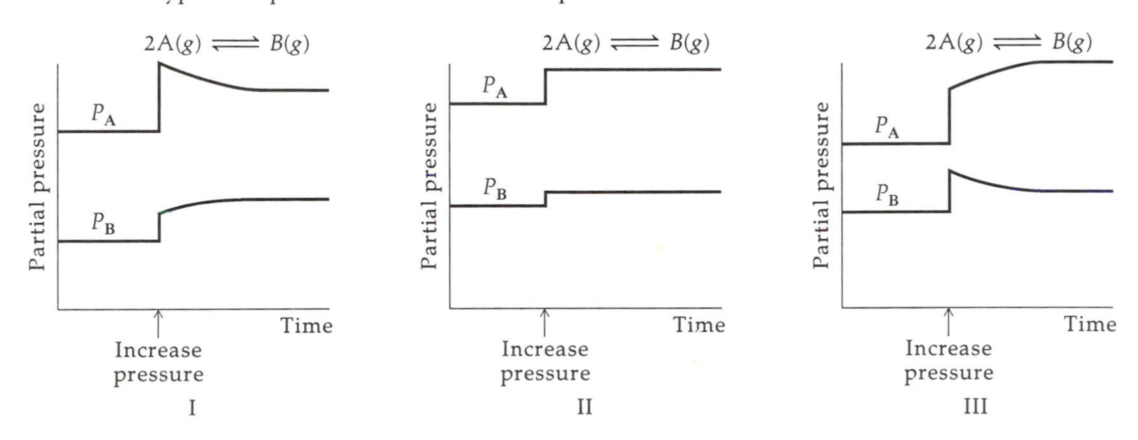

These results are derived rigorously in the text and can be summarized as follows.

An increase in pressure on an equilibrium at constant temperature, accomplished by a decrease in volume, results in a shift of the equilibrium so as to form fewer gas molecules.

A decrease in pressure on an equilibrium at constant temperature, accomplished by an increase in volume, results in a shift of the equilibrium so as to form more gas molecules.

EXAMPLE 1 The effect of changing the pressure on an equilibrium

What is the effect of increasing the pressure on the equilibrium shown? Assume there is no change in temperature.

$$2H_2(g) + CO(g) \rightleftharpoons CH_3OH(g)$$

SOLUTION An increase in pressure on an equilibrium at constant temperature, accomplished by a decrease in volume, results in a shift of the equilibrium so as to form fewer gas molecules. For this equilibrium, an increase in pressure causes some H_2 to react with some CO to form CH_3OH, resulting in fewer gas molecules overall.

EXERCISE What is the effect of decreasing the pressure at constant temperature on the equilibrium shown?

$$COCl_2(g) \rightleftharpoons CO(g) + Cl_2(g)$$

ANSWER: Some $COCl_2$ decomposes to form CO and Cl_2.

EXAMPLE 2 The effect of changing the pressure on an equilibrium

Describe the effect of increasing the pressure at constant temperature on the equilibrium shown.

$$H_2(g) + I_2(g) \rightleftharpoons 2HI(g)$$

SOLUTION Because the equilibrium has the same number of moles of gas on both sides, changing the pressure has no effect.

EXERCISE Describe the effect of decreasing the pressure at constant temperature on the following equilibrium.

$$H_2(g) + CO_2(g) \rightleftharpoons H_2O(g) + CO(g)$$

ANSWER: No effect

EXAMPLE 3 The effect of changing the pressure on an equilibrium

What is the effect of decreasing the pressure on the following equilibrium? Assume the temperature is constant.

$$2C(s) + O_2(g) \rightleftharpoons 2CO(g)$$

SOLUTION Decreasing the pressure on an equilibrium causes the equilibrium to shift in such a way as to form more *gas* molecules. In this case, decreasing the pressure causes some C and some O_2 to react to form CO.

EXERCISE Describe the effect of increasing the pressure on the following equilibrium at constant temperature.

$$CaCO_3(s) \rightleftharpoons CaO(s) + CO_2(g)$$

ANSWER: Some CO_2 will combine with CaO to form $CaCO_3$.

PITFALL The effect of addition of an inert gas

When an inert gas is added to a gaseous reaction mixture, the total pressures of the gaseous mixture increases; however, there is no effect on the equilibrium. We have considered changing pressure only by decreasing the volume of the reaction vessel at constant temperature.

13.8 THE EFFECT OF TEMPERATURE

KEY CONCEPT The change in equilibrium as temperature is changed

Changing the temperature affects the forward and reverse rates of reaction in an equilibrium and therefore affects the equilibrium. The overall effect of a temperature change on an equilibrium is given in the following table.

Temperature change	Exothermic reaction	Endothermic reaction
increase	more reactants form	more products form
decrease	more products form	more reactants form

EXAMPLE The effect of changing the temperature on an equilibrium

What is the effect of increasing the temperature on the equilibrium shown? Assume constant pressure.

$$2HI(g) \rightleftharpoons H_2(g) + I_2(g) \qquad \Delta H = -52.96 \text{ kJ}$$

SOLUTION Because the reaction is exothermic, increasing the temperature results in the formation of reactants. Hence, increasing the temperature causes some I_2 to react with some H_2 to form HI.

EXERCISE What is the effect of increasing the temperature on the following equilibrium? Assume the pressure is constant.

$$CaCO_3(s) \rightleftharpoons CaO(s) + CO_2(g) \qquad \Delta H = +178.5 \text{ kJ}$$

ANSWER: Some $CaCO_3$ decomposes to form CaO and CO_2.

PITFALL The effect of adding a catalyst

Addition of a catalyst has no effect on the position of an equilibrium. The catalyst does, however, speed up attainment of equilibrium.

DESCRIPTIVE CHEMISTRY TO REMEMBER

- **Nitric oxide (nitrogen monoxide),** NO, is an air pollutant generated in the internal combustion engine.
- **Phosgene,** $COCl_2$, is a poisonous gas.
- The **Haber–Bosch process** for the synthesis of ammonia is based on the use of an iron catalyst and high N_2 and H_2 pressure.

CHEMICAL EQUATIONS TO KNOW

- In the Haber process, nitrogen and hydrogen are in equilibrium with ammonia.

$$3H_2(g) + N_2(g) \rightleftharpoons 2NH_3(g)$$

- The production of sulfur trioxide (used in the production of sulfuric acid) involves an equilibrium with sulfur dioxide and oxygen.

$$2SO_2(g) + O_2(g) \rightleftharpoons 2SO_3(g)$$

- Hydrogen iodide can exist in equilibrium with hydrogen and iodine.

$$H_2(g) + I_2(g) \rightleftharpoons 2HI(g)$$

- Hydrogen chloride can exist in equilibrium with hydrogen and chlorine.

$$H_2(g) + Cl_2(g) \rightleftharpoons 2HCl(g)$$

- Nitric oxide (nitrogen monoxide) can exist in equilibrium with nitrogen and oxygen.

$$N_2(g) + O_2(g) \rightleftharpoons 2NO(g)$$

- Nitrous oxide (dinitrogen monoxide) can also exist in equilibrium with nitrogen and oxygen.

$$2N_2(g) + O_2(g) \rightleftharpoons 2N_2O(g)$$

- At a high temperature, water can exist in equilibrium with hydrogen and oxygen.

$$2H_2(g) + O_2(g) \rightleftharpoons 2H_2O(g)$$

- Ozone can exist in equilibrium with oxygen.

$$2O_3(g) \rightleftharpoons 3O_2(g)$$

- Gaseous phosphorus pentachloride can exist in equilibrium with phosphorus trichloride and chlorine.

$$PCl_5(g) \rightleftharpoons PCl_3(g) + Cl_2(g)$$

- Nitrosyl chloride can exist in equilibrium with nitric oxide and chlorine.

$$2NOCl(g) \rightleftharpoons 2NO(g) + Cl_2(g)$$

- In the Mond process, nickel and carbon monoxide are in equilibrium with nickel tetracarbonyl.

$$Ni(s) + 4CO(g) \rightleftharpoons Ni(CO)_4(g)$$

- Dinitrogen tetroxide can exist in equilibrium with nitrogen dioxide.

$$N_2O_4(g) \rightleftharpoons 2NO_2(g)$$

- Gaseous methanol can exist in equilibrium with carbon monoxide and hydrogen at high temperature.

$$CH_3OH(g) \rightleftharpoons 2H_2(g) + CO(g)$$

MATHEMATICAL EQUATIONS TO KNOW AND UNDERSTAND

$K_P = K_c(RT)^{\Delta n}$	relationship between K_c and K_P
When $Q_c > K_c$,	reactants have a tendency to form
When $Q_c = K_c$,	the reaction is at equilibrium
When $Q_c < K_c$,	products have a tendency to form
$K_c = \dfrac{k}{k'}$	for an elementary step in which k is the forward rate constant and k' is the rate constant for the reverse reaction
$x = \dfrac{-b \pm \sqrt{b^2 - 4ac}}{2a}$	quadratic formula

SELF-TEST EXERCISES

The description of chemical equilibrium

1. When N_2O_4 initially at 1.0 atm is placed in a flask at 298 K and allowed to come to equilibrium, the equilibrium partial pressure of N_2O_4 is 0.82 atm. What is the equilibrium partial pressure of NO_2?

$$N_2O_4(g) \rightleftharpoons 2NO_2(g)$$

(a) 0.18 atm (b) 0.36 atm (c) 0.82 atm (d) 0.41 atm (e) 1.64 atm

2. When NOCl, initially at 0.500 atm, is placed in a flask and allowed to come to equilibrium, the equilibrium partial pressure of Cl_2 is 0.201 atm. What is the equilibrium partial pressure of NOCl?

$$2NOCl(g) \rightleftharpoons 2NO(g) + Cl_2(g)$$

(a) 0.402 atm (b) 0.299 atm (c) 0.101 atm (d) 0.098 atm (e) 0.201 atm

3. What is K_c for the equilibrium shown if the concentrations at equilibrium are $[NO_2] = 0.041$ M, $[CO] = 0.033$ M, $[NO] = 1.8$ M, and $[CO_2] = 2.0$ M?

$$NO_2(g) + CO(g) \rightleftharpoons NO(g) + CO_2(g)$$

(a) 3.8×10^{-4} (b) 2.9 (c) 8.2×10^3 (d) 2.7×10^3 (e) 3.7

4. What is K_c for the equilibrium shown if the concentrations at equilibrium are $[CO] = 0.045$ M, $[Cl_2] = 0.045$ M, and $[COCl_2] = 0.50$ M?

$$COCl_2(g) \rightleftharpoons CO(g) + Cl_2(g)$$

(a) 9.0×10^{-2} M (b) 1.1×10^2 M (c) 4.1×10^{-3} M
(d) 2.5×10^2 M (e) 8.1×10^{-3} M

5. What is K_P for the equilibrium shown if the equilibrium partial pressures are $P_{Cl_2} = 2.2$ atm, $P_{H_2O} = 0.85$ atm, $P_{HCl} = 6.4$ atm, and $P_{O_2} = 1.6$ atm?

$$2Cl_2(g) + 2H_2O(g) \rightleftharpoons 4HCl(g) + O_2(g)$$

(a) 19 atm (b) 5.3×10^{-2} atm (c) 6.4×10^{-2} atm
(d) 1.3×10^{-3} atm (e) 7.7×10^2 atm

6. For which K_c for the hypothetical equilibrium shown does the equilibrium most strongly tend toward products?

$$A(g) + B(g) \rightleftharpoons C(g) + D(g)$$

(a) 0.015 (b) 8.1×10^{-3} (c) 1.0 (d) 290 (e) 3.2×10^3

7. What is Q_c for the equilibrium if $[Br_2] = 0.26$ M, $[NO] = 0.88$ M, and $[NOBr] = 0.33$ M?

$$2NOBr(g) \rightleftharpoons 2NO(g) + Br_2(g) \qquad K_c = 4.1 \times 10^{-4}\ M$$

(a) 1.8 M (b) 0.69 M (c) 1.4 M (d) 2.1 M (e) 0.34 M

8. For question 7, in what direction is there a tendency for the reaction to shift?
(a) to the left (b) to the right (c) neither direction

9. The equilibrium given occurs through an elementary bimolecular forward reaction and an elementary bimolecular reverse reaction. At 800 K, $K_c = 4.7$ and k' (the rate constant for the reverse reaction) is 0.26 L/mol·s. What is k, the rate constant for the forward reaction?

$$A + B \rightleftharpoons C + D$$

(a) 0.26 L/mol·s (b) 1.2 L/mol·s (c) 3.9 L/mol·s
(d) 18 L/mol·s (e) 4.9 L/mol·s

10. $K_c = 4.1 \times 10^8\ M^{-2}$ at 298 K for the equilibrium $N_2(g) + 3H_2(g) \rightleftharpoons 2NH_3(g)$. What is K_P?
(a) $1.7 \times 10^7\ atm^{-2}$ (b) $1.0 \times 10^{10}\ atm^{-2}$ (c) $6.9 \times 10^5\ atm^{-2}$
(d) $2.5 \times 10^{11}\ atm^{-2}$ (e) $5.2 \times 10^{11}\ atm^{-2}$

11. $K_P = 47.9$ atm at 127°C for the equilibrium shown. What is K_c at this temperature?

$$N_2O_4(g) \rightleftharpoons 2NO_2(g)$$

(a) 1.57×10^3 M (b) 5.17×10^4 M (c) 6.18×10^3 M
(d) 4.44×10^{-2} M (e) 1.46 M

12. What is P_{O_2} in the equilibrium shown if $P_{SO_3} = 2.5$ atm and $P_{SO_2} = 8.2 \times 10^{-3}$ atm?

$$2SO_2(g) + O_2(g) \rightleftharpoons 2SO_3(g) \qquad K_P = 3.0 \times 10^4\ atm^{-1}$$

(a) 9.1×10^6 atm (b) 5.2×10^5 atm (c) 3.1 atm (d) 98 atm (e) 4.2 atm

13. For a hypothetical equilibrium, what is [B] when $[A] = 2.5$ M and $[C] = 0.50$ M?

$$A(g) + 3B(g) \rightleftharpoons 2C(g) \qquad K_c = 6.4 \times 10^2\ M^{-2}$$

(a) 0.013 M (b) 0.054 M (c) 0.043 M (d) 1.6×10^{-4} M (e) 0.18 M

14. What is the equilibrium expression for the equilibrium shown?

$$2HgO(s) \rightleftharpoons 2Hg(s) + O_2(g)$$

(a) $[Hg]^2[O_2]$ (b) $[O_2]$ (c) $[O_2][Hg]^2/[HgO]^2$
(d) $[O_2]/[HgO]^2$ (e) $[Hg]^2/[HgO]^2$

15. What is the decomposition vapor pressure of $CuSO_4 \cdot 5H_2O(s)$ at 600 K, when $K_P = 2.94 \times 10^{-7}\ atm^5$ for the equilibrium shown?

$$CuSO_4 \cdot 5H_2O(s) \rightleftharpoons CuSO_4(s) + 5H_2O(g)$$

(a) 2.94×10^{-7} atm (b) 1.0 atm (c) 0.66 atm
(d) 0.049 atm (e) 5.4×10^{-4} atm

Equilibrium calculations

16. NO, initially at 5.0 M, is allowed to come to equilibrium. What is the equilibrium concentration of NO? (The answer is very sensitive to how and where you round off.)

$$2NO(g) \rightleftharpoons N_2(g) + O_2(g) \qquad K_c = 20$$

(a) 0.90 M (b) 0.50 M (c) 1.6 M (d) 2.2 M (e) 2.8 M

17. $K_c = 0.082$ M for the equilibrium shown. What is the equilibrium concentration of Cl_2 if $COCl_2$, initially at 0.50 M, is allowed to come to equilibrium?

$$COCl_2(g) \rightleftharpoons CO(g) + Cl_2(g)$$

(a) 0.17 M (b) 0.34 M (c) 0.20 M (d) 0.26 M (e) 0.44 M

18. Assume the temperature is such that $K_P = 0.54$ atm for the equilibrium shown. What is the equilibrium partial pressure of PCl_5 if PCl_5, initially at 1.20 atm, is allowed to come to equilibrium at this temperature?

$$PCl_5(g) \rightleftharpoons PCl_3(g) + Cl_2(g)$$

(a) 0.58 atm (b) 0.80 atm (c) 0.54 atm (d) 0.62 atm (e) 0.40 atm

19. $K_c = 2.5 \times 10^{-9}$ M^2 at 298 K for the equilibrium shown. What is the equilibrium concentration of H_2 if NH_3, initially at 25 mM, is allowed to come to equilibrium at 298 K?

$$2NH_3(g) \rightleftharpoons N_2(g) + 3H_2(g)$$

(a) 6.3×10^{-8} mM (b) 0.49 mM (c) 1.5 mM
(d) 1.9 mM (e) 2.1×10^{-3} mM

20. Assume the temperature is such that $K_P = 3.1 \times 10^{-8}$ atm for the equilibrium shown. What is the equilibrium partial pressure of S_2 if H_2S, initially at 0.50 atm, and H_2, initially at 0.010 atm, are allowed to come to equilibrium at this temperature?

$$2H_2S(g) \rightleftharpoons 2H_2(g) + S_2(g)$$

(a) 8.8×10^{-3} atm (b) 1.3×10^{-3} atm (c) 7.8×10^{-5} atm
(d) 1.6×10^{-6} atm (e) 2.8×10^{-4} atm

The response of equilibria to changes in the conditions

21. Which of the following causes the equilibrium shown to shift to the right?

$$3H_2(g) + N_2(g) \rightleftharpoons 2NH_3(g)$$

(a) addition of NH_3 (b) addition of N_2 (c) removal of H_2 (d) removal of N_2

22. Removal of H_2 from the equilibrium shown shifts the equilibrium

$$H_2(g) + I_2(g) \rightleftharpoons 2HI(g)$$

(a) to the left. (b) to the right. (c) neither to the left or right.

23. Addition of Fe(*s*) to the equilibrium shown shifts the equilibrium

$$2Fe_2O_3(s) + 3C(s) \rightleftharpoons 4Fe(s) + 3CO_2(g)$$

(a) to the left. (b) to the right. (c) neither to the left or right.

24. Increasing the pressure on the equilibrium shown shifts the equilibrium

$$Br_2(g) + 3F_2(g) \rightleftharpoons 2BrF_3(g)$$

(a) to the left. (b) to the right. (c) neither to the left or right.

25. Decreasing the pressure on the equilibrium shown shifts the equilibrium

$$Xe(g) + F_2(g) \rightleftharpoons XeF_2(g)$$

(a) to the left. (b) to the right. (c) neither to the left or right.

26. Decreasing the pressure on the equilibrium shown shifts the equilibrium

$$H_2(g) + Br_2(g) \rightleftharpoons 2HBr(g)$$

(a) to the left. (b) to the right. (c) neither to the left or right.

27. Increasing the temperature on the equilibrium shown shifts the equilibrium

$$2NO(g) + O_2(g) \rightleftharpoons 2NO_2(g) \qquad \Delta H = -114 \text{ kJ}$$

(a) to the left. (b) to the right. (c) neither to the left or right.

28. For which of the following changes does K_c also change?
(a) change in pressure (b) addition of a reagent
(c) change in temperature (d) none of these

Descriptive chemistry

29. What compound is formed in the Mond process?
(a) CO_2 (b) $Ni(CO)_4$ (c) $NiCl_4$ (d) NiO (e) SO_2

30. Gaseous methanol (CH_3OH) can exist in equilibrium with carbon monoxide (CO) and
(a) H_2 (b) H_2O (c) CH_4 (d) H_2O_2 (e) OH^-

31. In a hot automobile engine, N_2 is in equilibrium with
(a) N_2O (b) NO (c) NO_2 (d) N_2O_4 (e) NO_3

32. PCl_5 can exist in equilibrium with
(a) P_4 and Cl_2 (b) PCl_3 and P (c) PCl_3 (d) PCl_3 and Cl_2

33. The Haber process is used for the synthesis of
(a) NH_3 (b) esters (c) CH_3OH (d) NO (e) carbohydrates

14 ACIDS AND BASES

THE DEFINITIONS OF ACIDS AND BASES

KEY WORDS Define or explain each of the following terms in a written sentence or two.

amphiprotic
autoionization
Brønsted acid
Brønsted base
conjugate acid
conjugate base
hydronium ion
proton transfer

14.1 BRØNSTED ACIDS AND BASES

KEY CONCEPT A Brønsted acids and bases

The Brønsted–Lowry theory of acids and bases focuses on the transfer of an H^+ ion (a proton). In any reaction involving Brønsted acids and bases, we look for what the proton does in order to understand what chemistry is happening. With this in mind, we can make the following definitions. A **Brønsted acid** is a compound that donates a proton (H^+) in a chemical reaction; a **Brønsted base** is a compound that accepts a proton (H^+) in a chemical reaction. When a Brønsted acid donates a proton, a Brønsted base accepts it, so any chemical reaction involving Brønsted acids and bases must contain both; you can't have one without the other. As an example, consider the **ionization** of HNO_3 in water:

$$HNO_3(aq) + H_2O(l) \longrightarrow H_3O^+(aq) + NO_3^-(aq)$$

In this reaction, HNO_3 donates a proton and H_2O accepts it. HNO_3 is a Brønsted acid (it donates a proton), and H_2O is a Brønsted base (it accepts a proton). The fact that water is the solvent does not affect our view of the reaction; it is simply a reactant.

EXAMPLE 1 Recognizing a Brønsted acid and Brønsted base in a chemical reaction

Identify the Brønsted acid and Brønsted base in the reaction

$$NH_3 + CH_3^- \rightleftharpoons NH_2^- + CH_4$$

SOLUTION The substance that donates a proton is the Brønsted acid, the one that accepts it is the Brønsted base. In this reaction, NH_3 donates a proton (and becomes NH_2^-) and CH_3^- accepts a proton (and becomes CH_4). NH_3 acts as a Brønsted acid, and CH_3^- acts as a Brønsted base.

EXERCISE Identify the Brønsted acid and Brønsted base in the reaction

$$HS^-(aq) + H_2O(l) \rightleftharpoons H_2S(aq) + OH^-(aq)$$

ANSWER: HS^-, base; H_2O, acid

EXAMPLE 2 Predicting the products of a Brønsted acid–base reaction

Predict the products of the following Brønsted acid–base reaction:

$$HCO_3^-(aq) + OH^-(aq) \rightleftharpoons ?$$

SOLUTION A Brønsted acid–base reaction involves proton transfer, and the first hurdle here is to determine which of the two hydrogens in the reactants transfers as a proton. We must call on our knowledge that OH^- often accepts a proton to form water to conclude that it is most likely the Brønsted base in this reaction. Thus, we conclude that HCO_3^- is the Brønsted acid and donates a proton (forming CO_3^{2-}) and that OH^- is the Brønsted base and accepts the proton (forming H_2O). The correct equation is

$$HCO_3^-(aq) + OH^-(aq) \rightleftharpoons CO_3^{2-}(aq) + H_2O(l)$$

EXERCISE Predict the products of the Brønsted acid–base reaction:

$$HSO_4^-(aq) + OH^-(aq) \rightleftharpoons ?$$

ANSWER: $SO_4^{2-}(aq) + H_2O(l)$

KEY CONCEPT B Neutralization in Brønsted acid–base chemistry

In many solvents, the net ionic equation for acid–base neutralization is the reaction between a cationic Brønsted acid and an anionic Brønsted base to form neutral solvent molecules.

$$\text{Brønsted acid (cation)} + \text{Brønsted acid (anion)} \rightleftharpoons \text{solvent}$$

In water, this becomes

$$H_3O^+(aq) + OH^-(aq) \rightleftharpoons 2H_2O(l)$$

This neutralization reaction is a Brønsted acid–base reaction. H_3O^+ acts as a Brønsted acid by donating a proton to become H_2O, while OH^- acts as a Brønsted base by accepting the proton to also become H_2O.

EXAMPLE Brønsted acid–base neutralization in nonaqueous solvents

Write the net ionic equation for the Brønsted acid–base neutralization reaction that occurs in liquid hydrogen fluoride (HF).

SOLUTION A Brønsted acid–base neutralization in a solvent involves transfer of a proton from a cationic acid to an anionic base to give neutral solvent molecules. The acid is a solvent molecule that has acquired a proton, in this case H_2F^+, and the base is a solvent molecule that has lost a proton, in this case F^-. The neutralization reaction is

$$H_2F^+ + F^- \rightleftharpoons 2HF(l)$$

EXERCISE Write the neutralization reaction for the net ionic reaction that occurs in liquid sulfuric acid (H_2SO_4).

ANSWER: $H_3SO_4^+ + HSO_4^- \rightleftharpoons 2H_2SO_4(l)$

PITFALL The difference between a liquid solvent and an aqueous solution

The nonaqueous solvent neutralizations just described occur in (nearly) pure liquid solvent and not in aqueous solutions. The reactions shown would not occur if any water were present.

14.2 BRØNSTED ACID–BASE EQUILIBRIA

KEY CONCEPT A Brønsted equilibria

Proton-transfer reactions are extremely fast and facile. It is not suprising, therefore, to find that most Brønsted acid–base reactions actually exist as dynamic equilibria in which the proton rapidly moves back and forth between two Brønsted bases. For example, consider the ionization of aqueous hydrocyanic acid (HCN).

$$HCN(aq) + H_2O(l) \rightleftharpoons H_3O^+(aq) + CN^-(aq)$$

In this equilibrium, products form by transfer of a proton from HCN to H_2O; reactants reform by proton transfer from H_3O^+ to CN^-. For the forward reaction, HCN is a Brønsted acid and H_2O a Brønsted base; for the reverse reaction, H_3O^+ is a Brønsted acid and CN^- a Brønsted base. When a Brønsted acid loses its proton, the base formed is called the conjugate base of the acid; when a base accepts a proton to become an acid, the acid is called the conjugate acid of the base. CN^- is the conjugate base of HCN, and HCN is the conjugate acid of CN^-. Also, H_3O^+ is the conjugate acid of H_2O, and H_2O the conjugate base of H_3O^+. If we consider the reverse reaction, it becomes evident that H_3O^+ is an acid, as suggested, because it loses a proton; similarly, CN^- is a base because it gains a proton. As the following examples indicate, a conjugate acid and base differ only in the presence of an H^+.

Forward reaction	Reverse reaction	Conjugate acid–base pair
HCN (acid)	CN^- (base)	HCN, CN^-
H_2O (base)	H_3O^+ (acid)	H_3O^+, H_2O

EXAMPLE Recognizing conjugate acid–base pairs

Identify the conjugate acid–base pairs in the equilibrium

$$H_2S(aq) + NH_3(aq) \rightleftharpoons HS^-(aq) + NH_4{}^+(aq)$$

SOLUTION Members of a conjugate acid–base pair differ only in the presence of an H^+. In this reaction, H_2S donates a proton and becomes HS^-. H_2S and HS^- differ only in the presence of an H^+ and are a conjugate acid–base pair. The proton is accepted by NH_3, which then forms $NH_4{}^+$. NH_3 and $NH_4{}^+$ are a conjugate acid–base pair; they differ only in the presence of a proton.

EXERCISE Identify the conjugate acid–base pairs in the eqilibrium

$$CN^-(aq) + H_2O(l) \rightleftharpoons HCN(aq) + OH^-(aq)$$

ANSWER: CN^- and HCN; H_2O and OH^-

KEY CONCEPT B Autoionization

An **amphiprotic** compound is one that can donate a proton (to a proton acceptor) or accept a proton (from a proton donor). Such a compound can act as a Brønsted acid (when it donates a proton) or as a Brønsted base (when it accepts a proton). Amphiprotic solvents are common and undergo **autoionization**. In the autoionization process, one molecule of the solvent acts as a Brønsted acid and donates a proton, while another molecule of the solvent acts as a Brønsted base and accepts the proton. As with most Brønsted acid–base reactions, autoionization results in an equilibrium. For water, the autoionization equilibrium is

$$H_2O(l) + H_2O(l) \rightleftharpoons OH^-(aq) + H_3O^+(aq)$$

EXAMPLE Writing an autoionization equilibrium

Write the autoionization equilibrium for the amphiprotic solvent sulfuric acid, H_2SO_4.

SOLUTION Autoionization of an amphiprotic solvent occurs when one solvent molecule donates a proton to a second solvent molecule. For H_2SO_4, this is

$$H_2SO_4 + H_2SO_4 \rightleftharpoons H_3SO_4^+ + HSO_4^-$$

EXERCISE Write the autoionization equilibrium for the amphiprotic solvent HF.

ANSWER: $HF + HF \rightleftharpoons H_2F^+ + F^-$

KEY CONCEPT C The water autoionization constant

The autoionization of water is a dynamic equilibrium and is described by an equilibrium constant, K_w. The concentration of water does not appear in K_w, so we get

$$H_2O(l) + H_2O(l) \rightleftharpoons H_3O^+(aq) + OH^-(aq)$$
$$K_w = [H_3O^+][OH^-]$$

At 25°C, the experimentally determined value of K_w is $1.0 \times 10^{-14}\ M^2$. This means that at 25°C, the concentration of H_3O^+ multiplied by the concentration of OH^- *must* equal 1.0×10^{-14}, no matter what the source of the ions:

$$[H_3O^+][OH^-] = 1.0 \times 10^{-14}\ M^2 \qquad (25°C)$$

In pure water, the concentration of OH^- must be the same as that of H_3O^+, so we get

$$\text{Pure water: } [H_3O^+] = 1.0 \times 10^{-7}\ M \qquad [OH^-] = 1.0 \times 10^{-7}\ M$$

EXAMPLE Using K_w

An acid solution at 25°C is made up so that $[H_3O^+] = 3.4 \times 10^{-5}$ M. What is the hydroxide ion concentration in this solution? Is the hydroxide ion concentration higher or lower than that in pure water?

SOLUTION In any aqueous solution at 25°C, $[H_3O^+][OH^-] = 1.0 \times 10^{-14}\ M^2$. We are given an H_3O^+ concentration of 3.4×10^{-5}; thus,

$$[H_3O^+][OH^-] = 1.0 \times 10^{-14}\ M^2$$
$$(3.4 \times 10^{-5}\ M)\ [OH^-] = 1.0 \times 10^{-14}\ M^2$$
$$[OH^-] = \frac{1.0 \times 10^{-14}\ M^2}{3.4 \times 10^{-5}\ M}$$
$$= 2.9 \times 10^{-10}\ M$$

This OH^- concentration is lower than that of pure water because 2.9×10^{-10} M is less than 1.0×10^{-7} M. This result is expected because an increase in H_3O^+ concentration forces a decrease in the OH^- concentration.

EXERCISE What is the H_3O^+ concentration in a solution of a base in which $[OH^-] = 6.6 \times 10^{-5}$ M? Is the H_3O^+ concentration higher or lower than that in pure water?

ANSWER: 1.5×10^{-10} M; lower

SOLUTIONS OF STRONG ACIDS AND BASES

KEY WORDS Define or explain each of the following terms in a written sentence or two.

analyte
pH
pH curve
stoichiometric point
strong acid
strong base
titrant
universal indicator paper

14.3 STRONG ACIDS AND BASES

KEY CONCEPT Weak and strong acids and bases

An acid is classified as strong or weak depending on the percentage of the acid that ionizes when it is dissolved in water.

1. A **strong acid** is 100% ionized in aqueous solution at moderate concentrations. For example, when HCl is dissolved in water, all of the HCl molecules ionize into H_3O^+ and Cl^- ions. Because there is no detectable unionized acid, ionization constants are not defined for strong acids.

2. A **weak acid** is only partially ionized in aqueous solution at moderate concentrations. For example, when CH_3COOH is dissolved in water, most of the CH_3COOH molecules do not ionize. Only a few percent ionize into H_3O^+ and CH_3COO^- ions.

A base is classified as strong or weak depending on the percentage of the base that ionizes when it is dissolved in water.

3. A **strong base** is defined as a substance in which a reaction strongly favors formation of OH^-. Practically, there are two types of strong bases. One type is an ionic hydroxide that ionizes 100% in aqueous solution to produce relatively large amounts of $OH^-(aq)$; examples are NaOH, KOH, and $Ca(OH)_2$. The second type is an ionic compound that contains an anion that reacts completely with water to form $OH^-(aq)$; examples are Na_2O, KNH_2, and CaO.

$$\textit{First type:} \quad NaOH(s) \xrightarrow{100\%} Na^+(aq) + OH^-(aq)$$

$$\textit{Second type:} \quad O^{2-}(aq) + H_2O(l) \xrightarrow{100\%} 2OH^-(aq)$$

4. A **weak base** is only partially ionized in aqueous solution at moderate concentrations. For example, when NH_3 is dissolved in water, most of the NH_3 molecules do not react with water to form OH^- and NH_4^+. Only a few percent ionize into NH_4^+ and OH^- ions.

The terms strong and weak should be distinguished from **concentrated** and **dilute,** which describe how much solute is dissolved in a solution.

EXAMPLE 1 Using the terms weak, strong, concentrated, and dilute

Select from the four terms *weak, strong, concentrated,* and *dilute* to describe a 10 M solution of KOH.

SOLUTION We recognize that KOH is an ionic compound and, therefore, completely ionized in solution. This means it is a strong base. From the discussion in the text, it is evident that a 10 M solution of any solute would be considered a concentrated solution. Thus, 10 M KOH is a *concentrated* solution of a *strong* base.

EXERCISE Select from the terms *weak, strong, concentrated,* and *dilute* to describe a 0.30 M solution of HCl.

ANSWER: Dilute solution of a strong acid

EXAMPLE 2 Concentrations of the ions in a strong acid

What are the concentrations of H_3O^+, OH^-, and Cl^- in a 0.040 M solution of HCl?

SOLUTION HCl is a strong acid (see Table 14.4) and, therefore, undergoes 100% ionization. One good way to envision what happens in such a situation is to construct a table showing the concentrations of species before and after ionization. Because the concentration of H_2O does not change, it is not included in the table.

	$HCl(aq)$ $\xrightarrow{100\%}$	$H_3O^+(aq)$ +	$Cl^-(aq)$
before ionization	0.040 M	0 M	0 M
change upon 100% ionization	−0.040 M	+0.040 M	+0.040 M
after ionization	0 M	0.040 M	0.040 M

The changes that occur are determined by the stoichiometry of the ionization. Because one mole of H_3O^+ and one mole of Cl^- are formed from one mole of HCl, when 0.040 M HCl ionizes, 0.040 M H_3O^+ and 0.040 M Cl^- form. In answer to the question, the table supplies the concentrations of H_3O^+ and Cl^-. To determine $[OH^-]$, we use

$$[H_3O^+][OH^-] = 1.0 \times 10^{-14}\ M^2$$

We substitute $[H_3O^+] = 0.040$ M and solve for $[OH^-]$:

$$[H_3O^+][OH^-] = 1.0 \times 10^{-14}\ M^2$$

$$[H_3O^+] = \frac{1.0 \times 10^{-14}\ M^2}{[OH^-]}$$

$$= \frac{1.0 \times 10^{-14}\ M^2}{0.040\ M}$$

$$= 2.5 \times 10^{-13}\ M$$

EXERCISE What are the concentrations of H_3O^+, OH^-, and Ba^{2+} in a 0.0035 M solution of $Ba(OH)_2$? (Hint: Remember that when 1 mol $Ba(OH)_2$ ionizes, it forms 2 mol OH^-.)

ANSWER: $[OH^-] = 0.0070$ M; $[Ba^{2+}] = 0.0035$ M; and $[H_3O^+] = 2.9 \times 10^{-12}$ M

EXAMPLE 3 Relative H_3O^+ and OH^- concentrations in acid and base solutions

Which solution contains a higher concentration of OH^- ions, 0.10 M NaOH or 0.10 M NH_2NH_2? (Hint: Remember that hydrazine, NH_2NH_2, is a weak base, like NH_3; see Table 14.6.)

SOLUTION NaOH is a strong base that completely ionizes, so a 0.10 M solution of NaOH contains 0.10 M Na^+ ion and 0.10 M OH^- ion. NH_2NH_2 is a weak base; so in solution most of the NH_2NH_2 molecules remain unionized and very little $NH_2NH_3^+$ and OH^- are formed. Although we, as yet, have no way to determine the concentration of OH^- in a 0.10 M solution of NH_2NH_2 it is less than 0.10 M. For solutions of equal molarity, NaOH has a higher concentration of OH^- ions.

EXERCISE Which solution has a higher concentration of H_3O^+ ions, 0.20 M HBr or 0.20 M HF?

ANSWER: 0.20 M HBr

14.4 HYDROGEN ION CONCENTRATION AND pH

KEY CONCEPT A The pH of Solutions

The acidity of a solution is often reported by using the **pH** of the Solution. The pH is defined like the earlier p functions we encountered:

$$pH = -\log [H_3O^+]$$

This definition leads to the following at 25°C:

Acid solution:	$[H_3O^+] > 1.0 \times 10^{-7}$ M	$pH < 7.00$
Neutral solution:	$[H_3O^+] = 1.0 \times 10^{-7}$ M	$pH = 7.00$
Basic solution:	$[H_3O^+] < 1.0 \times 10^{-7}$ M	$pH > 7.00$

The pH of a solution can be approximately measured with **universal indicator paper,** which turns different colors when different pH solutions are placed on it. More accurate measurements are made with an instrument called a pH meter. Use of the pH is entirely a matter of convenience and conveys no more information than the hydronium ion concentration $[H_3O^+]$ itself.

EXAMPLE 1 Interpreting the pH

A typical hydronium ion concentration of the juices from a dill pickle is 3.3×10^{-2} M. To what pH does this correspond? Is it acidic, basic, or neutral?

SOLUTION The definition of pH is

$$pH = -\log [H_3O^+]$$

We are given $[H_3O^+] = 3.3 \times 10^{-2}$ M. Substituting into the definition of pH gives

$$\begin{aligned} pH &= -\log(3.3 \times 10^{-2}) \\ &= -(-1.48) \\ &= 1.48 \end{aligned}$$

Because the pH is less than 7.00, the solution is acidic.

EXERCISE What is the pH of a sample of human milk that has a hydronium ion concentration of 7.9×10^{-8} M? Is it acidic, basic, or neutral?

ANSWER: 7.10; barely basic

EXAMPLE 2 Calculating $[OH^-]$ from the pH

What is the hydroxide ion concentration of a sample of wine at 25°C with pH = 2.80? Is the wine acidic, basic, or neutral?

SOLUTION First, because the pH is less than 7.00, the wine is acidic. For the other part of the question, we can solve for $[OH^-]$ by calculating $[H_3O^+]$ from the pH and using the relationship

$$[H_3O^+][OH^-] = 1.0 \times 10^{-14}\ M^2$$

With the definitions antilog $z = 10^z$ for any z and antilog $(\log v) = v$ for any v, we take the antilog of both sides of the equation for pH and get

$$\begin{aligned} pH &= -\log[H_3O^+] \\ [H_3O^+] &= 10^{-pH} \\ &= 10^{-2.80} \\ &= 1.6 \times 10^{-3}\ M \end{aligned}$$

Now, we substitute this value of $[H_3O^+]$ into the equation $[H_3O^+][OH^-] = 1.0 \times 10^{-14}\ M^2$.

$$\begin{aligned} (1.6 \times 10^{-3}\ M)[OH^-] &= 1.0 \times 10^{-14}\ M^2 \\ [OH^-] &= 6.3 \times 10^{-12}\ M \end{aligned}$$

EXERCISE What is the hydroxide ion concentration of a sample of milk of magnesia, with pH = 10.47. Is it acidic, basic, or neutral?

ANSWER: 2.95×10^{-4} M; basic

KEY CONCEPT B The pH of strong acids and strong bases

A strong acid is one that undergoes 100% ionization. The hydronium ion (H_3O^+) concentration of a strong monoprotic acid is usually the same as the molarity of the acid itself. For instance, in 0.20 M $HCl(aq)$, we can construct the following table of values before and after ionization. (A reaction arrow ($\rightarrow$) is used rather than an equilibrium sign ($\rightleftharpoons$) because the reaction forms 100% products.)

	$HCl(aq)$	+	$H_2O(l)$	$\longrightarrow$	$H_3O^+(aq)$	+	$Cl^-(aq)$
concentration before 100% ionization	0.20 M				0		0
change that accompanies 100% ionization	−0.20 M				+0.20 M		+0.20 M
concentration after 100% ionization	0				0.20 M		0.20 M

Because all of the acid ionizes, the $[H_3O^+]$ concentration is the same as the starting HCl concentration. A laboratory reagent bottle labeled 0.20 M HCl actually contains no detectable HCl molecules! It contains, instead, 0.20 M H_3O^+ ion and 0.20 M Cl^- ion.

A strong base is one that undergoes 100% ionization. The hydroxide ion concentration can be calculated directly from the molarity of the base and the formula of the base. This is illustrated for a 0.20 M solution of the strong base $Ca(OH)_2$:

	$Ca(OH)_2(aq)$ $\longrightarrow$	$Ca^{2+}(aq)$ +	$2OH^-(aq)$
concentration before 100% ionization	0.20 M	0	0
change that accompanies 100% ionization	−0.20 M	+0.20 M	+0.40 M
concentration after 100% ionization	0	0.20 M	0.40 M

In setting up this table, we have used the fact that 1 mol $Ca(OH)_2$ yields 2 mol OH^- when it ionizes, so the increase in OH^- concentration is +0.40 (2 × 0.20).

In summary,

For a solution of monoprotic strong acid and molarity Y: $[H_3O^+] = Y$ M
For a solution of strong base with formula MOH and molarity Y: $[OH^-] = Y$ M
For a solution of strong base with formula $M(OH)_2$ and molarity Y: $[OH^-] = 2Y$ M

A convenient p function that is used in solutions is the pOH:

$$pOH = -\log [OH^-]$$

Because $[H_3O^+][OH^-] = 1.0 \times 10^{-14}$,

$$pH + pOH = 14.00$$

EXAMPLE Calculating the pH and pOH of a strong acid and strong base

What is the pH and pOH of a 0.025 M HNO_3 solution?

SOLUTION Nitric acid (HNO_3) is a strong monoprotic acid (see Table 14.4 of the text). For any strong monoprotic acid, the concentration of hydronium ion is the same as the concentration of the acid itself. So,

$$[H_3O^+] = 0.025 \text{ M}$$

Using the definition of pH results in the desired answer:

$$\begin{aligned} pH &= -\log [H_3O^+] \\ &= -\log (0.025) \\ &= 1.60 \end{aligned}$$

The pOH is calculated as follows:

$$\begin{aligned} pH + pOH &= 14.00 \\ 1.60 + pOH &= 14.00 \\ pOH &= 14.00 - 1.60 \\ &= 12.40 \end{aligned}$$

EXERCISE Calculate the pH and pOH of a 0.0050 M $Sr(OH)_2$ solution.

ANSWER: pH = 11.70; pOH = 2.30

PITFALL The initial concentration of H_3O^+ and OH^-

In these calculations, the initial concentrations of H_3O^+ and OH^- are assumed to be zero. In actuality, the autoionization of water gives starting concentrations of 1.0×10^{-7} M for each. However, the amount of H_3O^+ formed by a typical concentration of a strong acid completely overwhelms the tiny amount initially present; so assuming that the initial concentration is 0 M gives the correct answer. Similarly, the huge amount of OH^- formed by a typical concentration of a strong base is so large that it overwhelms the small amount of OH^- initially present, and it is acceptable to approximate the initial OH^- concentration as 0 M.

14.5 MIXTURES OF STRONG ACIDS AND BASES

KEY CONCEPT A Calculating pH when known amounts of acid and base are mixed

When a strong acid and strong base are mixed, the following reaction occurs.

$$H_3O^+(aq) + OH^-(aq) \longrightarrow 2H_2O(aq)$$

Because 1 mol of H_3O^+ reacts with 1 mol of OH^-, the excess H_3O^+ (or OH^-) can be calculated as shown in the table that follows. The pH of the solution that results depends on the relative amounts of acid and base that are mixed.

Condition	Calculation of moles of excess species	pH
moles H_3O^+ > moles OH^-	excess moles H_3O^+ = moles H_3O^+ added − moles OH^- added	pH < 7 (solution is acidic)
moles H_3O^+ = moles OH^-	neither H_3O^+ or OH^- is in excess	pH = 7 (solution is neutral)
moles H_3O^+ < moles OH^-	excess moles OH^- = moles OH^- added − moles H_3O^+ added	pH > 7 (solution is basic)

To calculate the concentration of H_3O^+ (or OH^-) we must be careful to use the total volume of solution, which is usually approximated as the sum of the volume of acid plus the volume of base; $V_{total} = V_{acid} + V_{base}$.

If moles H_3O^+ > moles OH^-, then $[H_3O^+] = \dfrac{\text{excess moles } H_3O^+}{V_{total}}$

If moles OH^- > moles H_3O^+, then $[OH^-] = \dfrac{\text{excess moles } OH^-}{V_{total}}$ and $[H_3O^+] = \dfrac{1.0 \times 10^{-14}\ M^2}{[OH^-]}$

EXAMPLE Calculation of the pH of a mixture of strong acid and strong base

What is the pH of the solution that results when 10.0 mL of 0.300 M HCl is mixed with 19.0 mL 0.150 M NaOH?

SOLUTION We must first calculate the moles of H_3O^+ and the moles of OH^- that are mixed, using the relationship moles = molarity × volume(in liters); we recognize that the molarity of H_3O^+ is the same as that of HCl because HCl is 100% ionized, and that the molarity of OH^- is the same as that of NaOH because NaOH is 100% ionized.

$$\text{moles } H_3O^+ = 0.300 \frac{\text{mol}}{\not{L}} \times 0.0100\not{L} = 0.00300 \text{ mol} \qquad \text{moles } OH^- = 0.150 \frac{\text{mol}}{\not{L}} \times 0.0190\not{L} = 0.00285 \text{ mol}$$

Because there are more moles of acid than base, we next calculate the excess moles of acid:

$$\begin{aligned} \text{excess moles } H_3O^+ &= \text{moles } H_3O^+ \text{ added} - \text{moles } OH^- \text{ added} \\ &= 0.00300 \text{ mol} - 0.00285 \text{ mol} \\ &= 0.00015 \text{ mol} \end{aligned}$$

It is now possible to calculate the molarity of H_3O^+, using the fact that the total volume is 29.0 mL:

$$V_{total} = 10.0 \text{ mL} + 19.0 \text{ mL}$$
$$= 29.0 \text{ mL}$$
$$= 0.0290 \text{ L}$$

Thus,

$$[H_3O^+] = \frac{\text{excess moles } H_3O^+}{V_{total}}$$
$$= \frac{0.00015 \text{ mol}}{0.0290 \text{ L}}$$
$$= 0.0052 \text{ M}$$

We can now calculate the pH:

$$\text{pH} = -\log [H_3O^+]$$
$$= -\log (0.0052)$$
$$= -(-2.28)$$
$$= 2.28$$

EXERCISE Now, 2.0 mL more of the 0.150 M NaOH (for a total of 21.0 mL) is added to the mixture described in the preceding example. What is the new pH? (Hint: When excess OH^- is added, $[OH^-]$ is calculated first; from this value, $[H_3O^+]$ and the pH are then calculated.)

ANSWER: pH = 11.68

KEY CONCEPT B Calculating the amount of acid (or base) needed for neutralization

For neutralization to occur, the moles of H_3O^+ added must equal the moles of OH^- added. We use the relationship moles = molarity × volume(in liters) to calculate the amount of acid needed to neutralize a given amount of base or the amount of base needed to neutralize a given amount of acid.

EXAMPLE The amount of base (or acid) needed to neutralize a given amount of acid (or base)

How much 0.0315 M $Ca(OH)_2$ is required to neutralize 5.00 mL of 0.115 M HNO_3?

SOLUTION We start by calculating the moles of H_3O^+ provided by the nitric acid; in doing so, we recognize that 1 mol of HNO_3 provides 1 mol of H_3O^+.

$$\text{moles } H_3O^+ \ (= \text{moles } HNO_3) = 0.115 \frac{\text{mol}}{\cancel{\text{L}}} \times 0.00500 \ \cancel{\text{L}}$$
$$= 0.000575 \text{ mol}$$

We next recall that 1 mol of OH^- is required to react with 1 mol of H_3O^+, so

$$\text{moles } OH^- \text{ required} = 0.000575 \text{ mol } OH^-$$

The moles of $Ca(OH)_2$ required are calculated by using the fact that 1 mol of $Ca(OH)_2$ supplies 2 mol of OH^-.

$$\text{moles } Ca(OH)_2 = 0.000575 \ \cancel{\text{mol } OH^-} \times \frac{1 \text{ mol } Ca(OH)_2}{2 \ \cancel{\text{mol } OH^-}}$$
$$= 0.000288 \text{ mol } Ca(OH)_2$$

We finally calculate the volume of calcium hydroxide required using moles = molarity × volume(in liters); we have moles of $Ca(OH)_2$ = 0.000288 mol and molarity = 0.0315 M.

$$\text{moles} = \text{molarity} \times \text{volume(in liters)}$$

$$0.000288 \text{ mol } Ca(OH)_2 = 0.0315 \frac{\text{mol } Ca(OH)_2}{\text{L}} \times \text{volume(in liters)}$$

$$\text{volume(in liters)} = \frac{0.000288 \cancel{\text{mol}}}{0.0315 \frac{\cancel{\text{mol}}}{\text{L}}}$$

$$= 0.00914 \text{ L}$$

$$= 9.14 \text{ mL}$$

EXERCISE What volume (in mL) of 0.205 M HCl is required to neutralize 25.0 mL of 0.255 M $Ba(OH)_2$?

ANSWER: 62.2 mL

14.6 pH CURVES FOR STRONG ACID–STRONG BASE TITRATIONS

KEY CONCEPT Titration of a strong base with a strong acid

A strong base is titrated with a strong acid by adding to the base a carefully measured volume of acid of known concentration. An important parameter in a titration is the pH at various points, because knowledge of the pH helps in choosing the correct indicator for the titration. We shall explore how the pH changes when the concentration of the base is known in advance. For the titration of NaOH(*aq*) with HCl(*aq*), the net reaction is

$$OH^-(aq) + H_3O^+(aq) \longrightarrow 2H_2O(l)$$

We should note that the stoichiometry of the reaction tells us that for every 1 mmol of acid added (giving 1 mmol of H_3O^+), 1 mmol of base reacts. Typically, we want to know the pH at four different stages of the titration:

1. before the addition of any acid
2. after addition of some acid but before the stoichiometric point is reached
3. at the stoichiometric point (the point at which just enough acid has been added to react with the base present at the start)
4. after the stoichiometric point (i.e., when more acid is added after the stoichiometric point has been reached)

At all points of the titration, the total volume equals the sum of the initial volume of base plus the volume of acid added. The total volume of solution increases as the titration is done; the volume of added acid must be taken into account.

EXAMPLE Calculating the pH at various points in a titration

10.00 mL of a 0.210 M NaOH(*aq*) solution is titrated with 0.0911 M HCl(*aq*). What is the pH of the titration mixture (a) before addition of any acid, (b) after addition of 12.00 mL HCl, (c) at the stoichiometric point (after addition of 23.05 mL HCl), and (d) after addition of a total of 30.00 mL HCl?

SOLUTION Throughout the solution to this problem, we shall use V_{total} to represent the total volume of titration mixture, in milliliters.

$$V_{total} = 10.00 + V_{acid}$$

where V_{acid} = the total volume of acid added and 10.00 mL is the original volume of base given in the problem.

(a) Before addition of any acid, the OH^- ion concentration is the same as that of the NaOH in solution:

$$[OH^-] = 0.210 \text{ M}$$
$$pOH = -\log [OH^-] = -\log (0.210)$$
$$= +0.68$$
$$pH + pOH = 14.00$$
$$pH = 14.00 - 0.68$$
$$= 13.32$$

(b) After addition of 12.00 mL HCl, $V_{total} = 10.00 \text{ mL} + 12.00 \text{ mL} = 22.00 \text{ mL}$. The pH at this point is determined by the millimoles of *unreacted* OH^- left in solution. This, in turn, is equal to the initial millimoles of OH^- minus the millimoles of OH^- that have reacted as a result of addition of acid:

$$\text{Millimoles unreacted } OH^- = \text{initial mmol } OH^- - \text{mmol } OH^- \text{ reacted by addition of acid}$$

$$\text{Initial millimoles } OH^- = 10.00 \cancel{\text{mL}} \times 0.210 \frac{\text{mmol}}{\cancel{\text{mL}}} = 2.10 \text{ mmol}$$

$$\text{Millimoles } OH^- \text{ reacted} = 12.00 \cancel{\text{mL}} \times 0.0911 \frac{\text{mmol}}{\cancel{\text{mL}}} = 1.09 \text{ mmol}$$

$$\text{Millimoles unreacted } OH^- = 2.10 \text{ mmol} - 1.09 \text{ mmol} = 1.01 \text{ mmol}$$

The molarity of OH^- equals the millimoles OH^- divided by the total volume of solution:

$$[OH^-] = \frac{1.01 \text{ mmol}}{22.00 \text{ mL}} = 0.0459 \text{ M}$$

From this, we calculate the pOH and the pH:

$$pOH = -\log [OH^-] = -\log (0.0459)$$
$$= 1.34$$
$$pH + pOH = 14.00$$
$$pH + 1.34 = 14.00$$
$$pH = 14.00 - 1.34$$
$$= 12.66$$

(c) At the stoichiometric point, the amount of acid added (23.05 mL, 2.10 mmol) is exactly that required to react with the original amount of base (10.00 mL, 2.10 mmol) present. In addition, neither Na^+ (from the NaOH) nor Cl^- (from HCl) affects the pH of the solution. Thus, there is no excess OH^- or H_3O^+ present in the titration mixture, only the amount that forms from the autoionization of water. The mixture (an NaCl solution) is neutral, with pH = 7.00.

(d) The total volume of the solution after addition of 30.00 mL HCl is 40.00 mL. ($V_{total} = 10.00 \text{ mL} + 30.00 \text{ mL}$.) At this point, a total of 2.73 mmol of acid have been added. The original 2.10 mmol of base have reacted with 2.10 mmol of this acid; and there is an extra 0.63 mmol (2.73 − 2.10 = 0.63) of acid. This extra acid now determines the pH of the solution:

$$[H_3O^+] = \frac{0.63 \text{ mmol}}{40.00 \text{ mL}}$$
$$= 0.016 \text{ M}$$
$$pH = -\log [H_3O^+]$$
$$= 1.80$$

EXERCISE 20.00 mL of a 0.0650 M NaOH(*aq*) solution is titrated with 0.1044 M HCl(*aq*). What is the pH of the reaction mixture (a) before addition of any acid, (b) after addition of 6.00 mL HCl, (c) at the stoichiometric point (after addition of 12.45 mL HCl), and (d) after addition of a total of 20.00 mL HCl?

ANSWER: (a) 12.81; (b) 12.38; (c) 7.00; (d) 1.71

EQUILIBRIA IN SOLUTIONS OF ACIDS AND BASES

KEY WORDS Define or explain each of the following terms in a written sentence or two.

acid ionization constant
base ionization constant
polyprotic acid
weak acid
weak base

14.7 IONIZATION CONSTANTS

KEY CONCEPT A Acid and base ionization constants

The ionization of an acid in water and of a base in water can both be described by equilibrium constants. The symbol K_a is used for acid ionization constants, and K_b for base ionization constants. With the rule that the concentration of water is never used in the ionization equilibrium expression when water is the solvent, the following can be written:

Ionization of the acid HBrO:

$$HBrO(aq) + H_2O(l) \rightleftharpoons H_3O^+(aq) + BrO^-(aq) \qquad K_a = \frac{[H_3O^+][BrO^-]}{[HBrO]}$$

Ionization of the base triethylamine:

$$(C_2H_5)_3N(aq) + H_2O(l) \rightleftharpoons (C_2H_5)_3NH^+(aq) + OH^-(aq) \qquad K_b = \frac{[(C_2H_5)_3NH^+][OH^-]}{[(C_2H_5)_3N]}$$

The size of the ionization constant gives us the relative strength of an acid or base. A larger ionization constant means that relatively more products are formed. An acid with a larger ionization constant is therefore a stronger proton donor and a stronger acid than one with a smaller ionization constant. Similarly, a base with a larger ionization constant is a stronger proton acceptor and a stronger base than one with a smaller ionization constant. The larger the ionization constant the stronger the acid (or base).

EXAMPLE Using equilibrium constants to gauge the relative strengths of acids and bases

Two acids, formic acid (HCOOH) and hydrobromous acid (HBrO), and their ionization constants are given. Write the ionization equilibrium for each, the equilibrium expression for each, and state which is the stronger acid.

$$HCOOH: K_a = 1.8 \times 10^{-4}\ M \qquad HBrO: K_a = 2.0 \times 10^{-9}\ M$$

SOLUTION The ionization equilibrium and equilibrium expression for each are given (we must remember that the concentration of water is not included in the ionization equilibrium expression when water is a solvent):

$$HCOOH(aq) + H_2O(l) \rightleftharpoons H_3O^+(aq) + HCOO^-(aq) \qquad K_a = \frac{[H_3O^+][HCOO^-]}{[HCOOH]}$$

$$HBrO(aq) + H_2O(l) \rightleftharpoons H_3O^+(aq) + BrO^-(aq) \qquad K_a = \frac{[H_3O^+][BrO^-]}{[HBrO]}$$

The acid with the larger equilibrium constant is the stronger of the two. Since 1.8×10^{-4} is greater than 2.0×10^{-9}, formic acid is a stronger acid than hydrobromous acid.

EXERCISE Write the ionization equilibria and equilibrium expressions for the bases pyridine (C_5H_5N, $K_b = 1.8 \times 10^{-9}$ M) and ammonia (NH_3, $K_b = 1.8 \times 10^{-5}$ M). State which is the stronger base.

ANSWER: $C_5H_5N(aq) + H_2O(l) \rightleftharpoons C_5H_5NH^+(aq) + OH^-(aq)$

$$K_b = \frac{[C_5H_5NH^+][OH^-]}{[C_5H_5N]}$$

$NH_3(aq) + H_2O(l) \rightleftharpoons NH_4^+(aq) + OH^-(aq)$

$$K_b = \frac{[NH_4^+][OH^-]}{[NH_3]}$$

Ammonia is the stronger base.

KEY CONCEPT B pK_a, pK_b, and pK_w

The so-called p function is quite common in the sciences. For any property or value,

$$p(\text{anything}) = -\log_{10}(\text{anything})$$

If we want to determine the p of some value, we first take the logarithm (to the base 10) of the value and then change the sign of the result. Thus,

$$pK_w = -\log K_w$$
$$pK_a = -\log K_a$$
$$pK_b = -\log K_b$$

For most calculators, the logarithm to the base 10 of a number is found by entering the number and pushing the log button. A few examples are

$$pK_w = -\log(1.0 \times 10^{-14}) = -(-14.00) = 14.00$$
$$pK_a(HCOOH) = -\log(1.8 \times 10^{-4}) = -(-3.74) = 3.74$$
$$pK_b(C_5H_5N) = -\log(1.8 \times 10^{-9}) = -(-8.74) = 8.74$$

Because $K_a \times K_b = K_w$ for the ionization constant of an acid (K_a) and the ionization constant of its conjugate base (K_b), the following relation holds:

$$pK_a(\text{conjugate acid}) + pK_b(\text{conjugate base}) = 14.00$$

EXAMPLE Calculating the pK_b of a conjugate base

What is the pK_a of the acid $C_6H_5NH^+$?

SOLUTION $C_6H_5NH^+$ is the conjugate acid of C_6H_5N (pyridine). The pK_b of pyridine is 8.74. Thus,

$$pK_a(C_6H_5NH^+) + pK_b(C_6H_5N) = 14.00$$
$$pK_a(C_6H_5NH^+) + 8.74 = 14.00$$
$$pK_a(C_6H_5NH^+) = 5.26$$

EXERCISE What is the pK_b of ClO^-? Use Table 14.6 of the text.

ANSWER: 6.47

KEY CONCEPT C The strength of conjugate acids and bases

If an acid HA is a strong acid, this means it easily donates H^+. Thus, A^- does not hold on to the H^+ very well; this means that A^- is a weak base. Similarly, if HA is a weak acid, it does not readily donate H^+. Thus, A^- holds on to the H^+ very strongly; this means that A^- is a strong base. Thus, for a conjugate acid–base pair:

- A strong conjugate acid has a weak conjugate base.
- A weak conjugate acid has a strong conjugate base.

The terms *weak* and *strong* are qualitative. We can quantify this idea by noting that because $pK_a + pK_b = 14.00$ for a conjugate acid–base pair:

- If a conjugate acid has a small pK_a (is a strong acid), the conjugate base must have a large pK_b (is a weak base).
- If a conjugate acid has a large pK_a (is a weak acid), the conjugate base must have a small pK_b (is a strong base).

EXAMPLE Judging the strength of conjugate acids and bases

Sulfurous acid, H_2SO_3, is a stronger acid than nitrous acid, HNO_2. Which is the stronger base, HSO_3^- or NO_2^-? Do not refer to Table 14.6 of the text.

SOLUTION Because H_2SO_3 is a stronger acid than HNO_2, NO_2^- holds on to protons more strongly than HSO_3^-; therefore, NO_2^- is a stronger proton acceptor than HSO_3^-, which means that NO_2^- is a stronger base.

EXERCISE The trimethylammonium ion $(CH_3)_3NH^+$ is a stronger acid than the dimethylammonium ion $(CH_3)_2NH_2^+$. Write the formula of the weaker conjugate base of these two acids. Do not refer to Table 14.6 of the text.

ANSWER: $(CH_3)_3N$

14.8 WEAK ACIDS AND BASES

KEY CONCEPT Limiting strengths in water

Some acids are so strong that when they are placed in water, they completely donate their protons to water to form H_3O^+:

$$HA(aq) + H_2O(l) \xrightarrow{100\%} H_3O^+(aq) + A^-(aq)$$

In fact, this will occur for any acid HA in which H_2O is a better proton acceptor than A^-. Thus, the proton-accepting ability of water determines whether an acid (in aqueous solution) is a strong acid or a weak acid. If H_2O is a better proton acceptor than A^-, then the acid is a strong acid; if H_2O is a weaker proton acceptor than A^-, then the acid will not donate all its protons to H_2O, and HA is a weak acid.

A similar argument can be used with bases. Some bases are so strong that when they are placed in water, they are able to completely strip protons from water to form OH^-:

$$B(aq) + H_2O(l) \xrightarrow{100\%} BH^+(aq) + OH^-(aq)$$

In fact, this will occur for any base B for which BH^+ is a weaker proton donor than H_2O. Thus, the proton-donating ability of water determines whether a base (in aqueous solution) is a strong base or a weak base. If BH^+ is a weaker proton donor than H_2O, then B is a strong base; if BH^+ is a stronger proton donor than H_2O, then B will not accept protons so strongly from H_2O, and B is a weak base.

The important point here is that the properties of the solvent play an important role in determining the strengths of acids and bases dissolved in the solvent. In water, for instance, all strong acids form H_3O^+ and all strong bases form OH^-. Water, in a sense, makes all strong acids appear the same (as H_3O^+ formers) and all strong bases appear the same (as OH^- formers), so this effect is called the **leveling effect** of water.

EXAMPLE Determining whether an acid or base is weak or strong

When acid HA is placed in water, it undergoes 100% ionization and completely forms H_3O^+ and A^-; when acid HA′ is placed in water, it also undergoes 100% ionization. Because both behave the same in water, is it correct to conclude that HA and HA′ have the same acid strength?

SOLUTION No. The fact that both acids behave the same in water shows that they are both strong enough acids to donate their protons completely to H_2O. However, it is possible to imagine a solvent (Sol) that is not as good a proton acceptor as water; in such a solvent, the acids are not 100% ionized, and therefore, their proton-donating abilities are differentiated:

$$HA + Sol \underset{}{\overset{100\%}{\rightleftharpoons}} SolH^+ + A^-$$

$$HA' + Sol \underset{}{\overset{5\%}{\rightleftharpoons}} SolH^+ + A'^-$$

EXERCISE Suppose you have two bases, B and B′, which are both strong bases in water, and you would like to see if they have different base strengths in another solvent. Would you choose a solvent whose protonated form is a better proton donator than protonated water or one that is worse?

ANSWER: Worse

14.9 THE STRUCTURE AND STRENGTHS OF ACIDS

KEY CONCEPT The strength of binary and oxoacids

In binary acids (HX), two features dominate the determination of acid strength: the polarity of the HX bond and the strength of the HX bond:

- Along a period, the more polar the H—X bond, the stronger the acid.
- In a group, the weaker the H—X bond, the stronger the acid.

In oxoacids, the two features that dominate the determination of acid strength are the oxidation number of the central atom and the electronegativity of the central atom:

- In different oxoacids that have the same central atom but a different number of oxygen atoms surrounding the central atom, the higher the oxidation number of the central atom, the stronger the acid. (It should be remembered that the more oxygen atoms around the central atom, the higher its oxidation number.)
- In different oxoacids that have the same number of oxygen atoms surrounding the central atom, the higher the electronegativity of the central atom, the stronger the acid.

EXAMPLE Predicting the strengths of acids

Which compound is the stronger acid, PH_3 or H_2S?

SOLUTION Because phosphorus and sulfur are in the same period, we expect that the polarity of the H—X bond will largely determine acid strengths. Because sulfur is more electronegative than phosphorus, the H—S bond will be more polar than the H—P bond; thus, we (correctly) predict that H_2S is the stronger acid. In fact, PH_3 is neither an acid nor a base.

EXERCISE $HBrO_2$ has not yet been discovered. When (and if) it is, predict whether it should be a stronger or weaker acid than $HClO_2$?

ANSWER: Weaker

14.10 THE pH OF WEAK ACIDS AND BASES

KEY CONCEPT Calculating the pH of weak acids and bases

The pH of a weak acid is calculated from the equilibrium H_3O^+ ion concentration and the definition $pH = -\log [H_3O^+]$. The pH of a weak base is calculated from the equilibrium OH^- ion concentration and the relations $pOH = -\log [OH^-]$ and $pH + pOH = 14.00$. For any base in which we know $[OH^-]$, we may equivalently use the relations $[H_3O^+][OH^-] = 1.0 \times 10^{-14}$ and $pH = -\log [H_3O^+]$ to determine the pH. In any case, if we are given the initial concentration of acid or base, it is necessary first to solve an equilibrium problem to get the equilibrium concentration of H_3O^+ or OH^-

▼ EXAMPLE 1 Calculating the pH

What is the pH of a 0.050 M solution of hypochlorous acid, HOCl? $K_a = 3.0 \times 10^{-5}$ M.

SOLUTION We first recognize that when 1 mol of HClO ionizes, it forms 1 mol of H_3O^+ and 1 mol of ClO^-. Thus, if we represent the amount of HClO that ionizes by $-x$, the amount of H_3O^+ that forms is $+x$, and the amount of ClO^- that forms is $+x$. The table of concentrations and changes is

	$HClO(aq)$ + $H_2O(l)$ $\rightleftharpoons$	$H_3O^+(aq)$ +	$ClO^-(aq)$
initial concentration	0.050 M	0	0
change to reach equilibrium	$-x$	$+x$	$+x$
equilibrium concentration	0.050 M $- x$	x	x

In setting up the table, we have assumed that the initial 1.0×10^{-7} M concentration of H_3O^+ (from the autoionization of water) is so small relative to the amount of H_3O^+ formed by ionization of the acid that we may consider the initial concentration of H_3O^+ to be 0. The equilibrium expression is

$$\frac{[H_3O^+][ClO^-]}{[HClO]} = K_a$$

Substituting the equilibrium concentrations and the value of K_a gives

$$\frac{(x)(x)}{0.050\ \text{M} - x} = 3.0 \times 10^{-5}\ \text{M}$$

We now assume that x is much smaller than 0.050 ($x \ll 0.050$), so $0.050 - x = 0.050$.

$$\frac{x^2}{0.050\ \text{M}} = 3.0 \times 10^{-5}\ \text{M}$$

$$x^2 = 1.5 \times 10^{-6}\ \text{M}^2$$

$$x = 1.2 \times 10^{-3}\ \text{M}$$

Before proceeding, we check to see that the assumptions made regarding the size of x are appropriate. Because x is approximately 10,000 times larger than 1.0×10^{-7}, neglecting 1.0×10^{-7} relative to x (i.e., setting the original concentration of H_3O^+ equal to 0 in the table) is justified. Second, x is 2.4% of 0.050:

$$\frac{1.2 \times 10^{-3}}{0.050} \times 100 = 2.4\%$$

As a rule of thumb, as long as x is not greater than 5% of the original concentration of 0.050, the approximation is justified (see the next pitfall).

Recognizing that $[H_3O^+] = x$, we get

$$[H_3O^+] = 1.2 \times 10^{-3}\ \text{M}$$

$$\text{pH} = -\log [H_3O^+]$$

$$= -\log (1.2 \times 10^{-3})$$

$$= -(-2.92)$$

$$= 2.92$$

EXERCISE What is the pH of a 0.33 M solution of HF(aq)? $K_a(\text{HF}) = 3.5 \times 10^{-4}$ M.

▲ ANSWER: 1.98

PITFALL When are approximations about one number being smaller than another correct?

We often make approximations such as

$$x \text{ is much smaller than } 0.23 \text{, so that}$$
$$0.23 - x = 0.23$$

How do we judge whether such an approximation is justified? In general, calculations involving equilibria as we do them may be as much as 5–10% off the correct values because of effects that we do not take into account. Thus, as long as any approximations do not introduce an error greater than 5%, the approximation is justified. Practically, if we say a "large value" is much greater than a "small value", so that

$$\text{Large value} - \text{small value} = \text{large value}$$

the approximation is justified if

$$\frac{\text{Small value}}{\text{Large value}} \times 100 < 5\%$$

EXAMPLE 2 Calculating the pH

What is the pH of a 0.60 M $(CH_3)_3N(aq)$ solution? $K_b\{(CH_3)_3N\} = 6.5 \times 10^{-5}$ M.

SOLUTION The equilibrium and table of concentration values for this problem are

	$(CH_3)_3N(aq)$ + $H_2O(l)$ $\rightleftharpoons$	$(CH_3)_3NH^+(aq)$ +	$OH^-(aq)$
initial concentration	0.60 M	0	0
change to reach equilibrium	$-x$	$+x$	$+x$
equilibrium concentration	0.60 M $- x$	x	x

As we did earlier with H_3O^+, we assume the amount of OH^- formed by the autoionization of water is too small to be of any importance, and we consider the initial concentration of OH^- to be 0 M. The equilibrium expression is

$$\frac{[(CH_3)_3NH^+][OH^-]}{[(CH_3)_3N]} = K_b$$

Substituting the equilibrium values and the value of K_b gives

$$\frac{(x)(x)}{0.60 \text{ M} - x} = 6.5 \times 10^{-5} \text{ M}$$

We now assume that x is much smaller than 0.60 ($x \ll 0.60$) so that $0.60 - x = 0.60$.

$$\frac{x^2}{0.60 \text{ M}} = 6.5 \times 10^{-5} \text{ M}$$
$$x^2 = 3.9 \times 10^{-5} \text{ M}^2$$
$$x = 6.2 \times 10^{-3} \text{ M}$$

x is large with respect to 1.0×10^{-7} and small with respect to 0.60 (x is 1.0% of 0.60), so our approximations are acceptable. Remembering that $x = [OH^-]$, we get

$$[OH^-] = 6.2 \times 10^{-3}$$
$$\text{pOH} = -\log [OH^-]$$
$$= -\log 6.2 \times 10^{-3}$$
$$= -(-2.21)$$
$$= +2.21$$

We finally calculate the pH:

$$\begin{aligned} pH + pOH &= 14.00 \\ pH + 2.21 &= 14.00 \\ pH &= 14.00 - 2.21 \\ &= 11.79 \end{aligned}$$

EXERCISE What is the pH of a 0.080 M $NH_2NH_2(aq)$ solution? pK_b $(NH_2NH_2) = 1.7 \times 10^{-6}$ M.

ANSWER: 10.57

14.11 POLYPROTIC ACIDS AND BASES

KEY CONCEPT A Polyprotic acids

Many Brønsted acids are able to donate more than one proton. Such acids are called polyprotic (many proton) acids. A **diprotic acid** is one that can donate two protons; carbonic acid H_2CO_3, sulfuric acid H_2SO_4, and oxalic acid HOOCCOOH are examples of diprotic acids. Polyprotic acids ionize in a stepwise fashion. First one proton ionizes, then a second, then a third, and so on. Each ionization step has a smaller equilibrium constant than the previous one. For instance, for carbonic acid,

$$H_2CO_3(aq) + H_2O(l) \rightleftharpoons H_3O^+(aq) + HCO_3^-(aq) \qquad K_{a1} = 4.3 \times 10^{-7}\ M$$

$$HCO_3^-(aq) + H_2O(l) \rightleftharpoons H_3O^+(aq) + CO_3^{2-}(aq) \qquad K_{a2} = 5.6 \times 10^{-11}\ M$$

Because of the large difference in K_{a1} and K_{a2} for most diprotic acids, the following approximations may be used when doing equilibrium calculations for them. For a diprotic acid H_2A:

1. The $H_3O^+(aq)$ and $HA^-(aq)$ equilibrium concentrations are determined by the first ionization step only. Thus, to determine the pH, we need only consider the first ionization step.
2. $[H_3O^+] = [HA^-]$ at equilibrium. This conclusion is a direct result of approximation 1.
3. $[A^{2-}] = K_{a2}$. This conclusion is a direct result of approximation 2.

One important exception to these rules occurs for H_2SO_4, in which the first ionization step is 100% complete. The second step has a relatively large equilibrium constant and does contribute H_3O^+ ions; it must be considered in calculating the pH. Also, for any acid, these approximations will not give the correct results if the concentration of acid is extremely low; in such a case, a more rigorous approach must be used.

EXAMPLE Calculating the pH of a sulfuric acid (H_2SO_4) solution

What is the pH of a 0.020 M $H_2SO_4(aq)$ solution?

SOLUTION The first ionization of $H_2SO_4(aq)$ is 100% complete. For 0.020 M H_2SO_4, this leads to 0.020 M $H_3O^+(aq)$ and 0.020 M $HSO_4^-(aq)$. In the second ionization step, the 0.020 M HSO_4^- ionizes with $K_{a2} = 0.012$ M. The equilibrium and table of concentration values are

	$HSO_4^-(aq)$ + $H_2O(l)$ ⇌	$H_3O^+(aq)$ +	$SO_4^{2-}(aq)$
initial concentration	0.020 M	0.020 M	0
change to reach equilibrium	$-x$	$+x$	$+x$
equilibrium concentration	0.020 M $- x$	0.020 M $+ x$	x

There is a new feature in this table. One of the products (H_3O^+) starts out with a nonzero concentration. The first ionization produces 0.020 M H_3O^+, which is present when the second ionization occurs. The equilibrium expression is

$$\frac{[H_3O^+][SO_4{}^{2-}]}{[HSO_4{}^-]} = K_{a2}$$

Substituting the equilibrium concentrations and the value of K_{a2} gives

$$\frac{(0.020\text{ M} + x)(x)}{(0.020\text{ M} - x)} = 0.012$$

For this example, we cannot make the assumption that x is much smaller than 0.020. Because K_{a2} is relatively large, considerable ionization occurs, and x will turn out to be a significant fraction of 0.020. We must solve for x without any approximations. Multiplying both sides of the equation by $(0.020 - x)$ gives

$$(0.020 + x)(x) = 0.012(0.020 - x)$$
$$0.020x + x^2 = 2.4 \times 10^{-4} - 0.012x$$

Subtracting $(2.4 \times 10^{-4} - 0.012x)$ from both sides of the equation gives

$$0.020x + x^2 - (2.4 \times 10^{-4} - 0.012x) = 0$$
$$x^2 + 0.032x - 2.4 \times 10^{-4} = 0$$

The equation is now in the form $ax^2 + bx + c = 0$, and the quadratic formula may be used to solve it:

$$x = \frac{-b \pm \sqrt{b^2 - 4ac}}{2a}$$

For $a = 1$, $b = 0.032$, and $c = -2.4 \times 10^{-4}$, we get

$$x = \frac{-0.032 \pm \sqrt{0.032^2 - (4)(1)(-2.4 \times 10^{-4})}}{2(1)}$$
$$= \frac{-0.032 \pm \sqrt{0.00102 + 0.00096}}{2}$$
$$= \frac{-0.032 \pm \sqrt{0.00198}}{2}$$
$$= \frac{-0.032 \pm 0.044}{2}$$

The two roots are

$$x = \frac{-0.032 - 0.044}{2} = -0.038 \qquad x = \frac{-0.032 + 0.044}{2} = 0.0060$$

The negative root would correspond to a negative molarity, which is impossible. This root has no physical meaning and is ignored. Thus, $x = 0.0060$, and from the table of concentrations,

$$[H_3O^+] = 0.020\text{ M} + x$$
$$= 0.020\text{ M} + 0.0060\text{ M}$$
$$= 0.026\text{ M}$$
$$\text{pH} = -\log[H_3O^+]$$
$$= -\log(0.026)$$
$$= 1.59$$

EXERCISE What is the pH of a 0.0040 M $H_2SO_4(aq)$ solution?

ANSWER: 2.19

KEY CONCEPT B Carbonic acid

Carbonic acid (H_2CO_3) is a diprotic acid with the following ionization equilibria and acid ionization constant:

$$H_2CO_3(aq) + H_2O(l) \rightleftharpoons H_3O^+(aq) + HCO_3^-(aq) \qquad K_{a1} = 4.3 \times 10^{-7}\ \mathrm{M}$$
$$HCO_3^-(aq) + H_2O(l) \rightleftharpoons H_3O^+(aq) + CO_3^{2-}(aq) \qquad K_{a2} = 5.6 \times 10^{-11}\ \mathrm{M}$$

When CO_2 dissolves in water, some of it reacts to form carbonic acid:

$$CO_2(aq) + H_2O(l) \rightleftharpoons H_2CO_3(aq)$$

Because approximately 1 molecule in every 480 undergoes this conversion, we can write the following equilibrium and ionization constant:

$$CO_2(aq) + 2H_2O(l) \rightleftharpoons H_3O^+(aq) + HCO_3^-(aq) \qquad K_{a1} = 2.0 \times 10^{-4}\ \mathrm{M}$$

EXAMPLE Calculating pH

Calculate the pH, HCO_3^- concentration, and CO_3^{2-} concentration of a 0.030 M solution of CO_2 in water at 25°C.

SOLUTION The equilibrium and table of concentration values for this problem is

	$CO_2(aq)$	$+ 2H_2O(l) \rightleftharpoons$	$H_3O^+(aq)$	$+ HCO_3^-(aq)$
initial concentration	0.030 M		0	0
change to reach equilibrium	$-x$		$+x$	$+x$
equilibrium concentration	0.030 M $- x$		x	x

The equilibrium expression is

$$\frac{[H_3O^+][HCO_3^-]}{[CO_2]} = K_{a1}$$

Substituting the equilibrium values from the table and the value of K_{a1} gives

$$\frac{(x)(x)}{0.030\ \mathrm{M} - x} = 2.0 \times 10^{-4}\ \mathrm{M}$$

We assume that x is much smaller than 0.030 ($x \ll 0.030$), so that $0.030 - x = 0.030$.

$$\frac{x^2}{0.030\ \mathrm{M}} = 2.0 \times 10^{-4}\ \mathrm{M}$$
$$x^2 = (0.030\ \mathrm{M})(2.0 \times 10^{-4}\ \mathrm{M})$$
$$= 6.0 \times 10^{-6}\ \mathrm{M}^2$$
$$x = 2.4 \times 10^{-3}\ \mathrm{M}$$

This value of x equals the H_3O^+ concentration and the HCO_3^- concentration:

$$[H_3O^+] = 2.4 \times 10^{-3}\ \mathrm{M}$$
$$\mathrm{pH} = -\log [H_3O^+]$$
$$= -\log (2.4 \times 10^{-3})$$
$$= -(-2.62)$$
$$= 2.62$$

To calculate the concentration of CO_3^{2-}, we consider the second ionization step:

$$HCO_3^-(aq) + H_2O(l) \rightleftharpoons H_3O^+(aq) + CO_3^{2-}(aq) \qquad K_{a2} = 5.6 \times 10^{-11}\text{ M}$$

The equilibrium expression for this ionization is

$$\frac{[H_3O^+][CO_3^{2-}]}{[HCO_3^-]} = K_{a2} = 5.6 \times 10^{-11}\text{ M}$$

Before the second ionization, $[HCO_3^-] = 2.4 \times 10^{-3}$ M and $[H_3O^+] = 2.4 \times 10^{-3}$ M. Because K_{a2} is very small, we assume that very little HCO_3^- ionizes and therefore that very little H_3O^+ is formed in the second ionization step. Thus, the concentrations of HCO_3^- and H_3O^+ do not change as a result of the second ionization; they stay equal to each other. The equilibrium expression for the second ionization then becomes

$$\frac{[CO_3^{2-}](2.4 \times 10^{-3}\text{ M})}{(2.4 \times 10^{-3}\text{ M})} = 5.6 \times 10^{-11}\text{ M}$$

$$[CO_3^{2-}] = 5.6 \times 10^{-11}\text{ M} \qquad (= K_{a2})$$

This reasoning is used for many diprotic acids and leads to the conclusion that for many diprotic acids H_2A, the equilibrium value of $[A^{2-}]$ equals K_{a2}.

EXERCISE What is the pH, HS^- concentration, and S^{2-} concentration of a 0.10 M $H_2S(aq)$ solution? $K_{a1} = 9.1 \times 10^{-8}$ M, $K_{a2} = 4.8 \times 10^{-11}$.

ANSWER: pH = 4.02; $[HS^-] = 9.5 \times 10^{-5}$ M; $[S^{2-}] = 4.8 \times 10^{-11}$ M

DESCRIPTIVE CHEMISTRY TO REMEMBER

- **Universal indicator paper** turns different colors at different pH values.
- The United States Environmental Protection Agency (EPA) defines waste as *corrosive* if its **pH** is either lower than 3.0 or higher than 12.5.
- The natural acidity of a stream is generally due to the presence of **carbonic acid** (H_2CO_3), **hydrogen phosphate** ions ($H_2PO_4^-$), or acids from plant tissue degradation.
- The smell of dead fish is due to **amines** (RNH_2).
- All **carboxylic acids** (acids containing the COOH group) are weak acids in water.
- **Ethanol,** CH_3CH_2OH, is the alcohol produced from the fermentation of grains such as wheat, barley, and rye. Its pK_a is 16, so it is not considered an acid.

CHEMICAL EQUATIONS TO KNOW

- According to Brønsted–Lowry theory, all acid–base neutralizations in aqueous solution are summarized by the reaction between hydronium ion and hydroxide ion to form water.

$$H_3O^+(aq) + OH^-(aq) \longrightarrow 2H_2O(l)$$

- The amide ion (NH_2^-) is a strong base and completely reacts with water to form hydroxide ion.

$$NH_2^-(aq) + H_2O(l) \longrightarrow NH_3(aq) + OH^-(aq)$$

- The hydride ion (H^-) is a strong base and completely reacts with water to form hydroxide ion.

$$H^-(aq) + H_2O(l) \longrightarrow H_2(g) + OH^-(aq)$$

- The oxide ion (O^{2-}) is a strong base and completely reacts with water to form hydroxide ion.

$$O^{2-}(aq) + H_2O(l) \longrightarrow 2OH^-(aq)$$

- Liquid ammonia is amphiprotic and undergoes autoionization.

$$2NH_3(l) \rightleftharpoons NH_4^-(am) + NH_2^-(am) \qquad pK_{am} = 33$$

- Acetic acid is amphiprotic and undergoes autoionization.

$$2CH_3COOH(l) \rightleftharpoons CH_3C(OH)_2^+(ac) + CH_3CO_2^-(ac) \qquad pK_{ac} = 12.6$$

- Adding acid to hydrogen carbonate ion shifts the following equilibrium to the left.

$$H_2CO_3(aq) + H_2O(l) \rightleftharpoons H_3O^+(aq) + HCO_3^-(aq)$$

The carbonic acid thus formed can then decompose to water and carbon dioxide.

$$H_2CO_3(aq) \rightleftharpoons H_2O(l) + CO_2(g)$$

MATHEMATICAL EQUATIONS TO KNOW AND UNDERSTAND

$pK_a = -\log K_a$ | $pK_a + pK_b = 14.00$

$pK_b = -\log K_b$ | $pOH = -\log [OH^-]$

$pH = -\log [H_3O^+]$ | $pH + pOH = 14.00$

SELF-TEST EXERCISES

The definitions of acids and bases

1. What is the Brønsted acid in the reaction $H_2S(g) + 2NH_3(g) \rightleftharpoons (NH_4)_2S(s)$?
(a) H_2S (b) NH_3 (c) NH_4^+ (d) S^{2-}

2. Which of the following can act both as a Brønsted acid and a Brønsted base?
(a) CO_2^{2-} (b) CH_4 (c) H_2O (d) H_2 (e) NH_4^+

3. What are the products of the Brønsted acid–base reaction: $HCN(aq) + H_2O(l) \rightleftharpoons ?$
(a) H_2CN^+, OH^- (b) $HCNO, H_2$ (c) CN^-, H_3O^+
(d) H_2, O^{2-} (e) CO_2, NO_2, H_3O^+

4. What are the products of thc Brønsted acid–base reaction $HS^-(aq) + OH^-(aq) \rightleftharpoons ?$
(a) H_2S, O^{2-} (b) S^{2-}, H_2O (c) H_2O_2, S_8 (d) SO_2, H_2O (e) H_2SO_4

5. What is the formula of the characteristic cationic acid formed in liquid (not aqueous) acetic acid, CH_3CO_2H?
(a) H_3O^+ (b) H^+ (c) $CH_3CO_2H^+$ (d) $H_9O_4^+$ (e) $CH_3CO_2H_2^+$

6. Which of the following is a conjugate acid–base pair in the following equilibrium?

$$F^-(aq) + H_2O(l) \rightleftharpoons HF(aq) + OH^-(aq)$$

(a) F^-, HF (b) HF, H_2O (c) F^-, H_2O (d) OH^-, HF (e) OH^-, F^-

7. Which of the following is the conjugate acid of NH_3?
(a) NH_2^- (b) NH_4^+ (c) HCl (d) H_3O^+ (e) H_2O

8. Which of the following is an amphiprotic solvent?
(a) CCl_4 (b) $POCl_3$ (c) H_2SO_4 (d) SO_2 (e) molten NaCl

9. Two of the equilibria shown are autoionizations. Which one is not?
(a) $H_2O + H_2O \rightleftharpoons H_2O_2 + H_2$ (b) $H_2O + H_2O \rightleftharpoons H_3O^+ + OH^-$
(c) $NH_3 + NH_3 \rightleftharpoons NH_4^+ + NH_2^-$

10. What is $[OH^-]$ in an aqueous solution with $[H_3O^+] = 4.7 \times 10^{-2}$ M?
(a) 1.0×10^{-7} M (b) 4.7×10^{-16} M (c) 3.3×10^{-5} M
(d) 2.1×10^{-13} M (e) 6.8×10^{-12} M

11. What is $[H_3O^+]$ in an aqueous solution with $[OH^-] = 3.8 \times 10^{-6}$ M?
(a) 6.1×10^{-5} M (b) 1.0×10^{-7} M (c) 8.1×10^{-11} M
(d) 3.8×10^{-8} M (e) 2.6×10^{-9} M

12. The autoionization constant, K_w, for pure water at 40°C is 2.9×10^{-14} M? What is $[H_3O^+]$ in pure water at 40°C?
(a) 1.0×10^{-7} M (b) 3.5×10^{-7} M (c) 1.7×10^{-7} M
(d) 2.9×10^{-7} M (e) 6.1×10^{-7} M

Solutions of strong acids and bases

13. A 0.0050 M solution of H_2SO_4 is a _______ solution of a _______ acid.
(a) dilute, weak (b) dilute, strong (c) concentrated, strong (d) concentrated, weak

14. What is the hydronium ion concentration of a 0.0052 M nitric acid (HNO_3) solution?
(a) 1.9×10^{-12} M (b) 1.0×10^{-7} M (c) 0.0104 M (d) 0.0052 M (e) 0.0026 M

15. What is $[H_3O^+]$ in a 0.025 M sodium hydroxide (NaOH) solution?
(a) 0.025 M (b) 4.0×10^{-13} M (c) 0.10 M
(d) 1.9×10^{-12} M (e) 1.0×10^{-7} M

16. What is the pH of lemon juice with a hydronium ion concentration of 2.40×10^{-3} M?
(a) 11.38 (b) 3.46 (c) 2.62 (d) 7.00 (e) 8.61

17. What is the pH of seawater that has a hydroxide ion concentration of 1.4×10^{-5} M?
(a) 5.22 (b) 4.85 (c) 8.78 (d) 9.15 (e) 7.00

18. Household ammonia typically has a pH of about 12.0 which classifies it as
(a) neutral. (b) acidic. (c) basic.

19. What is the hydronium ion concentration of a solution with pH = 4.40?
(a) 4.0×10^{-5} M (b) 1.0×10^{-7} M (c) 2.6×10^{-10} M
(d) 6.1×10^{-6} M (e) 2.6×10^{4} M

20. What is the hydroxide ion concentration of a solution with pH = 5.80?
(a) 6.3×10^{-9} M (b) 4.0×10^{-5} M (c) 1.0×10^{-7} M
(d) 1.6×10^{-6} M (e) 6.6×10^{-7} M

21. Which pH corresponds to the solution with the highest hydroxide ion concentration?
(a) 6.24 (b) 10.16 (c) 2.68 (d) 7.00

22. What is the pH of a 2.5×10^{-4} M KOH solution?
(a) 9.35 (b) 4.65 (c) 3.60 (d) 7.00 (e) 10.40

23. What is the pH of a 0.00183 M $HClO_4$ solution?
(a) 11.26 (b) 7.00 (c) −2.74 (d) 2.74 (e) −11.26

24. What is the pH of the solution that results when 10.0 mL of 0.0835 M HCl and 10.0 mL of 0.0850 M NaOH are mixed?
(a) 10.81 (b) 1.07 (c) 7.00 (d) 12.93 (e) 3.12

25. What is the pH of the solution that results when 12.5 mL of 0.0962 M HNO_3 and 5.75 mL of 0.102 M $Ba(OH)_2$ are mixed?
(a) 11.20 (b) 7.00 (c) 2.80 (d) 6.11 (e) 7.89

26. How many milliliters of 0.0250 M NaOH are required to neutralize 10.0 mL of 0.108 M HBr?
(a) 12.6 (b) 43.2 (c) 8.44 (d) 21.1 (e) 2.31

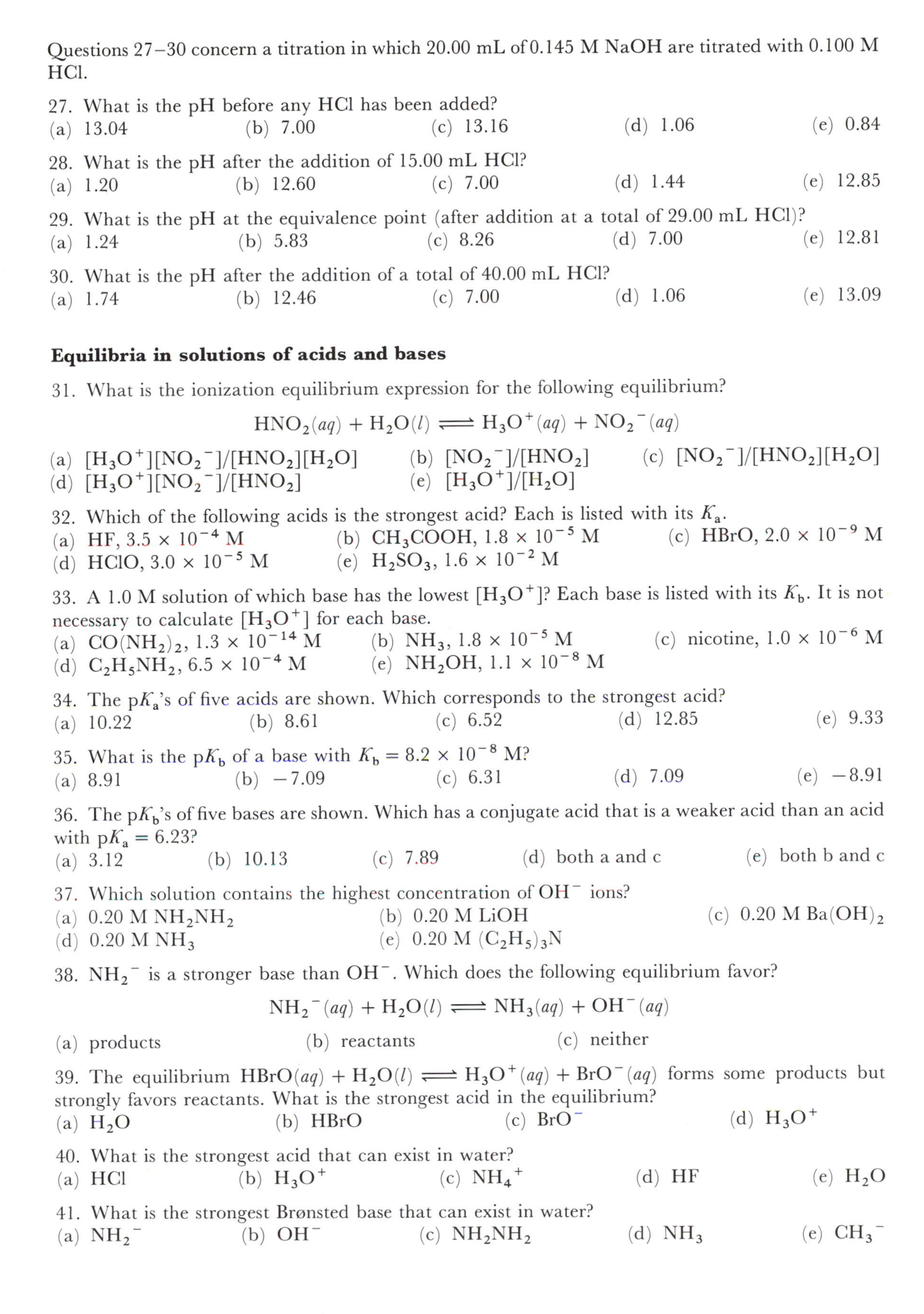

Questions 27–30 concern a titration in which 20.00 mL of 0.145 M NaOH are titrated with 0.100 M HCl.

27. What is the pH before any HCl has been added?
(a) 13.04 (b) 7.00 (c) 13.16 (d) 1.06 (e) 0.84

28. What is the pH after the addition of 15.00 mL HCl?
(a) 1.20 (b) 12.60 (c) 7.00 (d) 1.44 (e) 12.85

29. What is the pH at the equivalence point (after addition at a total of 29.00 mL HCl)?
(a) 1.24 (b) 5.83 (c) 8.26 (d) 7.00 (e) 12.81

30. What is the pH after the addition of a total of 40.00 mL HCl?
(a) 1.74 (b) 12.46 (c) 7.00 (d) 1.06 (e) 13.09

Equilibria in solutions of acids and bases

31. What is the ionization equilibrium expression for the following equilibrium?

$$HNO_2(aq) + H_2O(l) \rightleftharpoons H_3O^+(aq) + NO_2^-(aq)$$

(a) $[H_3O^+][NO_2^-]/[HNO_2][H_2O]$ (b) $[NO_2^-]/[HNO_2]$ (c) $[NO_2^-]/[HNO_2][H_2O]$
(d) $[H_3O^+][NO_2^-]/[HNO_2]$ (e) $[H_3O^+]/[H_2O]$

32. Which of the following acids is the strongest acid? Each is listed with its K_a.
(a) HF, 3.5×10^{-4} M (b) CH_3COOH, 1.8×10^{-5} M (c) HBrO, 2.0×10^{-9} M
(d) HClO, 3.0×10^{-5} M (e) H_2SO_3, 1.6×10^{-2} M

33. A 1.0 M solution of which base has the lowest $[H_3O^+]$? Each base is listed with its K_b. It is not necessary to calculate $[H_3O^+]$ for each base.
(a) $CO(NH_2)_2$, 1.3×10^{-14} M (b) NH_3, 1.8×10^{-5} M (c) nicotine, 1.0×10^{-6} M
(d) $C_2H_5NH_2$, 6.5×10^{-4} M (e) NH_2OH, 1.1×10^{-8} M

34. The pK_a's of five acids are shown. Which corresponds to the strongest acid?
(a) 10.22 (b) 8.61 (c) 6.52 (d) 12.85 (e) 9.33

35. What is the pK_b of a base with $K_b = 8.2 \times 10^{-8}$ M?
(a) 8.91 (b) −7.09 (c) 6.31 (d) 7.09 (e) −8.91

36. The pK_b's of five bases are shown. Which has a conjugate acid that is a weaker acid than an acid with pK_a = 6.23?
(a) 3.12 (b) 10.13 (c) 7.89 (d) both a and c (e) both b and c

37. Which solution contains the highest concentration of OH^- ions?
(a) 0.20 M NH_2NH_2 (b) 0.20 M LiOH (c) 0.20 M $Ba(OH)_2$
(d) 0.20 M NH_3 (e) 0.20 M $(C_2H_5)_3N$

38. NH_2^- is a stronger base than OH^-. Which does the following equilibrium favor?

$$NH_2^-(aq) + H_2O(l) \rightleftharpoons NH_3(aq) + OH^-(aq)$$

(a) products (b) reactants (c) neither

39. The equilibrium $HBrO(aq) + H_2O(l) \rightleftharpoons H_3O^+(aq) + BrO^-(aq)$ forms some products but strongly favors reactants. What is the strongest acid in the equilibrium?
(a) H_2O (b) HBrO (c) BrO^- (d) H_3O^+

40. What is the strongest acid that can exist in water?
(a) HCl (b) H_3O^+ (c) NH_4^+ (d) HF (e) H_2O

41. What is the strongest Brønsted base that can exist in water?
(a) NH_2^- (b) OH^- (c) NH_2NH_2 (d) NH_3 (e) CH_3^-

42. Ethylamine, $C_2H_5NH_2$, is a base. What is the equilibrium that occurs when it is placed in water?
(a) $C_2H_5NH_2(aq) + H_2O(l) \rightleftharpoons H_3O^+(aq) + C_2H_5NH^-(aq)$
(b) $C_2H_5NH_2(aq) \rightleftharpoons NH_2^-(aq) + C_2H_5^+(aq)$
(c) $C_2H_5NH_2(aq) + H_2O(l) \rightleftharpoons OH^-(aq) + C_2H_5NH_3^+(aq)$
(d) $C_2H_5NH_2(aq) + H_2O(l) \rightleftharpoons OH^-(aq) + C_2H_5NH^+(aq) + H_2(g)$

43. Which is the strongest acid? X refers to the atom other than H in each compound.

	CH_4	NH_3	H_2O_2
H—X bond enthalpy	440 kJ/mol	356 kJ/mol	365 kJ/mol
Electronegativity of X	2.1	3.0	3.5

(a) H_2O_2 (b) CH_4 (c) NH_3

44. Which is predicted to be the strongest acid?
(a) $HMnO_4$ (b) H_3PO_4 (c) $HTcO_4$ (d) H_2TiO_4 (e) HIO_4

45. Which of the following is the strongest acid?
(a) HClO (b) $HClO_2$ (c) $HClO_3$ (d) $HClO_4$

46. Without referring to any references, predict which is the strongest carboxylic acid.
(a) $CHCl_2COOH$ (b) $CH_2ClCOOH$ (c) CH_3COOH (d) CCl_3COOH

47. What is the hydronium ion concentration of a 0.30 M HBrO solution?
(a) 4.5×10^{-5} M (b) 2.4×10^{-5} M (c) 1.0×10^{-7} M
(d) 6.0×10^{-10} M (e) 8.1×10^{-4} M

48. What is the pH of a 0.50 M HCN solution?
(a) 7.00 (b) 9.68 (c) 5.55 (d) 4.80 (e) 2.43

49. What is the pH of a 2.0 M hydrazine solution?
(a) 10.04 (b) 7.00 (c) 2.74 (d) 11.27 (e) 8.26

50. What is the pH of a 0.25 M ammonia solution?
(a) 11.32 (b) 9.55 (c) 7.00 (d) 1.64 (e) 8.47

51. What is the hydronium ion concentration of a 0.40 M iodic acid solution?
(a) 0.35 M (b) 0.19 M (c) 0.26 M (d) 0.40 M (e) 0.068 M

52. What is the pH of a 0.0010 M $H_2SO_4(aq)$ solution? $K_{a2} = 0.012$.
(a) 3.00 (b) 2.73 (c) 1.54 (d) 1.74 (e) 2.22

53. What is the pH of a 0.0025 M H_2CO_3 solution?
(a) 4.49 (b) 9.51 (c) 7.00 (d) 6.37 (e) 7.63

54. What is the concentration of A^{2-} in a 0.20 M solution of H_2A, where H_2A is an unknown acid with $pK_{a1} = 4.22$, $pK_{a2} = 8.15$.
(a) 3.5×10^{-3} M (b) 6.0×10^{-5} M (c) 8.1×10^{-10} M
(d) 7.1×10^{-9} M (e) 3.8×10^{-5} M

Descriptive Chemistry

55. Which of the following contributes to the natural acidity of streams?
(a) HCl (b) H_2SO_3 (c) $HClO_3$ (d) H_2CO_3 (e) HF

56. Which group is characteristic of carboxylic acids?
(a) $—NH_2$ (b) —OH (c) —COOH (d) $—PO_3$ (e) $—CH_3$

57. What species is formed when O_2^- is placed in water?
(a) O_2 (b) H_2O (c) OH^- (d) H_2O_2
(e) None of these; O_2^- simply dissolves.

58. Which reaction summarizes all acid–base reactions in water?
(a) $H_3O^+(aq) + OH^-(aq) \longrightarrow 2H_2O(l)$
(b) $NH_2^-(aq) + H_2O(l) \longrightarrow NH_3(aq) + OH^-(aq)$
(c) $H^-(aq) + H_2O(l) \longrightarrow H_2(g) + OH^-(aq)$
(d) $O^{2-}(aq) + H_2O(l) \longrightarrow 2OH^-(aq)$
(e) $H_3O^+(aq) + NH_2^-(aq) \longrightarrow NH_3(aq) + H_2O(l)$

15 ACIDS, BASES, AND SALTS

SALTS AS ACIDS AND BASES

KEY WORDS Define or explain each of the following terms in a written sentence or two.

common-ion effect mixed solution

15.1 IONS AS ACIDS AND BASES

KEY CONCEPT A Cations as acids

One way for a salt to change the pH of a solution is through the reaction of its cation. There are two common types of cation reactions that change the pH. In one, a cation such as a protonated amine reacts with water to form hydronium ion—for example,

$$NH_2NH_3{}^+(aq) + H_2O(l) \rightleftharpoons NH_2NH_2(aq) + H_3O^+(aq)$$
$$(C_2H_5)_3NH^+(aq) + H_2O(l) \rightleftharpoons (C_2H_5)_3N(aq) + H_3O^+(aq)$$

These are acid ionizations; they result in formation of H_3O^+. The pK_a of each can be calculated from the pK_b of the unprotonated amine, as listed in Table 14.6 of the text. In a second type of reaction, a hydrated metal cation loses a proton in an ionization:

$$[Al(H_2O)_6]^{3+}(aq) + H_2O(l) \rightleftharpoons [Al(H_2O)_5(OH)]^{2+}(aq) + H_3O^+(aq)$$
$$[Fe(H_2O)_6]^{2+}(aq) + H_2O(l) \rightleftharpoons [Fe(H_2O)_5(OH)]^{+}(aq) + H_3O^+(aq)$$

All metals cations except those from Group I, those from Group II, and those with +1 charge undergo such reactions. Table 15.1 of the text lists the pK_a values of some of these cations. Both types of equilibria increase the acidity of a solution by increasing the concentration of $H_3O^+(aq)$.

EXAMPLE 1 Predicting whether a salt solution is acidic, basic, or neutral

Predict whether an aqueous solution of hydrazinium bromide, NH_2NH_3Br, is acidic basic, or neutral.

SOLUTION We first must decide what ions are formed when the salt is dissolved in water. In this case the salt is made of $NH_2NH_3{}^+$ and Br^- ions. Br^- is the conjugate base of the strong acid HBr and is, therefore, a very weak base; it will not affect the pH of the solution. However, the hydrazinium ion, $NH_2NH_3^+$, possesses a proton that will ionize; that is, because hydrazine is a weak base, its conjugate acid is relatively strong.

$$NH_2NH_3{}^+(aq) + H_2O(l) \rightleftharpoons NH_2NH_2(aq) + H_3O^+(aq)$$

A solution of NH_2NH_3Br, is, therefore, acidic.

EXERCISE Predict whether an aqueous solution of methylammonium nitrate, $CH_3NH_3NO_3$, is acidic, basic, or neutral. Write any equilibria involved.

ANSWER: Acidic; $CH_3NH_3^+(aq) + H_2O(l) \rightleftharpoons CH_3NH_2(aq) + H_3O^+(aq)$

EXAMPLE 2 Predicting whether a salt solution is acidic, basic, or neutral

Is a salt solution of sodium bicarbonate ($NaHCO_3$) acidic, basic, or neutral? $pK_a(HCO_3^-) = 10.25$, $pK_a(H_2CO_3) = 3.69$.

SOLUTION Na^+ has no effect on the pH of the solution, but we must consider two possible equilibria for the hydrogen carbonate ion. In the first, it acts as an Brønsted acid; in the second, as a Brønsted base.

$$HCO_3^-(aq) + H_2O(l) \rightleftharpoons H_3O^+(aq) + CO_3^{2-}(aq) \qquad pK_a = 10.25$$
$$HCO_3^-(aq) + H_2O(l) \rightleftharpoons H_2CO_3(aq) + OH^-(aq) \qquad pK_b = 10.32$$

If the first equilibrium predominates, the solution is acidic because H_3O^+ is formed in this equilibrium. If the second predominates, the solution is basic because OH^- is formed. In this case, the equilibrium constants for both equilibria are almost the same, so neither equilibrium prevails, and the solution is neutral. (The pK_b of the second equilibrium is calculated using $pK_a + pK_b = 14.00$, and the pK_a of carbonic acid, 3.69.)

EXERCISE Predict whether a solution of $NaHSO_4$ is acidic, basic, or neutral. $pK_a(HSO_4^-) = 1.92$. Write any equilibria involved.

ANSWER: Acidic; $HSO_4^-(aq) + H_2O(l) \rightleftharpoons H_3O^+(aq) + SO_4^{2-}(aq)$

EXAMPLE 3 Calculating the pH of a salt solution

What is the predicted pH of a 0.20 M $CH_3NH_3^+Cl^-(aq)$ solution? $pK_b(CH_3NH_2) = 3.44$.

SOLUTION Cl^- is a very weak Brønsted base and has no effect on the pH of the solution, so we focus on the methylammonium ion, $CH_3NH_3^+$. It participates in the equilibrium

$$CH_3NH_3^+(aq) + H_2O(l) \rightleftharpoons H_3O^+(aq) + CH_3NH_2(aq)$$

We calculate K_a for the equilibrium using the given value of $pK_b(CH_3NH_2)$:

$$pK_a(CH_3NH_3^+) + pK_b(CH_3NH_2) = 14.00$$
$$pK_a(CH_3NH_3^+) + 3.44 = 14.00$$
$$pK_a(CH_3NH_3^+) = 14.00 - 3.44 = 10.56$$
$$K_a = 10^{-10.56} = 2.8 \times 10^{-11}$$

The table of concentrations is

	$CH_3NH_3^+(aq)$	+ $H_2O(l)$ $\rightleftharpoons$	$H_3O^+(aq)$	+ $CH_3NH_2(aq)$
initial concentration	0.20 M		0	0
change to reach equilibrium	$-x$		$+x$	$+x$
equilibrium concentration	0.20 M $- x$		x	x

The initial concentration of H_3O^+ is zero because we assume the 1×10^{-7} M concentration of H_3O^+ formed by the autoionization of water is small enough to be neglected relative to the large amount of H_3O^+ formed in the equilibrium. The initial concentration of $CH_3NH_3^+$ (0.20 M) originates with the complete ionization of the 0.20 M CH_3NH_3Cl. Substituting the equilibrium values from the table into the equilibrium expression gives

$$\frac{[H_3O^+][CH_3NH_2]}{[CH_3NH_3^+]} = 2.8 \times 10^{-11}\ M$$
$$\frac{(x)(x)}{0.20\ M - x} = 2.8 \times 10^{-11}\ M$$

We now assume that x is so much smaller than 0.20, so that $0.20 - x = 0.20$. Thus

$$\frac{x^2}{0.20\ M} = 2.8 \times 10^{-11}\ M$$
$$x^2 = 0.20 \times 2.8 \times 10^{-11}\ M^2 = 5.6 \times 10^{-12}\ M^2$$
$$x = 2.4 \times 10^{-6}\ M$$

We note that x is very small relative to 0.20. It is also fairly large relative to 1×10^{-7} (24 times as large), so the approximations used are acceptable. x equals the equilibrium hydronium concentration, so we can determine the pH:

$$[H_3O^+] = 2.4 \times 10^{-6}\ M$$
$$pH = -\log [H_3O^+]$$
$$= -\log (2.4 \times 10^{-6})$$
$$= 5.63$$

EXERCISE What is the pH of a 0.085 M solution of $NH_2NH_3^+Cl^-$? $pK_b(NH_2NH_2) = 5.77$.

ANSWER: 4.65

EXAMPLE 4 Calculating the pH of a salt solution

What is the predicted pH of a 0.23 M $Co(NO_3)_2$ solution? $pK_a(Co^{2+}) = 8.89$. Assume only one proton is lost from the hydrated ion.

SOLUTION The NO_3^- ion is a very weak Brønsted base and it does not affect the pH of the solution, so we focus on the cobalt(II) ion (Co^{2+}). We do not know how many waters of hydration solvate the Co^{2+} ion, but this information is not required because we are told that only one proton ionizes from the hydrated ion. If there are m waters of hydration, the equilibrium and table of values for this problem are as follows.

	$Co(H_2O)_m^{2+}(aq)$ + $H_2O(l)$ $\rightleftharpoons$	$H_3O^+(aq)$ +	$Co(H_2O)_{m-1}(OH)^+(aq)$
initial concentration	0.23 M	0	0
change to equilibrium	$-x$	$+x$	x
equilibrium concentration	0.23 M $- x$	x	x

Thus,

$$\frac{[H_3O^+][Co(H_2O)_{m-1}(OH)^+]}{[Co(H_2O)_m^{2+}]} = K_a = 10^{-8.89}$$
$$= 1.3 \times 10^{-9}\ M$$

With the approximation that x is much smaller than 0.23, so that $0.23 - x = 0.23$, we get

$$\frac{x^2}{0.23\ M} = 1.3 \times 10^{-9}\ M$$
$$x^2 = 3.0 \times 10^{-10}\ M^2$$
$$x = 1.7 \times 10^{-5}\ M$$
$$[H_3O^+] = 1.7 \times 10^{-5}\ M$$
$$pH = -\log(1.7 \times 10^{-5})$$
$$= 4.76$$

EXERCISE What is the pH of a 0.30 M $Ni(NO_3)_2$ solution. $pK_a(Ni^{2+}) = 10.60$. Assume only one proton is lost from the hydrated ion.

ANSWER: 5.56

KEY CONCEPT B Anions as bases

An anion that is part of a salt may react with water to form OH^- ion and, therefore, increase the pH of a solution. If the anion is the conjugate base of a weak acid, the equilibrium may result in surprisingly large amounts of OH^-.

EXAMPLE 1 Predicting whether a salt solution is acidic, basic, or neutral

Predict whether a solution of sodium cyanide (NaCN) is acidic, basic, or neutral.

SOLUTION Na^+ cannot act as either a Brønsted base or Brønsted acid; it does not affect the acidity of the solution. However, CN^- is the conjugate base of the weak acid HCN ($pK_a = 9.31$), so CN^- reacts with H_2O to form HCN and OH^-. That is, the following equilibrium results in the formation of OH^- ion; a solution of NaCN is a basic.

$$CN^-(aq) + H_2O(l) \rightleftharpoons HCN(aq) + OH^-(aq)$$

EXERCISE Predict whether a solution of $Ca(NO_3)_2$ is acidic, basic, or neutral. Write any equilibria involved.

ANSWER: Neutral; no equilibrium occurs.

EXAMPLE 2 Calculating the pH of a salt solution

What is the pH of a 0.088 M potassium hypochlorite (KClO) solution? $pK_a = 7.53$ for hypochlorous acid, HClO.

SOLUTION As always in this type of problem, the first task is to write the equilibrium. Potassium ion (K^+) does not act as a Brønsted acid, so it does not affect the pH of the solution. We therefore focus on the hypochlorite ion. It participates in the equilibrium shown; the table of concentration values for this problem is also included.

	$ClO^-(aq)$ + $H_2O(l)$ $\rightleftharpoons$	$HClO(aq)$ +	$OH^-(aq)$
initial concentration	0.088 M	0	0
change to reach equilibrium	$-x$	$+x$	$+x$
equilibrium concentration	0.088 M $- x$	x	x

We calculate K_b for this equilibrium:

$$pK_b(ClO^-) + pK_a(HClO) = 14.00$$
$$pK_b(ClO^-) + 7.53 = 14.00$$
$$pK_b(ClO^-) = 6.47$$
$$K_b = 10^{-6.47}$$
$$= 3.4 \times 10^{-7}\ M$$

The equilibrium expression is

$$\frac{[HClO][OH^-]}{[ClO^-]} = 3.4 \times 10^{-7}\ M$$

Substituting in the equilibrium concentrations from the table of values gives

$$\frac{(x)(x)}{0.088\ M - x} = 3.4 \times 10^{-7}\ M$$

We assume that 0.088 is much greater than x, so that $0.088 - x = 0.088$.

$$\frac{x^2}{0.088\ M} = 3.4 \times 10^{-7}\ M$$
$$x^2 = 3.0 \times 10^{-8}\ M^2$$
$$x = 1.7 \times 10^{-4}\ M$$

The parameter x corresponds to the OH^- ion concentration. To calculate the pH in the most convenient way, we use

$$\begin{aligned} pH + pOH &= 14.00 \\ pOH &= -\log [OH^-]. \\ &= -\log (x) \\ &= -\log(1.7 \times 10^{-4}) \\ &= 3.76 \\ pH + 3.76 &= 14.00 \\ pH &= 10.24 \end{aligned}$$

The pH corresponds to a basic solution. This result is expected because the equilibrium forms OH^- as one of the products.

EXERCISE What is the pH of a 0.088 M potassium hypobromite (KBrO) solution? $pK_a(HBrO) = 8.69$.

ANSWER: 10.82

15.2 MIXED SOLUTIONS

KEY CONCEPT A The Common-ion effect

According to Le Chatelier's principle, addition of any reactant (that is not a pure liquid or solid) forces an equilibrium to shift toward products; in similar fashion, addition of any product (that is not a pure liquid or solid) forces the equilibrium to shift toward reactants. For instance, the addition of a soluble salt that contains cyanide ion (such as NaCN) to a solution of hydrocyanic acid (HCN) should force the acid equilibrium to the left; this lowers the concentration of H_3O^+ and, therefore, increases the pH.

$$HCN(aq) + H_2O(l) \rightleftharpoons H_3O^+(aq) + CN^-(aq)$$

Addition of such a salt illustrates the **common-ion effect.** Addition of an ion common to one already in solution shifts an equilibrium in one direction or another. As another example of the common-ion effect, addition of a salt containing ammonium ion (such as NH_4Cl), to a solution of ammonia shifts the base equilibrium to the left; this decreases the concentration of OH^- and, therefore, decreases the pH.

$$NH_3(aq) + H_2O(l) \rightleftharpoons OH^-(aq) + NH_4{}^+(aq)$$

EXAMPLE Using the common-ion effect to change the pH

What salt can be added to a hydrofluoric acid solution to increase the pH?

SOLUTION In any problem involving an equilibrium, we should first write the equilibrium involed:

$$HF(aq) + H_2O(l) \rightleftharpoons H_3O^+(aq) + F^-(aq)$$

To increase the pH, we must decrease the H_3O^+ concentration. In principle, this can be accomplished by removing HF from the solution or by adding F^- to it; both would cause the equilibrium to shift toward reactants, thereby lowering the H_3O^+ concentration. Practically, there is no way that addition of a salt can remove HF from solution. However, addition of any soluble fluoride, such as NaF, adds F^- and accomplishes an increase in pH.

EXERCISE Name a salt that could be added to a solution of hydrazine to decrease the pH.

ANSWER: Hydrazinium chloride, NH_2NH_3Cl (or any salt containing the hydrazinium ion, $NH_2NH_3^+$

KEY CONCEPT B The pH of mixed solutions

For our purposes (at this point), a mixed solution is a solution of a weak acid or weak base that also contains a common ion. To calculate the pH of such a solution, we must do a standard equilibrium calculation using a table of concentrations. Because a common ion is present, one of the ions formed from the equilibrium will have a nonzero concentration before equilibrium.

EXAMPLE Calculating the pH of a mixed solution

What is the pH of a solution that is 0.25 M in C_6H_5COOH and 0.15 M in NaC_6H_5COO? $pK_a(C_6H_5COOH) = 4.19$.

SOLUTION Sodium benzoate, like any other sodium salt, ionizes completely in aqueous solution, so a solution that is 0.15 M in NaC_6H_5COO actually contains 0.15 M $Na^+(aq)$ and 0.15 M $C_6H_5COO^-(aq)$. Our solution is a mixed solution of a weak acid and its conjugate base as a common ion. The table of concentrations follows.

	$C_6H_5COOH(aq)$ + $H_2O(l)$ $\rightleftharpoons$	$H_3O^+(aq)$ +	$C_6H_5COO^-(aq)$
before equilibrium	0.25 M	0 M	0.15 M
change to reach equilibrium	$-x$	$+x$	$+x$
after equilibrium	0.25 M $- x$	x	0.15 M $+ x$

Because $pK_a = 4.19$, $K_a = 10^{-4.19} = 6.5 \times 10^{-5}$ M. The equilibrium expression is

$$\frac{[H_3O^+][C_6H_5COO^-]}{[C_6H_5COOH]} = K_a$$

Substituting the equilibrium values from the table gives

$$\frac{(x)(0.15\ \text{M} + x)}{(0.25\ \text{M} - x)} = 6.5 \times 10^{-5}\ \text{M}$$

We assume $x \ll 0.15$ so that $0.15 + x = 0.15$, and $x \ll 0.25$ so that $0.25 - x = 0.25$. Thus,

$$\frac{x(0.15\ \cancel{\text{M}})}{(0.25\ \cancel{\text{M}})} = 6.5 \times 10^{-5}\ \text{M}$$

$$x = \frac{(0.25)(6.5 \times 10^{-5})}{0.15}\ \text{M}$$

$$= 1.1 \times 10^{-4}\ \text{M}$$

We note that because $0.25 > 1.1 \times 10^{-4}$ and $0.15 > 1.1 \times 10^{-4}$, our assumptions regarding the small size of x are justified. Also, since $1.1 \times 10^{-4} > 1.0 \times 10^{-7}$, the relatively small amount of hydronium formed by the autoionization of water can be ignored. Because x equals the hydronium ion concentration at equilibrium,

$$[H_3O^+] = 1.1 \times 10^{-4}\ \text{M}$$

$$\text{pH} = -\log [H_3O^+]$$

$$= -\log (1.1 \times 10^{-4})$$

$$= -(-3.96)$$

$$= 3.96$$

EXERCISE What is the pH of a solution that is 0.55 M in HF and 0.75 M in NaF? $pK_a(HF) = 3.75$.

ANSWER: 3.88

KEY CONCEPT C The dilution of mixed solutions

When two solutions are combined, the volume of the resulting solution is (approximated as) the sum of the volumes of the two original solutions. Because the number of moles of solute is not affected by combining the two solutions and, at the same time, the volume changes, the molarity of any solute is changed. This, in turn, affects the equilibrium concentrations of the species in the solution. In setting up a table of concentrations, the new volume must be taken into account in determining the concentrations before equilibrium.

EXAMPLE Calculating the pH of a diluted solution

What is the pH of the solution formed by mixing 250 mL of 0.80 M NH_4Cl with 250 mL of 0.40 M NH_3? $pK_b(NH_3) = 4.75$.

SOLUTION By mixing the two solutions, we end up with a mixed solution of a base (NH_3) and its conjugate acid (NH_4^+); the equilibrium involved is

$$NH_3(aq) + H_2O(l) \rightleftharpoons OH^-(aq) + NH_4^+(aq)$$

Because $pK_b = 4.75$, $K_b = 10^{-4.75} = 1.8 \times 10^{-5}$ M. To solve the equilibrium problem that evolves, we first need the original concentrations of NH_3 and NH_4Cl. We start by calculating the number of moles of each solute, using the fact that molarity × volume (in liters) = moles.

$$\text{Moles } NH_3 = 0.40 \frac{\text{mol}}{\cancel{\text{L}}} \times 0.250 \cancel{\text{L}} = 0.10 \text{ mol}$$

$$\text{Moles } NH_4Cl = 0.80 \frac{\text{mol}}{\cancel{\text{L}}} \times 0.250 \cancel{\text{L}} = 0.20 \text{ mol}$$

The volume of the final solution is 500 mL (0.500 L). We now calculate the molarity of the solutes in the final solution by dividing the moles of each solute by the volume of solution.

$$NH_3\text{:} \quad \text{molarity} = \frac{0.10 \text{ mol}}{0.500 \text{ L}} = 0.20 \text{ M}$$

$$NH_4Cl\text{:} \quad \text{molarity} = \frac{0.20 \text{ mol}}{0.500 \text{ L}} = 0.40 \text{ M}$$

With this information, we can construct the table of concentrations before and after equilibrium. In the table, we recognize that the concentration of NH_4^+ is the same as that of NH_4Cl because NH_4Cl ionizes completely.

	$NH_3(aq) + H_2O \rightleftharpoons$	$NH_4^+(aq) +$	$OH^-(aq)$
before equilibrium	0.20 M	0.40 M	0
change to reach equilibrium	$-x$	$+x$	$+x$
after equilibrium	0.20 M $- x$	0.40 M $+ x$	x

Substituting the equilibrium concentrations into the equilibrium expression, we get

$$\frac{[NH_4^+][OH^-]}{[NH_3]} = K_b$$

$$\frac{(0.40 \text{ M} + x)(x)}{(0.20 \text{ M} - x)} = 1.8 \times 10^{-5} \text{ M}$$

We now assume that $x \ll 0.40$ so that $0.40 + x = 0.40$, and that $x \ll 0.20$ so that $0.20 - x = 0.20$. Thus,

$$\frac{(x)(0.40 \text{ M})}{(0.20 \text{ M})} = 1.8 \times 10^{-5} \text{ M}$$

$$x = \frac{(0.20)(1.8 \times 10^{-5} \text{ M})}{0.40}$$

$$= 9.0 \times 10^{-6} \text{ M}$$

We note that because $9.0 \times 10^{-6} < 0.40$ and $9.0 \times 10^{-6} < 0.20$, our assumption of the smallness of x is proper. In addition, because $9.0 \times 10^{-6} > 1.0 \times 10^{-7}$, our ignoring the amount of OH^- formed from the autoionization of water (as we did by setting the initial concentration of OH^- to 0), is also appropriate. Thus,

$$[OH^-] = 9.0 \times 10^{-6}$$

$$pOH = -\log [OH^-]$$

$$= -(-5.05)$$

$$= 5.05$$

$$pH = 14.00 - 5.05$$

$$= 8.95$$

EXERCISE What is the pH of the solution prepared by mixing 450 mL of 1.07 M $C_2H_5NH_2$ and 150 mL of 2.40 M $C_2H_5NH_3Cl$? pK_b ($C_2H_5NH_2$) = 3.19.

ANSWER: 10.82

TITRATIONS AND pH CURVES

KEY WORDS Define or explain each of the following terms in a written sentence or two.

buffer   buffer capacity   Henderson–Hasselbach equation   indicator

15.3 THE VARIATION OF pH DURING A TITRATION

KEY CONCEPT pH changes during a titration

The anion of a weak acid is a strong Brønsted base. As a weak acid is titrated, the anion is liberated from the acid and may affect the pH of the titration mixture. For instance, F^- ion is liberated as HF is titrated; the effect of the F^- on the pH of the solution must be taken into account. (For a strong acid, such as HCl, the anion liberated is not a strong base; it has no effect on the pH.) We are, again, interested in the pH at four stages of the titration:

1. Before the addition of any base; the calculation is the same as that for the pH of a weak acid.
2. After addition of some base but before the stoichiometric point is reached; this calculation is the same as that for the pH of a mixed solution.
3. At the stoichiometric point; this calculation is the same as that of the pH of a salt.
4. After the stoichiometric point; this calculation is the same as that for the dilution of a base.

(The calculations for the titration of a weak base with a strong acid are similar to those for the titration of a weak acid with a strong base. An extra exercise is included for practice.)

EXAMPLE The pH during the titration of a weak acid with a strong base

Calculate the pH at the following points for the titration of 25.00 mL of 0.100 M HOCl with 0.150 M NaOH (a) before addition of any base, (b) after addition of 10.00 mL of base, (c) at the stoichiometric point (16.33 mL of base total), and (d) after addition of 5.00 mL of base beyond the stoichiometric point. K_a(HOCl) = 3.0×10^{-8} M, $pK_a = 7.53$.

SOLUTION First, we prepare a table showing the various amounts (in millimoles) of each of the chemical species involved at each stage of the titration after the added base reacts with acid but *before equilibrium*. To do this we recognize from the equation for the titration,

$$HOCl(aq) + OH^-(aq) \longrightarrow H_2O(l) + OCl^-(aq)$$

that 1 mmol NaOH reacts with 1 mmol HOCl to produce 1 mmol OCl^-. Also, the millimoles of OH^- are the same as the millimoles of NaOH, because NaOH is 100% ionized. (The millimoles of NaOH are calculated using the relation millimole solute = milliliter solution × molarity.) There are 2.50 mmol HOCl at the start of the titration (25.00 mL × 0.100 M = 2.50 mmol).

NaOH added, mL	OH^- added, mmol	HOCl left after reaction with OH^-, mmol	OCl^- formed after reaction, mmol	OH^- left after reaction, mmol	V_{total}
0.00	0.00	2.50	0.00	0.00	25.00
10.00	1.50	1.00	1.50	0.00	35.00
16.33	2.50	0.00	2.50	0.00	41.33
21.33	3.20	—	—	0.70	46.33

After the equivalence point (16.33 mL), there is no HOCl left to react, so two entries are left blank. The first row of entries applies to the situation before any NaOH is added, so no reaction is possible at this point, and the words *after reaction* in the column headings should be ignored. Before proceeding, you should confirm all of the entries in the table!

We also need K_b for this problem:

$$pK_b + pK_a = 14.00$$
$$pK_b = 14.00 - 7.53$$
$$= 6.47$$
$$K_b = 10^{-6.47}$$
$$= 3.4 \times 10^{-7}$$

The preceding table shows amounts before equilibrium. The calculations that follow take into account any equilibrium process that occurs at each stage of the titration.

(a) Before addition of any base, the problem is the same as that of determining the pH of a 0.100 M solution of HOCl. The equilibrium and table of values are as follows.

	$HOCl(aq)$ + $H_2O(l)$ $\rightleftharpoons$	$H_3O^+(aq)$ +	$OCl^-(aq)$
initial concentration	0.100 M	0	0
change to reach equilibrium	$-x$	$+x$	$+x$
equilibrium concentration	$0.100\ M - x$	x	x

Substituting the equilibrium values into the equilibrium expression gives

$$\frac{x^2}{0.100\ M - x} = 3.0 \times 10^{-8}\ M$$

If we assume that x is small and that $0.100 - x = 0.100$, then

$$x^2 = 3.0 \times 10^{-9}\ M^2$$
$$[H_3O^+] = x = 5.5 \times 10^{-5}\ M$$
$$pH = 4.26$$

(b) The table at the very start of the problem indicates that, after addition of 10.00 mL of NaOH, the solution consists of a mixture of a weak acid (HOCl) and its conjugate base (OCl^-); so it is a mixed solution. We calculate the pH using a table of concentrations. However, our first task is to calculate the initial concentrations of the two species involved in the equilibrium, HOCl and OCl^-.

$$[HOCl] = \frac{1.00 \text{ mmol}}{35.00 \text{ mL}} = 0.0286 \text{ M} \qquad [OCl^-] = \frac{1.50 \text{ mmol}}{35.00 \text{ mL}} = 0.0429 \text{ M}$$

Now we can set up the table of concentrations.

	$HOCl(aq)$ + $H_2O(l)$	$\rightleftharpoons$ $H_3O^+(aq)$ +	$OCl^-(aq)$
before equilibrium	0.0286 M	0 M	0.0429 M
change to reach equilibrium	$-x$	$+x$	$+x$
after equilibrium	$0.0286 \text{ M} - x$	x	$0.0429 \text{ M} + x$

Substituting the equilibrium values into the equilibrium expression gives

$$\frac{(x)(0.0429 \text{ M} + x)}{(0.0286 \text{ M} - x)} = 3.0 \times 10^{-8} \text{ M}$$

We now assume that $x \ll 0.0286$ so that $0.0286 - x = 0.0286$, and that $x \ll 0.0429$ so that $0.0429 + x = 0.0429$. Thus,

$$\frac{(x)(0.0429 \text{ M})}{0.0286 \text{ M}} = 3.0 \times 10^{-8} \text{ M}$$

$$x = \frac{(0.0286 \text{ M})(3.0 \times 10^{-8} \text{ M})}{0.0429 \text{ M}}$$

$$= 2.0 \times 10^{-8} \text{ M}$$

Because $x = [H_3O^+]$,

$$\begin{aligned} pH &= -\log [H_3O^+] \\ &= -\log (2.0 \times 10^{-8}) \\ &= -(-7.70) \\ &= 7.70 \end{aligned}$$

(c) At the stoichiometric point, exactly enough base has been added to react with the original amount of acid present (2.50 mmol). A solution of Na^+ ion and ClO^- ion remains. The pH is calculated in the same way as for a solution of the salt NaClO. Na^+ does not effect the pH, so we focus on the OCl^- ion.

$$\begin{aligned} [OCl^-] &= \frac{\text{millimoles } OCl^-}{\text{milliliters solution}} \\ &= \frac{2.50 \text{ mmol}}{41.67 \text{ mL}} \\ &= 0.0600 \text{ M} \end{aligned}$$

The equilibrium and table of concentration values are

	$OCl^-(aq)$ + $H_2O(l)$	$\rightleftharpoons$ $HOCl(aq)$ +	$OH^-(aq)$
initial concentration	0.0600 M	0	0
change to reach equilibrium	$-x$	$+x$	$+x$
equilibrium concentrations	$0.0600 \text{ M} - x$	x	x

The equilibrium expression is

$$\frac{[HOCl][OH^-]}{[OCl^-]} = K_b$$

Substituting in the equilibrium concentrations gives

$$\frac{(x)(x)}{0.0600 \text{ M} - x} = 3.4 \times 10^{-7} \text{ M}$$

We assume that $x \ll 0.0600$ so that $0.0600 - x = 0.0600$.

$$\frac{x^2}{0.0600 \text{ M}} = 3.4 \times 10^{-7} \text{ M}$$

$$x^2 = 2.0 \times 10^{-8} \text{ M}^2$$

$$[OH^-] = x = 1.4 \times 10^{-4} \text{ M}$$

$$\text{pOH} = -\log [OH^-] = -\log(1.4 \times 10^{-4})$$

$$= 3.85$$

$$\text{pH} = 14.00 - \text{pOH}$$

$$= 10.15$$

(d) After the stoichiometric point, the additional amount of strong base OH^- added completely overwhelms the small amount of OH^- formed by the ionization of OCl^- in water. The pH of the solution is determined by the additional OH^- added. From the table constructed at the start of the whole problem, we have

$$[OH^-] = \frac{0.70 \text{ mmol}}{46.33 \text{ mL}}$$

$$= 0.015 \text{ M}$$

$$\text{pOH} = 1.82$$

$$\text{pH} = 12.18$$

EXERCISE 1 Calculate the pH at the following points for the titration of 25.00 mL of 0.200 M benzoic acid (C_6H_5COOH) with 0.150 M NaOH: (a) before the addition of any NaOH, (b) after addition of 10.00 mL of NaOH, (c) at the stoichiometric point (33.33 mL of NaOH), and (d) after addition of 5.00 mL of NaOH beyond the stoichiometric point. $K_a(C_6H_5COOH) = 6.5 \times 10^{-5}$ M, $pK_a = 4.19$

ANSWER: (a) 2.44; (b) 3.81; (c) 8.56; (d) 12.20

EXERCISE 2 Calculate the pH at the following points for the titration of 50.00 mL of 0.100 M NH_2NH_2 with 0.200 M HCl: (a) before the addition of any acid, (b) after addition of 15.00 mL of HCl, (c) at the stoichiometric point (25.00 mL of HCl added), (d) after addition of 5.00 mL of acid past the stoichiometric point. $K_b(NH_2NH_2) = 1.7 \times 10^{-6}$ M, $pK_b = 5.77$.

ANSWER: (a) 10.61; (b) 8.05; (c) 4.70; (d) 1.90

15.4 INDICATORS AS WEAK ACIDS

KEY CONCEPT Acid–base indicators

An **acid–base indicator** is a compound that exhibits different colors in different pH ranges. A familiar indicator is phenolphthalein, which is colorless in an acid solution but pink above pH 8.2. Indicators are used to tell when to stop a titration. The choice of a specific indicator depends on the pH at the stoichiometric point and how fast the pH changes around the stoichiometric point. The *ideal* choice is one for which the pH at the stoichiometric point of the titration equals the pK of the indicator, however, if the pH $\pm$ 1 equals the pK of the indicator, the indicator will usually be suitable. Practically, this means that an indicator is chosen for which the pH at the stoichiometric point is in the middle of the range of pH values over which the indicator changes color (Table 15-3).

pH at stoichiometric point $\pm 1 = pK_{\text{ind}}$

pH at stoichiometric point = pH at midpoint of range over which indicator changes color

For instance, in a titration in which the pH at the stoichiometric point is 5.2, an excellent choice for an indicator is methyl red, which changes color in the pH range 4.8–6.0. The midpoint of this range is 5.4, which is close to 5.2.

EXAMPLE Choice of an indicator

Which indicator from Table 15.3 of the text would be best for a titration with pH 9.2 at the stoichiometric point?

SOLUTION We try to choose an indicator in which the midpoint of the pH range for the color change of the indicator is the same as the pH at the equivalence point of the titration. From the table, we find that thymol blue and phenolphthalein both look promising.

Thymol blue	pH range = 9.0–9.6	midpoint of range = 9.3
Phenolphthalein	pH range = 8.2–10.0	midpoint of range = 9.1

The midpoints of both are both quite close to the pH of 9.2 at the stoichiometric point. Either indicator would be suitable.

EXERCISE What indicator from Table 15.3 of the text would be best for a titration with pH 5.5 at the stoichiometric point?

ANSWER: Methyl red is the only choice from the table.

15.5 BUFFER SOLUTIONS

KEY CONCEPT A Buffers and buffer action

A **buffer** is a solution that does not change pH very dramatically when either acid or base is added in moderate amounts. A buffer can be prepared by mixing a weak acid with a salt of the weak acid or by mixing a weak base with a salt of the weak base. In the first case (weak acid + salt of weak acid), the buffer contains a weak acid and its conjugate base. In the second case (weak base + salt of weak base), the buffer contains a weak base and its conjugate acid. For example

Weak acid +	salt	gives	weak acid +	conjugate base
CH_3COOH	CH_3COONa		CH_3COOH	CH_3COO^-
Weak base +	salt	gives	weak base +	conjugate acid
NH_3	NH_4Cl		NH_3	NH_4^+

We can see how a buffer prevents drastic changes in pH by considering the equilibrium involved. For instance, for an acetic acid–sodium acetate buffer, the equilibrium is

$$CH_3COOH(aq) + H_2O(l) \rightleftharpoons CH_3COO^-(aq) + H_3O^+(aq)$$

This is a mixed solution, with relatively large amounts of both CH_3COOH and CH_3COO^- present. If a moderate amount of an acid is added, protons are mostly transferred to the base CH_3COO^- instead of to H_2O; very little H_3O^+ is formed as a result of the addition and the pH does not change very much. That is, after addition of an acid such as HCl, we have

$$HCl(aq) + CH_3COO^-(aq) \longrightarrow CH_3COOH(aq) + Cl^-(aq) \qquad \text{predominant reaction}$$
$$HCl(aq) + H_2O(l) \longrightarrow H_3O^+(aq) + Cl^-(aq) \qquad \text{unimportant reaction}$$

On addition of a moderate amount of base, the base reacts with CH_3COOH instead of forming free OH^- ions, so again the pH does not change very much. For example, if NaOH is added, we have

$$OH^-(aq) + CH_3COOH(aq) \longrightarrow CH_3COO^-(aq) + H_2O(l) \qquad \text{predominant reaction}$$
$$\text{Unreacted } OH^-(aq) \qquad \text{unimportant amount}$$

EXAMPLE Understanding buffers

Will a solution of hydrofluoric acid (HF) and sodium fluoride (NaF) act as a buffer? If it does, what reaction occurs when a small amount of strong acid (such as HCl) is added?

SOLUTION A solution of a weak acid and a salt of the acid will act as a buffer. Because HF is a weak acid and NaF is a salt of HF, the mixture will act as a buffer. When HCl is added, it will transfer protons mostly to F^- instead of to H_2O, so the H_3O^+ concentration will not change very much.

$$HCl(aq) + F^-(aq) \longrightarrow Cl^-(aq) + HF(aq) \quad \text{predominant reaction}$$
$$HCl(aq) + H_2O(aq) \longrightarrow H_3O^+(aq) + Cl^-(aq) \quad \text{unimportant reaction}$$

EXERCISE What reaction occurs in the HF–NaF buffer when a strong base is added?

ANSWER: $OH^-(aq) + HF(aq) \longrightarrow H_2O(aq) + F^-(aq)$

KEY CONCEPT B The pH of a buffer

The pH of a buffer solution can be calculated with the **Henderson–Hasselbalch equation,**

$$\text{pH} = \text{p}K_a + \log\frac{[\text{conjugate base}]}{[\text{conjugate acid}]}$$

The $\text{p}K_a$ is that of the conjugate acid. The concentrations of conjugate acid and conjugate base used usually correspond to those *before* equilibrium occurs. Using the concentration before equilibrium is tantamount to assuming that x (the change in concentrations to reach equilibrium) is small with regard to the initial concentrations of both acid and base. In this regard, we should emphasize that using the Henderson–Hasselbalch equation is completely equivalent to setting up a table of concentrations and solving an equilibrium problem.

EXAMPLE Calculating the pH of a buffer

What is the pH of the buffer prepared by making up a solution that is 0.50 M acetic acid (CH_3COOH) and 0.40 M sodium acetate (CH_3COONa)? $\text{p}K_a(CH_3COOH) = 4.75$

SOLUTION The pH of a buffer can be calculated with the Henderson–Hasselbalch equation. In this problem, CH_3COOH is the acid and CH_3COO^- is the base. We recognize that sodium acetate is 100% dissociated in solution, so the concentration of acetate is the same as that of the sodium acetate. Thus,

$$[CH_3COOH] = 0.50\text{ M} \qquad [CH_3COO^-] = 0.40\text{ M}$$

$$\begin{aligned}\text{pH} &= 4.75 + \log\frac{[CH_3COO^-]}{[CH_3COOH]}\\ &= 4.75 + \log\left(\frac{0.40\text{ M}}{0.50\text{ M}}\right)\\ &= 4.75 - 0.10\\ &= 4.65\end{aligned}$$

If we round off to one significant figure as the text does, the answer is pH 5. (You should now solve this problem as an equilibrium problem using a table of concentrations and assure that the answer comes out the same.)

EXERCISE What is the pH of the buffer prepared by making up a solution that is 0.30 M ammonia (NH_3) and 0.20 M ammonium chloride (NH_4Cl)? $\text{p}K_b(NH_3) = 4.75$.

ANSWER: 4.82 or, to one significant figure, 5

KEY CONCEPT C Buffer capacity

A buffer operates by providing a weak acid to react with added base and a weak base to react with added acid. If enough base is added to a buffer to react with all of the weak acid, the buffer is no longer able to protect against the addition of more base. If enough acid is added to a buffer to react with all of the weak base, the buffer is no longer able to protect against the addition of more acid. We conclude that any buffer has a limit to its buffering ability. As a working rule of thumb, when the ratio of conjugate base to conjugate acid in the buffer is less than 0.1 or more than 10, we assume that the buffer's capacity has been exceeded.

$$\frac{[\text{conjugate base}]}{[\text{conjugate acid}]} \geq 10$$ buffer no longer protects against base; not enough acid is present

$$10 > \frac{[\text{conjugate base}]}{[\text{conjugate acid}]} > 0.1$$ buffer protects against both acid and base

$$\frac{[\text{conjugate base}]}{[\text{conjugate acid}]} \leq 0.1$$ buffer no longer protects against acid; not enough base is present

These limits define what is called the **buffer capacity** of the buffer. If we substitute these ratios into the Henderson–Haselbalch equation (as upper and lower bounds, so that the inequalities become equal signs), we arrive at the following:

$$\frac{[\text{conjugate base}]}{[\text{conjugate acid}]} = 10 \qquad \text{pH} = \text{p}K_a + \log 10 = \text{p}K_a + 1$$

$$\frac{[\text{conjugate base}]}{[\text{conjugate acid}]} = 0.1 \qquad \text{pH} = \text{p}K_a + \log 0.1 = \text{p}K_a - 1$$

These results indicate that a buffer works best in the pH range from $\text{p}K_a - 1$ to $\text{p}K_a + 1$. So, for instance, a benzoic acid–sodium benzoate buffer, for which pK_a(benzoic acid) = 4.19, buffers in the pH range 3.19 (4.19 − 1) to 5.19 (4.19 + 1).

EXAMPLE Estimating the buffer capacity of a buffer

An acetic acid–sodium acetate buffer is prepared by making up a solution that is 0.050 M in acetic acid and 0.050 M in sodium acetate. How many milliliters of 1.0 M NaOH can be added to 250 mL of this buffer before its buffer capacity is exceeded?

SOLUTION The reaction that occurs when NaOH is added to this buffer is

$$CH_3COOH(aq) + OH^-(aq) \rightleftharpoons CH_3COO^-(aq) + H_2O(l)$$

As NaOH is added, the CH_3COOH reacts (so its concentration decreases) and CH_3COO^- is formed (so its concentration increases). Thus, the ratio [conjugate base]/[conjugate acid] gets larger. The buffer capacity will be exceeded when the ratio gets large enough that

$$\frac{[\text{conjugate base}]}{[\text{conjugate acid}]} = 10$$

To determine when this ratio is reached, we first calculate the millimoles of acetic acid (CH_3COOH) and acetate ion (CH_3COO^-) in the 250 mL of buffer:

$$\text{Millimoles } CH_3COOH = 250 \text{ mL} \times 0.050 \text{ M} = 12.5 \text{ mmol}$$
$$\text{Millimoles } CH_3COO^- = 250 \text{ mL} \times 0.050 \text{ M} = 12.5 \text{ mmol}$$

When x mmol NaOH is added, x mmol CH_3COOH reacts and x mmol CH_3COO^- is formed. So after addition of x mmol NaOH,

$$\text{Millimoles } CH_3COOH = 12.5 \text{ mmol} - x$$
$$\text{Millimoles } CH_3COO^- = 12.5 \text{ mmol} + x$$

Because both CH_3COOH and CH_3COO^- are in the same solution, the volumes cancel in the ratio of concentrations; this is,

$$\frac{[\text{conjugate base}]}{[\text{conjugate acid}]} = \frac{\text{millimoles conjugate base}}{\text{millimoles conjugate acid}} = \frac{\text{millimoles } CH_3COO^-}{\text{millimoles } CH_3COOH}$$

For our problem, the buffer capacity is exceeded when

$$\frac{12.5 \text{ mmol} + x}{12.5 \text{ mmol} - x} = 10$$

$$12.5 \text{ mmol} + x = 10(12.5 \text{ mmol} - x)$$

$$12.5 \text{ mmol} + x = 125 \text{ mmol} - 10x$$

$$10x + x = 125 \text{ mmol} - 12.5 \text{ mmol}$$

$$11x = 112.5 \text{ mmol}$$

$$x = \frac{112.5 \text{ mmol}}{11}$$

$$= 10.2 \text{ mmol}$$

This is the number of millimoles of NaOH that just exceeds the buffer capacity. Because the concentration of NaOH being added is 1.0 M, we have

$$\text{Milliliters NaOH} = \frac{10.2 \text{ mmol}}{1.0 \text{ mmol/mL}}$$

$$= 10.2 \text{ mL} = 10 \text{ mL}$$

EXERCISE An ammonia–ammonium chloride buffer is prepared by making up a solution that is 0.40 M in ammonia (NH_3) and 0.35 M in ammonium chloride (NH_4Cl). How many milliliters of 2.0 M HCl can be added to 100 mL of this buffer before its buffer capacity is exceeded?

▲ ANSWER: 17 mL

SOLUBILITY EQUILIBRIA

KEY WORDS Define or explain each of the following terms in a written sentence or two.

dissolution
formation constant
ion product
qualitative analysis
selective precipitation
solubility product

15.6 THE SOLUBILITY CONSTANT

KEY CONCEPT A The solubility constant

We learned earlier that a sparingly soluble salt may participate in a dynamic equilibrium in which the solid salt is in equilibrium with dissolved salt in saturated solution. For instance, for $CaSO_4$,

$$CaSO_4(s) \rightleftharpoons Ca^{2+}(aq) + SO_4^{2-}(aq)$$

We can write an equilibrium expression for this type of equilibrium in the standard way. Because the concentration of a pure solid does not appear in the equilibrium expression, we get

$$K_s = [Ca^{2+}][SO_4^{2-}]$$

The subscript s stands for solubility constant, which is the name given to the equilibrium constants that describe this type of equilibrium. The solubility of a sparingly soluble salt can be estimated from the solubility constant, and the solubility constant estimated from the solubility, as illustrated in the examples that follow.

EXAMPLE 1 Estimating solubility

Estimate the solubility of $BaCO_3$ at 25°C. $K_s = 8.1 \times 10^{-9}$ M^2.

SOLUTION We first construct a table of concentration values just as we have done for other equilibria. An upper case P denotes the solid precipitate in the mixture; it is not actually a concentration

	$BaCO_3(s) \rightleftharpoons$	$Ba^{2+}(aq)$ +	$CO_3^{2-}(aq)$
initial concentrations	P	0	0
change to reach equilibrium	$-S$	$+S$	$+S$
equilibrium concentrations	$P - S$	S	S

We place a very significant interpretation on S. As can be seen from the table of concentration values, it is the amount of $BaCO_3$ that dissolves, expressed in mol/L (M); S is the solubility. The solubility constant expression is

$$[Ba^{2+}][CO_3^{2-}] = K_s$$

Substituting values from the table and the value of K_s gives

$$(S)(S) = 8.1 \times 10^{-9}\ M^2$$
$$S = 9.0 \times 10^{-5}\ M$$

The solubility of $BaCO_3$ is 9.0×10^{-5} mol/L, or 9.0×10^{-5} M.

EXERCISE What is the solubility of CuCl? $K_s = 1.0 \times 10^{-6}$ M^2?

ANSWER: 1.0×10^{-3} M.

EXAMPLE 2 Estimating solubility

According to the value of K_s, what is the solubility of Cu_2S? $K_s = 2.0 \times 10^{-47}$ M^3.

SOLUTION In the table of concentration values for this equilibrium, which follows, we are careful to remember that 2 mol Cu^+ ion is formed for every 1 mol of Cu_2S that dissolves.

	$Cu_2S(s) \rightleftharpoons$	$2Cu^+(aq)$ +	$S^{2-}(aq)$
initial concentration	P	0	0
change to reach equilibrium	$-S$	$+2S$	$+S$
equilibrium concentration	$P - S$	$2S$	S

S has the same interpretation as in the previous problem. It represents the amount of Cu_2S that dissolves and is its solubility in moles per liter. The solubility expression is

$$[Cu^+]^2[S^{2-}] = K_s$$

Substituting values from the concentration table and the solubility constant gives

$$(2S)^2(S) = 2.0 \times 10^{-47}\ M^3$$
$$4S^3 = 2.0 \times 10^{-47}\ M^3$$
$$S^3 = 5.0 \times 10^{-48}\ M^3$$
$$S = (5.0 \times 10^{-48}\ M^3)^{1/3}$$

The superscript $\frac{1}{3}$ indicates that the cube root of a number should be taken. To get the cube root of 5.0×10^{-48} on calculators with standard arithmetic notation, do the following:

Enter 5.0×10^{-48}
Depress the Y^x key
Enter 0.33333333
Depress the = key

The result is the solubility of Cu_2S:

$$S = 1.7 \times 10^{-16}\ \text{M}$$

EXERCISE Estimate the solubility of $Fe(OH)_2$? $K_s = 1.6 \times 10^{-14}\ \text{M}^3$.

ANSWER: 1.6×10^{-5} M

EXAMPLE 3 Estimating the solubility constant

The solubility of lead(II) bromide ($PbBr_2$) at 20°C is 0.84 g/100 mL water. Estimate K_s for $PbBr_2$.

SOLUTION The equilibrium and table of concentration values are

	$PbBr_2(s)$ $\rightleftharpoons$	$Pb^{2+}(aq)$ +	$2Br^-(aq)$
initial concentration	P	0	0
change to reach equilibrium	$-S$	$+S$	$+2S$
equilibrium concentration	$P - S$	S	$2S$

The equilibrium expression is

$$[Pb^{2+}][Br^-]^2 = K_s$$

Substituting the values from the table of concentration values gives

$$(S)(2S)^2 = K_s$$
$$4S^3 = K_s$$

We now need the value of S to calculate K_s. The molar mass of $PbBr_2$ is 367 g/mol. Thus,

$$0.84\ \text{g PbBr}_2 \times \frac{1\ \text{mol PbBr}_2}{367\ \text{g}} = 2.3 \times 10^{-3}\ \text{mol PbBr}_2$$

We assume that when 0.84 g $PbBr_2$ dissolves in 100 mL water, 100 mL of solution results. That is, the amount of $PbBr_2$ is so small that the volume of solution is the same as the volume of solvent. 100 mL is the same as 0.100 L of solution, so

$$S = \frac{2.3 \times 10^{-3}\ \text{mol}}{0.10\ \text{L}}$$
$$= 2.3 \times 10^{-2}\ \text{M}$$

Substituting this value of S into the equilibrium expression gives

$$K_s = 4S^3$$
$$= 4(2.3 \times 10^{-2}\ \text{M})^3$$
$$= 4.9 \times 10^{-5}\ \text{M}^3$$

This value is a little different from that in Table 15.5 of the text, because the value in the text table is determined by a more accurate method.

EXERCISE The solubility of silver carbonate (Ag_2CO_3) is 3×10^{-3} g/100 mL water. (a) Estimate K_s for silver carbonate and (b) compare your result with the value of $6.2 \times 10^{-12}\ \text{M}^3$ given in Table 15.5 of the text.

ANSWER: (a) $5 \times 10^{-12}\ \text{M}^3$; (b) close to value in Table 15.5

KEY CONCEPT B The common-ion effect

According to Le Chatelier's principle, addition of any product (that is not a pure solid or pure liquid) to a dynamic equilibrium forces the equilibrium to shift toward reactants. For instance, for the equilibrium

$$CuI(s) \rightleftharpoons Cu^{+}(aq) + I^{-}(aq)$$

addition of Cu^{+} ion (by the addition of $CuNO_3$, for example) would cause precipitation of some CuI. Addition of I^{-} ion (by the addition of NaI, for example) would also cause precipitation of some CuI. Thus, less CuI is dissolved in the presence of an external source of Cu^{+} ion or I^{-} ion than in pure water. The lowering of the solubility of a sparingly soluble salt by the presence of an ion in the salt is due to the common-ion effect we encountered earlier. We can quantitatively calculate solubilities in the presence of a common ion, as shown in the next two examples.

EXAMPLE 1 Calculating solubility

Calculate the solubility of $CaSO_4$ in 0.10 M $Na_2SO_4(aq)$. $K_s(CaSO_4) = 2.4 \times 10^{-5}$ M^2.

SOLUTION We first note that 0.10 M Na_2SO_4 is fully dissociated and the concentration of SO_4^{2-} is, therefore, 0.10 M. The equilibrium and table of concentration values then become:

	$CaSO_4(s)$ $\rightleftharpoons$	$Ca^{2+}(aq)$ +	$SO_4^{2-}(aq)$
initial concentration	P	0	0.10 M
change to reach equilibrium	$-S$	$+S$	+S
equilibrium concentration	$P - S$	S	0.10 M + S

The equilibrium expression is

$$[Ca^{2+}][SO_4^{2-}] = K_s$$

Substituting the equilibrium concentrations and value of K_s gives

$$(S)(0.10 \text{ M} + S) = 2.4 \times 10^{-5} \text{ M}^2$$

We now assume that S is very small compared with 0.10, so 0.10 M + S = 0.10 M.

$$(S)(0.10 \text{ M}) = 2.4 \times 10^{-5} \text{ M}^2$$

$$S = 2.4 \times 10^{-4} \text{ M}$$

This value of S is very much smaller than 0.10, so the approximation made is justified. Because S is the solubility (it is the amount of solid that dissolves), the solubility of $CaSO_4$ in 0.10 M Na_2SO_4 is 2.4×10^{-4} mol/L (M). (In pure water, the solubility is 5.0×10^{-3} M, about five times larger.)

EXERCISE What is the solubility of CuI in 0.30 M NaI? $K_s(CuI) = 1.0 \times 10^{-6}$ M^2.

ANSWER: 3.3×10^{-6} M

EXAMPLE 2 Calculating solubility

What is the solubility of $BaF_2(s)$ in 0.20 M NaF? $K_s(BaF_2) = 1.7 \times 10^{-6}$ M^3.

SOLUTION In 0.20 M NaF, $[F^{-}]$ = 0.20 M. The equilibrium and table of values follows:

	$BaF_2(s)$ $\rightleftharpoons$	$Ba^{2+}(aq)$ +	$2F^{-}(aq)$
initial concentration	P	0	0.20 M
change to reach equilibrium	$-S$	$+S$	$+2S$
equilibrium concentration	$P - S$	S	0.20 M + $2S$

In setting up this table, we should note that the initial F^- concentration originates with NaF and does not come from the dissolving of BaF_2. Thus, we do *not* multiply 0.20 M by 2; there is only one mole of F^- ions per mole of KF. The equilibrium expression is

$$[Ba^{2+}][F^-]^2 = K_s$$

Substituting the equilibrium values and value of K_s gives

$$(S)(0.20 \text{ M} + 2S)^2 = 1.7 \times 10^{-6} \text{ M}^3$$

We assume that $2S$ is much smaller than 0.20, so $0.20 \text{ M} + 2S = 0.20 \text{ M}$.

$$(S)(0.20 \text{ M})^2 = 1.7 \times 10^{-6} \text{ M}^3$$
$$S = 4.3 \times 10^{-5} \text{ M}$$

The solubility of $BaF_2(s)$ in 0.20 M NaF is S, 4.3×10^{-5} mol/L.

EXERCISE What is the solubility of PbI_2 in 0.50 M NaI? $K_s(PbI_2) = 1.4 \times 10^{-8} \text{ M}^3$.

ANSWER: 5.6×10^{-8} M

15.7 PRECIPITATION REACTIONS AND QUALITATIVE ANALYSIS

KEY CONCEPT Precipitation reactions and analysis

Qualitative analysis involves identifying the cations and/or anions in a unknown solution, often through precipitation reactions that give precipitates with characteristic and well-known colors. One important aspect of this process is being able to predict whether a precipitate will form. To do so, we use Q_s, the expression that has the same form as K_s but uses the actual concentrations of ions in a given solution at some moment; these do not have to be the equilibrium concentrations. We then compare Q_s with K_s for the compound we want to precipitate.

- When $Q_s > K_s$, the compound precipitates.
- When $Q_s = K_s$, a saturated solution has been made and, most likely, no visible precipitate forms.
- When $Q_s < K_s$, no precipitate forms (and part of any precipitate already present dissolves).

EXAMPLE 1 Predicting if a precipitate forms

Does $CaCO_3$ precipitate if a solution which is initially 2.0×10^{-4} M in Ca^{2+} is made 4.0×10^{-4} M in Na_2CO_3? $K_s(CaCO_3) = 8.7 \times 10^{-9} \text{ M}^2$.

SOLUTION The size of Q_s relative to K_s determines if a precipitate forms. For $CaCO_3$,

$$CaCO_3(s) \rightleftharpoons Ca^{2+}(aq) + CO_3^{2-}(aq)$$
$$Q_s = [Ca^{2+}][CO_3^{2-}]$$
$$= (2.0 \times 10^{-4} \text{ M})(4.0 \times 10^{-4} \text{ M})$$
$$= 8.0 \times 10^{-8} \text{ M}^2$$

Because $8.0 \times 10^{-8} \text{ M}^2 > 8.7 \times 10^{-9} \text{ M}^2$, $Q_s > K_s$ and $CaCO_3$ precipitates.

EXERCISE Does CuS precipitate if a solution that is 5×10^{-8} M in Cu^{2+} has its pH adjusted so that the S^{2-} ion concentration is 6×10^{-20} M? $K_s(CuS) = 8.5 \times 10^{-45} \text{ M}^2$.

ANSWER: Yes

EXAMPLE 2 Predicting if a precipitate forms

Assume we add 0.050 mL (about 1 drop) of 8×10^{-2} M NaF to 1.00 mL of a solution that is 0.010 M in Ba^{2+}, Pb^{2+}, and Ca^{2+}. Which ions precipitate as fluorides? $K_s(BaF_2) = 1.7 \times 10^{-6} \text{ M}^2$, $K_s(PbF_2) = 3.7 \times 10^{-8} \text{ M}^2$, and $K_s(CaF_2) = 4.0 \times 10^{-11} \text{ M}^2$.

SOLUTION The value of Q_s for each salt determines if it precipitates. To calculate Q_s we must determine the concentration of each ion after mixing of the two solutions. We do this using the relationship that describes the molarity change for dilution of a solution, $M_1V_1 = M_2V_2$. In this case, the original molarity and volume for each metal are $M_1 = 0.010$ M and $V_1 = 1.00$ mL. The final volume V_2 is 1.05 mL (1.00 mL + 0.050 mL = 1.05 mL). Thus,

$$M_1V_1 = M_2V_2$$
$$M_2 = \frac{M_1V_1}{V_2}$$
$$= \frac{(0.010\ \text{M})(1.00\ \cancel{\text{mL}})}{1.05\ \cancel{\text{mL}}}$$
$$= 0.0095\ \text{M}$$

For the F^- ion, $M_1 = 8 \times 10^{-2}$, $V_1 = 0.050$ mL, and $V_2 = 1.05$ mL. We calculate

$$M_2 = \frac{(8 \times 10^{-2}\ \text{M})(0.050\ \cancel{\text{mL}})}{1.05\ \cancel{\text{mL}}}$$
$$= 4 \times 10^{-3}\ \text{M}$$

It is now possible to calculate Q_s for all of the salts. Because the concentration of each metal is the same, we get

$$Q_s = [M^{2+}][F^-]^2$$

where the symbol M^{2+} represents a metal cation. Substituting the values previously calculated gives

$$Q_s = (0.0095\ \text{M})(4 \times 10^{-3}\ \text{M})^2$$
$$= 2 \times 10^{-7}\ \text{M}^3$$

By comparing Q_s with K_s for each salt, we finally determine which ions precipitate:

BaF_2: $2 \times 10^{-7} < 1.7 \times 10^{-6}$, so $Q_s < K_s$ and BaF_2 does not precipitate.
PbF_2: $2 \times 10^{-7} > 3.7 \times 10^{-8}$, so $Q_s > K_s$ and PbF_2 precipitates.
CaF_2: $2 \times 10^{-7} > 4.0 \times 10^{-11}$, so $Q_s > K_s$ and BaF_2 precipitates.

EXERCISE Does adding another 0.45 mL NaF (for a total of 0.50 mL) in the preceding problem succeed in precipitating BaF_2?

ANSWER: Yes

15.8 DISSOLUTION OF PRECIPITATES

KEY CONCEPT A Dissolution by removal of an ion

It is possible to dissolve a sparingly soluble solid MX, by removing the anion X^- from solution:

$$MX(s) \rightleftharpoons M^+(aq) + X^-(aq)$$

According to Le Chatlier's principle, if $X^-(aq)$ is removed from solution, the equilibrium will shift toward products. This means that $MX(s)$ will dissolve. Four situations in which this technique is used are the following:

1. When the solid is a hydroxide such as $Fe(OH)_3$, addition of an acid removes OH^- from solution (because H_3O^+ reacts with OH^-), and so the hydroxide dissolves.

   $H_3O^+(aq) + OH^-(aq) \rightleftharpoons H_2O(l)$ Acid removes OH^- from solution
   $Fe(OH)_3(s) \rightleftharpoons Fe^{3+}(aq) + 3OH^-(aq)$ Equilibrium shifts to right and $Fe(OH)_3$ dissolves

2. When the solid has a basic anion, such as F^-, CO_3^{2-}, SO_3^{2-}, or S^{2-}, addition of an acid removes the anion from solution (because H_3O^+ reacts with any basic anion), and so the salt dissolves.

For example, with zinc carbonate:

$2H_3O^+(aq) + CO_3^{2-}(aq) \rightleftharpoons 3H_2O(l) + CO_2(aq)$ Acid removes CO_3^{2-} from solution

$ZnCO_3(aq) \rightleftharpoons Zn^{2+}(aq) + CO_3^{2-}(aq)$ Equilibrium shifts to right and $ZnCO_3$ dissolves

3. It is sometimes possible to change the oxidation number of an ion in a salt in order to remove the ion from solution. This shifts the solubility equilibrium toward products and, therefore, dissolves the salt. For example, with insoluble sulfides (such as CuS), we can oxidize the sulfide ion to solid sulfur (which precipitates out of solution) and thereby dissolve the sulfide.

$3S^{2-}(aq) + 8H^+(aq) + 2NO_3^-(aq) \longrightarrow 3S(s) + 2NO(g) + 4H_2O(l)$ Oxidation removes S^{2-}

$CuS(s) \rightleftharpoons Cu^{2+}(aq) + S^{2-}(aq)$ Equilibrium shifts to right and CuS dissolves

4. When it is possible to form a complex ion of one of the ions in a salt, the ion will be removed from solution and the salt will dissolve. For instance, with AgCl, ammonia can be used to remove Ag^+ from solution so that the AgCl dissolves.

$Ag^+(aq) + 2NH_3(aq) \rightleftharpoons Ag(NH_3)_2^+$ Ag^+ is removed from solution

$AgCl(s) \rightleftharpoons Ag^+(aq) + Cl^-(aq)$ Equilibrium shifts to right and AgCl dissolves

EXAMPLE Dissolving a precipitate by removal of an ion from solution

Develop a method to dissolve the insoluble salt cobalt(II) cyanide, $Co(CN)_2$. Write the reactions involved.

SOLUTION The cyanide ion, which is the anion of a weak acid, is a base. If we add an acid (such as HCl) to the solution, the cyanide ion will react with the acid and be removed from solution. Thus, $Co(CN)_2$ will dissolve.

$H_3O^+(aq) + CN^-(aq) \rightleftharpoons H_2O(l) + HCN(aq)$ Acid removes CN^- from solution

$Co(CN)_2(aq) \rightleftharpoons Co^{2+}(aq) + 2CN^-(aq)$ Equilibrium shifts to right and $Co(CN)_2$ dissolves

EXERCISE Develop a method to dissolve the insoluble salt gold(III) bromide, $AuBr_3$. Write the reactions involved.

ANSWER: Remove Au^{3+} from solution by adding CN^- (typically as KCN) to form the complex $Au(CN)_4^-$.

$Au^{3+}(aq) + CN^-(aq) \rightleftharpoons Au(CN)_4^-(aq)$ Au^{3+} is removed from solution

$AuBr_3(s) \rightleftharpoons Au^{3+}(aq) + 3Br^-(aq)$ Equilibrium shifts to right and $AuBr_3$ dissolves

KEY CONCEPT B The effect of pH on the solubilities of compounds

The solubility of any salt that contains a Brønsted base is dependent on the pH of the solution. Let's consider $CaSO_4$ as an example, since it contains the Brønsted base SO_4^{2-}:

$$CaSO_4(s) \rightleftharpoons Ca^{2+}(aq) + SO_4^{2-}(aq)$$

SO_4^{2-} participates in the equilibrium

$$SO_4^{2-}(aq) + H_3O^+(aq) \rightleftharpoons HSO_4^-(aq) + H_2O(l)$$

As the H_3O^+ concentration is increased by the addition of acid, SO_4^{2-} reacts to form HSO_4^-. As the SO_4^{2-} reacts and is withdrawn from solution, the equilibrium involving $CaSO_4$ shifts to the right, and more $CaSO_4$ dissolves. Thus, as the pH is decreased, $CaSO_4$ solubility increases. Conversely, as the pH increases, the second equilibrium shifts toward reactants, forming SO_4^{2-}. This causes the solubility equilibrium for $CaSO_4$ to shift toward $CaSO_4(s)$, reflecting a decrease in solubility. In summary,

$$\text{pH decrease} \longrightarrow [H_3O^+] \text{ increase} \longrightarrow [SO_4^{2-}] \text{ decrease} \longrightarrow \text{higher } CaSO_4 \text{ solubility}$$
$$\text{pH increase} \longrightarrow [H_3O^+] \text{ decrease} \longrightarrow [SO_4^{2-}] \text{ increase} \longrightarrow \text{lower } CaSO_4 \text{ solubility}$$

EXAMPLE Effect of pH on the solubility of a salt

What happens to the solubility of ZnS as the pH is decreased?

SOLUTION The two important equilibria involved are

$$ZnS(s) \rightleftharpoons Zn^{2+}(aq) + S^{2-}(aq)$$

and

$$S^{2-}(aq) + H_3O^+(aq) \rightleftharpoons HS^-(aq) + H_2O(l)$$

A decrease in pH means that the concentration of H_3O^+ increases. This forces the second equilibrium to the right, which decreases the S^{2-} concentration. This in turn drives the first equilibrium to the right, which increases the solubility of ZnS. Thus, as the pH decreases, the solubility of ZnS increases.

EXERCISE What happens to the solubility of AgCl as the pH is increased?

ANSWER: Nothing; Cl^- is too weak a Brønsted base for any effect.

KEY CONCEPT C Formation constants

We now consider how to quantitatively account for the solubility of a compound that contains an ion that forms a complex. Let us assume we want to know the solubility of AgCl in the presence of ammonia. The two equilibria involved are

$$AgCl(s) \rightleftharpoons Ag^+(aq) + Cl^-(aq) \qquad K_s = [Ag^+][Cl^-]$$

$$Ag^+(aq) + 2NH_3(aq) \rightleftharpoons Ag(NH_3)_2^+(aq) \qquad K_f = \frac{[Ag(NH_3)_2^+]}{[Ag^+][NH_3]^2}$$

K_f is the **formation constant** of the complex $Ag(NH_3)_2^+$. Adding these two chemical equations allows us to write the equilibrium that shows how the AgCl dissolves in ammonia. In addition, when two chemical equilibria are added, K for the new equilibrium equals the product of the two original equilibrium constants. Thus,

$$AgCl(s) + 2NH_3(aq) \rightleftharpoons Ag(NH_3)_2^+(aq) + Cl^-(aq) \qquad K = K_s \times K_f = \frac{[Ag(NH_3)_2^+][Cl^-]}{[NH_3]^2}$$

It is now possible to solve an equilibrium problem by constructing a table of concentrations and using the new equilibrium constant.

EXAMPLE Calculating the solubility of AgCl in the presence of ammonia

What is the solubility of AgCl in 0.0050 M NH_3?

SOLUTION We set up a table of equilibrium concentrations, using the symbol P to represent a solid precipitate. We note that the initial concentration of ammonia is 0.0050 M, and we use the symbol S to represent the solubility of AgCl.

	$AgCl(s)$	+	**$2NH_3(aq)$**	$\rightleftharpoons$	**$Ag(NH_3)_2^+(aq)$**	+	**$Cl^-(aq)$**
before equilibrium	P		0.0050 M		0 M		0 M
change to reach equilibrium	$-S$		$-2S$		$+S$		$+S$
after equilibrium	$P - S$		$0.0050\text{ M} - 2S$		S		S

We substitute the equilibrium values into the equilibrium expression and solve for S. Because $K_s = 1.6 \times 10^{-10}$ and $K_f = 1.6 \times 10^7$, we get

$$\frac{[Ag(NH_3)_2^+][Cl^-]}{[NH_3]^2} = K_s \times K_f$$

$$\frac{(S)(S)}{(0.0050\ M - 2S)^2} = (1.6 \times 10^{-10})(1.6 \times 10^7)$$

$$\frac{S^2}{(0.0050\ M - 2S)^2} = 2.6 \times 10^3$$

We now take the square root of both sides of the equation

$$\frac{S}{0.0050\ M - 2S} = 5.1 \times 10^{-2}$$

To solve for S, we multiply both sides of the equation by $0.0050\ M - 2S$ and then collect all terms with S on the left side of the equation:

$$S = (5.1 \times 10^{-2})(0.0050\ M - 2S)$$

$$S = 2.6 \times 10^{-4}\ M - 0.10S$$

$$1.10S = 2.6 \times 10^{-4}\ M$$

$$S = \frac{2.6 \times 10^{-4}\ M}{1.10}$$

$$= 2.4 \times 10^{-4}\ M$$

Because S stands for the solubility of AgCl, the answer is 2.4×10^{-4} M.

EXERCISE What is the solubility of AgBr in 0.0025 M KCN?

ANSWER: 5.0×10^{-5} M

DESCRIPTIVE CHEMISTRY TO REMEMBER

- All **metal cations,** except those of Group I, Group II, and those with +1 charge, produce acid solutions.
- **River water** may contain the following ions: CO_3^{2-}, NO_3^-, PO_4^{3-}, Cl^-, Ca^{2+}, Na^+, NH_4^+, Fe^{3+}, Fe^{2+}, Al^{3+}, and Mg^{2+}.
- **Formic acid** (HCOOH) is the acid present in ant venom.
- An **indicator** is a weak Brønsted acid that has one color in acid form (HIn) and another color in basic form (In^-).
- Blood is maintained at pH = 7.4 by **buffer action**.
- Formation of the **complex** $Au(CN)_2^-$ is used to extract gold from ores.
- **Aqua regia** is a 3:1 mixture of HCl and HNO_3. It is unusual in its ability to dissolve gold.
- **Qualitative analysis** is the analysis of a sample to determine the elements it contains.

CHEMICAL EQUATIONS TO KNOW

- An example of a metal ion acting as an acid is illustrated by the behavior of chromium(III) in aqueous solution.

$$Cr(H_2O)_6^{3+}(aq) \rightleftharpoons Cr(H_2O)_5OH^{2+}(aq) + H_3O^+(aq)$$

- The ammonium ion is acidic.

$$NH_4^+(aq) \rightleftharpoons NH_3(aq) + H_3O^+(aq)$$

- The fluoride ion is a base.

$$F^-(aq) + H_2O(l) \rightleftharpoons HF(aq) + OH^-(aq)$$

- Adding acid to sulfite ion (SO_3^{2-}) results in formation of sulfur dioxide (SO_2).

$$2H_3O^+(aq) + SO_3^{2-}(aq) \longrightarrow SO_2(aq) + 3H_2O(l)$$

- Nitric acid can oxidize sulfide ion to elemental sulfur.

$$3S^{2-}(aq) + 8H^+(aq) + 2NO_3^-(aq) \longrightarrow 3S(s) + 2NO(g) + 4H_2O(l)$$

- Formation of the complex ion $Ag(S_2O_3)_2^{3-}$ is used in photographic processing to remove silver halide from exposed film after it has been developed.

$$Ag^+(aq) + 2S_2O_3^{2-}(aq) \rightleftharpoons Ag(S_2O_3)_2^{3-}(aq)$$

- Formation of a gold cyanide complex is used to dissolve gold by encouraging oxidation of the gold.

$$4Au(s) + 8CN^-(aq) + O_2(g) + 2H_2O(l) \longrightarrow 4Au(CN)_2^-(aq) + 4OH^-(aq)$$

MATHEMATICAL EQUATIONS TO KNOW AND UNDERSTAND

$$pH = pK_a + \log \frac{[A^-]}{[HA]}$$ Henderson–Hasselbalch equation

SELF-TEST EXERCISES

Ions as acids and bases

1. What is the pH of a 0.25 M solution of NH_4Br?
(a) 7.00 (b) 0.62 (c) 13.41 (d) 9.06 (e) 4.93

2. What is the pH of a 0.14 M solution of morphine hydrochloride? Morphine hydrochloride may be represented as $RNH_3^+Cl^-$, where R is a large organic part of the molecule. pK_b (morphine) = 5.79.
(a) 0.90 (b) 4.53 (c) 9.46 (d) 13.21 (e) 7.00

3. For equal concentrations of each cation, which one results in the most acidic solution?
(a) NH_4^+, $pK_b(NH_3) = 4.75$ (b) $CH_3NH_3^+$, $pK_b(CH_3NH_2) = 3.44$
(c) $(C_2H_5)_3NH^+$, $pK_b[(C_2H_5)_3N] = 2.99$

4. What is the pH of a 0.12 M solution of $Co(NO_3)_2$? $pK_a(Co^{2+}) = 8.89$. Assume only one proton is lost from the hydrated ion.
(a) 4.91 (b) 9.09 (c) 0.94 (d) 7.00 (e) 13.06

5. What is the pH of a 0.25 M solution of potassium formate, NaHCOO? $pK_a(HCOOH) = 3.75$.
(a) 8.56 (b) 2.29 (c) 11.71 (d) 7.00 (e) 5.44

6. What is the pH of a solution of 0.050 M calcium acetate $Ca(CH_3COO)_2$? $pK_a(CH_3COOH) = 4.75$.
(a) 10.32 (b) 5.28 (c) 8.72 (d) 3.74 (e) 7.66

7. What is the pH of a solution that is 0.55 M in hypochlorous acid (HClO) and 0.35 M in potassium hypochlorite? $pK_a(HClO) = 7.53$.
(a) 7.91 (b) 7.33 (c) 7.70 (d) 7.00 (e) 7.58

8. What is the pH of a solution that is 0.10 M in hydrazinium chloride (NH_2NH_3Cl) and 0.080 M in hydrazine (NH_2NH_2)? $pK_b(NH_2NH_2) = 5.77$.
(a) 8.13 (b) 5.49 (c) 5.67 (d) 5.92 (e) 8.33

9. A solution made up with equimolar concentrations of a weak acid and its conjugate base has a pH of 6.05. What is the pK_a for the acid?
(a) 5.95 (b) 6.35 (c) 6.05 (d) need more information to tell

10. What is the percentage anilinium ion ($C_6H_5NH_3^+$) deprotonated in a 0.030 M solution of anilinium chloride ($C_6H_5NH_3Cl$)? $K_a(C_6H_5NH_3^+) = 2.3 \times 10^{-5}$ M.
(a) 0.077% (b) 1.4% (c) 0.82% (d) 0.23% (e) 2.7%

11. What is the percentage hypobromite ion (BrO^-) protonated in a 0.16 M solution of sodium hypobromite? $K_a(HBrO) = 3.0 \times 10^{-5}$ M.
(a) 0.019% (b) 3.0% (c) 4.6×10^{-3}%
(d) 2.1×10^{-7}% (e) 0.033%

12. What is the pH of the solution prepared by mixing 25.0 mL of 0.024 M HF with 15.0 mL of 0.15 M NaF? $K_a(HF) = 3.5 \times 10^{-4}$ M.
(a) 6.41 (b) 7.00 (c) 7.59 (d) 4.03 (e) 9.97

13. What is the pH of the solution prepared by mixing 25.0 mL of 0.0040 M HCl with 60.0 mL of 0.025 M NaCl?
(a) 7.00 (b) 2.93 (c) 8.44 (d) 11.12 (e) 6.14

Titrations and pH curves

Questions 14–17 concern a titration in which 20.00 mL of 0.15 M benzoic acid (C_6H_5COOH) are titrated with 0.10 M NaOH. p$K_a(C_6H_5COOH) = 4.19$.

14. What is the pH before any base is added?
(a) 1.46 (b) 2.51 (c) 4.22 (d) 7.00 (e) 11.50

15. What is the pH after addition of 20.00 mL NaOH?
(a) 7.00 (b) 4.26 (c) 4.49 (d) 3.96 (e) 7.30

16. What is the pH at the stoichiometric point (30.00 mL NaOH)?
(a) 8.36 (b) 8.43 (c) 5.52 (d) 7.00 (e) 6.29

17. What is the pH after addition of 10.00 mL NaOH past the stoichiometric point?
(a) 12.22 (b) 1.84 (c) 7.00 (d) 10.67 (e) 3.45

Questions 18–21 concern a titration in which 15.00 mL of 0.120 M morphine are titrated with 0.100 M HCl. pK_b (morphine) = 5.79.

18. What is the pH before the addition of any acid?
(a) 3.46 (b) 0.94 (c) 13.18 (d) 10.65 (e) 7.00

19. What is the pH after addition of 5.00 mL HCl?
(a) 7.79 (b) 7.00 (c) 6.20 (d) 8.24 (e) 8.62

20. What is the pH at the stoichiometric point (18.00 mL HCl added)?
(a) 3.00 (b) 0.35 (c) 4.74 (d) 9.76 (e) 11.04

21. What is the pH after addition of 5.00 mL HCl beyond the stoichiometric point?
(a) 1.88 (b) 7.00 (c) 5.15 (d) 3.12 (e) 12.11

22. Which indicator of the following five would be the best to use for a titration with pH 7.4 at the stoichiometric point? Each indicator is shown with the pH range in which it changes color.
(a) methyl red (4.8–6.0) (b) bromthymol blue (6.0–7.6) (c) thymol blue (9.0–9.6)
(d) phenol red (6.8–8.0) (e) phenolphthalein (8.2–10.0)

23. What is the pH of a formic acid–sodium formate buffer that is 0.80 M in formic acid (HCOOH) and 0.50 M in sodium formate (HCOONa)?
(a) 4.14 (b) 4.00 (c) 4.39 (d) 3.55 (e) 3.81

24. What is the pH of a trimethylammonium chloride–trimethylamine buffer that is 0.20 M in trimethylammonium chloride [$(CH_3)_3NHCl$] and 0.25 M in trimethylamine [$(CH_3)_3N$]? $pK_b[(CH_3)_3N] = 4.19$.
(a) 9.91 (b) 9.88 (c) 9.65 (d) 10.00 (e) 9.71

25. How many milliliters of 2.0 M HCl can be added to 100 mL of a buffer that is 0.20 M in ammonia and 0.20 M in ammonium chloride before the buffer capacity is exceeded?
(a) 8.2 (b) 6.0 (c) 9.3 (d) 5.3 (e) 7.5

26. How many milliliters of 1.5 M NaOH can be added to 250 mL of a buffer that is 0.050 M in acetic acid and 0.040 M in sodium acetate before the buffer capacity is exceeded?
(a) 12 (b) 7.0 (c) 16 (d) 5.0 (e) 10

Solubility equilibria

27. What is the solubility of iron(II) sulfide (FeS) in mol/L? $K_s = 6.3 \times 10^{-18}\ M^2$.
(a) 4.0×10^{-35} M (b) 6.3×10^{-18} M (c) 1.2×10^{-17} M
(d) 2.5×10^{-9} M (e) 6.3×10^{-10} M

28. What is the solubility of copper(II) iodate [$Cu(IO_3)_2$] in mol/L? $K_s = 1.4 \times 10^{-7}\ M^3$.
(a) 3.8×10^{-4} M (b) 1.9×10^{-4} M (c) 1.4×10^{-7} M
(d) 5.2×10^{-3} M (e) 3.3×10^{-3} M

29. What is the solubility of bismuth(III) sulfide (Bi_2S_3) in mol/L? $K_s = 1.0 \times 10^{-97}\ M^5$.
(a) 2.9×10^{-33} M (b) 1.6×10^{-20} M (c) 3.2×10^{-49} M
(d) 4.0×10^{-20} M (e) 2.0×10^{-98} M

30. The solubility of magnesium carbonate ($MgCO_3$) at 20°C is 1×10^{-2} g/100 g water. Estimate K_{sp} for magnesium carbonate.
(a) $4 \times 10^{-4}\ M^2$ (b) $1 \times 10^{-1}\ M^2$ (c) $5 \times 10^{-7}\ M^2$
(d) $1 \times 10^{-6}\ M^2$ (e) $7 \times 10^{-9}\ M^2$

31. The solubility of calcium fluoride (CaF_2) at 20°C is 2×10^{-3} g/100 g water. Estimate K_s from this information.
(a) $7 \times 10^{-11}\ M^3$ (b) $7 \times 10^{-8}\ M^3$ (c) $4 \times 10^{-4}\ M^3$
(d) $2 \times 10^{-3}\ M^3$ (e) $7 \times 10^{-14}\ M^3$

32. What is the solubility of AgCl in 0.20 M NaCl? $K_s(AgCl) = 1.6 \times 10^{-10}\ M^2$.
(a) 1.3×10^{-5} M (b) 8.0×10^{-10} M (c) 3.5×10^{-10} M
(d) 1.6×10^{-10} M (e) 3.2×10^{-11} M

33. What is the solubility of BaF_2 in 0.25 M NaF? $K_s(BaF_2) = 1.7 \times 10^{-6}\ M^3$.
(a) 1.7×10^{-6} M (b) 2.7×10^{-5} M (c) 6.8×10^{-6} M
(d) 1.3×10^{-3} M (e) 4.4×10^{-7} M

34. What is the solubility of Sb_2S_3 in a solution in which $[S^{2-}] = 5.0 \times 10^{-3}$ M? $K_s = 1.7 \times 10^{-93}\ M^5$.
(a) 5.6×10^{-20} M (b) 2.2×10^{-21} M (c) 5.8×10^{-44} M
(d) 3.4×10^{-94} M (e) 2.8×10^{-19} M

35. What is the solubility of $Al(OH)_3(s)$ when the pH = 5.12? $K_s = 1.0 \times 10^{-33}\ M^4$.
(a) 2.5×10^{-9} M (b) 1.2×10^{-7} M (c) 8.1×10^{-5} M
(d) 5.2×10^{-8} M (e) 4.4×10^{-7} M

36. What is the solubility of $Al(OH)_3$ when the solution pH equals 10.55?
(a) 1.4×10^{-2} M (b) 2.0×10^{-9} M (c) 2.8×10^{-12} M
(d) 3.4×10^{-8} M (e) 2.2×10^{-23} M

37. What happens to the solubility of PbS when the pH is increased?
(a) Solubility increases. (b) Solubility decreases. (c) Solubility remains the same.

38. What happens to the solubility of $PbCl_2$ when the pH is decreased?
(a) Solubility increases. (b) Solubility decreases. (c) Solubility remains the same.

39. A solution is prepared that is 2×10^{-4} M in Ag^+ and 5×10^{-9} M in Br^-. Does AgBr precipitate? $K_s(AgBr) = 7.7 \times 10^{-13}$ M^2.
(a) Yes (b) No

40. 0.20 mL of 6×10^{-3} M Na_2CO_3 is added to 1.0 mL of a solution that is 5×10^{-5} M in Ag^+, Mg^{2+}, and Ba^{2+}. Which ions precipitate? $K_s(Ag_2CO_3) = 6.2 \times 10^{-12}$ M^3, $K_s(MgCO_3) = 1.0 \times 10^{-5}$ M^2, $K_s(BaCO_3) = 8.1 \times 10^{-9}$ M^2.
(a) Ag^+ only (b) Mg^{2+} only (c) Ba^{2+} only
(d) Mg^{2+} and Ba^{2+} (e) All precipitate.

41. What is the solubility of AgCl(*s*) in 0.25 M NH_3? $K_f\{[Ag(NH_3)_2]^+\} = 1.6 \times 10^7$ M^{-2}, $K_s(AgCl) = 1.6 \times 10^{-10}$ M^2.
(a) 0.25 M (b) 1.6×10^7 M (c) 4.0×10^{-3} M (d) 0.051 M (e) 0.013 M

42. Which reagent could be used to dissolve copper(II) hydroxide, $Cu(OH)_2(s)$?
(a) CO_2 (b) HCl (c) NaCl (d) NH_3 (e) KNO_3

43. Which reagent could be used to dissolve radium carbonate ($RaCO_3$)?
(a) CaO (b) $CaCl_2$ (c) NH_3 (d) KCl (e) HNO_3

44. Based on the information in Table 15.7, which solid will dissolve in KCN?
(a) WO_3 (b) $CaCO_3$ (c) HgO (d) $Fe(OH)_2$ (e) Ir_2O_3

45. Given the equilibrium constants for equilibriums I and II, what is the equilibrium constant for equilibrium III?

$$\begin{array}{lrcll} \text{I:} & AgI(s) & \rightleftharpoons & Ag^+(aq) + I^-(aq) & K_s = 1.5 \times 10^{-16} \\ \text{II:} & Ag^+(aq) + 2CN^-(aq) & \rightleftharpoons & Ag(CN)_2^- & K_f = 5.6 \times 10^8 \\ \text{III:} & AgI(s) + 2CN^-(aq) & \rightleftharpoons & Ag(CN)_2^- + I^-(aq) & K = ? \end{array}$$

(a) $K = 7.1 \times 10^{-10}$ (b) $K = 1.4 \times 10^9$ (c) $K = 3.7 \times 10^{24}$
(d) $K = 8.4 \times 10^{-8}$ (e) $K = 2.7 \times 10^{-25}$

Descriptive Chemistry

46. Which of the following cations would you expect to be acidic when dissolved in water?
(a) Fe^{3+} (b) Mg^{2+} (c) Na^+ (d) Ag^+ (e) Ba^{2+}

47. All but one of the following ions is found in river water. Which one is not?
(a) Ca^{2+} (b) Al^{3+} (c) Cl^- (d) CO_3^{2-} (e) S^{2-}

48. Aqua regia is a 1:3 mixture of
(a) HNO_3 and NaCl. (b) HNO_3 and HCl. (c) NaOH and $NaNO_3$.
(d) Au and Ag. (e) H_2O and H_2O_2.

49. Which ion is basic?
(a) NH_4^+ (b) O^{2-} (c) Cl^- (d) K^+ (e) NO_3^-

50. What products are formed when aluminum hydroxide, $Al(OH)_3$, is heated?
(a) Al and H_2O (b) Al_2O_3 and H_2O (c) Al, H_2, and O_2 (d) Al_2O_3 and H_2

51. What products are formed when acid is added to $SO_3^{2-}(aq)$?
(a) S_8, H_2, and O_2 (b) SO_4^{2-} and H_2 (c) SO_3 and H_2 (d) SO_2 and H_2O

16 THERMODYNAMICS AND EQUILIBRIUM

THE FIRST LAW OF THERMODYNAMICS

KEY WORDS Define or explain each of the following terms in a written sentence or two.

enthalpy
first law of thermodynamics
internal energy
thermodynamics
work

16.1 HEAT, WORK, AND ENERGY

KEY CONCEPT A Internal Energy

In chemical systems we are concerned with the energy of a collection of particles (atoms, molecules, or formula units). The total energy of all the particles in a sample is called the **internal energy** (U) of the sample; it is a state property. If heat is added to a system or work is done on the system, the internal energy increases; if heat is removed from the system or the system does work on the surroundings, the internal energy decreases:

$$\underset{\Delta U}{\text{Change in internal energy}} = \underset{q}{\text{heat}} + \underset{w}{\text{work}}$$

The sign conventions we use in this equation are shown in the table.

Heat change	Sign of heat and/or work	Internal energy
heat leaves system	−	decreases
heat added to system	+	increases
work done on system	+	increases
work done by system	−	decreases

EXAMPLE Calculating a change in internal energy

A chemical reaction is run in such a way that 844 J of heat is evolved and, at the same time, 118 J of work is done on the surroundings by the chemical system. What is the change in the internal energy of the system?

SOLUTION When heat leaves the system, it is assigned a negative sign, and when work is done by the system, it is assigned a negative sign. Thus, for our problem,

$$q = -844\text{ J} \qquad w = -118\text{ J}$$

The change in internal energy is given by

$$\Delta U = q + w$$

We now substitute the given values of q and w:

$$\begin{aligned}\Delta U &= -844\text{ J} + (-118\text{ J}) \\ &= -962\text{ J}\end{aligned}$$

EXERCISE A chemical reaction is run in such a way that 1.55 kJ of heat is evolved and, at the same time, 328 J of work is done on the chemical system. What is the change in the internal energy of the system?

ANSWER: −1.23 kJ

KEY CONCEPT B Expansion work

Work is defined as distance times force, where the distance is the distance a system moves and force is the force opposing the movement. The units of work are the same as the units of energy, $kg \cdot m^2/s^2$, which shows that work and energy are equivalent. One $kg \cdot m^2/s^2$ is a joule (J), the SI unit of energy. For a gas in a cylinder with a movable piston, work is involved whenever the gas expands or is compressed. The work of expansion or compression is

$$\begin{aligned} \text{Work} &= -\text{pressure} \times \text{change in volume} \\ &= -P \times \Delta V \end{aligned}$$

When the system does work on the surroundings (by expansion of the gas), the work is a negative quantity and will come out negative using the relationship just given. When the surroundings do work on the system (through a compression), the work is positive.

EXAMPLE 1 Calculating the work in the compression or expansion of a gas

A gas trapped in a cylinder expands from 2.33 L to 3.57 L against an external pressure of 1.12 atm. How much work (in joules) is associated with this change?

SOLUTION The work done in the expansion or compression of gas is

$$w = -P \times \Delta V$$

ΔV is 1.24 L (3.57 L − 2.33 L), and the work is done against 1.12 atm pressure, so P is 1.12 atm. Thus,

$$\begin{aligned} w &= -P \times \Delta V \\ &= -1.12 \text{ atm} \times 1.24 \text{ L} \\ &= -1.39 \text{ L} \cdot \text{atm} \end{aligned}$$

This answer must be converted to joules. Because 1 L·atm = 101 J, we have

$$-1.39 \cancel{\text{L atm}} \times \frac{101 \text{ J}}{\cancel{\text{L atm}}} = -140 \text{ J}$$

The answer is negative, as expected, because the system (the trapped gas) does work on the surroundings (the atmosphere) in an expansion.

EXERCISE Calculate the work done when a gas is compressed from 40.3 to 35.6 L against a pressure of 0.57 atm.

ANSWER: 270 J

EXAMPLE 2 Calculating the work accompanying a chemical reaction

Calculate the work done when the following reaction is run at standard temperature and pressure (STP).

$$CaCO_3(s) \longrightarrow CaO(s) + CO_2(g)$$

SOLUTION In this reaction, 1 mol of gas is formed from a solid at STP, so the change in volume, because of the formation of the gas, is 22.4 L. We assume that the volumes of the solids involved are so small that they can be ignored. At STP, the pressure is 1 atm (exactly); so the gas formed must expand into the atmosphere against 1 atm pressure. The result is converted to joules with the relation 1 L·atm = 101 J.

$$\begin{aligned} w &= -P \times \Delta V \\ &= -1 \text{ atm} \times 22.4 \text{ L} \\ &= -22.4 \text{ L atm} \end{aligned}$$

$$\begin{aligned} -22.4 \cancel{\text{L atm}} \times \frac{101 \text{ J}}{\cancel{\text{L atm}}} &= -2.26 \times 10^3 \text{ J} \\ &= -2.26 \text{ kJ} \end{aligned}$$

EXERCISE Calculate the work done when the reaction shown is run at STP.

$$3H_2(g) + N_2(g) \longrightarrow 2NH_3(g)$$

ANSWER: 4.52 kJ

KEY CONCEPT C Measuring a change in internal energy

The change in internal energy is $\Delta U = q + w$, with $w = -P \times \Delta V$. When a process occurs in a closed container, it is impossible for the volume to change. In such a circumstance, $\Delta V = 0$ and $\Delta U = q$. Because this relationship holds for constant volume processes only, the heat transfer is given the symbol q_V, where the subscript connotes a constant volume process.

Constant volume: $\Delta U = q_V$ — Internal energy change is determined by transfer of heat only.

EXAMPLE Understanding constant volume processes

A chemical reaction run in an open beaker releases 1.23 kJ of heat. The same reaction run in a closed, stainless steel cylinder releases 1.25 kJ. What is ΔU for the reaction?

SOLUTION The heat transfer that occurs when a reaction is run at constant volume equals the change in internal energy. When the reaction is run in an open beaker, the volume may change; when it is run in a closed container; the volume cannot change. Hence,

$$\Delta U = q_V$$

Because heat is released, $q_V = -1.25$ kJ and

$$\Delta U = -1.25 \text{ kJ}$$

EXERCISE When the following reaction is run in a sealed bomb calorimeter, 145 J of heat is absorbed. When it is run in an open beaker, 145 J of heat is also absorbed. Explain why the heat transfer is the same in both circumstances. What is ΔU for the reaction?

$$Ca^{2+}(aq) + SO_4{}^{2-}(aq) \longrightarrow CaSO_4(s)$$

ANSWER: In this particular case, no gases are involved; so no change in volume occurs, and $\Delta V = 0$ no matter how the reaction is run. Thus, the heat transfer in an open beaker is the same as the heat transfer in a sealed container: $\Delta U = +145$ J

16.2 Enthalpy

KEY CONCEPT Enthalpy and constant pressure processes

The change in enthalpy is the heat transfer that occurs when a reaction is run at constant pressure (as in an open container). If we use the symbol q_P to indicate the heat transfer at constant pressure, then $\Delta H = q_P$. Because the volume is allowed to change in such a situation, the heat transfer that occurs comes from two possible sources: (a) any change in internal energy that occurs and (b) any work that occurs. Thus, the enthalpy change is determined by both changes in internal energy and work done.

Constant pressure: $\Delta H = q_P$ — Enthalpy change is determined by transfer of heat; the heat accounts for changes in internal energy and any work that must be accomplished.

EXAMPLE Using the definition of enthalpy

A chemical reaction run in an open beaker releases 1.23 kJ of heat. The same reaction run in a closed stainless steel cylinder releases 1.25 kJ. What is ΔH for the reaction?

SOLUTION The change in enthalpy for a chemical reaction is the heat released or absorbed when the reaction is run at constant pressure. When a reaction is run in an open container, the pressure is constant at atmospheric pressure, and the heat released or absorbed is ΔH. Because in this example heat is released, a negative sign is associated with the change. Thus, $\Delta H = -1.23$ kJ. (The difference of 0.02 kJ is the work used to make room in the atmosphere for the products to form.)

EXERCISE A reaction run at constant volume absorbs 3.55 kJ of heat from the surroundings. A separate experiment indicates that the work that accompanies the reaction when it is run at constant pressure is -0.08 kJ. What is ΔH for the reaction?

ANSWER: 3.47 kJ

THE DIRECTION OF SPONTANEOUS CHANGE

KEY WORDS Define or explain each of the following terms in a written sentence or two.

entropy	spontaneous change	standard reaction entropy
second law of thermodynamics	standard molar entropy	Trouton's rule

16.3 ENTROPY AND SPONTANEOUS CHANGE

KEY CONCEPT A Disorder and spontaneous change

Most processes always occur in one direction. For example:

1. When a hot piece of metal is placed in cool water, the metal cools and the water warms until both are at the same temperature. When a piece of metal at some temperature is placed in water at that same temperature, there is never a change in which the metal warms and the water cools.
2. When a bottle of gas is opened in a room, the gas molecules spontaneously spread throughout the room. Never will the spread-out molecules suddenly rush back and gather inside the bottle.
3. When an ice cube is placed in a beaker of water at 25°C, the ice melts and the water cools. A beaker of water at the cooler temperature never suddenly warms a little as an ice cube spontaneously forms in the water.

These are examples of **spontaneous processes,** which are processes that have a tendency to occur without outside intervention. In each example, as in all spontaneous processes, the direction of change is one that leads from order to disorder. By disorder, we mean that there are more choices for the location of molecules or energy. As an analogy, imagine that you take a handful of marbles out of a bag and throw them on the floor; the marbles have more freedom to be in different locations as they come to rest on the floor than when they are confined in the bag. It is this "choice" of location that defines the notion of disorder. In molecular systems, it is the change from order to disorder that is actually the driving force that makes spontaneous processes occur.

In example 1, the thermal energy originally confined to the atoms of the hot metal disperses throughout the water as chaotic thermal motion of the water molecules; because the energy has spread over more quantum levels, an increase in disorder has occurred. In example 2, the molecules originally contained in a small volume disperse and occupy a greater volume; because the molecules in the greater volume have more opportunity to move from place to place and be in different locations, an increase in disorder has occurred. In example 3, both effects occur; thermal energy becomes dispersed over more quantum levels and molecules spread from a smaller to larger volume. Thus, when energy spreads out among more molecules (and hence more quantum levels) as chaotic thermal motion (examples 1 and 3), or when molecules disperse from a smaller to a larger volume (examples 2 and 3), disorder increases.

EXAMPLE Spontaneous processes

Explain how the melting ice in example 3 results in an increase in the disorder of matter.

SOLUTION When the ice melts, the water molecules originally trapped at one site in the solid have access to the entire volume of the liquid. This results in an increase in the disorder of the water molecules.

EXERCISE When a solid dissolves in a solvent, is there an increase or decrease in molecular disorder?

ANSWER: Increase

KEY CONCEPT B Entropy as a measure of disorder

The entropy (S) of a sample is a measure of the disorder (both energy disorder and molecular disorder) of the sample. The greater the entropy, the greater the disorder.

EXAMPLE Qualitative prediction of the change in entropy

Is there an increase, decrease, or no change in entropy (as measured by molecular disorder, neglecting energy disorder) in the chemical reaction

$$2SO_2(g) + O_2(g) \longrightarrow 2SO_3(g)$$

SOLUTION There are 3 mol of gaseous reactants and 2 mol of gaseous products. When the component atoms are contained in 3 mol of gas, they are more spread out then when they are contained in 2 mol of gas. The reactants are more disordered than the products, so there is a decrease in disorder in the chemical reaction. There is a decrease in entropy for the reaction ($\Delta S < 0$), if energy disorder is ignored.

EXERCISE Is there an increase, decrease, or no change in entropy (as measured by molecular disorder) in the chemical reaction

$$2Na(s) + 2H_2O(l) \longrightarrow 2Na^+(aq) + 2OH^-(aq) + H_2(g)$$

ANSWER: Increase; $\Delta S > 0$

KEY CONCEPT C The Boltzmann formula for the entropy

The Boltzmann formula

$$S = k \ln W$$

allows us to advance from the notion of disorder to a precise calculated value of the entropy of a system. In the formula, k is the **Boltzmann constant** (1.38×10^{-23} J/K) and W (for molecular disorder) is the number of different ways of arranging the molecules in the system without changing the total energy of the system. For N molecules, in which each molecule can have P positions,

$$W = P^N$$

Substituting this into the Boltzmann formula gives

$$\begin{aligned} S &= k \ln P^N \\ &= kN \ln P \end{aligned}$$

For 1 mol of material, $N = 6.022 \times 10^{23}$ and

$$\begin{aligned} S &= (1.38 \times 10^{-23}\ \text{J/K})(6.02 \times 10^{23}) \ln P \\ &= (8.31\ \text{J/K}) \ln P \end{aligned}$$

A perfect crystal is one for which $W = 1$. Thus, the entropy of a perfect crystal at 0 K is zero. At $T = 0$ K, there is no chaotic thermal motion, so there is no energy disorder. If $W = 1$ there is no matter disorder. Thus, the total entropy of the perfect crystal is zero at 0 K $\{S = k \ln W = k \ln 1 = k(0) = 0\}$.

EXAMPLE Estimating the entropy of a crystal at 0 K

Calculate the entropy of 1 mol of crystalline $CFCl_3$ at 0 K according to the Boltzmann formula.

SOLUTION $CFCl_3$ is a tetrahedral molecule. There are four possible orientations of the molecule in a crystal, depending on which of the four equivalent tetrahedral positions is occupied by the single F. There should not be much difference in the energy of the four orientations because the halogens F and Cl are similar. We can, therefore, use the Boltzmann formula with $W = 4$. For 1 mol of substance,

$$\begin{aligned} S &= (8.31\ \text{J/K}) \ln 4 \\ &= (8.31\ \text{J/K})(1.39) \\ &= 11.5\ \text{J/K} \end{aligned}$$

EXERCISE Use the Boltzmann formula to estimate at 0 K the entropy of 1 mol of crystalline SOF_4. This is a trigonal bipyramidal molecule.

ANSWER: 13.4 J/K

KEY CONCEPT D Standard molar entropies

The standard molar entropy of a substance is the entropy per mole of the pure substance at 1 atm pressure. Some features of standard molar entropies are apparent from a examination of Appendix 2A of the text.

1. As the temperature of a substance increases, its entropy increases.
2. The entropy of a substance depends on the physical state of the substance:

$$S(\text{gas}) > S(\text{liquid}) > S(\text{solid})$$

3. Larger, more complex substances tend to have a greater entropy than smaller, simpler substances (for the same physical state at the same temperature).
4. All entropies at 298 K are positive. (At 0 K, the entropy of all substances is zero or near zero, and as the temperature increases, the entropy increases.)

EXAMPLE Calculating the entropy of a physical change

One form of SiO_2 (tridymite) has $S° = 11.2$ J/K·mol at 298 K. Another form of SiO_2 (quartz) has $S° = 10.00$ J/K·mol.
(a) What is the entropy change when 100 g of quartz changes to 100 g of tridymite at 298 K? (b) One of these two forms is crystalline and the other is amorphous. Which is crystalline?

SOLUTION (a) The process is

$$SiO_2(\text{quartz}) \longrightarrow SiO_2(\text{tridymite})$$

If 1 mol of SiO_2 undergoes this change, the entropy change is

$$\begin{aligned} \Delta S° &= S°(\text{tridymite}) - S°(\text{quartz}) \\ &= 11.2\ \text{J/K·mol} - 10.00\ \text{J/K·mol} \\ &= 1.2\ \text{J/K·mol} \end{aligned}$$

The moles of SiO_2 in 100 g are calculated from its molar mass, 60.08 g/mol.

$$\text{Moles } SiO_2 = 100 \not{g} \times \frac{1 \text{ mol } SiO_2}{60.08 \not{g}}$$
$$= 1.66 \text{ mol } SiO_2$$

So, for 100 g of SiO_2 (1.66 mol),

$$1.66 \not{\text{mol}}\ SiO_2 \times 1.2 \frac{\text{J}}{\not{\text{mol}} \cdot \text{K}} = 2.0 \text{ J/K}$$

(b) An amorphous substance is more disordered than a crystalline substance. More disorder implies a higher molar entropy. Thus, tridymite, with the higher standard molar entropy, must be the amorphous form and quartz, with the lower standard molar entropy, must be the crystalline form.

EXERCISE What is the entropy change when 50.0 g of rhombic sulfur ($S° = 254.4$ J/K·mol) change to monoclinic sulfur ($S° = 260.8$ J/K·mol)? Which form of sulfur is more disordered?

$$S_8(\text{rhombic}) \longrightarrow S_8(\text{monoclinic})$$

ANSWER: +1.3 J/K; monoclinic

KEY CONCEPT E Standard reaction entropies

The entropies in Table 16.1 (and Appendix 2A) of the text can be used to calculate the change in entropy for a reaction at standard conditions.

$$\underset{\text{Pure, 1 atm pressure}}{\text{Reactants at standard conditions}} \longrightarrow \underset{\text{Pure, 1 atm pressure}}{\text{products at standard conditions}}$$

$$\Delta S° = S°(\text{products}) - S°(\text{reactants})$$

This calculation is done in precisely the same manner as calculations of reaction enthalpy from $\Delta H_f°$ values. As before, we must remember that each entropy in the table is a per-mole quantity, so the number of moles of each reactant and product in the chemical equation must be taken into account.

EXAMPLE Calculating a standard reaction entropy

Calculate the standard reaction entropy for the following reaction and comment on the sign of the result.

$$2H_2(g) + CO(g) \longrightarrow CH_3OH(l) \qquad \Delta S° = ?$$

SOLUTION $\Delta S° = S°(\text{products}) - S°(\text{reactants})$. We first retrieve the value of $S°$ for each reactant and product from Appendix 2A or Table 16.1 of the text.

$$\overset{130.7 \text{ J/K·mol}}{2H_2(g)} + \overset{197.7 \text{ J/K·mol}}{CO(g)} \longrightarrow \overset{126.8 \text{ J/K·mol}}{CH_3OH(l)}$$

Substituting into the equation for calculating the standard reaction entropy gives

$$\begin{aligned}\Delta S° &= S°(\text{products}) - S°(\text{reactants}) \\ &= [1 \text{ mol } CH_3OH \times S°\{CH_3OH(l)\}] - [1 \text{ mol } CO \times S°\{CO(g)\} + 2 \text{ mol } H_2 \times S°\{H_2(g)\}] \\ &= \left[1 \not{\text{mol}\ CH_3OH} \times 126.8 \frac{\text{J}}{\not{\text{mol}} \cdot \text{K}}\right] - \left[\left(1 \not{\text{mol}\ CO} \times 197.7 \frac{\text{J}}{\not{\text{mol}} \cdot \text{K}}\right) + \left(2 \not{\text{mol}\ H_2} \times 130.7 \frac{\text{J}}{\not{\text{mol}} \cdot \text{K}}\right)\right] \\ &= [126.8 \text{ J/K}] - [459.1 \text{ J/K}] \\ &= -332.3 \text{ J/K}\end{aligned}$$

In this reaction, 3 mol of a gas become 1 mol of a liquid; so a decrease in disorder and a decrease in entropy occur. This is reflected in the negative sign of $\Delta S°$.

EXERCISE First predict the sign of the entropy change and then calculate the standard reaction entropy for the reaction

$$2SO_2(g) + O_2(g) \longrightarrow 2SO_3(g)$$

ANSWER: Negative; -188.1 J/K

PITFALL Elements have finite values of $S°$

Unlike the values of the standard enthalpies of formation, the standard molar entropies of the elements are not zero. Do not forget to include them in calculations of standard reaction entropies.

16.4 THE ENTROPY CHANGE IN THE SURROUNDINGS

KEY CONCEPT A The importance of the entropy change of the surroundings

For a process to be a spontaneous process, there must be an increase in the total entropy of the system and the surroundings. An increase in the entropy of either the system or surroundings can offset a decrease in entropy of the other.

$$\text{Spontaneous process: } \begin{pmatrix}\text{change in entropy}\\ \text{of system}\end{pmatrix} + \begin{pmatrix}\text{change in entropy}\\ \text{of surroundings}\end{pmatrix} > 0$$

When this occurs, the overall disorder of the system plus surroundings increases. We have already seen how to calculate the change in entropy of a reaction system. For a chemical reaction or change of state, the entropy change of the system is what we earlier called the entropy of reaction. The entropy change of the surroundings is calculated with the following relationship:

$$\text{Change in entropy of surroundings} = \frac{\text{heat transfer to or from surroundings}}{T \text{ (in kelvins) of surroundings}}$$

$$\Delta S_{\text{surr}} = \frac{\text{heat transfer to or from surroundings}}{T_{\text{surr}}}$$

For both endothermic and exothermic reactions, the heat flow (from the surroundings' point of view) has a sign opposite to that of the reaction enthalpy. This leads to

$$\Delta S_{\text{surr}} = \frac{-\Delta H}{T_{\text{surr}}}$$

Heat flow into the surroundings increases the entropy of the surroundings because the chaotic thermal motion of the molecules in the surroundings increases, whereas heat flow out of the surroundings decreases its entropy.

Exothermic reaction ➡ $\Delta H < 0$ ➡ heat flow into surroundings ➡ disorder in surroundings increases ➡ $\Delta S_{\text{surr}} > 0$

Endothermic reaction ➡ $\Delta H > 0$ ➡ heat flow out of surroundings ➡ disorder in surroundings decreases ➡ $\Delta S_{\text{surr}} < 0$

EXAMPLE Calculation of the total entropy change of a reaction

Calculate the standard entropy change at 298 K of the system and of the surroundings for the reaction

$$3Fe(s) + 2O_2(g) \longrightarrow Fe_3O_4(s)$$

Is this reaction (which is a major one that occurs when your car rusts) a spontaneous process?

SOLUTION We anticipate the need to calculate $\Delta S°$ and $\Delta H°$ for the reaction by assembling a table of enthalpies of formations and standard molar entropies.

	$3Fe(s)$ +	$2O_2(g)$	$\longrightarrow Fe_3O_4(s)$
$\Delta H_f°$ (per mole)	0	0	−1118 kJ
$S°$ (per mole)	27.28 J/K	205.14 J/K	146.4 J/K

The entropy change for the system (the standard reaction entropy) is

$$\begin{aligned}\Delta S° &= S°(\text{products}) - S°(\text{reactants})\\ &= \left[(1\ \cancel{\text{mol Fe}_3\text{O}_4}) \times \left(146.4\,\frac{\text{J}}{\cancel{\text{mol}}\cdot\text{K}}\right)\right]\\ &\quad - \left[\left(2\ \cancel{\text{mol O}_2} \times 205.14\,\frac{\text{J}}{\cancel{\text{mol}}\cdot\text{K}}\right) + (3\ \cancel{\text{mol Fe}}) \times \left(27.28\,\frac{\text{J}}{\cancel{\text{mol}}\cdot\text{K}}\right)\right]\\ &= [146.4\text{ J/K}] - [492.12\text{ J/K}]\\ &= -345.7\text{ J/K}\end{aligned}$$

To calculate the entropy change in the surroundings, we need the standard enthalpy of reaction:

$$\begin{aligned}\Delta H° &= \left[(1\ \cancel{\text{mol Fe}_3\text{O}_4}) \times \left(-1118\,\frac{\text{kJ}}{\cancel{\text{mol}}}\right)\right] - \left[(2\ \cancel{\text{mol O}_2}) \times \left(0\,\frac{\text{kJ}}{\cancel{\text{mol}}}\right) + (3\ \cancel{\text{mol Fe}}) \times \left(0\,\frac{\text{kJ}}{\cancel{\text{mol}}}\right)\right]\\ &= -1118\text{ kJ} = -1.118 \times 10^6\text{ J}\end{aligned}$$

The entropy change in the surroundings is

$$\begin{aligned}\Delta S_{surr} &= \frac{-\Delta H°}{T_{surr}}\\ &= \frac{-(-1.118 \times 10^6\text{ J})}{298\text{ K}}\\ &= +3.75 \times 10^3\text{ J/K}\end{aligned}$$

The total entropy change for the reaction is

$$\begin{aligned}\Delta S°(\text{system + surroundings}) &= -345.7\text{ J/K} + 3.75 \times 10^3\text{ J/K}\\ &= +3.40 \times 10^3\text{ J/K}\end{aligned}$$

The large positive change in entropy indicates the reaction is a spontaneous reaction.

EXERCISE Calculate the total standard entropy change at 298 K for the reaction shown and state whether it is a spontaneous reaction.

$$CuSO_4\cdot 5H_2O(s) \longrightarrow CuSO_4(s) + 5H_2O(g)$$

ANSWER: −251 J/K; not a spontaneous reaction

KEY CONCEPT B Systems at equilibrium

There is no net change for a system at equilibrium, so the total entropy change for a system at equilibrium is zero.

$$\text{System at equilibrium: } \begin{pmatrix}\text{entropy change}\\ \text{of system}\end{pmatrix} + \begin{pmatrix}\text{entropy change of}\\ \text{surroundings}\end{pmatrix} = 0$$

$$\Delta S \qquad\qquad \left(\frac{-\Delta H}{T_{surr}}\right)$$

Two examples of equilibria occur at the boiling and freezing points of a substance:

Boiling/condensation point	pure liquid in equilibrium with vapor at 1 atm (standard conditions)	$\Delta S^\circ_{vap} - \dfrac{\Delta H^\circ_{vap}}{T_b} = 0$
Melting/freezing point	pure liquid in equilibrium with pure solid both at 1 atm (standard conditions)	$\Delta S^\circ_{fus} - \dfrac{\Delta H^\circ_{fus}}{T_m} = 0$

We make the assumption here that the surroundings are at the same temperature as the sample, so T_{surr} becomes T_m for the melting/freezing equilibrium and T_b for the boiling/condensation process. Also, the values of ΔS° and ΔH° should correspond to the temperature of the equilibrium; but for temperatures near 25°C, the values of ΔS° and ΔH° at 25°C can be used as an approximation.

▼ **EXAMPLE Calculating the molar entropy of a phase change**

What is the molar entropy of vaporization for $CCl_4(g)$, which has a molar enthalpy of vaporization of 30.00 kJ/mol and boils at 349.9 K?

SOLUTION The molar entropy of vaporization can be calculated with the formula

$$\Delta S^\circ_{vap} - \frac{\Delta H^\circ_{vap}}{T_b} = 0$$

This rearranges to

$$\Delta S^\circ_{vap} = \frac{\Delta H^\circ_{vap}}{T_b}$$

Substituting the given data gives

$$\Delta S^\circ_{vap} = \frac{30.00 \times 10^3 \text{ J/mol}}{349.9 \text{ K}}$$
$$= 85.7 \text{ J/K}\cdot\text{mol}$$

EXERCISE The molar heat of melting of O_2 is 444 J/mol and its melting temperature is −218.8°C. What is ΔS for the melting of O_2?

▲ ANSWER: 8.17 J/K·mol

16.5 THE SECOND LAW

KEY CONCEPT The second law

The **second law of thermodynamics** formalizes the proposition we made earlier regarding the total change in entropy of a spontaneous change. If we define the "universe" as the system and surroundings combined, the second law of thermodynamics states the following: a spontaneous change is accompanied by an increase in the total entropy of the universe. An exothermic reaction increases the entropy of the surroundings by increasing the chaotic thermal motion of the surroundings. Thus, even if the entropy of the system decreases as the reactants become products, this decrease may be offset by an increase in the entropy of the surroundings, and the change may still be a spontaneous change. An endothermic reaction decreases the entropy of the surroundings by decreasing the chaotic thermal motion of the surroundings. This decrease may be offset by an increase in the entropy change of the system as the reactants become products, and the change may still be a spontaneous change. For chemical reactions, the entropy change of the system consists of the entropy change, $\Delta S^\circ = S^\circ(\text{products}) - S^\circ(\text{reactants})$. Very often the entropy change of the system is relatively small, so the entropy change of the surroundings determines whether the change is a spontaneous change. That is, it is often the exothermicity or endothermicity of a reaction that determines if the change is a spontaneous change.

EXAMPLE 1 Qualitative use of the second law

The following reaction is exothermic. Without detailed calculations, comment on whether it is a spontaneous reaction.

$$2C_4H_{10}(l) + 13O_2(g) \longrightarrow 8CO_2(g) + 10H_2O(g)$$

SOLUTION Because the reaction is exothermic, the entropy change in the surroundings is positive and contributes toward a spontaneous change. In addition, the reaction involves forming 18 mol of gas from 10 mol of gas and 2 mol of liquid, so the entropy change of the system is almost certainly positive. This also contributes to making the reaction a spontaneous change. The entropy effects in the surroundings and in the system are positive, so the reaction is spontaneous. In summary,

$\Delta S_{surr} > 0$ and $\Delta S_{system} > 0$ so the process is spontaneous.

EXERCISE The reaction below is endothermic. Without detailed calculations, comment on whether it is a spontaneous reaction.

$$3H_2O(g) + 2CO_2(g) \longrightarrow 3O_2(g) + C_2H_5OH(l)$$

ANSWER: $\Delta S_{surr} < 0$ and $\Delta S_{system} < 0$; not a spontaneous reaction

EXAMPLE 2 Qualitative use of the second law

The following reaction is endothermic. Without detailed calculations, comment on whether it is a spontaneous reaction.

$$SiO_2(s) + 2C(s) \longrightarrow Si(s) + 2CO(g)$$

SOLUTION An endothermic reaction decreases the entropy of the surroundings, which contributes to a nonspontaneous change. In the reaction, 2 mol of gas and 1 mol of solid are formed from 3 mol of solid, so the entropy change of the system is most likely positive. The two factors have opposite effects, so whether the reaction is spontaneous depends on which effect is larger.

$\Delta S_{surr} < 0$ and $\Delta S_{system} > 0$ opposite effects; cannot tell if reaction is spontaneous without detailed calculation

EXERCISE The following reaction is exothermic. Without detailed calculations, comment on whether it is a spontaneous reaction.

$$C_2H_4(g) + Cl_2(g) \longrightarrow C_2H_4Cl_2(l)$$

ANSWER: $\Delta S_{surr} > 0$ and $\Delta S_{system} < 0$; cannot tell if reaction is a spontaneous process without detailed calculations

FREE ENERGY

KEY WORDS

Define or explain each of the following terms in a written sentence or two.

Gibbs free energy
standard free energy of formation
standard free energy of reaction
thermodynamically unstable compound

16.6 FOCUSING ON THE SYSTEM

KEY CONCEPT A The free energy change of the system

Whether a process is spontaneous depends on the total entropy change of the process. The **Gibbs free energy** change (ΔG) of the system is defined in such a way that it depends directly on the total entropy change:

$$\Delta G = -T\Delta S_{total}$$

Here, ΔS_{total} is the entropy change of the universe:

$$\Delta S_{total} = \Delta S_{surr} + \Delta S_{system}$$

Because any spontaneous change has a positive ΔS_{total}, it must also have a negative ΔG.

$$\text{Spontaneous change: } \Delta S_{total} > 0$$
$$\Delta G < 0$$

Notice that ΔG does not have the subscript "total" because the free energy change is for the system only. The free energy change of the system "keeps track of" the entropy change of the universe. Any spontaneous change occurs so as to maximize the entropy of the universe and minimize the free energy of the system. From the definition of the Gibbs free energy, $G = H - TS$, its change during a process is

$$\Delta G_{system} = \Delta H_{system} - T\Delta S_{system}$$

We write this simply as

$$\Delta G = \Delta H - T\Delta S$$

It must be remembered that ΔS now refers to the change in the *system* only. In earlier sections, we used this symbol to indicate the entropy change of the universe.

EXAMPLE Calculating the free energy change

A reaction has $\Delta H = +36.2$ kJ and $\Delta S = 115$ J/K at 25°C. Calculate ΔG and state whether the reaction is spontaneous.

SOLUTION The free energy change is given by

$$\Delta G = \Delta H - T\Delta S$$

Substituting the given values

$$\Delta H = 36.2 \text{ kJ} = 3.62 \times 10^4 \text{ J}$$
$$T = 25 + 273 = 298 \text{ K}$$
$$\Delta S = 115 \text{ J/K}$$

gives

$$\Delta G = 3.62 \times 10^4 \text{ J} - (298 \text{ K})\left(115 \frac{\text{J}}{\text{K}}\right)$$
$$= 3.62 \times 10^4 \text{ J} - 3.43 \times 10^4 \text{ J}$$
$$= 0.19 \times 10^4 \text{ J}$$
$$= 1.9 \times 10^3 \text{ J}$$

The reaction is not spontaneous because ΔG is positive.

EXERCISE What is ΔG for the reaction in the example at 100°C. (Assume ΔH and ΔS do not change in going from 25 to 100°C.) Is the reaction spontaneous at 100°C?

ANSWER: −6.70 kJ; yes

PITFALL The proper units for $\Delta G = \Delta H - T\Delta S$

Because values of ΔH are typically expressed in kilojoules and values of ΔS are usually expressed in joules per kelvin (not kilojoules per kelvin), the units of either ΔH or ΔS must be adjusted so that the two are consistent. In the preceding example, for instance, the units of ΔH are changed from kilojoules to joules. The temperature must be expressed in kelvins when using the relation $\Delta G = \Delta H - T\Delta S$.

KEY CONCEPT B Factors that determine the sign of ΔG

The factors that determine the sign of ΔG are (a) the sign of ΔH, (b) the sign of ΔS, and (c) the temperature. An exothermic process has $\Delta H < 0$. Because $\Delta G = \Delta H - T\Delta S$, a negative ΔH contributes to making ΔG negative. An endothermic process, however, has $\Delta H > 0$, and the positive ΔH contributes to making ΔG positive. Because of the minus sign in $\Delta G = \Delta H - T\Delta S$, a positive ΔS tends to make ΔG negative whereas a negative ΔS tends to make ΔG positive. The effect of temperature is illustrated in the previous example and the accompanying exercise. At 25°C the reaction described is not spontaneous, but at 100°C it is. The temperature is a weighting factor for ΔS; the higher the temperature, the more important is ΔS in determining the sign of ΔG. These ideas are summarized in the following table.

Sign of ΔH	Sign of ΔS	Temperature	Sign of ΔG	Nature of change
−	+	all	−	spontaneous process
+	−	all	+	not a spontaneous process
−	+	low	−	spontaneous process
		high	+	not a spontaneous process
+	−	low	+	not a spontaneous process
		high	−	spontaneous process

It is possible for ΔG to be zero. In such a circumstance, the system is at equilibrium. When a system is at equilibrium, the change represented by the chemical equation is not a spontaneous process, nor is the reverse process a spontaneous process.

EXAMPLE Calculating the minimum temperature needed for decomposition

Calculate the minimum temperature in degrees Celsius at which the decomposition of $CuSO_4(s)$ to $CuO(s)$ + $SO_3(g)$ is spontaneous. Data from Appendix 2A of the text will prove helpful.

$$CuSO_4(s) \longrightarrow CuO(s) + SO_3(g)$$

SOLUTION The decomposition is spontaneous when $\Delta G < 0$ for the reaction. A reaction will occur (or try to occur) only when it is a spontaneous process. Thus, the minimum temperature needed for the decomposition is given by the condition $\Delta G < 0$. We will calculate the temperature for the condition $\Delta G = 0$;

$$\Delta G = \Delta H - T\Delta S$$
$$0 = \Delta H - T\Delta S$$
$$T = \frac{\Delta H}{\Delta S}$$

From the enthalpies of formation for the compounds in the reaction, we have

$$\Delta H^\circ = \Delta H_f^\circ(\text{products}) - \Delta H_f^\circ(\text{reactants})$$

$$\begin{aligned}\Delta H^\circ &= \{1 \text{ mol CuO}(s) \times \Delta H_f^\circ[\text{CuO}(s)] + 1 \text{ mol SO}_3 \times \Delta H_f^\circ[\text{SO}_3(g)]\} \\ &\quad - \{1 \text{ mol CuSO}_4 \times \Delta H_f^\circ[\text{CuSO}_4(s)]\} \\ &= \left\{1 \cancel{\text{mol CuO}} \times \left(-157.3 \frac{\text{kJ}}{\cancel{\text{mol}}}\right) + 1 \cancel{\text{mol}} \text{ SO}_3 \times \left(-395.72 \frac{\text{kJ}}{\cancel{\text{mol}}}\right)\right\} \\ &\quad - \left\{1 \text{ mol } \cancel{\text{CuSO}_4} \times \left(-771.36 \frac{\text{kJ}}{\cancel{\text{mol}}}\right)\right\} \\ &= \{-553.02 \text{ kJ}\} - \{-771.36 \text{ kJ}\} \\ &= +218.3 \text{ kJ} = 2.183 \times 10^5 \text{ J}\end{aligned}$$

From the standard molar entropies, we have

$$\begin{aligned}\Delta S^\circ &= S^\circ(\text{products}) - S^\circ(\text{reactants}) \\ &= S^\circ\{1 \text{ mol CuO}(s) \times S^\circ(\text{CuO}) + 1 \text{ mol SO}_3(g) \times S^\circ(\text{SO}_3)\} - \{1 \text{ mol CuSO}_4(s) \times S^\circ(\text{CuSO}_4(s))\} \\ &= \left\{1 \cancel{\text{mol CuO}} \times 42.63 \frac{\text{J}}{\cancel{\text{mol}} \cdot \text{K}} + 1 \cancel{\text{mol SO}_3} \times 256.76 \frac{\text{J}}{\cancel{\text{mol}} \cdot \text{K}}\right\} - \left\{1 \cancel{\text{mol CuSO}_4} \times 109.0 \frac{\text{J}}{\cancel{\text{mol}} \cdot \text{K}}\right\} \\ &= \{299.39 \text{ J/K}\} - \{109.0 \text{ J/K}\} \\ &= 190.4 \text{ J/K}\end{aligned}$$

We now assume that ΔH and ΔS do not change significantly with temperature, so we can use the values of ΔH° and ΔS° determined at 25°C (from the data in Appendix 2A) at other temperatures.

$$\begin{aligned}T &= \frac{\Delta H}{\Delta S} \\ &= \frac{2.183 \times 10^5 \cancel{\text{J}}}{190.4 \cancel{\text{J}}/\text{K}} \\ &= 1147 \text{ K} \\ T(^\circ\text{C}) &= 1147 - 273 \\ &= 874^\circ\text{C}\end{aligned}$$

This temperature is calculated from the condition $\Delta G = 0$. Any temperature above this makes the decomposition a spontaneous process.

EXERCISE Calculate the minimum temperature in °C at which the decomposition of $CuSO_4 \cdot 5H_2O(s)$ to $CuSO_4(s)$ and $H_2O(g)$ is a spontaneous process.

$$\text{CuSO}_4 \cdot 5\text{H}_2\text{O}(s) \longrightarrow \text{CuSO}_4(s) + 5\text{H}_2\text{O}(g)$$

ANSWER: 125°C

KEY CONCEPT C Standard free energies of formation

If a reaction is run under standard conditions, that is, with pure reactants and products at 1 atm pressure, the free energy change associated with the reaction is called the **standard free energy:**

$$\text{Reactants (pure, 1 atm)} \longrightarrow \text{products (pure, 1 atm)}$$
$$\Delta G = \text{standard free energy } (\Delta G^\circ)$$

The **standard free energy of formation** of a compound is the standard reaction free energy for synthesis of the compound from its elements:

$$\text{Elements (pure, 1 atm)} \longrightarrow \text{compound (pure, 1 atm)}$$
$$\Delta G = \text{standard free energy of formation of compound } (\Delta G_f^\circ)$$

A table of standard free energies of formation is given in Appendix 2A of the text. A compound is classified as **thermodynamically unstable** if its standard free energy of formation is positive. This means that there is a *tendency* for the compound to decompose into its elements. A **thermodynamically stable** compound is one with a negative ΔG_f°, because the standard reaction free energy for decomposition of the compound into its elements is then positive, and the decomposition is not a spontaneous process.

EXAMPLE Calculating the standard free energy of formation

Calculate the standard free energy of formation of $CO_2(g)$ at 25°C from its standard enthalpy of formation and standard molar entropy. Is $CO_2(g)$ thermodynamically stable or thermodynamically unstable?

SOLUTION The standard free energy of formation is the free energy change associated with the following reaction, where it is understood that all products and reactants are pure and at 1 atm pressure:

$$\text{C(graphite)} + O_2(g) \longrightarrow CO_2(g) \qquad \Delta G_f^\circ = ?$$

To calculate ΔG_f° we use the following steps:

1. $\Delta G_f^\circ = \Delta H_f^\circ - T\Delta S^\circ$

2. $\Delta H_f^\circ = \Delta H_f^\circ(\text{products}) - \Delta H_f^\circ(\text{reactants})$

$$= \left\{1\ \cancel{\text{mol CO}_2} \times \left(-393.51\ \frac{\text{kJ}}{\cancel{\text{mol}}}\right)\right\} - \left\{1\ \cancel{\text{mol C}} \times \left(0\ \frac{\text{kJ}}{\cancel{\text{mol}}}\right) + 1\ \cancel{\text{mol O}_2} \times \left(0\ \frac{\text{kJ}}{\cancel{\text{mol}}}\right)\right\}$$
$$= -393.51\ \text{kJ}$$
$$= -3.9351 \times 10^5\ \text{J}$$

3. $\Delta S^\circ = S^\circ(\text{products}) - S^\circ(\text{reactants})$

$$= \left\{1\ \cancel{\text{mol CO}_2} \times \left(213.74\ \frac{\text{J}}{\text{K}\cdot\cancel{\text{mol}}}\right)\right\} - \left\{1\ \cancel{\text{mol C}} \times \left(5.740\ \frac{\text{J}}{\text{K}\cdot\cancel{\text{mol}}}\right) + 1\ \cancel{\text{mol O}_2} \times \left(205.1\ \frac{\text{J}}{\text{K}\cdot\cancel{\text{mol}}}\right)\right\}$$
$$= +2.9\ \text{J/K}$$

Substituting into the required equation gives

$$\begin{aligned}\Delta G^\circ &= \Delta H^\circ - T\Delta S^\circ \\ &= -3.9351 \times 10^5\ \text{J} - (298\ \cancel{\text{K}})(2.9\ \text{J}/\cancel{\text{K}}) \\ &= -3.9351 \times 10^5\ \text{J} - 8.6 \times 10^2\ \text{J} \\ &= -3.9437 \times 10^5\ \text{J} \\ &= -394.37\ \text{kJ}\end{aligned}$$

The answer -394.37 kJ is the standard free energy of formation for 1 mol of CO_2, so $\Delta G_f^\circ(CO_2, g) = -394.37$ kJ/mol. Because $CO_2(g)$ has a negative standard free energy of formation, it is thermodynamically stable.

EXERCISE Calculate the standard free energy of formation of $CCl_4(l)$ at 25°C from its standard enthalpy of formation and standard molar entropy. Is $CCl_4(l)$ a thermodynamically stable compound?

ANSWER: -65.2 kJ/mol; yes

16.7 SPONTANEOUS REACTIONS

KEY CONCEPT Using standard free energies of formation

Standard free energies of formation can be used to calculate a standard reaction free energy, in a manner analogous to using standard enthalpies of formation to calculate standard reaction enthalpies and standard molar entropies to calculate standard reaction entropies.

$$\Delta G^\circ = \Delta G_f^\circ(\text{products}) - \Delta G_f^\circ(\text{reactants})$$

In using standard free energies of formation, we must remember to take into account the number of moles of reactants and products expressed in the balanced equation.

▼ **EXAMPLE Calculating the free energy of formation**

Is the production of butane gas [$C_4H_{10}(g)$] at 298 K from the reaction of methane [$CH_4(g)$] with ethyne [$C_2H_2(g)$] a spontaneous reaction?

SOLUTION To answer this question, we must calculate ΔG for the reaction

$$2CH_4(g) + C_2H_2(g) \longrightarrow C_4H_{10}(g)$$

We assume all of the gases are at a partial pressure of 1 atm so that we can use ΔG° to answer the question.

$$\begin{aligned}\Delta G^\circ &= \Delta G_f^\circ(\text{products}) - \Delta G_f^\circ(\text{reactants}) \\ &= \left\{1 \cancel{\text{mol } C_4H_{10}} \times \left(-17.03 \frac{\text{kJ}}{\cancel{\text{mol}}}\right)\right\} \\ &\quad - \left\{2 \cancel{\text{mol } CH_4} \times \left(-50.72 \frac{\text{kJ}}{\cancel{\text{mol}}}\right) + 1 \cancel{\text{mol } C_2H_2} \times \left(209.20 \frac{\text{kJ}}{\cancel{\text{mol}}}\right)\right\} \\ &= \{-17.03 \text{ kJ}\} - \{107.76 \text{ kJ}\} \\ &= -124.79 \text{ kJ}\end{aligned}$$

Because $\Delta G^\circ < 0$, the production of butane from the reaction of methane and ethyne at the specified conditions (1 atm pressure) is a spontaneous process. This does not mean it is easy to do so! We may still have to devise a way to make the reaction occur at a reasonable rate. The negative ΔG° tells us the reaction is a spontaneous process and tends to occur, but does not guarantee it will occur at a finite rate.

EXERCISE An entrepreneur asks you to invest your life savings in a brand new energy-producing device that depends on using sand (SiO_2) as a fuel in the following reaction

$$2SiO_2(s) + N_2(g) \longrightarrow 2Si(s) + N_2O_4(g)$$

With the high availability of sand and atmospheric nitrogen, not to mention the lucrative sales of silicon to the semiconductor industry, the entrepreneur guarantees you will get rich. How much money should you invest?

ANSWER: None. The reaction has a positive $\Delta G(+1811 \text{ kJ})$, so it does not occur as a spontaneous process. The scheme will not work. ▲

EQUILIBRIA

16.8–16.9 FREE ENERGY AND COMPOSITION; THE EQUILIBRIUM CONSTANT

KEY CONCEPT Equilibrium constants and ΔG

For any equilibrium, ΔG is related to K/Q where K is the equilibrium constant and Q is the mass action expression:

$$\Delta G = -RT \ln (K/Q)$$

When all the reactants and products are in their standard states for a reaction, then $Q = 1$ and $\Delta G = \Delta G^\circ$.

$$\text{Reactants (pure, 1 atm)} \rightleftharpoons \text{products (pure, 1 atm)}$$

$$\Delta G^\circ = -RT \ln K$$

This applies to any of the equilibria we have encountered up to now. However, there are some rules regarding how gases are handled:

1. For any equilibrium involving gases only, K_P should be used.
2. For any equilibrium involving a nondissolved gas, the concentration of the gas must be expressed as a partial pressure and not as a molar concentration.
3. Gas partial pressures must be expressed in atmospheres and not in torrs.

EXAMPLE 1 Calculating an equilibrium constant from $\Delta G°$

Use the data in Appendix 2A of the text to calculate the equilibrium constant K_P at 25°C for the equilibrium

$$2SO_2(g) + O_2(g) \rightleftharpoons 2SO_3(g)$$

SOLUTION For any equilibrium,

$$\Delta G° = -RT \ln K_P$$

We use the standard free energies of formation of SO_3(-371.06 kJ/mol) and SO_2 (-300.19 kJ/mol) to calculate $\Delta G°$:

$$\begin{aligned}\Delta G° &= \Delta G_f°(\text{products}) - \Delta G_f°(\text{reactants})\\ &= \left\{2\text{ mol SO}_3 \times \left(-371.06\,\frac{\text{kJ}}{\text{mol}}\right)\right\} - \left\{2\text{ mol SO}_2 \times \left(-300.19\,\frac{\text{kJ}}{\text{mol}}\right)\right\}\\ &= \{-742.12\text{ kJ}\} - \{-600.38\text{ kJ}\}\\ &= -742.12\text{ kJ} + 600.38\text{ kJ}\\ &= -141.74\text{ kJ}\end{aligned}$$

Thus, at 25°C (298 K),

$$\begin{aligned}-141.74\text{ kJ} &= -RT\ln K_P\\ &= -\left(8.314\,\frac{\text{J}}{\text{K}}\right)(298\text{ K})\ln K_P\\ &= -2.48\text{ kJ}\ln K_P\end{aligned}$$

Dividing both sides of the equation by 2.48 kJ gives

$$\begin{aligned}\ln K_P &= \frac{141.74\text{ kJ}}{2.48\text{ kJ}}\\ &= 57.2\end{aligned}$$

This gives

$$\begin{aligned}K_P &= e^{57.2}\\ &= 7 \times 10^{24}\end{aligned}$$

(The value for this answer is very sensitive to how and where you round off in the calculation.)

EXERCISE Calculate K_P at 25°C for the equilibrium shown from the free energies of formation of the substances involved.

$$H_2(g) + I_2(g) \rightleftharpoons 2HI(g)$$

ANSWER: 0.254

EXAMPLE 2 Calculating a vapor pressure from $\Delta G°$

Calculate the vapor pressure of chloroform ($CHCl_3$) at 25°C from the standard free energies of formation of liquid chloroform (-71.84 kJ/mol) and gaseous chloroform (-70.12 kJ/mol).

SOLUTION A vapor pressure is measured when the vapor and liquid are in equilibrium with each other.

$$CHCl_3(l) \rightleftharpoons CHCl_3(g)$$

The equilibrium constant for this equilibrium is

$$K_P = P_{CHCl_3}$$
$$P_{CHCl_3} = \text{partial pressure of } CHCl_3 = \text{vapor pressure of } CHCl_3$$

Thus, if we determine the value of the equilibrium constant, we shall have the vapor pressure of chloroform. We use the relationship between K_P and ΔG°:

$$\Delta G^\circ = -RT \ln K_P$$

First, we calculate ΔG° from the standard free energies of formation:

$$\begin{aligned}\Delta G^\circ &= \Delta G_f^\circ(\text{products}) - \Delta G_f^\circ(\text{reactants}) \\ &= \left\{1 \cancel{\text{mol } CHCl_3}(g) \times \left(-70.12 \frac{\text{kJ}}{\cancel{\text{mol}}}\right)\right\} - \left\{1 \cancel{\text{mol } CHCl_3}(l) \times \left(-71.84 \frac{\text{kJ}}{\cancel{\text{mol}}}\right)\right\} \\ &= \{-70.12 \text{ kJ}\} - \{-71.84 \text{ kJ}\} \\ &= -70.12 \text{ kJ} + 71.84 \text{ kJ} \\ &= +1.72 \text{ kJ}\end{aligned}$$

We now substitute this value into the required equation, in which we use the fact that $RT = 2.48$ kJ at 298 K:

$$\begin{aligned}+1.72 \cancel{\text{kJ}} &= -2.48 \cancel{\text{kJ}} \ln K_P \\ \ln K_P &= -\frac{1.72}{2.48} = -0.694 \\ K_P &= e^{-0.694} \\ &= 0.500 \text{ atm}\end{aligned}$$

Because of the fact that the standard states refer to 1 atm pressure, K_P comes out in atmospheres. Therefore, the vapor pressure of chloroform at 25°C is 0.500 atm, or 380 Torr.

EXERCISE The vapor pressure of methanol at 25°C is 124 Torr. The standard free energy of formation of liquid methanol at 25°C is −166.22 kJ/mol. What is the standard free energy of formation of gaseous methanol at 25°C? (Remember that pressure must be expressed in atmospheres.)

ANSWER: −162 kJ/mol

CHEMICAL EQUATIONS TO KNOW

- The oxidation of ammonia is part of the Ostwald process for making nitric acid.

$$4NH_3(g) + 5O_2(g) \longrightarrow 4NO(g) + 6H_2O(g)$$

- Pure liquid hydrogen peroxide has a tendency to decompose into water and oxygen.

$$2H_2O_2(l) \longrightarrow 2H_2O(l) + O_2(g)$$

- The reaction between barium hydroxide and ammonium thiocyanate is endothermic but spontaneous.

$$Ba(OH)_2 \cdot 8H_2O(s) + 2NH_4SCN(s) \longrightarrow Ba(SCN)_2(aq) + 2NH_3(g) + 10H_2O(l)$$

- Carbon is used to reduce iron(III) oxide in one of the reactions in a blast furnace.

$$2Fe_2O_3(s) + 3C(s) \longrightarrow 4Fe(s) + 3CO_2(g)$$

MATHEMATICAL EQUATIONS TO KNOW AND UNDERSTAND

$\Delta U = q + w$	change in internal energy
$w = -P \times \Delta V$	expansion work
$\Delta U = q_V$	heat at constant volume
$\Delta H = q_P$	heat at constant pressure
$S = k \ln W$	entropy of a crystal
$\Delta S_{surr} = \dfrac{-\Delta H}{T}$	entropy change in surroundings caused by transfer of heat
$\Delta G = \Delta H - T\Delta S$	free energy change
$\Delta G° = \Delta H° - T\Delta S°$	free energy change at standard conditions
$\Delta G = -RT \ln \dfrac{K}{Q}$	reaction free energy at arbitrary composition
$\Delta G° = -RT \ln K$	free energy relation to any equilibrium constant

SELF-TEST EXERCISES

The first law of thermodynamics

1. A reaction is run during which 1.23 kJ of heat is released and the reaction system does 235 J of work on the surroundings. What is ΔU for this process?
(a) 10×10^3 J (b) -10×10^3 J (c) 1.57 kJ (d) -1.47 kJ (e) 1.23 kJ

2. Under what conditions is it possible to create or destroy energy?
(a) constant pressure (b) constant volume (c) constant temperature
(d) in isolated systems (e) energy cannot be created or destroyed

3. What is the work done when 2 mol of benzene (C_6H_6) is combusted at STP?

$$2C_6H_6(l) + 15O_2(g) \longrightarrow 12CO_2(g) + 6H_2O(g)$$

(a) 40.7 kJ (b) 6.79 kJ (c) 3.0×10^2J (d) 11.6 kJ (e) 1.8 kJ

4. A reaction run at constant volume liberates 6.24 kJ; when run at constant pressure, 6.44 kJ is liberated. What is ΔH for the reaction?
(a) -0.20 kJ (b) -6.44 kJ (c) $+0.20$ kJ (d) -12.68 kJ (e) 6.24 kJ

5. What a reaction is run at constant volume, 8.69 kJ of heat is absorbed; when run at constant pressure, 8.15 kJ is absorbed. What is ΔU for the reaction?
(a) -16.84 kJ (b) -0.54 kJ (c) -4.35 kJ (d) 8.69 kJ (e) 8.15 kJ

The direction of spontaneous change

6. Which process results in an increase in energy disorder?
(a) A deck of cards is shuffled.
(b) A beaker of hot water cools in a cold room.
(c) A piece of copper at 25°C is dropped into water at 25°C.
(d) Ideal gas molecules spread throughout a room.

7. When ideal gas molecules escape from a small volume and disperse throughout a room,
(a) the disorder of energy increases. (b) the disorder of energy decreases.
(c) the disorder of matter decreases. (d) the disorder of matter increases.
(e) there is no change in disorder.

8. Which of the following reactions has a positive entropy change? (Do not calculate the standard reaction entropy for each.)
(a) $CaO(s) + H_2O(g) \longrightarrow Ca(OH)_2(s)$
(b) $2NO_2(g) \longrightarrow N_2O_4(g)$
(c) $CuSO_4(s) + 5H_2O(l) \longrightarrow CuSO_4 \cdot 5H_2O(s)$
(d) $PCl_5(s) \longrightarrow PCl_3(l) + Cl_2(g)$
(e) $H_2O(g) \longrightarrow H_2O(l)$

9. What molar entropy at 0 K for a IOF_5 crystal is predicted by the Boltzmann formula?
(a) 0 J/K (b) 14.9 J/K (c) 5.76 J/K (d) 9.13 J/K (e) 13.4 J/K

10. What is the molar entropy at 0 K for the trigonal bipyramidal compound SOF_4?
(a) 0 J/K (b) 14.9 J/K (c) 5.76 J/K (d) 9.13 J/K (e) 13.4 J/K

11. What is the predicted molar entropy at 0 K for CO_2?
(a) 0 J/K (b) 14.9 J/K (c) 5.76 J/K (d) 9.13 J/K (e) 13.4 J/K

12. Without referring to any tables, predict which has the greatest standard molar entropy at 298 K?
(a) $H_2O(l)$ (b) $Ca(s)$ (c) $Ca(OH)_2(s)$ (d) $CH_4(g)$ (e) $C_2H_4(g)$

13. Predict which has the greatest standard molar entropy at the given temperature.
(a) $H_2O(l, 300\ K)$ (b) $H_2O(s, 260\ K)$ (c) $H_2O(g, 450\ K)$
(d) $H_2O(l, 400\ K)$ (e) $H_2O(g, 385\ K)$

14. Calculate the standard reaction entropy for the reaction

$$2H_2O(l) + O_2(g) \longrightarrow 2H_2O_2(l)$$

(a) −125.8 J/K (b) −165.6 J/K (c) 564.2 J/K
(d) −30.2 J/K (e) +39.7 J/K

15. Calculate the standard reaction entropy for the reaction

$$2HCl(g) + F_2(g) \longrightarrow 2HF(g) + Cl_2(g)$$

(a) −26.26 J/K (b) −35.11 J/K (c) −5.97 J/K
(d) 7.16 J/K (e) +3.19 J/K

16. What is the entropy change in the surroundings at 298 K for a reaction for which $\Delta H = -32.2$ kJ?
(a) −108 J/K (b) −32.2 KJ/K (c) 9.60 J/K
(d) +32.2 J/K (e) 108 J/K

17. An endothermic reaction ______ the entropy of the surroundings by ______ the chaotic thermal motion in the surroundings.
(a) increases, increasing (b) increases, decreasing
(c) decreases, decreasing (d) decreases, increasing

18. For a system at equilibrium, the total entropy change in the system plus surroundings
(a) is zero. (b) is positive.
(c) is negative. (d) depends on the temperature.

19. What is the sign of the entropy change for the following process?

$$H_2O(l, 25°C) \longrightarrow H_2O(g, 1\ atm, 25°C)$$

(a) zero (b) positive (c) negative

20. What is the molar entropy of melting of CCl_4? Its enthalpy of melting is 2.47 kJ/mol and its melting point is 250 K.
(a) 618 kJ/K (b) 2.47 kJ/K (c) 15.3 J/K (d) 9.88 J/K (e) 21.2 J/K

21. What is the boiling point of methanol? Its enthalpy of vaporization is 35.27 kJ/mol and its entropy of vaporization is 104.6 J/K.
(a) 297.6 K (b) 337.2 K (c) 481.2 K (d) 368.9 K (e) 412.6 K

22. A hydrogen-bonded liquid is typically ______ disorderly than a non-hydrogen-bonded liquid and therefore has a ______ entropy of vaporization than a non-hydrogen-bonded liquid.
(a) less, higher (b) less, lower (c) more, higher (d) more, lower

23. The reaction shown is exothermic:

$$H_2(g) + O_2(g) \longrightarrow H_2O_2(l)$$

Which statement is correct? (Answer without making detailed calculations.)
(a) The reaction is a spontaneous process.
(b) The reaction is not a spontaneous process.
(c) It is impossible to judge whether it is a spontaneous process without calculations.

24. The reaction shown is exothermic:

$$2CO(g) \longrightarrow 2C(s) + O_2(g)$$

Which statement is correct? (Answer without making detailed calculations.)
(a) The reaction is a spontaneous process.
(b) The reaction is not a spontaneous process.
(c) It is impossible to judge whether it is a spontaneous process without calculations.

Free energy

25. Calculate ΔG at 100°C for a reaction for which $\Delta H = -35.1$ kJ and $\Delta S = -225$ J/K. Also state whether the reaction is spontaneous or not.
(a) +57.6 kJ, spontaneous (b) +48.8 kJ, spontaneous
(c) +48.8 kJ, not spontaneous (d) −119.0 kJ, spontaneous
(e) +57.6 kJ, not spontaneous

26. A change with a positive change in total entropy has a ______ change in free energy and ______ a spontaneous change.
(a) negative, is (b) positive, is (c) negative, is not
(d) positive, is not (e) zero, is

27. A reaction that is spontaneous at all temperatures must have a ______ reaction enthalpy and ______ reaction entropy.
(a) positive, positive (b) negative, negative
(c) positive, negative (d) negative, positive

28. A reaction with $\Delta H = 361$ kJ and $\Delta S = +121$ J/K is spontaneous at
(a) low temperatures only. (b) high temperatures only. (c) all temperatures.

29. What is the minimum temperature at which a reaction with $\Delta H = 182$ kJ and $\Delta S = 203$ J/K is a spontaneous process?
(a) 1115 K (b) 897 K (c) 0.897 K (d) 1.12 K (e) 36.9 K

30. Which of the compounds shown is the most thermodynamically stable? Each is shown with its standard free energy of formation.
(a) $CaCl_2(s)$, −748.1 kJ/mol (b) $HBr(g)$, −53.45 kJ/mol (c) $SO_2(g)$, −300.19 kJ/mol
(d) $H_2O(l)$, −237.13 kJ/mol (e) $NH_3(g)$, +328.1 kJ/mol

31. Does the following reaction tend to occur at 25°C? $\Delta G_f^\circ(PbO) = -188.89$ kJ/mol.

$$Pb(s) + H_2O(g) \longrightarrow PbO(s) + H_2(g)$$

(a) yes (b) no

32. What is $\Delta G°$ for the following reaction? $\Delta G_f^\circ(NaHCO_3) = -851.9$ kJ/mol, $\Delta G_f^\circ(Na_2CO_3) = -1048.2$ kJ/mol.

$$2NaHCO_3(s) \longrightarrow Na_2CO_3(s) + CO_2(g) + H_2O(g)$$

(a) -1256.2 kJ (b) 32.7 kJ (c) 884.6 kJ (d) -3374.9 kJ (e) -2523.0 kJ

Equilibria

33. What is the calculated equilibrium constant, at 25°C, for the equilibrium shown? $\Delta G_f^\circ(SO_2) = -300.19$ kJ/mol, $\Delta G_f^\circ(SO_3) = -371.06$ kJ/mol.

$$2SO_2(g) + O_2(g) \rightleftharpoons 2SO_3(g)$$

(a) 2.5×10^{-25} atm^{-1} (b) 2.6×10^{12} atm^{-1} (c) 3.8×10^{-13} atm^{-1}
(d) 7.0×10^{24} atm^{-1} (e) 1.03 atm^{-1}

34. What is the equilibrium constant, at 25°C, for

$$PCl_5(g) \rightleftharpoons PCl_3(g) + Cl_2(g)$$

(a) 1.1×10^{13} atm (b) 9.4×10^{-14} atm (c) 3.1×10^{-7} atm
(d) 3.3×10^{6} atm (e) 4.2×10^{3} atm

35. What is the vapor pressure of $POCl_3(l)$ at 25°C? The standard free energy of formation of the liquid is -520.9 kJ/mol and that of the gas is -513.0 kJ/mol.
(a) 31.4 Torr (b) 24.2 Torr (c) 15.8 Torr (d) 81.6 Torr (e) 2.2 Torr

Descriptive chemistry

36. What products are formed when hydrogen peroxide (H_2O_2) decomposes?
(a) H_2 and O_2 (b) H_2O and O_2 (c) H_2 and NaOH (d) H_2 and O_3

37. Which of the following reactions is both spontaneous and endothermic?
(a) $2Fe_2O_3(s) + 3C(s) \longrightarrow 4Fe(s) + 3CO_2(g)$
(b) $Pb(s) + H_2O(g) \longrightarrow PbO(s) + H_2(g)$
(c) $4NH_3(g) + 5O_2(g) \longrightarrow 4NO(g) + 6H_2O(g)$
(d) $Ba(OH)_2 \cdot 8H_2O(s) + 2NH_4SCN(s) \longrightarrow Ba(SCN)_2(aq) + 2NH_3(g) + 10H_2O(l)$

17 ELECTROCHEMISTRY

ELECTROCHEMICAL CELLS

KEY WORDS Define or explain each of the following terms in a written sentence or two.

anode
cathode
cell potential
electrode
electrolyte
galvanic cell
primary cell
redox couple
salt bridge
secondary cell

17.1 GALVANIC CELLS

KEY CONCEPT A Redox review and introduction to galvanic cells

A redox reaction is one that involves a change in oxidation number of one or more reactants, usually through electron transfer. Oxidation is loss of electrons, and reduction is gain of electrons. (Remember, "LEO the lion goes GER"; LEO for Lose Electrons Oxidation and GER for Gain Electrons Reduction.) The substance that is oxidized is the reducing agent, and the substance reduced is the oxidizing agent. For example, in the reaction

$$Zn(s) + Cu^{2+}(aq) \longrightarrow Zn^{2+}(aq) + Cu(s)$$

(a) $Zn(s)$ loses two electrons (to become Zn^{2+}) and is oxidized; $Zn(s)$ is the reducing agent (it reduces Cu^{2+}). (b) Cu^{2+} gains two electrons [to become $Cu(s)$] and is reduced; Cu^{2+} is the oxidizing agent [it oxidizes $Zn(s)$]. This reaction is spontaneous ($\Delta G < 0$); that is, the electrons move from Zn to Cu^{2+} without any outside intervention. If a piece of zinc is dropped into a $CuSO_4$ solution, the electron transfer takes place where the Cu^{2+} ions bump into the solid Zn. However, the reaction can be set up so that the electrons are forced to flow through a wire in order to get from the Zn to the Cu^{2+} ions. This is accomplished by separating the Zn metal from the aqueous Cu^{2+}, as shown in the figure. The electrons flowing through a wire constitute an electric current.

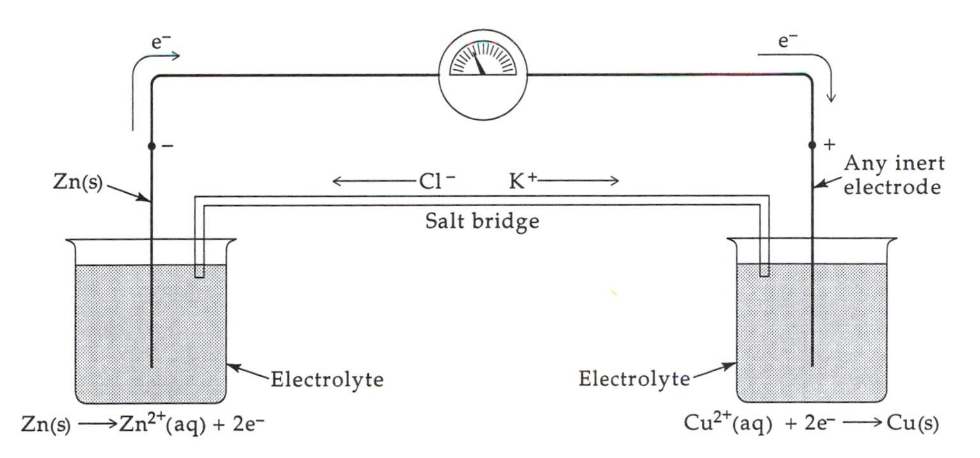

The salt bridge is a hollow glass tube filled with a gelled electrolyte such as KCl; it provides negative charge to the beaker on the left (in the form of Cl^- ions) to make up for the loss of negative charge as electrons flow; it also provides positive charge to the right-hand beaker (in the form of K^+ ions) to balance the excess negative charge created in that beaker by the flow of electrons. Because electrons flow *from* the left-hand electrode, it appears to be a source of electrons and has a negative charge; the electrons flow *toward* the right-hand electrode, so it is given a positive charge. A reaction must be spontaneous to provide a current. If it is not spontaneous, no reaction occurs and no current is generated.

EXAMPLE Identifying electron transfer in a redox equation

For the reaction

$$Mg(s) + 2H^+(aq) \longrightarrow Mg^{2+}(aq) + H_2(g)$$

identify the species oxidized, the species reduced, the oxidizing agent, and the reducing agent. This reaction is spontaneous; could it be used to generate an electric curent?

SOLUTION Mg(*s*) loses two electrons when it becomes Mg^{2+}, so Mg(*s*) is oxidized. Mg(*s*) must be the reducing agent (it reduces H^+). H^+ gains electrons to become H_2, so H^+ is reduced. Because H^+ is reduced, it must be the oxidizing agent (it oxidizes Mg). The reaction could be set up so that the electrons flow through an external wire. Because it is a spontaneous reaction, it would provide an electric current.

EXERCISE For the reaction

$$Cu(s) + Hg^{2+}(aq) \longrightarrow Cu^{2+}(aq) + Hg(l)$$

identify the species oxidized, the species reduced, the oxidizing agent, and the reducing agent. This reaction is spontaneous; could it be used to generate an electric current?

ANSWER: Cu(*s*), oxidized, reducing agent; Hg^{2+}, reduced, oxidizing agent. It could be used to generate a current.

KEY CONCEPT B Redox couples and electrode notation

A **galvanic cell** is a cell that uses a spontaneous chemical reaction to produce an electric current. Most galvanic cells contain a conducting liquid **electrolyte** (either a solution or a melt). Two **electrodes** dip into the electrolyte; these serve as a conduit for electrons and give the generated current access to an external circuit outside the cell. The electrodes may also be chemical reactants. The cell works as follows:

1. Oxidation occurs at one of the electrodes (called the **anode**).
2. The electrons released by the oxidation are shunted out of the cell to an external circuit (where they can be used to do some form of work) and then back into the cell by way of the second electrode.
3. At the second electrode (called the **cathode**), reduction occurs by gain of electrons.

In a cell using the reaction

$$Zn(s) + Cu^{2+}(aq) \longrightarrow Zn^{2+}(aq) + Cu(s)$$

the anode and cathode reactions are

Anode:	$Zn(s) \longrightarrow Zn^{2+}(aq) + 2e^-$	Oxidation
Cathode:	$2e^- + Cu^{2+}(aq) \longrightarrow Cu(s)$	Reduction

A **redox couple** consists of the two forms of a substance (oxidized form and reduced form) that appear in the half-reaction for the substance, written in the form

Oxidized form/reduced form

For the preceding two half-reactions, we have

$$\text{Zn couple: } Zn^{2+}/Zn$$
$$\text{Cu couple: } Cu^{2+}/Cu$$

A **cell diagram** is a shorthand way of writing the essential chemistry and components of an electrochemical cell. It is written as

$$\text{Anode compartment} \,\|\, \text{cathode compartment}$$

The double vertical bar represents the salt bridge. Each compartment is written to show the electrodes and any chemicals present. The electrode is written first for the anode compartment and last for the cathode compartment, so the electrodes are always the first and last symbols in the cell diagram. A single vertical line indicates an area of contact between two different cell components. Two chemicals present in a single solution are separated by a comma.

EXAMPLE Writing cell diagrams

A cell is constructed by dipping a tin electrode into a tin(II) chloride ($SnCl_2$) solution in one beaker and a platinum electrode into a silver nitrate ($AgNO_3$) solution in another beaker. A salt bridge is used. What is the cell diagram for this cell? The reaction is

$$Sn(s) + 2Ag^{+}(aq) \longrightarrow 2Ag(s) + Sn^{2+}(aq)$$

SOLUTION The two half-reactions are

$$\begin{array}{lrl} \text{Anode:} & Sn(s) & \longrightarrow Sn^{2+}(aq) + 2e^{-} \\ \text{Cathode:} & 2Ag^{+}(aq) + 2e^{-} & \longrightarrow 2Ag(s) \end{array}$$

The tin anode is a reactant. The cell diagram is

$$Sn(s)\,|\,Sn^{2+}(aq)\,\|\,Ag^{+}(aq)\,|\,Ag(s)\,|\,Pt(s)$$

The solid silver metal that forms in the reaction plates the cathode; thus, there is an area of contact between $Ag(s)$ and $Pt(s)$, as indicated in the cell diagram. In such a diagram, the anode is the first substance listed [a piece of tin, $Sn(s)$, in our example], and the cathode is the last substance listed [a piece of platinum, $Pt(s)$, in our example].

EXERCISE Write the cell diagram for a cell that uses the reaction

$$Mg(s) + 2H^{+}(aq) \longrightarrow H_2(g) + Mg^{2+}(aq)$$

A platinum electrode is used with the H^{+}/H_2 couple, and a salt bridge is used.

ANSWER: $Mg(s)\,|\,Mg^{2+}(aq)\,\|\,H^{+}(aq)\,|\,H_2(g)\,|\,Pt(s)$

PITFALL What is (and isn't) in a cell diagram?

A cell diagram gives the essential components and chemicals in a cell but does not tell us how the cell is actually constructed. Building a working chemical cell is a subtle art. Each cell has its own idiosyncracies and problems that must be considered, and many different cell designs exist.

KEY CONCEPT C Cell potential

The **cell potential**, which is measured in units of volts (V), is a measure of how "hard" the electrons in the external circuit of the cell are pushed from the anode to the cathode. The SI unit of volts is defined such that when 1 coulomb (C) of charge moves from the anode to the cathode, with a potential difference of 1 volt between the two electrodes, 1 joule of energy is released. A joule is a volt · coulomb; that is, 1 joule = 1 volt × 1 coulomb.

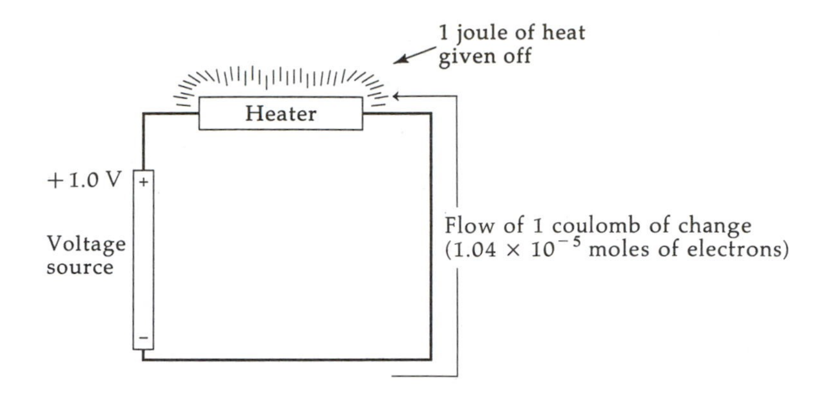

17.2 PRACTICAL CELLS

See Descriptive Chemistry To Remember in this chapter.

ELECTROCHEMISTRY AND THERMODYNAMICS

KEY WORDS Define or explain each of the following terms in a written sentence or two.

corrosion
electrochemical series
electrode potential
Faraday constant
glass electrode
passivation
sacrificial electrode
standard cell potential
standard electrode potential
standard hydrogen electrode

17.3 CELL POTENTIAL AND REACTION FREE ENERGY

KEY CONCEPT A Cell potential and free energy

For a cell operating at any conditions of concentration, there exists a relationship between the cell potential E(cell) and ΔG for the reaction:

$$\Delta G = -nFE(\text{cell}) \quad \text{(all conditions)}$$

where n is the number of electrons transferred for the reaction as written and F is the **Faraday constant**, which gives the charge in coulombs associated with a mole of electrons (96,485 C/mol e^-). For a cell in which the reactants and products are all at standard conditions (1 atm pressure, pure), E(cell), becomes $E°$(cell) and ΔG becomes $\Delta G°$.

$$\Delta G° = -nFE°(\text{cell})$$

$E°$(cell) is called the **standard cell potential.** These relationships give the expected correlations between ΔG and E(cell).

1. A cell works only for a spontaneous process ($\Delta G < 0$), and the cell voltage is always positive ($E(\text{cell}) > 0$). The minus sign in the equation $\Delta G = -nFE(\text{cell})$ assures the difference in the signs of ΔG and E(cell).
2. At equilibrium $\Delta G = 0$. Not net reaction occurs, so E(cell) also must equal zero. The equation predicts this.
3. A large negative ΔG means the reaction has a very strong tendency to occur. This implies that E(cell) should be large and positive, as predicted by $\Delta G = -nFE(\text{cell})$.

When using $\Delta G = -nFE(\text{cell})$, recall that $1\ J = 1\ C{\cdot}V$.

EXAMPLE 1 Calculating ΔG from E(cell)

E(cell) = 0.25 V for the following reaction. What is ΔG for this reaction?

$$Cd(s) + Pb^{2+}(aq, 0.020\ M) \longrightarrow Cd^{2+}(aq, 0.10\ M) + Pb(s)$$

SOLUTION The relationship between ΔG and E(cell) is

$$\Delta G = -nFE(\text{cell})$$

$Cd(s)$ loses two electrons when it changes to Cd^{2+}, so $n = 2$. [This is confirmed by the fact that Pb^{2+} gains two electrons as it changes to $Pb(s)$.] Thus,

$$\begin{aligned} n &= 2\ \text{mol e}^- \\ E(\text{cell}) &= +0.25\ \text{V} \\ \Delta G &= -(2\ \cancel{\text{mol e}^-})\left(9.65 \times 10^4 \frac{\text{C}}{\cancel{\text{mol e}^-}}\right)(0.25\ \text{V}) \\ &= -4.8 \times 10^4\ \text{C}\cdot\text{V} \\ &= -4.8 \times 10^4\ \text{J} \\ &= -48\ \text{kJ} \end{aligned}$$

EXERCISE What is ΔG for the reaction in the preceding example when $[Pb^{2+}] = 1.0$ M and $[Cd^{2+}] = 1.0$ M? For these conditions (standard conditions), $\Delta G = \Delta G^\circ$, $E(\text{cell}) = E^\circ(\text{cell})$ and the measured cell potential is 0.27 V.

ANSWER: -52 kJ

EXAMPLE 2 Calculating E(cell) from ΔG

$\Delta G = -718$ kJ at 25°C for the following reaction. What is E(cell) for this reaction?

$$2Al(s) + 3Fe^{2+}(aq, 2.0\ M) \rightleftharpoons 2Al^{3+}(aq, 0.030\ M) + 3Fe(s)$$

SOLUTION The equation that relates ΔG to E(cell) is

$$\Delta G = -nFE(\text{cell})$$

$$E(\text{cell}) = \frac{\Delta G}{-nF}$$

In the chemical equation, two $Al(s)$ atoms lose six electrons to become two Al^{3+} ions, so $n = 6$. [This is confirmed by noting that three Fe^{2+} ions must gain six electrons to become three $Fe(s)$ atoms.] Because 1 V = 1 J/C, it is convenient to express ΔG in joules rather than kilojoules. Thus, we have

$$\begin{aligned} \Delta G &= -7.18 \times 10^5\ \text{J} \\ n &= 6 \\ E(\text{cell}) &= \frac{-7.18 \times 10^5\ \text{J}}{-(6\ \cancel{\text{mol e}^-})(9.65 \times 10^4\ \text{C}/\cancel{\text{mol e}^-})} \\ &= +1.24\ \text{J/C} \\ &= 1.24\ \text{V} \end{aligned}$$

EXERCISE At standard conditions, $\Delta G^\circ = -706$ kJ for the reaction in the preceding example. What is E°(cell)?

ANSWER: $+1.22$ V

KEY CONCEPT B Electrode potentials

Just as a cell reaction can be thought of as the result of two half-reactions, a cell potential can be thought of as the sum of the two half-reaction potentials. The half-reaction potentials are called **electrode potentials** or, if the cell is at standard conditions, **standard electrode potentials**. A table of standard electrode potentials can be constructed by assigning a value of 0 V exactly to the electrode potential of the **standard hydrogen electrode (SHE)** and comparing other electrode potentials to the SHE.

Standard hydrogen electrode half-reaction: $2H^+(aq, 1\text{ M}) + 2e^- \longrightarrow H_2(g, 1\text{ atm})$
$E° = 0$ V(exactly)

Any half-reaction may be written either as a reduction process or an oxidation process. The electrode potential for an oxidation half-reaction is called an **oxidation potential** and that for a reduction half-reaction a **reduction potential**. For instance,

$$Zn(s) \longrightarrow Zn^{2+}(aq) + 2e^- \qquad E° = \text{oxidation potential} = +0.76\text{ V}$$
$$Zn^{2+}(aq) + 2e^- \longrightarrow Zn(s) \qquad E° = \text{reduction potential} = -0.76\text{ V}$$

As illustrated for the Zn^{2+}/Zn couple, reversing a half-reaction changes the sign of the electrode potential. Thus, for any couple,

$$\text{Oxidation potential} = -\text{reduction potential}$$

It is quite possible for one of the electrode potentials in a working cell to be negative if the sum of the two electrode potentials for the cell is positive. If the calculated cell potential, E(cell), for a proposed cell is negative, the cell will not work as proposed. The reverse reaction will have a positive cell potential, so a cell using the reverse reaction will function.

EXAMPLE 1 Calculating $E°$(cell) from electrode potentials

What is $E°$(cell) for a proposed cell that uses the following reaction? Will the cell function at standard conditions?

$$Ti(s) + Fe^{2+}(aq) \longrightarrow Fe(s) + Ti^{2+}(aq)$$

SOLUTION To calculate $E°$(cell), we must first determine which half-reactions are involved and what their standard electrode potentials are. The two half-reactions are

Anode: $Ti(s) \longrightarrow Ti^{2+}(aq) + 2e^-$

Cathode: $Fe^{2+}(aq) + 2e^- \longrightarrow Fe(s)$

The relevant reduction potentials given in Appendix 2B of the text are

Ti^{2+}/Ti couple: $Ti^{2+}(aq) + 2e^- \longrightarrow Ti(s)$ $\quad E° = -1.63$ V

Fe^{2+}/Fe couple: $Fe^{2+}(aq) + 2e^- \longrightarrow Fe(s)$ $\quad E° = -0.44$ V

To get the correct cell voltage, we must write each half-reaction as it functions in the overall cell reaction and sum the electrode potentials, $E°(\text{cell}) = E°(\text{anode}) + E°(\text{cathode})$.

	Reaction	
Anode:	$Ti(s) \longrightarrow Ti^{2+}(aq) + 2e^-$	$E° = +1.63$ V
Cathode:	$Fe^{2+}(aq) + 2e^- \longrightarrow Fe(s)$	$E° = -0.44$ V
	$Ti(s) + Fe^{2+}(aq) \longrightarrow Fe(s) + Ti^{2+}(aq)$	$E°(\text{cell}) = 1.19$ V

Note that the Ti^{2+}/Ti couple appears as an oxidation half-reaction, so its $E°$ (+1.63 V) is the negative of the standard reduction potential (−1.63 V). Because the proposed cell has a positive standard cell potential (+1.19 V), it will function at standard conditions.

EXERCISE What is E°(cell) for a proposed cell that uses the following reaction? Will the cell function at standard conditions?

$$Ni(s) + V^{2+}(aq) \longrightarrow Ni^{2+}(aq) + V(s)$$

ANSWER: −0.96 V; no

EXAMPLE 2 Predicting E°(cell) for a proposed cell

What is E°(cell) for a proposed cell that uses the following reaction? Will the cell function at standard conditions?

$$2In(s) + 6H_2O(l) \longrightarrow 2In^{3+}(aq) + 3H_2(g) + 6OH^-(aq)$$

SOLUTION The two half-reactions involved in the cell reaction are

$$\begin{array}{ll} \text{Cathode:} & 2H_2O(l) + 2e^- \longrightarrow H_2(g) + 2OH^-(aq) \\ \text{Anode:} & In(s) \longrightarrow In^{3+}(aq) + 3e^- \end{array}$$

The standard reduction potentials for these half-reactions are

$$\begin{array}{lll} H_2O/H_2 \text{ couple:} & 2H_2O(l) + 2e^- \longrightarrow H_2(g) + 2OH^-(aq) & E^\circ = -0.83\text{ V} \\ In^{3+}/In \text{ couple:} & In^{3+}(aq) + 3e^- \longrightarrow In(s) & E^\circ = -0.34\text{ V} \end{array}$$

To get the standard cell potential, we add the half-reactions and sum the standard electrode potentials for the reactions as written. Remember that multiplying the half-reactions by a number (so that electrons cancel when the half-reactions are added) does not effect the standard electrode potentials. Reversing a half-reaction changes the sign of the standard electrode potential. Thus, we use 3 × the H_2O/H_2 couple written as reduction (reaction 1) and 2 × In^{3+}/In written as oxidation (reaction 2):

$$\begin{array}{rll} 6H_2O(l) + 6e^- \longrightarrow 3H_2(g) + 6OH^-(aq) & E^\circ = -0.83\text{ V} & (1) \\ 2In(s) \longrightarrow 2In^{3+}(aq) + 6e^- & E^\circ = +0.34\text{ V} & (2) \\ \hline 6H_2O(l) + 2In(s) \longrightarrow 3H_2(g) + 6OH^-(aq) + 2In^{3+}(aq) & E^\circ(\text{cell}) = -0.49\text{ V} & \end{array}$$

The standard cell potential for the proposed cell is negative, so the cell would *not* function at standard conditions. A cell constructed to take advantage of the reverse reaction would function, however, with a cell potential of 0.49 V.

EXERCISE A fuel cell is proposed that would take advantage of the following reaction. What is E°(cell) for the proposed cell? Would it function at standard conditions?

$$2H_2(g) + O_2(g) \longrightarrow 2H_2O(l)$$

ANSWER: 1.23 V yes

17.4 THE ELECTROCHEMICAL SERIES

KEY CONCEPT A The electrochemical series

The half-reactions in Table 17.3 of text are all written in the form

Oxidized form of substance + electron ⟶ reduced form of substance

The oxidized form may act as an oxidizing agent in a reaction (it takes electrons from another substance), in which case it is convenient to think of the half-reaction as

Oxidizing agent + electrons ⟶ reduced form

The substances at the left of Table 17.3 of the text act as oxidizing agents. Because the half-reactions high in the table have very positive reduction potentials, these half-reactions tend to go very strongly.

The oxidizing agents high in the table (such as F_2 and Ce^{4+}) are therefore very strong oxidizing agents. Any oxidizing agent from a couple high in the table can oxidize the reduced form of any couple lower in the table. For example, F_2 can oxidize Au. In a similar fashion, the reduced form may act as a reducing agent in a reaction (it gives electrons to another substance), in which case it is convenient to think of the reverse of the half-reaction as

$$\text{Reducing agent} \longrightarrow \text{oxidized form} + \text{electrons}$$

Reducing agents low in the table (such as Li and K) have very large oxidation potentials for this reverse reaction, so they reduce other substances very strongly. They are good reducing agents. The reverse of a half-reaction low in the table, when combined with a half-reaction higher in the table, has a positive $E°$(cell). For example, Li can reduce Na^+.

The table allows us (a) to predict which proposed oxidations and reductions from the half-reactions in the table are possible, and (b) to rank the relative oxidizing power of the oxidizing agents and the relative reducing power of the reducing agents. Table 17.3 of the text is an **electrochemical series**, which is a list of reduction half-reactions, with the best oxidizing agents (on the left of the half-reactions) at the top of the list and the best reducing agents (on the right of the half-reactions) at the bottom. Table 17.3 is required for the examples and exercises that follow.

EXAMPLE 1 Relative oxidizing strengths

Is it possible to use Br_2 to oxidize Ag?

SOLUTION Oxidizing agents higher in the electrochemical series will oxidize the *reduced form* of couples lower in the series. Br_2 is above Ag, so Br_2 should oxidize Ag. We show the two reduction half-reactions involved to stress the point that the oxidizing agent is on the left and the substance oxidized on the right, as they appear in the table.

$$\begin{array}{llll} (Br_2,\ \text{higher}) & Br_2 + 2e^- \longrightarrow 2Br^- & \\ & Ag^+ + e^- \longrightarrow Ag & (Ag,\ \text{lower}) \end{array}$$

We can check the result by determining $E°$(cell) for the proposed reaction in the usual way. The positive value obtained for the cell potential confirms that the reaction should proceed spontaneously.

$$\begin{array}{lr} Br_2 + 2e^- \longrightarrow 2Br^- & E° = +1.09\text{ V} \\ 2Ag \longrightarrow 2Ag^+ + 2e^- & E° = -0.80\text{ V} \\ \hline Br_2 + 2Ag \longrightarrow 2Ag^+ \rightarrow 2Br^- & E°(\text{cell}) = +0.29\text{ V} \end{array}$$

EXERCISE It is possible to oxidize Fe to Fe^{3+} with Pb?

ANSWER: No

EXAMPLE 2 Relative reducing strengths

A cell is proposed that would use Ag to reduce I_2 for the cell reaction. Predict whether the cell will work.

SOLUTION Ag (the reducing agent in the Ag^+/Ag couple) is higher than I_2 (the oxidized form in the I_2/I^- couple) in the electrochemical series. A reducing agent higher in the table will not reduce the oxidized form of a substance lower in the table, so we predict the cell will not work. The reduction half-reactions, as they appear in Table 17.3, are shown here to stress that, in the table, the reducing agent is on the right of a half-reaction and the species to be reduced on the left of a half-reaction.

$$\begin{array}{lll} & Ag^+ + e^- \longrightarrow Ag & (Ag,\ \text{higher}) \\ (I_2,\ \text{lower}) & I_2 + 2e^- \longrightarrow 2I^- & \end{array}$$

To confirm the prediction, we can calculate $E°$(cell) in the normal way. The negative standard cell potential confirms that the reaction will not work at standard conditions.

$$
\begin{array}{lr}
2Ag \longrightarrow 2Ag^+ + 2e^- & E° = -0.80 \text{ V} \\
I_2 + 2e^- \longrightarrow 2I^- & E° = +0.54 \text{ V} \\
\hline
I_2 + 2Ag \longrightarrow 2Ag^+ + 2I^- & E°(\text{cell}) = -0.26 \text{ V}
\end{array}
$$

EXERCISE Can Zn be used to reduce Fe^{3+} to Fe?

ANSWER: Yes

KEY CONCEPT B The reaction of metals with acids

The reaction of a metal with acid can be represented by

$$\text{Metal}(s) + 2H^+(aq) \longrightarrow \text{metal cation}(aq) + H_2(g)$$

The two half-reactions are

$$
\begin{array}{ll}
\text{Metal}(s) \longrightarrow \text{metal cation}(aq) + \text{electron(s)} & E° \text{ depends on metal} \\
2H^+(aq) + 2e^- \longrightarrow H_2(g) & E° = 0 \text{ V}
\end{array}
$$

The oxidation potential of the metal must be positive for the reaction to occur at standard conditions. Only couples below hydrogen in the electrochemical series have positive oxidation potentials (because those below hydrogen have negative reduction potentials). The only metals that react with acids are those in couples below hydrogen (such as Fe, Pb, Sn, and Zn).

EXAMPLE Predicting if a metal should react with an acid

Should H_2 form if a piece of tin is dropped into water?

SOLUTION The reaction we are considering is

$$Sn(s) + 2H^+(aq) \longrightarrow Sn^{2+}(aq) + H_2(g)$$

$Sn(s)$ is part of the Sn^{2+}/Sn couple. The reduction half-reaction for this couple is below that for the H^+/H_2 couple in Table 17.3 of the text, so Sn should react as proposed. It will reduce H^+ to H_2.

EXERCISE Should copper react with an acid to form H_2?

ANSWER: No

17.5 The Dependence of Cell Potential on Concentration

KEY CONCEPT A The equilibrium constant of a reaction

The equilibrium constant of a reaction is related to the standard cell potential of the reaction through the relation

$$\ln K = \frac{nFE°(\text{cell})}{RT}$$

At 25°C, this becomes

$$\ln K = \frac{nE°(\text{cell})}{0.02569 \text{ V}}$$

This relation gives us a method for calculating the equilibrium constant for any reaction that can be written as the sum of two half-reactions and for which $E°$(cell) is known.

EXAMPLE 1 Calculating an equilibrium constant

What is the equilibrium constant at 25°C for the reaction

$$Fe^{3+}(aq) + Cu^{+}(aq) \longrightarrow Fe^{2+}(aq) + Cu^{2+}(aq)$$

SOLUTION We first determine E°(cell) for the reaction from the standard reduction potentials in Table 17.3 of the text.

$$\begin{array}{rcl|r} Fe^{3+}(aq) + e^- & \longrightarrow & Fe^{2+}(aq) & E^\circ = +0.77\text{ V} \\ Cu^{+}(aq) & \longrightarrow & Cu^{2+}(aq) + e^- & E^\circ = -0.15\text{ V} \\ \hline Fe^{3+}(aq) + Cu^{+}(aq) & \longrightarrow & Fe^{2+}(aq) + Cu^{2+}(aq) & E^\circ(\text{cell}) = +0.62\text{ V} \end{array}$$

From the half-reactions, $n = 1$. Substituting into the equation for K gives

$$\ln K = \frac{(1)(+0.62\text{ V})}{0.02569\text{ V}}$$
$$= 24$$
$$K = e^{24}$$
$$= 3 \times 10^{10} \quad \text{(no units)}$$

EXERCISE What is the equilibrium constant for the reaction

$$Ni^{2+}(aq) + 2In(s) \longrightarrow Ni(s) + 2In^{+}(aq)$$

ANSWER: 9×10^{-4} M

EXAMPLE 2 Calculating a solubility product

Use standard cell potentials to calculate K_s for AgI at 25°C.

SOLUTION Because the equilibrium involved is not a redox reaction, we first have to formulate two half-reactions that add up to the overall equilibrium for the dissolution of AgI(s):

$$AgI(s) \rightleftharpoons Ag^{+}(aq) + I^{-}(aq)$$

This can be done with the following two half-reactions (which are shown with their electrode potentials):

$$\begin{array}{rcl|r} Ag(s) & \longrightarrow & Ag^{+}(aq) + e^- & E^\circ = -0.80\text{ V} \\ AgI(s) + e^- & \longrightarrow & Ag(s) + I^{-}(aq) & E^\circ = -0.15\text{ V} \\ \hline AgI(s) & \longrightarrow & Ag^{+}(aq) + I^{-}(aq) & E^\circ(\text{cell}) = -0.95\text{ V} \end{array}$$

From the half-reactions, $n = 1$. Thus,

$$\ln K = \frac{nE^\circ(\text{cell})}{0.02569\text{ V}}$$
$$= \frac{(1)(-0.95\text{ V})}{0.02569\text{ V}}$$
$$= -37$$
$$K = e^{-37}$$
$$K_s = 9 \times 10^{-17}\text{ M}^2$$

EXERCISE Use standard cell potentials to calculate K_s for $Cd(OH)_2$ at 25°C.

ANSWER: 1×10^{-14} M^3

KEY CONCEPT B The dependence of cell potential on concentrations

The **Nernst equation** relates the concentrations of the substances in a cell to the measured cell potential. If E(cell) is the cell potential and E°(cell) the standard cell potential,

$$E(\text{cell}) = E^\circ(\text{cell}) - \frac{RT}{nF} \ln Q$$

or, at 25°C,

$$E(\text{cell}) = E^\circ(\text{cell}) - \frac{0.02569 \text{ V}}{n} \ln Q$$

Q, the mass action quotient we encountered in Chapters 15 and 16, is an expression that resembles the equilibrium expression but uses the actual concentrations present at some time (not necessarily the equilibrium concentrations); n is the number of electrons transferred for the reaction as written. By using an electrode for which Q depends on the hydronium ion concentration, we can directly relate the cell potential to the pH. For instance, as shown in the text, when $n = 2$ and the hydronium ion is a product of the cell reaction used,

$$E(\text{cell}) = E + 0.0592 \text{ pH}$$

The value of E depends on the details of the cell.

EXAMPLE Calculating E(cell) for a cell at nonstandard conditions

Use the Nernst equation to calculate the cell potential at 25°C for the reaction

$$\text{Fe}(s) + 2\text{Ag}^+(aq, 0.050 \text{ M}) \longrightarrow \text{Fe}^{2+}(aq, 2.0 \text{ M}) + 2\text{Ag}(s)$$

SOLUTION E°(cell) is calculated with the standard electrode potentials of the two half-reactions involved:

$$\begin{array}{rcll} \text{Fe}(s) & \longrightarrow & \text{Fe}^{2+}(aq) + 2\text{e}^- & E^\circ = +0.44 \text{ V} \\ 2\text{Ag}^+(aq) + 2\text{e}^- & \longrightarrow & 2\text{Ag}(s) & E^\circ = +0.80 \text{ V} \\ \hline \text{Fe}(s) + 2\text{Ag}^+(aq) & \longrightarrow & \text{Fe}^{2+}(aq) + 2\text{Ag}(s) & E^\circ(\text{cell}) = \quad 1.24 \text{ V} \end{array}$$

Q has the same form as the equilibrium expression:

$$Q = \frac{[\text{Fe}^{2+}]}{[\text{Ag}^+]^2}$$

Because $[\text{Fe}^{2+}] = 2.0$ M and $[\text{Ag}^+] = 0.050$ M,

$$Q = \frac{(2.0)}{(0.050)^2} = 8.0 \times 10^2$$

Substituting $n = 2$ and the values calculated above into the Nernst equation gives

$$\begin{aligned} E(\text{cell}) &= 1.24 \text{ V} - \frac{0.02569 \text{ V}}{2} \ln 8.0 \times 10^2 \\ &= 1.24 \text{ V} - \frac{0.02569 \text{ V}}{2} (6.68) \\ &= 1.24 \text{ V} - 0.0859 \text{ V} \\ &= 1.15 \text{ V} \end{aligned}$$

EXERCISE Calculate E(cell) at 25°C for the reaction

$$\text{Sn}^{4+}(aq, 2.0 \text{ M}) + 2\text{In}^{2+}(aq, 2.0 \text{ M}) \longrightarrow \text{Sn}^{2+}(aq, 0.10 \text{ M}) + 2\text{In}^{3+}(aq, 0.10 \text{ M})$$

ANSWER: 0.76 V

ELECTROLYSIS

KEY WORDS Define or explain each of the following terms in a written sentence or two.

Down's process | electroplating | overpotential

17.6 THE POTENTIAL NEEDED FOR ELECTROLYSIS

KEY CONCEPT A Electrolysis and the voltage needed for electrolysis

In the electrolysis process, an electric current is used to bring about a chemical change. In an electrolysis cell, as in galvanic cells, reduction occurs at the cathode and oxidation at the anode. The cathode must be connected to a source of electrons and the anode to a sink for electrons.

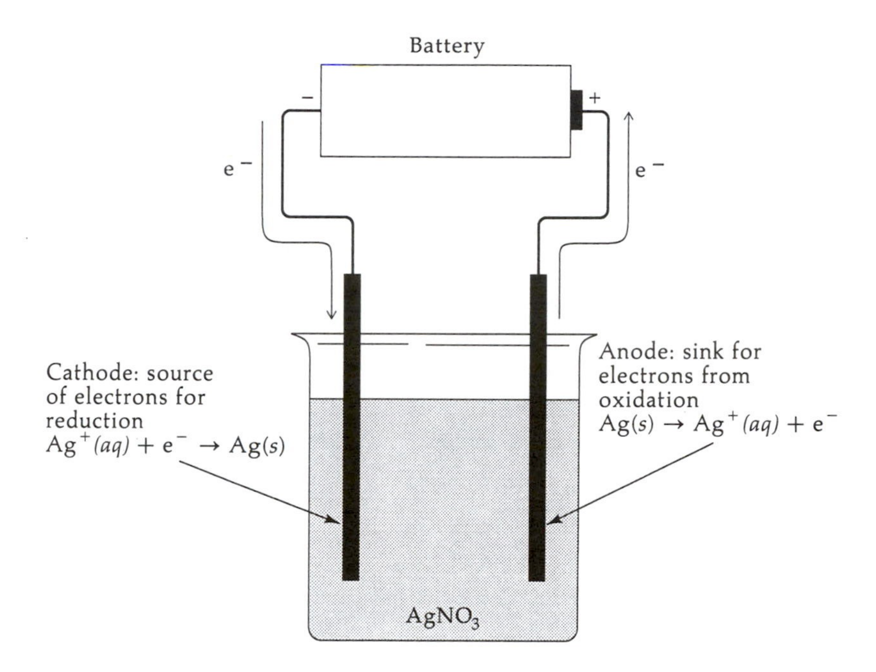

In this electrolysis cell, oxidation of $Ag(s)$ occurs at a silver anode, and aqueous Ag^+ ion is reduced (and plated out) at the cathode. Because no spontaneous chemical reaction occurs, both electrodes can be placed in the same solution.

Electrolysis forces a reaction to run in a direction opposite to its spontaneous direction. For instance, let us assume concentrations are such that E(cell) for the following reaction is +0.15 V:

$$Ag(s) + Fe^{3+}(aq) \longrightarrow Fe^{2+}(aq) + Ag^+(aq) \qquad E(\text{cell}) = +0.15\text{ V}$$

If we want to force this reaction to run in the reverse direction in an electrolysis cell, we must apply a minimum voltage of 0.15 V to the cell:

$$Fe^{2+}(aq) + Ag^+(aq) \xrightarrow{E(\text{applied}) = 0.15\text{ V}} Ag(s) + Fe^{3+}(aq)$$

In practice, the voltage calculated in this manner is not enough to cause an electrolysis to occur. For complex reasons, a higher applied voltage must often be used, especially when a gas is produced. This excess voltage, above and beyond the theoretical minimum voltage, is called an **overvoltage.** Reported overvoltages usually refer to the reaction at a single electrode half-reaction. Overvoltages depend on a number of factors; two important ones are the electrode material and the substance produced at the electrode.

EXAMPLE 1 Calculating the minimum voltage required for an electrolysis

What is the minimum voltage needed to run the following reaction in an electrolysis cell at standard conditions at 25°C?

$$2Cu^{2+}(aq) + 2H_2O(l) \longrightarrow 2Cu(s) + 4H^+(aq) + O_2(g)$$

SOLUTION An electrolysis forces a reaction to run in the direction opposite to that of the spontaneous process. The minimum voltage needed for the electrolysis equals the cell potential generated by the reverse reaction. Thus, we calculate E(cell) at standard conditions at 25°C for the reverse of the electrolysis reaction:

$$\begin{array}{rcl|r} 2Cu(s) & \longrightarrow & 2Cu^{2+}(aq) + 4e^- & E° = -0.34\text{ V} \\ 4H^+(aq) + O_2(g) + 4e^- & \longrightarrow & 2H_2O(l) & E° = +1.23\text{ V} \\ \hline 4H^+(aq) + O_2(g) + 2Cu(s) & \longrightarrow & 2Cu^{2+}(aq) + 2H_2O(l) & E°(\text{cell}) = 0.89\text{ V} \end{array}$$

As required, this reaction is the reverse of the electrolysis reaction. Because it has a cell potential of 0.89 V, a minimum of 0.89 V is required to force the electrolysis reaction to occur:

$$2Cu^{2+}(aq) + 2H_2O(l) \xrightarrow{E(\text{applied}) = 0.89\text{ V}} 2Cu(s) + 4H^+(aq) + O_2(g)$$

EXERCISE What is the minimum voltage required to run the following reaction in an electrolysis cell at standard conditions at 25°C?

$$4Ag^+(aq) + 2H_2O(l) \longrightarrow 4Ag(s) + 4H^+(aq) + O_2(g)$$

ANSWER: 0.43 V

EXAMPLE 2 Calculating the voltage needed for an electrolysis

What is the minimum voltage needed to run the electrolysis in the preceding example at 25°C when $[Cu^{2+}] = 0.10$ M and $[H^+] = 1.0 \times 10^{-7}$ M? Assume the pressure of O_2 remains at (exactly) 1 atm.

SOLUTION To determine the minimum voltage for an electrolysis, we must calculate E(cell) for the reverse spontaneous process:

$$2Cu(s) + 4H^+(aq, 1.0 \times 10^{-7}\text{ M}) + O_2(g, 1\text{ atm}) \longrightarrow 2Cu^{2+}(aq, 0.10\text{ M}) + 2H_2O(l)$$

For this reaction, $n = 4$ and $E°(\text{cell}) = 0.89$ V (see preceding example), and

$$\begin{aligned} Q &= \frac{[Cu^{2+}]^2}{[H^+]^4 P_{O_2}} \\ &= \frac{(0.10)^2}{(1.0 \times 10^{-7})^4(1)} \\ &= 1.0 \times 10^{26} \end{aligned}$$

The Nernst equation now gives

$$\begin{aligned} E(\text{cell}) &= E°(\text{cell}) - \frac{0.02569\text{ V}}{n} \ln Q \\ &= 0.89\text{ V} - \frac{0.02569\text{ V}}{4} \ln(1.0 \times 10^{26}) \\ &= 1.24\text{ V} - \frac{0.02569\text{ V}}{4}(59.87) \\ &= 0.51\text{ V} \end{aligned}$$

Because $E(\text{cell}) = 0.51$ V for the spontaneous reverse process, 0.51 V is the minimum voltage that must be applied to get the original electrolysis reaction to occur:

$$2Cu^{2+}(aq, 0.10\text{ M}) + 2H_2O(l) \xrightarrow{E(\text{applied}) = 0.51\text{ V}} 2Cu(s) + 4H^+(aq, 1.0 \times 10^{-7}\text{ M}) + O_2(g, 1\text{ atm})$$

EXERCISE What is the minimum voltage needed to run the electrolysis in the preceding exercise at 25°C when $[Ag^+] = 0.20$ M and $[H^+] = 1.0 \times 10^{-7}$ M. Assume the pressure of O_2 is exactly 1 atm.

ANSWER: 0.06 V

EXAMPLE 3 Calculating the voltage needed for electrolysis with overpotential

Assume the overpotential for formation of O_2 in the preceding example is 0.66 V and that the overpotential for formation of Ag is negligible. What is the actual voltage needed to get the electrolysis to work?

SOLUTION The overpotential is the voltage above and beyond the minimum theoretical voltage that must be applied for an electrolysis cell to function. Thus, 0.66 V must be added to the calculated 0.51 V. A total of 1.17 V (0.66 V + 0.51 V = 1.17 V) is required.

EXERCISE Assume the overpotential for formation of O_2 in the cell in the previous exercise is 0.53 V and that for formation of Ag is negligible. (The overpotential for O_2 is different from that in the example because we assume a different electrode material is used.) What is the actual voltage needed to get the electrolysis to work?

ANSWER: 0.59 V

KEY CONCEPT B Competing oxidations and reductions

When more than one reducible species is present, there is a choice regarding which is reduced during the electrolysis. The species with the most positive reduction potential is most easily reduced and is reduced first. If more than one oxidizable species is present, the one with the most positive oxidation potential is oxidized first.

Water abounds in aqueous solutions, so the possibility exists that water will be oxidized and/or reduced during an electrolysis. The reduction and oxidation potentials of neutral water (pH = 7) are

$$\text{Reduction:} \quad 2H_2O(l) + 2e^- \longrightarrow H_2(g) + 2OH^-(aq) \qquad E^\circ = -0.42\ \text{V}$$

$$\text{Oxidation:} \quad 2H_2O(l) \longrightarrow 4e^- + 4H^+(aq) + O_2(g) \qquad E^\circ = -0.81\ \text{V}$$

If we ignore overvoltage effects, any reducible species with a reduction potential more positive than −0.42 V (e.g., +0.61 V, −0.31 V) is reduced before water, but water is reduced before a species with a reduction potential less positive than −0.42 V (e.g., −0.96 V, −1.11 V). A species with an oxidation potential more positive than −0.81 V (e.g., +1.16 V, −0.62 V) is oxidized before water, whereas water is oxidized before a species with an oxidation potential less positive than −0.82 V (e.g., −1.03 V, −1.86 V).

EXAMPLE Predicting the product of an electrolysis

What products form when an aqueous solution of $MgBr_2$ is electrolyzed? Assume standard state conditions for $MgBr_2$ and no overvoltages.

SOLUTION The first task is to recognize all the possibilities for oxidation and reduction. Mg^{2+} is not easily oxidized, but it can be reduced. Water can also be reduced. The relevant half-reactions and reduction potentials are

$$Mg^{2+}(aq) + 2e^- \longrightarrow Mg(s) \qquad E^\circ = -2.36\ \text{V}$$

$$2H_2O(l) + 2e^- \longrightarrow H_2(g) + 2OH^-(aq) \qquad E^\circ = -0.42\ \text{V}$$

Because water has a more positive reduction potential than Mg^{2+} ion (−0.42 V > −2.36 V), the water is reduced first. Br^- is not easily reduced, but it can be oxidized. Water can also be oxidized. The relevant half-reactions and oxidation potentials are

$$2Br^-(aq) \longrightarrow Br_2(l) + 2e^- \qquad E^\circ = -1.09\ \text{V}$$

$$2H_2O(l) \longrightarrow 4e^- + 4H^+(aq) + O_2(g) \qquad E^\circ = -0.81\ \text{V}$$

Because H_2O has a more positive oxidation potential than Br^- ($-0.81\ V > -1.09\ V$), H_2O is oxidized first. If we neglect overpotential, the overall electrolysis reaction and required voltage are

$$2H_2O(l) \xrightarrow{E(\text{applied}) = 1.23\ V} 2H_2(g) + O_2(g)$$

The E(applied) is calculated with the techniques of the previous section ($0.42\ V + 0.81\ V = 1.23\ V$).

EXERCISE What products form when an aqueous solution of CuF_2 is electrolyzed. Write the electrolysis reaction.

ANSWER: $Cu(s)$, $O_2(g)$, H^+;
$$2Cu^{2+}(aq) + 2H_2O(l) \longrightarrow 2Cu(s) + 4H^+(aq) + O_2(g)$$

PITFALL Use of $E°$ to predict what gets oxidized or reduced

Often the $E°$ values used in predicting products are based on the assumption that standard state concentrations are present. If concentrations are different from standard state concentrations, $E°$ values may not be reliable predictors of what gets oxidized or reduced. Also, we must consider overpotentials in order to determine the actual reactions that occur in an electrolysis.

17.7 THE EXTENT OF ELECTROLYSIS

KEY CONCEPT The extent of electrolysis

The amount of product formed during an electrolysis depends on three factors: (a) the current used, (b) how long the electrolysis runs, and (c) the moles of electrons required per mole of product in the relevant half-reaction. Because 1 ampere = 1 coulomb/second, the current multiplied by time gives the total charge that flows through an electrolytic cell:

$$\text{Current (in amperes, A)} \times \text{time (in seconds, s)} = \text{charge (in coulombs, C)}$$

The charge is related to the moles of electrons that flow through the cell by the Faraday constant,

$$1\ \text{mol}\ e^- = 96{,}485\ C$$

The moles of electrons can be treated like a reactant or product in a half-reaction and, as such, can be related to the moles of other substances in the half-reaction. For instance, in the reduction of Al^{3+},

$$Al^{3+} + 3e^- \longrightarrow Al$$

the production of 3 mol of electrons by an electric current results in the production of 1 mol of Al. The connections and relations among the various quantities are shown in the following figure:

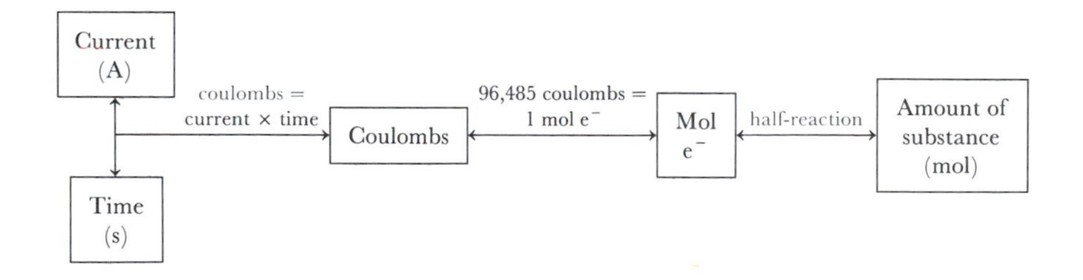

EXAMPLE 1 Predicting the amount of product formed in an electrolysis

How many grams of Al are produced by a 25-A current running for 2.0 h? The cathode reaction is $Al^{3+}(l) + 3e^- \rightarrow Al(s)$.

SOLUTION The amount of Al produced depends on the number of moles of electrons that flow through the electrolysis cell, which, in turn, depends on the total charge that flows. Using the fact that 1 A = 1 C/s, we get the total charge:

$$25\,\frac{\text{C}}{\cancel{\text{s}}} \times 2.0\,\cancel{\text{h}} \times \frac{3600\,\cancel{\text{s}}}{1\,\cancel{\text{h}}} = 1.8 \times 10^5\ \text{C}$$

The fact that 1 mol of electrons carries 96,485 C of charge allows us to calculate the moles of electrons that carry 1.8×10^5 C of charge:

$$1.8 \times 10^5\,\cancel{\text{C}} \times \frac{1\ \text{mol e}^-}{96{,}485\,\cancel{\text{C}}} = 1.9\ \text{mol e}^-$$

The grams of aluminum produced can now be calculated. The half-reaction indicates that 3 mol of electrons are required to produce 1 mol of aluminum:

$$1.9\,\cancel{\text{mol e}^-} \times \frac{1\,\cancel{\text{mol Al}}}{3\,\cancel{\text{mol e}^-}} \times \frac{26.98\ \text{g Al}}{1\,\cancel{\text{mol Al}}} = 17\ \text{g Al}$$

EXERCISE How many grams of $Cu(s)$ are produced from $Cu^{2+}(aq)$ by a 15.0-A current running for 24.0 h?

ANSWER: 427 g

EXAMPLE 2 Calculating the time required for an electrolysis

How much time (in seconds) is required to produce 150 g Na fron a NaCl melt by a 25-A current?

SOLUTION The reduction involved is $Na^+(l) + e^- \rightarrow Na(s)$. The moles of electrons needed to produce 150 grams is

$$150\,\cancel{\text{g Na}} \times \frac{1\,\cancel{\text{mol Na}}}{23.0\,\cancel{\text{g Na}}} \times \frac{1\ \text{mol e}^-}{1\,\cancel{\text{mol Na}}} = 6.52\ \text{mol e}^-$$

With this result, we can determine the total charge that must flow:

$$6.52\,\cancel{\text{mol e}^-} \times \frac{96{,}485\ \text{C}}{1\,\cancel{\text{mol e}^-}} = 6.29 \times 10^5\ \text{C}$$

Because the current is 25 A, the charge is delivered at the rate of 25 C/s:

$$6.29 \times 10^5\,\cancel{\text{C}} \times \frac{1\text{s}}{25\,\cancel{\text{C}}} = 2.5 \times 10^4\ \text{s}$$

EXERCISE How much time (in seconds) is required to plate 1.20 kg of Cr from Cr^{3+} with a 125-A current?

ANSWER: 5.34×10^4 s

DESCRIPTIVE CHEMISTRY TO REMEMBER

- A **primary cell** is a sealed cell that is not designed to be recharged. A **secondary cell** is a cell that is designed to be rechargeable. A **fuel cell** is a primary cell that operates on a continually renewed supply of fuel.

Some practical cells

- **Leclanché** cell (standard dry cell), a primary cell: Zn anode, MnO_2 cathode, electrolyte contains NH_4Cl, cell potential = 1.5 V.

  Anode relation: $4NH_4^+(aq) + Zn(s) \longrightarrow [Zn(NH_3)_4]^{2+}(aq) + 4\,H^+(aq)$

  Cathode reaction: $MnO_2(s) + H_2O(l) + e^- \longrightarrow MnO(OH)(s) + OH^-(aq)$

- **Alkaline** cell, a primary cell: Zn anode, MnO_2 cathode, electrolyte contains KOH, cell potential = 1.5 V.

$$\text{Anode reaction:} \quad Zn(s) + 2OH^-(aq) \longrightarrow Zn(OH)_2(s) + 2e^-$$
$$\text{Cathode reaction:} \quad MnO_2(s) + H_2O(l) + e^- \longrightarrow MnO(OH)(s) + OH^-(aq)$$

- **Mercury** cell, a primary cell: Zn anode, HgO cathode, electrolyte contains KOH, cell potential = 1.3 V.

$$\text{Anode reaction:} \quad Zn(s) + 2OH^-(aq) \longrightarrow Zn(OH)_2(s) + 2e^-$$
$$\text{Cathode reaction:} \quad HgO(s) + H_2O(l) + 2e^- \longrightarrow Hg(l) + 2OH^-(aq)$$

- **Lead–acid** cell, a secondary cell: Pb anode, PbO_2 cathode, electrolyte contains $H_2SO_4(aq)$, cell potential = 2 V.

$$\text{Anode reaction:} \quad Pb(s) + SO_4^{2-}(aq) \longrightarrow PbSO_4(s) + 2e^-$$
$$\text{Cathode reaction:} \quad PbO_2(s) + 4H^+(aq) + SO_4^{2-}(aq) + 2e^- \longrightarrow PbSO_4(s) + 2H_2O(l)$$

- **Nickel–cadmium** cell (nicad cell), a secondary cell: Cd anode, $Ni(OH)_3$ cathode, electrolyte contains $OH^-(aq)$, cell potential = 1.4 V.

$$\text{Anode reaction:} \quad 2OH^-(aq) + Cd(s) \longrightarrow Cd(OH)_2(aq) + 2e^-$$
$$\text{Cathode reaction:} \quad Ni(OH)_3(s) + e^- \longrightarrow Ni(OH)_2(s) + OH^-(aq)$$

- **Passivation** is the formation on a metal of a surface film that protects the metal from further reaction. As an example, aluminum and zinc exposed to air both form oxide coats that prevent further oxidation.
- **Corrosion** is the oxidation of metals by water (H_2O/H_2, OH^- couple) or oxygen (O_2, H^+/H_2O couple).
- In the **Downs process,** molten NaCl is electrolyzed to produce sodium metal.
- **Electroplating** is a process in which the object to be plated is made the cathode in an electrolytic cell, and a surface coating of a metal is deposited on the object as metal ions in solution are reduced. Chromium plating for automobile parts is done by electrolysis.
- Attempts to electroplate metallic **chromium** from $Cr^{3+}(aq)$ solutions have been unsuccessful because the hydrated ion $[Cr(H_2O)_6]^{3+}$ is so stable that it cannot be reduced.

CHEMICAL EQUATIONS TO KNOW

- In the Daniel cell, $Cu^{2+}(aq)$ is reduced to $Cu(s)$, and $Zn(s)$ is oxidized to $Zn^{2+}(aq)$.

$$Zn(s) + Cu^{2+}(aq) \longrightarrow Cu(s) + Zn^{2+}(aq)$$

- The overall reaction in the familiar commercial dry cell is

$$Zn(s) + 2MnO_2(s) + 4NH_4^+(aq) + 2OH^-(aq) \longrightarrow [Zn(NH_3)_4]^{2+}(aq) + 2H_2O(l) + 2MnO(OH)(s)$$

- The overall reaction in a mercury cell is

$$Zn(s) + HgO(s) + H_2O(l) \longrightarrow Zn(OH)_2(s) + Hg(l)$$

- The overall reaction in a lead–acid cell (during discharge) is

$$Pb(s) + PbO_2(s) + 2H_2SO_4(aq) \longrightarrow 2PbSO_4(s) + 2H_2O(l)$$

- In the electrolysis of aqueous NaCl, Cl_2 is produced at the anode and OH^- at the cathode. If these products are allowed to mix, hypochlorite ion is produced.

$$Cl_2(g) + 2OH^-(aq) \longrightarrow 2ClO^-(aq) + H_2(g)$$

- The electroplating of chromium, a process that uses a lot of electricity because six electrons are required for every atom of chromium, uses the reduction half-reaction

$$CrO_3(aq) + 6H^+(aq) + 6e^- \longrightarrow Cr(s) + 3H_2O(l)$$

MATHEMATICAL EQUATIONS TO KNOW AND UNDERSTAND

$\Delta G = -nFE(\text{cell})$	free energy of electrochemical cell
$\Delta G^\circ = -nFE^\circ(\text{cell})$	standard free energy of electrochemical cell
$E^\circ(\text{cell}) = E^\circ(\text{anode}) + E^\circ(\text{cathode})$	standard potential of cell
$\Delta G^\circ = -RT \ln K$	relation of equilibrium constant to standard free energy
$nFE^\circ(\text{cell}) = RT \ln K$	relation of equilibrium constant to standard cell potential
$\Delta G = \Delta G^\circ + RT \ln Q$	free energy of chemical process at any conditions
$E(\text{cell}) = E^\circ(\text{cell}) + \dfrac{RT}{nF} \ln Q$	cell potential of cell at any conditions

Coulombs delivered
= current in amperes × time in seconds

SELF-TEST EXERCISES

Electrochemical cells

1. The oxidizing agent in the reaction $2Na(s) + 2H_2O(l) \longrightarrow 2NaOH(aq) + H_2(g)$ is
(a) NaOH (b) H_2O (c) H_2 (d) Na

2. What species is reduced in the following reaction?

$$2Cr(s) + 3MnO_2(s) + 12H^+(aq) \longrightarrow 2Cr^{3+}(aq) + 3Mn^{2+}(aq) + 6H_2O(l)$$

(a) Cr (b) MnO_2 (c) Cr^{3+} (d) H^+ (e) H_2O

3. For the following cell reaction, what is produced at the cathode?

$$Zn(s) + Pb^{2+}(aq) \longrightarrow Zn^{2+}(aq) + Pb(s)$$

(a) Zn (b) Pb^{2+} (c) Zn^{2+} (d) Pb

4. What is the cell diagram for the cell that uses the following reaction; the Fe^{3+}/Fe^{2+} couple uses a platinum electrode.

$$2Fe^{3+}(aq) + Cu(s) \longrightarrow Cu^{2+}(aq) + 2Fe^{2+}(aq)$$

(a) $Cu(s) | Cu^{2+}(aq) || Fe^{3+}(aq) | Pt(s)$
(b) $Pt(s) | Fe^{3+}(aq), Fe^{2+}(aq) || Cu^{2+}(aq) | Cu(s)$
(c) $Pt(s) | Fe^{3+}(aq) || Cu^{2+}(aq) | Pt(s)$
(d) $Pt(s) | Cu(s) | Cu^{2+}(aq) || Fe^{3+}(aq), Fe^{2+}(aq) | Cu(s)$
(e) $Cu(s) | Cu^{2+}(aq) || Fe^{3+}(aq), Fe^{2+}(aq) | Pt(s)$

5. What is the anode reaction for the cell represented by the cell diagram that follows?

$$Pt(s) | H_2(g) | HCl(aq) || Hg_2Cl_2(s) | Hg(l)$$

(a) $H_2(g) \longrightarrow 2H^+(aq) + 2e^-$
(b) $Pt(s) \longrightarrow Pt^{2+}(aq) + 2e^-$
(c) $2Hg(l) + 2Cl^-(aq) \longrightarrow Hg_2Cl_2(s) + 2e^-$
(d) $Hg_2Cl_2(s) + 2e^- \longrightarrow 2Hg(l) + 2Cl^-(aq)$
(e) $2HCl(aq) + 2e^- \longrightarrow H_2(g) + 2Cl^-(aq)$

6. Which of the following is a reasonable cathode reaction for a cell using the given nonredox reaction?

$$MnCO_3(s) \rightleftharpoons Mn^{2+}(aq) + CO_3{}^{2-}(aq)$$

(a) $Mn(s) \longrightarrow Mn^{2+}(aq) + 2e^-$
(b) $2CO_3{}^{2-}(aq) \longrightarrow 2C(s) + 3O_2(g) + 4e^-$
(c) $Mn^{2+}(aq) + 2e^- \longrightarrow Mn(s)$
(d) $MnCO_3(s) + 2e^- \longrightarrow Mn(s) + CO_3{}^{2-}(aq)$

7. Which of the following is a reasonable cathode reaction for a cell using the given reaction?

$$CuS(s) \longrightarrow Cu^{2+}(aq) + S^{2-}(aq)$$

(a) $CuS(s) \longrightarrow Cu^{2+}(aq) + S^{2-}(aq) + 2e^-$
(b) $Cu^{2+}(aq) + 2e^- \longrightarrow Cu(s)$
(c) $CuS(s) + 2e^- \longrightarrow Cu(s) + S^{2-}(aq)$
(d) $Cu(s) + S^{2-}(aq) \longrightarrow CuS(s) + 2e^-$

8. 1 joule = which of the following?
(a) 1 volt/coulomb
(b) 1 coulomb/volt
(c) 1 volt · coulomb
(d) 1×10^{-5} volt · coulomb

9. A cell uses an oxidation half-reaction that is a strongly spontaneous process and a reduction half-reaction that is not spontaneous. What is a reasonable estimate for the cell potential?
(a) +2.5 V (b) −1.3 V (c) 0.0 V (d) +0.15 V (e) −0.18 V

Thermodynamics and electrochemistry

10. What is ΔG for the reaction $Mg(s) + 2H^+(aq) \longrightarrow Mg^{2+}(aq) + H_2(g)$ if it is run under conditions such that E(cell) = +1.65 V?
(a) +159 kJ (b) +318 kJ (c) 0 kJ (d) −159 kJ (e) −318 kJ

11. What is ΔG for the reaction $2Ce(s) + 3Mg^{2+}(aq) \longrightarrow 3Mg(s) + 2Ce^{3+}(aq)$ if it is run under conditions such that E(cell) = +0.562 V?
(a) −325 kJ (b) +325 kJ (c) −163 kJ (d) +163 kJ (e) −108 kJ

12. What is E(cell) for the reaction $Zn(s) + Cu^{2+}(aq) \longrightarrow Zn^{2+}(aq) + Cu(s)$ if it is run under conditions such that $\Delta G = -532$ kJ.
(a) +1.10 V (b) −3.16 V (c) −0.03 V (d) +2.76 V (e) +5.51 V

13. Assume E(cell) = 1.20 V for A $\longrightarrow$ B. What is E(cell) for the reaction 2A $\longrightarrow$ 2B?
(a) 0.30 V (b) 4.80 V (c) 0.60 V (d) 2.40 V (e) 1.20 V

14. What is $E°$(cell) for $Cd(s) + Sn^{2+}(aq) \longrightarrow Sn(s) + Cd^{2+}(aq)$?
(a) 0.0 V (b) 0.26 V (c) −0.40 V (d) −0.55 V (e) +0.14 V

15. What is $E°$(cell) for $2Na(s) + 2H_2O(l) \longrightarrow 2NaOH(aq) + H_2(g)$?
(a) 2.71 V (b) −0.83 V (c) 1.88 V (d) −3.54 V (e) +1.62 V

16. What is $E°$(cell) for $Au(s) + 3Ag^+(aq) \longrightarrow Au^{3+}(aq) + 3Ag(s)$?
(a) +0.20 V (b) −1.40 V (c) −0.60 V (d) +1.00 V (e) +0.80 V

17. What is $E°$(cell) for the cell $Pt(s) | Fe^{3+}(aq), Fe^{2+}(aq) || Cu^+(aq) | Cu(s)$?
(a) −0.27 V (b) +0.83 V (c) 0.25 V (d) 1.29 V (e) +1.01 V

18. Which of the following is the strongest oxidizing agent? (Consult Table 17.3 of the text.)
(a) Cl_2 (b) H_2O (c) Pb (d) Li (e) Ag^+

19. Which of the following is the strongest reducing agent? (Consult Table 17.3.)
(a) Ca (b) Al^{3+} (c) H_2 (d) Cu (e) Br_2

20. Which of the following will oxidize H_2O? (Consult Table 17.3.)
(a) K (b) Au^+ (c) Ag^+ (d) Fe^{3+} (e) Pb

21. Which of the following will reduce Pb^{2+}? (Consult Appendix 2B of the text.)
(a) Cl_2 (b) Zn (c) Ca^{2+} (d) Ag (e) Fe^{3+}

22. Which metal reacts with an acid to form H_2? (Consult Table 17.3.)
(a) Mg (b) Cu (c) Ag (d) Au

23. What is K at 25°C for the reaction $Cl_2(g) + 2F^-(aq) \longrightarrow F_2(g) + 2Cl^-(aq)$?
(a) 7.8×10^{-52} (b) 8.7×10^{-18} (c) 1.2×10^{17} (d) 7.6×10^{-35} (e) 2.6×10^{15}

24. Calculate K_s for $Hg_2Cl_2(s)$ using standard cell potentials.
(a) 3.7×10^{17} M^3 (b) 1.6×10^{-9} M^3 (c) 6.1×10^{8} M^3 (d) 2.7×10^{-18} M^3 (e) 8.6×10^{-5}M^3

25. What is E(cell) for the following reaction?

$$2Fe^{3+}(aq, 0.20 \text{ M}) + Cu(s) \longrightarrow Cu^{2+}(aq, 0.10 \text{ M}) + 2Fe^{2+}(aq, 0.10 \text{ M})$$

(a) 0.37 V (b) 0.43 V (c) 0.48 V (d) 0.55 V (e) 0.31 V

Electrolysis

26. Neglecting overpotentials, what voltage must we apply to run the electrolysis reaction $2Na^+(aq) + 2I^-(aq) \rightarrow 2Na(s) + I_2(s)$ at standard conditions at 25°C?
(a) 0.54 V (b) 2.71 V (c) 1.67 V (d) 3.25 V (e) 2.17 V

27. Neglecting overpotentials, what voltage must we apply to run the following electrolysis reaction at 25°C?

$$2I^-(aq, 0.20 \text{ M}) + 2H_2O(l) \longrightarrow I_2(s) + H_2(g, 1 \text{ atm}) + 2OH^-(aq, 5.0 \times 10^{-3} \text{ M})$$

(a) 1.17 V (b) 1.37 V (c) 1.46 V (d) 1.28 V (e) 1.15 V

28. What voltage must be applied to run the following electrolysis at 25°C? Assume the overpotential for production of $O_2(g)$ is 0.48 V and that for production of $Zn(s)$ is negligible.

$$2Zn^{2+}(aq, 0.50 \text{ M}) + 2H_2O(l) \longrightarrow 2Zn(s) + O_2(g, 1 \text{ atm}) + 4H^+(aq, 0.20 \text{ M})$$

(a) 0.76 V (b) 1.93 V (c) 1.96 V (d) 1.23 V (e) 2.44 V

29. What products should be formed during the electrolysis of aqueous $NiBr_2$? Assume no overpotentials are present and standard concentrations of solute are used.
(a) Br_2, H_2, OH^- (b) Ni, O_2, H^+ (c) Br_2, O_2, H^+ (d) Ni, H_2, OH^- (e) O_2, H_2

30. An aqueous solution containing Ag^+, Cu^{2+}, and Zn^{2+} is electrolyzed. What product should form at the cathode?
(a) Ag (b) Cu (c) Zn (d) H_2, OH^- (e) O_2, H^+

31. How many grams of aluminium can be plated out of a Al^{3+} solution with a current of 100 A running for 12.0 h?
(a) 0.112 (b) 154 (c) 402 (d) 1.21×10^3 (e) 0.335

32. Using a 75.0-A current, how long should it take to plate out 5.00 g of nickel from a $NiCl_2$ solution?
(a) 124 s (b) 62.0 s (c) 31.0 s (d) 496 s (e) 219 s

Descriptive chemistry

33. What is the anode reaction in a dry cell?
(a) $MnO_2(s) + H_2O(l) + e^- \longrightarrow MnO(OH)(s) + OH^-(aq)$
(b) $4NH_4^+(aq) + Zn(s) \longrightarrow [Zn(NH_3)_4]^{2+}(aq) + 4H^+(aq) + 2e^-$
(c) $Pb(s) + SO_4^{2-}(aq) \longrightarrow PbSO_4(s) + 2e^-$
(d) $2H^+(aq) + 2e^- \longrightarrow H_2(g)$
(e) $Cd(s) + 2OH^-(aq) \longrightarrow Cd(OH)_2(s) + 2e^-$

34. The anode and cathode reactions in the lead–acid battery both yield the same product under discharge. What is it?
(a) Pb (b) PbO_2 (c) H_2O (d) $PbSO_4$ (e) H^+

35. Which of the following is the cell that uses a constantly renewed supply of fuel to generate its current?
(a) primary cell (b) nicad cell (c) secondary cell (d) fuel cell (e) lead–acid cell

18 HYDROGEN AND THE s-BLOCK ELEMENTS

HYDROGEN

KEY WORDS Define or explain each of the following terms in a written sentence or two.

hydrogenation
hydrometallurgical extraction
interstitial hydride
metallic hydride
reforming reaction
saline hydride
shift reaction
synthesis gas
water-splitting reactions

18.1 THE ELEMENT HYDROGEN

KEY CONCEPT A The uses and manufacture of hydrogen

About 3×10^8 kg of hydrogen per year are used in a variety of industrial processes in the United States.

Uses	Chemical equations
rocket fuel	$2H_2(l) + O_2(l) \longrightarrow 2H_2O(g)$
conversion to ammonia	$3H_2(g) + N_2(g) \longrightarrow 2NH_3(g)$
conversion to methanol	$2H_2(g) + CO(g) \longrightarrow CH_3OH(l)$
hydrometallurgical extraction	$H_2(g) + Cu^{2+}(aq) \longrightarrow Cu(s) + 2H^+$ (example)
hydrogenation	$H_2(g) + —C{=}C— \longrightarrow —CH—CH—$

The principal manufacturing method for hydrogen employs two reactions, the reforming reaction followed by the shift reaction:

Reforming reaction: $CH_4(g) + H_2O(g) \longrightarrow CO(g) + 3H_2(g)$
Shift reaction: $CO(g) + H_2O(g) \longrightarrow CO_2(g) + H_2(g)$

Two typical laboratory preparations are (a) reaction of an active metal with an acid and (b) reaction of a hydride with water:

Active metal with acid: $Zn(s) + 2HCl(aq) \longrightarrow ZnCl_2(aq) + H_2(g)$
Hydride with water: $CaH_2(s) + 2H_2O(l) \longrightarrow Ca(OH)_2(aq) + 2H_2(g)$

EXAMPLE Calculating the mass of the production of H_2

How many kilograms of H_2 can be produced from 10.0 kg of CH_4 and excess H_2O?

SOLUTION This synthesis involves the reforming reaction and the shift reaction in a standard calculation of the grams of product that result from a given amount of reactant. In this case, methane (CH_4) is the limiting reagent, so we first calculate the moles of methane with its molar mass of 16.04 g/mol:

$$10.0\ \text{kg} \times \frac{1000\ \text{g}}{1\ \text{kg}} \times \frac{1\ \text{mol}}{16.04\ \text{g}} = 623\ \text{mol}\ CH_4$$

We now use the balanced equation to calculate the moles CO and moles H_2 produced in the reforming reaction:

$$623\ \cancel{\text{mol CH}_4} \times \frac{1\ \text{mol CO}}{1\ \cancel{\text{mol CH}_4}} = 623\ \text{mol CO}$$

$$623\ \cancel{\text{mol CH}_4} \times \frac{3\ \text{mol H}_2}{1\ \cancel{\text{mol CH}_4}} = 1.87 \times 10^3\ \text{mol H}_2$$

The CO is converted to H_2 in the shift reaction, so we calculate the moles of H_2 produced:

$$623\ \cancel{\text{mol CO}} \times \frac{1\ \text{mol H}_2}{1\ \cancel{\text{mol CO}}} = 623\ \text{mol H}_2$$

The total of 2.49×10^3 mol H_2 produced (1.87×10^3 mol + 623 mol = 2.49×10^3 mol) is converted to kilograms to get the desired answer:

$$2.49 \times 10^3\ \cancel{\text{mol H}_2} \times \frac{2.02\ \cancel{\text{g}}}{1\ \cancel{\text{mol H}_2}} \times \frac{1\ \text{kg}}{1000\ \cancel{\text{g}}} = 5.03\ \text{kg}$$

EXERCISE Which laboratory preparation of H_2 gives the higher yield of hydrogen per gram of the more expensive starting material (Zn, CaH_2)? Try to do this exercise without detailed calculations.

ANSWER: Reaction of CaH_2 with H_2O

KEY CONCEPT B The special characteristics of hydrogen

Hydrogen is so unique that it is not thought of as part of any group or block of elements. Two special characteristics make its chemistry unique.

1. The small size of the atom, which allows it to hydrogen bond.
2. The exceptionally small size of the cation (H^+), which makes it an exceptionally strong Lewis acid; the small size also leads to extremely fast proton transfer in Brønsted acid–base reactions.

Another unique aspect of hydrogen is the relative masses of its three isotopes.

Name	Symbol	Nucleus	Mass, g/mol
protium	^1H	1 proton	1.01
deuterium	^2H	1 proton + 1 neutron	2.02
tritium*	^3H	1 proton + 2 neutrons	3.03

* Radioactive

No other element has an isotope with twice the mass of a lighter isotope *and* an isotope with three times the mass of the ligher one. Also, hydrogen is the only element whose isotopes are commonly called by their own names (protium, deuterium, and tritium).

EXAMPLE Predicting the solubility of hydrogen-containing compounds

Account for the high solubility of propanol, $CH_3CH_2CH_2OH$, in water even though propane, $CH_3CH_2CH_3$, is virtually insoluble in water.

SOLUTION Propane is a nonpolar molecule because of its low bond polarities and symmetric structure, so it doesn't dissolve in highly polar solvents like water. Propanol, with its polar —OH group, is able to undergo hydrogen bonding with the water solvent and dissolves easily.

EXERCISE Why are proton transfer reactions so fast?

ANSWER: The small size of H^+ allows it to move from place to place without colliding with or bumping nearby atoms.

18.2 IMPORTANT COMPOUNDS OF HYDROGEN

KEY CONCEPT Compounds of hydrogen

We consider binary hydrides only, compounds in which hydrogen combines with only one other element.

Saline hydrides. Hydrogen forms ionic hydrides (also known as saline hydrides) when it combines with a strongly electropositive metal. They are formed by heating the element with H_2. The hydrides all contain the hydride ion, H^-.

$$2K(s) + H_2(g) \longrightarrow 2KH(s) \qquad (K^+ \text{ and } H^- \text{ ions combined})$$
$$Ca(s) + H_2(g) \longrightarrow CaH_2(s) \qquad (Ca^{2+} \text{ and } H^- \text{ ions combined})$$

The ionic hydrides are powerful reducing agents. They lose electrons easily.

$$2H^-(aq) \longrightarrow 2e^- + H_2(g) \qquad E^\circ = 2.25 \text{ V}$$

The large positive electrode potential indicates an extremely favorable process. H^- cannot exist in water because its oxidation potential (+2.25 V), when added to the reduction potential of neutral water (−0.42 V), gives a positive cell potential:

$$\begin{array}{rcll} 2H^-(aq) & \longrightarrow & 2e^- + H_2(g) & E^\circ = 2.25 \text{ V} \\ 2H_2O(l) + 2e^- & \longrightarrow & 2OH^-(aq) + H_2(g) & E^\circ = -0.42 \text{ V} \\ \hline 2H^-(aq) + 2H_2O(l) & \longrightarrow & 2OH^-(aq) + 2H_2(g) & E^\circ = +1.83 \text{ V} \end{array}$$

From the Brønsted acid–base point of view, H^- is a stronger base than OH^-, so H^- forms OH^- when it combines with water.

Molecular compounds of hydrogen. The common covalent hydrides are compounds in which hydrogen is combined with a nonmetal; examples are H_2O, NH_3, H_2S, C_2H_2, and PH_3. They are synthesized with a variety of techniques, as shown by the following examples.

Direct combination of elements: $H_2(g) + Cl_2(g) \xrightarrow{\text{light}} 2HCl(g)$

Protonation of Brønsted base S^{2-}: $S^{2-}(aq) + 2HCl(aq) \longrightarrow H_2S(g) + 2Cl^-(aq)$

C_2^{2-}: $C_2^{2-}(s) + 2H_2O(l) \longrightarrow C_2H_2(g) + 2OH^-(aq)$

Br^-: $KBr(aq) + H_3PO_4(aq) \longrightarrow HBr(g) + KH_2PO_4(aq)$

Deprotonation of Brønsted acid: $PH_4^+(aq) + OH^-(aq) \longrightarrow PH_3(g) + H_2O(l)$

The polymeric hydrides aluminum hydride and beryllium hydride also are covalent hydrides.

Interstitial hydrides. Heating certain *d*-block metals in the presence of hydrogen (H_2) results in the formation of hydrides in which the hydride ions occupy holes or gaps between the metal atoms in the structure.

$$2Cu(s) + H_2(g) \longrightarrow 2CuH(s) \qquad \text{(an interstitial hydride)}$$

EXAMPLE Synthesis of a covalent hydride

Suggest a method for the preparation of ammonia (NH_3) that uses protonation of a Brønsted base.

SOLUTION One approach to this problem is to decide which Brønsted base becomes ammonia when the base is protonated. Removal of a proton from NH_3 leaves NH_2^-, so protonation of NH_2^- results in formation of ammonia:

$$NH_2^-(aq) + H_2O(l) \longrightarrow NH_3(g) + OH^-(aq)$$

EXERCISE Suggest a method for the preparation of ammonia that uses deprotonation of a Brønsted acid.

ANSWER: $NH_4^+(aq) + OH^-(aq) \longrightarrow NH_3(g) + H_2O(l)$

GROUP I: ALKALI METALS

KEY WORDS Define or explain each of the terms with a written sentence or two.

metal–ammonia solutions Solvay process

18.3 GROUP I ELEMENTS

KEY CONCEPT The Group I elements: Properties and production

All the alkali metals (Group I) are very easily oxidized and are therefore never found in nature in their elemental form (with oxidation state 0). The elements can be formed by electrolysis of a molten salt; for instance, sodium is produced by the electrolysis of NaCl:

$$2NaCl(l) \xrightarrow{\text{electrolysis}} 2Na(s) + Cl_2(g)$$

They also can be made by reduction with a powerful reducing agent, such as sodium, which can be used to produce potassium:

$$KCl(l) + Na(g) \xrightarrow{750\text{ K}} NaCl(s) + K(g)$$

Ionization energies. For many properties, lithium is anomalous relative to the rest of the group. First ionization energies are low for all of the alkali metals, but the second ionization energies are extremely large because they involve breaking up a noble-gas core. The ease of removing the first electron results in the chemistry of the alkali metals being dominated by formation of ions with a +1 charge. Ionization energies decrease as we go down the group because the size of the atoms increases.

Reduction potentials. All the reduction potentials of the alkali-metal cations are very negative, which indicates that the metals themselves are easily oxidized and are therefore strong reducing agents. Lithium's reduction potential is especially large, which can be attributed to the fact that the small Li^+ cation is strongly hydrated. The strong hydration is very exothermic, so formation of the cation is thermodynamically favored (i.e., $\Delta G < 0$). This, in turn, leads to a very positive oxidation potential (and very negative reduction potential).

Reaction with water and ammonia. All the alkali metals react with water, with the speed of the reaction increasing down the group.

$2Li(s) + 2H_2O(l) \longrightarrow 2LiOH(aq) + H_2(g)$	slow
$2Na(s) + 2H_2O(l) \longrightarrow 2NaOH(aq) + H_2(g)$	fast; H_2 may ignite (dangerous)
$2K(s) + 2H_2O(l) \longrightarrow 2KOH(aq) + H_2(g)$	very fast; H_2 ignites (dangerous)
$2Cs(s) + 2H_2O(l) \longrightarrow 2CsOH(aq) + H_2(g)$	dangerously explosive
$2Rb(s) + 2H_2O(l) \longrightarrow 2RbOH(aq) + H_2(g)$	dangerously explosive

The reaction with lithium is anomalously slow because the small lithium atom is bound tightly in the solid, and also because lithium, relative to the other Group I elements, has a high ionization potential and high sublimation energy.

Reaction with nonmetals. All the alkali metals react directly with most of the nonmetals (other than the noble gases). Lithium, in anomalous fashion, is the only alkali metal that burns in nitrogen. The main reactions for burning in air are

Li:	$6Li(s) + N_2(g) \longrightarrow 2Li_3N(s)$	lithium nitride formed
	$4Li(s) + O_2(g) \longrightarrow 2Li_2O(s)$	lithium oxide formed
Na:	$2Na(s) + O_2(g) \longrightarrow Na_2O_2(s)$	sodium peroxide formed
K:	$K(s) + O_2(g) \longrightarrow KO_2(s)$	potassium superoxide formed
Cs:	$Cs(s) + O_2(g) \longrightarrow CsO_2(s)$	cesium superoxide formed
Rb:	$Rb(s) + O_2(g) \longrightarrow RbO_2(s)$	rubidium superoxide formed

The oxides of Na, K, Cs, and Rb are produced by decomposition of the respective carbonates:

$$Na_2CO_3(s) \longrightarrow Na_2O(s) + CO_2(g)$$
$$K_2CO_3(s) \longrightarrow K_2O(s) + CO_2(g)$$
$$Cs_2CO_3(s) \longrightarrow Cs_2O(s) + CO_2(g)$$
$$Rb_2CO_3(s) \longrightarrow Rb_2O(s) + CO_2(g)$$

Group I superoxides can be used to purify air by reacting with the unwanted CO_2 and producing the desired O_2. For instance,

Potassium superoxide:	$4KO_2(s) + 2CO_2(g) \longrightarrow 2K_2CO_3(s) + 3O_2(g)$
Sodium superoxide:	$4NaO_2(s) + 2CO_2(g) \longrightarrow 2Na_2CO_3(s) + 3O_2(g)$

EXAMPLE Cohesive forces in the alkali metals

Based on the speed of reaction with water, which metal is held together by stronger cohesive forces, Na or K?

SOLUTION Lithium reacts relatively slowly with water because the lithium atoms are held very tightly together in the metal and so removing them from the metal takes a lot of energy. This results in a high activation energy (and slow speed) for the reaction. Because sodium reacts more slowly than potassium with water, we conclude that the reaction with sodium has a higher activation energy because the sodium atoms are held more tightly in the solid than potassium atoms. In other words, sodium is held together by stronger cohesive forces.

EXERCISE What is the oxidation state of oxygen in the oxide ion, the peroxide ion, and the superoxide ion?

ANSWER: Oxide, -2; peroxide, -1; superoxide, $-\frac{1}{2}$

18.4 IMPORTANT GROUP I COMPOUNDS

KEY CONCEPT A Important alkali metal compounds

Lithium compounds. The small size and high polarizing power of the Li^+ cation results in a notable tendency toward covalent bonding for lithium. It also results in a strong ion–dipole interaction, so lithium salts are often hydrated.

Sodium chloride, NaCl (table salt). Sodium chloride is obtained by mining or evaporation of brine (seawater). NaCl is used for a vast number of industrial processes such as the production of NaOH, Na_2SO_4, and Na_2CO_3. It is also a necessary nutritional mineral and is used to season food. For a variety of reasons, Na^+ ions are found in seawater at a much higher concentration than K^+ ions, even though the two ions have similar abundances on earth.

Sodium hydroxide, NaOH (caustic soda). Sodium hydroxide is produced in large quantities by the electrolysis of aqueous NaCl. It is used, among other things, as an inexpensive starting material for the production of other sodium salts.

$$2NaCl(aq) + 2H_2O(l) \xrightarrow{\text{electrolysis}} 2NaOH(aq) + Cl_2(g) + H_2(g)$$

Sodium sulfate, Na_2SO_4 (salt cake); $Na_2SO_4 \cdot 10H_2O$ (Glauber's salt). Sodium sulfate is mined or produced by reaction of NaCl with sulfuric acid. Na_2SO_4 is used in paper making and as a substitute for phosphates in detergents.

$$H_2SO_4(aq) + 2NaCl(s) \longrightarrow Na_2SO_4(aq) + 2HCl(g)$$

Sodium carbonate, Na_2CO_3 (soda ash); $Na_2CO_3 \cdot 10H_2O$ (washing soda). Sodium carbonate is mined or produced by the Solvay process, which uses the overall reaction

$$2NaCl(aq) + CaCO_3(s) \longrightarrow Na_2CO_3(s) + CaCl_2(s)$$

It is used in large amounts in glass making as a source of sodium oxide, which is produced in a Lewis acid–base decomposition:

$$Na_2CO_3(s) \longrightarrow Na_2O(s) + CO_2(g)$$

$$CO_3^{2-} \longrightarrow \underset{\text{Lewis base}}{O^{2-}} + \underset{\text{Lewis acid}}{CO_2}$$

Washing soda, which is quite soluble, was formerly used to precipitate Mg^{2+} and Ca^{2+} from hard water:

$$Ca^{2+}(aq) + CO_3^{2-}(aq) \longrightarrow CaCO_3(s)$$

$$Mg^{2+}(aq) + CO_3^{2-}(aq) \longrightarrow MgCO_3(s)$$

It also provides a basic environment for aqueous solutions:

$$CO_3^{2-}(aq) + H_2O(l) \longrightarrow HCO_3^-(aq) + OH^-(aq)$$

Potassium compounds. Potassium is obtained (in the form of K^+ ions) principally through mining of KCl and $KCl \cdot MgCl_2 \cdot 6H_2O$ (a mixed salt). Potassium nitrate (KNO_3) is obtained by mixing KCl with $NaNO_3$ in an aqueous solution. Potassium nitrate is more soluble than sodium chloride; so NaCl can be crystallized out of a solution containing Na^+, K^+, Cl^-, and NO_3^- ions, and potassium nitrate then obtained by evaporation of the remaining solution. Potassium nitrate releases O_2 when heated:

$$2KNO_3(s) \longrightarrow 2KNO_2(s) + O_2(g)$$

Thus, KNO_3 is a good oxidizing agent (when heated). It is the oxidizing agent in gunpowder, which uses the two reactions:

$$2KNO_3(s) + 4C(s) \longrightarrow K_2CO_3(s) + 3CO(g) + N_2(g)$$

$$2KNO_3(s) + 2S(s) \longrightarrow K_2SO_4(s) + SO_2(g) + N_2(g)$$

The rapid formation of six moles of gas from a solid results in a rapid increase in volume, which we call an *explosion*.

▼ **EXAMPLE The role of reagents from Group I elements**

What chemical role does NaCl play (a) in the production of NaOH and (b) in the production of Na_2SO_4?

SOLUTION (a) In the manufacture of NaOH, NaCl makes the water an electrolyte and provides the Na^+ ion in the NaOH. (The hydroxide ion originates with the water and is produced by the reduction of H_2O.)

$$2NaCl(aq) + 2H_2O(l) \xrightarrow{\text{electrolysis}} 2NaOH(aq) + Cl_2(g) + H_2(g)$$

(b) In the manufacture of Na_2SO_4, the chloride ion in NaCl acts as a Brønsted base that removes protons from sulfuric acid. It also provides the Na^+ ion for the product.

$$H_2SO_4(aq) + 2NaCl(s) \longrightarrow Na_2SO_4(aq) + 2HCl(g)$$

EXERCISE In the following reaction, which occurs when black gunpowder explodes, state what is oxidized, what is reduced, what the oxidizing agent is, and what the reducing agent is.

$$2KNO_3(s) + 4C(s) \longrightarrow K_2CO_3(s) + 3CO(g) + N_2(g)$$

ANSWER: C, oxidized; N, reduced; KNO_3, oxidizing agent; $C(s)$, reducing agent ▲

KEY CONCEPT B The Solvay process

The Solvay process is used to produce sodium carbonate (Na_2CO_3) from NaCl in locales where the carbonate is not available as a mineral resource. The overall reaction is

$$2NaCl(aq) + CaCO_3(s) \longrightarrow Na_2CO_3(s) + CaCl_2(s)$$

The process is actually a multistep process that uses the following reactions:

$$CaCO_3(s) \longrightarrow CaO(s) + CO_2(g) \quad (1)$$
$$CO_2(g) + H_2O(l) \longrightarrow H_2CO_3(aq) \quad (2)$$
$$H_2CO_3(aq) + NH_3(aq) \longrightarrow NH_4^+(aq) + HCO_3^-(aq) \quad (3)$$
$$NaCl(aq) + NH_4^+(aq) + HCO_3^-(aq) \longrightarrow NaHCO_3(s) + NH_4Cl(aq) \quad (4)$$
$$2NaHCO_3(s) \longrightarrow Na_2CO_3(s) + CO_2(g) + H_2O(g) \quad (5)$$

In the process, the CO_2 produced in reaction 5 is reused in reaction 2; and ammonia, which is very expensive, is recovered as follows:

$$CaO(s) + H_2O(l) \longrightarrow Ca(OH)_2(aq) \quad \text{(CaO from reaction 1)} \quad (6)$$
$$2NH_4^+(aq) + Ca(OH)_2(aq) \longrightarrow 2NH_3(g) + 2H_2O(l) + Ca^{2+}(aq) \quad (NH_4^+ \text{ from reaction 3}) \quad (7)$$

▼ **EXAMPLE The reactions of the Solvay process**

Which of the reactions in the Solvay process are Brønsted acid–base reactions?

SOLUTION A Brønsted acid–base reaction involves a proton transfer. Reactions 3, 6, and 7 involve proton transfer and are therefore Brønsted acid–base reactions.

EXERCISE Which of the reactions in the Solvay process is a Lewis acid–base reaction?

ANSWER: Reactions 1 and 5 ▲

GROUP II: ALKALINE EARTH METALS

KEY WORDS Define or explain each of the following terms in a written sentence or two.

alkaline earth metals   lime   slaked lime

18.5 GROUP II ELEMENTS

KEY CONCEPT The elements: Production, properties, and uses

All the alkaline earths (Group II elements) are too reactive to be found in ores as pure metals with oxidation state 0. The elements calcium, strontium, and barium are called the *alkaline earth metals.*

Production. Beryllium occurs in ores mainly as beryl ($3BeO \cdot Al_2O_3 \cdot 6SiO_2$) and is obtained by converting beryl to the chloride and electrolyzing the chloride.

$$BeCl_2(l) \longrightarrow Be(s) + Cl_2(g)$$

Magnesium occurs as the mixed carbonate dolomite ($CaCO_3 \cdot MgCO_3$) and is obtained either by electrolytic reduction or chemical reduction. The electrolytic reduction starts with seawater (which contains Mg^{2+}) and uses the following steps:

Precipitate $Mg(OH)_2$:	$Mg^{2+}(aq) + 2OH^-(aq) \longrightarrow Mg(OH)_2(s)$	(1)
Synthesize $MgCl_2$:	$Mg(OH)_2(s) + 2HCl(aq) \longrightarrow MgCl_2(s) + 2H_2O(l)$	(2)
Electrolyze to Mg:	$MgCl_2(l) \longrightarrow Mg(s) + Cl_2(g)$	(3)

Calcium, strontium, and barium are found mostly in the sea as dolomite ($CaCO_3 \cdot MgCO_3$) and in limestone ($CaCO_3$), strontianite ($SrSO_4$), and barite ($BaSO_4$). All three are obtained by either electrolytic or chemical reduction. Two examples are

Electrolytic reduction for Ca:	$CaCl_2(l) \longrightarrow Ca(l) + Cl_2(g)$
Chemical reduction for Ba:	$3BaO(s) + 2Al(s) \longrightarrow Al_2O_3(s) + 3Ba(s)$

Properties. The ionization of two electrons from the alkaline earth ns^2 configuration (leaving an inert-gas core) requires relatively little energy, but the lattice enthalpies of the Group 2 ionic solids are rather large. Thus, the elements (with the exception of beryllium) generally exist in their compounds as ions with +2 charge. They always have +2 oxidation number in their compounds. Because of its high polarizing power, beryllium has a tendency toward covalency; it also has a diagonal relationship with aluminum. Thus, Be_2O_3 is amphoteric. Beryllium is extremely toxic, both as a metal and in its compounds.

All the Group II cations have strongly negative reduction potentials. The metals therefore have large positive oxidation potentials. This means they are oxidized easily and are powerful reducing agents. All except beryllium reduce water:

$Be(s) + 2H_2O(l) \longrightarrow$ no reaction	
$Mg(s) + 2H_2O(l) \longrightarrow Mg(OH)_2(aq) + H_2(g)$	reacts with hot water
$Ca(s) + 2H_2O(l) \longrightarrow Ca(OH)_2(aq) + H_2(g)$	reacts with cold water
$Sr(s) + 2H_2O(l) \longrightarrow Sr(OH)_2(aq) + H_2(g)$	reacts with cold water
$Ba(s) + 2H_2O(l) \longrightarrow Ba(OH)_2(aq) + H_2(g)$	reacts with cold water

Even though the reaction of Be and Mg with water has a high positive standard potential E(cell), both metals are passivated by an oxide coat, so the reaction does not occur with beryllium at all and occurs with magnesium only if hot water is used. This indicates that $Be(s)$ is more strongly passivated than $Mg(s)$.

Uses. Some of the uses of the Group II elements are shown in the following table.

Element	Uses
Be	construction of missiles, satellites, x-ray tube windows, nonsparking tools
Mg	construction of airplanes; in fireworks and incendiary devices
Ca	as a getter in the production of steel (to remove unwanted oxides)
Ba	as a getter in vacuum tube construction (to remove O_2)

EXAMPLE Properties of Group II compounds

Suggest a reason, based on the lattice enthalpies of their respective oxides, why Be(s) is more strongly passivated than Mg(s).

SOLUTION The lattice enthalpy of BeO is 4293 kJ/mol; that of MgO is 3889 kJ/mol. This difference is largely due to the smaller size of the Be^{2+} ion relative to the Mg^{2+} ion. The higher lattice enthalpy means that it is more difficult for water to chemically break through the passivating BeO coat on Be than through the MgO coat on Mg.

EXERCISE Beryllium expresses amphoteric behavior by dissolving in both acid and base. Write the reaction for the dissolution of Be(s) in NaOH(aq) and in HCl(aq).

ANSWER: $Be(s) + 2OH^-(aq) + 2H_2O(l) \longrightarrow [Be(OH)_4]^{2-}(aq) + H_2(g)$; $Be(s) + 2HCl(aq) \longrightarrow BeCl_2(aq) + H_2(g)$

18.6 IMPORTANT GROUP II COMPOUNDS

KEY CONCEPT Important Group II compounds

Beryllium compounds. The properties of beryllium compounds are mostly the result of the small size and high polarizing power of the Be^{2+} cation. The small size limits the coordination number of the cation to 4, so tetrahedral BeX_4 units occur in many beryllium compounds. BeH_2 and $BeCl_2$ are polymeric, with neighboring Be atoms held together by bonding molecular orbitals in a Be–H–Be and Be–Cl–Be group, respectively. The chloride is produced by reaction 1, which follows; the hydrated Be^{2+} ion is acidic (reaction 2).

$$C(s) + BeO(s) + Cl_2(g) \longrightarrow BeCl_2(g) + CO(g) \quad (1)$$

$$[Be(H_2O)_4]^{2+}(aq) + H_2O(l) \longrightarrow [Be(H_2O)_3OH]^+(aq) + H_3O^+(aq) \quad (2)$$

Magnesium oxide, MgO. Burning magnesium in air yields both the nitride and the oxide:

$$2Mg(s) + O_2(g) \longrightarrow 2MgO(s)$$

$$3Mg(s) + N_2(g) \longrightarrow Mg_3N_2(s)$$

Pure oxide is formed by heating and decomposition of the carbonate. The oxide, with its small Mg^{2+} and O^{2-} ions, has a very high lattice enthalpy. It is extremely stable and melts at the very high temperature of 2800°C.

$$MgCO_3(s) \longrightarrow MgO(s) + CO_2(g)$$

Magnesium hydroxide, $Mg(OH)_2$ (milk of magnesia). Magnesium hydroxide is not very soluble and is a mild base. It is used as an antacid to neutralize stomach acid (HCl). The $MgCl_2$ formed acts as a purgative.

$$Mg(OH)_2(s) + 2HCl(aq) \longrightarrow MgCl_2(aq) + 2H_2O(l)$$

Chlorophyll. Chlorophyll contains magnesium in the form of +2 ions. In the structure, four nitrogen atoms on a single large organic part of the molecule act as Lewis bases and bond to the Mg^{2+}, which acts as a Lewis acid. Chlorophyll is one of the key compounds in the photosynthetic process in green plants. It captures light from the sun and funnels it (as free energy) into the photosynthesis reaction,

$$6CO_2(g) + 6H_2O(l) \longrightarrow C_6H_{12}O_6(aq) + 6O_2(g)$$

Calcium carbonate. $CaCO_3$ (calcite, argonite, chalk, limestone, and marble are forms of $CaCO_3$). The calcium in hard water originates with calcium carbonate. Calcium carbonate itself is insoluble. However, Ca^{2+} ions dissolve in a two-stage process. First, atmospheric CO_2 dissolves to form carbonic acid. When the now acidic water comes into contact with mineral deposits of $CaCO_3$ (such as limestone), the $CaCO_3$ dissolves:

$$CO_2(g) + H_2O(l) \longrightarrow H_2CO_3(aq)$$
$$CaCO_3(s) + H_2CO_3(aq) \longrightarrow Ca(HCO_3)_2(aq)$$

The calcium hydrogen carbonate is soluble in water and dissolution occurs readily, resulting in a solution of Ca^{2+} and HCO_3^- ions. When the water is heated in domestic water systems, CO_2 is driven off and some carbonic acid decomposes:

$$H_2CO_3(aq) \longrightarrow CO_2(g) + H_2O(l)$$

This lowers the concentration of H_2CO_3, which, in turn, shifts the equilibrium below to the right and results in the production of carbonate ion (CO_3^{2-}). The CO_3^{2-} thus formed reacts with the Ca^{2+} already present precipitating $CaCO_3$ which clogs hot water pipes and boilers.

$$HCO_3^-(aq) + HCO_3^-(aq) \rightleftharpoons CO_3^{2-}(aq) + H_2CO_3(aq)$$
$$Ca^{2+}(aq) + CO_3^{2-}(aq) \longrightarrow CaCO_3(s)$$

Calcium oxide and hydroxide, CaO (lime or quicklime); $Ca(OH)_2$ (slaked lime). Heating $CaCO_3$ produces CaO:

$$CaCO_3(s) \longrightarrow CaO(s) + CO_2(g)$$

Addition of water to CaO (slaking) produces $Ca(OH)_2$ in a strongly exothermic Lewis acid–base reaction;

$$CaO(s) + H_2O(l) \longrightarrow Ca(OH)_2(s)$$

Aqueous $Ca(OH)_2$ (limewater) is used to test for carbon dioxide. If an unknown gas containing CO_2 is bubbled through limewater, a visible precipitate forms, indicating the presence of the CO_2:

$$CO_2(g) + Ca(OH)_2(aq) \longrightarrow CaCO_3(s) + H_2O(l)$$

Calcium oxide is used industrially in ironmaking where, in a Lewis acid–base reaction, it is added to the melt in a blast furnace to remove unwanted SiO_2 from iron ore:

$$CaO(s) + SiO_2(s) \longrightarrow CaSiO_3(l)$$
$$\underset{\text{Lewis base}}{O^{2-}} + \underset{\text{Lewis acid}}{SiO_2} \longrightarrow SiO_3^{2-}$$

Addition of CaO to water lowers the concentration of Ca^{2+} ions from hard water through the following reactions. Some of the Ca^{2+} in the last reaction comes from the added CaO.

$$CaO(s) + H_2O(l) \longrightarrow Ca^{2+}(aq) + 2OH^-(aq)$$
$$OH^-(aq) + HCO_3^-(aq) \longrightarrow CO_3^{2-}(aq) + H_2O(l)$$
$$Ca^{2+}(aq) + CO_3^{2-}(aq) \longrightarrow CaCO_3(s)$$

Heating of $CaO(s)$ with carbon results in the formation of calcium carbide (more properly called calcium acetylide):

$$CaO(s) + 3C(s) \longrightarrow CaC_2(s) + CO(g)$$

The acetylide ion (C_2^{2-}) is a stronger Brønsted base than OH^-, so it reacts with water to form ethyne (acetylene, C_2H_2). This sequence of reactions provides a route from inorganic carbon $\{C(s)\}$ to an organic compound (C_2H_2).

$$CaC_2(s) + 2H_2O(l) \longrightarrow C_2H_2(g) + Ca(OH)_2(aq)$$

Structural calcium. Calcium compounds are rigid because of the large electrostatic attraction between the small Ca^{2+} ion and anions. This rigidity makes calcium compounds useful as structural components. For example, common mortar contains $Ca(OH)_2$, which sets to a hard mass through the reaction with atmospheric CO_2:

$$Ca(OH)_2(s) + CO_2(g) \longrightarrow CaCO_3(s) + H_2O(l)$$

Tooth enamel is hydroxyapatite, a calcium-containing mineral $\{Ca_5(PO_4)_3OH\}$. Tooth decay occurs when certain bacteria that thrive in the mouth produce acids that dissolve the enamel:

$$Ca_5(PO_4)_3OH(s) + 4H_3O^+(aq) \longrightarrow 5Ca^{2+}(aq) + 3HPO_4^{2-}(aq) + 5H_2O(l)$$

Fluoroapatite $\{Ca_5(PO_4)_3F\}$, in which a fluoride ion replaces the hydroxide ion in hydroxyapatite, is more resistant to this type of attack:

$$Ca_5(PO_4)_3OH(s) + F^-(aq) \longrightarrow Ca_5(PO_4)_3F(s) + OH^-(aq)$$

EXAMPLE The properties of Group II compounds

Use a thermodynamic argument, suggest a reason why $CaCO_3$ must be heated to make it decompose to CaO and CO_2.

SOLUTION The reaction is $CaCO_3(s) \longrightarrow CaO(s) + CO_2(g)$. We shall consider the factors that contribute to ΔG ($= \Delta H - T\Delta S$) for the reaction. Because the net process in the decomposition is the breaking of a C—O bond, ΔH must be positive. The formation of a gas from a solid results in a large positive ΔS. Thus, we have

$$\Delta H > 0 \qquad \text{and} \qquad \Delta S > 0$$

A reaction with positive ΔH and positive ΔS is spontaneous only at relatively high temperatures, so $CaCO_3$ must be heated for it to decompose.

EXERCISE Suggest two reasons why fluoroapatite is more resistant to acid attack than hydroxyapatite.

ANSWER: (a) OH^- is a stronger base than F^-, so OH^- reacts more readily with acid; (b) F^- is smaller than OH^-, so F^- experiences stronger interactions with its cationic neighbors and holds the enamel together in a tighter, stronger structure.

SELF-TEST EXERCISES

Hydrogen

1. What is the electron configuration of the hydride ion?
(a) $1s^1$ (b) $1s^2$ (c) $1s^22s^1$ (d) $1s^22s^2$ (e) no electrons

2. What is the role of $H_2(g)$ in the Haber process, $N_2(g) + 3H_2(g) \longrightarrow 2NH_3(g)$?
(a) Brønsted acid (b) Brønsted base (c) oxidizing agent
(d) reducing agent (e) catalyst

3. How many liters of $H_2(g)$ are required to produce 50.0 L of $NH_3(g)$? Assume all volumes are measured at STP.

$$N_2(g) + 3H_2(g) \longrightarrow 2NH_3(g)$$

(a) 75.0 (b) 150 (c) 33.3 (d) 50.0 (e) 100

4. In the reaction of the hydride ion with water, H^- acts as all of the following except one. Which is the exception?

$$H^-(aq) + H_2O(l) \longrightarrow OH^-(aq) + H_2(g)$$

(a) reducing agent (b) electrolyte (c) Brønsted base
(d) Lewis base (e) substance oxidized

5. Which of the following is a covalent hydride?
(a) C_2H_2 (b) CaH_2 (c) H_2 (d) NaH (e) CuH

6. What Brønsted base must be protonated to form CH_4?
(a) C_2H_2 (b) CO_2 (c) CH_5^+ (d) CH_3^- (e) CH_3Cl

7. Which of the following is a polymeric hydride?
(a) CaH_2 (b) NaH (c) AlH_3 (d) CuH (e) C_2H_6

8. Which metal is not active enough to produce H_2 by reaction with water?
(a) Zn (b) Cu (c) Na (d) Mg

9. Which of the isotopes of hydrogen is/are radioactive?
(a) protium (b) deuterium (c) tritium (d) a and b (e) all of these

10. Which of the following covalent hydrides has the highest heat of vaporization?
(a) H_2O (b) C_2H_2 (c) HCl (d) CH_4 (e) HBr

11. What is the volume, at STP, of the 3×10^8 kg H_2 used in the United States each year?
(a) 3×10^8 L (b) 6×10^8 L (c) 8×10^{10} L (d) 3×10^{12} L (e) 6×10^{12} L

Group I: Alkali metals

12. The large negative reduction potentials for the alkali metal cations are due to
(a) large ΔH_{sub} for the metals (b) positive ionization energies for the metal.
(c) large hydration energy for the cation. (d) positive electron affinities for the metals.

13. Calculate the enthalpy for the half-cell oxidation of $Li(s)$, using the following data.

Sublimation energy of $Li(s)$ = 161 kJ/mol
Ionization energy of $Li(g)$ = 520.3 kJ/mol
Hydration energy of $Li^+(g)$ = −520 kJ/mol

(a) +1201 kJ/mol (b) −161 kJ/mol (c) 161 kJ/mol
(d) −1201 kJ/mol (e) −879 kJ/mol

14. Which of the following reacts with water least violently?
(a) Na (b) Li (c) Cs (d) Rb (e) K

15. What products are formed when Na reacts with water?
(a) NaH, O_2 (b) NaH, H_2O_2 (c) NaOH, H_2 (d) NaOH, O_2 (e) NaOH, H_2O_2

16. What unusual species is formed when potassium metal is dissolved in liquid ammonia? (*am* = solvated by ammonia molecules.)
(a) $e^-(am)$ (b) $H_2(g)$ (c) $NH_2^+(am)$ (d) $NH_4^+(am)$ (e) $NH_3(g)$

17. Only one of the alkali metals forms a nitride when burned in air. Which one?
(a) Na (b) Cs (c) Rb (d) Li (e) K

18. What is the major product when sodium is burned in air?
(a) Na_2O_2 (b) Na_3N (c) Na_2O (d) NaO_2 (e) $NaNO_3$

19. What is the major product when potassium is burned in air?
(a) K_2O_2 (b) K_3N (c) K_2O (d) KNO_3 (e) KO_2

20. What is the oxidation state of oxygen in the superoxide ion?
(a) -1 (b) -2 (c) $+1$ (d) $-\frac{1}{2}$ (e) 0

21. Which of the following is washing soda?
(a) NaCl (b) $Na_2CO_3 \cdot 10H_2O$ (c) Na_2SO_4 (d) NaOH (e) $NaNO_3$

22. Which of the alkali metals exhibits significant covalency in its compounds?
(a) Li (b) Na (c) K (d) Cs (e) Rb

23. Which of the following reactions is used commercially to produce Na_2SO_4?
(a) $2Na(s) + K_2SO_4(l) \longrightarrow Na_2SO_4(s) + 2K(s)$
(b) $(NH_4)_2SO_4(aq) + 2NaOH(aq) \longrightarrow Na_2SO_4(aq) + 2NH_3(g) + 2H_2O(l)$
(c) $H_2SO_4(aq) + 2NaCl(s) \longrightarrow Na_2SO_4(aq) + 2HCl(g)$
(d) $K_2SO_4(s) + 2NaCl(aq) \longrightarrow Na_2SO_4(aq) + 2KCl(aq)$

24. How many kilograms of sodium can be produced from the electrolysis of molten NaCl by a 2.50×10^4 A current running for 24.0 h?
(a) 199 (b) 225 (c) 386 (d) 515 (e) 418

25. Which of the following is a common oxidizing agent?
(a) KCl (b) KNO_3 (c) K_2SO_4 (d) K_2CO_3 (e) K

26. Which of the reactions used in the Solvay process is a hydrolysis?
(a) $CaCO_3(s) \longrightarrow CaO(s) + CO_2(g)$
(b) $CO_2(g) + H_2O(l) \longrightarrow H_2CO_3(aq)$
(c) $H_2CO_3(aq) + NH_3(aq) \longrightarrow NH_4^+(aq) + HCO_3^-(aq)$
(d) $NaCl(aq) + NH_4^+(aq) + HCO_3^-(aq) \longrightarrow NaHCO_3(s) + NH_4Cl(aq)$
(e) $2NaHCO_3(s) \longrightarrow Na_2CO_3(s) + CO_2(g) + H_2O(g)$

Group II: Alkaline earths

27. Which alkaline earth shares a diagonal relationship with aluminum?
(a) Mg (b) Ca (c) Sr (d) Be (e) Ba

28. Only one of the alkaline earths forms a significant amount of the nitride when it is burned in air. Which one?
(a) Ba (b) Sr (c) Be (d) Ca (e) Mg

29. All the alkaline earths but one react with water. Which one doesn't?
(a) Mg (b) Be (c) Ca (d) Sr (e) Ba

30. What products are formed when barium reacts with water?
(a) $Ba(OH)_2$, H_2 (b) BaH_2, O_2 (c) $Ba(OH)_2$, O_2 (d) BaO, H_2O (e) BaO, H_2

31. What is the oxidizing agent when barium reacts with water?
(a) Ba (b) H_2O (c) Ba^{2+} (d) OH^- (e) O_2

32. Which of the following compounds exhibits substantial covalency?
(a) BaO (b) $CaCl_2$ (c) SrO (d) $Sr(OH)_2$ (e) $BeCl_2$

33. What is the most common structural unit for beryllium?
(a) BeX_4 tetrahedra (b) BeX_6 octahedra (c) Be–Be dimers (d) BeX dimers

34. Which of the following alkaline earths is most strongly passivated by an oxide coat?
(a) Sr (b) Ca (c) Be (d) Ba (e) Mg

35. Which reaction is typically used to obtain pure MgO?
(a) $MgCO_3(s) \longrightarrow MgO(s) + CO_2(g)$
(b) $2Mg(s) + O_2(g) \longrightarrow 2MgO(s)$
(c) $MgSO_4(s) \longrightarrow MgO(s) + SO_3(g)$
(d) $Mg(OH)_2(s) \longrightarrow MgO(s) + H_2O(g)$

36. Limestone is
(a) $MgCO_3$ (b) $CaCl_2$ (c) CaO (d) $CaCO_3$ (e) CaC_2

37. What reaction is used to detect CO_2 bubbled into limewater?
(a) $CO_2(g) + Ca(OH)_2(aq) \longrightarrow CaCO_3(s) + H_2O(l)$
(b) $CO_2(g) + H_2O(l) \longrightarrow H_2CO_3(aq)$
(c) $6CO_2(g) + 6H_2O(l) \longrightarrow C_6H_{12}O_6(aq) + 6O_2(g)$
(d) $CO_2(g) + CaO(s) \longrightarrow CaCO_3(s)$

38. How many liters of $C_2H_2(g)$ (measured at STP) can be produced from 100 g of CaO?
(a) 1.78 (b) 39.9 (c) 250 (d) 61.1 (e) 18.5

39. Hydroxyapatite $[Ca_5(PO_4)_3OH]$ is the structural material in
(a) mortar. (b) tooth enamel. (c) marble. (d) plant stems.

40. At what temperature does the decomposition of $MgCO_3$ into MgO and CO_2 become thermodynamically favorable?

	$MgCO_3(s)$	**$MgO(s)$**	**$CO_2(g)$**
ΔH_f°, kJ/mol	−1095.8	−601.7	−393.51
S°, J/mol·K	65.6	26.94	213.74

(a) 575 K (b) 1230 K (c) 2110 K (d) 3610 K (e) 4440 K

19 The p-BLOCK ELEMENTS: I

The *p*-block elements are Groups III–VIII. Helium is in the same area of the periodic table as the *p*-block elements but is not actually a *p*-block element, because its electron configuration is $1s^2$.

GROUP III: BORON AND ALUMINUM

KEY WORDS Define or explain each of the following terms in a written sentence or two.

anhydride
Baeyer process
boranes
borohydrides
Hall process
thermite reaction

19.1–19.3 THE GROUP III ELEMENTS, OXIDES, AND IMPORTANT COMPOUNDS

KEY CONCEPT A Elemental boron and aluminum

Boron is a nonmetal and as such forms acidic oxides and hydroxides. In almost all of its compounds, it has a +3 oxidation state. It is mined as the minerals borax ($Na_2B_4O_7 \cdot 10H_2O$) and kernite ($Na_2B_4O_7 \cdot 7H_2O$). Extraction of elemental boron involves conversion to boron oxide followed by reduction with magnesium:

$$B_2O_3(s) + 3Mg(s) \longrightarrow 2B(s) + 3MgO(s)$$

Elemental boron has several allotropes. It is used to harden steels; it is also used to produce fibers to strengthen plastics. Chemically, boron is quite inert.

Aluminum is an amphoteric metal. It reacts with both bases and nonoxidizing acids:

$$2Al(s) + 6H_3O^+(aq) \longrightarrow 2Al^{3+}(aq) + 3H_2(g) + 6H_2O(l)$$
$$2Al(s) + 2OH^-(aq) + 6H_2O(l) \longrightarrow 2[Al(OH)_4]^-(aq) + 3H_2(g)$$

Oxidizing acids lead to the formation of a passivating oxide film on aluminum surfaces that prevents further reaction. Aluminum has a +3 oxidation state in almost all of its compounds. It is mined as bauxite, an impure ore containing Al_2O_3; the aluminum is extracted with the electrolytic **Hall process:**

$$2Al_2O_3(s) + 3C(s) \longrightarrow 4Al(s) + 3CO_2(g)$$

The element is a strong, light metal and an excellent conductor. Its low density, wide availability, and resistance to corrosion (because of a passivating oxide film) makes it widely used in structural applications where low density and high strength are required. Chemically, it is frequently used in a class of reactions called **thermite reactions,** in which it reacts with a metal oxide to produce the free metal:

$$Fe_2O_3(s) + 2Al(s) \longrightarrow 2Fe(l) + Al_2O_3(s)$$

Aluminum oxide is a strong reducing agent and is used to reduce some oxide ores for which the more common reducing agents won't work.

EXAMPLE Aluminum and boron extraction

Assume an ore is found which is 8.0% boron. How many grams of Mg are required to extract all of the boron from 2.4 kg of the ore?

SOLUTION The reaction is $B_2O_3(s) + 3Mg(s) \longrightarrow 2B(s) + 3MgO(s)$. We first calculate the amount of boron that is expected from the percentage in the ore. Because 8.0% translates to a fractional amount of boron in the ore of 0.080 g of boron to 1 g of ore, we get

$$0.080 \frac{\text{g B}}{\cancel{\text{g ore}}} \times (2.4 \times 10^3 \cancel{\text{g ore}}) = 1.9 \times 10^2 \text{ g B}$$

We now use the coefficients in the chemical equation to calculate the amount of magnesium required to produce 1.9×10^2 g B,

$$1.9 \times 10^2 \cancel{\text{g B}} \times \frac{1 \cancel{\text{mol B}}}{10.81 \cancel{\text{g B}}} \times \frac{3 \cancel{\text{mol Mg}}}{2 \cancel{\text{mol B}}} \times \frac{24.31 \text{ g Mg}}{1 \cancel{\text{mol Mg}}} = 6.4 \times 10^2 \text{ g Mg}$$

EXERCISE How many kilograms of aluminum can be produced electrolytically by a 1.0×10^4 A current running for 24 h? 1 F = 96,485 C.

ANSWER: 81 kg

KEY CONCEPT B The compounds of boron

Boric acid, $B(OH)_3$. Boric acid is used as a mild antiseptic. It is a weak monoprotic acid. On heating it dehydrates to its anhydride, B_2O_3:

$$B(OH)_3(aq) + 2H_2O(l) \longrightarrow [B(OH)_4]^-(aq) + H_3O^+(aq)$$
$$2B(OH)_3(s) \longrightarrow B_2O_3(s) + 3H_2O(g)$$

Boron carbide, $B_{12}C_3$. When boron is heated strongly with graphite, it forms boron carbide:

$$2B(s) + 3C(s) \longrightarrow B_{12}C_3(s)$$

Boron nitride, BN. When boron is heated strongly with ammonia, boron nitride is formed:

$$12B(s) + 2NH_3(g) \longrightarrow 2BN(s) + 3H_2(g)$$

Boron halides. The boron halides are made either by direct reaction of the elements or from the oxide. Two reactions that use the oxide are

$$B_2O_3(s) + 3CaF_2(s) + 3H_2SO_4(l) \longrightarrow 2BF_3(g) + 3CaSO_4(s) + 3H_2O(l)$$
$$B_2O_3(s) + 3C(s) + 3Cl_2(g) \longrightarrow 2BCl_3(g) + 3CO(g)$$

All of the halides are electron deficient, with an incomplete octet on boron. The molecules are planar triangular and, because of the incomplete octet, are strong Lewis acids.

Boranes. Boranes are compounds of boron with hydrogen, such as B_2H_6 and $B_{10}H_{14}$. Diborane (B_2H_6), which is extremely reactive, is produced by the reaction of sodium borohydride with boron trifluoride. Boranes are electron-deficient compounds with three center, two-electron B—H—B bonds.

$$4BF_3 + 3BH_4^- \longrightarrow 3BF_4^- + 2B_2H_6(g)$$

Borohydrides (compounds containing BH_4^-). Sodium borohydride, a common borohydride, is produced by the reaction of sodium hydride on boron trichloride. Borohydride ion is an effective and important reducing agent.

$$4NaH + BCl_3 \longrightarrow NaBH_4 + 3NaCl$$

▼ EXAMPLE The reactions of boron compounds

How many milliliters of 0.200 M NaOH are required to neutralize 0.535 g $B(OH)_3$?

SOLUTION The reaction involved is $B(OH)_3(aq) + OH^-(aq) \longrightarrow [B(OH)_4]^-(aq)$. We first calculate the moles of $B(OH)_3$ present, using the molar mass of $B(OH)_3$, 61.83 g/mol:

$$0.535\ \cancel{g} \times \frac{1\text{mol } B(OH)_3}{61.83\ \cancel{g}} = 8.65 \times 10^{-3} \text{ mol } B(OH)_3$$

The chemical equation tells us that 1 mol of $B(OH)_3$ requires 1 mol $OH^-(aq)$. This allows us to calculate the moles NaOH required:

$$8.65 \times 10^{-3}\ \cancel{\text{mol } B(OH)_3} \times \frac{1 \text{ mol NaOH}}{1\ \cancel{\text{mol } B(OH)_3}} = 8.65 \times 10^{-3} \text{ mol NaOH}$$

From the molarity of the NaOH and the moles required, we can calculate the volume of NaOH required:

$$8.65 \times 10^{-3}\ \cancel{\text{mol NaOH}} \times \frac{1 \text{ L}}{0.200\ \cancel{\text{mol NaOH}}} \times \frac{1000 \text{ mL}}{1 \text{ L}} = 43.3 \text{ mL}$$

EXERCISE In a synthesis of BN from B and NH_3, 0.25 L of H_2 (measured at STP) are formed. How many grams of BN form at the same time?

▲ ANSWER: 0.18 g BN

KEY CONCEPT C The compounds of aluminum

Alumina (Al_2O_3). Alumina is the common name for the various aluminum oxides with the formula Al_2O_3. It is made by the **Baeyer process** by acidifying a solution of $[Al(OH)_4]^-$ with CO_2 to get $Al(OH)_3$ and then heating the $Al(OH)_3$ to get alumina:

$$[Al(OH)_4]^-(aq) + CO_2(g) \longrightarrow HCO_3^-(aq) + Al(OH)_3(s)$$
$$2Al(OH)_3(s) \longrightarrow Al_2O_3(s) + 3H_2O(g)$$

The aluminum oxide called γ-alumina is a moderately reactive form of alumina. It dissolves in acids to give salts with the Al^{3+} ion:

$$Al_2O_3(s) + 6H^+(aq) \longrightarrow 2Al^{3+}(aq) + 3H_2O(l)$$

However, α-alumina is an extremely hard, unreactive form of alumina. It dissolves in alkalis to give the aluminate ion, $[Al(OH)_4]^-$:

$$Al_2O_3(s) + 2OH^-(aq) + 3H_2O(l) \longrightarrow 2[Al(OH)_4]^-(aq)$$

Mixing Al^{3+} with aluminate ion results in the formation of a flocculent precipitate of $Al(OH)_3$, which is used to capture impurities in water.

$$Al^{3+}(aq) + 3[Al(OH)_4]^-(aq) \longrightarrow 4Al(OH)_3(s)$$

Aluminum trichloride, ($AlCl_3$). Aluminum trichloride is formed as an ionic compound by the reaction of chlorine with aluminum or of alumina with chlorine in the presence of carbon.

$$2Al(s) + 3Cl_2(g) \longrightarrow 2AlCl_3(s)$$
$$Al_2O_3(s) + 3Cl_2(g) + 3C(s) \longrightarrow 2AlCl_3(s) + 3CO(g)$$

Heating the ionic $AlCl_3$ to its melting point of 192°C results in a rearrangement to form a molecular liquid of Al_2Cl_6 molecules. Molecular Al_2Cl_6 consists of units in which an unshared pair of electrons on chlorine (a Lewis base) is donated to a vacant orbital on an adjacent aluminum atom (a Lewis acid). Aluminum halides react with water in a highly exothermic reaction. The white solid lithium aluminum hydride, an important reducing agent in organic chemistry, is prepared from aluminum chloride:

$$4LiH + AlCl_3 \longrightarrow LiAlH_4 + 3LiCl$$

EXAMPLE Physical properties of boron and aluminum compounds

Explain why the boiling points of the boron halides indicate that London forces are dominant in the intermolecular attractions of these compounds. Why are London forces the dominant forces? The boiling points are

$$BF_3: \ -101°C \qquad BCl_3: \ -107°C \qquad BBr_3: \ 91.3°C \qquad BI_3: \ 210°C$$

SOLUTION Overall, the boiling points increase from BF_3 to BI_3 (even though there is a slight reversal of the trend at BCl_3). Thus, as the compounds get heavier, the boiling points increase. This behavior is typical of situations in which London forces dominate because the strength of London forces increases as the number of electrons increases and a compound becomes more polarizable. The only other foces that might affect the boiling points are dipole–dipole interactions; however, the boron halides are triangular planar molecules, and the individual bond dipoles cancel to give a net dipole moment of zero, so no dipole–dipole attractions exist in these compounds.

EXERCISE Explain why the ionic $AlCl_3$ expands greatly when it changes to molecular Al_2Cl_6.

ANSWER: The strong ionic bonds in $AlCl_3$ hold the ions together more effectively than the weak London forces hold Al_2Cl_6 molecules together.

GROUP IV: CARBON AND SILICON

KEY WORDS Define or explain each of the following terms in a written sentence or two.

amphiboles
cements
ceramics
covalent carbides
glass
interstitial carbides
saline carbides
silicones

19.4–19.6 GROUP IV ELEMENTS, OXIDES, AND IMPORTANT COMPOUNDS

KEY CONCEPT A Elemental carbon and silicon and other Group IV elements

All Group IV elements have an s^2p^2 valence electron configuration. Carbon and silicon at the top of Group IV are nonmetals, whereas tin and lead at the bottom are metals. Both tin and lead, however, show amphoteric behavior, indicating some nonmetal character is present. For instance, tin reacts with both acids and alkalies:

$$Sn(s) + 2HCl(\text{conc}, aq) \longrightarrow SnCl_2(aq) + H_2(g)$$
$$Sn(s) + 2OH^-(aq) + 4H_2O(l) \longrightarrow [Sn(OH)_6]^{2-}(aq) + 2H_2(g)$$

Carbon. Solid carbon exists in two common allotropic forms, graphite and diamond. Graphite is the thermodynamically most stable form at normal conditions. Graphite consists of flat planes of carbon atoms; in diamond, however, each carbon atom is tetrahedrally bonded to four others. These differences make graphite soft and electrically conducting and make diamond hard and nonconducting.

Silicon. Silicon occurs naturally as silicates (compounds containing SiO_4^{4-}) and as silicon dioxide. Quartz, quartzite, and sand are forms of SiO_2. Pure silicon, which finds wide use in the semiconductor industry, is obtained from quartzite in a three-step process:

$$SiO_2(s) + 2C(s) \longrightarrow Si(s) + 2CO(g) \qquad \text{(impure Si formed)}$$
$$Si(s) + 2Cl_2(g) \longrightarrow SiCl_4(l)$$
$$SiCl_4(l) + 2H_2(g) \longrightarrow Si(s) + 4HCl(g) \qquad \text{(pure Si formed)}$$

Germanium. Germanium is used mainly in the semiconductor industry.

Tin. Tin is obtained from its ore cassiterite (SnO_2) through reduction with carbon in a process that must be done carefully to avoid contamination with iron. Tin is used mainly for plating.

$$SnO_2(s) + C(s) \longrightarrow Sn(l) + CO_2(g)$$

Lead. Lead is obtained from its ore galena (PbS) and is very dense and chemically inert. Its high density makes it useful as a radiation shield. Lead forms passivating oxide, chloride, and sulfate films. Because lead is very toxic, its use in plumbing may result in some lead contamination of domestic water supplies.

EXAMPLE The reactions of the Group IV elements

Why does the iron in a tin-plated iron can corrode in preference to the tin coat?

SOLUTION Corrosion is an oxidation process. The oxidation of iron is thermodynamically favored over the oxidation of tine:

$$Fe(s) \longrightarrow Fe^{2+}(aq) + 2e^- \qquad E^\circ = +0.44 \text{ V}$$
$$Sn(s) \longrightarrow Sn^{2+}(aq) + 2e^- \qquad E^\circ = +0.14 \text{ V}$$

Thus, if moisture gains access to the iron and permits electron conduction to a cathode reaction, the iron will be oxidized first.

EXERCISE How many liters of H_2 (measured at STP) are produced when 10.0 g Sn is placed in 10.00 mL of hot concentrated HCl (which is approximately 12 M HCl).

ANSWER: 1.3 L

KEY CONCEPT B The compounds of carbon

Carbon monoxide, (CO). Carbon monoxide is produced when an organic compound or carbon itself is burned in a limited supply of oxygen. Commercially it is produced as **synthesis gas** (reaction 1). It is produced in the laboratory by dehydrating formic acid (reaction 2).

$$CH_4(g) + H_2O(g) \longrightarrow CO(g) + 3H_2(g) \qquad (1)$$
$$HCOOH(aq) \longrightarrow CO(g) + H_2O(l) \qquad (2)$$

Carbon monoxide is very toxic and a moderately good Lewis base as a result of the presence of lone pairs. For example, in a typical reaction, CO reacts with a *d*-block metal or ion:

$$Ni(s) + 4CO(g) \longrightarrow [Ni(CO)_4](l)$$

The toxicity of CO is partially related to its Lewis-base character because it attaches more strongly than oxygen to the iron in hemoglobin and prevents the blood from transporting oxygen. Carbon monoxide is a reducing agent.

Carbon dioxide (CO_2) *and carbonates* (CO_3^{2-}). Carbon dioxide is a weak Lewis acid, as is evidenced by its slight reaction with water:

$$CO_2(aq) + H_2O(l) \rightleftharpoons H_2CO_3(aq)$$

Carbonate ion is formed as a product of this equilibrium. When carbonate salts are heated, carbon dioxide and an oxide are formed. For example,

$$CaCO_3(s) \longrightarrow CaO(s) + CO_2(g)$$

Carbonates and hydrogen carbonates (salts containing HCO_3^-) evolve carbon dioxide when treated with acid. The Lewis acid H^+ reacts with oxygen as O^{2-} to form $OH^-(aq)$:

$$CO_3^{2-}(aq) + H^+(aq) \longrightarrow CO_2(g) + OH^-(aq)$$

The reaction of a hydrogen carbonate with acid is used in baking to form CO_2 and thereby cause doughs to rise:

$$2NaHCO_3(s) + Ca(H_2PO_4)_2(s) \longrightarrow Na_2HPO_4(s) + 2CO_2(g) + 2H_2O(g) + CaHPO_4(s)$$

Carbides. Carbides are compounds containing the carbide ion (C^{4-}). [Some compounds containing the acetylide ion (C_2^{2-}) are commonly called carbides even though they are not formally carbides.] Group I and II metals form **saline carbides** when their oxides are heated with carbon; examples are CaC_2, Li_2C_2 and Be_2C. **Covalent carbides** are carbides formed by metals with electronegativities similar to that of carbon, such as silicon and boron. Silicon carbide, SiC, known as carborundum, is an extremely hard, covalent carbide. It is produced by the reaction of silicon dioxide with carbon:

$$SiO_2(s) + 3C(s) \longrightarrow SiC(s) + 2CO(g)$$

In **interstitial carbides,** carbon atoms lie in holes in close-packed arrays of *d*-block metal atoms. Bonds between the carbon atoms and metal atoms stabilize the metal lattice and help to hold the metal atoms together. Examples are Cr_3C and Fe_3C.

Carbon halides. All Group IV elements form tetrachlorides. Carbon tetrachloride (CCl_4) is an important industrial solvent, but its toxicity limits its use. Chlorofluorocarbons, known as Freons, are used as solvents and as refrigerants. It is accepted that some Freons cause damage to the stratospheric ozone layer. Freons are produced by successive replacement of chlorine in chlorinated hydrocarbons such as carbon tetrachloride, which itself is formed by chlorination of methane:

$$CH_4(g) + 4Cl_2(g) \longrightarrow CCl_4(l) + 4HCl(g)$$
$$CCl_4(l) + HF(l) \longrightarrow CFCl_3(g) + HCl(g)$$

Cyanides (compounds containing CN^-). The cyanide ion is quite stable. The parent acid of the cyanides, HCN, is made by the reaction

$$2CH_4(g) + 2NH_3(g) + 3O_2(g) \longrightarrow 2HCN(g) + 6H_2O(g)$$

Cyanides are very strong Lewis bases, as is illustrated by the formation of the extremely stable ferrocyanide ion:

$$Fe^{2+}(aq) + 6CN^-(aq) \longrightarrow [Fe(CN)_6]^{4-}(aq)$$

The cyanide ion is highly toxic to the central nervous system; it also disrupts transfer of oxygen to cellular tissues.

EXAMPLE The structure and reactions of carbon compounds

Use standard enthalpies of formation, bond dissociation energies, and the following reaction to estimate the bond dissociation energy of carbon monoxide. Is your result consistent with the Lewis structure of CO?

$$CH_4(g) + H_2O(g) \longrightarrow CO(g) + 3H_2(g)$$

Standard enthalpies of formation are $CH_4(g)$, -74.8 kJ/mol; $H_2O(g)$, -241.8 kJ/mol; and $CO(g)$, -110.53 kJ/mol. Bond dissociation energies are C—H, 435 kJ/mol; O—H, 431 kJ/mol; and H—H, 436 kJ/mol.

SOLUTION We first calculate the reaction enthalpy using the enthalpies of formation:

$$\Delta H^\circ = [1 \text{ mol} \times \Delta H_f^\circ(CO(g))] - [1 \text{ mol} \times \Delta H_f^\circ(H_2O(g)) + 1 \text{ mol} \times \Delta H_f^\circ(CH_4(g))]$$
$$= [1 \cancel{\text{mol}} \times -110.53 \text{ kJ}/\cancel{\text{mol}}] - [(1 \cancel{\text{mol}} \times -241.8 \text{ kJ}/\cancel{\text{mol}}) + (1 \cancel{\text{mol}} \times -74.8 \text{ kJ}/\cancel{\text{mol}})]$$
$$= +206.1 \text{ kJ}$$

The reaction enthalpy is approximately equal to the energy required to break the bonds in CH_4 and H_2O minus the energy released when bonds are made in CO and H_2:

$$\Delta H^\circ = [(4 \times \text{C—H bond energy}) + (2 \times \text{O—H bond energy})]$$
$$- [\text{CO bond energy} + (3 \times \text{H—H bond energy}]$$
$$= [(4 \times 435 \text{ kJ}) + (2 \times 431 \text{ kJ})] - [(\text{CO bond energy}) + (3 \times 436 \text{ kJ})]$$

We substitute the value of ΔH° and solve for the CO bond energy:

$$206.1 \text{ kJ} = 1294 \text{ kJ} - \text{CO bond energy.}$$
$$\text{CO bond energy} = 1088 \text{ kJ}$$

This extremely high bond energy is consistent with the Lewis structure of CO, which indicates that CO is a triple bonded molecule, :C≡O:.

EXERCISE Explain, on the basis of its Lewis structure, why CN^- is a Lewis base.

ANSWER: The Lewis structure is $:C{\equiv}N:^-$. The unshared pairs make it a Lewis base. (In most of its reactions as a Lewis base, the unshared pair on carbon is donated.)

KEY CONCEPT C The compounds of silicon

Silicon dioxide (SiO_2, or silica). Silica occurs in pure form naturally only as quartz and sand. Silica is very stable, but can be attacked by HF (reaction 3). It is also attacked by the Lewis base OH^- in molten NaOH and by O^{2-} in molten Na_2CO_3 (reaction 4).

$$SiO_2(s) + 6HF(aq) \longrightarrow [SiF_6]^{2-}(aq) + 2H_3O^+(aq) \quad (3)$$
$$SiO_2(s) + 2Na_2CO_3(l) \longrightarrow Na_4SiO_4(l) + 2CO_2(g) \quad (4)$$

Silicates (compounds containing ${SiO_4}^{4-}$ units). The compound sodium silicate is formed in reaction 4. **Orthosilicates** are the simplest silicates and contain isolated ${SiO_4}^{4-}$ ions. Na_4SiO_4 and $ZrSiO_4$ are examples. The **pyroxenes** consist of SiO_4 chains in which two corner atoms are shared by two neighboring units, so that the "average" unit is a ${SiO_3}^{2-}$ unit. Cations placed along the chain in a more or less random manner provide electrical neutrality. $NaAl(SiO_3)_2$ and $CaMg(SiO_3)_2$ are examples. In **amphiboles** the chains of silicate ions are linked to form a cross-linked ladderlike structure containing ${Si_4O_{11}}^{6-}$ units. Tremolite, $Ca_2Mg_5(Si_4O_{11})_2(OH)_2$, is an example. Almost all amphiboles contain hydroxide ions attached to the metal. Sheets containing ${Si_2O_5}^{2-}$ units, with cations lying between the sheets and linking them together, are also known. $Mg_3(Si_2O_5)_2(OH)_2$ is an example. Other structural types of silicon oxides exist.

Aluminosilicates and ceramics. Replacing a Si^{4+} ion with an Al^{3+} ion (plus extra cations to make up for the charge difference) results in the formation of **aluminosilicates.** Mica, of which one form is $KMg_3(Si_3AlO_{10})(OH)_2$ is an example. The extra cations hold the sheets of tetrahedra together, which makes this type of material rather hard. Feldspar is a silicate material in which more than half the silicon is replaced by aluminum. Leaching of the cations in a feldspar results in clay. Heating clays to drive out the water trapped between the sheets of aluminosilicate tetrahedra results in formation of **ceramics. Cements** are produced by melting aluminosilicates and then allowing them to solidify. Cements are one of the components used in making concrete.

Silicones. Silicones are made of long —O—Si—O— chains with the remaining silicon bonding positions occupied by organic groups. They possess both hydrophobic and hydropilic properties, which makes them useful as fabric waterproofers and in biological applications, as surgical and cosmetic implants.

Glasses. Heating silica (SiO_2) above its melting point and cooling it slowly results in the formation of fused silica, a **glass.** Addition of various metal ions results in what is popularly called "glass." Heating the silica causes rupture of many Si—O bonds, which enables the metal ions to form ionic bonds with some of the oxygen atoms in the silica tetrahedra. Thus, covalent Si—O bonds are replaced by metal–oxygen ionic bonds. The nature of the glass depends on the metal added. Addition of Na^+ and Ca^{2+} (12% Na_2O and 12% CaO) results in **soda-lime glass,** the common glass used for windows and bottles. Reduction of the proportions of Na^+ and Ca^{2+} and addition of 16% B_2O_3 result in **borosilicate glass.** This type of glass, such as Pyrex, is more resistant to heat shock than soda-lime glass. Photochromic sunglasses, which darken in sunlight, contain added Ag^+ and Cu^+ ions. Sunlight causes reduction of the silver ions and darkening of the glass (reaction 5). Removal of the sunlight allows the process to reverse (reaction 6) because the reduction potential (in the glass) of the Ag^+/Ag couple is less than that of the Cu^{2+}/Cu^+ couple.

$$Ag^+(s) + Cu^+(s) \longrightarrow Ag(s) + Cu^{2+}(s) \tag{5}$$

$$Ag(s) + Cu^{2+}(s) \longrightarrow Ag^+(s) + Cu^+(s) \tag{6}$$

Tin and lead oxides. Tin(II) oxide and tin(IV) oxide are both amphoteric. Thermodynamically, tin(IV) oxide is the more stable form. Lead(II) oxide may be red or yellow, depending on how it is prepared; it is amphoteric and a strong oxidizing agent. Unlike tin, in which the +4 oxidation state is more stable for the oxide, lead(II) oxide is thermodynamically more stable than lead(IV) oxide. The relative ease of interconversion between lead(II) and lead(IV) is the basis of the use of the lead–acid storage battery.

▼ EXAMPLE The structure of silicon compounds

Use VSEPR theory to confirm that the silicate ion is tetrahedral.

SOLUTION Silicon donates 4 valence electrons and the four oxygens donate 24 valence electrons ($4 \times 6 = 24$). The −4 charge requires 4 additional electrons, so the total number of valence electrons is 32 ($4 + 24 + 4 = 32$). The Lewis structure is

$$\left[\begin{array}{ccc} & :\ddot{O}: & \\ & | & \\ :\ddot{O}- & Si & -\ddot{O}: \\ & | & \\ & :\ddot{O}: & \end{array}\right]^{4-}$$

There are four electron pairs surrounding silicon; thus they take on a tetrahedral shape and the ion itself is tetrahedral.

EXERCISE Explain why silicon tetrachloride reacts with water and carbon tetrachloride does not.

$$SiCl_4(l) + 2H_2O(l) \longrightarrow SiO_2(s) + 4HCl(aq)$$

$$CCl_4(l) + 2H_2O(l) \longrightarrow \text{no reaction}$$

ANSWER: Silicon is larger than carbon and can expand its valence electron shell beyond an octet to accommodate five bonds during reaction as the Lewis base H_2O attacks and Cl^- leaves.

GROUP V: NITROGEN AND PHOSPHORUS

KEY WORDS Define or explain each of the following terms in a written sentence or two

Harber–Bosch process
nitrogen fixing
Ostwald process
polyphosphoric acids

19.7–19.9 THE GROUP V ELEMENTS AND IMPORTANT COMPOUNDS

KEY CONCEPT A Nitrogen, phosphorus, and the other Group V elements

The elements nitrogen and phosphorus at the top of Group V are nonmetals, whereas bismuth at the bottom is a metal. Nitrogen differs substantially from its congeners in its high electronegativity, small size, ability to form multiple bonds, and lack of usable *d*-orbitals. Nitrogen displays oxidation numbers from -3 to $+5$. Elemental nitrogen is prepared by fractional distillation of liquid air. The extremely strong triple bond in N_2 makes it very stable. **Nitrogen fixation** is the process in which N_2 is converted to compounds usable by plants. The **Haber–Bosch process** is a major industrial nitrogen-fixation technique. Bacteria found in the root nodules of legumes also fix N_2.

Phosphorus is prepared from phosphate rock such as calcium phosphate:

$$2Ca_3(PO_4)_2(s) + 6SiO_2(s) + 10C(s) \longrightarrow P_4(g) + 6CaSiO_3(l) + 10CO(g)$$

One of the two common allotropic forms of phosphorus, white phosphorus, is a highly reactive, extremely toxic material. It is a molecular solid consisting of P_4 tetrahedra. It bursts into flame spontaneously on contact with air. Red phosphorus, the second common allotrope, likely consists of linked P_4 tetrahedra. It is less reactive than white phosphorus and is used in match heads because it can be ignited by friction.

EXAMPLE Nitrogen and phosphorus

Nitrogen-fixation bacteria in the root nodules of legumes convert N_2 to NH_3. How would you classify this reaction (e.g., acid–base, redox, decomposition, precipitation, or other)?

SOLUTION The reaction is a redox reaction because the oxidation state of nitrogen changes from 0 to -3.

EXERCISE How many kilograms of phosphorus are present in 1×10^3 kg $Ca_3(PO_4)_2$?

ANSWER: 2×10^2 kg

KEY CONCEPT B Compounds of nitrogen

Nitrogen has a rich and complex chemistry, with a wide range of oxidation states, as indicated in the following discussion.

Ammonia (NH_3: oxidation number $= -3$). Ammonia is produced industrially by the **Haber–Bosch** process. The ammonia produced is the starting point for the production of many nitrogen-containing compounds and is used as a fertilizer. Ammonia is a weak Brønsted base, and when neutralized with acid, it forms ammonium salts. For instance,

$$HCl(aq) + NH_3(aq) \longrightarrow NH_4Cl(aq)$$

The ammonium ion is a weak Brønsted acid:

$$NH_4^+(aq) + H_2O(l) \longrightarrow NH_3(aq) + H_3O^+(aq)$$

Ammonia is a Lewis base, as is evidenced by its reaction with aqueous copper(II) ion:

$$4NH_3(aq) + Cu^{2+}(aq) \longrightarrow [Cu(NH_3)_4]^{2+}(aq)$$

In this reaction, unshared pairs on nitrogen move into unoccupied orbitals on Cu^{2+}. Ammonium salts containing an oxidizing anion (such as nitrate) can be oxidized in a self-oxidation:

$$NH_4NO_3(s) \longrightarrow N_2O(g) + 2H_2O(g)$$
$$2NH_4NO_3(s) \longrightarrow 2N_2(g) + O_2(g) + 4H_2O(g)$$

Hydrazine (N_2H_4: oxidation number $= -2$). Hydrazine is produced by the gentle oxidation of ammonia:

$$2NH_3(aq) + OCl^-(aq) \longrightarrow N_2H_4(aq) + H_2O(l)$$

Hydrazine is a dangerously explosive, oily liquid. A mixture of methylhydrazine (CH_3NHNH_2) and liquid N_2O_4 is used as a rocket fuel because these two liquids ignite on contact and produce a large volume of gas:

$$4CH_3NHNH_2(l) + 5N_2O_4(l) \longrightarrow 9N_2(g) + 12H_2O(g) + 4CO_2(g)$$

Dinitrogen oxide (N_2O: oxidation number $= +1$). N_2O (laughing gas) is the oxide of nitrogen with the lowest oxidation number for nitrogen. It is formed by the gentle heating of ammonium nitrate:

$$NH_4NO_3(s) \longrightarrow N_2O(g) + 2H_2O(g)$$

N_2O is fairly unreactive. It is toxic if inhaled in large amounts.

Nitric oxide (NO: oxidation number $= +2$). NO is manufactured industrially by the catalytic oxidation of NH_3:

$$4NH_3(g) + 5O_2(g) \longrightarrow 4NO(g) + 6H_2O(g)$$

The endothermic formation of NO from the oxidation of N_2 occurs readily at the high temperatures that occur in automobile engines and turbine engine exhausts:

$$N_2(g) + O_2(g) \longrightarrow 2NO(g)$$

NO is oxidized further to NO_2 on exposure to air. Through the formation of NO_2, nitric oxide contributes to smog, acid rain, and destruction of the stratospheric ozone layer.

Nitrogen dioxide (NO_2: oxidation number $= +4$). NO_2, like NO, is paramagnetic. It disproportionates in water to form nitric acid:

$$3NO_2(s) + H_2O(l) \longrightarrow 2HNO_3(aq) + NO(g)$$

It is prepared in the laboratory by heating lead(II) nitrate:

$$2Pb(NO_3)_2(s) \longrightarrow 4NO_2(g) + 2PbO(s) + O_2(g)$$

It exists in equilibrium with its dimer, dinitrogen tetroxide:

$$2NO_2(g) \rightleftharpoons N_2O_4(g)$$

Nitrous acid (HNO_2: oxidation number $= +3$). Nitrous acid can be produced by mixing its anhydride, dinitrogen trioxide, with water:

$$N_2O_3(g) + H_2O(l) \longrightarrow 2HNO_2(aq)$$

Nitrous acid has not been isolated as a pure compound, but it is used in aqueous solution. It is a weak acid ($pK_a = 3.4$).

Nitric acid (HNO_3: oxidation number = +5). Nitric acid, a widely used industrial and laboratory acid, is produced by the three-step **Ostwald process:**

$$4NH_3(g) + 2O_2(g) \longrightarrow 4NO(g) + 6H_2(g)$$
$$2NO(g) + O_2(g) \longrightarrow 2NO_2(g)$$
$$3NO_2(g) + H_2O(l) \longrightarrow 2HNO_3(aq) + NO(g)$$

The anhydride of nitric acid is dinitrogen pentoxide:

$$N_2O_5(s) + H_2O(l) \longrightarrow 2HNO_3(l)$$

Dinitrogen pentoxide consists of NO_2^+ cations and NO_3^- anions. Nitrogen has its highest oxidation state in nitric acid, and the acid is therefore an excellent oxidizing agent. A mixture of 1 part by volume HNO_3 with 3 parts by volume HCl is called **aqua regia.** It is one of the few substances that reacts with gold. Nitrate ion oxidizes the gold while chloride ion encourages the reaction by complexing the gold(III) ion formed:

$$Au(s) + NO_3^-(aq) + 4Cl^-(aq) + 4H^+(aq) \longrightarrow [AuCl_4]^-(aq) + NO(g) + 2H_2O(l)$$

When neutralized, nitric acid forms a nitrate salt. Heavy metal nitrate salts decompose into NO_2 and an oxide when heated, whereas lighter metal salts form a nitrite salt and O_2:

$$2Pb(NO_3)_2(s) \longrightarrow 2PbO(s) + 4NO_2(g) + O_2(g)$$
$$2KNO_3(s) \longrightarrow 2KNO_2(s) + O_2(g)$$

EXAMPLE Reactions of nitrogen compounds

What is the change in volume (measured at STP) when 100 g of methylhydrazine react with excess dinitrogen tetroxide? Assume the liquids have zero volume.

$$4CH_3NHNH_2(l) + 5N_2O_4(l) \longrightarrow 9N_2(g) + 12H_2O(g) + 4CO_2(g)$$

SOLUTION We first use the molar mass of methylhydrazine (46.07 g/mol) to calculate the moles of methylhydrazine in 100 g:

$$100\,\cancel{g} \times \frac{1\text{ mol } CH_3NHNH_2}{46.07\,\cancel{g}} = 2.17\text{ mol } CH_3NHNH_2$$

We now use the chemical equation, which tells us that 4 mol of methylhydrazine yields 25 mol of gas (9 + 12 + 4 = 25), to calculate the moles of gas formed:

$$2.17\,\cancel{\text{mol } CH_3NHNH_2} \times \frac{25\text{ mol gas}}{4\,\cancel{\text{mol } CH_3NHNH_2}} = 13.6\text{ mol gas}$$

Finally, by using the molar mass of an ideal gas at STP ($V_m = 22.4$ L/mol) we determine the answer:

$$13.6\,\cancel{\text{mol gas}} \times 22.4\text{ L}/\cancel{\text{mol}} = 305\text{ L}$$

EXERCISE In the potentially explosive reaction of ammonium nitrate shown, what is oxidized and what is reduced?

$$2NH_4NO_3(s) \longrightarrow 2N_2(g) + O_2(g) + 4H_2O(g)$$

ANSWER: N in NH_4^+ is oxidized from oxidation number −3 to 0. N in NO_3^- is reduced from oxidation number +5 to 0. Oxygen in NO_3^- is oxidized from oxidation number −2 to 0.

KEY CONCEPT C The compounds of phosphorus

Phosphine (PH_3). Phosphine is formed by deprotonation of phosphonium ion by a suitable base, such as OH^-. It is an extremely poisonous gas and is neither an acid nor a base.

$$PH_4^+(aq) + OH^-(aq) \longrightarrow H_2O(l) + PH_3(g)$$

Phosphorus halides. Phosphorus trichloride and phosphorus pentachloride are the two most important halides of phosphorus. PCl_3 is used as the starting point for the production of a number of important phosphorus-containing compounds. It is a molecular liquid of pyramidal PCl_3 molecules. It reacts with water to form phosphorus acid:

$$PCl_3(l) + 3H_2O(l) \longrightarrow H_3PO_3(aq) + 3HCl(aq)$$

Phosphorus pentachloride reacts violently with water to form phosphoric acid:

$$PCl_5(s) + 4H_2O(l) \longrightarrow H_3PO_4(aq) + 5HCl(aq)$$

In the solid, PCl_5 is an ionic compound consisting of $[PCl_4]^+$ and $[PCl_6]^-$ ions. On vaporization, it forms trigonal bipyramidal PCl_5 molecules.

Phosphorus(III) oxide and phosphorus acid (P_4O_6 and H_3PO_3). Phosphorus(III) oxide is prepared by the burning of white phosphorus in a limited supply of air. P_4O_6 molecules consist of P_4 tetrahedra, with an oxygen atom lying on each edge of the tetrahedron, between two phosphorus atoms. A tetrahedron has six edges, hence the formula P_4O_6. Phosphorus(III) oxide is the anhydride of phosphorus acid. The structure of phosphorus acid is unusual in that one of the hydrogen atoms is attached to the central phosphorus atom.

Phosphorus(V) oxide and phosphoric acid (P_4O_{10} and H_3PO_4). Phosphorus(V) oxide is prepared by burning phosphorus in excess oxygen. The molecule is the same as that of P_4O_6 (described earlier) but with an additional oxygen attached to each phosphorus at the apices of the tetrahedron. P_4O_{10} is the anyhydride of phosphoric acid. Phosphorus(V) oxide traps and reacts with water with great efficiency and is widely used as a drying agent. It must be used carefully, however, because the reaction with water can be violent. Phosphoric acid is prepared commercially by the action of sulfuric acid on phosphate rock [$Ca_3(PO_4)_2$] or fluoroapatite [$Ca_5F(PO_4)_3$]:

$$Ca_3(PO_4)_2(s) + 3H_2SO_4(l) \longrightarrow 2H_3PO_4(l) + 3CaSO_4$$

$$Ca_5F(PO_4)_3(s) + 5H_2SO_4(l) + 10H_2O(l) \longrightarrow 3H_3PO_4(l) + 5CaSO_4{\cdot}2H_2O(s) + HF(g)$$

Phosphoric acid is a triprotic acid. It is only moderately oxidizing, even though its phosphorus is in a high oxidation state. Phosphoric acid is the parent acid of phosphate salts (salts containing the tetrahedral phosphate ion, PO_4^{3-}). Phosphates are generally not very soluble, which makes them appropriate structural material for bones and teeth.

Polyphosphates. Polyphosphates are compounds that contain linked PO_4^{3-} tetrahedra. The simplest such structure is the pyrophosphate ion ($P_2O_7^{4-}$), in which two phosphate ions are linked by an oxygen atom through —P—O—P— bonds, as shown in the following figure.

$$\left[\begin{array}{ccccccccc} & & :\ddot{O} & & & & :\ddot{O} & & \\ & & \| & & & & \| & & \\ :\ddot{\underset{..}{O}} & - & P & - & \ddot{\underset{..}{O}} & - & P & - & \ddot{\underset{..}{O}}: \\ & & | & & & & | & & \\ & & :\underset{..}{O}: & & & & :\underset{..}{O}: & & \end{array}\right]^{4-}$$

More complicated structures formed by longer chains or rings also exist. The biochemically most important polyphosphate is undoubtedly adenosine triphosphate (ATP), which contains three phosphate tetrahedra linked by —P—O—P— bonds. The hydrolysis of ATP to adenosine diphosphate (ADP) by the rupture of an O—P bond releases energy that is used by cells to drive the biochemical reactions in the cell:

$$ATP + H_2O \longrightarrow ADP + HPO_4^{2-} + 41\ kJ$$

It is not an exaggeration to call ATP the "fuel of life," because it provides the energy for many of the biochemical reactions of cells.

SELF-TEST EXERCISES

Group III: Boron and aluminum

1. One of the following does not properly describe elemental boron. Which one?
(a) metal (b) acidic oxide (c) several allotropes (d) hardens steel (e) very inert

2. How many grams of boron are there in 1.00 kg of its ore kernite, $Na_2B_4O_7 \cdot 7H_2O$?
(a) 327 (b) 266 (c) 132 (d) 33.0 (e) 429

3. The common oxidation state of aluminum in its compounds is
(a) +2 (b) +1 (c) +3 (d) 0 (e) −4

4. What is the time (in hours) required to electrolytically reduce 1.0 metric ton of aluminum with a current of 1.0×10^4 A? 1 metric ton = 1000 kg
(a) 3.0×10^2 h (b) 1.0×10^2 h (c) 1.0×10^{-2} h (d) 1.0×10^6 h (e) 1.0×10^4 h

5. What is the anhydride of boric acid, $B(OH)_3$?
(a) BN (b) B_2O_3 (c) B_2H_6 (d) $B_{12}C_3$ (e) BF_3

6. The pK_a of boric acid is 9.00. What is the pH of a 0.020 M solution of boric acid?

$$B(OH)_3(aq) + 2H_2O(l) \rightleftharpoons [B(OH)_4]^-(aq) + H_3O^+(aq)$$

(a) 1.78 (b) 7.00 (c) 3.14 (d) 8.65 (e) 5.35

7. Which structure best represents the Al_2Cl_6 molecule?

```
(a)      Cl   Cl                 (b) Cl              Cl
         |    |                        \              |
    Cl — Al — Al — Cl                   Al — Cl — Al — Cl
         |    |                        /              |
         Cl   Cl                     Cl              Cl

(c)        Cl — Cl                (d) Cl      Cl      Cl
          /        \                   \    /  \    /
    Al — Cl         Cl — Al             Al      Al
          \        /                   /    \  /    \
           Cl — Cl                    Cl      Cl      Cl
```

8. The solubility product of $Al(OH)_3$ is approximately 1×10^{-33} M^4. What is the solubility of $Al(OH)_3$ in g/100 mL in a solution with pH = 5.44?
(a) 2×10^{-22} (b) 2×10^{-21} (c) 2×10^{-16} (d) 2×10^{-5} (e) 4×10^{-7}

Group IV: Carbon and silicon

9. Graphite possesses one metallike property. What is the property?
(a) lustrousness (b) ductility (c) forms positive ions
(d) electrical conductivity (e) forms basic oxides

10. It takes extremely high pressures to convert graphite to diamond. Given that the density of graphite is 2.22 g/cm^3, what is a reasonable density for diamond?
(a) 1.98 gm/cm^3 (b) 2.22 gm/cm^3 (c) 3.51 gm/cm^3

11. What process is used to obtain ultrapure silicon?
(a) fractional distillation (b) filtration (c) column chromatography
(d) paper chromatography (e) zone refining

12. 1.00 kg SiO_2 is processed and results in the production of 126 g of ultrapure silicon. What is the percentage yield?
(a) 26.9% (b) 12.6% (c) 18.7% (d) 44.9% (e) 35.2%

13. Which carbon species is a reducing agent?
(a) CO (b) CCl_4 (c) CO_2 (d) CN^- (e) CH_4

14. The boiling points of three carbon compounds are CO_2, −78°C; CS_2, 46.2°C; and CSe_2, 125.1°C. What is the dominant force that determines the boiling points of these compounds?
(a) ion–dipole (b) London (c) dipole–dipole (d) ion–ion

15. What is the structural unit for silicates?
(a) SiO_2 linear units (b) SiO_2 angular units (c) SiO_4 tetrahedra
(d) SiO_4 square planar units (e) SiF_6 octahedra

16. How many milligrams of silica can be produced by reacting 10.0 mL of 0.200 M HCl with 1.00 g sodium silicate?

$$4HCl(aq) + Na_4SiO_4(s) \longrightarrow SiO_2(s) + 4NaCl(aq) + 2H_2O(l)$$

(a) 266 (b) 41.2 (c) 129 (d) 326 (e) 30.0

Group V: Nitrogen and phosphorus

17. One of the following does not describe N. Which one?
(a) high electronegativity (b) highly reactive (c) forms multiple bonds
(d) small size (e) no usable *d* orbitals

18. On a mole basis, air is 78% N_2. What is the partial pressure of N_2 in a 8.5-L flask filled with air at 325 K? There is a total of 0.544 mol of air in the flask. $R = 0.0821$ L·atm/mol·K.
(a) 3.1 atm (b) 1.3 atm (c) 2.4 atm (d) 1.7 atm (e) 3.6 atm

19. What is the P—P—P bond angle in P_4?
(a) 90° (b) 120° (c) 45° (d) 60° (e) 180°

20. What partial pressure (in torr) of P_4 develops when 10.0 g $Ca_3(PO_4)_2$ are completely reacted as shown at 800°C in a 5.0-L closed vessel?

$$2Ca_3(PO_4)_2(s) + 6SiO_2(s) + 10C(s) \longrightarrow P_4(g) + 6CaSiO_3(s) + 10CO(g)$$

(a) 444 Torr (b) 677 Torr (c) 326 Torr (d) 918 Torr (e) 215 Torr

21. What is the pH of a 0.15 M solution of nitrous acid? $pK_a = 3.47$

$$HNO_2(aq) + H_2O(l) \rightleftharpoons H_3O^+(aq) + NO_2^-(aq)$$

(a) 3.40 (b) 2.11 (c) 1.45 (d) 1.71 (e) 4.38

22. What is the oxidation number of phosphorus in the pyrophosphate ion, $P_2O_7^{4-}$?
(a) +3 (b) +9 (c) +5 (d) −6 (e) −2

20 THE p-BLOCK ELEMENTS: II

Here we consider the elements in Groups VI–VIII, which, with the exception of polonium, are all nonmetals. There is great variation in the chemical properties of these elements, from the highly reactive halogens to the generally inert noble gases.

GROUP VI: OXYGEN AND SULFUR

KEY WORDS Define or explain each of the following terms in a written sentence or two.

catenate
chalcogens
chemiluminescence
clathrate
Claus process
colloidal solution
contact process
Frasch process
hydrolyze
polysulfane

20.1 GROUP VI ELEMENTS

KEY CONCEPT A Elemental oxygen and sulfur

Oxygen is an important industrial product that is used extensively in the steel industry. It is a colorless, odorless, paramagnetic gas. The paramagnetism of O_2 is accounted for by molecular orbital theory, but a satisfactory Lewis structure for O_2 cannot be drawn. Elemental oxygen exists in two allotropic forms, O_2 and O_3 (ozone). Ozone is produced by an electric discharge through O_2; its pungent odor can be detected near sparking electrical equipment and after lightning strikes. The following table gives some properties of O_2 and O_3.

Property	O_2	O_3
boiling point	−183°C	−112°C
liquid color	pale blue	dark blue
atmosphere	23% by mass	forms stratospheric ozone layer
magnetism	paramagnetic	diamagnetic

Sulfur is found in ores and as elemental sulfur. The **Frasch process** is used to mine elemental sulfur by melting the sulfur (mp = 165°C) with superheated water and forcing it to the surface with compressed air. The **Claus process** recovers sulfur from H_2S in a process that involves two reactions:

$$2H_2S(g) + 3O_2(g) \longrightarrow 2SO_2(g) + 2H_2O(l)$$
$$2H_2S(g) + SO_2(g) \longrightarrow 3S(s) + 2H_2O(l)$$

Sulfur is used primarily to produce sulfuric acid and to vulcanize rubber. Vulcanization increases the toughness of rubber by introducing cross links between the natural rubber polymer chains. Elemental sulfur exists in its most stable allotropic form as $S_8(s)$ rings, but for simplicity it is often represented as $S(s)$.

EXAMPLE The structure of sulfur and oxygen

Use VSEPR theory to predict the shape of ozone and the hybridization of the central oxygen atom.

SOLUTION The Lewis structure of O_3 (actually, one of two equivalent resonance structures) has 18 valence electrons:

$$:\ddot{O}=\ddot{O}-\ddot{\underset{\cdot\cdot}{O}}:$$

The central atom has three electron pairs around it (remember that a double bond counts as one electron pair.) Three pairs of electrons result in a trigonal orientation for the electron pairs and sp^2 hybridization. Because one of the electron pairs is a lone pair, the molecule has a bent or angular shape.

EXERCISE In the second reaction associated with the Frasch process, which atom is oxidized and which atom is reduced?

$$2H_2S(g) + SO_2(g) \longrightarrow 3S(s) + 2H_2O(l)$$

ANSWER: S in H_2S is oxidized, and S in SO_2 is reduced.

KEY CONCEPT B Trends and differences in Group VI elements

There is a gradual change from nonmetallic behavior at the top of Group VI to metallic behavior at the bottom, as in Groups IV and V. Other changes occur in going down the group. Sulfur, selenium, and tellurium are able to form relatively long chains (**catenate**) whereas oxygen is not. Sulfur and oxygen are both quite reactive and combine with most of the other elements.

EXAMPLE Trends in Group VI

What trend in boiling points would you expect for the diatomic molecules X_2 for the first four Group VI elements.

SOLUTION For a homonuclear diatomic molecule, the only intermolecular forces are London forces. As the molecular masses of the molecules increase, the London forces increase and boiling points increase.

EXERCISE The boiling points of the Group VI elements increase from oxygen to tellurium, but the boiling point of polonium is lower than that of tellurium. Suggest a reason for the reversal in the trend.

ANSWER: The type of force that determines the boiling point changes from London forces (for O through Te) to metallic bonding (for Po).

20.2 COMPOUNDS OF GROUP VI ELEMENTS AND HYDROGEN

KEY CONCEPT A Water

Due to its unique set of physical and chemical properties, water is one of the most remarkable compounds known. Water for domestic water supplies must undergo many types of purification. High-purity water for special applications is obtained by distillation and/or ion exchange. Oxygen contamination in water can be reduced to a low level by reduction with hydrazine:

$$O_2(aq) + N_2H_4(aq) \longrightarrow 2H_2O(l) + N_2(aq)$$

Water has a higher boiling point than expected because of the extensive hydrogen bonding between H_2O molecules. Hydrogen bonding also causes a more open structure and, therefore, a lower density for the solid than for the liquid. With its high polarity, water is an excellent solvent for ionic compounds. Chemically, it can both donate and accept protons (it is amphiprotic); so it is both a Brønsted acid and a Brønsted base. It can act as an oxidizing agent and a reducing agent. With its unshared electron pairs on oxygen, water can also act as a Lewis base. Thus, it can form complexes with transition metals and can hydrolyze certain substances (such as PCl_5). In its hydrolysis reactions, a bond between oxygen and another element is formed. Hydrolysis reactions can occur with or without a change in oxidation number. Oxides of metals are bases because the oxide ion is a stronger base than OH^-, and oxide ion immediately reacts with water to form OH^-. Some of the reactivities of water are as follows:

As a Brønsted base: $CH_3COOH(aq) + H_2O(l) \longrightarrow CH_3CO_2^-(aq) + H_3O^+(aq)$

As a Brønsted acid: $H_2O(l) + NH_3(aq) \longrightarrow NH_4^+(aq) + OH^-(aq)$

As an oxidizing agent: $2Na(s) + 2H_2O(l) \longrightarrow 2NaOH(aq) + H_2(g)$

As a reducing agent: $2H_2O(l) + 2F_2(g) \longrightarrow 4HF(aq) + O_2(g)$

Hydrolysis without a change in oxidation state: $PCl_5(s) + 4H_2O(l) \longrightarrow H_3PO_4(aq) + 5HCl(aq)$

Hydrolysis with a change in oxidation state (of Cl in this case): $Cl_2(g) + H_2O(l) \longrightarrow ClOH(aq) + HCl(aq)$

EXAMPLE The reactions of water

Describe how water can be considered a Lewis base in its reaction with PCl_5. What is the Lewis acid?

SOLUTION The reaction is

$$PCl_5(s) + 4H_2O(l) \longrightarrow H_3PO_4(aq) + 5HCl(aq)$$

In this reaction, the unshared electron pairs on the oxygens move into empty orbitals on the phosphorus atoms to form new P—O bonds as P—Cl bonds are broken. Thus, water is the Lewis base, and PCl_5 is the Lewis acid.

EXERCISE What is the reduction potential at 25°C for the following half-reaction at pH = 11.0? $E° = -0.83$ V. Assume that the partial pressure of H_2 is 1 atm exactly.

$$2H_2O(l) + 2e^- \longrightarrow 2OH^-(aq) + H_2(g)$$

ANSWER: −0.65 V

KEY CONCEPT B Hydrogen peroxide (H_2O_2)

H_2O_2 is a highly reactive, pale blue liquid. Many of its physical properties are similar to those of water; one exception is its substantially higher density (1.44 g/mL). Chemically, hydrogen peroxide is more acidic than water ($pK_a = 11.75$). It is a strong oxidizing agent, but it can also function as a reducing agent.

As an oxidizing agent: $2Fe^{2+}(aq) + H_2O_2(aq) + 2H^+(aq) \longrightarrow 2Fe^{3+}(aq) + 2H_2O(l)$

As a reducing agent: $Cl_2(g) + H_2O_2(aq) + 2OH^-(aq) \longrightarrow 2Cl^-(aq) + 2H_2O(l) + O_2(g)$

The O—O bond in H_2O_2 is very weak. Oxygen has a −1 oxidation state in H_2O_2.

EXAMPLE The physical properties of hydrogen peroxide

The boiling point of hydrogen peroxide is very different from that of water. Predict whether it is higher or lower than water's, and give the reason for your prediction.

SOLUTION H_2O_2 has two oxygens, so it seems reasonable to expect that it may be more extensively hydrogen bonded than H_2O, which has only one oxygen. The more extensive hydrogen bonding should lead to a higher boiling point; thus, we predict that H_2O_2has a higher boiling point than water. This prediction is correct; the boiling point of H_2O_2 is 152°C.

EXERCISE Explain why H_2O_2 is more dense than H_2O.

ANSWER: More extensive hydrogen bonding holds the molecules closer together.

20.3 IMPORTANT COMPOUNDS OF SULFUR

KEY CONCEPT A Sulfur dioxide and sulfites

Sulfur burns in air to form SO_2, which is a poisonous gas. Volcanic activity, the combustion of fuels contaminated with sulfur, and the oxidation of H_2S are the major sources of the SO_2 found in the atmosphere. The equation for the oxidation of H_2S is

$$2H_2S(g) + 3O_2(g) \longrightarrow 2SO_2(g) + 2H_2(g)$$

SO_2 is an acidic oxide and forms sulfurous acid when it reacts with water:

$$SO_2(g) + H_2O(l) \longrightarrow H_2SO_3(aq)$$

In aqueous solutions, H_2SO_3 exists as a mixture of $SO_2{\cdot}7H_2O$ in equilibrium with H_3O^+ and SO_3^{2-}, with few or no H_2SO_3 molecules present. In $SO_2{\cdot}7H_2O$, each SO_2 molecule is located in a cage formed by surrounding water molecules in a structure called a **clathrate.** SO_2 and sulfite ion (SO_3^{2-}) can act as oxidizing or reducing agents:

SO_3^{2-} as an oxidizing agent: $4HSO_3^-(aq) + 2HS^-(aq) \longrightarrow 3S_2O_3^{2-}(aq) + 3H_2O(l)$

SO_3^{2-} as an reducing agent: $SO_3^{2-}(aq) + S(s) \longrightarrow S_2O_3^{2-}(aq)$

EXAMPLE The structure and reactions of SO_2

What is $\Delta G°$ for the reaction

$$2H_2S(g) + 3O_2(g) \longrightarrow 2SO_2(g) + 2H_2O(g)$$

Would you expect the H_2S that escapes into the atmosphere to stay as H_2S or to form SO_2?

SOLUTION For $\Delta G°$, we use the standard free energies of formation of the reactants and products:

$$\begin{aligned}\Delta G° &= [\{2 \text{ mol} \times \Delta G_f°(H_2O(g))\} + \{2 \text{ mol} \times \Delta G_f°(SO_2(g))\}] \\ &\quad - [\{2 \text{ mol} \times \Delta G_f°(H_2S(g))\} + \{3 \text{ mol} \times \Delta G_f°(O_2(g))\}] \\ &= [\{2 \text{ mol} \times (-228.57 \text{ kJ/mol})\} + \{2 \text{ mol} \times (-300.19 \text{ kJ/mol})\}] \\ &\quad - [\{2 \text{ mol} \times (-33.56 \text{ kJ/mol})\} + \{3 \text{ mol} \times (0 \text{ kJ/mol})\}] \\ &= [-1057.52 \text{ kJ}] - [-67.12 \text{ kJ}] \\ &= -990.40 \text{ kJ}\end{aligned}$$

The large negative free energy of reaction indicates that the products are thermodynamically strongly favored, so we would expect H_2S to form SO_2.

EXERCISE Use VSEPR theory to predict the shape of the SO_3^{2-} ion.

ANSWER: Pyramidal

KEY CONCEPT B Sulfur trioxide and the sulfates

Sulfur dioxide can be oxidized directly to sulfur trioxide by reaction with O_2:

$$2SO_2(g) + O_2(g) \longrightarrow 2SO_3(g)$$

The uncatalyzed reaction is slow in the atmosphere, but it can be catalyzed by the presence of metal cations in droplets of water or by certain surfaces. Indirect pathways also exist in the atmosphere for the conversion of SO_2 to SO_3. SO_3 is a triangular planar molecule and forms S_3O_9 trimers in the solid. The **contact process** for the production of H_2SO_4 involves the production of SO_3, which, by dissolving in "oleum" (98% H_2SO_4) and reacting with water, is converted to H_2SO_4.

$$S(g) + O_2(g) \xrightarrow{1000°C} SO_2(g)$$

$$2SO_2(g) + O_2(g) \xrightarrow{500°C,\, V_2O_5} 2SO_3(g)$$

Sulfuric acid is a strong Brønsted acid, a powerful dehydrating agent, and a mild oxidizing agent. In water, its first ionization is complete:

$$H_2SO_4(aq) + H_2O(l) \longrightarrow HSO_4^-(aq) + H_3O^+(aq)$$

HSO_4^- is a weak acid with $pK_a = 1.92$:

$$HSO_4^-(aq) + H_2O(l) \rightleftharpoons SO_4^{2-}(aq) + H_3O^+(aq)$$

As a dehydrating agent, sulfuric acid is able to extract water from a variety of compounds, including sucrose ($C_{12}H_{22}O_{11}$) and formic acid (HCOOH):

$$C_{12}H_{22}O_{11}(s) \xrightarrow{\text{Conc. } H_2SO_4} 12C(s) + 11H_2O(l)$$

$$HCOOH(l) \xrightarrow{\text{Conc. } H_2SO_4} CO(g) + H_2O(l)$$

Due to the large negative enthalpy of solution involved, mixing sulfuric acid with water can cause violent splashing; so for reasons of safety, sulfuric acid is always carefully added to water, not the water to acid. The sulfate ion is a mild oxidizing agent:

$$4H^+(aq) + 2e^- + SO_4^{2-}(aq) \longrightarrow SO_2(aq) + 2H_2O(l) \qquad \Delta E° = 0.17 \text{ V}$$

EXAMPLE The reactions of sulfuric acid

Mixing sulfuric acid with solid sodium chloride results in the formation of dry hydrogen chloride. Write the equation for the reaction, state what type of reaction it is, and suggest why the hydrogen chloride is dry.

SOLUTION Because HCl is one of the products formed by the reaction of H_2SO_4 with NaCl, it is reasonable to assume that the Na^+ and SO_4^{2-} that remain form sodium sulfate. The equation is

$$H_2SO_4(l) + 2NaCl(s) \longrightarrow 2HCl(g) + Na_2SO_4(s)$$

The reaction is a Lewis acid–base reaction in which the Lewis base Cl^- reacts with the Lewis acid H^+ to form HCl, and the salt Na_2SO_4 forms from the remaining ions. The HCl is dry because sulfuric acid is a strong dehydrating agent and removes any stray water present.

EXERCISE Pure sulfuric acid shows very high conductivity. Explain why.

ANSWER: It undergoes the self-ionization $2H_2SO_4(l) \rightleftharpoons HSO_4^- + H_3SO_4^+$; the ions formed conduct an electric current.

GROUP VII: THE HALOGENS

KEY WORDS Define or explain each of the following terms in a written sentence or two.

halogens　　　　interhalogens

20.4 THE GROUP VII ELEMENTS

KEY CONCEPT A Group VII elements

The group VII elements F, Cl, Br, I, and As are called the halogens.

Fluorine is notable for being the most electronegative element. Elemental fluorine is obtained by the electrolysis of an anhydrous, molten KF–HF mixture. Fluorine is highly reactive and highly oxidizing; it is used in the production of SF_6, teflon, and UF_6 (for isotopic enrichment).

Chlorine is obtained by the electrolysis of molten or aqueous NaCl. It is a pale green-yellow gas that is used in a number of industrial processes.

Bromine, a corrosive, reddish-brown liquid of Br_2 molecules, is produced by the oxidation of Br^- by Cl_2. Bromine has a number of industrial uses, including organic chemical synthesis and photographic processes.

$$Cl_2(g) + 2Br^-(aq) \longrightarrow Br_2(l) + 2Cl^-(aq)$$

Iodine occurs as I^- in seawater and as an impurity in Chile saltpeter. The element is produced by the oxidation of I^- with Cl_2. It is a blue-black, lustrous solid of I_2 molecules. I_2 is slightly soluble in water, but it dissolves well in iodide solutions since it reacts with aqueous I^- to form a complex:

$$I_2(aq) + I^-(aq) \longrightarrow I_3^-(aq)$$

Because iodine is an essential trace element in human nutrition, iodides are often added to table salt. This "iodized salt" can prevent iodine deficiency, which leads to goiter (enlargement of the thyroid gland).

EXAMPLE Reactions and properties of the halogens

Comment on why it is possible to use Cl_2 to produce Br_2 from Br^-. What is $E°$ for the reaction?

SOLUTION The reaction is an oxidation–reduction reaction. Using the reduction potentials for the two half-reactions that are involved allows us to calculate $E°$ for the reaction:

$$\begin{array}{rcl} Cl_2(g) + 2e^- & \longrightarrow & 2Cl^-(aq) \qquad E°_{red} = 1.36\text{ V} \\ Br^-(aq) & \longrightarrow & Br_2(l) + 2e^- \qquad E°_{ox} = -1.07\text{ V} \\ \hline Cl_2(g) + 2Br^-(aq) & \longrightarrow & Br_2(l) + 2Cl^-(aq) \qquad E° = +0.29\text{ V} \end{array}$$

Because the standard reaction potential is positive, the reaction is thermodynamically favored and can be used to produce Br_2.

EXERCISE Is the lustrous appearance of I_2 consistent with its position in Group VII?

ANSWER: Yes. A lustrous appearance is a metallic property; going down a group results in a increase in metallic properties. I_2 is a nonmetal, but it has a metal like appearance.

KEY CONCEPT B Trends and differences

The halogens show smooth variations in the properties of the elements and their compounds, with some exceptions for fluorine. Many of their physical properties are determined by London forces. Fluorine displays a -1 oxidation number in its compounds, whereas the other halogens typically show -1, 0, 1, 3, 5, and 7 in their compounds. Because of fluorine's small size, high electronegativity, and lack of usable *d* orbitals, fluorine compounds are somewhat unusual. For instance, fluorine tends to bring out high oxidation numbers in elements to which it is bonded (e.g., $+7$ for I in IF_7). The small size of the fluoride ion results in high lattice enthalpies for fluorides, which makes them less soluble than other halides. The low bond energy of F_2 and high strength of fluorine's bonds with other elements result in highly exothermic reactions when F_2 reacts. For instance, the reaction of F_2 with CH_4 is more exothermic than the reaction of Cl_2 with CH_4:

$$F_2(g) + CH_4(g) \longrightarrow CH_3F(g) + HF(g) \qquad \Delta H^\circ = -434 \text{ kJ}$$
$$Cl_2(g) + CH_4 \longrightarrow CH_3Cl(g) + HCl(g) \qquad \Delta H^\circ = -98 \text{ kJ}$$

The strong bonds formed by fluorine make many of its compounds relatively inert. Fluorine's ability to form hydrogen bonds results in relatively high melting points, boiling points, and enthalpies of vaporization for many of its compounds.

▼ EXAMPLE The properties of covalent halogens

HCl, HBr, and HI are all strong acids whereas HF is a weak acid. Explain.

SOLUTION The HF bond is exceptionally strong (fluorine forms strong bonds in its compounds), so it does not ionize nearly as easily as HCl, HBr, and HI.

EXERCISE Fluoridation of domestic water supplies can be accomplished by adding NaF to the water to get about 1 ppm (part per million) by mass of F^- ion. This means that for every million grams of water, one gram of fluoride ion is present. How many grams of NaF must be added per gallon of water to get to 1 ppm of F^-? 1 qt = 0.9463 L.

▲ ANSWER: 0.008 g

20.5 HALIDES

KEY CONCEPT The halides

The hydrogen halides (HX, X = halogen) can be prepared by the direct reaction of the elements or by the action of a nonvolatile acid on a metal halide:

Direct reaction of elements: $$Cl_2 + H_2(g) \longrightarrow 2HCl(g)$$

Acid + metal halide: $$CaF_2(s) + H_2SO_4(aq, \text{conc}) \longrightarrow Ca(HSO_4)_2(aq) + 2HF(g)$$
$$KI(s) + H_3PO_4(aq) \longrightarrow KH_2PO_4(aq) + HI(g)$$

All hydrogen halides are colorless, pungent gases above 19°C. HF has the unusual property of attacking and dissolving silica glasses and is used for glass etching. In this reaction, F^- acts as a Lewis base and replaces oxygen atoms in the silicate structure:

$$6HF(aq) + SiO_2(s) \longrightarrow H_2SiF_6(aq) + 2H_2O(l)$$

In the etching process, the disruption of the glass structure at the surface of the glass causes the glass to become opaque.

Metal halides may be formed by direct reaction, by the reaction of a halogen with a metal oxide in the presence of a reducing agent, and in aqueous solution by neutralization or precipitation.

Direct reaction:	$Cu(s) + Cl_2(g) \longrightarrow CuCl_2(s)$
Halogen + metal oxide:	$Cr_2O_3(s) + 3C(s) + 3Cl_2(g) \longrightarrow 2CrCl_3(s) + 3CO(g)$
Precipitation:	$BaCl_2(aq) + 2KF(aq) \longrightarrow BaF_2(s) + 2KCl(aq)$

EXAMPLE Physical and chemical properties of halides

Predict the trend in enthalpies of vaporization for the hydrogen halides.

SOLUTION Enthalpies of vaporization are determined largely by intermolecular attractions. There are three intermolecular attractions to consider: hydrogen bonding, dipole–dipole attractions, and London forces. The presence of hydrogen bonding in HF makes its enthalpy of vaporization the highest of the halogen halides. Dipole–dipole attractions decrease slightly in going down the group because the electronegatives of the halogens decrease slightly from 3.0 for Cl to 2.5 for I. However, London forces increase substantially because there is a significant increase in the number of electrons in going from HCl to HI. The increase in London forces should result in an increase in enthalpies of vaporization in going from HCl to HI.

EXERCISE Why doesn't HCl dissolve glass like HF does?

ANSWER: The Si—F bond that forms when HF reacts with SiO_2 is very strong whereas the Si—Cl bond that would form with HCl is not.

20.6 HALOGEN OXIDES AND OXOACIDS

KEY CONCEPT Halogen oxides and oxoacids

The *hypohalous acids* (HOX, halogen oxidation number = +1) are prepared by the direct reaction of a halogen with water, and the corresponding hypohalite salts, are prepared by the reaction of a halogen with aqueous alkali:

Hypochorous acid:	$Cl_2(g) + H_2O(l) \longrightarrow HOCl(aq) + HCl(aq)$
Sodium hypochlorite:	$Cl_2(g) + NaOH(aq) \longrightarrow NaOCl(aq) + HCl(aq)$

Hypochlorites oxidize organic material in water by producing oxygen in aqueous solution; the O_2 readily oxidizes organic materials. The O_2 production occurs in a two-step process, the net result of which is

$$2OCl^-(aq) \longrightarrow 2Cl^-(aq) + O_2(g)$$

Chlorates (ClO_3^-, chlorine oxidation number = +5) are prepared by the reaction of chlorine with hot aqueous alkali:

$$3Cl_2(g) + 6OH^-(aq, hot) \longrightarrow ClO_3^-(aq) + 5Cl^-(aq) + 3H_2O(l)$$

Chlorates decompose when heated, with the identity of the final product depending on the presence of a catalyst:

$$4KClO_3(l) \xrightarrow{\Delta} 3KClO_4(s) + KCl(s)$$
$$2KClO_3(l) \xrightarrow{\Delta, MnO_2} 2KCl(s) + 3O_2(g)$$

Chlorates are good oxidizing agents. They are also used to produce the important compound ClO_2, frequently through reduction with SO_2:

$$2NaClO_3(aq) + SO_2(g) + H_2SO_4(aq) \longrightarrow 2NaHSO_4(aq) + 2ClO_2(g)$$

Chlorine dioxide (ClO_2, chlorine oxidation number = +4) is an odd-electron, paramagnetic, yellow-green gas that is used to bleach paper pulp. It is highly reactive and may explode violently if the circumstances are right.

Perchlorates (salts containing ClO_4^-, chlorine oxidation number = +7) are prepared by the electrolysis of aqueous chlorates. The half-reaction involved is

$$ClO_3^-(aq) + H_2O(l) \longrightarrow ClO_4^-(aq) + 2H^+(aq) + 2e^-$$

Perchlorates and perchloric acid ($HClO_4$) are powerful oxidizing agents; perchloric acid in contact with small amounts of organic matter may explode.

EXAMPLE The reactions and structure of halogen oxides

The hypohalous ions ClO^-, BrO^-, and IO^- all undergo the disproportion reaction

$$3XO^-(aq) \longrightarrow XO_3^-(aq) + 2X^-(aq)$$

The reaction is slowest with hypochlorite, faster with hypobromate, and fastest with hypoiodate. Suggest a reason for this order of reactivities.

SOLUTION The reaction is a Lewis acid–base reaction in which oxygen atoms form new bonds with the halogen atoms and the old halogen–oxygen bonds break. As the size of the central halogen atom increases, there is more room for the electrons on the oxygen atom to approach the halogen and form a bond, so the reaction becomes faster.

EXERCISE Predict the relative strengths of the following acids in water: HOCl, HOBr, and HOI.

ANSWER: Strongest HOCl > HOBr > HOI weakest

GROUP VIII: THE NOBLE GASES

KEY WORDS Define or explain each of the following terms in a written sentence or two.

cryogenics | fluorinating agent | superfluid

20.7 GROUP VIII ELEMENTS

KEY CONCEPT Group VIII elements

The Group VIII elements are He, Ne, Ar, Xe, Kr, and Ra. Until 1962 no compounds of these so-called inert gases had been synthesized. Now that synthesis of some compounds has been accomplished, the group is called the *noble gases*. All have close-shell configurations and are chemically quite unreactive. All occur as monatomic gases.

Helium is the second most abundant element in the universe. It is formed in the earth through the emission of alpha particles (helium nuclei), which capture two electrons to become helium atoms. Its low density and lack of flammability (unlike hydrogen) make it an ideal gas for lighter-than-air aircraft. Helium has the lowest boiling point of any known substance (4.2 K) and finds extensive use as a **cryogenic** coolant. (Cryogenics is the science of low temperatures.) It cannot be solidified at normal pressure. Below 1 K, it becomes a **superfluid,** a fluid that has zero viscosity and, therefore, no resistance to flow.

Neon, argon, krypton, and *xenon* find specialized uses as cryogenic fluids and in lighting applications. *Radon* is highly radioactive. It is formed as a natural product of radioactive decay in the earth and seeps out of the ground in some areas of the United States (and elsewhere). In those areas, there is the possibility that it may find its way into houses, where it may pose a health threat because of its radioactivity.

EXAMPLE The physical properties of the noble gases

The phase diagram of helium follows. At what temperature and pressure are the liquid, solid, and vapor at equilibrium?

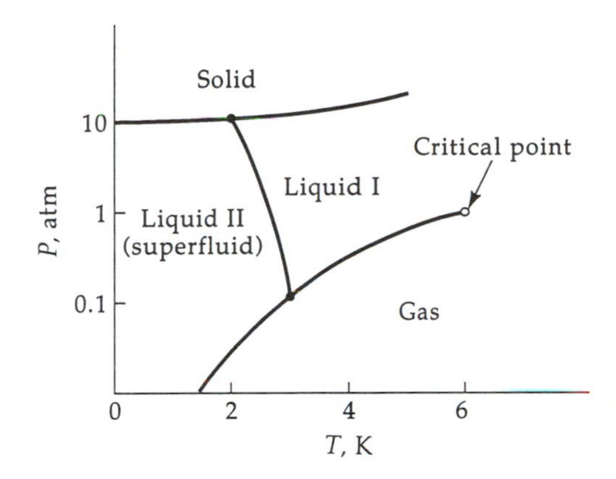

SOLUTION One of the many unique features of helium is its lack of a standard triple point. There is no temperature and pressure at which the liquid, solid, and vapor coexist at equilibrium.

EXERCISE Explain why helium does not form a solid at normal pressures.

ANSWER: The intermolecular London forces are so weak that they cannot bring the atoms together to form the solid.

20.8 COMPOUNDS OF THE NOBLE GASES

KEY CONCEPT The compounds of the noble gases

Helium, neon, and argon form no stable compounds. Krypton forms one stable neutral compound, KrF_2. Xenon displays a rich chemistry involving, for the most part, Xe—O and Xe—F bonds. Because of radon's radioactivity, its chemistry is largely unexplored. Compounds with relatively unstable Xe—N, Xe—C, and Kr—N bonds have been reported. Xenon displays oxidation states of +2, +4, +6, and +8. Direct reaction of fluorine with xenon at high temperature results in the formation of XeF_2 (Xe oxidation number = +2), XeF_4 (Xe oxidation number = +4), and XeF_6 (Xe oxidation number = +6). All three compounds are covalent molecules in the gas phase. They are powerful **fluorinating agents** (reagents that attach fluorine to other substances). Xenon trioxide can be synthesized by the hydrolysis of xenon tetrafluoride:

$$6XeF_4(s) + 12H_2O(l) \longrightarrow 2XeO_3(aq) + 4Xe(g) + 3O_2(g) + 24HF(aq)$$

The trioxide is the anhydride of xenic acid (H_2XeO_4); in aqueous alkali, the acid forms $HXeO_4^-$, which further disproportionates to the perxenate ion (XeO_6^{4-}, Xe oxidation state = +8).

$$2HXeO_4^-(aq) + 2OH^-(aq) \longrightarrow XeO_6^{4-}(aq) + Xe(g) + O_2(g) + 2H_2O(l)$$

EXAMPLE The prediction of oxidizing ability

Would you expect the perxenate ion to be a good oxidizing agent? Why?

SOLUTION We would predict it to be a good oxidizing agent because the xenon in perxenate has a very high oxidation number (+8) and a strong tendency to gain electrons to get a lower oxidation number.

EXERCISE Suggest why XeF_4 is such a good fluorinating agent in reactions such as

$$\text{Compound} + XeF_4(s) \longrightarrow \text{compound—F} + Xe(g)$$

ANSWER: When XeF_4 gives up its fluorine atoms, it forms the very stable $Xe(g)$. The formation of a stable product helps to make a reaction occur.

SELF-TEST EXERCISES

Group VI: Oxygen and sulfur

1. According to molecular orbital theory, what is the bond order of O_2?
(a) 1/2 (b) 0 (c) 2 (d) 3/2 (e) 1

2. Use that fact that O_2 is 23% by mass of the atmosphere to estimate its partial pressure when atmospheric pressure is 1.0 atm. Assume the remaining atmosphere gas is N_2.
(a) 0.16 atm (b) 0.21 atm (c) 1.0 atm (d) 0.23 atm (e) 0.32 atm

3. From the boiling points of O_2 (−183°C) and O_3 (−112°C), predict the boiling point of S_2. (Because of chemical reactions, this boiling point has not yet been measured.)
(a) −254°C (b) −41°C (c) −183°C (d) −112°C (e) −148°C

4. How many kilograms of solid sulfur can be produced from 1.0 kg of H_2S?
(a) 1.4 (b) 0.82 (c) 0.65 (d) 0.94 (e) 0.79

5. Which of the Group VI elements is best described as a nonmetal with some metallic characteristics?
(a) O (b) S (c) Se (d) Te (e) Po

6. Which of the Group VI elements has the least ability to catenate?
(a) O (b) Se (c) Po (d) Te (e) S

7. All of the following phrases but one describe water. Which does not describe water?
(a) extensively hydrogen bonded as a liquid (b) amphiprotic (c) low polarity
(d) oxidizing agent (e) Lewis base

8. The entropy of vaporization of $H_2S(l)$ at its boiling point is 87.7 J/mol·K. What is a reasonable estimate of the entropy of vaporization of water?
(a) 109 J/mol·K (b) 87.7 J/mol·K (c) 66.4 J/mol·K

9. What is the role of water in the following reaction?

$$PCl_5(s) + 4H_2O(l) \longrightarrow H_3PO_4(aq) + 5HCl(aq)$$

(a) Brønsted base (b) reducing agent (c) oxidizing agent
(d) Lewis base (e) solvent

10. What is the pH of a 3.0% by mass solution of H_2O_2? Assume the solution has a density of 1.0 g/mL. The pK_a of H_2O_2 is 11.75.
(a) 11.86 (b) 5.90 (c) 8.43 (d) 2.35 (e) 7.00

11. What is a clathrate?
(a) a compound in which molecules are trapped in a cage
(b) an oxygen-containing compound that is amphiprotic
(c) a Lewis base that is also an oxidizing agent
(d) a strongly hydrated covalent compound

12. Sulfur trioxide is the anhydride of which acid?
(a) H_2SO_3 (b) $H_2S_2O_6$ (c) H_2S (d) $H_2S_2O_3$ (e) H_2SO_4

13. The high boiling point of H_2SO_4 is largely due to
(a) London forces. (b) dipole–dipole attractions.
(c) self-ionization and ion–ion attractions. (d) hydrogen bonding.

14. The ionization energy of SF_6 is high because of the
(a) low electronegativity of sulfur. (b) highly symmetrical octahedral shape.
(c) high electronegativity of fluorine. (d) large number of unshared electron pairs.
(e) expanded octet of sulfur.

Group VII: The halogens

15. Which of the halogens is found in nature in its elemental form?
(a) fluorine (b) chlorine (c) bromine (d) iodine (e) none of these

16. What is the hybridization of the central iodine atom in I_3^-?
(a) sp^2 (b) sp^3 (c) sp (d) sp^3d^2 (e) sp^3d

17. From the trends in Group VII, predict which of the following does not describe astatine. (Due to its short half-life, the properties of At are not well known.)
(a) lustrous (b) colorless (c) solid
(d) exists as an At_2 molecule (e) forms -1 ion

18. Which compound has the highest lattice enthalpy?
(a) NaCl (b) NaF (c) NaI (d) NaBr

19. How many liters of HF(g) measured at STP can be produced from 25.0 g of CaF_2?
(a) 22.4 L (b) 28.6 L (c) 14.3 L (d) 7.15 L (e) 11.2 L

20. Which hydrogen halide has the highest enthalpy of vaporization?
(a) HI (b) HBr (c) HCl (d) HF

21. Chlorine has oxidation numbers ranging from -1 to $+7$. With what oxidation state is it most likely to participate in disproportionation reactions?
(a) $+1$ (b) -1 (c) $+7$ (d) $+3$

22. What products are formed when $KClO_3$ is heated with no catalyst present?
(a) $KClO_4$, KCl (b) $KClO_2$, KCl (c) KCl, O_2
(d) KCl, ClO_2 (e) No reaction occurs.

23. Perchloric acid is prepared by the aqueous electrolytic oxidation of
(a) ClF_3 (b) ClO_2 (c) Cl_2 (d) ClO_3^- (e) HCl

Group VIII: The noble gases

24. Which noble gas has the highest boiling point?
(a) Kr (b) Ne (c) Ar (d) Xe (e) Rn

25. A superfluid is a liquid
(a) that has zero viscosity. (b) with unusually strong hydrogen bonds.
(c) that will not vaporize or solidify. (d) that can dissolve any known solid.
(e) that reacts with almost all known substances.

26. Three noble gases have been observed to form stable, neutral compounds. Which noble gas is least likely to form such compounds?
(a) He (b) Ne (c) Ar (d) Xe (e) Kr

27. Xenon fluorides are prepared by heating Xe with
(a) HF (b) NaF (c) IF_7 (d) F_2 (e) PtF_4

28. Xenon has been found to form stable bonds with only one element besides fluorine. What element is it?
(a) chlorine (b) carbon (c) oxygen (d) sodium (e) hydrogen

29. All of the following but one are known xenon species. Which one has not been synthesized?
(a) XeO_6^{4-} (b) $HXeO_4^-$ (c) XeO_2 (d) XeO_3

21 THE *d*-BLOCK ELEMENTS

THE *d*-BLOCK ELEMENTS AND THEIR COMPOUNDS

KEY WORDS Define or explain each of the following terms in a written sentence or two.

coinage metals	lanthanide contraction	sandwich compound
ferromagnetism	metallocenes	titanates
hydrometallurgical processes	piezoelectric	transition metal

21.1 TRENDS IN PROPERTIES

KEY CONCEPT A Trends in physical properties

The members of the *d* block of the periodic table are called the **transition metals.** An important aspect of their chemistry is that their *d*-electrons are more strongly bound than their valence *s* electrons. The elements in the last row of the *d* block (from La through Hg) also have chemical and physical properties that are strongly affected by the presence of underlying *f* electrons. The lack of shielding from *d* electrons causes the metallic radii to decrease slightly in going from left to right and the first ionization energies to increase slightly. The trend is not uniform, however. Iron has the smallest radius and highest ionization energy; electron–electron repulsions in doubly occupied *d* orbitals increase the atomic radii and decrease the ionization energies after iron. The trends are approximately the same for the second and third rows of the *d* block. However, the **lanthanide contraction** decreases the metallic radii and increases the ionization energies of the third-row elements, so they are similar to the values in the second row. The lanthanide contraction also results in high densities for the third row of the *d* block because the relatively heavy elements have small radii and correspondingly small volumes.

▼ **EXAMPLE Predicting the radius of a *d*-block element**

The metallic radius of Ni is 124 pm and that of the element directly below it, Pd, is 137 pm. Predict the radius of Pt, the element below Pd.

SOLUTION Owing to the lanthanide contraction, the metallic radius of each of the elements in the third row of the *d* block is very close to that of the element directly above. In fact, from tantalum through mercury, the metallic radius is no more than 3 pm greater than the element above. Lanthanum and hafnium are still only 9 pm and 7 pm greater, respectively, than the elements directly above. We thus predict that the radius of Pt is between 138 and 140 pm, no more than 3 pm greater than that of Pd. (The actual radius is 139 pm.)

EXERCISE The metallic radius of Cu is 128 pm and that of the element directly below it, Ag, is 144 pm. Predict the radius of Au, the element below Ag.

▲ ANSWER: Predicted between 145 and 147 pm; actual 146 pm

KEY CONCEPT B Trends in chemical properties

The *d* block elements display a wide range of oxidation numbers. Most of the elements have only one or two important oxidation numbers—that is, the oxidation numbers that appear in most of their compounds. Elements with oxidation numbers high in their range tend to be good oxidizing agents;

for instance, MnO_4^-, in which Mn has a +7 oxidation number, is a common and important oxidizing agent. Compounds that have elements with oxidation numbers in the middle of their range tend to disproportionate; for example, Cu^+ tends to disproportionate to Cu^0 and Cu^{2+}. Compounds with elements having oxidation numbers low in their range tend to be good reducing agents; for example, Fe^{2+} is a commonly used reducing agent. Most *d* metal oxides are basic, but there is a shift toward acidic oxides as the metal oxidation number increases. The elements on the left of the *d* block resemble their *s* block neighbors whereas those on the right resemble their *p* block neighbors.

EXAMPLE Predicting the chemical properties of the *d*-block elements.

Predict whether V_2O_5 (an orange solid) is a good oxidizing agent, a good reducing agent, or neither. Will it dissolve best in acid or base?

SOLUTION Vanadium has a +5 oxidation number in V_2O_5. According to Figure 21.6 of the text, this is the highest oxidation number that vanadium attains. We thus predict it will be a good oxidizing agent and that it will tend to be an acidic oxide. Because it is an acidic oxide, it will dissolve best in base.

EXERCISE Which of the following is most likely to disproportionate: $Mn^{3+}(aq)$, $Sc^{3+}(aq)$, or $ReO_4^-(aq)$?

ANSWER: $Mn^{3+}(aq)$

21.2 SCANDIUM THROUGH NICKEL

KEY CONCEPT A The elements scandium through manganese

Scandium is highly reactive and has few commercial uses. It exists only with a +3 oxidation number, and its aqueous ion $[Sc(H_2O)_6]^{3+}$ is fairly acidic.

Titanium is a light, strong metal that is passivated by an oxide coat. It appears in its ore with its highest oxidation number (+4) and, as might be expected, is difficult to remove from its ore. The +4 oxidation number is the most stable oxidation number for titanium and appears in its most important compound, TiO_2. This nontoxic compound is used as a pigment in paints. Titanium(IV) forms a series of compounds containing the titanate ion, TiO_3^{2-}. Barium titanate ($BaTiO_3$) is **piezoelectric**, which means that it develops an electric potential when it is mechanically stressed.

Vanadium is a soft, silver-gray metal used in the production of iron alloys. Its most important compound is V_2O_5, which is used as a catalyst in the contact process for sulfuric acid. It forms many different colored compounds and thus finds use in glazes in the ceramics industry.

Chromium is a bright corrosion-resistant metal that forms compounds with many different colors. It is used in the production of stainless steel, as a coating for magnetic tapes (as CrO_2), and for chromium plating. The yellow solid sodium chromate (Na_2CrO_4), in which chromium has a +6 oxidation number, is one of chromium's most useful compounds. Placing the chromate ion in acid solution converts it to the dichromate ion ($Cr_2O_7^{2-}$) with no change in oxidation number:

$$2CrO_4^{2-}(aq) + 2H^+(aq) \longrightarrow Cr_2O_7^{2-}(aq) + H_2O(l)$$

Dichromates are common laboratory oxidizing agents; they are also corrosive and toxic, so they must be handled with care.

Manganese is a gray metal. Because of its lack of resistance to corrosion, it is rarely used alone. It is an important component of many alloys, however. In steel, it removes traces of sulfur by forming a sulfide that is not incorporated into the final steel produced. It is also used to make alloys of nonferrous metals. Manganese displays the oxidation numbers +1 through +7 in its compounds. The most stable oxidation number is +2, but +4 and +7 are also important. MnO_2 takes part in the cathode reaction in the Lelanche dry cell. MnO_2 is also the starting point for the production of many manganese compounds. Potassium permanganate, which contains Mn(VII) in the permanganate ion (MnO_4^-) is an extremely versatile laboratory oxidizing agent in acid solution. (It decomposes in basic solution.) Permanganate ion is reduced to Mn^{2+} when it is used as an oxidizing agent in acid solution:

$$MnO_4^-(aq) + 8H^+(aq) + 5e^- \longrightarrow Mn^{2+}(aq) + 4H_2O(l)$$

EXAMPLE The reactions and properties of some *d*-block elements

Is an aqueous solution of Cr^{3+} likely to be acidic, or neutral?

SOLUTION Cr^{3+} is a small, highly charged ion. As with Al^{3+} and Ti^{3+}, we would expect Cr^{3+} to form a highly hydrated ion in which electrons in the solvating water molecules are polarized, thus making the ion acidic. The equilibrium follows; in fact, Cr^{3+} has a pK_a of 4:

$$[Cr(H_2O)_6]^{3+}(aq) + H_2O(l) \rightleftharpoons [Cr(H_2O)_5(OH)]^{2+}(aq) + H_3O^+(aq)$$

EXERCISE $TiCl_4$ melts at $-23°C$ and boils at $136°C$. Is its bonding covalent or ionic?

ANSWER: Covalent

KEY CONCEPT B The elements iron through nickel

Iron is an important structural metal. It is extracted from its ore in a blast furnace. Calcium oxide (from the decomposition of limestone in the furnace charge) is used to remove silicon, aluminum, and phosphorus impurities from the ore through the reactions

$$CaO(s) + SiO_2(s) \longrightarrow CaSiO_3(l)$$
$$CaO(s) + Al_2O_3(s) \longrightarrow Ca(AlO_2)_2(l)$$
$$6CaO(s) + P_4O_{10}(s) \longrightarrow 2Ca_3(PO_4)_2(l)$$

The first stage in making steel (an alloy of iron) is to lower the carbon content of the pig iron that comes from a blast furnace and to remove the remaining impurities. This is done in a modern version of the **Bessemer converter** using the **basic oxygen process,** in which oxygen and powdered limestone are forced through the molten metal. Iron is not a particularly hard metal. Alloying it makes it harder and improves its resistance to corrosion. Iron, its oxides, and many of its alloys display the property of **ferromagnetism**, the ability to be permanently magnetized. This is accomplished by forcing the magnetic fields of many of the electrons in the iron atoms to become aligned in the same direction, resulting in a large overall magnetic field.

Iron corrodes in moist air. With acids that have nonoxidizing anions (such as HCl), it reacts to form H_2 and a variety of colored salts. Aqueous solutions of Fe^{3+} are acidic:

$$[Fe(H_2O)_6]^{3+}(aq) + H_2O(l) \rightleftharpoons [Fe(H_2O)_5OH]^{2+}(aq) + H_3O^+(aq)$$

Lowering the pH of such a solution nearly to zero forces this equilibrium to the left and tends to stabilize the solvated Fe^{3+} ion. Iron forms a number of compounds in which it has a zero oxidation number; $Fe(CO)_5$ is an example. **Metallocenes** (also known as **sandwich compounds**), in which a metal ion is bonded between two planar cyclopentadienide ion rings, were first made with Fe^{2+} as the bonded ion. Iron is an essential dietary element because it is part of the hemoglobin complex that is responsible for oxygen transport.

Cobalt is used mainly in iron alloys. However, as a component of vitamin B_{12}, it also is an essential dietary element.

Nickel is a hard, silver-white metal that is used mainly in the production of alloys; its most stable oxidation number is +2. Nickel is found mostly as a sulfide ore (NiS). Roasting this ore converts it to NiO. In the **Mond process**, NiO is reduced to Ni with hydrogen; the impure product thus produced is converted to $Ni(CO)_4$ (nickel carbonyl), and the impurities are removed. Pure nickel is then recovered from the nickel carbonyl. Nickel(III) is reduced to nickel(II) in the operation of nickel-cadmium (nicad) batteries.

EXAMPLE The properties of iron and nickel

Explain why iron pentacarbonyl, $Fe(CO)_5$, has such a low boiling point (103°C).

SOLUTION $Fe(CO)_5$ is made up of neutral CO molecules bonded to a neutral Fe atom. No ions are involved. Thus, the intermolecular interactions in this compound do not involve ion–ion interactions. We have not learned how to predict the shape of this compound, but its low boiling point is an indication that it has the highly symmetric trigonal-bipyramidal shape and, therefore, no net dipole moment. This means that the only form of intermolecular attractions present are weak London forces, and the boiling point of the compound is correspondingly low.

EXERCISE How many liters of $CO(g)$ (measured at 50°C and 1.00 atm pressure) are required to convert 125 g Ni to $Ni(CO)_4$ in the Mond process?

$$Ni(s) + 4CO(g) \longrightarrow Ni(CO)_4(g)$$

ANSWER: 226 L

21.3 COPPER, ZINC, AND THEIR CONGENERS

KEY CONCEPT The elements copper through mercury

The elements Cu, Ag, and Au, (with the configuration $d^{10}s^1$) and Zn, Cd, and Hg (with the configuration $d^{10}s^2$) are known as the **coinage metals**. All are relatively unreactive.

Copper ore is separated from unwanted rock using **froth flotation**. The metal is then extracted from its ore using either the **pyrometallurgical process** or the **hydrometallurgical process**. Copper is an excellent electrical conductor and forms useful alloys such as bronze (when alloyed with tin) and brass (when alloyed with zinc). Copper metal does not displace H^+ from acids, but it may be oxidized by acids (such as HNO_3) that have anions that are good oxidizing agents. Copper corrodes in moist air to form $Cu_2(OH)_2CO_3(s)$, a green compound that is apparent on aged decorative copper trim:

$$2Cu(s) + H_2O(l) + O_2(g) + CO_2(g) \longrightarrow Cu_2(OH)_2CO_3(s)$$

Copper forms compounds with an oxidation number of +1, but in water Cu(I) tends to disproportionate to $Cu(s)$ and Cu(II).

Silver lies above H^+ in the electrochemical series (below H^+ in the activity series), so it does not produce H_2 in acid solutions. It is oxidized by nitric acid in a reaction that is reminiscent of copper's reaction with nitric acid:

$$3Ag(s) + 4H^+(aq) + NO_3^-(aq) \longrightarrow 3Ag^+(aq) + NO(g) + 2H_2O(l)$$
$$3Cu(s) + 8H^+(aq) + 2NO_3^-(aq) \longrightarrow 3Cu^{2+}(aq) + 2NO(g) + 4H_2O(l)$$

Aqueous Ag(I), unlike aqueous Cu(I), does not disproportionate, and almost all silver compounds have silver with a +1 oxidation number. AgF and $AgNO_3$ are the only common soluble silver salts. Silver is used extensively in the photographic industry.

Gold is extremely inert. Its most common oxidation number in compounds is +3. Aqueous Au(I) tends to disproportionate into Au(*s*) and Au(III).

Zinc occurs mainly as ZnS. It is obtained from its ore by froth flotation followed by smelting with coke. It is a silvery, reactive amphoteric metal. The oxidation number of zinc in most of its compounds is +2. It reacts with acids to form Zn^{2+} and with alkalis to form $[Zn(OH)_4]^{2-}$:

$$Zn(s) + 2H^+(aq) \longrightarrow Zn^{2+}(aq) + H_2(g)$$
$$Zn(s) + 2OH^-(aq) + 2H_2O(l) \longrightarrow [Zn(OH)_4]^{2-} + H_2(g)$$

Cadmium is similar to zinc but more metallic, and its oxides are more basic than zinc's. Cadmium salts are extremely toxic. The oxidation number of cadmium in most of its compounds is +2.

Mercury occurs as HgS (cinnabar). It is obtained from its ore by froth flotation followed by roasting in air. The volatile metal distills off the roasting mixture as Hg(*g*). It is then condensed to Hg(*l*). Mercury lies above hydrogen in the electrochemical series, so it does not form H_2 when placed in acids. It can be oxidized by nitric acid, in a reaction that is very similar to those we saw earlier for copper and silver:

$$3Hg(l) + 8H^+(aq) + 2NO_3^-(aq) \longrightarrow 3Hg^{2+}(aq) + 2NO(g) + 4H_2O(l)$$

Mercury is found in its compounds with oxidation numbers +1 and +2. The +1 ion is unusual; it is a covalently bonded, diatomic cation $Hg{-}Hg^{2+}$, written as $Hg_2{}^{2+}$. In a reaction that is often used to test chemically for the presence of mercury, Hg_2Cl_2 disproportionates when mixed with aqueous ammonia:

$$Hg_2{}^{2+}(aq) \longrightarrow Hg(l) + Hg^{2+}(aq)$$

Mercury vapors and salts are both toxic.

EXAMPLE The coinage metals

Explain why gold is less reactive than silver.

SOLUTION Because of the lanthanide contraction, gold and silver have the same metallic radii (144 pm). However, gold has 32 more protons in the nucleus (which result in an additional +32 nuclear charge), so its outer electrons are much more tightly bound than silver's. Gold's valence electrons are, therefore, much less easily lost than silver's, so it is much less reactive.

EXERCISE Why don't Zn, Cd, and Hg form compounds with a +3 oxidation number?

ANSWER: The high, poorly shielded nuclear charge holds the *d* electrons too tightly.

COMPLEXES OF THE *d*-BLOCK ELEMENTS

KEY WORDS Define or explain each of the following terms in a written sentence or two.

chiral
coordination sphere
enantiomer
ligand
optical isomer
optically active
plane-polarized light
racemic mixture
stereoisomer
structural isomer

21.4 THE STRUCTURES OF COMPLEXES

KEY CONCEPT The structures of complexes

The Lewis acid in a complex is always a metal cation or metal atom and is frequently referred to as the *central metal ion* or *central metal atom.* The Lewis base, which bonds to the central metal ion or atom, is called a **ligand.** The term **coordination compound** refers to a complex or compound that contains a complex. The ligands that are directly bonded to the central ion are collectively referred to as the **coordination sphere.** Two common and important types of complexes are those that have six ligands in their coordination sphere (a *six-coordinate* complex) and those with four ligands in their coordination sphere (a *four-coordinate* complex). Six-coordinate complexes are usually octahedral, whereas four coordinate complexes are either tetrahedral or square planar.

Ligands are called **monodentate** (one-toothed) if they form only one bond to a central ion in a complex. A **bidentate ligand** can simultaneously form two bonds to the central ion, and a **polydentate ligand** can simultaneously form two or more bonds. A **chelate** compound is a complex that contains a polydentate ligand that forms a ring with the central ion as part of the ring. Polydentate ligands that form stable chelates are called *chelating agents.*

A complex is named by naming the ligands in alphabetical order with a prefix to indicate the number of each ligand, followed by the name of the central metal, including its oxidation state in roman numerals in parentheses. If the whole complex is neutral or positively charged, the metal name used is the same as the element name. If the complex is negatively charged, the metal is named by adding the suffix *-ate* to the stem name of the metal. Complicated ligand names are placed in parentheses, with the prefix *bis-* (rather than di-) to indicate two of them are present or *tris-* (instead of tri-) to indicate three are present. Finally, an ionic coordination compound is named like any other ionic compound, with the name of the cation first, followed by the name of the anion.

▼ **EXAMPLE Naming complexes**

(a) Name $[TiCl_2F_4]^{2-}$, and (b) write the formula of tetraaquaethylenediamminechromium(IV). State the shape of both.

SOLUTION (a) This complex has four fluoride ions and two chloride ions as ligands. Each ligand has a -1 charge, so the ligands contribute a charge of -6 to the complex. Because the total charge of the complex is -2, the titanium must have a $+4$ charge and a $+4$ oxidation number. Because the complex is negatively charged, it has an *-ate* suffix and is named dichlorotetrafluorotitanate(IV) ion. Because there are six ligands, it is a six-coordinate complex and is likely octahedral shaped. (b) This complex has four water molecules (*tetraaqua*) and one ethylenediamine (en) molecule as ligands. All the ligands are neutral, so the total charge on the complex is the same as that of the central metal ion, $+4$. The formula is $[Cr(en)(H_2O)_4]^{4+}$. Because ethylenediamine is a bidentate ligand, the complex is a six-coordinate complex and has an octahedral shape.

EXERCISE (a) Name the complex $[Ni(H_2O)_2(NH_3)_4]^{2+}$, and (b) write the structural formula of monocarbonylpentacyanomolybdate(IV)

▲ ANSWER: (a) Tetraamminediaquanickel(II); (b) $[Mo(CN)_5(CO)]^-$

21.5 ISOMERISM

KEY CONCEPT Isomerism

Isomers are different compounds that have the same molecular formula. There are two broad classes of isomers. **Structural isomers** are isomers in which one or more atoms are linked to different neighbors in different isomers. **Stereoisomers** are isomers in which all the atoms are linked to the same neighbors in different isomers, but the arrangement of atoms in space differs for different isomers. Two types of stereoisomerism are geometrical isomerism and optical isomerism. In **geometrical isomers,** the ligands are arranged such that neighboring ligands are different for the different isomers. The *cis-trans* isomerism that occurs for octahedral complexes of the form MX_4Y_2 is a type of geometrical isomerism. In the *trans* isomer, the two Y ligands are across from each other; in the *cis* isomer, the two Y ligands are next to each other:

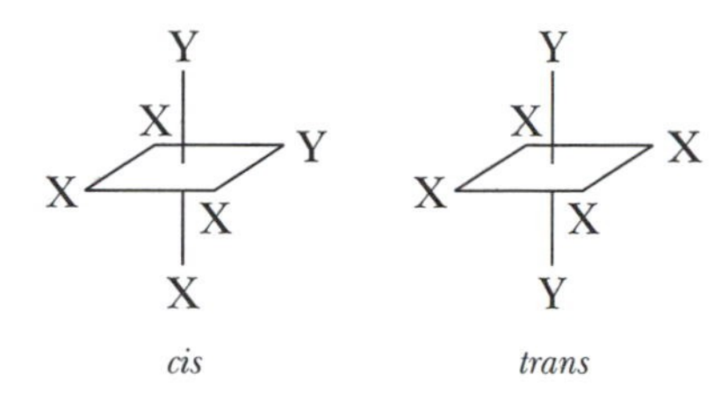

In **optical isomerism,** a molecule or ion and its mirror image are not the same. To understand the relationship of a mirror image to the object making the image when optical isomerism occurs, place your left hand next to your right hand, palm to palm. One hand appears as a mirror image of the other. But are your left and right hands the same? Try putting a left-hand glove on your right hand; clearly, they are not the same. The relationship of a right hand to a left hand is the same as the relationship between two optical isomers. They are mirror images of each other, but different. Optical isomers are not geometric isomers. A molecule that is different from its mirror image is said to be **chiral.** We should be aware that in many cases a molecule is the same as its mirror image and is, therefore, not chiral. Each of the two types of a chiral molecule (the molecule and its mirror image) is called an **enantiomer.** Enantiomers have identical properties except that they rotate the plane of **plane-polarized light** in opposite but equal directions. A molecule or ion that rotates the plane of plane-polarized light is said to be **optically active.** The **(+)-enantiomer** rotates the plane clockwise (as seen by an observer the light beam is aimed at) and the **(−)-enantiomer** rotates the plane counterclockwise by exactly the same amount. When both enantiomers are present in equal concentrations in a solution, polarized light is unaffected because every molecule that accomplishes a clockwise rotation has a mate that undoes that rotation with an equal counterclockwise rotation. Such an mixture of (+)- and (−)-enantiomers is called a **racemic mixture.** Biological syntheses of chiral molecules almost always result in the production of only one enantiomer.

▼ EXAMPLE The chirality of complexes

Is the following octahedral complex chiral?

SOLUTION If the mirror image of the complex is different from the complex, the complex is chiral. Otherwise, it is not chiral. To find out the relationship between the complex and its mirror image, we draw the complex (a) and its mirror image (b) in the following figure. Now, we reorient the mirror image (on paper) to see if it is identical to the original complex. In this case, a 90° clockwise rotation (looking down from the top of the complex) results in (c). A point-by-point comparison indicates that (a) and (c) are identical. Thus, the complex and its mirror image are identical and the complex is not chiral.

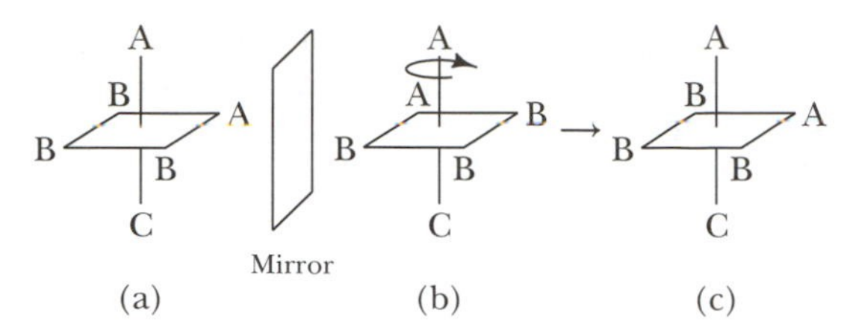

EXERCISE Is the following octahedral complex chiral?

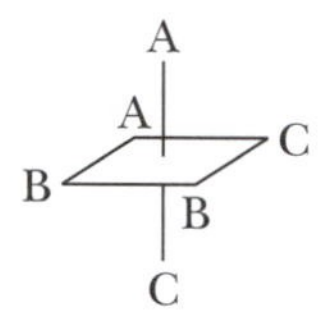

ANSWER: Yes

CRYSTAL FIELD THEORY

KEY WORDS Define or explain each of the following terms in a written sentence or two.

complementary color
diamagnetic
e orbital
high-spin complex
ligand field splitting
low-spin complex
paramagnetic
spectrochemical series
strong field ligand
t orbital
weak-field ligand

21.6 THE EFFECTS OF LIGANDS ON *d* ELECTRONS

KEY CONCEPT The effects of ligands on *d* electrons

The starting point of **crystal field theory** is the assumption that the ligands in a complex have bonded to and oriented around the central metal atom. We then ask what effects the ligands have on the central metal *d* electrons. The ligands are often negatively charged or, at the very least, have a large electron cloud directed at the central atom. In an octahedral complex, two of the five *d* orbitals point directly toward one or more ligands, and three of the five point between the ligands. Electrons in the two orbitals that point directly at the ligands experience repulsions from the electrons on the ligands and have a higher energy than they would if they were uniformly dispersed around the nucleus. These higher energy orbitals are called ***e* orbitals.** Electrons in the three orbitals that point between the ligands have a lower energy than they would if they were uniformly dispersed around the nucleus. These lower energy orbitals are called ***t* orbitals.** This results in the following orbital diagram.

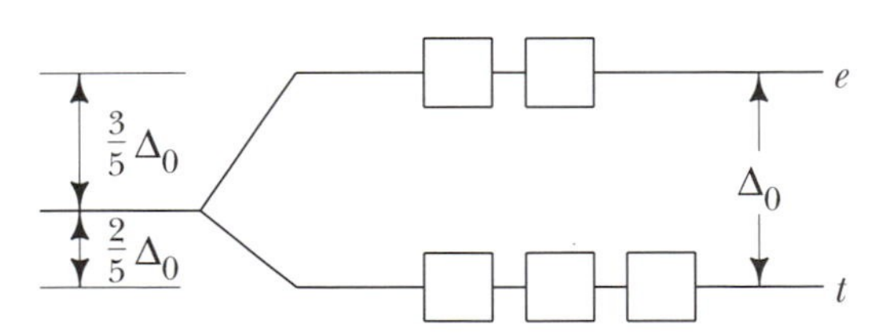

The difference in energy between the two sets of orbitals is called the **ligand field splitting,** symbolized Δ_0. Electrons in the lower energy orbitals can absorb a photon with energy $E = \Delta_0$, which often results in a colored complex. Using the relationship $E_{photon} = h\nu$, we find that the wavelength λ of the photon is related to Δ_0 through the relationship

$$\Delta_0 = hc/\lambda$$

Some ligands, called **weak-field ligands,** produce a small Δ_0 when they form a complex with a metal. Others, called **strong-field ligands,** result in a large Δ_0. The **spectrochemical series** is a list of ligands ordered according to the value of Δ_0 they typically cause in a complex. The color of complexes is caused by the absorption of specific wavelengths of light corresponding to Δ_0. Let's assume that white light, a mixture of photons of all colors with wavelengths ranging from about 420 nm (violet) to about 700 nm (red), impinges on a complex. If the complex absorbs photons of only one color, the transmitted light will no longer be white but will manifest the fact that one of the colors of light has been absorbed. The following table (and the color wheel in Figure 21.28 of the text) indicate the relationship between the light absorbed and the color of the complex (the light transmitted). The wavelengths of absorbed light are approximate.

Color of light absorbed	Color of complex
violet (420 nm)	yellow
blue (460 nm)	orange
blue-green (490 nm)	red
green (520 nm)	purple
yellow-green (560 nm)	red-violet
yellow (570 nm)	violet
orange (600 nm)	blue
red (700 nm)	green

The corresponding colors in the two columns are called **complementary colors.** However, we must be careful in interpreting the colors of complexes because complexes usually absorb a wide range of colors, not a single wavelength.

▼ EXAMPLE The absorption of light by complexes

$[Co(NH_3)_6]^{3+}$ displays an absorption maximum at 437 nm. What is Δ_0 (in kJ/mol) for the complex? What color is the complex?

SOLUTION We use the relationship $\Delta_0 = hc/\lambda$ to calculate Δ_0. Substituting the values $h = 6.63 \times 10^{-34}$ J·s, $c = 3.00 \times 10^8$ m/s and $\lambda = 437$ nm (437×10^{-9} m) gives

$$\Delta_0 = \frac{(6.63 \times 10^{-34}\ \text{J}\cdot\cancel{\text{s}})(3.00 \times 10^8\ \cancel{\text{m}/\text{s}})}{437 \times 10^{-9}\ \cancel{\text{m}}}$$
$$= 4.55 \times 10^{-19}\ \text{J}$$

This is the energy absorbed by a single complex that absorbs a single photon. For a mole of the complex, we must multiply this by Avogadro's number, 6.02×10^{23}:

$$\Delta_0\ (\text{kJ/mol}) = (4.55 \times 10^{-19}\ \text{J})(6.02 \times 10^{23})$$
$$= 2.74 \times 10^5\ \text{J/mol}$$
$$= 274\ \text{kJ/mol}$$

From the preceding table, a complex that absorbs light at 437 nm is yellow-orange.

EXERCISE $[CrF_6]^{3-}$ displays an absorption maximum at 658 nm. What is Δ_0 (in kJ/mol) for the complex? What color is the complex?

▲ ANSWER: 182 kJ/mol; blue-green

21.7 THE ELECTRONIC STRUCTURES OF MANY-ELECTRON COMPLEXES

KEY CONCEPT The electronic structure of many-electron complexes

To build up the valence-electron configuration of the central metal electrons in the *t* and *e* orbitals of a complex, we must add electrons to the orbitals so as to achieve the lowest possible energy. Each orbital can hold a maximum of two electrons, and the two electrons in an orbital must have paired spins. Hund's rule applies, but it must be applied properly: When Δ_0 is large, the low-energy *t* orbitals fill before any electrons enter the *e* orbitals. Hund's rule is applied to the lower *t* orbitals first and then, after they fill, to the upper *e* orbitals. When Δ_0 is small, electrons enter all five orbitals (following Hund's rule) before any pairing of electrons begins. The filling of orbitals for octahedral complexes is as follows:

	Large Δ_o		Small Δ_o	
	t-orbitals	*e*-orbitals	*t*-orbitals	*e*-orbitals
d^1	[↑] [ ] [ ]	[ ] [ ]	[↑] [ ] [ ]	[ ] [ ]
d^2	[↑] [↑] [ ]	[ ] [ ]	[↑] [↑] [ ]	[ ] [ ]
d^3	[↑] [↑] [↑]	[ ] [ ]	[↑] [↑] [↑]	[ ] [ ]
d^4	[↑↓] [↑] [↑]	[ ] [ ]	[↑] [↑] [↑]	[↑] [ ]
d^5	[↑↓] [↑↓] [↑]	[ ] [ ]	[↑] [↑] [↑]	[↑] [↑]
d^6	[↑↓] [↑↓] [↑↓]	[ ] [ ]	[↑↓] [↑] [↑]	[↑] [↑]
d^7	[↑↓] [↑↓] [↑↓]	[↑] [ ]	[↑↓] [↑↓] [↑]	[↑] [↑]
d^8	[↑↓] [↑↓] [↑↓]	[↑] [↑]	[↑↓] [↑↓] [↑↓]	[↑] [↑]
d^9	[↑↓] [↑↓] [↑↓]	[↑↓] [↑]	[↑↓] [↑↓] [↑↓]	[↑↓] [↑]
d^{10}	[↑↓] [↑↓] [↑↓]	[↑↓] [↑↓]	[↑↓] [↑↓] [↑↓]	[↑↓] [↑↓]

It should be noted that for the d^1, d^2, d^3, d^8, d^9, and d^{10} configurations, the size of Δ_0 does not affect the electron configuration. But for the d^4, d^5, d^6, and d^7 configurations, different electron configurations are obtained for the different sizes of Δ_0. For octahedral d^4 through d^7 complexes, the configurations obtained with a large Δ_0 minimize the number of unpaired electrons and are called **low-spin complexes.** Conversely, octahedral d^4 through d^7 complexes with a small Δ_0 have a maximum of unpaired electrons and are called **high-spin complexes.** Unpaired electrons make a species **paramagnetic,** and as we can see from the configurations, many types of complexes can have unpaired electrons and are, therefore, paramagnetic. Species with no unpaired electrons are said to be **diamagnetic.** Some of the electron configurations shown would result in diamagnetic complexes.

EXAMPLE Predicting the properties of a complex

Predict whether $[Co(CN)_6]^{3-}$ is a high-spin or low-spin complex. How many unpaired electrons does it possess? Is it paramagnetic or diamagnetic?

SOLUTION We must first determine how many *d* electrons are present in the complex. Because each CN^- contributes a -1 charge to the complex, the cobalt must have a $+3$ charge. A neutral cobalt atom has an $[Ar]3d^74s^2$ configuration, so Co^{3+} possesses a d^6 configuration. CN^- is very high in the spectrochemical series; thus, we expect a very large Δ_O. This, in turn, leads to a low-spin complex. A low-spin, d^6 complex has six electrons in the three low-energy *t* orbitals; so all the electrons are paired and the complex is diamagnetic.

EXERCISE Predict whether $[Fe(SCN)_6]^{3-}$ is a high-spin or low-spin complex. How many electrons does it possess? Is it paramagnetic or diamagnetic?

ANSWER: High-spin, d^5 complex; five unpaired electrons; paramagnetic

SELF-TEST EXERCISES

The *d*-block elements and their compounds

1. What electron configuration should lead to the largest metallic radius?
(a) $[Ar]3d^14s^2$ (b) $[Kr]4d^15s^2$ (c) $[Kr]4d^55s^1$ (d) $[Ar]3d^{10}4s^2$ (e) $[Kr]4d^75s^1$

2. Which element is expected to have a metallic radius nearest to that of Pd?
(a) Ni (b) Pt (c) Ag (d) Ru (e) Co

3. Five transition metal species are shown along with the range of oxidation numbers the metal displays in its compounds. Which species is likely to be the best oxidizing agent?
(a) Fe^{2+} ($+2 \longrightarrow +6$) (b) MnO_2 ($+1 \longrightarrow +6$) (c) V_2O_5 ($+1 \longrightarrow +5$)
(d) NiO ($+2 \longrightarrow +4$) (e) Sc^{3+} ($+3$ only)

4. Which oxide of vanadium is most acidic?
(a) V_2O_3 (b) VO (c) VO_2 (d) V_2O_5

5. Judging by electron configuration and position in the periodic table, which of the following would you expect to be, overall, the most reactive?
(a) Sc (b) Ti (c) V (d) Mn (e) Fe

6. The Sc^{3+} ion is
(a) basic. (b) neutral. (c) acidic.

7. Which compound of chromium is a common laboratory oxidizing agent?
(a) $CrCl_3$ (b) $K_4[Cr(CN)_6]$ (c) Cr_2O_3 (d) $KCrOF_4$ (e) $K_2Cr_2O_7$

8. In its compounds manganese displays oxidation states ranging from $+1$ through $+7$. Which manganese species is most likely to disproportionate?
(a) Mn(*s*) (b) MnO_4^{2-} (c) Mn^{3+} (d) MnO_4^-

9. A ferromagnetic material is one that
(a) can permanently magnetize iron. (b) can be permanently magnetized.
(c) can be chemically reduced by Fe^{2+}. (d) can be chemically oxidized by Fe^{3+}.
(e) has no unpaired electrons.

10. Which element is biologically important in oxygen transport in mammals?
(a) Zn (b) Fe (c) Se (d) Cr (e) Au

11. What products are obtained when NiS is roasted in air?
(a) Ni, SO_2 (b) Ni, H_2S (c) $NiSO_4$ (d) NiO_2, SO_3 (e) NiO, SO_2

12. What is the oxidation state of nickel in $Ni(CO)_4$?
(a) 1 (b) 3 (c) 0 (d) 4 (e) 2

13. The low reactivity of the coinage metals is due to
(a) the high shielding ability of *s* electrons.
(b) their large number of oxidation states.
(c) the poor shielding ability *d* electrons.
(d) their high electrical conductivity.
(e) the lanthanide contraction.

14. How many milliliters of H_2 (measured at STP) are required to reduce the Cu^{2+} found in 100 mL of 0.250 M $CuCl_2$?

$$Cu^{2+}(aq) + H_2(g) \longrightarrow Cu(s) + 2H^+(aq)$$

(a) 422 mL (b) 220 mL (c) 900 mL (d) 1.23×10^4 mL (e) 560 mL

15. All of the statements that follow except one describe gold. Which one does not describe gold?
(a) reduces H^+ to form H_2 (b) excellent conductor (c) yellow color
(d) dissolves in aqua regia (e) forms complexes

16. What is the formula of mercury(I) chloride?
(a) HgCl (b) Hg_2Cl_2 (c) Hg_2Cl (d) $HgCl_2$

Complexes of the *d*-block elements

17. A complex that has a small K_f but for which the rate of decomposition is very slow is
(a) stable and labile.
(b) unstable and labile.
(c) stable and nonlabile.
(d) unstable and nonlabile.

18. All of the following but one describe complexes in general. Which one does not?
(a) colored (b) used in analysis (c) magnetic
(d) biologically irrelevant (e) undergo substitution reactions

19. Which of the following can act as a ligand?
(a) H_2O (b) Fe (c) Ca^{2+} (d) C_2H_2 (e) CH_4

20. What is the name of $[CoCl(NH_3)_5]^{2+}$?
(a) pentaammoniamonochlorocobaltate
(b) pentaammoniamonochlorocobalt(II)
(c) pentamminechlorocobalt(II)
(d) pentamminechlorocobalt(III)

21. What is the formula (including the charge) of aquatribromoplatinate(II) ion?
(a) $[PtBr_3(H_2O)]^-$ (b) $[Pt(Br_2)_3(H_2O)]^{2+}$ (c) $[PtBr_2(H_2O)]^0$
(d) $[PtBr_3(H_2O)_3]^-$ (e) $[PtBr_3(H_2O)]^{2+}$

22. Which of the following represents a *cis* isomer?

(a) B; B, B, A, A; B
(b) A; B, B, B, B; A
(c) B; B, A, A, B; B

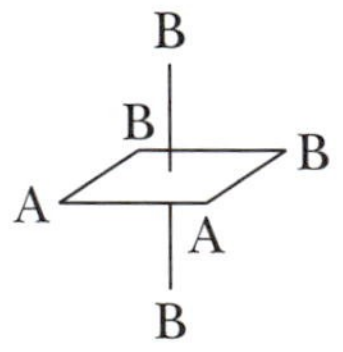

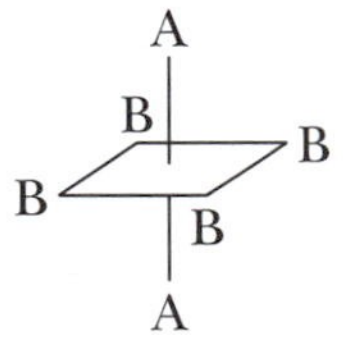

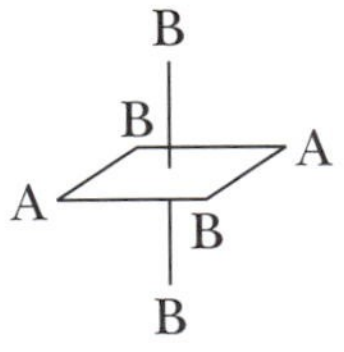

23. Which of the following is chiral?

(a) C; B, B, A, A; C
(b) C; C, B, A, B; A
(c) C; B, A, A, B; C

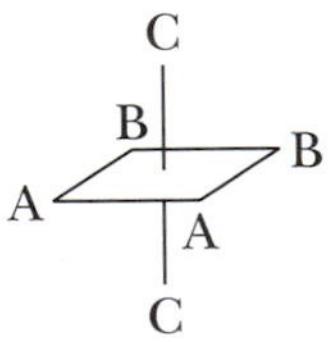

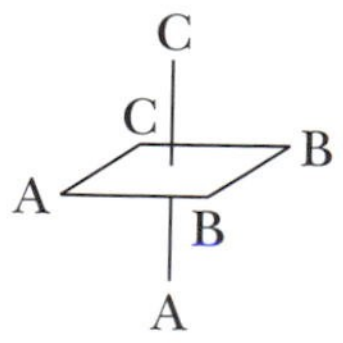

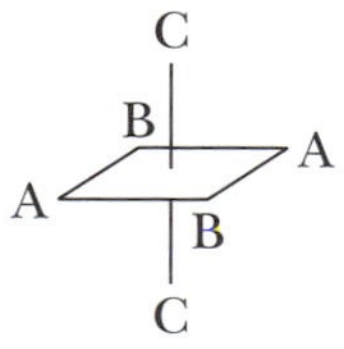

24. Which is not optically active?
(a) a 1:1 mixture of (+)- and (−)-enantiomer
(b) a solution of (+)-enantiomer
(c) a solution of (−)-enantiomer
(d) a 1:2 mixture of (+)- and (−)-enantiomer
(e) a 2:1 mixture of (+)- and (−)-enantiomer

Crystal field theory

25. What is Δ_0 (in kJ/mol) for a complex that absorbs light at 425 nm?
(a) 321 kJ/mol (b) 4.68×10^{-9} kJ/mol (c) 3.11×10^{-19} kJ/mol
(d) 355 kJ/mol (e) 282 kJ/mol

26. What color would you expect for a complex with $\Delta_0 = 222$ kJ/mol?
(a) red (b) green-yellow (c) red-violet (d) green (e) red-orange

27. Which of the following has the largest Δ_0
(a) $[Fe(H_2O)_6]^{3+}$ (b) $[Fe(SCN)_6]^{4-}$ (c) $[Co(CN)_6]^{3-}$
(d) $[AuCl_4]^-$ (e) $[MnF_6]^{2-}$

28. Which ligand is most likely to result in the formation of a high-spin complex?
(a) SCN^- (b) CN^- (c) CO (d) H_2O (e) NH_3

29. How many unpaired electrons are present in $[Fe(CN)_6]^{3-}$?
(a) 0 (b) 1 (c) 3 (d) 4 (e) 5

30. Which type of complex is diamagnetic?
(a) d^6 (high spin) (b) d^5 (high spin) (c) d^3
(d) d^6 (low spin) (e) d^4 (low spin)

22 NUCLEAR CHEMISTRY

NUCLEAR STABILITY

KEY WORDS Define or explain each of the following terms in a written sentence or two.

α particle
β particle
daughter nucleus
magic numbers
nucleon
nucleosynthesis
nuclide
positron
radioactive element
radioactive series

22.1 NUCLEAR STRUCTURE AND NUCLEAR RADIATION

KEY CONCEPT A Recognizing isotopes and nuclides

Isotopes are atoms with the same atomic number but different mass numbers. For instance, carbon-12 and carbon-14 both have an atomic number of 6, but they have different mass numbers, so both are isotopes of carbon. A **nuclide** is a specific isotope of an element: carbon-12 is a nuclide, carbon-14 is a nuclide, and vanadium-54 is a nuclide. Approximately 850 nuclides are known. A common way of representing a nuclide is with the symbol ${}^{A}_{Z}\text{E}$ where E is the element symbol of the nuclide, A is the mass number, and Z is the atomic number. (The symbolism is also used for fundamental particles, in which case Z is the charge on the particle.) Because the mass number equals the total number of neutrons and protons (the total number of **nucleons**), $A - Z$ equals the number of neutrons in the nucleus of the nuclide. The symbol ${}^{A}\text{E}$ is also used to represent a nuclide because writing the atomic number is redundant; once the element symbol is known, the atomic number is known.

EXAMPLE Determining the number of nucleons in a nuclide

How many neutrons, protons, and nucleons are there in an atom of sodium-25. What is the symbol for this nuclide?

SOLUTION The number of protons is determined by the identity of the element. Sodium has an atomic number of 11, so it has 11 protons ($Z = 11$). The mass number of this nuclide is 25. This equals the sum of the number of neutrons plus protons, so the number of neutrons is $25 - 11$ or 14. The protons and neutrons are the nucleons in the atom, so there are 25 nucleons. The symbol for the nuclide is ${}^{25}_{11}\text{Na}$.

EXERCISE How many neutrons, protons, and nucleons are there in an atom of rubidium-96? What is the symbol for the nuclide?

ANSWER: 59 neutrons, 37 protons, 96 nucleons; ${}^{96}_{37}\text{Rb}$

KEY CONCEPT B Radiation from nuclei

The term *radiation* refers to particles and/or photons that are emitted from the nuclei of atoms. (X-rays are an exception because the extranuclear electrons are involved in their production.) There are four major types of radiation. Three of them (alpha, beta, and positron) occur through the emission of a particle from the nucleus. The last (gamma) corresponds to the emission of a high-energy photon from a nucleus in an excited state. The different radiations were first characterized by their behavior as they travel between electrically charged plates of opposite charge.

Positively charged particles are attracted by the negative plate and repelled by the positive plate, whereas negatively charged particles are attracted by the positive plate and repelled by the negative one. Table 22.1 of the text lists some properties of various forms of radiation. It should be noted that physicists often use the term *electron* and the symbol β^- for what we call beta emission.

EXAMPLE Distinguishing the different types of radiactive emissions

Comment on the relative direction and extent of deflection experienced by α particles and β particles as they travel between electrically charged plates of opposite charge.

SOLUTION Alpha particles are relatively massive (mass number = 4) and have a +2 charge. They are attracted by the negative plate and repelled by the positive plate, but they are not deflected much because of their large mass. Beta particles have very little mass (mass number = 0) and a negative charge. They are attracted by the positive plate and repelled by the negative plate; they undergo relatively large deflections because of their low mass.

EXERCISE Comment on the direction and extent of deflection of a γ-ray photon as it passes between electrically charged plates of opposite charge.

ANSWER: No deflection occurs.

KEY CONCEPT C Nuclear disintegration

The emission of a particle (other than a neutron) from the nucleus of an atom results in a change in the atomic number of the nuclide involved. The new nuclide that results is called the **daughter nucleus,** or daughter nuclide, and the conversion of one element into another is called **nuclear transmutation.** The daughter nucleus is usually formed in an excited state and emits one or more γ-ray photons as it relaxes to its ground state.

EXAMPLE 1 Recognizing a nuclear transmutation

Does the equation that follows represent a nuclear transmutation? Explain your answer.

$$^{239}_{93}\text{Np} \longrightarrow {}^{239}_{94}\text{Pu} + \beta$$

SOLUTION The equation shows the conversion of a neptunium-239 nucleus into a plutonium-239 nucleus through the emission of a β particle (or 1 mol neptunium-239 into 1 mol plutonium-239 through the emission of 1 mol of β particles). Because the equation represents the conversion of one element into another, it is a nuclear transmutation.

EXERCISE What is the daughter nuclide in the transmutation equation given in the example?

ANSWER: Plutonium-239

EXAMPLE 2 Understanding and calculating the wavelength of γ radiation

The nuclear disintegration in the preceding example includes the emission of a γ-ray photon of energy 8.88×10^{-14} J. What is the origin of the γ radiation?

SOLUTION The plutonium nucleus is formed in an excited state (represented in the following equation by an asterisk). When it relaxes to the ground state, it emits a γ-ray photon:

$$^{239}_{94}\text{Pu}^* \longrightarrow {}^{239}_{94}\text{Pu} + \gamma$$

EXERCISE What is the wavelength, in picometers, of the γ-ray photon emitted? The energy of the photon is given in this example; $h = 6.63 \times 10^{-34}$ J·s; $c = 3.00 \times 10^8$ m/s.

ANSWER: 2.24 pm

22.2 THE IDENTITIES OF DAUGHTER NUCLIDES

KEY CONCEPT Alpha and beta disintegration, electron capture, and positron emission

The three radioactive particle emissions and electron capture result in different types of change in the nucleus that undergoes transmutation. These are summarized in the table.

Event	Change in number of protons	Change in number of neutrons	Change in mass number
α emission	decrease by 2	decrease by 2	decrease by 4
β emission	increase by 1	decrease by 1	no change
positron emission	decrease by 1	increase by 1	no change
electron capture	decrease by 1	increase by 1	no change

When the original nuclide and the type of reaction are known, the correct product of a nuclear reaction can always be determined by writing a balanced nuclear equation. A balanced nuclear equation is one in which the sum of the mass numbers on the left equals the sum of the mass numbers on the right, and the sum of charges on the left equals the sum of the charges on the right. Practically, this means that the sum of the superscripts must be the same on the left as on the right and that the sum of the subscripts must be the same on the left as on the right. For instance, for the equation

$${}^{133}_{53}\text{I} \longrightarrow {}^{0}_{-1}\text{e} + {}^{133}_{54}\text{Xe}$$

Mass balance: $133 = 133 + 0$ Charge balance: $53 = -1 + 54$

It should be recognized that the electron emitted as a β particle is emitted from the nucleus; it is not one of the extranuclear electrons. However, electron capture involves the capture of one of the atom's extranuclear *s* electrons by the nucleus.

EXAMPLE 1 Identifying the element formed by α decay

What element is formed when polonium-210 undergoes α decay?

SOLUTION We first formulate the problem by recognizing that, for α decay, an α particle is one of the products of the nuclear reaction. We thus have

$${}^{210}_{84}\text{Po} \longrightarrow {}^{A}_{Z}\text{E} + {}^{4}_{2}\text{He}$$

To achieve mass balance, the sum of the superscripts on the left side of the equation must equal the sum of the superscripts on the right side. Thus, $A = 206$ $(210 = 206 + 4)$. Similarly, for charge balance the sum of the subscripts on the left side of the equation must equal the sum of the subscripts on the right. Thus, $Z = 82$ $(84 = 82 + 2)$. The daughter nuclide formed is lead-206:

$${}^{210}_{84}\text{Po} \longrightarrow {}^{206}_{82}\text{Pb} + {}^{4}_{2}\text{He}$$

EXERCISE What element is formed when neptunium-237 undergoes α decay?

ANSWER: Protactinium-233

EXAMPLE 2 Identifying the element formed by β decay

What daughter nuclide is formed when actinium-227 undergoes β decay?

SOLUTION We formulate the problem by recognizing that a β particle (${}^{0}_{-1}\text{e}$) is one of the products of the nuclear reaction:

$${}^{227}_{89}\text{Ac} \longrightarrow {}^{A}_{Z}\text{E} + {}^{0}_{-1}\text{e}$$

Mass balance requires that $A = 227$ and charge balance requires that $Z = 90$ ($89 = 90 - 1$). Thus the daughter nuclide is thorium-227:

$$^{227}_{89}\text{Ac} \longrightarrow {}^{227}_{90}\text{Th} + {}^{0}_{-1}\text{e}$$

EXERCISE What daughter nuclide is formed when tantalum-186 undergoes β decay?

ANSWER: Tungsten-186

EXAMPLE 3 Identifying the element formed by electron capture

What daughter nuclide is formed when iridium-186 undergoes electron capture?

SOLUTION We formulate the problem by recognizing that for electron capture, an electron is one of the reactants. We must determine E in the equation

$$^{186}_{77}\text{Ir} + {}^{0}_{-1}\text{e} \longrightarrow {}^{A}_{Z}\text{E}$$

Mass balance requires that $186 + 0 = A$, so $A = 186$. Charge balance requires that $77 + (-1) = Z$, so $Z = 76$. The daughter nuclide is osmium-186:

$$^{186}_{77}\text{Ir} + {}^{0}_{-1}\text{e} \longrightarrow {}^{186}_{76}\text{Os}$$

EXERCISE What daughter nuclide results from electron capture by xenon-116?

ANSWER: Iodine-116

EXAMPLE 4 Identifying the nuclide formed by positron emission

What daughter nuclide is formed when bismuth-192 undergoes positron emission?

SOLUTION We formulate the problem by recognizing that a positron (${}^{0}_{1}\text{e}$) must be one of the products of the nuclear reaction when positron emission occurs. The problem is to determine E in the equation

$$^{192}_{83}\text{Bi} \longrightarrow {}^{A}_{Z}\text{E} + {}^{0}_{1}\text{e}$$

Mass balance requires that $192 = A + 0$, so $A = 192$. Charge balance requires that $83 = Z + 1$, so $Z = 82$. The daughter nuclide is lead-192:

$$^{192}_{83}\text{Bi} \longrightarrow {}^{192}_{82}\text{Pb} + {}^{0}_{1}\text{e}$$

EXERCISE What daughter nuclide results from positron emission from tin-111?

ANSWER: Indium-111

EXAMPLE 5 Identifying the type of transmutation that occurs

What process occurs when platinum-197 forms the daughter nuclide gold-197?

SOLUTION In this case we know the original nuclide and the daughter nuclide and must determine E in the equation

$$^{197}_{78}\text{Pt} \longrightarrow {}^{A}_{Z}\text{E} + {}^{197}_{79}\text{Au}$$

Mass balance requires that $197 = A + 197$, so $A = 0$. Charge balance requires that $78 = Z + 79$, so $Z = -1$. E is a β particle, so the process that occurs is β emission. Another approach is to realize that the process in which the number of protons increases by 1 and the mass number stays the same is β emission.

EXERCISE What process transmutes strontium-83 to rubidium-83?

ANSWER: Electron capture or positron decay

22.3 THE PATTERN OF NUCLEAR STABILITY

KEY CONCEPT The band of stability and sea of instability

The underlying reasons that explain why some nuclides are stable and others are radioactive are not well understood; however, an important empirical observation allows us to organize our thinking. If we plot all of the stable (nonradioactive) nuclei, with atomic number on the x axis and mass number on the y axis, we find that the stable nuclei lie in a narrow band (the **band of stability**) that ends at $Z = 83$. The inference that can be made from the existence of this band is that the relative number of protons and neutrons helps to determine the stability of nuclei. A nucleus may be unstable because it has too many protons (is proton rich) or too many neutrons (is neutron rich). The fact that the band of stability ends at $Z = 83$ informs us that nuclei with more than 83 protons are unstable, suggesting that to be stable a nucleus cannot be too massive. The nuclides that lie outside the band of stability are said to be in the **sea of instability.** In Figure 22.9 of the text, nuclides to the upper left of the band of stability are neutron rich, those to the lower right are proton rich, and those to the right of $Z = 83$ are too massive to be stable. Nuclides that are too massive *tend* to decay by α emission; those that are neutron rich, by β emission; and those that are proton rich, by electron capture or positron emission.

Type of nuclide	Most common type of decay	Result of decay
too massive	α decay	mass number decreases by 4
neutron rich	β decay	Z increases, number of neutrons decreases
proton rich	electron capture or positron emission	Z decreases, number of neutrons increases

EXAMPLE 1 Predicting the mode of decay

What is a likely mode of decay for $^{236}_{92}U$?

SOLUTION For nuclides with $Z > 83$, α decay is a common and likely mode of decay. Without information to the contrary, we assume that uranium-236 undergoes α decay.

EXERCISE Which of the following is most likely to undergo α decay?

$$^{242}_{96}Pu \qquad ^{188}_{79}Au \qquad ^{165}_{69}Tm$$

ANSWER: Plutonium-242

EXAMPLE 2 Predicting the mode of decay

What is a likely mode of decay for $^{176}_{79}Au$? (Use Figure 22.9 of the text for help.)

SOLUTION Plotting this nuclide on Figure 22.9 of the text shows that it lies below the band of stability. Nuclides in this area of the plot have too many protons relative to the number of neutrons present; that is, if the number of protons is decreased, the point for the nuclide moves toward the band of stability. (Check this for yourself!) Nuclides that are proton rich may decay by either electron capture or positron emission. In a sample of gold-176, some nuclei will decay by positron emission, some by electron capture.

EXERCISE What is a likely mode of decay for $^{201}_{78}Pt$?

ANSWER: β emission

PITFALL Do all nuclides with $Z > 83$ undergo α emission?

Not all nuclides with $Z > 83$ undergo α emission as their mode of decay. Even though α emission is quite common for massive nuclides, it is not the only mode of decay. For example, $^{245}_{94}Pu$ and $^{237}_{92}U$ both decay entirely by β emission.

22.4 NUCLEOSYNTHESIS

KEY CONCEPT Neutron and charged-particle induced transmutation

Nucleosynthesis, as the word implies, is the formation (synthesis) of elements. Because neutrons have zero charge, they can approach and enter a positively charged nucleus with relative ease. A neutron entering a nucleus does not result in a new element but forms a new isotope of the original element. However, the nuclide formed frequently undergoes radioactive decay that results in a new element. As an example, bombarding iron-54 with a neutron results in the emission of a proton and the formation of manganese-54:

$$^{54}_{26}Fe + ^{1}_{0}n \longrightarrow ^{54}_{25}Mn + ^{1}_{1}H$$

A nucleosynthesis reaction is frequently written in a shortened form, for example:

Starting nuclide (bombarding particle, particle emitted) new nuclide

$$^{54}_{26}Fe(^{1}_{0}n,\ ^{1}_{1}H)^{54}_{25}Mn$$

High-energy charged particles may also be used to accomplish nucleosynthesis reactions.

EXAMPLE 1 Neutron induced transmutation

When $^{238}_{92}U$ is bombarded with a neutron, the product formed undergoes β decay. What is the final product that results? Write the nuclear equation in the conventional shortened form.

SOLUTION The reactants in this nuclear reaction are uranium-238 and a neutron. The products are a β particle and the unknown product E:

$$^{238}_{92}U + ^{1}_{0}n \longrightarrow ^{0}_{-1}\beta + ^{A}_{Z}E$$

Mass number balance requires that $238 + 1 = 0 + A$, so $A = 239$. Charge balance requires that $92 + 0 = -1 + Z$, so $Z = 93$. E is neptunium-239, $^{239}_{93}Np$. The starting nuclide is uranium-238, the bombarding particle is a neutron, the particle emitted is a β particle, and the nuclide formed is neptunium-239. The shortened form of the nuclear equation is

$$^{238}_{92}U(^{1}_{0}n,\ ^{0}_{-1}\beta)^{239}_{93}Np$$

EXERCISE What nuclide is produced in the atmosphere through the bombardment of nitrogen-14 by high-energy neutrons? A proton ($^{1}_{1}H$) is emitted as part of the overall nuclear reaction. Also, write the nucleosynthesis equation in the conventional shortened form.

ANSWER: $^{14}_{6}C$; $^{14}_{7}N(^{1}_{0}n,^{1}_{1}H)^{14}_{6}C$

EXAMPLE 2 Charged-particle induced transmutation

What is E in the following nucleosynthesis?

$$^{27}_{13}Al(^{4}_{2}He,^{1}_{0}n)^{A}_{Z}E$$

SOLUTION Mass number balance requires that $27 + 4 = 1 + A$, so $A = 30$. Charge balance requires $13 + 2 = 0 + Z$, so $Z = 15$. Therefore, E is phosphorus-30.

EXERCISE What is the bombarding particle in the following nucleosynthesis?

$$^{246}_{96}\text{Cm}(\text{E},5^{1}_{0}\text{n})^{254}_{102}\text{No}$$

ANSWER: $^{13}_{6}C$

RADIOACTIVITY

KEY WORDS Define or explain each of the following terms in a written sentence or two.

activity
becquerel
curie
decay constant
dose
dose equivalent
rad
relative biological effectiveness
rem
scintillation counter

22.5 MEASURING RADIOACTIVITY

KEY CONCEPT A Differences in penetrating power

The penetrating power of a radioactive particle depends on (a) its charge, (b) its mass, and (c) its energy. An α particle, with its high $+2$ charge and large mass, interacts strongly with any matter it enters and does not penetrate very far. A sheet of aluminum will stop all but the most energetic α particles. Even in air, a typical α particle may travel only a few centimeters or tens of centimeters before stopping. However, the powerful interaction of α particles with matter makes them particularly dangerous if they are able to reach radiation-sensitive internal organs or bone marrow. A β particle, with its -1 charge and small mass, interacts less effectively with matter and is able to penetrate deeper into a substance than is an α particle of equal energy. For particles of equal energy, a β particle penetrates matter 500 to 1000 times as far as an α particle. A γ-ray photon has zero mass and zero charge and, therefore, interacts far less with matter than an α particle or β particle of equal energy; γ-ray photons are thus very penetrating. Fallout shelters are constructed with thick concrete walls in order to stop highly penetrating γ radiation.

EXAMPLE Explaining penetrating power

Predict the penetrating power of a high-energy proton relative to that of an α particle of equal energy.

SOLUTION A proton has a $+1$ charge, which is less than the $+2$ charge on an α particle. We predict that for particles of equal energy, a proton is more penetrating than an α particle.

EXERCISE How does the penetrating power of a positron compare with that of a β particle?

ANSWER: They are about the same.

KEY CONCEPT B The activity of radioactive sources

The activity of a radioactive source is measured by the number of nuclear disintegrations that occur per second. The number of disintegrations per second can be measured by a **Geiger counter** or **scintillation counter.** A source that undergoes 3.7×10^{10} nuclear disintegrations per second is called a **1-curie** (Ci) source. In the SI system, the unit of activity is the **becquerel** (Bq); one becquerel is equivalent to one disintegration per second (dis/s).

$$1 \text{ Ci} = 3.7 \times 10^{10} \text{ dis/s}$$

$$1 \text{ Bq} = 1 \text{ dis/s}$$

EXAMPLE Calculation of the activity of a source

How many disintegrations per second occur in a 40.0 Ci cobalt-60 radioactive source?

SOLUTION A 1-Ci source undergoes 3.7×10^{10} dis/s. We can use this definition as a conversion factor to answer the problem:

$$40.0 \text{ Ci} \times \frac{3.7 \times 10^{10} \text{ dis/s}}{1 \text{ Ci}} = 1.5 \times 10^{12} \text{ dis/s}$$

EXERCISE A sample of a radioactive source undergoes 3.3×10^{11} dis/s. What is the activity of the source, in curies and in becquerels?

ANSWER: 8.9 Ci; 3.3×10^{11} Bq

KEY CONCEPT C Doses of radioactivity

The radiation dose is the energy deposited in a sample per unit of mass. One standard unit for measuring radiation dose is the rad (*r*adiation *a*bsorbed *d*ose). One **rad** corresponds to exactly 10^{-2} J of energy deposited in a kilogram of sample:

$$1 \text{ rad} = 10^{-2} \text{ J/kg}$$

A dose of 500 rad from γ radiation is enough to cause death in 50% of the humans that receive this dose. This is equivalent to only 5 J of energy per kilogram of tissue. Alpha radiation is approximately 20 times as effective per rad as β and γ radiation in causing biological damage. To account for the difference in biological damage that different types of radiation cause, each type of radiation is assigned a **relative biological effectiveness, Q.** Beta and γ radiation both have $Q = 1$, whereas α radiation has $Q = 20$. (These numbers are approximate because Q also depends on the tissue being irradiated.) The dose in rads times Q is a measure of the expected biological effect of any radiation and is called the **dose equivalent.** The unit for the dose equivalent is the **rem** (*r*adiation *e*quivalent *m*an).

$$\text{Rem} = \text{rad} \times Q$$

EXAMPLE Doses from radiation

What is the dose equivalent (in rem) when 2.5 J of energy is delivered to 0.85 kg of tissue by α radiation?

SOLUTION We first calculate the energy delivered per kilogram of tissue, by dividing the total energy delivered by the mass of the tissue in kilograms:

$$\begin{aligned} \text{Energy per kg} &= 2.5 \text{ J/0.85 kg} \\ &= 2.9 \text{ J/kg} \end{aligned}$$

Now we use the fact that 1 rad = 10^{-2} J/kg to calculate the dose in rads:

$$\begin{aligned} \text{Dose in rad} &= 2.9 \text{ J/kg} \times \frac{1 \text{ rad}}{10^{-2} \text{ J/kg}} \\ &= 2.9 \times 10^{2} \text{ rad} \end{aligned}$$

Finally, we use $Q = 20$ for α radiation to calculate the equivalent dose in rems:

$$\begin{aligned} \text{Rem} &= \text{rad} \times Q \\ &= 2.9 \times 10^{2} \times 20 \text{ rem} \\ &= 5.8 \times 10^{3} \text{ rem} \end{aligned}$$

EXERCISE A 58-kg person is bathed in β radiation. Because β radiation is only moderately penetrating, only 2.0 kg of the person's tissue is irradiated, with a total dose received of 23 rem. What is the total energy absorbed by the tissue?

ANSWER: 0.46 J

22.6 THE RATE OF NUCLEAR DISINTEGRATION

KEY CONCEPT A The law of radioactive decay

The radioactive decay of a radioactive nuclide follows a first-order rate law. If N equals the number of radioactive nuclei present at any time, the rate of decay is proportional to N:

$$\text{Rate of decay} = kN$$

In addition, if N_0 is the number of radioactive nuclei at time = 0, and N is the number of radioactive nuclei at time = t, then the familiar first-order integrated rate law is obtained:

$$\ln\left(\frac{N_0}{N}\right) = kt$$

where k is the first-order rate constant associated with the decay and is called the **decay constant.** Like any first-order rate constant, it has units of 1/time. The half-life of a radioactive element is defined as the time it takes for one-half of a sample to undergo radioactive decay. The following bar graph shows the decay of a radioactive sample through three half-lives. Note that the mass of the radioactive sample (shaded bars) decreases by half for every half-life. At the same time, the mass of the daughter nuclides increases, as shown by the unshaded bars. On the scale of this experiment, mass is conserved. The sum of the masses of the radioactive sample and the daughter nuclides is always 8 g.

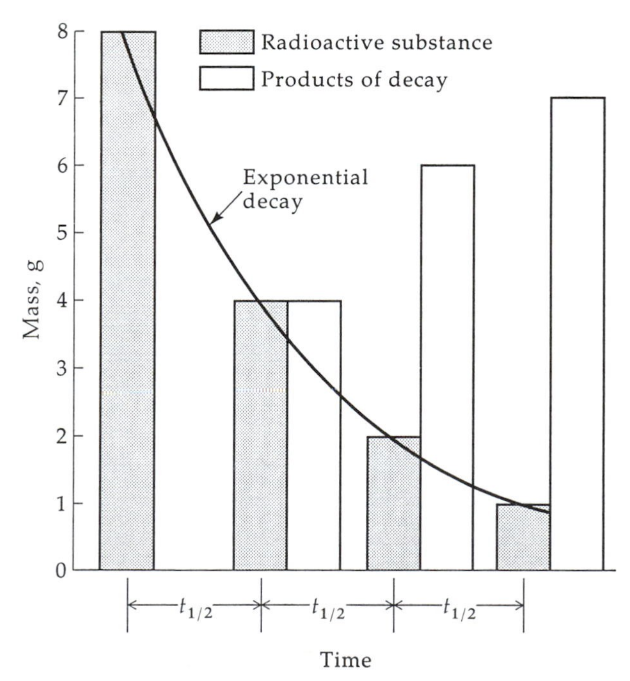

The half-life of an element is related to the decay constant through the relation

$$t_{1/2} = \frac{0.693}{k}$$

EXAMPLE 1 First-order decay kinetics

The decay constant for the decay of rhodium-99 is 0.043 day^{-1}. How many grams of rhodium-99 remain if a 15-g sample is allowed to decay for exactly 10 days?

SOLUTION Because the mass of a sample is proportional to the number of nuclei present, we can write the integrated rate law as

$$\ln\left(\frac{m_0}{m}\right) = kt$$

where m_0 is the mass of the radioactive sample present at time = 0 and m is the mass present at time = t. Substituting $k = 0.043\ \text{day}^{-1}$ and $t = 10$ day into the equation gives

$$\ln\left(\frac{m_0}{m}\right) = 0.043\ \text{day}^{-1} \times 10\ \text{day}$$
$$= 0.43$$

We next take the exponential of both sides of the equation. Recalling that $e^{\ln y} = y$ and calculating $e^{0.43} =$ 1.54 give

$$\frac{m_0}{m} = 1.54$$

We finally use the fact that $m_0 = 15$ g to calculate the desired result:

$$\frac{15\ \text{g}}{m} = 1.54$$
$$m = 9.7\ \text{g}$$

EXERCISE The decay constant for the decay of lead-209 is 0.21 h^{-1}. How many grams of lead-209 remain if a 2.6-kg sample is allowed to decay for 3.0×10^2 min?

ANSWER: 0.91 kg

EXAMPLE 2 Calculating half-lifes

What is the half-life of an element with $k = 0.387\ (\text{ms})^{-1}$?

SOLUTION The relationship between the half-life of an element and the decay constant for its decay is

$$t_{1/2} = 0.693/k$$

Substituting in $k = 0.387\ (\text{ms})^{-1}$ gives

$$t_{1/2} = \frac{0.693}{0.387\ (\text{ms})^{-1}}$$
$$= 1.79\ \text{ms}$$

EXERCISE What is the half-life of an element that has a first-order decay constant of 0.020 s^{-1}

ANSWER: 35 s

EXAMPLE 3 Using the first-order integrated rate law for radioactive decay

A 26-mg sample of thorium-215 is allowed to decay until 20 mg remain. How long was the sample allowed to decay? The half-life of thorium-215 is 1.2 s.

SOLUTION The integrated rate law for radioactive decay using the mass of a sample is

$$\ln\left(\frac{m_0}{m}\right) = kt$$

We first calculate k using the relation

$$k = \frac{0.693}{t_{1/2}} = \frac{0.693}{1.2\ s} = 0.58\ s^{-1}$$

Substituting $m = 20$ mg, $m_0 = 26$ mg, and $k = 0.58\ s^{-1}$ into the integrated rate law gives

$$\ln\left(\frac{26\ \cancel{mg}}{20\ \cancel{mg}}\right) = 0.58\ s^{-1} \times t$$
$$\ln(1.3) = 0.58\ s^{-1} \times t$$
$$0.26 = 0.58\ s^{-1} \times t$$
$$t = \frac{0.26}{0.58\ s^{-1}} = 0.45\ s$$

EXERCISE How much time is required for the decay of a sample of rhenium-192 to 0.20% of its original mass? $t_{1/2} = 16$ s.

ANSWER: 143 s

KEY CONCEPT B Radiocarbon dating

A small percentage of the carbon atoms in atmospheric CO_2 is radioactive carbon-14, which has a half-life of 5730 years(y). Plants and animals incorporate carbon-14 into their tissue, plants through photosynthesis and animals by being part of the food web. The carbon-14 present in any living organism decays at its normal rate, but the organism takes in fresh carbon-14 as it lives. Thus, a type of dynamic equilibrium is reached for a live organism, in which the relative amount of carbon-14 in the organism is the same as that in atmospheric CO_2. However, when the organism dies, the incorporated carbon-14 continues to decay, but it is no longer replenished with fresh carbon-14. The relative amount of carbon-14 found in a sample of the dead organism's tissue is a measure of the time since the organism died. For instance, if a sample has exactly one-half of the radioactive carbon it originally carried when it was alive, we would be safe in concluding that the organism died 5730 y (one half-life) previously. The technique of determining the "age" (actually, the time since the organism died) of a formerly living sample by the amount of carbon-14 it contains is called **radiocarbon dating.**

EXAMPLE Interpreting radiocarbon dating

A sample of charcoal from a hearth discovered in a acrheological dig undergoes 6100 carbon-14 disintegrations per gram of carbon in a 24-h period. Carbon from a modern source undergoes 920 disintegrations per hour (dis/h) per gram of carbon. How old is the piece of charcoal; that is, how long ago did the tree from which the charcoal was formed die?

SOLUTION We assume that the modern carbon source is an accurate reflection of the carbon-14 present at the time the tree was alive. This permits the calculation of the number of disintegrations per gram of carbon that would have occurred in a 24-h period when the tree was alive:

$$920\ \frac{\text{dis}}{\cancel{h}} \times 24\ h = 2.2 \times 10^4\ \text{dis} \qquad \text{(in 24 h)}$$

Because the number of disintegrations is proportional to the number of carbon-14 nuclei present, we can write the first-order integrated rate law as

$$\ln\left(\frac{dis_0}{dis}\right) = kt$$

where dis = the number of disintegrations in a certain time interval at time = t and dis_0 = the number of disintegrations in the same time interval at time = 0. Also, $k = 0.693(t_{1/2})^{-1} = 1.21 \times 10^{-4}\ \text{y}^{-1}$. Substituting $dis = 6100$ and $dis_0 = 2.2 \times 10^4$ into the integrated rate equation along with the value of k just calculated gives

$$\ln\left(\frac{2.2 \times 10^4\ \cancel{\text{dis}}}{6100\ \cancel{\text{dis}}}\right) = 1.21 \times 10^{-4}\ \text{y}^{-1} \times t$$

$$\ln(3.6) = 1.21 \times 10^{-4}\ \text{y}^{-1} \times t$$

$$1.3 = 1.21 \times 10^{-4}\ \text{y}^{-1} \times t$$

$$t = \frac{1.3}{1.21 \times 10^{-4}\ \text{y}^{-1}}$$

$$= 1.1 \times 10^4\ \text{y}$$

The tree died approximately 11,000 years ago.

EXERCISE A sample of wood from an ancient boat undergoes 2800 carbon-14 disintegrations in a 30-h period. How old is the piece of wood?

ANSWER: 1.9×10^4 yr

NUCLEAR ENERGY

KEY WORDS Define or explain each of the following terms in a written sentence or two.

breeder reactor
critical mass
fallout
fissile
fissionable
induced nuclear fission
mass–energy conversion
moderator
nuclear reactor
plasma
supercritical mass
thermonuclear explosion

22.7 MASS–ENERGY CONVERSION

KEY CONCEPT Nuclear Binding Energy

The nuclear binding energy of a nuclide is the energy released when the neutrons and protons in the nucleus combine to form the nucleus. The binding energy is a measure of the stability of the nucleus. For instance, for sulfur-35, which contains 16 protons and 19 neutrons,

$$16\ \text{protons} + 19\ \text{neutrons} \longrightarrow \text{sulfur-35 nucleus} \qquad \text{energy released} = \text{binding energy}$$

A nuclide with a high binding energy is more stable than one with a low binding energy. The nuclear binding energy E_{bind} is given by

$$E_{\text{bind}} = \Delta m \times c^2$$

where c is the speed of light (3.00×10^8 m/s) and Δm is the difference in mass between the nucleus and the separated neutrons and protons that make up the nucleus; in actual use, the mass of the hydrogen atom is used instead of the mass of the proton.

$$\Delta m = (\text{mass of separated neutrons + protons}) - (\text{mass of nucleus})$$

$$= \begin{pmatrix}\text{mass of neutrons + mass of hydrogen} \\ \text{atoms equal to number of protons}\end{pmatrix} - (\text{mass of nucleus})$$

EXAMPLE Calculating the binding energy of a nuclide

What is the binding energy (in kJ/mol) of nitrogen-14, which has an atomic mass of 14.0031 u? The mass of a hydrogen atom is 1.0078 u, and that of a neutron 1.0087 u.

SOLUTION The binding energy is given by

$$E_{bind} = \Delta m \times c^2$$

Nitrogen has atomic number 7, so it contains 7 protons. The isotope of mass number 14 contains 7 neutrons ($14 - 7 = 7$). Thus,

$$\begin{aligned}\Delta m &= [(7 \times \text{mass of hydrogen atom}) + (7 \times \text{mass of neutron})] - (\text{mass of nitrogen-14 nucleus}) \\ &= [(7 \times 1.0078\ \text{u}) + (7 \times 1.0087\ \text{u})] - (14.0031\ \text{u}) \\ &= 0.1124\ \text{u}\end{aligned}$$

We convert this mass difference from atomic mass units to kilograms so that the calculated energy will be expressed in joules. The conversion factor is $1\ \text{u} = 1.661 \times 10^{-27}$ kg.

$$0.1124\ \cancel{\text{u}} \times \frac{1.661 \times 10^{-27}\ \text{kg}}{1\ \cancel{\text{u}}} = 1.867 \times 10^{-28}\ \text{kg}$$

Substituting this value into the original equation gives the binding energy for one atom:

$$\begin{aligned}E_{bind} &= \Delta m \times c^2 \\ &= (1.867 \times 10^{-28}\ \text{kg})(3.00 \times 10^{8}\ \text{m/s})^2 \\ &= (1.867 \times 10^{-28}\ \text{kg})(9.00 \times 10^{16}\ \text{m}^2/\text{s}^2) \\ &= 1.681 \times 10^{-11}\ \text{J} \\ &= 1.681 \times 10^{-14}\ \text{kJ}\end{aligned}$$

Avogadro's number is used to convert to kJ/mol; the binding energy of nitrogen-14 is 1.012×10^{10} kJ/mol:

$$1.681 \times 10^{-14}\ \frac{\text{kJ}}{\cancel{\text{atom}}} \times 6.022 \times 10^{23}\ \frac{\cancel{\text{atom}}}{\text{mol}} = 1.012 \times 10^{10}\ \text{kJ/mol}$$

EXERCISE What is the binding energy, in kJ/mol, of copper-62, which has an atomic mass of 61.9326 u?

ANSWER: 5.228×10^{10} kJ/mol

22.8 NUCLEAR FISSION

KEY CONCEPT A Types of nuclear fission

Fission is a process in which a nucleus breaks up into smaller nuclei. The breakup is accompanied by the release of a large amount of energy. **Spontaneous nuclear fission** occurs when a nucleus breaks up without first absorbing a particle. When bombarding a nucleus with a neutron causes the nucleus to break up into smaller nuclei, **induced nuclear fission** is said to occur, as shown by the following reaction:

$$^{1}_{0}\text{n} + ^{235}_{92}\text{U} \longrightarrow ^{131}_{50}\text{Sn} + ^{103}_{42}\text{Mo} + 2\,^{1}_{0}\text{n}$$

A nuclide that can undergo induced nuclear fission is called a **fissionable** nuclide. A nuclide that can undergo fission when it is bombarded by a slow-moving neutron is called a **fissile nuclide.** Uranium-235, plutonium-239, and uranium-233 are examples of fissile nuclides. A self-sustaining fission reaction can occur in a sample if the fission reaction produces neutrons that are captured by nuclei in the sample, which then undergo fission, producing more neutrons, and so on. If the sample is too small or is inappropriately shaped, many neutrons will escape instead of hitting nuclei and a sustained fission reaction does not occur. For a large sample of the proper shape, few neutrons escape and a sustained nuclear reaction can occur. A sample of the proper shape and size to sustain fission is said to possess a **critical mass.** A sample with a **supercritical mass** is large enough not only to sustain a fission reaction but also to explode. In a nuclear reactor, fission is assisted by the **moderator**, a substance that slows neutrons so they can collide more effectively with nuclei. The fission is controlled both by the geometry of the sample and by **control rods** made of a substance that absorbs neutrons and prevents a runaway nuclear reaction. The difference between a fission explosion and controlled fission is schematically indicated in the figure. Each × represents a fission event. The neutrons from the fission are shown but the products are not.

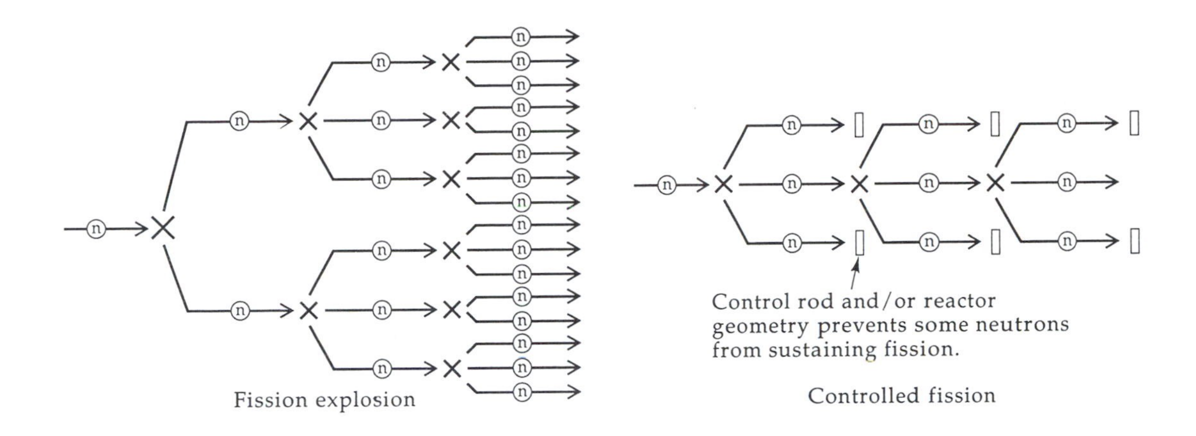

▼ EXAMPLE Writing a nuclear fission reaction

What is the missing product E in the following fission reaction?

$$^{1}_{0}n + ^{235}_{92}U \longrightarrow ^{A}_{Z}E + ^{139}_{56}Ba + 3^{1}_{0}n$$

SOLUTION For mass number balance, the sum of the superscripts on the left must equal the sum of the superscripts on the right. Thus, $1 + 235 = A + 139 + 3$, and $A = 94$. For charge balance, the sum of the subscripts on the left must equal the sum of the subscripts on the right; thus, $0 + 92 = Z + 56 + 0$, so $Z = 36$. The missing product is krypton-94, $^{94}_{36}Kr$.

EXERCISE What is the missing product in the following nuclear fission reaction?

$$^{1}_{0}n + ^{235}_{92}U \longrightarrow ^{A}_{Z}E + ^{137}_{52}Te + 2^{1}_{0}n$$

ANSWER: Zirconium-97, $^{97}_{40}Zr$ ▲

KEY CONCEPT B The energy released during fission

The energy released during a fission reaction is reflected in the difference in the masses of the reactants and products through the equation $E = c^2\Delta m$. For this situation, E is the energy released, and Δm is the difference in the masses of the products and reactants,

$$\Delta m = (\text{mass of reactants}) - (\text{mass of products})$$

▼ EXAMPLE Calculating the heat released during fission

What is the energy released, in kJ/mole uranium-235, in the fission reaction

$$^{1}_{0}n + ^{235}_{92}U \longrightarrow ^{97}_{40}Zr + ^{137}_{52}Te + 2^{1}_{0}n$$

The masses of the various species are neutron, 1.0087 u; uranium-235, 235.0439 u; zirconium-97, 96.9110 u; and tellurium-137, 136.9254 u.

SOLUTION The energy released is given by

$$E = c^2 \Delta m$$

$$\begin{aligned}\Delta m &= (m_{\text{neutron}} + m_{\text{uranium}-235}) - (m_{\text{zirconium}-97} + m_{\text{tellurium}-137} + 2\, m_{\text{neutron}}) \\ &= (1.0087 \text{ u} + 235.0439 \text{ u}) - (96.9110 \text{ u} + 136.9254 \text{ u} + 2 \times 1.0087 \text{ u}) \\ &= (236.0526 \text{ u}) - (235.8538 \text{ u}) \\ &= 0.1988 \text{ u}\end{aligned}$$

We convert the mass to kilograms and at the same time calculate the energy:

$$\begin{aligned}E &= (0.1988 \text{ u} \times 1.661 \times 10^{-27} \text{ kg/u}) \times (3.00 \times 10^{8} \text{ m/s})^2 \\ &= 2.97 \times 10^{-11} \text{ J} \\ &= 2.97 \times 10^{-14} \text{ kJ}\end{aligned}$$

This is the energy released when one atom of uranium-235 undergoes fission. For a mole of uranium-235, we get

$$\begin{aligned}E \text{ (per mole)} &= 2.97 \times 10^{-14} \frac{\text{kJ}}{\text{atom}} \times 6.02 \times 10^{23} \frac{\text{atom}}{\text{mol}} \\ &= 1.79 \times 10^{10} \text{ kJ/mol}\end{aligned}$$

EXERCISE What is the energy released (in kJ/mol of uranium-235) during the following fission:

$${}^{1}_{0}\text{n} + {}^{235}_{92}\text{U} \longrightarrow {}^{90}_{38}\text{Sr} + {}^{143}_{54}\text{Xe} + 3\,{}^{1}_{0}\text{n}$$

The masses of the nuclides involved are strontium-90, 90.9102 u, and xenon-143, 142.9296 u. The preceding example problem has additional required masses.

ANSWER: 1.68×10^{10} kJ/mol

22.9 NUCLEAR FUSION

KEY CONCEPT The energy released during fusion

In the nuclear fission process, a heavy nucleus disintegrates into lighter nuclei with the release of energy. In the **nuclear fusion** process, lighter nuclei are fused together into a heavier nucleus, also with the release of energy. In both cases, the binding energy of the products is more than that of the reactants, so there is a release of energy when the process occurs. The energy released during fusion is reflected in the difference in the masses of the reactants and products and is calculated in the same way as for fission.

EXAMPLE Calculating the energy released during fusion

Calculate the energy released when 1.00 g ^{2}H (deuterium) reacts in the fusion reaction

$${}^{2}_{1}\text{H} + {}^{2}_{1}\text{H} \longrightarrow {}^{3}_{2}\text{He} + {}^{1}_{0}\text{n}$$

The masses of the species involved are ${}^{2}_{1}$H, 2.0140 u; ${}^{3}_{2}$He, 3.01603 u; and ${}^{1}_{0}$n, 1.0087 u.

SOLUTION The energy released is given by the relationship

$$\begin{aligned}E &= c^2 \Delta m \\ \Delta m &= \text{(mass of reactants)} - \text{(mass of products)}\end{aligned}$$

For our problem,

$$\begin{aligned}\Delta m &= (2 \times \text{mass deuterium}) - (\text{mass helium-3} + \text{mass neutron}) \\ &= (2 \times 2.0140 \text{ u}) - (3.01603 \text{ u} + 1.0087 \text{ u}) \\ &= 0.0033 \text{ u}\end{aligned}$$

We convert this to kilograms and solve for E simultaneously:

$$\begin{aligned} E &= (3.00 \times 10^{8} \text{ m/s})^{2} \times 0.0033 \cancel{u} \times 1.661 \times 10^{-27} \text{ kg}/\cancel{u} \\ &= 4.9 \times 10^{-13} \text{ J} \\ &= 4.9 \times 10^{-16} \text{ kJ} \end{aligned}$$

This result is for the fusion of two atoms of deuterium. The energy released per atom is half this amount or 2.5×10^{-16} kJ. For 1.00 g of deuterium, we get

$$1.00 \cancel{\text{g } ^{2}\text{H}} \times \frac{1 \cancel{\text{mol } ^{2}\text{H}}}{2.0140 \cancel{\text{g}}} \times 6.02 \times 10^{23} \frac{\text{atom}}{\cancel{\text{mol}}} = 2.99 \times 10^{23} \text{ atom}$$

This is the number of deuterium atoms that undergo fusion. Multiplying this number by the energy released per atom gives us the answer to the problem:

$$2.99 \times 10^{23} \cancel{\text{atom}} \times 2.5 \times 10^{-16} \frac{\text{kJ}}{\cancel{\text{atom}}} = 7.5 \times 10^{7} \text{ kJ}$$

Exercise Calculate the energy released when 1 mol ^{2}H fuses with 1 mol ^{3}H (tritium) in the fusion reaction

$$^{2}_{1}\text{H} + ^{3}_{1}\text{H} \longrightarrow ^{4}_{2}\text{He} + ^{1}_{0}\text{n}$$

The masses of the species not given in the example are $^{3}_{1}H$, 3.01605 u, and $^{4}_{2}He$, 4.00260 u

ANSWER: 1.69×10^{9} kJ

22.10 CHEMICAL ASPECTS OF NUCLEAR POWER

KEY CONCEPT Nuclear fuel and nuclear waste

Uranium is mined as the ore **pitchblende** (UO_2), which contains the desired fissile uranium-235 as 0.7% of the uranium present. Extraction is the process of converting the uranium in the ore to a desired chemical form. A common extraction process involves converting the UO_2 to UF_6. The 0.7% concentration of uranium-235 is not high enough to be useful as a fuel, so the percentage of uranium-235 must be increased. The process of increasing the percentage of uranium-235 in a sample is called **isotopic enrichment,** and the reason for producing UF_6 is to use it in the enrichment process. The enrichment process uses the difference in masses of $^{235}UF_6$ (mass = 349.03 u) and $^{238}UF_6$ (mass = 352.04 u). In the gas diffusion process, the difference in masses leads to a difference in rates of diffusion and the possibility of enrichment.

$$\begin{aligned} \frac{\text{Rate of diffusion of } ^{235}\text{UF}_6}{\text{Rate of diffusion of } ^{238}\text{UF}_6} &= \sqrt{\frac{\text{mass of } ^{238}\text{UF}_6}{\text{mass of } ^{235}\text{UF}_6}} \\ &= \sqrt{\frac{352.04 \cancel{u}}{349.03 \cancel{u}}} \\ &= 1.0043 \end{aligned}$$

The enriched UF_6 is converted back to UO_2 or to uranium metal for use as a fuel. After the fuel has undergone fission for some time, the concentration of ^{235}U becomes too low for further fission. The spent fuel contains highly radiaoctive and unwanted products of fission as well as some remaining ^{235}U and the ^{239}Pu formed during fission. The uranium and plutonium are separated from the unwanted products using a **solvent extraction** process. This process uses the relatively high solubility of uranium and plutonium oxides in a solvent that does not dissolve the unwanted products. The recovered uranium and plutonium are recycled for use as fuels. It is hoped that a way can be discovered to store the unwanted, highly radioactive products for many thousands of years until they are no longer dangerously radioactive. One idea for long-term storage that has been extensively studied is underground storage in deep mines. One of the problems with underground storage is **leaching,** the process by which underground water slowly dissolves the stored wastes and disperses it to deep underground water sources.

EXAMPLE The uranium fuel cycle

The complete process of using uranium as a nuclear fuel involves four steps that together are referred to as the uranium fuel cycle. What are the four steps?

SOLUTION These steps are

1. mining,
2. extraction
3. enrichment
4. recycling

EXERCISE Why is it desirable to extract uranium and plutonium from spent nuclear fuel?

ANSWER: They can be recycled and reused as fuel.

DESCRIPTIVE CHEMISTRY TO REMEMBER

- The three natural **radioactive series** are

  Uranium-238 series: starts at uranium-238 and ends at lead-206
  Uranium-235 series: starts at uranium-235 and ends at lead-207
  Thorium-232 series: starts at thorium-232 and ends at lead-208

- The typical annual dose of radiation received by a person in the United States is **0.2 rem.** The actual figure depends on where the individual lives.
- In a **breeder reactor** neutrons are not moderated (slowed down). Some of the high-energy neutrons are used to convert ^{238}U, which is not fissile, to ^{239}Pu, which is fissile.
- The **purex process** for separating uranium and plutonium from the unwanted products of fission uses the preferential solubility of the oxides of uranium and plutonium in a mixture of kerosene (80%) and tritertiarybutyl phosphate (20%) for the separation.

CHEMICAL EQUATIONS TO KNOW

- The first step in one extraction scheme for uranium is oxidation of the UO_2 in the ore from a +4 oxidation state to a +6 oxidation state.

$$3UO_2(s) + 8HNO_3(aq) \longrightarrow 3UO_2(NO_3)_2(aq) + 2NO(g) + 4H_2O(l)$$

- After purification, the uranium(VI) nitrate is decomposed to uranium(VI) oxide by heating to 600°C.

$$UO_2(NO_3)_2(s) \longrightarrow UO_3(s) + NO(g) + NO_2(g) + O_2(g)$$

- The $UO_3(s)$ is reduced to uranium(IV) fluoride.

$$UO_3(s) + H_2(g) \longrightarrow UO_2(s) + H_2O(g)$$
$$UO_2(s) + 4HF(aq) \longrightarrow UF_4(s) + 2H_2O(l)$$

- The UF_4 may be reduced to uranium metal for various uses or oxidized to UF_6 for enrichment.

$$UF_4(s) + 2Mg(s) \longrightarrow U(s) + 2MgF_2(s)$$
$$UF_4(s) + F_2(g) \xrightarrow{450°C} UF_6(s)$$

- Enriched uranium as the hexafluoride can be reduced to an oxide for direct use or as a starting material for further reduction.

$$UF_6(s) + H_2(g) + 2H_2O(g) \longrightarrow UO_2(s) + 6HF(g)$$

MATHEMATICAL EQUATIONS TO KNOW AND UNDERSTAND

$\text{Rem} = \text{rad} \times Q$	definition of rem
$\text{Rate of decay} = k \times N$	rate of radioactive decay
$\ln\left(\frac{N_0}{N}\right) = kt$	integrated rate law for radioactive decay
$t_{1/2} = 0.693/k$	relation of half-life to decay constant
$E = c^2\Delta m$	mass-energy relation

SELF-TEST EXERCISES

Nuclear stability

1. What particle is deflected most as it moves between two oppositely charged plates?
(a) neutron (b) $^{4}_{2}\text{He}$ (c) γ (d) $^{12}_{6}\text{C}$ (e) β^+

2. What is the energy of a γ-ray photon with $\lambda = 2.65$ pm? $h = 6.63 \times 10^{-34}$ J · s; $c = 3.00 \times 10^8$ m/s.
(a) 6.54×10^{-8} J (b) 7.51×10^{-14} J (c) 1.76×10^{-45} J
(d) 7.95×10^{-4} J (e) 1.16×10^{-12} J

3. What product is formed when ^{47}K undergoes beta decay?
(a) ^{47}Ar (b) ^{47}Ca (c) ^{46}Ca (d) ^{46}K (e) ^{48}K

4. What product is formed when ^{72}As undergoes positron emission?
(a) ^{72}Ge (b) ^{73}As (c) ^{74}As (d) ^{73}Ge (e) ^{72}Se

5. What product is formed when ^{226}Th undergoes alpha decay?
(a) ^{224}Rn (b) ^{224}Ra (c) ^{222}Ra (d) ^{222}Rn (e) ^{230}Th

6. What product is formed when ^{160}Er undergoes electron capture?
(a) ^{159}Ho (b) ^{159}Er (c) ^{160}Ho (d) ^{161}Yb (e) ^{160}Tm

7. What is a likely mode of decay for ^{241}Am?
(a) positron emission (b) electron capture (c) beta emission (d) alpha emission

8. What is a likely mode of decay for ^{184}Tl?
(a) electron capture (b) positron emission (c) alpha emission
(d) gamma emission (e) both electron capture and positron emission

9. What is a likely mode of decay for a light element that is neutron rich?
(a) positron emission (b) electron capture (c) gamma emission
(d) beta emission (e) both positron emission and electron capture

10. What is the product of the following nucleosynthesis $^{239}_{94}\text{Pu}(\alpha, \text{n})\text{E}$?
(a) $^{242}_{96}\text{Cm}$ (b) $^{241}_{96}\text{Cm}$ (c) $^{243}_{96}\text{Cm}$ (d) $^{235}_{92}\text{U}$ (e) $^{236}_{92}\text{U}$

11. Bombarding $^{239}_{94}\text{Pu}$ with a single neutron results in beta emisson and formation of
(a) $^{238}_{95}\text{Am}$ (b) $^{239}_{93}\text{Np}$ (c) $^{240}_{93}\text{Np}$ (d) $^{240}_{94}\text{Pu}$ (e) $^{240}_{95}\text{Am}$

Radioactivity

12. Which of the following types of radioactivity has the highest penetrating power?
(a) alpha (b) beta (c) gamma (d) positron

13. What is the activity of a radioactive source that undergoes 5.5×10^{13} dis/s?
(a) 1.5 kCi (b) 6.7 mCi (c) 20 Ci (d) 49 mCi (e) 7.9×10^{2} Ci

14. What is the dose in rem when 70 kg of tissue receives the equivalent of 6.2 J of alpha radiation? 1 rad = 10^{-2} J/kg; $Q = 20$ for alpha radiation.
(a) 18 mrem (b) 87 rem (c) 1.2 rem (d) 1.8×10^{2} rem (e) 5.6 rem

15. How many grams of potassium-44 remain if a 25-g sample is allowed to decay for 1.0 h? The half-life of potassium-44 is 22.4 min.
(a) 0.44 g (b) 1.6×10^{2} g (c) 3.1 g (d) 3.9 g (e) 2.6 g

16. If 13.2 g of a radioactive sample remains after the sample ages for 12.0 min, how much sample was originally present? The half-life of the substance is 5.40 min.
(a) 122 g (b) 2.82 g (c) 1.43 g (d) 61.6 g (e) 166 g

17. A sample of wood from an ancient boat is found to undergo 5200 carbon-14 disintegrations per gram of carbon in an 18-h period. How old is the boat? Current samples of carbon decay at the rate of 920 dis/h. The half-life of carbon-14 is 5730 y.
(a) 2.6×10^{4} y (b) 1.8×10^{3} y (c) 2.2×10^{3} y
(d) 4.6×10^{3} y (e) 9.6×10^{3} y

Nuclear power

18. What is the role of a moderator in a nuclear reactor?
(a) slows down neutrons (b) captures neutrons (c) disperses heat from reactor core
(d) emits electrons (e) encourages proper cooling of core

19. What is the missing product in the following nuclear fission reaction?

$${}^{1}_{0}\text{n} + {}^{235}_{92}\text{U} \longrightarrow ? + {}^{89}_{37}\text{Rb} + 3\,{}^{1}_{0}\text{n}$$

(a) ${}^{144}_{55}\text{Cs}$ (b) ${}^{146}_{55}\text{Cs}$ (c) ${}^{143}_{55}\text{Cs}$ (d) ${}^{144}_{54}\text{Xe}$ (e) ${}^{146}_{54}\text{Xe}$

20. What is the binding energy, in kJ/mol, of ${}^{34}_{15}\text{Si}$? The mass of silicon-34 is 33.9764 u, that of a hydrogen atom is 1.0078 u, and that of a neutron is 1.0087 u.
(a) 3.18×10^{11} (b) 5.17×10^{11} (c) 4.59×10^{-11}
(d) 6.54×10^{10} (e) 2.76×10^{10}

21. How much energy (in kJ/mol of uranium-235) is released in the following fission reaction? The masses of the various species involved are neutron, 1.0087 u; uranium-235, 235.0439 u; lathanum-147, 146.9281 u; and bromine-87, 86.9207 u.

$${}^{1}_{0}\text{n} + {}^{235}_{92}\text{U} \longrightarrow {}^{147}_{57}\text{La} + {}^{87}_{35}\text{Br} + 2\,{}^{1}_{0}\text{n}$$

(a) 2.57×10^{10} (b) 8.31×10^{10} (c) 1.68×10^{10}
(d) 7.40×10^{10} (e) 5.45×10^{10}

22. How much energy (in kilojoules) is released when 1.00 g ${}^{2}\text{H}$ undergoes fusion? The masses of the species involved are ${}^{2}\text{H}$, 2.0140 u; ${}^{3}\text{H}$, 3.01603 u; ${}^{4}\text{He}$, 4.00260 u; and n, 1.0087 u. The reaction is ${}^{2}_{1}\text{H} + {}^{3}_{1}\text{H} \longrightarrow {}^{4}_{2}\text{He} + {}^{1}_{0}\text{n}$.
(a) 1.69×10^{9} (b) 8.36×10^{8} (c) 4.28×10^{10}
(d) 8.36×10^{11} (e) 1.69×10^{12}

23. UO_3 is converted to UF_6 in a four-step process. How many grams of UF_6 can be produced from 100 g UO_3? It is not necessary to do a calculation for each step.
(a) 81.3 g (b) 66.8 g (c) 91.4 g (d) 100 g (e) 123 g

Descriptive chemistry

24. What compound is used in the enrichment of uranium-235 by the gas diffusion method?
(a) UO_2 (b) UF_6 (c) UF_4 (d) UO_3 (e) U

25. The oxidation of UO_2 by HNO_3 results in the formation of which uranium compound?
(a) UO_3 (b) U (c) $UO_2(NO_3)_2$ (d) no reaction

26. One of the following nuclides is not at the head of a natural radioactive series. Which one?
(a) uranium-238 (b) uranium-235 (c) plutonium-239 (d) thorium-232

27. What is the annual dose of radiation received by an average U.S. citizen?
(a) 2 rem (b) 0.2 rem (c) 500 rem (d) 5 krem

28. What is the product when UF_4 is reduced by reaction with Mg?
(a) UF_6 (b) UO_2 (c) UO_3 (d) U (e) $UO_2(NO_3)_2$

23 THE HYDROCARBONS

The branch of chemistry that deals with most of the compounds of carbon is called **organic chemistry.** The compounds, such as CH_4 (methane) and $C_{12}H_{22}O_{11}$ (sucrose), are called **organic compounds.** A few compounds of carbon, such as CO_2, CO, and the carbonates, are considered inorganic rather than organic. The **hydrocarbons** are the binary compounds of carbon and hydrogen; examples are CH_4, C_2H_6 (ethane), and C_6H_6 (benzene). Two broad classes of hydrocarbons are the **saturated hydrocarbons,** which are hydrocarbons with no carbon–carbon multiple bonds, and **unsaturated hydrocarbons,** which are those with one or more carbon–carbon multiple bonds.

THE ALKANES

23.1–23.3 ISOMERISM, NOMENCLATURE, AND PROPERTIES

KEY WORDS Define or explain each of the following terms in a written sentence or two.

alkane
alkyl group
conformation
cycloalkane
homologous series
isomerization
methylene group
structural isomerism
substituent
substitution reaction

KEY CONCEPT A Alkanes and isomerism

A **homologous series** of compounds is a family of compounds in which a CH_2 group (a **methylene group**) is added to a member of the family to generate the next member. The **alkanes** are a homologous series of compounds that have the molecular formula C_nH_{2n+2}. The first three members of the alkane family are CH_4 ($n = 1$), C_2H_6 ($n = 2$), and C_3H_8 (propane, $n = 3$). A carbon atom in a typical alkane has sp^3 hybridization and a tetrahedral shape, and is bonded to four other atoms. The carbon–carbon bonds in the alkanes are all single bonds; rotation around these bonds occurs easily at room temperature. **Isomerism** is the occurrence of two different componds with the same molecular formula. **Structural isomerism** is a type of isomerism in which at least one atom in one isomer is bonded to different neighbors than in another isomer. For instance, there are two compounds with the formula C_4H_{10} ($n = 4$):

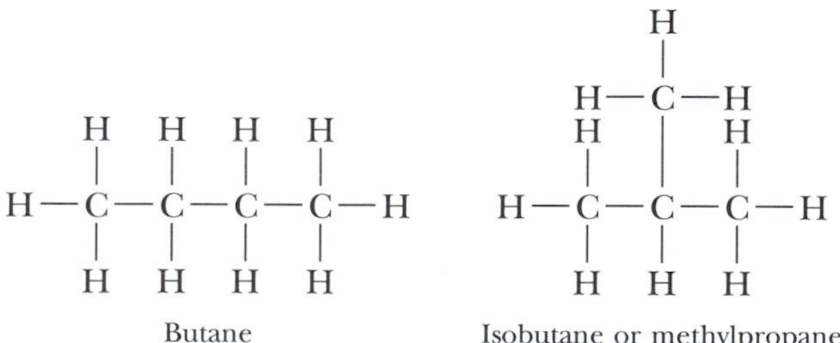

Butane   Isobutane or methylpropane

The central carbon in methylpropane is bonded to three other carbons and one hydrogen, as neighbors; there is no carbon with these same neighbors in butane, so methylpropane and butane are structural isomers. Another way of indicating these structures is

$$CH_3CH_2CH_2CH_3 \qquad CH_3CH(CH_3)CH_3$$

Butane   Methylpropane

The parentheses around the CH_3 in the formula for methylpropane indicate the CH_3 group is bonded to the preceding carbon atom. A third way of writing the formula for butane is $CH_3(CH_2)_2CH_3$. In this case, the parentheses are used to indicate how many methylene groups in a row are in the formula.

▼ EXAMPLE Isomerism in the alkanes

Without referring to the text, draw all the structural isomers that have the formula C_5H_{12}.

SOLUTION A good method for drawing structural isomers is to first draw the carbon "skeletons" and then add the required hydrogen atoms. We start with the five carbons bonded in a straight chain and then move one or more of the carbons to get all the other structural isomers:

$$
\mathrm{C{-}C{-}C{-}C{-}C}
\qquad
\begin{array}{c}
\mathrm{C} \\ | \\ \mathrm{C{-}C{-}C{-}C}
\end{array}
\qquad
\begin{array}{c}
\mathrm{C} \\ | \\ \mathrm{C{-}C{-}C} \\ | \\ \mathrm{C}
\end{array}
$$

You may wonder why we do not include

$$
\begin{array}{c}
\mathrm{C} \\ | \\ \mathrm{C{-}C{-}C{-}C}
\end{array}
\quad \text{and} \quad
\begin{array}{c}
\mathrm{C{-}C{-}C{-}C} \\ | \\ \mathrm{C}
\end{array}
$$

But if you make a list of the carbon atoms and the neighbors they have, you will find that these structures represent the same isomer as the center structure in the first figure. Molecular models (constructed from Styrofoam balls and toothpicks, or gum drops and toothpicks, or a commercial molecular-model set) also will show that the three skeletons match up atom-to-atom. Thus, the only possible arrangements for the five carbons of an alkane are those in the first figure. Any others you can draw are simply replicas of one of these three, oriented differently on the paper. To complete the drawings, we add in the hydrogen atoms, remembering there must be a total of 12 hydrogens and that every carbon atom is bonded to 4 other atoms:

$$
\begin{array}{ccccccccccc}
 & & \mathrm{H} & & \mathrm{H} & & \mathrm{H} & & \mathrm{H} & & \mathrm{H} \\
 & & | & & | & & | & & | & & | \\
\mathrm{H} & - & \mathrm{C} & - & \mathrm{C} & - & \mathrm{C} & - & \mathrm{C} & - & \mathrm{C}{-}\mathrm{H} \\
 & & | & & | & & | & & | & & | \\
 & & \mathrm{H} & & \mathrm{H} & & \mathrm{H} & & \mathrm{H} & & \mathrm{H}
\end{array}
\qquad
\begin{array}{c}
\mathrm{H} \\ | \\ \mathrm{H{-}C{-}H} \\
\mathrm{H}\quad\mathrm{H}\quad | \quad\mathrm{H} \\
|\quad\;|\quad\;|\quad\;| \\
\mathrm{H{-}C{-}C{-}C{-}C{-}H} \\
|\quad\;|\quad\;|\quad\;| \\
\mathrm{H}\quad\mathrm{H}\quad\mathrm{H}\quad\mathrm{H}
\end{array}
\qquad
\begin{array}{c}
\mathrm{H} \\ | \\ \mathrm{H{-}C{-}H} \\
\mathrm{H}\quad | \quad\mathrm{H} \\
|\quad\;|\quad\;| \\
\mathrm{H{-}C{-}C{-}C{-}H} \\
|\quad\;|\quad\;| \\
\mathrm{H}\quad | \quad\mathrm{H} \\
\mathrm{H{-}C{-}H} \\ | \\ \mathrm{H}
\end{array}
$$

EXERCISE The formulas for four alkanes with formula C_6H_{14} are shown in the figure. Which are distinct isomers and which represent the same compound?

$$
\underset{\text{I}}{\begin{array}{r}
\mathrm{CH_3}\quad \\ |\quad\;\; \\ \mathrm{CH_3CH_2CH_2CHCH_3}
\end{array}}
\qquad
\underset{\text{II}}{\begin{array}{c}
\mathrm{CH_3} \\ | \\ \mathrm{CH_3CH_2C{-}CH_3} \\ | \\ \mathrm{CH_3}
\end{array}}
\qquad
\underset{\text{III}}{\begin{array}{l}
\quad\;\;\mathrm{CH_3} \\ \quad\;\; | \\ \mathrm{CH_3{-}CHCH_2CH_2CH_3}
\end{array}}
$$

$$
\underset{\text{IV}}{\begin{array}{c}
\mathrm{CH_3CH_2CHCH_2CH_3} \\ | \\ \mathrm{CH_3}
\end{array}}
$$

▲ ANSWER: I, II, and IV are distinct; I and III are the same.

KEY CONCEPT B Alkane nomenclature

If the carbon atoms in an alkane are bonded in a row, so that the formula is of the form $CH_3(CH_2)_mCH_3$, the compound is named according to the number of carbon atoms present. Compounds of this type are called **unbranched alkanes.** Table 23.1 of the text gives examples of such compounds. All are named with a stem name and the suffix *-ane*. A **branched alkane** has one or more carbons that are not in a row of carbon atoms and thus form **side chains.** Butane is an unbranched alkane, and methylpropane is a branched alkane with a CH_3 side chain. When naming the branched alkanes, we treat side chains as **substituents,** which are atoms or groups of atoms that have been substituted for a hydrogen atom. To name a branched alkane, we follow these steps:

Step 1. Identify the longest chain of carbon atoms in the molecule.
Step 2. Number the carbon atoms on the longest chain so that the carbon atoms with substituents have the lowest possible numbers.
Step 3. Identify each substituent and the number of the carbon atom it is located on.
Step 4. Name the compound by listing the substituents in alphabetical order, with the numbered location of each substituent preceding its name; the name(s) of the substituent(s) are followed by the parent name of the alkane, which is determined by the number of carbon atoms in the longest chain. For instance, the octane with a methyl group at the number 3 carbon would be 3-methyloctane.
Step 5. If two or more of the same substituents are present (such as two methyl groups), a Greek prefix, such as *di-*, *tri-*, or *tetra-* is attached to the name of the group; and the numbers of the carbon atoms to which the groups are attached, separated by commas, are included in the name. For instance, if our octane had a methyl group at carbon 4 as well as at carbon 3, its name would be 3,4-dimethyloctane.
Step 6. Numbers in a name are always separated from letters by a hyphen; numbers are separated from each other by commas.

EXAMPLE 1 Naming unbranched alkanes

What is the name of the alkane $CH_3(CH_2)_6CH_3$?

SOLUTION The alkane contains a total of eight carbon atoms, so the stem of the name is obtained by combining the prefix for eight (*oct-*) with the alkane suffix *-ane* to get the name *octane*.

EXERCISE What is the name of the alkane $CH_3(CH_2)_7CH_3$?

ANSWER: Nonane

EXAMPLE 2 Naming a branched alkane

Name the alkane

$$CH_3-CH_2-CH_2-\overset{\displaystyle CH_3}{\underset{\displaystyle CH_3}{\overset{|}{\underset{|}{C}}}}-CH_3$$

SOLUTION The first step is to number the carbons of the longest carbon chain so that the substituents appear on the lowest numbered carbons:

$$\underset{5}{CH_3}-\underset{4}{CH_2}-\underset{3}{CH_2}-\overset{\displaystyle CH_3}{\underset{\displaystyle CH_3}{\overset{|}{\underset{2|}{C}}}}-\underset{1}{CH_3}$$

The longest carbon chain contains five carbons, so the compound is a pentane. There are two methyl groups on carbon number 2. Combining all this information gives the name 2,2-dimethylpentane.

EXERCISE Name the alkane

$$
\begin{array}{ccccccccccc}
 & & & & CH_3 & & CH_2CH_3 & & & & \\
 & & & & | & & | & & & & \\
CH_3 & - & CH_2 & - & CH & - & CH & - & CH & - & CH_3 \\
 & & & & & & & & | & & \\
 & & & & & & & & CH_3 & &
\end{array}
$$

ANSWER: 3-Ethyl-2,4-dimethylhexane

KEY CONCEPT C The properties of alkanes

Since the only important intermolecular attractions in the alkanes are London forces, the alkanes show a smooth variation in many physical properties as alkane size increases. In branched alkanes, the side chains prevent the molecules from getting as close to each other as unbranched molecules of the same size might, so the branched molecules experience lower molecular attractions and have physical properties that reflect these smaller attractions. Chemically, the strong C—H and C—C bonds of the alkanes make them relatively unreactive. Three important types of reactions that the alkanes do undergo are oxidation, isomerization, and subsitution reactions. A typical oxidation is the complete combustion to carbon dioxide and water in excess oxygen; this is always a highly exothermic reaction and is the basis for using the alkanes as fuels. The catalytic **reforming** process used by oil refineries accomplishes **isomerization,** the conversion of one isomer to another. One of the reactions that occurs is the conversion of unbranched alkanes to branched alkanes. For example, butane is converted to methylpropane;

$$
\begin{array}{ccccccccccccc}
 & & & & & & & & & & CH_3 & & \\
 & & & & & & & & & & | & & \\
CH_3 & - & CH_2 & - & CH_2 & - & CH_3 & \longrightarrow & CH_3 & - & CH & - & CH_3
\end{array}
$$

In a **substitution reaction,** a group or atom is substituted for another group or atom in the original molecule. The substitution of a chlorine atom for a hydrogen atom through a radical chain reaction is an example of an alkane substitution reaction. For example, in methane the reaction is

$$CH_4(g) + Cl_2(g) \longrightarrow CH_3Cl(g) + HCl(g)$$

Many such reactions are difficult to control, and multiple substitutions may occur. For the chlorination of methane it is not unusual to get CH_2Cl_2, $CHCl_3$, and CCl_4 as products as well as CH_3Cl.

EXAMPLE The physical and chemical properties of the alkanes

Arrange the following alkanes in order of increasing boiling points.

$CH_3CH_2CH_3$	$CH_3CH_2CH_2CH_3$	$CH_3CH(CH_3)CH_3$	$CH_3CH_2CH_2CH_2CH_3$
Propane	Butane	Methylpropane	Pentane

SOLUTION For the alkanes, the boiling points increase as the size (molecular weight) of the compound increases. The smallest is propane, so it has the lowest boiling point; the largest is pentane, so it has the highest boiling point. Butane and methylpropane, the intermediate compounds, are both the same size (4 carbons); but branched compounds experience smaller intermolecular attractions than unbranched compounds of the same size, so methylpropane has a lower boiling point than butane. The order is

Lowest boiling point propane $<$ methylpropane $<$ butane $<$ pentane highest boiling point

EXERCISE Write the formulas and names of all of the substitution products from the chlorination of ethane.

ANSWER: CH_3CH_2Cl (chloroethane); CH_2ClCH_2Cl (1,2-dichloroethane); $CHCl_2CH_3$ (1,1-dichloroethane); $CH_2ClCHCl_2$ (1,1,2-trichloroethane); CH_3CCl_3 (1,1,1-trichloroethane); CH_2ClCCl_3 (1,1,1,2-tetrachloroethane); $CH_2Cl_2CHCl_2$ (1,1,2,2-tetrachloroethane); $CHCl_2CCl_3$ (1,1,1,2,2-pentachloroethane); and CCl_3CCl_3 (hexachloroethane; numbers are not used in this name because there is no ambiguity regarding the location of the chlorine atoms without them).

THE ALKENES AND THE ALKYNES

23.4–23.7 ALKENE NOMENCLATURE, THE CARBON–CARBON DOUBLE BOND, AND POLYMERIZATION; ALKYNES

KEY WORDS Define or explain each of the following terms in a written sentence or two.

addition reaction
alkene
alkyne
cis-trans isomers
dehydrogenation reaction
elimination reaction
geometrical isomers
hydrohalogenation reaction
polymer
Ziegler–Natta catalyst

KEY CONCEPT A The alkenes and alkene nomenclature

The alkenes are a homologous series that contain a carbon–carbon double bond and have a molecular formula of the type C_nH_{2n}. C_2H_4 (ethene, also called ethylene) and $CH_2{=}CHCH_3$ (propene) are the two simplest alkenes. Alkenes can be produced from alkanes or substituted alkanes by an **elimination reaction;** in this reaction a small molecule such as water, HCl, HBr, or H_2 is driven out of (eliminated from) the reactant molecule, leaving a double bond. The atoms or groups removed are usually on adjacent carbon atoms in the reactant molecule. Catalytic **dehydrogenation** eliminates H_2:

$$CH_3CH_3 \longrightarrow CH_2{=}CH_2 + H_2$$

In a **dehydrohalogenation** reaction, a hydrohalide, such as HCl, HBr, or HI, is eliminated from an alkane on which the halogen has been substituted:

$$CH_3{-}\underset{\displaystyle |}{\overset{Br}{}}\!\!CH{-}\overset{H}{\underset{}{}}\!CH_2 + OH^- \longrightarrow CH_3CH{=}CH_2 + Br^- + H_2O$$

Alkenes are named using the stem name of the corresponding alkane with a number that specifies the location of the double bond. The number is obtained by numbering the longest carbon chain so that the double bond is associated with the lowest numbered carbon possible and assigning to the double bond the lower number of the two double-bonded carbons:

$$\overset{5}{C}H_3\overset{4}{C}H_2\overset{3}{C}H{=}\overset{2}{C}H\overset{1}{C}H_3$$

2-Pentene (not 3-pentene)

Other substituents are named as for the alkanes, with a number specifying the location of the group and an appropriate prefix to denote how many groups of each kind are present. Numbering the longest chain so that the double bond has the lowest number takes precedence over keeping the substituent numbers low.

EXAMPLE 1 Naming an alkene

Name the alkene

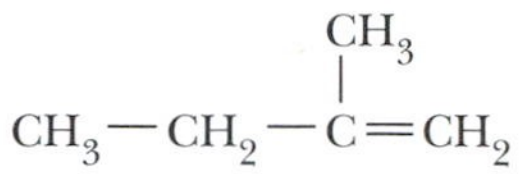

$$\begin{array}{c} CH_3 \\ | \\ CH_3-CH_2-C=CH_2 \end{array}$$

SOLUTION The carbon chain is numbered so that the carbons associated with the double bond have the lowest possible numbers. The double bond is assigned the number 1 because this is the lower of the numbers of the two double-bonded carbons, and the methyl group is at position 2. The name is 2-methyl-1-butene.

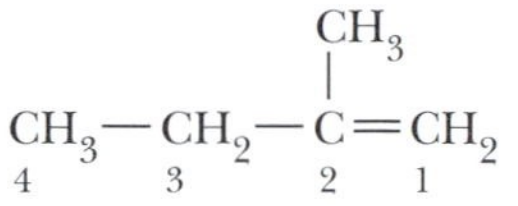

$$\begin{array}{ccccccc} & & & & CH_3 & & \\ & & & & | & & \\ CH_3 & - & CH_2 & - & C & = & CH_2 \\ 4 & & 3 & & 2 & & 1 \end{array}$$

EXERCISE Name the following alkene.

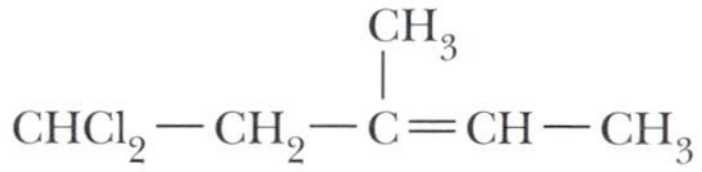

$$\begin{array}{c} CH_3 \\ | \\ CHCl_2-CH_2-C=CH-CH_3 \end{array}$$

ANSWER: 5,5-Dichloro-3-methyl-2-pentene. (Note that a low number for the double bond takes precedence over low numbers for the substituents.)

EXAMPLE 2 Elimination reactions

Name the alkene formed when butane undergoes dehydrogenation.

SOLUTION Two products are possible since H_2 can be eliminated from the two central carbons to give 2-butene or from an end carbon and its neighbor to give 1-butene:

$$\underset{\text{2-Butene}}{CH_3CH=CHCH_3} \qquad \underset{\text{1-Butene}}{CH_2=CHCH_2CH_3}$$

EXERCISE What product is formed when 1-bromobutane is heated in an organic solvent containing OH^-?

ANSWER: 1-Butene (only)

KEY CONCEPT B The properties of the carbon–carbon double bond

Rotation around a carbon–carbon double bond does not occur to any appreciable extent at normal temperatures. The lack of flexibility introduced into alkenes because of the rigidity of the double bond prevents alkene molecules from packing as close to each other as the corresponding alkanes (which are more flexible), so the intermolecular attractions in the alkenes are smaller. The physical properties of the alkenes reflect this difference. The effective lack of rotation around the double bond also makes possible a type of isomerism called **geometrical isomerism.** Geometric isomers are compounds with the same molecular formulas and with atoms bonded to the same neighbors, but with different arrangements of atoms in space. As an example, for 2-butene, two geometric isomers are possible:

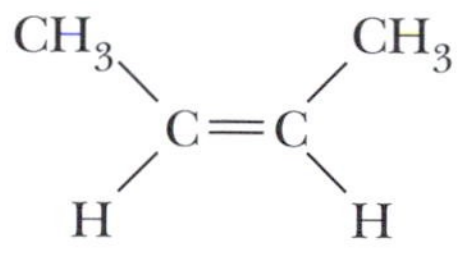

$$\underset{\textit{cis}\text{-2-Butene}}{\begin{array}{ccc} CH_3 & & CH_3 \\ & \diagdown\, C=C \,\diagup & \\ H & \diagup \quad \diagdown & H \end{array}} \qquad \underset{\textit{trans}\text{-2-Butene}}{\begin{array}{ccc} CH_3 & & H \\ & \diagdown\, C=C \,\diagup & \\ H & \diagup \quad \diagdown & CH_3 \end{array}}$$

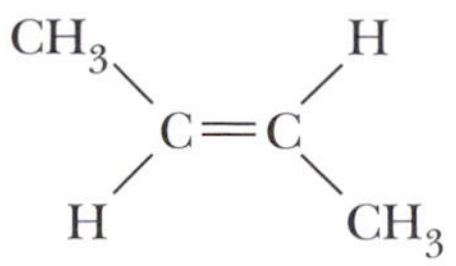

In the trans isomer, the two methyl groups are across the double bond from each other (hence the designation *trans*). In the cis isomer, they are on the same side of the double bond. Because rotation around the double bond does not occur under normal circumstances, the two isomers are distinct compounds with different chemical and physical properties. If enough energy is added to a pure sample of either compound, a **cis-trans isomerization** reaction may occur in which the energy excites rotation around the double bond and part of the sample is converted to the other isomer.

The characteristic reaction of the double bond is an **addition reaction,** a reaction in which a small molecule such as H_2O, HCl, HBr, Cl_2, Br_2, or H_2 adds to the double bond. In the addition, one part of the small molecule ends up on one carbon associated with the double bond and the rest of the small molecule ends up on the other double-bonded carbon. In the process, the double bond becomes a single bond. For example:

Hydrogenation (addition of H_2): $CH_2{=}CH_2 + H_2 \rightarrow CH_3CH_3$
Halogenation (addition of a halogen molecule): $CH_2{=}CH_2 + Cl_2 \rightarrow CH_2ClCH_2Cl$
Hydrohalogenation (addition of a hydrohalide): $CH_2{=}CH_2 + HBr \rightarrow CH_3CH_2Br$
Hydration (addition of water): $CH_2{=}CH_2 + H_2O \rightarrow CH_3CH_2OH$

EXAMPLE 1 Cis-trans isomerism

Is cis-trans isomerism possible for 1-butene?

SOLUTION We first draw the compound in a way that would show any possible cis-trans isomerism:

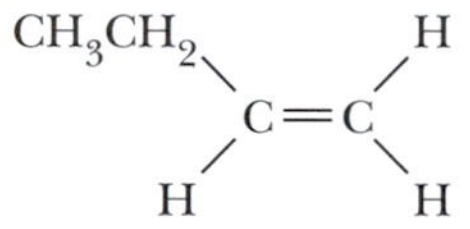

```
CH3CH2       H
      \     /
       C=C
      /     \
     H       H
```

As we can see, no difference results if we interchange the two hydrogens on the right side of the structure or if we interchange the ethyl group (CH_3CH_2) with the hydrogen beneath it. Thus, cis-trans isomerism is not possible, and 1-butene exists in only one form.

EXERCISE What is the name of the following compound?

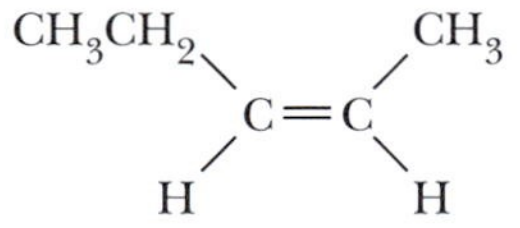

```
CH3CH2       CH3
      \     /
       C=C
      /     \
     H       H
```

ANSWER: *cis*-2-Pentene

EXAMPLE 2 Addition reactions

What product is formed when HBr adds to the double bond in ethene, $CH_2{=}CH_2$.

SOLUTION In this addition reaction, H adds to one of the double-bonded carbon atoms and Br adds to the other. The product is bromoethane:

$$CH_2{=}CH_2 + HBr \longrightarrow CH_2BrCH_3$$

EXERCISE What is the name of the product formed by the addition of Cl_2 to ethene?

ANSWER: 1,2-Dichloroethane

KEY CONCEPT C Alkene polymerization

Alkenes undergo a reaction called **addition polymerization,** in which one alkene adds to another alkene, then a third adds to the first two, then a fourth, then a fifth, and so on. This results in a long chain of linked units. The single alkene is called the **monomer;** the linked long-chain species formed from it is called a **polymer.** In this **polymerization reaction** the double bond is converted to a single bond (as in the previous addition reactions). The polymerization may occur through a free-radical chain mechanism in which, for example, a $CH_3{-}CHX\cdot$ radical adds to $CH_2{=}CHX$ to form a second, longer free radical, which then adds to another $CH_2{=}CHX$, and so on, to form a long polymer chain.

$$CH_3{-}\underset{}{\overset{X}{\overset{|}{C}}}H\cdot + CH_2{=}\overset{X}{\overset{|}{C}}H \longrightarrow CH_3{-}\overset{X}{\overset{|}{C}}H{-}CH_2{-}\overset{X}{\overset{|}{C}}H\cdot$$

$$CH_3{-}\overset{X}{\overset{|}{C}}H{-}CH_2{-}\overset{X}{\overset{|}{C}}H\cdot + CH_2{=}\overset{X}{\overset{|}{C}}H \longrightarrow CH_3{-}\overset{X}{\overset{|}{C}}H{-}CH_2{-}\overset{X}{\overset{|}{C}}H{-}CH_2{-}\overset{X}{\overset{|}{C}}H\cdot$$

Another important polymerization procedure uses a **Ziegler–Natta catalyst,** such as triethylaluminum, $Al(CH_2CH_3)_3$. This catalyst causes side groups to be spatially arranged in a regular fashion along the polymer chain in either an **isotactic polymer** or **syndiotactic polymer** (see Figure 23.12 of the text). This leads to a high-density material because the regular arrangement allows the polymer chains to pack closely together. In an **atactic** polymer, the side groups are randomly arranged and the polymer chains cannot pack as closely together; the material is of lower density. Rubber is a natural polymer made of isoprene $[CH_2{=}C(CH_3){-}C(CH_3){=}CH_2]$ monomers. Vulcanization adds S—S cross links to the natural polymer to strengthen it. A Ziegler–Natta catalyst is used to produce synthetic rubber; without this type of catalyst, a useless product results. By using other than isoprene monomers, chemists have produced a variety of synthetic rubbers, which are collectively called **elastomers.** If a mixture of monomer units is used, a **copolymer** is produced, which sometimes results in more desirable products.

EXAMPLE Polymerization

What side chain results when propene is polymerized using a free-radical chain reaction?

SOLUTION We imagine that we start with a propene unit that has been converted to a free radical, and we allow it to add to a second propene. We treat the methyl group as the X in the reaction written previously. Thus, the side group is a methyl group:

$$CH_3{-}\underset{CH_3}{\underset{|}{C}}H\cdot + CH_2{=}CH_2CH_3 \longrightarrow CH_3{-}\underset{CH_3}{\underset{|}{C}}H{-}CH_2{-}\underset{CH_3}{\underset{|}{C}}H{-}CH_2\cdot$$

EXERCISE Does rotation around carbon–carbon single bonds convert an isotactic polymer into a syndioactic polymer?

ANSWER: No

KEY CONCEPT D Alkynes

Alkynes are hydrocarbons that contain a carbon–carbon triple bond. They form a homologous series with formulas of the type C_nH_{2n-2}. The two simplest alkynes are $CH{\equiv}CH$ (ethyne) and $CH{\equiv}CCH_3$ (propyne). Alkynes are named using the stem name of the longest chain of carbons in the compound with the suffix *-yne*. Substituents are treated as in alkanes and alkenes. The alkynes can be produced by the dehydrohalogenation of dibromoalkanes in which the bromine atoms are on adjacent carbons. The reactions of alkynes are similar to those of alkenes, with addition across the triple bond to form a double-bonded compound followed by continued addition being the common mode of reaction.

EXAMPLE Naming alkynes and the reactions of alkynes

Name the alkyne shown and predict the products obtained by reacting it with HBr.

$$CH_3CH(CH_3)C{\equiv}CH$$

SOLUTION The longest carbon chain contains four carbons, with the triple bond between carbons 1 and 2 and a methyl group on carbon 3. The compound is thus 3-methyl-1-butyne. Addition of HBr across the triple bond gives two products, depending on whether Br ends up on carbon 1 or carbon 2:

$$\underset{\text{I}}{CH_3CH(CH_3)CBr{=}CH_2} \qquad \underset{\text{II}}{CH_3CH(CH_3)CH{=}CHBr}$$

Compound I, the product formed when Br adds to carbon 2, is 2-bromo-3-methyl-1-butene; it is the predominant product. Compound II, the product formed when Br adds to carbon 1, is 1-bromo-3-methyl-1-butene. It is usually impossible to stop the addition at this stage, and so a second HBr adds across the double bond. This leads to three products: 2,2-dibromo-3-methylbutane; 1,2-dibromo-3-methylbutane; and 1,1-dibromo-3-methylbutane.

EXERCISE What product(s) are formed when 2-pentyne is reacted with Br_2?

ANSWER: 2,3-Dibromo-2-pentene and 2,2,3,3-tetrabromopentane

AROMATIC HYDROCARBONS

23.8–23.9 NOMENCLATURE AND REACTIONS

KEY WORDS Define or explain each of the following terms in a written sentence or two.

aliphatic
arenes
aromatic
Friedel–Crafts alkylation
meta (*m*-)
ortho (*o*-)
para (*p*-)
polycyclic

KEY CONCEPT A Aromatic hydrocarbons and nomenclature

Aromatic compounds are compounds that are based on or contain a benzene (C_6H_6) ring:

```
          H
          |
          C
       ／    ＼
H—C    ◯    C—H
   |            |
H—C          C—H
       ＼    ／
          C
          |
          H
```

Such compounds are also called **arenes.** Even though they are unsaturated, they are quite unreactive; and when they react, they do so more by substitution than by addition. All carbon–carbon bonds in the benzene ring are equivalent. There are no isolated double bonds in benzene but rather a π system with delocalized orbitals that extend over all six carbon atoms. The delocalized π orbitals contain six electrons. There are also **polycyclic** arenes, which contain two or more benzene-type rings fused together. The conventional symbols for benzene and the polycyclic naphthalene and anthracene are

Benzene

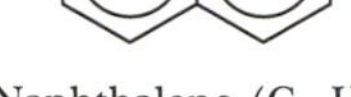

Naphthalene ($C_{10}H_8$)

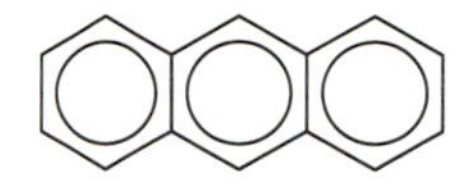

Anthracene ($C_{14}H_{10}$)

When a single substituent is present on a benzene ring, the compound is named by using the substituent name as a prefix in front of the word *benzene*. Thus, C_6H_5Cl is called chlorobenzene. If two of the same substituents are present, the common method of naming uses the prefix *ortho-* (abbreviated *o-*) to denote that the two substituents are on adjacent carbons, the prefix *meta-* (abbreviated *m-*) to denote that the two substituents are separated by one carbon, and the prefix *para-* (abbreviated *p-*) to denote that the two substituents are separated by two carbons (i.e., they are across the ring from each other). For example:

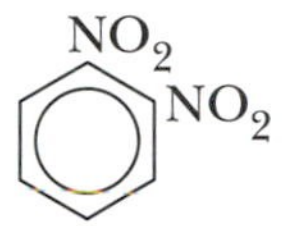

o-Dinitrobenzene
1,2-Dinitrobenzene

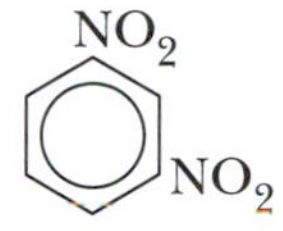

m-Dinitrobenzene
1,3-Dinitrobenzene

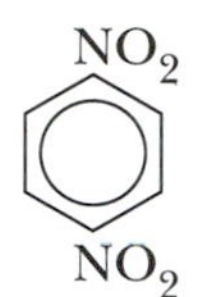

p-Dinitrobenzene
1,3-Dinitrobenzene

Systematic names are obtained by numbering the carbon atoms on the ring starting at one of the groups and naming the compound as a substituted benzene. Numbering is done so as to place substituents on the lowest numbered carbon atoms possible. The ortho, meta, and para labels are also used in some special circumstances, such as in naming substituted toluenes ($C_6H_5CH_3$ is toluene). Thus, 1-chloro-3-methylbenzene is also called *p*-chlorotoluene.

EXAMPLE Naming aromatic compounds

What is the name of the following compound?

Cl
Cl

SOLUTION The two chlorine atoms are on adjacent carbons. If we number the carbons of the ring, starting with one of the chlorines (which one is immaterial) and numbering so as to keep the numbers of the substituents as low as possible, the two chlorines are on carbons 1 and 2, and the compound is 1,2-dichlorobenzene. Using the ortho, meta, and para system and noting that the two chlorines are on adjacent carbon atoms, we name the compound *o*-dichlorobenzene.

EXERCISE Name the following compound.

ANSWER: 1-Bromo-3-chlorobenzene or *p*-bromochlorobenzene

KEY CONCEPT B Reactions of aromatic hydrocarbons

Reactions involving benzene rings are generally substitution reactions in which some atom or group replaces a hydrogen; the π system is not disrupted. For instance, in the presence of the catalyst $FeBr_3$, the reaction of benzene with Br_2 results in the formation of bromobenzene:

$$C_6H_6 + Br_2 \xrightarrow{FeBr_3} C_6H_5Br + HBr$$

Alkyl groups can be attached to a benzene ring using **Friedel–Crafts alkylation.** In this reaction, a chlorinated alkane is mixed with benzene (or a substituted benzene) in the presence of the catalyst $AlCl_3$. The alkyl group attached to the chlorine adds to the ring:

$$\text{Alkyl—Cl} + C_6H_6 \xrightarrow{AlCl_3} C_6H_5\text{—Alkyl}$$

When benzene, minus one hydrogen, is a substituent, it is called a **phenyl group** (C_6H_5). Thus, the compound $CH{\equiv}C(C_6H_5)$ is given the common name phenylacetylene. The benzene ring can be made to undergo addition under relatively harsh conditions.

EXAMPLE The reactions of aromatic hydrocarbons

Suggest a method for preparing ethylbenzene from ethene and benzene and any other inorganic reagents required.

SOLUTION We can add an ethyl group to benzene using chloroethane (CH_3CH_2Cl) in a Friedel–Crafts reaction. We do not have chloroethane as a starting reagent but can produce it by reacting the ethene we do have with HCl. The synthetic route is then

$$CH_2{=}CH_2 + HCl \longrightarrow CH_3CH_2Cl$$

$$CH_3CH_2Cl + C_6H_6 \xrightarrow{AlCl_3} C_6H_5CH_2CH_3$$

EXERCISE Suggest a route to prepare propylbenzene from propene and benzene.

ANSWER:
$$CH_3CH{=}CH_2 + Cl_2 \longrightarrow CH_3CHClCH_2Cl$$
$$CH_3CHClCH_2Cl \longrightarrow CH_3CH{=}CHCl + HCl$$
$$CH_3CH{=}CHCl + H_2 \longrightarrow CH_3CH_2CH_2Cl$$
$$CH_3CH_2CH_2Cl + C_6H_6 \longrightarrow CH_3CH_2CH_2(C_6H_5) + HCl$$

SELF-TEST EXERCISES

The alkanes

1. All but one of the following is a structural isomer of $CH_3CH_2CH(CH_3)CH_3$. Which one is not?
(a) $CH_3CH_2CH_2CH_3$ (b) $CH_3CH_2CH_2CH_2CH_3$ (c) $C(CH_3)_4$

2. What is the name of $CH_3(CH_2)_2CH_3$?
(a) pentane (b) 1-methylpentane (c) cyclopentane (d) hexane (e) butane

3. What is the name of $CH_3C(CH_3)_2CH_2CH_2CH_3$?
(a) dimethylpentane (b) 4,4-dimethylpentane (c) 4-dimethylpentane
(d) 2,2-dimethylpentane (e) heptane

4. Which alkane has the highest boiling point?
(a) 2,2-dimethylpentane (b) pentane (c) hexane
(d) 3-methylpentane (e) 2,2-dimethylbutane

5. What product is formed when CH_3CH_3 is mixed with chlorine and flashed with light?
(a) No reaction occurs. (b) CH_3CH_2Cl (c) CH_2ClCH_2Cl
(d) CH_3CHCl_2 (e) compounds (b), (c), and (d)

The alkenes and the alkynes

6. What compound is formed when 2-bromopentane undergoes dehydrohalogenation?
(a) 2,2-dibromopentane (b) 2-pentene (c) hexene
(d) pentane (e) 3-pentene

7. Name $CH_3CH(CH_3)CH_2CH{=}CH_2$.
(a) 4-methyl-2-pentene (b) 2-methyl-4-pentene (c) hexene
(d) 4-methyl-1-pentene (e) 2-methyl-5-pentene

8. Name the following compound.

$$\begin{array}{ccc} CH_3 & & H \\ & C{=}C & \\ H & & CH_2CH_3 \end{array}$$

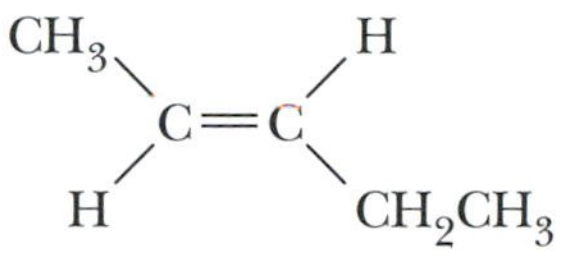

(a) *cis*-2-pentene (b) *trans*-2-pentene (c) *cis*-3-pentene
(d) *trans*-3-pentene (e) pentene

9. What product is formed when HI adds to ethene?
(a) iodoethane (b) 1,2-diiodoethane (c) 1,1-diiodoethane (d) hexaiodoethane

10. Which type of polymer has an irregular orientation of side chains.
(a) isotactic (b) syndiotactic (c) atactic

Aromatic hydrocarbons

11. Name the following compound:

CH_2CH_3 (benzene ring) Cl

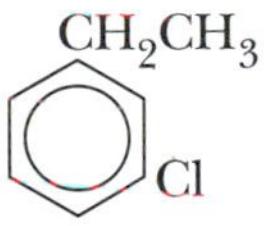

(a) *m*-chlorotoulene (b) chloroethylbenzene (c) xylene
(d) 3-chloro-1-ethylbenzene (e) 4-chloro-6-ethylbenzene

12. What product is formed when chloromethane is mixed with benzene in the presence of $AlCl_3$?
(a) hexane (b) chlorobenzene (c) 1,4-dichlorobenzene
(d) *p*-chlorotoluene (e) toluene

13. What product is formed when Br_2 is mixed with benzene in the presence of $FeBr_3$?
(a) 2,2-dibromohexane (b) 2,2-dibromohexene (c) bromobenzene
(d) cyclohexane (e) bromocyclohexane

14. What product is formed when benzene is reacted with H_2 at 200°C and 30 atm in the presence of a nickel catalyst?
(a) No reaction occurs. (b) cyclohexene (c) hexane
(d) cyclohexane

24 FUNCTIONAL GROUPS AND BIOMOLECULES

THE HYDROXYL GROUP

KEY WORDS Define or explain each of the following terms in a written sentence or two.

alcohol	inductive effect	resonance effect
crown ether	phenol	secondary alcohol
ether	primary alcohol	tertiary alcohol
functional group		

24.1 ALCOHOLS AND ETHERS

KEY CONCEPT Alcohols and ethers

An **alcohol** is an organic compound that contains an —OH (hydroxyl) group that is not attached to an aromatic ring or to a carbonyl group. The —OH group is an example of a **functional group,** which is a group of atoms or bonds that endow a molecule with a specific set of properties. Methanol (CH_3OH) is the simplest alcohol; it is obtained as a natural product from the distillation of wood or by the reaction of carbon monoxide with hydrogen gas at high pressure and temperature with ZnO as a catalyst. Ethanol (CH_3CH_2OH) is obtained as the result of the fermentation of sugars by yeast or by the acid-catalyzed hydration of ethene. Alcohols are classified as **primary, secondary,** or **tertiary** according to the number of hydrogen atoms attached to the carbon to which the —OH group is attached: RCH_2OH is a primary alcohol, R_2CHOH is a secondary alcohol, and R_3COH is a tertiary alcohol. Alcohols are named by using the stem name of the alkane and the suffix *-ol*. The position of the —OH group is indicated with a number; the longest carbon chain is numbered so as to put the —OH group at the lowest possible numbered carbon.

CH_3OH	$CH_3CH_2CH(OH)CH_3$	$CH_3CH_2CH_2CH_2OH$
Methanol	2-Butanol (not 3-butanol)	1-Butanol (not 4-butanol)

If the name of the alcohol is unambiguous without a number, no number is used (as in methanol and ethanol). A **diol** is an alcohol with two —OH groups, and a **triol** is one with three —OH groups. Diols and triols are more extensively hydrogen-bonded than alcohols with one —OH because the additional —OH groups provide more opportunities for hydrogen bonding; thus, diols and triols have higher boiling points, melting points, enthalpies of melting and vaporization, and viscosities than alcohols of similar molecular weight.

Alcohols are amphiprotic. For instance, ethanol is a weak Brønsted acid ($pK_a = 16$) and ionizes to form the **ethoxide ion,** $CH_3CH_2O^-(aq)$; ethanol can also accept a proton from an acid to form an **oxonium ion** (an ion of the form ROH_2^+). Primary alcohols can be converted to aldehydes (a compound that contains the —CHO group) by gentle oxidation ([O] = oxidation). Aldehydes themselves are easily oxidized and may convert further to carboxylic acids:

$$\underset{\text{Primary alcohol}}{R-\overset{\overset{\large OH}{|}}{C}-H} \xrightarrow{[O]} \underset{\text{Aldehyde}}{R-\overset{\overset{\large O}{\|}}{C}-H} \xrightarrow{[O]} \underset{\text{Carboxylic acid}}{R-\overset{\overset{\large O}{\|}}{C}-OH}$$

Oxidation of a secondary alcohol results in the production of a ketone (a compound that contains an $R_2C{=}O$ group):

$$\underset{\text{Secondary alcohol}}{R-\overset{\overset{\displaystyle OH}{|}}{C}-R} \xrightarrow{[O]} \underset{\text{Ketone}}{R-\overset{\overset{\displaystyle O}{\|}}{C}-R}$$

EXAMPLE 1 Naming alcohols

Name the following alcohols and classify each as primary, secondary, or tertiary:
(a) $CH_3CH(OH)CH(CH_3)CH_3$
(b) $CHCl_2CH_2CH_2OH$
(c) $CH_3CH_2C(CH_3)(OH)CH_3$

SOLUTION (a) The longest chain in this alcohol is four carbons long, so the alcohol is a butanol. The chain is numbered so that the —OH group has the lowest possible number; thus, the —OH is on carbon 2 and the methyl group is on carbon 3, and the correct name is 3-methyl-2-butanol. The —OH is attached to a carbon that is bonded to one hydrogen and two carbons, so this is a secondary alcohol. (b) This alcohol has the —OH group on carbon 1 and two chlorines on carbon 3. The longest chain of carbon atoms is three carbons long, so the compound is 3,3-dichloro-1-propanol. The —OH is attached to a carbon that is bonded to two hydrogens and one carbon, so this is a primary alcohol. (c) This alcohol has a four-carbon chain with an —OH and a —CH_3 attached to carbon 2. It is 2-methyl-2-butanol. The —OH is attached to a carbon that is bonded to three carbons, so this is a tertiary alcohol.

EXERCISE Name the following alcohols and classify each one as primary, secondary, or tertiary: (a) $CH_3CH_2CH_2OH$, (b) $CH_3CH(OH)CH_3$, and (c) $CH_3CH_2CH(OH)CH(CH_3)CH_3$.

ANSWER: (a) 1-Propanol, primary; (b) 2-propanol, secondary; (c) 2-methyl-3-pentanol, secondary

EXAMPLE 2 The reactions and properties of alcohols

Draw the structure of the compound obtained when 1-butanol undergoes gentle oxidation. What type of compound is formed?

SOLUTION 1-Butanol is a primary alcohol. When it undergoes gentle oxidation, two hydrogen atoms are removed to form an aldehyde:

$$CH_3CH_2CH_2CH_2OH \xrightarrow{[O]} CH_3CH_2CH_2\overset{\overset{\displaystyle O}{\|}}{C}H$$

EXERCISE What is the structure of the compound formed when 1-butanol undergoes vigorous oxidation? What type of compound is produced?

ANSWER: $CH_3CH_2CH_2\overset{\overset{\displaystyle O}{\|}}{C}OH$; a carboxylic acid

24.2 PHENOLS

KEY CONCEPT Phenols

A **phenol** is a compound that has an —OH group attached directly to a benzene ring. The simplest example, C_6H_5OH, is called *phenol*. Phenols are not alcohols; they have different chemical and physical properties from alcohols. Phenols are stronger acids than alcohols; for example, phenol has $pK_a = 9.89$. $C_6H_5O^-$ is the **phenoxide ion.** The relatively high acidity of phenol stems from the stability of the phenoxide ion, which is introduced by the delocalization of the negative charge. The synthesis of **phenolic resins,** which are extensively cross-linked polymers containing phenolic groups, takes advantage of the fact that the phenoxide ion is a rich source of electrons.

EXAMPLE The properties and reactions of phenols

Which is a stronger acid, methanol or 2,4,6-trichlorophenol?

SOLUTION Phenols are generally stronger acids than alcohols, so we predict that 2,4,6-trichlorophenol is the stronger acid.

EXERCISE Why is phenol more reactive in base than in acid?

ANSWER: In base, phenol exists as phenoxide ion; phenoxide ion is more reactive than phenol itself.

THE CARBONYL GROUP

KEY WORDS Define or explain each of the following terms in a written sentence or two.

acetal
aldehyde
carbohydrate
carbonyl group
cyanohydrin
furanose
hemiacetal
ketone
monosaccharide
pyranose
reducing sugar

24.3 ALDEHYDES AND KETONES

KEY CONCEPT Aldehydes and ketones

The **carbonyl group** (–C(=O)–) appears in many classes of compounds, including aldehydes and ketones. In **aldehydes,** at least one of the atoms bonded to the carbonyl carbon is a hydrogen. In a ketone, both of the atoms bonded to the carbonyl carbon are carbons:

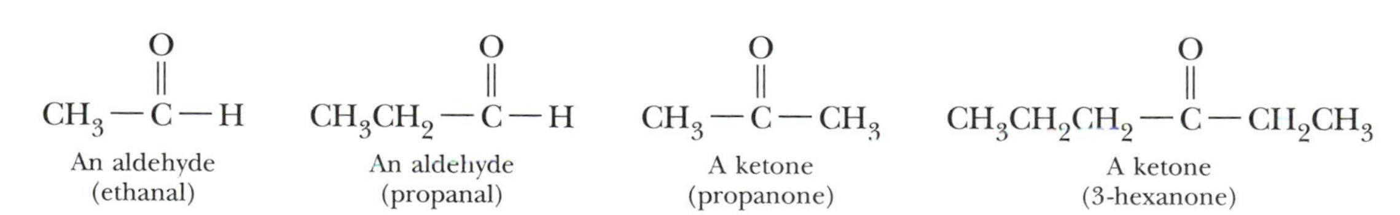

Primary alcohols become aldehydes by the loss of two hydrogen atoms. Secondary alcohols lead to ketones also by loss of two hydrogen atoms. Aldehydes are named systematically by using the stem name of the alkane with the suffix *-al* as for ethanal (from ethane) and propanal (from propane). If additional functional groups are present, the longest carbon chain containing the —CHO group is numbered with the carbon in the —CHO as carbon number 1. The systematic name of a ketone is obtained by changing the *-e* in the alkane to *-one*. The location of the C=O group is obtained by numbering the longest carbon chain containing the C=O so that the C=O group has the lowest possible number. Hence, in 3-hexanone the carbon in the C=O group is the number 3 carbon in the chain. Propanone has no number because the C=O group must be at the central carbon; if it were on either of the end carbons, an aldehyde structure would result.

Aldehydes are obtained by the gentle oxidation of primary alcohols. Ketones are obtained by the oxidation of secondary alcohols. The ease of oxidation of aldehydes is used as a basis for distinguishing aldehydes from ketones. When we place an aldehyde in Fehling's solution, which is a weakly oxidizing solution of Cu^{2+} and tartrate ion, a brick-red precipitate of Cu_2O forms. Ketones do not react with Fehling's solution. Similarly, aldehydes reduce Ag^+ in Tollen's reagent to a visible, solid Ag coat (a "silver mirror") on the surface of the glassware that the reaction is run in. Ketones do not react with Tollen's reagent.

In a reaction that is very important because it builds a new carbon–carbon bond, cyanide ion reacts with the carbonyl group in aldehydes and a few ketones to give a **cyanohydrin:**

$$\mathrm{R{-}\overset{\overset{\displaystyle O}{\|}}{C}{-}R + CN^- + H^+ \longrightarrow R{-}\underset{\underset{\displaystyle R}{|}}{\overset{\overset{\displaystyle OH}{|}}{C}}{-}C{\equiv}N}$$

A cyanohydrin

A cyanohydrin is a compound containing a —CN group and an —OH group attached to the same carbon. Aldehydes react with alcohols in the presence of dry HCl to form a **hemiacetal,** a compound with an —OH group and an —OR group attached to the same carbon. The hemiacetal may react with more alcohol to form an **acetal,** a compound with two —OR groups attached to the same carbon:

$$\mathrm{R{-}\overset{\overset{\displaystyle O}{\|}}{C}{-}H + R'OH \longrightarrow R{-}\underset{\underset{\displaystyle H}{|}}{\overset{\overset{\displaystyle OH}{|}}{C}}{-}OR' + R'OH \longrightarrow R{-}\underset{\underset{\displaystyle H}{|}}{\overset{\overset{\displaystyle OR'}{|}}{C}}{-}OR'}$$

An aldehyde — A hemiacetal — An acetal

An aldehyde with an —OH group (an alcohol function) attached four or five carbons away from the aldehyde function may form a cyclic hemiacetal.

EXAMPLE 1 Naming aldehydes and ketones

Name the following compounds:

(a) $CH_2ClCH_2CH_2CHO$

(b) $CH_3CH(CH_3)\overset{\overset{O}{\|}}{C}CH_3$

(c) $CH_3C(CH_3)_2CHO$

SOLUTION (a) Because the compound contains a —CHO group, it is an aldehyde. The four-carbon chain is numbered with a 1 on the carbon in the aldehyde group, so the chlorine is on carbon 4. The compound is 4-chlorobutanal (from butane). (b) The longest chain in the compound is four carbons long. The chain is numbered so that the C═O group contains the lowest numbered carbon possible; thus, the C═O is at carbon 2 and the methyl group is at carbon 3. The compound is 3-methylbutanone. (There is no need to use a number to designate the position of the carbonyl group because there is no choice regarding its location.) (c) This is a three-carbon aldehyde with two methyl groups on carbon 2; hence, it is 2,2-dimethylpropanal.

EXERCISE Name the following compounds:

(a) $CH_3CH(CH_3)CH(CH_3)CHO$

(b) $CH_3CH_2C(=O)CH(CH_3)CH_3$

(c) $CH(Br)_2CH_2CH_2C(=O)CH_3$

ANSWER: (a) 2,3-Dimethylbutanal, (b) 2-methyl-3-pentanone, (c) 5,5-dibromo-2-pentanone

EXAMPLE 2 The reactions of aldehydes and ketones

Compound A (an alcohol) is oxidized and the product B purified. B contains a carbonyl group, is not a carboxylic acid, and does not react with Fehling's solution or Tollen's reagent. What type of alcohol is A? Do you expect that B would react with an alcohol to form an hemiacetal?

SOLUTION Because B is the product of the oxidation of an alcohol and is not a carboxylic acid, we conclude that it is either an aldehyde or a ketone. Furthermore, because it does not reduce Tollen's reagent or Fehling's solution, it is not an aldehyde; it must, therefore, be a ketone. Ketones are produced by the oxidation of a secondary alcohol; thus, we conclude that A is a secondary alcohol. Ketones do not form hemiacetals, so we predict that B would not form a hemiacetal.

EXERCISE Two products are formed when 2-methyl-1-pentanol is oxidized. Product A forms a brick-red precipitate when it is mixed with Fehling's solution. Product B does not react with Tollen's reagent. Name product A, and identify the type of compound product B is.

ANSWER: A is 2-methylpentanal; B is a carboxylic acid.

24.4 CARBOHYDRATES

KEY CONCEPT Carbohydrates

Carbohydrates are a class of compounds that includes sugars and starches. Many have formulas of the form $C_m(H_2O)_n$. The common sugar glucose, $C_6H_{12}O_6$, is both an aldehyde and an alcohol, as can be seen from its structural formula:

$$H-\underset{H}{\overset{OH}{C}}-\underset{H}{\overset{OH}{C^*}}-\underset{H}{\overset{OH}{C}}-\underset{H}{\overset{OH}{C}}-\underset{H}{\overset{OH}{C}}-\overset{O}{\overset{\|}{C}}-H$$

The aldehyde portion of glucose can react with the starred carbon on the same molecule to produce a cyclic hemiacetal form of glucose called glucopyranose. This **pyranose** form of glucose has a six-membered ring with oxygen as one of the ring atoms. In aqueous solution, the cyclic hemiacetal form of glucose is in equilibrium with the open-chain aldehyde form. The open-chain form reacts with Fehling's solution and Tollen's reagent, so glucose is called a **reducing sugar. Fructose** (also $C_6H_{12}O_6$, an isomer of glucose) undergoes an intramolecular reaction like glucose to become a six-membered pyranose form; it also exists in a **furanose** form that contains a five-membered ring. Fructose does not contain an aldehyde group, but one of its hydroxy groups reduces Fehling's solution and Tollen's reagent, so fructose is also a reducing sugar. Sucrose ($C_{12}H_{22}O_{11}$) is formed by the reaction of a fructopyranose unit with a glucopyranose unit, so it is a **disaccharide** ("two sugar").

The individual sugar units are called **monosaccharides.** When several monosaccharides combine, a **polysaccharide** such as cellulose (an indigestible polysaccharide) and starch (a digestible polysaccharide) results.

EXAMPLE The properties and reactions of carbohydrates

The open-chain structure of fructose contains neither an aldehyde function nor an oxidizable ketone function. Explain why it gives positive Fehling's solution and Tollen's reagent tests.

SOLUTION Fehling's solution and Tollen's reagent both work by being reduced by a good reducing agent, that is, by a molecule that is easily oxidized. We know that alcohols are oxidizable, but they are usually not easily enough oxidized to give a positive Fehling's solution or Tollen's reagent test. However, because of the proximity of a carbonyl group, one of the alcohol functions on fructose is easily oxidizable; this makes fructose a reducing sugar.

EXERCISE Is sucrose a reducing sugar? Explain.

ANSWER Sucrose is a disaccharide made up of the cyclic forms of fructose and glucose. In sucrose the monosaccharides are not in equilibrium with their respective open-chain structures, so they cannot act as reducing agents; thus, sucrose is not a reducing sugar.

THE CARBOXYL GROUP

KEY WORDS Define or explain each of the following terms in a written sentence or two.

carboxyl group
carboxylate group
carboxylic acid
condensation polymer
condensation reaction
ester
fatty acid
lipid
polyester

24.5 CARBOXYLIC ACIDS

KEY CONCEPT Carboxylic acids

The **carboxyl group** contains a hydroxyl group and carbonyl group. It is often abbreviated as —COOH (even though it contains no oxygen–oxygen bond) or $—CO_2H$. Loss of the hydroxyl proton results in the formation of the resonance-stabilized **carboxylate group** ($—CO_2^-$) in which both oxygens are equivalent. **Carboxylic acids** are compounds that contain the carboxyl group. They are named systematically by changing the *-e* of the parent alkane to *-oic acid*. For many of the smaller carboxylic acids, common names (in parentheses for the acids that follow) are used instead of systematic names.

$$\mathrm{H-\overset{\overset{\displaystyle O}{\|}}{C}-OH} \qquad \mathrm{CH_3-\overset{\overset{\displaystyle O}{\|}}{C}-OH} \qquad \mathrm{CH_3CH_2CH_2-\overset{\overset{\displaystyle O}{\|}}{C}-OH}$$

Methanoic acid (formic acid) — Ethanoic acid (acetic acid) — Butanoic acid (butyric acid)

Carboxylic acids can be prepared by moderately vigorous oxidation of primary alcohols or aldehydes or, in some cases, by the direct oxidation of alkyl groups. Carboxylic acids are weak Brønsted acids. Their acidic property results partially from the withdrawal of the electron from the —OH group by the nearby carbonyl oxygen, which weakens the oxygen–hydrogen bond; this results in easier ionization of the —OH hydrogen than occurs in alcohols. Carboxylic acids with long hydrocarbon chains are called **fatty acids.**

EXAMPLE The properties of carboxylic acids

Acetic acid has a boiling point of 118°C whereas propanol, with the same molecular weight, boils at 97°C. Explain this difference.

SOLUTION Acetic acid contains two oxygens that can hydrogen bond whereas propanol contains only one. The more extensive hydrogen bonding in the acid results in a higher boiling point.

EXERCISE Which is more acidic, acetic acid or chloroacetic acid? Why?

ANSWER: Chloroacetic acid is more acidic because chlorine withdraws electrons from the OH group, weakening the O—H bond and making ionization easier.

24.6 ESTERS

KEY CONCEPT Esters

Esters are formed by the reaction of an alcohol with a carboxylic acid. If we keep track of the oxygen atoms involved by using ^{18}O as a tracer, we find that the oxygen originally in the alcohol becomes part of the ester:

$$R{-}\overset{\overset{\displaystyle O}{\|}}{C}{-}OH + R'^{18}OH \longrightarrow R{-}\overset{\overset{\displaystyle O}{\|}}{C}{-}^{18}OR' + H_2O$$

This reaction is a **condensation reaction**, a reaction in which two molecules combine to form a larger molecule with the elimination of a small molecule such as water. Fats and oils are made up largely of esters. Fats and oils are members of a class of compounds called **lipids,** which are the water insoluble (and nonpolar-solvent soluble) constituents of cells.

Polyesters are polymers produced by linking together carboxylic acids and alcohols in a long chain. They are examples of **condensation polymers,** which are polymers formed by a series of condensation reactions. In forming such a polymer, a di-acid is reacted with a diol, and the polymer is formed by alternate linking of acid and alcohol fragments.

EXAMPLE The properties and reactions of esters

Draw the structure of the ester formed by the reaction of acetic acid with ethanol.

SOLUTION An ester is formed structurally by removing the hydroxyl hydrogen from the alcohol and the hydroxyl group from the acid and bonding the alcohol oxygen to the carbonyl carbon. Removing the hydroxyl hydrogen from ethanol leaves the CH_3CH_2O— fragment. Removing the hydroxyl group from acetic acid leaves the $CH_3C{=}O$ fragment. Joining the alcohol oxygen to the carbonyl carbon gives the desired structure:

$$CH_3{-}\overset{\overset{\displaystyle O}{\|}}{C}{-}OCH_2CH_3$$

EXERCISE Explain why lipids are insoluble in water and soluble in nonpolar solvents.

ANSWER: Lipids are esters of acids with very long hydrocarbon chains that make the lipid insoluble in water and soluble in nonpolar solvents.

FUNCTIONAL GROUPS CONTAINING NITROGEN

KEY WORDS Define or explain each of the following terms in a written sentence or two.

α-helix
amide
amine
amino acid
base pair
codon
denaturation
essential amino acid
nucleic acid
nucleoside
nucleotide
polynucleotide
polypeptide chain
racemic mixture

24.7 AMINES

KEY CONCEPT Amines

Amines are compounds derived from ammonia, in which one or more of the hydrogens in ammonia has been replaced by an alkyl group. Some examples are

$$\underset{\text{Ethylamine}}{H-\overset{\overset{\displaystyle H}{|}}{N}-CH_2CH_3} \qquad \underset{\text{Dimethylamine}}{CH_3-\overset{\overset{\displaystyle H}{|}}{N}-CH_3} \qquad \underset{\text{Diethylmethylamiine}}{CH_3CH_2-\overset{\overset{\displaystyle CH_3}{|}}{N}-CH_2CH_3}$$

Amines are classified as **primary, secondary,** or **tertiary,** depending on the number of carbon atoms that are attached to the nitrogen: RNH_2 is a primary amine, R_2NH is a secondary amine, and R_3N is a tertiary amine. In the examples just shown, ethylamine is a primary amine, dimethylamine a secondary amine, and diethylmethylamine a tertiary amine.

The characteristic reactions of amines involve their properties as Lewis bases and Brønsted bases. Stabilization of the conjugate acids through electron donation by alkyl groups makes alkylamines stronger bases than ammonia. In arylamines, on the other hand, the base itself (rather than the conjugate acid) is stabilized through resonance with the ring, so the base is weaker than ammonia. The conjugate acid of an amine is called a **quaternary ammonium ion.** For example, the conjugate acid of dimethylamine is $(CH_3)_2NH_2^+$, a quaternary ammonium ion. Amines condense with carboxylic acids to form **amides,** compounds that contain the **amido group,** $-CO(NR_2)$. The R here may be an alkyl group or a hydrogen. Amides can also be synthesized by the reaction of an amine with an acyl chloride, $R-CO(Cl)$, or by heating the alkylammonium salt of a carboxylic acid. This last method is used to make **polyamides** such as nylon.

EXAMPLE Naming amines

Name the following amines and classify them as primary, secondary or tertiary: (a) $CH_3CH_2NH_2$, (b) $(CH_3CH_2)_2NCH_3$ and (c) $CH_3CH_2N(CH_2CH_2CH_3)_2$.

SOLUTION Amines are named by identifying the alkyl groups attached to nitrogen and listing them in alphabetical order followed by the word *amine.* The prefixes *di-* and *tri-* do not determine the alphabetical order. The classification of the amine depends on the number of alkyl groups attached to the nitrogen. (a) A single ethyl group is attached to the nitrogen; so the compound is ethylamine, and it is a primary amine. (b) Two ethyl groups and one methyl group are attached to the nitrogen; so the compound is diethylmethylamine, and it is a tertiary amine. (c) Two propyl groups and an ethyl group are attached to the nitrogen, so this ethyldipropylamine is a tertiary amine.

EXERCISE Name the following amines, and classify them as primary, secondary, or tertiary: (a) $(CH_3CH_2)_2NCH_2CH_2CH_3$, (b) $(CH_2CH_3)_2NH$, and (c) CH_3NH_2.

ANSWER: (a) Diethylpropylamine, tertiary; (b) diethylamine, secondary; (c) methylamine, primary

24.8 AMINO ACIDS

KEY CONCEPT Amino acids

An amino acid is a compound containing both a carboxylic acid group and an amine group. The biologically important amino acids have structures of the form

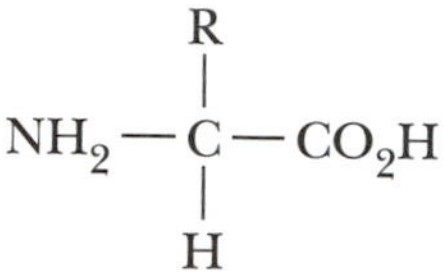

$$NH_2-\underset{\displaystyle H}{\overset{\displaystyle R}{\overset{|}{\underset{|}{C}}}}-CO_2H$$

A large number of amino acids are chiral and therefore display optical activity. A typical laboratory synthesis of an amino acid produces equal amounts of both L- and D-optical isomers; such a mixture is called a **racemic mixture.** It does not rotate the plane of polarized light (it is optically inactive). Amino acids found in nature are exclusively the L-enantiomer.

Amino acids are the building blocks of proteins. In proteins, the amino acids are linked by peptide bonds. Proteins are **polypeptides,** which are copolymers containing a mixture of approximately 20 amino acids incorporated into the chain. The component amino acid building blocks of a protein are often called amino acid **residues.** (Remember that part of the amino acid structure is lost when it is incorporated into the polypepetide; what is left is called the residue.)

The **primary structure** of a protein is the sequence of amino acids in the peptide chain. Two peptides with different sequences of amino acids have different primary structures. The three-dimensional shape that is assumed by the chain of amino acids is called the **secondary structure.** One of the two common secondary structures is the **α helix,** a structure in which the long chain twists to form a helical structure (like a spiral staircase). A second common secondary structure involves the cross-linking of chains by hydrogen bonding to form a **β-pleated sheet.** Hydrogen bonds between the N—H and C=O groups of the amino acid residues give rise to these structures. The **tertiary structure** of a polypeptide refers to how the α helices or β-pleated sheets twist and turn to determine the overall shape of the polypeptide. A good analogy is provided by taking a helical toy Slinky and twisting it back onto itself to give a specifically shaped result. Both hydrogen bonding and **disulfide links** (S—S links between sulfur atoms) help to determine the tertiary structure of polypeptides. In many proteins, polypeptide units with the appropriate tertiary structure come together to form a more massive unit that is the functioning protein. The way in which the units join determine the **quaternary structure** of the protein. Any modification of any of the structures primary, secondary, tertiary, or quaternary) usually results in the malfunction or dysfunction of the protein involved. The loss of one or more of the structural features of a protein is called **denaturation** of the protein. A denaturated protein cannot function as it should.

EXAMPLE The structure and properties of amino acids and polypeptides

Draw the structure of Ala-Gly-Gly, and locate the peptide linkages.

SOLUTION A peptide is formed through a condensation of the carboxylic-acid end of an amino acid with the amine end of another amino acid, accompanied by the elimination of a water molecule. The three amino acids we have are

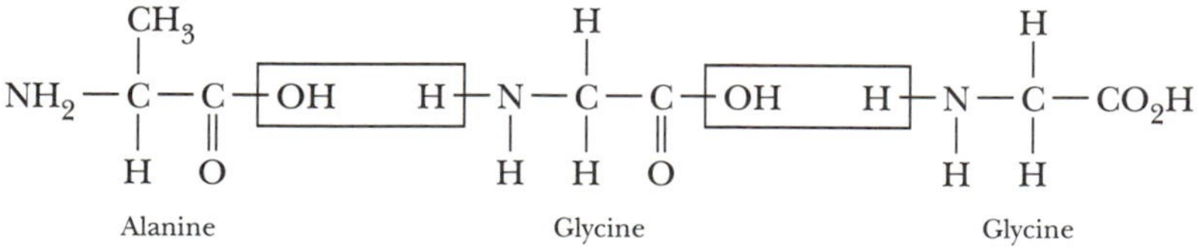

$$\underset{\text{Alanine}}{NH_2-\underset{\displaystyle H}{\overset{\displaystyle CH_3}{\overset{|}{\underset{|}{C}}}}-\underset{\displaystyle O}{\underset{\|}{C}}-\boxed{OH \quad H}-}\underset{\text{Glycine}}{\underset{\displaystyle H}{\underset{|}{N}}-\underset{\displaystyle H}{\overset{\displaystyle H}{\overset{|}{\underset{|}{C}}}}-\underset{\displaystyle O}{\underset{\|}{C}}-\boxed{OH \quad H}-}\underset{\text{Glycine}}{\underset{\displaystyle H}{\underset{|}{N}}-\underset{\displaystyle H}{\overset{\displaystyle H}{\overset{|}{\underset{|}{C}}}}-CO_2H}$$

The elimination of water (as indicated by the rectangles) and the formation of C—N bonds results in the formation of peptide linkages, which are shown in boldface in the following.

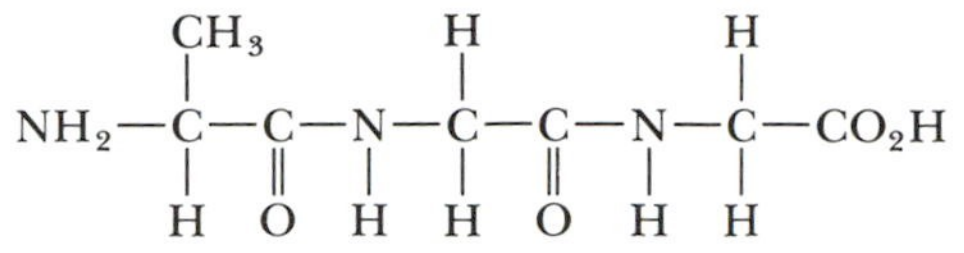

$$
\begin{array}{ccccccccccccccc}
 & & CH_3 & & & & & & H & & & & & & H \\
 & & | & & & & & & | & & & & & & | \\
NH_2 & - & C & - & C & - & N & - & C & - & C & - & N & - & C-CO_2H \\
 & & | & & \| & & | & & | & & \| & & | & & | \\
 & & H & & O & & H & & H & & O & & H & & H
\end{array}
$$

EXERCISE Let us assume that a protein is produced in which the primary structure is not what it should be. Which of the other three structures of the protein is likely to be affected.? Will the protein function properly?

ANSWER: If the primary structure of a protein chain is wrong, all of the other structures (secondary, tertiary, and quaternary) are likely to be adversely affected. The protein will not function properly.

24.9 DNA AND RNA

KEY CONCEPT DNA and RNA

The biological ribonucleic acids (RNA) and deoxyribonucleic acids (DNA) are polymers with structures (for both) that can be schematically represented as

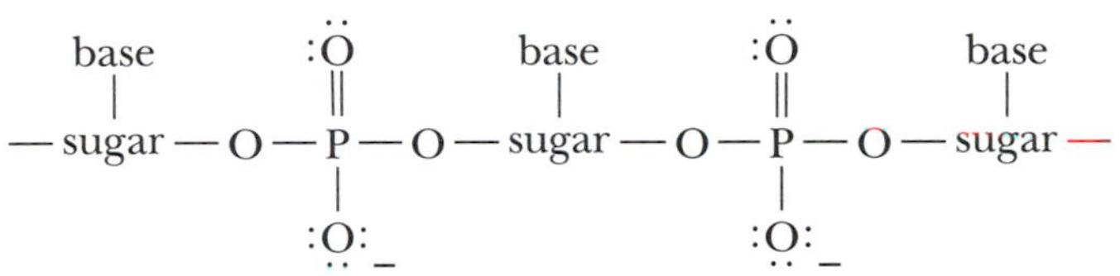

$$
\begin{array}{ccccccccccccc}
 & \text{base} & & & & :\ddot{O} & & & \text{base} & & & & :\ddot{O} & & & \text{base} & \\
 & | & & & & \| & & & | & & & & \| & & & | & \\
- & \text{sugar} & - & O & - & P & - & O & - \text{sugar} - & O & - & P & - & O & - & \text{sugar} & - \\
 & & & & & | & & & & & & | & & & & & \\
 & & & & & :\ddot{O}:^- & & & & & & :\ddot{O}:^- & & & & &
\end{array}
$$

The bases involved are adenine, guanine, cytosine, and thymine for DNA, and the first three and uracil for RNA. The sugars in both DNA and RNA are derived from ribose in its furanose form. A base-sugar unit is called a **nucleoside;** these are formed by a condensation between the base and ribose with the elimination of a water molecule. There are four nucleosides in DNA and four in RNA, corresponding to the four bases plus one sugar for each. A base-sugar-PO_4H_2 unit, formed by the condensation of a nucleoside with phosphoric acid, again with the elimination of a water molecule, is called a **nucleotide.** Again, four nucleotides are used in the construction of RNA and four in DNA. Nucleotides are the monomer units that make up DNA and RNA, and these two **polynucleotides** are copolymers of the different nucleotides. DNA itself is usually found as two chains intertwined to form a double helix. The double helix is held together by hydrogen bonding between adenine and thymine and between guanine and cytosine. Adenine and thymine fit together spatially to form hydrogen bonds, as do guanine and cytosine. Thus, they are called the AT (adenine-thymine) and GC (guanine-cytosine) **base pairs.** Other pairings are not as favorable as these and do not (or, at least, should not) occur.

Cellular replication of DNA is accomplished by separating the polynucleotide strands and building new partners for each single strand. Because the AT pairing and the GC pairing are faithfully reproduced, two new double-stranded DNA units are produced. The **genetic code** is based on the sequence of nucleotides in DNA. The basic unit of the code, called a **codon,** is a sequence of three nucleotides in DNA.

EXAMPLE RNA, DNA, and the genetic code

What bonds form the repetitive backbone structure of DNA?

SOLUTION The backbone bonding of DNA takes place when an oxygen in the phosphate fragment bonds to a carbon on ribose and a second carbon in ribose (two carbons away) bonds to the next phosphate fragment. That is, as the figure schematically shows, the string of phosphate groups and sugars, but not the bases, makes up the backbone.

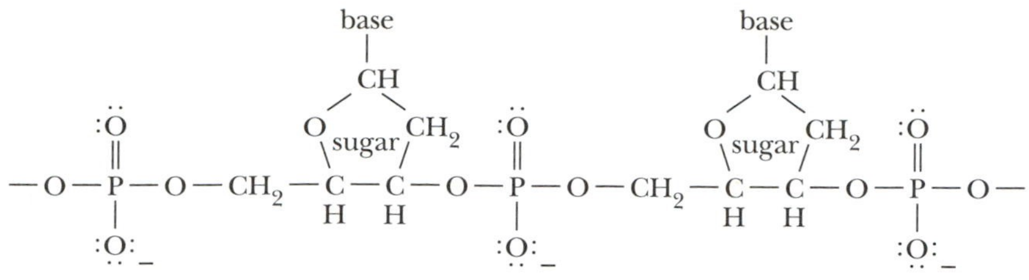

```
                    base                      base
                     |                         |
                     CH                        CH
                    /  \                      /  \
  :Ö               O    CH2   :Ö             O    CH2   :Ö
   ||               \sugar/    ||             \sugar/    ||
—O—P—O—CH2—C—C—O—P—O—CH2—C—C—O—P—O—
   |               H  H        |              H  H       |
  :O:⁻                        :O:⁻                      :O:⁻
```

EXERCISE What tripeptide sequence results from the DNA sequence GAGCATGTA? Assume that the code is read from left to right.

ANSWER: Gln-His-Val

SELF-TEST EXERCISES

The hydroxyl group

1. Which of the following is an Arrhenius base?
(a) CH_3OH (b) CH_3CH_2OH (c) C_6H_5OH
(d) $CH_3CH(OH)CH_3$ (e) none of these

2. Which of the following is a secondary alcohol?
(a) CH_3OH (b) $CH_3CH(OH)CH_3$ (c) $(CH_3)_3COH$
(d) CH_3CH_2OH (e) $CH_3CH_2CH_2OH$

3. The name of $CH_3CH(OH)CH(CH_3)CH_3$ is
(a) 2-methyl-3-butanol (b) 1-methyl-2-butanol (c) 3-methyl-2-butanol
(d) 2-methyl-1-butanol (e) methylbutanol

4. Which of the following forms in the first step of the acid dehydration of ethanol?
(a) ethoxide ion (b) carbonium ion (c) oxonium ion
(d) hydroxide ion (e) bromonium ion

5. What product is formed when 1-butanol is mixed with concentrated H_2SO_4 at high temperature?
(a) butane (b) 2-methylpropene (c) 1-butene (d) 3-methylpropane

6. What compound is formed by the gentle oxidation of 1-propanol?
(a) propanal (b) propanone (c) propanoic acid
(d) propane (e) 2-propanol

7. Which compound is the strongest acid?
(a) ethanol (b) 1-butanol (c) phenol (d) 2-butanol (e) 2-methyl-2-propanol

The carbonyl group

8. What is the name of the following compound?

$$CH_3\overset{\overset{\displaystyle O}{\|}}{C}HCH_2CH_3$$

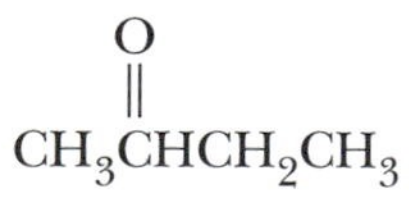

(a) butanol (b) butanal (c) butanone (d) 2-butanol (e) butene

9. What is the name of the following compound?

$$CH_3CH(CH_3)\overset{\overset{\displaystyle O}{\|}}{C}H$$

(a) 3-methyl-4-propanol (b) 2-methylpropanone (c) 2-methyl-1-propanone
(d) 3-methyl-4-propanone (e) 2-methylpropanal

10. The oxidation of 2-pentanol produces
(a) 2-pentanone. (b) pentanal. (c) 3-pentanone.
(d) 2-butanone. (e) butanal.

11. The gentle oxidation of an alcohol results in a compound that reduces Tollen's reagent. Which of the following may be the alcohol?
(a) 2-pentanol (b) 2-butanol (c) 1-propanol
(d) 2,2-dimethyl-2-pentanol (e) 3-pentanol

12. Which of the following represents a hemiacetal?

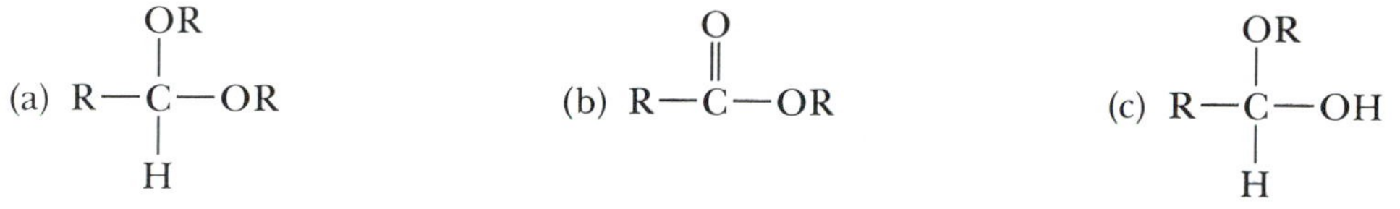
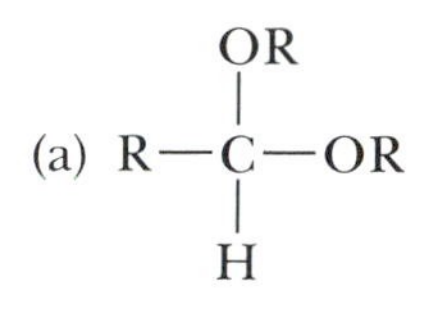

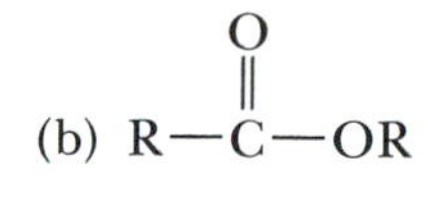

(c) R—C—OH (with OR above and H below)

13. All of the following but one describe glucose. Which one doesn't describe glucose?
(a) reducing sugar (b) polyalcohol (c) aldehyde
(d) disaccharide (e) has pyranose form

14. Which displays the structure of the carboxyl group?

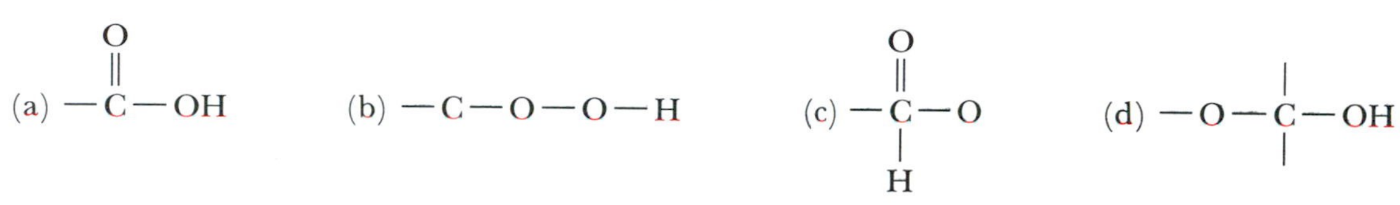

15. What is the name of $CH_3CH(CH_3)COOH$?
(a) 2-methylbutanoic acid (b) 2-methylpropanoic acid
(c) 2-methyl-3-propanoic acid (d) 1-methylpropanoic acid

16. Which order of acid strengths is correct? (SA = strongest acid, WA = weakest acid)
(a) SA mineral acid > phenol > typical carboxylic acid > alcohol WA
(b) SA mineral acid > typical carboxylic acid > phenol > alcohol WA
(c) SA mineral acid > phenol > alcohol > typical carboxylic acid WA
(d) SA typical carboxylic acid > mineral acid > phenol > alcohol WA

17. Which of the following represents an ester?

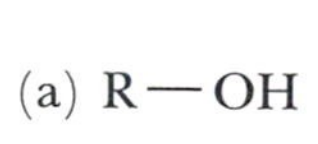

Functional groups containing nitrogen

18. Which of the following is a secondary amine?
(a) $(CH_3)_2NH$ (b) $CH_3CH_2N(CH_3)_2$ (c) NH_3 (d) CH_3NH_2 (e) $CH_3NH_3^+$

19. What is the name of $(CH_3)_2NCH_2CH_2CH_3$?
(a) tetramethylamine (b) dimethylpropylamine (c) methylmethylpropylamine
(d) butylamine

20. HBr is added to ethene and the product A is reacted with CN^- to get product B. B is completely reduced with lithium aluminum hydride to get C. What is C?
(a) 1-bromo-1-cyanoethane (b) ethylamine (c) diethylamine (d) cyanoethane
(e) propylamine

21. Delocalization of the lone electron pair on nitrogen in arylamines makes arylamines ______ bases than alkylamines.
(a) stronger (b) weaker

22. Which of the following amino acids is chiral?

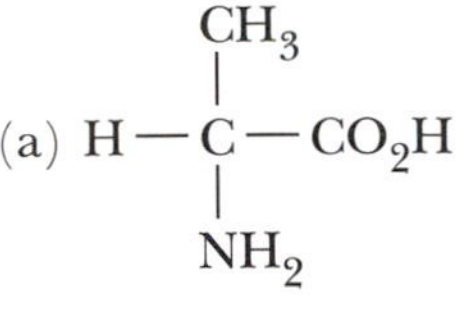

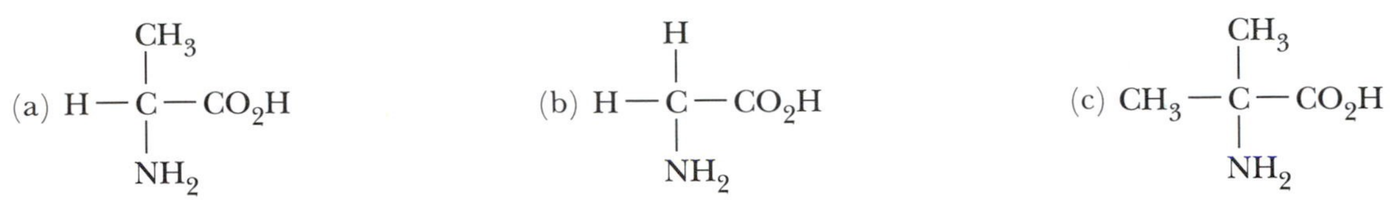

23. The spatial arranging of a polypeptide into an α helix or β-pleated sheet refers to which structure of a protein?
(a) primary (b) secondary (c) tertiary (d) quarternary

24. Which of the following represents a peptide linkage?

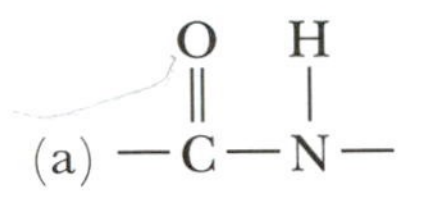

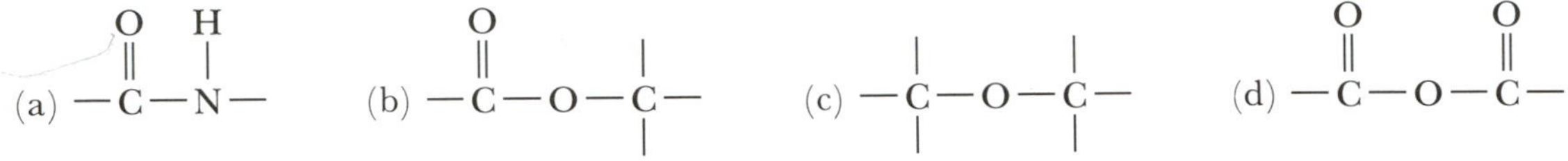

25. A nucleotide is a
(a) DNA base. (b) DNA base + sugar. (c) DNA base + phosphate fragment.
(d) DNA base + sugar + phosphate fragment.

26. The double helix of DNA is held together by
(a) S—S bridges. (b) AC and GT hydrogen bonds.
(c) peptide bonds. (d) AT and GC hydrogen bonds.

ANSWERS TO SELF-TEST EXERCISES

CHAPTER 1: 1b, 2e, 3a, 4c, 5a, 6e, 7d, 8b, 9d, 10c, 11d, 12a, 13c, 14e, 15b, 16d, 17a, 18a, 19c, 20e, 21d, 22b, 23b, 24c, 25e, 26d, 27d, 28b, 29a, 30e, 31c, 32e, 33c, 34c, 35a, 36a, 37a, 38e, 39e, 40d, 41a, 42d, 43c, 44b, 45b, 46d, 47c, 48a, 49d.

CHAPTER 2: 1e, 2c, 3c, 4a, 5a, 6b, 7e, 8c, 9c, 10c, 11e, 12e, 13c, 14b, 15b, 16c, 17a, 18d, 19c, 20b, 21a, 22a, 23b, 24e, 25b, 26d, 27b, 28b, 29d, 30c, 31c, 32b, 33a, 34e, 35b, 36b, 37b, 38a, 39d, 40c, 41b, 42a, 43c, 44c, 45d, 46e, 47b, 48b, 49b, 50c, 51d, 52b, 53d, 54e.

CHAPTER 3: 1c, 2d, 3c, 4a, 5d, 6c, 7b, 8b, 9b, 10e, 11e, 12a, 13e, 14c, 15e, 16b, 17b, 18a, 19d, 20b, 21c, 22a, 23a, 24b, 25c, 26d, 27d, 28e, 29b, 30b, 31a, 32c, 33b, 34a, 35d, 36c, 37d, 38e, 39b, 40d, 41b, 42b, 43a.

CHAPTER 4: 1c, 2e, 3a, 4b, 5b, 6c, 7a, 8a, 9d, 10a, 11b, 12c, 13e, 14e, 15c, 16b, 17c, 18d, 19e, 20b, 21c, 22a, 23e, 24c, 25d, 26e, 27d, 28a, 29e, 30b, 31d, 32e, 33d, 34a, 35e, 36b, 37d, 38d, 39c, 40b, 41d.

CHAPTER 5: 1d, 2a, 3e, 4c, 5b, 6b, 7c, 8e, 9b, 10a, 11b, 12b, 13b, 14e, 15d, 16d, 17e, 18a, 19c, 20a, 21b, 22b, 23b, 24b, 25e, 26c, 27a, 28c, 29d, 30b, 31a, 32b, 33d, 34c, 35d, 36e, 37c, 38d, 39e, 40a, 41b.

CHAPTER 6: 1a, 2b, 3c, 4d, 5b, 6e, 7e, 8c, 9b, 10c, 11a, 12c, 13e, 14c, 15a, 16a, 17b, 18d, 19c, 20e, 21d, 22d, 23d, 24a, 25a, 26e, 27b, 28c, 29c, 30b, 31c, 32c, 33a, 34a, 35c, 36b, 37b, 38d, 39c, 40a, 41c, 42e, 43b, 44c, 45b, 46a, 47c, 48e.

CHAPTER 7: 1c, 2c, 3d, 4b, 5a, 6a, 7d, 8e, 9c, 10c, 11b, 12e, 13e, 14d, 15a, 16b, 17a, 18a, 19a, 20e, 21b, 22b, 23d, 24a, 25e, 26c, 27c, 28e, 29e, 30a, 31a, 32c, 33b, 34d, 35a, 36a, 37c, 38d, 39e, 40b, 41a, 42d, 43c, 44b, 45d, 46a, 47b.

CHAPTER 8: 1e, 2a, 3b, 4c, 5e, 6a, 7d, 8d, 9c, 10e, 11b, 12e, 13d, 14c, 15e, 16c, 17b, 18a, 19a, 20d, 21d, 22c, 23a, 24c, 25c, 26b, 27d, 28e, 29e, 30a, 31d, 32c, 33a, 34b, 35b, 36b, 37e, 38b, 39a, 40d, 41e, 42c, 43a, 44a, 45e, 46e, 47c, 48b, 49d, 50e, 51a, 52a.

CHAPTER 9: 1a, 2d, 3e, 4c, 5d, 6b, 7a, 8b, 9e, 10c, 11b, 12b, 13d, 14e, 15b, 16c, 17b, 18e, 19b, 20c, 21d, 22a, 23e, 24b, 25c, 26b, 27b, 28e, 29b, 30d, 31e, 32a, 33a, 34e, 35b, 36b, 37e, 38d, 39a, 40b, 41d, 42b, 43d, 44b, 45c, 46b, 47c, 48a.

CHAPTER 10: 1d, 2a, 3c, 4b, 5d, 6e, 7c, 8d, 9c, 10c, 11b, 12e, 13b, 14b, 15d, 16a, 17b, 18a, 19b, 20e, 21c, 22e, 23b, 24b, 25e, 26b, 27a, 28d, 29e, 30c, 31a, 32b, 33b, 34e, 35a, 36d, 37e, 38b, 39b, 40a, 41c, 42b, 43a, 44d, 45b, 46c, 47d, 48a, 49e, 50b.

CHAPTER 11: 1b, 2c, 3e, 4b, 5a, 6e, 7d, 8a, 9e, 10d, 11c, 12e, 13e, 14a, 15e, 16a, 17a, 18c, 19d, 20b, 21d, 22d, 23b, 24d, 25a, 26b, 27e, 28e, 29a, 30c, 31d, 32a, 33a, 34c, 35c, 36b, 37e, 38c, 39a, 40d, 41e, 42b, 43c, 44b.

CHAPTER 12: 1a, 2e, 3c, 4b, 5b, 6d, 7b, 8c, 9b, 10e, 11a, 12a, 13d, 14a, 15e, 16e, 17e, 18b, 19d, 20c, 21b, 22a, 23b, 24b, 25d, 26b, 27d, 28c, 29c, 30a, 31a, 32d, 33c, 34d, 35c, 36e, 37d, 38d, 39b.

CHAPTER 13: 1b, 2d, 3d, 4c, 5e, 6e, 7a, 8a, 9b, 10c, 11e, 12c, 13b, 14b, 15d, 16b, 17a, 18d, 19c, 20c, 21b, 22a, 23c, 24b, 25a, 26c, 27a, 28c, 29b, 30a, 31b, 32d, 33a.

CHAPTER 14: 1a, 2c, 3c, 4b, 5e, 6a, 7b, 8c, 9a, 10d, 11e, 12c, 13b, 14d, 15b, 16c, 17d, 18c, 19a, 20a, 21b, 22e, 23d, 24a, 25c, 26b, 27c, 28b, 29d, 30a, 31d, 32e, 33d, 34c, 35d, 36a, 37c, 38a, 39d, 40b, 41b, 42c, 43a, 44e, 45d, 46d, 47b, 48d, 49d, 50a, 51b, 52b, 53a, 54d, 55d, 56c, 57c, 58a.

CHAPTER 15: 1e, 2b, 3a, 4a, 5a, 6c, 7b, 8a, 9c, 10e, 11c, 12d, 13b, 14b, 15c, 16b, 17a, 18d, 19e, 20c, 21a, 22d, 23d, 24a, 25a, 26b, 27d, 28e, 29b, 30d, 31a, 32b, 33b, 34c, 35e, 36a, 37a, 38c, 39a, 40c, 41e, 42b, 43e, 44d, 45d, 46a, 47e, 48b, 49b, 50b, 51d.

CHAPTER 16: 1d, 2e, 3b, 4b, 5d, 6b, 7d, 8d, 9b, 10e, 11a, 12e, 13c, 14a, 15c, 16e, 17c, 18a, 19b, 20d, 21b, 22a, 23c, 24b, 25c, 26a, 27d, 28b, 29b, 30a, 31b, 32b, 33d, 34c, 35a, 36b, 37d.

CHAPTER 17: 1b, 2b, 3d, 4e, 5a, 6d, 7c, 8c, 9d, 10e, 11a, 12d, 13e, 14b, 15c, 16c, 17c, 18a, 19a, 20b, 21b, 22a, 23a, 24d, 25c, 26d, 27d, 28e, 29b, 30a, 31c, 32e, 33b, 34d, 35d.

CHAPTER 18: 1b, 2d, 3a, 4b, 5a, 6d, 7c, 8b, 9c, 10a, 11d, 12d, 13c, 14b, 15c, 16a, 17d, 18a, 19e, 20d, 21b, 22a, 23c, 24d, 25b, 26b, 27d, 28e, 29b, 30a, 31b, 32e, 33a, 34c, 35a, 36d, 37a, 38b, 39b, 40a.

CHAPTER 19: 1a, 2c, 3c, 4a, 5b, 6e, 7d, 8e, 9d, 10c, 11e, 12a, 13a, 14b, 15c, 16e, 17b, 18b, 19d, 20e, 21b, 22c.

CHAPTER 20: 1c, 2b, 3b, 4d, 5d, 6a, 7c, 8a, 9d, 10b, 11a, 12e, 13d, 14c, 15e, 16e, 17b, 18b, 19c, 20d, 21a, 22a, 23d, 24e, 25a, 26a, 27d, 28c, 29c.

CHAPTER 21: 1b, 2b, 3c, 4d, 5a, 6c, 7e, 8c, 9b, 10b, 11e, 12c, 13c, 14e, 15a, 16b, 17d, 18d, 19a, 20d, 21a, 22a, 23b, 24a, 25e, 26c, 27c, 28a, 29b, 30d.

CHAPTER 22: 1e, 2b, 3b, 4a, 5c, 6c, 7d, 8e, 9d, 10a, 11e, 12c, 13a, 14d, 15d, 16d, 17e, 18a, 19a, 20e, 21c, 22b, 23e, 24b, 25c, 26c, 27b, 28d.

CHAPTER 23: 1a, 2e, 3d, 4c, 5e, 6b, 7d, 8b, 9a, 10c, 11d, 12e, 13c, 14d.

CHAPTER 24: 1e, 2b, 3c, 4c, 5c, 6a, 7c, 8c, 9e, 10a, 11c, 12c, 13d, 14a, 15b, 16b, 17d, 18a, 19b, 20e, 21b, 22a, 23b, 24a, 25d, 26d.